Recapturing a Future for Space Exploration

Life and Physical Sciences Research for a New Era

Committee for the Decadal Survey on Biological and Physical Sciences in Space

Space Studies Board

Aeronautics and Space Engineering Board

Division on Engineering and Physical Sciences

NATIONAL RESEARCH COUNCIL
OF THE NATIONAL ACADEMIES

THE NATIONAL ACADEMIES PRESS
Washington, D.C.
www.nap.edu

THE NATIONAL ACADEMIES PRESS **500 Fifth Street, N.W.** **Washington, DC 20001**

NOTICE: The project that is the subject of this report was approved by the Governing Board of the National Research Council, whose members are drawn from the councils of the National Academy of Sciences, the National Academy of Engineering, and the Institute of Medicine. The members of the committee responsible for the report were chosen for their special competences and with regard for appropriate balance.

This study is based on work supported by Contracts NNH05CC16C and NNH10CC48B between the National Academy of Sciences and the National Aeronautics and Space Administration. Any opinions, findings, conclusions, or recommendations expressed in this publication are those of the author(s) and do not necessarily reflect the views of the agency that provided support for the project.

International Standard Book Number-3: 978-0-309-16384-2
International Standard Book Number-0: 0-309-16384-6

Copies of this report are available free of charge from:

Space Studies Board
National Research Council
500 Fifth Street, N.W.
Washington, DC 20001

Additional copies of this report are available from the National Academies Press, 500 Fifth Street, N.W., Lockbox 285, Washington, DC 20055; (800) 624-2422 or (202) 334-3133 (in the Washington metropolitan area); Internet, http://www.nap.edu.

Printed in the United States of America

OTHER RECENT REPORTS OF THE SPACE STUDIES BOARD AND THE AERONAUTICS AND SPACE ENGINEERING BOARD

Assessment of Impediments to Interagency Collaboration on Space and Earth Science Missions (Space Studies Board [SSB], 2011)
Panel Reports—New Worlds, New Horizons in Astronomy and Astrophysics (BPA and SSB, 2011)
Vision and Voyages for Planetary Science in the Decade 2013-2022 [prepublication version] (SSB, 2011)

Advancing Aeronautical Safety: A Review of NASA's Aviation Safety-Related Research Programs (Aeronautics and Space Engineering Board [ASEB], 2010)
Capabilities for the Future: An Assessment of NASA Laboratories for Basic Research (Laboratory Assessments Board [LAB] with SSB and ASEB, 2010)
Controlling Cost Growth of NASA Earth and Space Science Missions (SSB, 2010)
Defending Planet Earth: Near-Earth-Object Surveys and Hazard Mitigation Strategies (SSB with ASEB, 2010)
An Enabling Foundation for NASA's Space and Earth Science Missions (SSB, 2010)
Life and Physical Sciences Research for a New Era of Space Exploration: An Interim Report (SSB with ASEB, 2010)
New Worlds, New Horizons in Astronomy and Astrophysics (Board on Physics and Astronomy [BPA] and SSB, 2010)
Report of the Panel on Implementing Recommendations from the New Worlds, New Horizons Decadal Survey [prepublication version] (BPA and SSB, 2010)
Revitalizing NASA's Suborbital Program: Advancing Science, Driving Innovation, and Developing a Workforce (SSB, 2010)

America's Future in Space: Aligning the Civil Space Program with National Needs (SSB with ASEB, 2009)
Approaches to Future Space Cooperation and Competition in a Globalizing World: Summary of a Workshop (SSB with ASEB, 2009)
An Assessment of NASA's National Aviation Operations Monitoring Service (ASEB, 2009)
Assessment of Planetary Protection Requirements for Mars Sample Return Missions (SSB, 2009)
Fostering Visions for the Future: A Review of the NASA Institute for Advanced Concepts (ASEB, 2009)
Near-Earth Object Surveys and Hazard Mitigation Strategies: Interim Report (SSB with ASEB, 2009)
A Performance Assessment of NASA's Heliophysics Program (SSB, 2009)
Radioisotope Power Systems: An Imperative for Maintaining U.S. Leadership in Space Exploration (SSB with ASEB, 2009)

Assessing the Research and Development Plan for the Next Generation Air Transportation System: Summary of a Workshop (ASEB, 2008)
A Constrained Space Exploration Technology Program: A Review of NASA's Exploration Technology Development Program (ASEB, 2008)
Ensuring the Climate Record from the NPOESS and GOES-R Spacecraft: Elements of a Strategy to Recover Measurement Capabilities Lost in Program Restructuring (SSB, 2008)
Launching Science: Science Opportunities Provided by NASA's Constellation System (SSB with ASEB, 2008)
Managing Space Radiation Risk in the New Era of Space Exploration (ASEB, 2008)
NASA Aeronautics Research: An Assessment (ASEB, 2008)
Opening New Frontiers in Space: Choices for the Next New Frontiers Announcement of Opportunity (SSB, 2008)
Review of NASA's Exploration Technology Development Program: An Interim Report (ASEB, 2008)
Science Opportunities Enabled by NASA's Constellation System: Interim Report (SSB with ASEB, 2008)
Severe Space Weather Events—Understanding Societal and Economic Impacts: A Workshop Report (SSB, 2008)
Space Science and the International Traffic in Arms Regulations: Summary of a Workshop (SSB, 2008)
United States Civil Space Policy: Summary of a Workshop (SSB with ASEB, 2008)
Wake Turbulence: An Obstacle to Increased Air Traffic Capacity (ASEB, 2008)

Limited copies of SSB reports are available free of charge from

Space Studies Board
National Research Council
The Keck Center of the National Academies
500 Fifth Street, N.W., Washington, DC 20001
(202) 334-4777/ssb@nas.edu
www.nationalacademies.org/ssb/ssb.html

COMMITTEE FOR THE DECADAL SURVEY ON BIOLOGICAL AND PHYSICAL SCIENCES IN SPACE

ELIZABETH R. CANTWELL, Lawrence Livermore National Laboratory, *Co-chair*
WENDY M. KOHRT, University of Colorado, Denver, *Co-chair*
LARS BERGLUND, University of California, Davis
NICHOLAS P. BIGELOW, University of Rochester
LEONARD H. CAVENY, Independent Consultant, Fort Washington, Maryland
VIJAY K. DHIR, University of California, Los Angeles
JOEL E. DIMSDALE, University of California, San Diego, School of Medicine
NIKOLAOS A. GATSONIS, Worcester Polytechnic Institute
SIMON GILROY, University of Wisconsin-Madison
BENJAMIN D. LEVINE, University of Texas Southwestern Medical Center at Dallas
RODOLFO R. LLINAS,* New York University Medical Center
KATHRYN V. LOGAN, Virginia Polytechnic Institute and State University
PHILIPPA MARRACK,† National Jewish Health
GABOR A. SOMORJAI, University of California, Berkeley
CHARLES M. TIPTON, University of Arizona
JOSE L. TORERO, University of Edinburgh, Scotland
ROBERT WEGENG, Pacific Northwest National Laboratory
GAYLE E. WOLOSCHAK, Northwestern University Feinberg School of Medicine

ANIMAL AND HUMAN BIOLOGY PANEL

KENNETH M. BALDWIN, University of California, Irvine, *Chair*
FRANÇOIS M. ABBOUD, University of Iowa, Roy J. and Lucille A. Carver College of Medicine
PETER R. CAVANAGH, University of Washington
V. REGGIE EDGERTON, University of California, Los Angeles
DONNA MURASKO, Drexel University
JOHN T. POTTS, JR., Massachusetts General Hospital
APRIL E. RONCA, Wake Forest University School of Medicine
CHARLES M. TIPTON, University of Arizona
CHARLES H. TURNER,‡ Indiana University-Purdue University, Indianapolis
JOHN B. WEST, University of California, San Diego

APPLIED PHYSICAL SCIENCES PANEL

PETER W. VOORHEES, Northwestern University, *Chair*
NIKOLAOS A. GATSONIS, Worcester Polytechnic Institute
RICHARD T. LAHEY, JR., Rensselaer Polytechnic Institute
RICHARD M. LUEPTOW, Northwestern University
JOHN J. MOORE, Colorado School of Mines
ELAINE S. ORAN, Naval Research Laboratory
AMY L. RECHENMACHER, University of Southern California
JAMES S. T'IEN, Case Western Reserve University
MARK M. WEISLOGEL, Portland State University

---

*Through mid-December 2009.
†Through mid-May 2010.
‡Deceased July 2010.

FUNDAMENTAL PHYSICAL SCIENCES PANEL

ROBERT V. DUNCAN, University of Missouri, *Chair*
NICHOLAS P. BIGELOW, University of Rochester
PAUL M. CHAIKIN, New York University
RONALD G. LARSON, University of Michigan, Ann Arbor
W. CARL LINEBERGER, University of Colorado, Boulder
RONALD WALSWORTH, Harvard University

HUMAN BEHAVIOR AND MENTAL HEALTH PANEL

THOMAS J. BALKIN, Walter Reed Army Institute of Research, *Chair*
JOEL E. DIMSDALE, University of California, San Diego, School of Medicine
NICK KANAS, University of California, San Francisco
GLORIA R. LEON, University of Minnesota, Minneapolis
LAWRENCE A. PALINKAS, University of Southern California
MRIGANKA SUR,[§] Massachusetts Institute of Technology

INTEGRATIVE AND TRANSLATIONAL RESEARCH FOR THE HUMAN SYSTEMS PANEL

JAMES A. PAWELCZYK, Pennsylvania State University, *Chair*
ALAN R. HARGENS, University of California, San Diego
ROBERT L. HELMREICH, University of Texas, Austin (retired)
JOANNE R. LUPTON, Texas A&M University, College Station
CHARLES M. OMAN, Massachusetts Institute of Technology
DAVID ROBERTSON, Vanderbilt University
SUZANNE M. SCHNEIDER, University of New Mexico
GAYLE E. WOLOSCHAK, Northwestern University Feinberg School of Medicine

PLANT AND MICROBIAL BIOLOGY PANEL

TERRI L. LOMAX, North Carolina State University, *Chair*
PAUL BLOUNT, University of Texas Southwestern Medical Center at Dallas
ROBERT J. FERL, University of Florida
SIMON GILROY, University of Wisconsin-Madison
E. PETER GREENBERG, University of Washington School of Medicine

TRANSLATION TO SPACE EXPLORATION SYSTEMS PANEL

JAMES P. BAGIAN, U.S. Air Force and University of Michigan, *Chair*
FREDERICK R. BEST, Texas A&M University, College Station
DAVID C. BYERS,[§] Independent Consultant, Torrance, California
LEONARD H. CAVENY, Independent Consultant, Fort Washington, Maryland
MICHAEL B. DUKE, Colorado School of Mines (retired)
JOHN P. KIZITO, North Carolina A&T State University
DAVID Y. KUSNIERKIEWICZ, Johns Hopkins University, Applied Physics Laboratory
E. THOMAS MAHEFKEY, JR., Heat Transfer Technology Consultants
DAVA J. NEWMAN, Massachusetts Institute of Technology

[§]Through mid-December 2009.

RICHARD J. ROBY, Combustion Science and Engineering, Inc.
GUILLERMO TROTTI, Trotti and Associates, Inc.
ALAN WILHITE, Georgia Institute of Technology

STAFF

SANDRA J. GRAHAM, Senior Program Officer, Space Studies Board, *Study Director*
ALAN C. ANGLEMAN, Senior Program Officer, Aeronautics and Space Engineering Board
IAN W. PRYKE, Senior Program Officer, Space Studies Board
ROBERT L. RIEMER,[¶] Senior Program Officer, Board on Physics and Astronomy
MAUREEN MELLODY, Program Officer, Aeronautics and Space Engineering Board
REGINA NORTH, Consultant, Washington, D.C.
CATHERINE A. GRUBER, Editor, Space Studies Board
LEWIS GROSWALD, Research Associate, Space Studies Board
DANIELLE JOHNSON-BLAND,[¶] Senior Program Assistant, Committee on Law and Justice
LAURA TOTH,[¶] Senior Program Assistant, National Materials Advisory Board
LINDA M. WALKER, Senior Program Assistant, Space Studies Board
ERIC WHITAKER,[¶] Senior Program Assistant, Computer Science and Telecommunications Board

MICHAEL H. MOLONEY, Director, Space Studies Board, and Director, Aeronautics and Space Engineering Board

[¶]Staff from other National Research Council boards who assisted with the survey.

SPACE STUDIES BOARD

Preface

In May 2009, the National Research Council's (NRC) Committee for the Decadal Survey on Biological and Physical Sciences in Space began a series of meetings initiated as a result of the following language in the explanatory statement accompanying the FY 2008 Omnibus Appropriations Act (P.L. 110-161):

> Achieving the goals of the Exploration Initiative will require a greater understanding of life and physical sciences phenomena in microgravity as well as in the partial gravity environments of the Moon and Mars. Therefore, the Administrator is directed to enter into an arrangement with the National Research Council to conduct a "decadal survey" of life and physical sciences research in microgravity and partial gravity to establish priorities for research for the 2010-020 decade.

In response to this language, a statement of task for an NRC study was developed in consultation with members of the life and physical sciences communities, NASA, and congressional staff. The guiding principle of the study was to set an agenda for research in the next decade that would use the unique characteristics of the space environment to address complex problems in the life and physical sciences, so as to deliver both new knowledge and practical benefits for humankind as it embarks on a new era of space exploration. Specifically, the decadal survey committee was asked to define research areas, recommend a research portfolio and a timeline for conducting that research, identify facility and platform requirements as appropriate, provide rationales for suggested program elements, define dependencies among research objectives, identify terrestrial benefits, and specify whether the results of the research would directly enable exploration or would produce fundamental new knowledge. The research areas identified were to be categorized as either (1) required to enable exploration missions or (2) enabled or facilitated because of exploration missions. The complete statement of task for the study is given in Appendix A of this report.

As one of its earliest tasks, the committee divided the broad spectrum of relevant disciplines and charge elements into seven broad focus areas and organized the following study panels to address each theme:

- Animal and Human Biology Panel,
- Applied Physical Sciences Panel,
- Fundamental Physical Sciences Panel,
- Human Behavior and Mental Health Panel,
- Integrative and Translational Research for Human Systems Panel,

- Plant and Microbial Biology Panel, and
- Translation to Space Exploration Systems Panel.

The level of integration between the committee and the panels, and among many of the panels, was exceptionally high throughout the study. In general, the panels had primary responsibility for gathering data on the status of the relevant research areas, performing analysis, and developing a chapter in their assigned theme area; the committee provided continuous direction and feedback to the panels, integrated the input of the panels, and developed the chapters that responded directly to the statement of task. The report is thus the product of the committee's and the panels' combined efforts. In addition to the expertise embodied in the committee and the panels, broad community input was provided to the study at town hall meetings held in conjunction with professional society meetings, in approximately 150 white papers submitted by individuals and teams from the community,* and through numerous briefings and direct exchanges. Based on these inputs and its own deliberations, the committee in conjunction with the panels reviewed those areas of research that seemed most promising and selected the research topics and themes that are discussed in this report.

Although decadal surveys are a long-standing tradition in many other fields supported by NASA, such as astronomy and planetary science, this report represents the first decadal survey of NASA's life and physical sciences programs. In preparing it, the committee and its panels drew heavily on a number of past NRC studies that looked at many of the disciplines represented in this study. Two such reports were considered foundation documents on which the current study has built but whose work it does not reproduce: The 1998 NRC report *A Strategy for Research in Space Biology and Medicine in the New Century* provided a detailed assessment of the impact of spaceflight on the minds and physiological systems of humans, as well as effects in plants and animals. The 2006 NRC report *Microgravity Research in Support of Technologies for the Human Exploration and Development of Space and Planetary Bodies* examined the new capabilities that NASA would have to develop in order to explore the solar system and identified the underlying physical processes on which these capabilities depended. Both reports describe the phenomenological changes that occur—in biological systems and physical processes—at a level of detail that the current report does not attempt to replicate. Readers interested in better understanding these processes are referred to the two earlier studies.

This full survey report was preceded by the committee's interim report, released in July 2010, titled *Life and Physical Sciences Research for a New Era of Space Exploration: An Interim Report* (http://www.nap.edu/catalog.php?record_id=12944). Drawing on preliminary analyses performed by the committee and its panels, that report provided guidance on near-term programmatic issues related to the organization and management of the life and physical sciences research enterprise at NASA. It also identified a number of broad topics that represent near-term opportunities for research on the International Space Station. The guidance in the interim report is incorporated into the more detailed examination of programmatic issues and research needs for all platforms in this full report.

*Available at http://www8.nationalacademies.org/SSBSurvey/PublicViewMicro.aspx.

Acknowledgment of Reviewers

This report has been reviewed in draft form by individuals chosen for their diverse perspectives and technical expertise, in accordance with procedures approved by the Report Review Committee of the National Research Council (NRC). The purpose of this independent review is to provide candid and critical comments that will assist the institution in making its published report as sound as possible and to ensure that the report meets institutional standards for objectivity, evidence, and responsiveness to the study charge. The review comments and draft manuscript remain confidential to protect the integrity of the deliberative process. We wish to thank the following individuals for their review of this report:

Andreas Acrivos, City College of the City University of New York;
Robert L. Ash, Old Dominion University;
Henry W. Brandhorst, Jr., Carbon-Free Energy, LLC;
Edward J. Britt, Pratt & Whitney Space Propulsion;
Jay C. Buckey, Jr., Dartmouth-Hitchcock Medical Center;
Jonathan B. Clark, National Space Biomedical Research Institute;
Michael E. Fisher, University of Maryland;
Lennard A. Fisk, University of Michigan;
F. Andrew Gaffney, Vanderbilt University School of Medicine;
Kurt Gibble, Pennsylvania State University;
Roger Hangarter, Indiana University;
Kathryn D. Held, Massachusetts General Hospital/Harvard Medical School;
Edward W. Hodgson, Jr., Hamilton Sundstrand Corporation;
George M. Homsy, University of British Columbia;
Mamoru Ishii, Purdue University;
David M. Klaus, University of Colorado, Boulder;
Richard H. Kohrs, NASA (retired);
Rodolfo R. Llinas, New York University Medical Center;
Jay S. Loeffler, Massachusetts General Hospital;
David E. Longnecker, Association of American Medical Colleges;
Robert Marcus, Independent Consultant and Eli Lilly and Company (retired);

Gail Martin, University of California, San Francisco;
Ralph Napolitano, Iowa State University;
Robert J. Naumann, University of Alabama, Huntsville;
Mary Jane Osborn, University of Connecticut Health Center;
William Paloski, University of Houston;
G. Kim Prisk, University of California, San Diego;
Emery I. Reeves, Independent Consultant and U.S. Air Force Academy (retired);
Danny A. Riley, Medical College of Wisconsin;
Gerald Sonnenfeld, Clemson University;
T. Peter Stein, University of Medicine and Dentistry of New Jersey;
Thomas G. Stoebe, University of Washington;
Pete Suedfeld, University of British Columbia (emeritus);
Peter B. Sunderland, University of Maryland;
George W. Swenson, Jr., University of Illinois;
Scott Tremaine, Institute for Advanced Study;
Russell Turner, Oregon State University;
Forman A. Williams, University of California, San Diego;
Eugene Wissler, University of Texas, Austin; and
A. Thomas Young, Lockheed Martin Corporation (retired).

Although the reviewers listed above have provided many constructive comments and suggestions, they were not asked to endorse the conclusions or recommendations, nor did they see the final draft of the report before its release. The review of this report was overseen by Martha P. Haynes, Cornell University, and Laurence R. Young, Massachusetts Institute of Technology. Appointed by the NRC, they were responsible for making certain that an independent examination of this report was carried out in accordance with institutional procedures and that all review comments were carefully considered. Responsibility for the final content of this report rests entirely with the authoring committee and the institution.

Contents

APPENDIXES

Summary

SCIENCE AND EXPLORATION

More than four decades have passed since a human first set foot on the Moon. Great strides have been made since in our understanding of what is required to support an enduring human presence in space, as evidenced by progressively more advanced orbiting human outposts, culminating in the current International Space Station (ISS). However, of the more than 500 humans who have so far ventured into space, most have gone only as far as near-Earth orbit, and none have traveled beyond the orbit of the Moon. Achieving humans' further progress into the solar system has proved far more difficult than imagined in the heady days of the Apollo missions, but the potential rewards remain substantial. Overcoming the challenges posed by risk and cost—and developing the technology and capabilities to make long space voyages feasible—is an achievable goal. Further, the scientific accomplishments required to meet this goal will bring a deeper understanding of the performance of people, animals, plants, microbes, materials, and engineered systems not only in the space environment but also on Earth, providing terrestrial benefits by advancing fundamental knowledge in these areas.

During its more than 50-year history, NASA's success in human space exploration has depended on the agency's ability to effectively address a wide range of biomedical, engineering, physical science, and related obstacles—an achievement made possible by NASA's strong and productive commitments to life and physical sciences research for human space exploration, and by its use of human space exploration infrastructures for scientific discovery.* This partnership of NASA with the research community reflects the original mandate from Congress in 1958 to promote science and technology, an endeavor that requires an active and vibrant research program. The committee acknowledges the many achievements of NASA, which are all the more remarkable given budgetary challenges and changing directions within the agency. In the past decade, however, a consequence of those challenges has been a life and physical sciences research program that was dramatically reduced in both scale and scope, with the result that the agency is poorly positioned to take full advantage of the scientific opportunities offered by the now fully equipped and staffed ISS laboratory, or to effectively pursue the scientific research needed to support the development of advanced human exploration capabilities.

Although its review has left it deeply concerned about the current state of NASA's life and physical sciences research, the Committee for the Decadal Survey on Biological and Physical Sciences in Space is nevertheless

*These programs' accomplishments are described in several National Research Council (NRC) reports—see for example, *Assessment of Directions in Microgravity and Physical Sciences Research at NASA* (The National Academies Press, Washington, D.C., 2003).

convinced that a focused science and engineering program can achieve successes that will bring the space community, the U.S. public, and policymakers to an understanding that we are ready for the next significant phase of human space exploration. The goal of this report is to lay out steps whereby NASA can reinvigorate its partnership with the life and physical sciences research community and develop a forward-looking portfolio of research that will provide the basis for recapturing the excitement and value of human spaceflight—thereby enabling the U.S. space program to deliver on new exploration initiatives that serve the nation, excite the public, and place the United States again at the forefront of space exploration for the global good. This report examines the fundamental science and technology that underpin developments whose payoffs for human exploration programs will be substantial, as the following examples illustrate:

- An effective countermeasures program to attenuate the adverse effects of the space environment on the health and performance capabilities of astronauts, a development that will make it possible to conduct prolonged human space exploration missions.
- A deeper understanding of the mechanistic role of gravity in the regulation of biological systems (e.g., mechanisms by which microgravity triggers the loss of bone mass or cardiovascular function)—understanding that will provide insights for strategies to optimize biological function during spaceflight as well as on Earth (e.g., slowing the loss of bone or cardiovascular function with aging).
- Game changers, such as architecture-altering systems involving on-orbit depots for cryogenic rocket fuels, an example of a revolutionary advance possible only with the scientific understanding required to make this Apollo-era notion a reality. As an example, for some lunar missions such a depot could produce major cost savings by enabling use of an Ares I type launch system rather than a much larger Ares V type system.
- The critical ability to collect or produce large amounts of water from a source such as the Moon or Mars, which requires a scientific understanding of how to retrieve and refine water-bearing materials from extremely cold, rugged regions under partial-gravity conditions. Once cost-effective production is available, water can be transported to either surface bases or orbit for use in the many exploration functions that require it. Major cost savings will result from using that water in a photovoltaic-powered electrolysis and cryogenics plant to produce liquid oxygen and hydrogen for propulsion.
- Advances stemming from research on fire retardants, fire suppression, fire sensors, and combustion in microgravity that provide the basis for a comprehensive fire-safety system, greatly reducing the likelihood of a catastrophic event.
- Regenerative fuel cells that can provide lunar surface power for the long eclipse period (14 days) at high rates (e.g., greater than tens of kilowatts). Research on low-mass tankage, thermal management, and fluid handling in low gravity is on track to achieve regenerative fuel cells with specific energy greater than two times that of advanced batteries.

In keeping with its charge, the committee developed recommendations for research fitting in either one or both of these two broad categories:

1. *Research that* ***enables*** *space exploration:* scientific research in the life and physical sciences that is needed to develop advanced exploration technologies and processes, particularly those that are profoundly affected by operation in a space environment.
2. *Research* ***enabled by*** *access to space:* scientific research in the life and physical sciences that takes advantage of unique aspects of the space environment to significantly advance fundamental scientific understanding.

The key research challenges, and the steps needed to craft a program of research capable of facilitating the progress of human exploration in space, are highlighted below and described in more detail in the body of the report. In the committee's view, these are steps that NASA will have to take in order to recapture a vision of space exploration that is achievable and that has inspired the country, and humanity, since the founding of NASA.

ESTABLISHING A SPACE LIFE AND PHYSICAL SCIENCES RESEARCH PROGRAM: PROGRAMMATIC ISSUES

Research in the complex environment of space requires a strong, flexible, and supportive programmatic structure. Also essential to a vibrant and ultimately successful life and physical sciences space research program is a partnership between NASA and the scientific community at large. The present program, however, has contracted to below critical mass and is perceived from outside NASA as lacking the stature within the agency and the commitment of resources to attract researchers or to accomplish real advances. For this program to effectively promote research to meet the national space exploration agenda, a number of issues will have to be addressed.

Administrative Oversight of Life and Physical Sciences Research

Currently, life and physical science endeavors have no clear institutional home at NASA. In the context of a programmatic home for an integrated research agenda, program leadership and execution are likely to be productive only if aggregated under a single management structure and housed in a NASA directorate or key organization that understands both the value of science and its potential application in future exploration missions. The committee concluded that:

- *Leadership with both true scientific gravitas and a sufficiently high level in the overall organizational structure at NASA is needed to ensure that there will be a "voice at the table" when the agency engages in difficult deliberations about prioritizing resources and engaging in new activities.*
- *The successful renewal of a life and physical sciences research program will depend on strong leadership with a unique authority over a dedicated and enduring research funding stream.*
- *It is important that the positioning of leadership within the agency allows the conduct of the necessary research programs as well as interactions, integration, and influence within the mission-planning elements that develop new exploration options.*

Elevating the Priority of Life and Physical Sciences Research in Space Exploration

It is of paramount importance that the life and physical sciences research portfolio supported by NASA, both extramurally and intramurally, receives appropriate attention within the agency and that its organizational structure is optimally designed to meet NASA's needs. The committee concluded that:

- *The success of future space exploration depends on life and physical sciences research being central to NASA's exploration mission and being embraced throughout the agency as an essential translational step in the execution of space exploration missions.*
- *A successful life and physical sciences program will depend on research being an integral component of spaceflight operations and on astronauts' participation in these endeavors being viewed as a component of each mission.*
- *The collection and analysis of a broad array of physiological and psychological data from astronauts before, during, and after a mission are necessary for advancing knowledge of the effects of the space environment on human health and for improving the safety of human space exploration. If there are legal concerns about implementing this approach, they could be addressed by the Department of Health and Human Services Secretary's Advisory Committee on Human Research Protections.*

Establishing a Stable and Sufficient Funding Base

A renewed funding base for fundamental and applied life and physical sciences research is essential for attracting the scientific community needed to meet the prioritized research objectives laid out in this report. Researchers must have a reasonable level of confidence in the sustainability of research funding if they are expected to focus their laboratories, staff, and students on research issues relevant to space exploration. The committee concluded that:

- *In accord with elevating the priority of life and physical sciences research, it is important that the budget to support research be sufficient, sustained, and appropriately balanced between intramural and extramural activities. As a general conclusion regarding the allocation of funds, an extramural budget should support an extramural research program sufficiently robust to ensure a stable community of scientists and engineers who are prepared to lead future space exploration research and train the next generation of scientists and engineers.*
- *Research productivity and efficiency will be enhanced if the historical collaborations of NASA with other sponsoring agencies, such as the National Institutes of Health, are sustained, strengthened, and expanded to include other agencies.*

Improving the Process for Solicitation and Review of High-Quality Research

Familiarity with, and the predictability of, the research solicitation process are critical to enabling researchers to plan and conduct activities in their laboratories that enable them to prepare high-quality research proposals. Regularity in frequency of solicitations, ideally multiple solicitations per year, would help to ensure that the community of investigators remains focused on life and physical science research areas relevant to the agency, thereby creating a sustainable research network. The committee concluded that:

- *Regularly issued solicitations for NASA-sponsored life and physical sciences research are necessary to attract investigators to research that enables or is enabled by space exploration. Effective solicitations should include broad research announcements to encourage a wide array of highly innovative applications, targeted research announcements to ensure that high-priority mission-oriented goals are met, and team research announcements that specifically foster multidisciplinary translational research.*
- *The legitimacy of NASA's peer-review systems for extramural and intramural research hinges on the assurance that the review process, including the actions taken by NASA as a result of review recommendations, is transparent and incorporates a clear rationale for prioritizing intramural and extramural investigations.*
- *The quality of NASA-supported research and its interactions with the scientific community would be enhanced by the assembly of a research advisory committee, composed of 10 to 15 independent life and physical scientists, to oversee and endorse the process by which intramural and extramural research projects are selected for support after peer review of their scientific merit. Such a committee would be charged with advising and making recommendations to the leadership of the life and physical sciences program on matters relating to research activities.*

Rejuvenating a Strong Pipeline of Intellectual Capital Through Training and Mentoring Programs

A critical number of investigators is required to sustain a healthy and productive scientific community. A strong pipeline of intellectual capital can be developed by modeling a training and mentoring program on other successful programs in the life and physical sciences. Building a program in life and physical sciences would benefit from ensuring that an adequate number of flight- and ground-based investigators are participating in research that will enable future space exploration. The committee concluded that:

- *Educational programs and training opportunities effectively expand the pool of graduate students, scientists, and engineers who will be prepared to improve the translational application of fundamental and applied life and physical sciences research to space exploration needs.*

Linking Science to Needed Mission Capabilities Through Multidisciplinary Translational Programs

Complex systems problems of the type that human exploration missions will increasingly encounter will need to be solved with integrated teams that are likely to include scientists from a number of disciplines, as well as engineers, mission analysts, and technology developers. The interplay between and among the life and physical sciences and engineering, along with a strong focus on cost-effectiveness, will require multidisciplinary approaches. Multidisciplinary translational programs can link the science to the gaps in mission capabilities through planned and enabled data collection mechanisms. The committee concluded that:

- *A long-term strategic plan to maximize team research opportunities and initiatives would accelerate the trajectory of research discoveries and improve the efficiency of translating those discoveries to solutions for the complex problems associated with space exploration.*
- *Improved central information networks would facilitate data sharing with and analysis by the life and physical science communities and would enhance the science results derived from flight opportunities.*

ESTABLISHING A LIFE AND PHYSICAL SCIENCES RESEARCH PROGRAM: AN INTEGRATED MICROGRAVITY RESEARCH PORTFOLIO

Areas of Highest-Priority Research

NASA has a strong and successful track record in human spaceflight made possible by a backbone of science and engineering accomplishments. Decisions regarding future space exploration, however, will require the generation and use of new knowledge in the life and physical sciences for successful implementation of any options chosen. Chapters 4 through 10 in this report identify and prioritize research questions important both to conducting successful space exploration and to increasing the fundamental understanding of physics and biology that is enabled by experimentation in the space environment. These two interconnected concepts—that science is enabled by access to space and that science enables future exploration missions—testify to the powerful complementarity of science and the human spaceflight endeavor. For example, the research recommended in this report addresses unanswered questions related to the health and welfare of humans undertaking extended space missions, to technologies needed to support such missions, and to logistical issues with potential impacts on the health of space travelers, such as ensuring adequate nutrition, protection against exposure to radiation, suitable thermoregulation, appropriate immune function, and attention to stress and behavioral factors. At the same time, progress in answering such questions will find broader applications as well.

It is not possible in this brief summary to describe or even adequately summarize the highest-priority research recommended by the committee. However, the recommendations selected (from a much larger body of discipline suggestions and recommendations) as having the highest overall priority for the coming decade are listed briefly as broad topics below. The committee considered these recommendations to be the minimal set called for in its charge to develop an integrated portfolio of research enabling and enabled by access to space and thus did not attempt to further prioritize among them. In addition, it recognized that further prioritization among these disparate topic areas will be possible only in the context of specific policy directions to be set by NASA and the nation. Nevertheless, the committee has provided tools and metrics that will allow NASA to carry out further prioritization (as summarized below in the section "Research Portfolio Implementation").

The recommended research portfolio is divided into the five disciplines areas and two integrative translational areas represented by the study panels that the committee directed. The extensive details (such as research timeframes and categorizations as enabling, enabled-by, or both) of the research recommended as having the highest priority are presented in Chapters 4 through 10 of the report, and much of this information is summarized in the research portfolio discussion in Chapter 13.

Plant and Microbial Biology

Plants and microbes evolved at Earth's gravity (1 *g*), and spaceflight represents a completely novel environment for these organisms. Understanding how they respond to these conditions holds great potential for advancing knowledge of how life operates on Earth. In addition, plants are important candidates for components of a biologically based life support system for prolonged spaceflight missions, and microbes play complex and essential roles in both positive and negative aspects of human health, in the potential for degradation of the crew environment through fouling of equipment, and in bioprocessing of the wastes of habitation in long-duration missions. The highest-priority research, focusing on these basic and applied aspects of plant and microbial biology, includes:

- Multigenerational studies of International Space Station microbial population dynamics;
- Plant and microbial growth and physiological responses; and
- Roles of microbial and plant systems in long-term life support systems.

Behavior and Mental Health

The unusual environmental, psychological, and social conditions of spaceflight missions limit and define the range of crew activities and trigger mental and behavioral adaptations. The adaptation processes include responses that result in variations in astronauts' mental and physical health, and strongly stress and affect crew performance, productivity, and well-being. It is important to develop new methods, and to improve current methods, for minimizing psychiatric and sociopsychological costs inherent in spaceflight missions, and to better understand issues related to the selection, training, and in-flight and post-flight support of astronaut crews. The highest-priority research includes:

- Mission-relevant performance measures;
- Long-duration mission simulations;
- Role of genetic, physiological, and psychological factors in resilience to stressors; and
- Team performance factors in isolated autonomous environments.

Animal and Human Biology

Human physiology is altered in both dramatic and subtle ways in the spaceflight environment. Many of these changes profoundly limit the ability of humans to explore space, yet also shed light on fundamental biological mechanisms of medical and scientific interest on Earth. The highest-priority research, focusing on both basic mechanisms and development of countermeasures, includes:

- Studies of bone preservation and bone-loss reversibility factors and countermeasures, including pharmaceutical therapies;
- In-flight animal studies of bone loss and pharmaceutical countermeasures;
- Mechanisms regulating skeletal muscle protein balance and turnover;
- Prototype exercise countermeasures for single and multiple systems;
- Patterns of muscle retrainment following spaceflight;
- Changes in vascular/interstitial pressures during long-duration space missions;
- Effects of prolonged reduced gravity on organism performance, capacity mechanisms, and orthostatic intolerance;
- Screening strategies for subclinical coronary heart disease;
- Aerosol deposition in the lungs of humans and animals in reduced gravity;
- T cell activation and mechanisms of immune system changes during spaceflight;
- Animal studies incorporating immunization challenges in space; and
- Studies of multigenerational functional and structural changes in rodents in space.

Crosscutting Issues for Humans in the Space Environment

Translating knowledge from laboratory discoveries to spaceflight conditions is a two-fold task involving horizontal integration (multidisciplinary and transdisciplinary) and vertical translation (interaction among basic, preclinical, and clinical scientists to translate fundamental discoveries into improvements in the health and well-being of crew members during and after their missions). To address the cumulative effect of a range of physiological and behavioral changes, an integrated research approach is warranted. The highest-priority crosscutting research issues include:

- Integrative, multisystem mechanisms of post-landing orthostatic intolerance;
- Countermeasure testing of artificial gravity;
- Decompression effects;
- Food, nutrition, and energy balance in astronauts;
- Continued studies of short- and long-term radiation effects in astronauts and animals;
- Cell studies of radiation toxicity endpoints;
- Gender differences in physiological effects of spaceflight; and
- Biophysical principles of thermal balance.

Fundamental Physical Sciences in Space

The fundamental physical sciences research at NASA has two overarching quests: (1) to discover and explore the laws governing matter, space, and time and (2) to discover and understand the organizing principles of complex systems from which structure and dynamics emerge. Space offers unique conditions in which to address important questions about the fundamental laws of nature, and it allows sensitivity in measurements beyond that of ground-based experiments in many areas. Research areas of highest priority are the following:

- Study of complex fluids and soft matter in the microgravity laboratory;
- Precision measurements of the fundamental forces and symmetries;
- Physics and applications of quantum gases (gases at very low temperatures where quantum effects dominate); and
- Behavior of matter near critical phase transition.

Applied Physical Sciences

Applied physical sciences research, especially in fluid physics, combustion, and materials science, is needed to address design challenges for many key exploration technologies. This research will enable new exploration capabilities and yield new insights into a broad range of physical phenomena in space and on Earth, particularly with regard to improved power generation, propulsion, life support, and safety. Applied physical sciences research topics of particular interest are as follows:

- Reduced-gravity multiphase flows, cryogenics, and heat transfer database development and modeling;
- Interfacial flows and phenomena in exploration systems;
- Dynamic granular material behavior and subsurface geotechnics;
- Strategies and methods for dust mitigation;
- Complex fluid physics in a reduced-gravity environment;
- Fire safety research to improve screening of materials in terms of flammability and fire suppression;
- Combustion processes and modeling;
- Materials synthesis and processing to control microstructures and properties;
- Advanced materials design and development for exploration; and
- Research on processes for in situ resource utilization.

Translation to Space Exploration Systems

The translation of research to space exploration systems includes identification of the technologies that enable exploration missions to the Moon, Mars, and elsewhere, as well as the research in life and physical sciences that is needed to develop these enabling technologies, processes, and capabilities. The highest-priority research areas to support objectives and operational systems in space exploration include:

- Two-phase flow and thermal management;
- Cryogenic fluid management;
- Mobility, rovers, and robotic systems;
- Dust mitigation systems;
- Radiation protection systems;
- Closed-loop life support systems;
- Thermoregulation technologies;
- Fire safety: materials standards and particle detectors;
- Fire suppression and post-fire strategies;
- Regenerative fuel cells;
- Energy conversion technologies;
- Fission surface power;
- Ascent and descent propulsion technologies;
- Space nuclear propulsion;
- Lunar water and oxygen extraction systems; and
- Planning for surface operations, including in situ resource utilization and surface habitats.

For each of the high-priority research areas identified above, the committee classified the research recommendations as enabling for future space exploration options, enabled by the environment of space that exploration missions will encounter, or both.

Research Portfolio Implementation

While the committee believes that any healthy, integrated program of life and physical sciences research will give consideration to the full set of recommended research areas discussed in this report—and will certainly incorporate the recommendations identified as having the highest priority by the committee and its panels—it fully recognizes that further prioritization and decisions on the relative timing of research support in various areas will be determined by future policy decisions. For example, and only as an illustration, a policy decision to send humans to Mars within the next few decades would elevate the priority of enabling research on dust mitigation systems, whereas a policy decision to focus primarily on advancing fundamental knowledge through the use of space would elevate the priority of critical phase transition studies. The committee therefore provided for future flexibility in the implementation of its recommended portfolio by mapping all of the high-priority research areas against the metrics used to select them. These eight overarching metrics, listed below with clarifying criteria (see also Table 13.3) added in parentheses, can be used as a basis for policy-related ordering of an integrated research portfolio. Examples of how this might be done are provided in the report.

- The extent to which the results of the research will reduce uncertainty about both the benefits and the risks of space exploration (*Positive Impact on Exploration Efforts, Improved Access to Data or to Samples, Risk Reduction*)
- The extent to which the results of the research will reduce the costs of space exploration (*Potential to Enhance Mission Options or to Reduce Mission Costs*)
- The extent to which the results of the research may lead to entirely new options for exploration missions (*Positive Impact on Exploration Efforts, Improved Access to Data or to Samples*)
- The extent to which the results of the research will fully or partially answer grand science challenges that the space environment provides a unique means to address (*Relative Impact Within Research Field*)

- The extent to which the results of the research are uniquely needed by NASA, as opposed to any other agencies (*Needs Unique to NASA Exploration Programs*)
- The extent to which the results of the research can be synergistic with other agencies' needs (*Research Programs That Could Be Dual-Use*)
- The extent to which the research must use the space environment to achieve useful knowledge (*Research Value of Using Reduced-Gravity Environment*)
- The extent to which the results of the research could lead to either faster or better solutions to terrestrial problems or to terrestrial economic benefit (*Ability to Translate Results to Terrestrial Needs*)

Facilities, Platforms, and the International Space Station

Facility and platform requirements are identified for each of the various areas of research discussed in this report. Free-flyers, suborbital spaceflights, parabolic aircraft, and drop towers are all important platforms, each offering unique advantages that might make them the optimal choice for certain experiments. Ground-based laboratory research is critically important in preparing most investigations for eventual flight, and there are some questions that can be addressed primarily through ground research. Eventually, access to lunar and planetary surfaces will make it possible to conduct critical studies in the partial-gravity regime and will enable test bed studies of systems that will have to operate in those environments. These facilities enable studies of the effects of various aspects of the space environment, including reduced gravity, increased radiation, vacuum and planetary atmospheres, and human isolation.

Typically, because of the cost and scarcity of the resource, spaceflight research is part of a continuum of efforts that extend from laboratories and analog environments on the ground, through other low-gravity platforms as needed and available, and eventually into extended-duration flight. Although research on the ISS is only one component of this endeavor, the capabilities provided by the ISS are vital to answering many of the most important research questions detailed in this report. The ISS provides a unique platform for research, and past NRC studies have noted the critical importance of its capabilities to support the goal of long-term human exploration in space.[†] These include the ability to perform experiments of extended duration, access to human subjects, the ability to continually revise experiment parameters based on previous results, the flexibility in experimental design provided by human operators, and the availability of sophisticated experimental facilities with significant power and data resources. The ISS is the only existing and available platform of its kind, and it is essential that its presence and dedication to research for the life and physical sciences be fully utilized in the decade ahead.

With the retirement of the space shuttle program in 2011, it will also be important for NASA to foster interactions with the commercial sector, particularly commercial flight providers, in a manner that addresses research needs, with attention to such issues as control of intellectual property, technology transfer, conflicts of interest, and data integrity.

Science Impact on Defining Space Exploration

Implicit in this report are integrative visions for the science advances necessary to underpin and enable revolutionary systems and bold exploration architectures for human space exploration. Impediments to revitalizing the U.S. space exploration agenda include costs, past inabilities to predict costs and schedule, and uncertainties about mission and crew risk. Research community leaders recognize their obligations to address those impediments. The starting point of much of space-related life sciences research is the reduction of risks to missions and crews. Thus, the recommended life sciences research portfolio centers on an integrated scientific pursuit to reduce the health hazards facing space explorers, while also advancing fundamental scientific discoveries. Similarly, revolutionary

---

[†]See, for example, National Research Council, *Review of NASA Plans for the International Space Station*, The National Academies Press, Washington, D.C., 2006.

and architecture-changing systems will be developed not simply by addressing technological barriers, but also by unlocking the unknowns of the fundamental physical behaviors and processes on which the development and operation of advanced space technologies will depend. This report is thus much more than a catalog of research recommendations; it specifies the scientific resources and tools to help in defining and developing with greater confidence the future of U.S. space exploration and scientific discovery.

1

Introduction

THE EXPLORATION IMPERATIVE

The Committee for the Decadal Survey on Biological and Physical Sciences in Space was tasked in this study to review the next 10 years' scientific challenges and opportunities for the U.S. space program as it advances its tradition of exploration and discovery in microgravity environments. The committee was instructed in its task to disregard considering specific budgetary recommendations and instead to focus on one enormously challenging question: What are the key scientific challenges that life and physical sciences research in space must address in the next 10 years? This review, or decadal survey, takes place in the context of the many remarkable achievements of the National Aeronautics and Space Administration (NASA) in exploring and studying space during the past 40 years. However, NASA's continued preeminence in this endeavor cannot be assumed, because other nations have both a keen interest, and the intent to make their own mark, in space. The European Space Agency, Russian Federal Space Agency, Japanese Aerospace Exploration Agency, Canadian Space Agency, and Chinese National Space Administration are among the many international agencies now working to make substantial progress in exploration. Furthermore, the continued leadership of the United States in space activities depends on clear goals for the research program and a consistent level of fiscal support for implementation of such programs.[1]

These are not new issues. The history of terrestrial exploration was also marked by a tension between nations' competition versus cooperation, by the financial costs of exploration versus pressing local economic needs, and by governmental decisions that either facilitated or hindered exploration. What is also evident from history is that exploration is inherently costly, dangerous, and unpredictable, requiring nations to make long-term fiscal and scientific investments in activities whose benefits are frequently unimagined or unrecognized for many years.

The responsibility for conducting a decadal survey and providing recommendations, priorities, and timetables was assumed by an NRC-appointed committee that subsequently appointed expert chairs and members to the following advisory panels:

- Animal and Human Biology Panel,
- Applied Physical Sciences Panel,
- Fundamental Physical Sciences Panel,
- Human Behavior and Mental Health Panel,
- Integrative and Translational Research for Human Systems Panel,

- Plant and Microbial Biology Panel, and
- Translation to Space Exploration Systems Panel.

After appointment, members received documents and hours of briefings from NASA, as well as from commercial and academic authorities on biological and physical science matters, pertaining to previous utilization of National Academies reports; NASA's research facilities, capabilities, procedures, and needs for exploration; lunar exploration and habitation; the International Space Station (ISS) as a platform for physical and biological research; the potential for and feasibility of commercial platforms; and research findings from space which, when combined with the extensive literature on microgravity and partial-gravity research, became the basis for recommendations, priorities, and timetables.

The report is divided into 13 chapters that summarize the deliberations of the committee and input of its seven panels of experts. Information, perspectives, and advice were obtained from the general public and experts in the field through town halls at professional society meetings and from solicited white papers from informed and concerned scientists. Various representatives of past and current NASA programs, experts from a range of disciplines, and speakers from private companies that are increasingly involved in space exploration all provided briefings to the panels and the committee.

Before presenting the findings of this decadal survey, the committee considered lessons learned from humankind's long experience with exploration and with the challenges of progressively broadening the frontiers of the known world. Throughout human history, exploration has driven some of our most inspiring achievements and profound discoveries. By discovering that the stars at night were distant points in a three-dimensional space, humans realized that Earth is not flat and that it is not the center of the universe. We gained the courage to travel over great distances and discovered new lands, new materials, and new resources. Moreover, the process of exploration resulted in new ways of thinking about the world and ourselves. For example, the quest to explore exacerbated the risk of incurring serious diseases, such as scurvy, which led to new ways of understanding health and illness (e.g., the importance of nutrition). Exploration inspired competition and government backing to develop new technologies, such as methods for accurate navigation (e.g., John Harrison's time pieces to derive accurate measures of longitude). The drive to explore continues today, and the key frontier of the future is space.

Multiple and varied platforms have and will continue to contribute to the acquisition of knowledge and to the enhancement of exploration. The ISS will play a pivotal role as a dedicated experimental laboratory for biological and physical research. Additionally, the promise that commercial transportation may one day provide operational platforms for microgravity and partial-gravity research should not be ignored.

Many believe that humanity is destined for a future in space. However, even with all of the resources of Earth, the enormous challenges of space voyages will not be overcome unless guided by significantly enhanced scientific research. In this context, it is interesting to review lessons learned from the history of Earth-based exploration.

Exploration has always been shaped by scientific and engineering research confronting new challenges. It took many years to learn how to deal with the destruction of wooden-hulled ships by Toredo worms; it took more years yet to learn that ships were jeopardized by galvanic interactions between the copper sheathing of hulls, which was effective against the worms, and ships' iron bolts. It is likely that similar unanticipated problems and solutions will be encountered in the hazardous environment of space, where materials are exposed to radiation, extreme temperatures, and microgravity.

In addition to the scientific understanding of materials and structures, exploration is shaped by the responses of crews to novel and severe environments. For instance, contested authority lines and crew frictions were the rule, not the exception, in early explorations, particularly when vessels sailed beyond the reach of their home governance. As humankind confronts unknown and extreme environments, it is only a matter of time before we encounter new disease pathogenesis. Thus, we can anticipate that new and unforeseen nutritional and radiation problems will pose major challenges for crew health, safety, and performance in space.

In the history of exploration, the importance of staging areas cannot be overestimated. Staging areas for navigation (e.g., the Canary Islands) were vital in acclimating sailors to voyages of increasing distances from "home." Such staging areas were also vital as trading depots where stockpiles for protracted voyages could be stored. Discovery of new food types and food preparation techniques, as well as new sources of water, were key to facilitating distant exploration. Similarly, experimentation with new navigational aids was facilitated in the relatively familiar environment of a staging station. With its large size and scientific focus, the ISS is likely to develop crucial experience as a type of proto-staging area.

It is extremely difficult to forecast the economic benefits of exploration. Indeed, the initial impetus for discovery of the New World was the quest for spice. While seeking spice, explorers discovered new continents. While looking for gold, they stumbled upon lands of unparalleled fertility. Way stations in space may offer similar uniquely valuable economic opportunities that are difficult to calculate in advance. The microgravity environment allows an entirely new way of crafting materials. A relatively safe environment close to Earth offers an opportunity (and revenues therefrom) for thousands of people to experience a microgravity environment. Furthermore, the selection criteria for space tourists will likely be less restrictive than those used for professional astronauts. As a result there will be opportunities for acquiring increasing amounts of medical insights in space.

Given that space voyages may eventually last for years and that an emergency return to Earth because of crew health or vessel material problems will be impractical, the safest way of venturing into space is to acquire experience with numerous missions of increasing complexity and duration. Such missions will reveal existing limitations as well as facilitate acquisition of new knowledge.

Throughout history, the exploration quest has demanded three things: that risks are balanced against the safety of the explorers, the tools are available to enable them to explore, and there are inspiring discoveries to be made. These remain key elements of a healthy space exploration program.

STUDY CONTEXT, CHALLENGES, AND ORGANIZATION

The obstacles faced by humanity in becoming a spacefaring species have been enormous. The United States has overcome many initial hurdles to deliver the lunar landings, the space shuttle, and, in partnership with other nations, the ISS. More than four decades after humans first traveled to the surface of the Moon, some 500 out of Earth's nearly 7 billion people have traveled into space. Yet, of those, only two dozen have traveled further than low Earth orbit and none have traveled beyond the orbit of the Moon. Despite tremendous advances in technology, human exploration of space remains a tremendously difficult, risky, and expensive endeavor.

Looking to the future, significant improvements are needed in spacecraft, life support systems, and space technologies to enhance and enable the human and robotic missions that NASA will conduct under the U.S. space exploration policy. The missions beyond low Earth orbit to and back from planetary bodies and beyond will involve a combination of environmental risk factors such as reduced gravity levels and increased exposure to radiation. Human explorers will require advanced life support systems and will be subjected to extended-duration confinement in close quarters. For extended-duration missions conducted at large distances from Earth and for which resupply will not be an option, technologies that are self-sustaining and/or adaptive will be necessary. These missions present multidisciplinary scientific and engineering challenges and opportunities for enabling research that are both fundamental and applied in nature. Meeting these scientific challenges will require an understanding of biological and physical processes, as well as their interaction, in the presence of partial-gravity and microgravity environments.

In the context of extraordinary advances in the life and physical sciences and with the realization that national policy decisions will continue to shift near-term exploration goals, the committee focused on surveying broadly and intensively the scientific issues necessary to advance knowledge in the next decade. Such a task is never easy; it relies on interpolation and extrapolation from existing knowledge sources and educated assumptions about new developments. The committee grappled with all of these issues as well as the thorny problem of how to organize the scientific efforts themselves procedurally so that they would flourish in the next decade.

ORGANIZATION OF THIS REPORT

As described in the Preface, the current report is the outcome of a highly integrated effort by the committee and its panels, drawing on extensive input from the scientific community. Nevertheless, important differences exist between the various chapters developed by the individual panels. Potential metrics were shared and discussed among the panels and the committee, and each panel selected, refined, and applied the specific metrics that were most appropriate for its own theme area. These metrics were later aggregated and synthesized into a common set of criteria against which all of the highest-priority recommendations were mapped. In organizing their chapter material, each panel worked from a common template, which they then revised to fit the demands of their subject material. Accordingly, each panel chapter (Chapters 4 through 10) contains a review of the current status of knowledge in the applicable disciplines and topics; an assessment of gaps that need to be addressed; recommendations to address these gaps; a selection of the recommendations that the authoring panel considers to be of the highest priority; and discussion of the time frame, facilities, and platforms needed to support the recommended research. Neither the order in which these chapters appear, nor their relative length, should be taken as an indication of the relative importance assigned to them. In general, topics with the greatest commonality were grouped together both in the report and within the chapters. Thus some chapters cover a much larger scientific "terrain" than others, some have a greater number of recent flight results to consider, and some with particularly technical topics require more detailed background discussion than others. Finally, each of these chapters was drafted by a different panel, and while all panels worked toward a similar format, their eventual approach to summarizing their material differed according to the unique characteristics and needs of the fields under review.

While the report recognizes the powerful advantages of the ISS in carrying out many types of critical research, the research was selected independently of the consideration of what platform should be used and whether that platform was available. Instead, platform needs were identified only after the research was selected.

The theme areas addressed by the seven panels were divided into five discipline areas and two integrative translational areas. In the latter, attention was given to the crosscutting science and technology issues for human survival in the space environment and to the translational research essential to the development of affordable space exploration systems. The concept of translational research has attracted considerable attention in recent years, in particular in the health sciences. Translational research seems important to almost everyone, and in the life sciences it has emerged as a priority of the National Institutes of Health. However, as pointed out by Woolf,[2] translational research means different things to different people. Over time, the definition of translational research has been subject to much debate. In the area of medicine, the concept of a translational continuum has emerged as stretching from basic science discovery through early and late stages of application development and then onto dissemination and adoption of those applications in clinical health practice. In other research areas, the concept of translational research has emerged as research linking fundamental to applied science. Many perceive it as research carrying a promise of being transformative rather than incremental. While defining translational research uniformly across life and physical sciences is challenging, the committee has, for the purpose of this report, defined it as "the effective translation of knowledge, mechanisms, and techniques generated in one scientific domain, such as basic or fundamental science, into new scientific tools or applications that advance research in another scientific domain, such as clinical or applied science."

As noted previously, final panel recommendations resulted from a joint process in which the advisory panels interacted with each other and with the survey committee. The resulting set of highest-priority research recommendations form the basis of the research portfolio laid out in Chapter 13. In addition, as per the committee's statement of task, budget considerations did not play a role in the selection of research, the setting of priorities, or the creation of an integrated portfolio. However, the committee was cognizant of the role that both budget and policy direction will play in implementing the recommended portfolio and has therefore provided guidance and examples in Chapter 13 for how NASA might approach a policy-related ordering of an integrated research portfolio.

In the process of discovery and analysis, certain themes arose repeatedly in discussions with the community and within the panels and committee. These themes related to the obstacles that would need to be overcome to enable the development of a successful research program. The results of those discussions and analysis, which are presented in Chapter 12, are considered by the committee and its panels to be at least equal in importance to the selection of research.

To provide context for the discussions in the report, the committee has also summarized, to the best of its ability, the available information on the research facilities that are, or are likely to become, available to the scientific community supporting NASA's life and physical sciences research. This information, along with a very general overview of NASA's program evolution in this area, is provided in the report's opening chapters.

REFERENCES

1. National Research Council. 2009. *America's Future in Space: Aligning the Civil Space Program with National Needs.* The National Academies Press, Washington, D.C.
2. Woolf, S.H. 2008. The meaning of translational research and why it matters. *Journal of the American Medical Association* 299:211-213.

2

Review of NASA's Program Evolution in the Life and Physical Sciences in Low-Gravity and Microgravity Environments

A brief synopsis of the evolution of the life and physical sciences program at NASA along with a sampling of its accomplishments is provided below. The committee's intention is not that this synopsis be comprehensive but rather that it set the stage for the chapters that follow by providing a general introduction and overview of the content, purpose, and history of NASA programs. This review is also intended to provide some indications of the potential for a vibrant life and physical sciences program at NASA and of the current status of this research endeavor within the agency.

ORIGIN OF THE RESEARCH PROGRAM

Since its establishment by Congress in 1958, NASA has recognized broad mandates to (1) extend the presence of humans in space while maintaining their health and safety and (2) provide the necessary platforms and resources to enable scientific inquiry on how the force of gravity affects physical phenomena and living organisms. It was quickly appreciated that the existing environment of space was sufficiently different from that of Earth that the design of spacecraft and space-ready technologies would be a significant challenge. Moreover, it was realized that the near-absence of gravity has a fundamentally unique effect on many biological and physical phenomena that cannot be investigated for any duration of time on Earth. Subsequently, it was understood that space provided opportunities for answering fundamental science questions whose impact would extend far beyond the mission needs of NASA, including results that could enable commercial applications.

One of the major challenges from the outset involved deciding where the organization and management of a microgravity life and physical sciences research mission would reside within the overall NASA administrative infrastructure, which was then being managed by scientists and engineers focused on rocket design and launch capability for space exploration. From the 1980s through 1992, the life and physical sciences were housed in the Office of Space Science and Applications (OSSA) alongside other sciences such as astrophysics and planetary exploration. Rather than maintaining the life and physical sciences research program as a partner to cognate science programs constituting NASA's research portfolio—e.g., astronomy, Earth science, astrophysics, and planetary science—the senior management at NASA came to view "microgravity sciences" as supportive of, and subordinate to, the agency's major space programs aimed at developing human spaceflight operations. Thus, the life and physical sciences research program was relocated within the infrastructure of spaceflight operations rather than being an equal partner with NASA's other science programs.

This decision generated the perception among many scientists outside the agency that, because microgravity life and physical sciences research had no representation at the highest administrative levels within NASA, research in this field was considered unimportant or peripheral to the agency. The acceptance of the life and physical sciences research program waned among many scientists, engineers, physicians, and clinicians working outside NASA. It is noteworthy that these early views of NASA microgravity sciences continue to the present.

EVOLUTION OF THE RESEARCH MISSION

By the 1970s, offices for life and microgravity sciences were established at NASA headquarters. In the life sciences, a director position was created and the research mission was formalized as three distinct programs: (1) gravitational biology—understanding the role of gravity in the development and evolution of life; (2) biomedical research—characterizing and removing the primary physiological and psychological obstacles to extending human spaceflight; and (3) operational medicine—developing medical and life support systems to enable human expansion beyond Earth and into the solar system. These three programs had support from specific NASA centers (viz., Johnson Space Center and Ames Research Center), and the three themes articulated by these initial programs continue to define the spaceflight life sciences research missions and focus. In comparison, the development of the microgravity physical sciences program was more diffuse, in part because of its evolution from specific technological research that was related to the development of spaceflight systems. Fluid physics was one of the earliest thrusts because of its relevance to flight systems such as those for propellant control. Broadly, the physical sciences research activities that emerged can be categorized as (1) fluid physics, (2) materials science (crystal growth, metallurgy, soft matter, etc.), (3) biotechnology science (electrophoresis, protein crystallography, etc.), (4) combustion science, and (5) fundamental physics.

The Apollo Program and Skylab

While orbital animal life sciences research began with the V2 rocket in 1948[1]—and the Mercury and Gemini missions can be said to have entailed research to answer some very fundamental questions about space environments and our ability to conduct science and exploration missions—it was not until the Apollo Flight Program (1968-1972) that dedicated microgravity life and physical sciences research began within NASA. In total, eleven crew missions were completed: two involved Earth-orbiting; two lunar orbiting; one lunar swing-by; and six Moon-landing missions. Significant information was obtained from these flights, which were conducted within the framework of the Operational Medicine Program. This program was focused primarily on documenting the physiological effects on crew astronauts during varying flight durations and ascertaining whether there were serious deleterious effects. In the physical sciences, early research focused on small "suitcase" experiments. Specifically, to experience microgravity, experiments could be performed only when the spacecraft was in free flight (no thrusters or engines activated) and within the extremely tight constraints on time and crew availability. Experiments on fluid flow, thermal transport, and electrophoresis of biological molecules were performed. In spite of these early milestones, minimal scientific data were obtained that were publishable in a peer-reviewed scientific journal. However, the Apollo microgravity program established a precedent for hypothesis-driven research in the microgravity sciences.

In the early 1970s, Skylab became the first U.S. space station. Four Skylab missions were conducted from May 1973 to February 1974, ranging in duration from 28 to 84 days. Much of the research in the physical sciences during those flights was targeted at the effects of low gravity on buoyancy-driven convective flow and on materials processing. The reduction in, or absence of, buoyancy in microgravity transforms combustion processes and is important to solidification and crystallization, among many processes. In the absence of buoyancy, many reactions are limited by diffusion, and not only do they change in dynamical character but the length scale of the process also grows and, overall, becomes simpler to understand. Specifically, combustion can become diffusion-limited, and lack of buoyancy transforms solidification of metals under welding conditions and most scenarios of crystal formation. Containerless processing experiments, which were expected to have potential for commercial application, were included but were not deemed successful. In parallel, a range of life sciences missions were also conducted on Skylab. These covered processes such as cellular development and plant growth, but a key component was to demonstrate that humans could live in space for relatively long durations while conducting a variety of tasks, including performance of extravehicular activities, and return to Earth without physiological deficits in performance capability. The outcomes of these missions revealed several critical findings, including the following: (1) there were clear-cut deficiencies in cardiovascular functions when exercise was performed by the astronauts at the same absolute intensity before and after the mission; (2) losses in muscle mass and strength as well as in bone density were documented; and (3) loss of movement fidelity, including post-flight balance instabilities, was observed (in the absence of vision abnormalities). It is now known that these alterations were hallmarks of prolonged exposure to the spaceflight environment. Even though the sample size was small, the observations were sufficiently robust to be reported in peer-reviewed scientific journals. Overall, the Skylab program laid the foundation for future microgravity studies of greater breadth and depth.

Dedicated Life Sciences Missions on the Shuttle

In 1978, NASA released a formal research announcement calling for research proposals for the laboratory facility of the space shuttle-based space transport system (STS). The unique feature of this announcement was that the research was to be rigidly peer-reviewed by two different panels: one focusing on scientific merit and the other examining the feasibility of conducting the study with the available facility and equipment infrastructure of the STS laboratory. A similar review model was used in the physical sciences. The first life sciences mission was set for the mid-1980s but was postponed until 1991 following the explosion of the space shuttle *Challenger,* after which shuttle missions were suspended until NASA deemed that it was safe to continue the shuttle program.

The access to space afforded by the STS missions provided for a broad portfolio of life sciences experimentation aimed at assessing the effects of microgravity and spaceflight on biological responses. Of particular note, however, are the three dedicated Space Life Sciences (SLS) missions that were flown in the decade of the 1990s: SLS-1 in 1991 lasted 9 days; SLS-2 in 1993 lasted 14 days; and Neurolab, a dedicated mission for the neurosciences in 1998, lasted 16 days. Within these three missions was a wide scope of experiments, ranging from plant and cell biology studies to complementary human and animal projects. The human studies were enhanced by NASA astronauts and payload scientists, who not only conducted the research as surrogate investigators but also served as the subjects for the composite human science package.

The research topics investigated across the three missions were broad, covering all the physiological systems discussed in this report. A unique feature of these missions was that laboratories for the animal studies were established at both the launch and the landing sites so that ground-based analyses could be conducted in close proximity to take-off and, especially, at landing to minimize physiological alterations occurring during the recovery period. In Neurolab, as with SLS-2, animal subjects were studied in detail during actual spaceflight. Some of the animal specimens were acquired during flight and then compared to animal samples obtained following landing. This was a major accomplishment because all animal studies prior to SLS-2 were performed during varying time intervals after landing, making it impossible to separate in the various experiments the effects of landing from the effects of the spaceflight environment itself.

An additional unique feature of the SLS and Neurolab missions was the synergy established within the international community of investigators and agencies. For example, in the Neurolab mission all of the ground-based research prior to flight was funded by several institutes within the National Institutes of Health, especially the National Institute of Neurological Disorders and Stroke. Investigators from the Japanese Aerospace Exploration Agency and the European Space Agency were also involved.

Although the life sciences program had its roots in issues of crew health and safety, fundamental biology also grew to be a substantive part of the program, particularly in the area of plant biology. For example, research on the effects of loss of convection on root zone hypoxia showed the impact of spaceflight on plant metabolism, and comprehensive gene expression studies revealed genome-wide effects of spaceflight on gene expression patterns.

In many ways, the dedicated flight program in space life sciences served as an important model of how flight-based research can be integrated across (1) project science disciplines, (2) national and international space research programs, and (3) national and international funding agencies. There have been repeated calls within the life sciences community to recapture the synergy that was present in the Neurolab mission.

Dedicated Physical Sciences Missions on the Shuttle

During the flight history of the STS, a range of physical sciences investigations were flown, largely in the context of Spacelab. Among these missions were U.S. Microgravity Laboratory (USML) missions USML-1 and USML-2, as well as Microgravity Science Lab (MSL-1) and several international collaborative missions. The topics of all these experiments mirrored the five physical sciences categories listed above. Among the fluid physics experiments were investigations of surface tension and thermocapillary-driven flow, which generated new insights into instabilities and oscillatory flow excitations. Multiphase fluid flow experiments and investigations of bubbles, droplets, and coalescence were also an active part of the program, as was work on complex fluids. In materials science, crystal growth was an active theme, with one goal being to grow large, homogeneous crystals containing an exceptionally low number of defects in alloys (e.g., ZnCdTe) as well as in organic compounds, most notably in protein crystals. Using the TEMPUS electromagnetic levitator, investigations of undercooled liquids led to the design of glass-forming metal alloys and metallic glasses that had significant commercial impact. Experiments also explored dendrite growth in the absence of convective heat transfer, research that involved materials science and fluid physics. Combustion was an active area driven by fundamental questions and the relevance to fire safety. Central to this research were investigations on flame propagation and extinction, combustion ignition and autoignition, smouldering, and droplet combustion that yielded significant results. Among these were elegant experiments on ball flames. During the Spacelab era, fundamental physics began to emerge as a significant thrust in microgravity research; four experiments were conducted on phase transitions and critical phenomena at both moderate and ultra-low temperatures. Their results included the most precise measurement to date of the superfluid critical point in liquid helium. Collectively, through the shuttle era, research in the physical sciences generated an impressive number of peer-reviewed publications, landmark measurements, and discoveries, all of which could be achieved only through access to space.[2]

An important element of the STS period of physical sciences research was the rise of discipline working groups. These groups formed advisory committees for NASA and provided an increasingly coherent interface between the scientific community, potential flight opportunities, and NASA's leadership. They organized conferences and workshops that brought NASA leaders and scientists together with university and industrial scientists and helped to develop a microgravity science community and heritage. Near the latter part of the STS period, the physical sciences research community had grown significantly to include Nobel laureates, members of the National Academies, and some of the best and brightest young scientists of the era. One objective of this report is to present a vision of a program that will recapture for NASA the strength and significance of the research portfolio during the STS period and the excellence of its participants.

CURRENT STATUS AND POTENTIAL OF THE LIFE AND PHYSICAL SCIENCES RESEARCH PROGRAMS

From the inception of the International Space Station (ISS) program, elaborate plans were made by NASA and its international partners, especially the Japanese Aerospace Exploration Agency and the European Space Agency, to outfit the ISS as a world-class research laboratory for undertaking cutting-edge research and to provide opportunities that would expand research in microgravity to periods longer than 6 months. This was to be a major leap for the United States, given that the longest missions on Spacelab were less than 20 days. One important goal was to establish a first-class animal facility on the ISS containing Advanced Animal Habitats capable of separately housing 6 rats or 8 to 10 mice that could be utilized for a variety of studies on the effects of long-duration microgravity, while using a centrifuge as a control modality to maintain homeostasis at different gravity loads up to the standard 1 *g*. Other planned ISS infrastructure goals included shared facilities for combustion and materials research, as well as a flexible low-temperature physics facility. Some of these facilities have been realized and have enabled important scientific milestones such as the achievement of plant growth from seed to seed on orbit. Over the ISS development period, the organization and prioritization of the life and physical sciences changed significantly. These changes included the establishment of the Office of Biological and Physical Research as NASA's fifth strategic enterprise and an increased focus on the Human Exploration and Development of Space program.

Unfortunately, over only a few years' time, nearly all non-medical research in this program was canceled owing to a lack of funds and budget reprioritizations focused on the Constellation program. This was a serious blow to the life and physical sciences program within NASA because, since the late 1990s, there had been little or no flight program of substance to expand on the knowledge base established throughout the prior decade. From 2003 to the present time, the budget for biological and fundamental microgravity sciences within NASA has been reduced by more than 90 percent from its prior level, with only modest protection via a congressional mandate, and little opportunity remains (as mandated to NASA) for conducting even ground-based research. This is true for U.S.-led fundamental and applied hypothesis-driven research initiatives, and the committee notes that the only human research being performed on the ISS is agency-sponsored mission operations research. Although the remaining components of the life and physical sciences program are providing important scientific information, as described in subsequent chapters of this report, the extramural research community in these fields of space science is not part of the equation. Potential approaches to addressing these and other organizational issues are discussed in Chapter 12.

REFERENCES

1. Grindeland, R.E., Ilyn, E.A., Holley, D.C., and Skidmore, M.G. 2005. International collaboration on Russian spacecraft and the case for free flyer biosatellites. Pp. 41-80 in *Experimentation with Animal Models in Space* (G. Sonnenfeld, ed.). Advances in Space Biology and Medicine, Volume 10. Elsevier, Amsterdam.
2. National Research Council. 2003. *Assessment of Directions in Microgravity and Physical Sciences Research at NASA.* The National Academies Press, Washington, D.C.

3

Conducting Microgravity Research: U.S. and International Facilities

This chapter is designed to provide detailed information on U.S. and international microgravity science and research facilities and accommodations, as well as their current status and availability. Like Chapter 2, it helps set the stage for the panel chapters that follow because it provides a reasonably complete and coherent picture of the research capabilities that are referred to frequently, but in a more fragmented manner, in the remainder of the report. This chapter covers major facilities for microgravity research. It is intended to help scientists and researchers understand microgravity research capabilities available nationally and internationally for specific areas of research. It includes hardware currently in development or proposed for future development and facilities that were canceled but considered high priority by the scientific community. The details provided in this chapter were obtained from various official international space agency sources describing hardware and capabilities. Units included in descriptions are retained as originally provided and therefore vary between standards.

Facilities and hardware designed to accommodate microgravity experiments can be divided into space- and ground-based, pressurized and nonpressurized, and automated and non-automated.

The International Space Station (ISS) is the sole space-based facility providing long-term laboratory modules for scientists worldwide to carry out pressurized and nonpressurized microgravity experiments. The ISS can accommodate a wide range of life and physical sciences research in its six dedicated laboratory modules and, in addition, provides external truss and exposed facility sites to accommodate external attached payloads for technology development, observational science, and other tests. There are also free-flyers* and satellites dedicated to physical and life sciences research, providing a platform for long-duration missions. However, free-flyers typically do not allow access to astronauts and are fully automated.

Ground-based facilities for microgravity physical sciences include parabolic flights, drop towers, and sounding rockets. In addition, space life science researchers require access to highly specialized ground facilities, for instance the NASA Space Radiation Laboratory (NSRL) at Brookhaven National Laboratory (BNL), and the National Aeronautics and Space Administration (NASA) bed rest facilities at Johnson Space Center (JSC).

This chapter is divided into four sections addressing the types of facilities involved in microgravity research and associated infrastructure: (1) facilities aboard the ISS, (2) global space transportation systems, (3) free-flyers, and (4) U.S. and international ground-based microgravity research facilities. While these descriptions of major, relevant facilities have been included in the report for the convenience of the reader, it should be kept in mind that

*Free-flyers have their own dedicated section titled "Free-Flyers" below in this chapter, following the "Global Space Transportation" section.

the selection of facilities and their capabilities continue to evolve. Readers are referred to NASA websites for the most current information on facilities and equipment for research.[†]

INTERNATIONAL SPACE STATION

The major intended purpose of the ISS is to provide an Earth-orbiting research facility that houses experiment payloads, distributes resource utilities, and supports permanent human habitation for conducting science and research experiments in a microgravity environment. It is expected to serve as a world-class orbiting national and international laboratory for conducting high-value scientific research and providing access to microgravity resources for major areas of science and technology development. The ISS sustains a habitable living and working environment in space for extended periods of time, and astronauts become not only the operators of experiments but also the subjects of space research. NASA has partnered with four other space agencies on the ISS Program: the Russian Federal Space Agency (Roscosmos), the Canadian Space Agency (CSA), the Japanese Aerospace Exploration Agency (JAXA), and the European Space Agency (ESA).

Based on international barters, the United States owns 50 percent of all science/experiment racks located in the ESA and JAXA laboratory modules. The U.S. principal investigators working with NASA have access to these facilities but not necessarily to facilities owned and operated exclusively by ISS partners. Exclusively owned Russian facilities are a good example of the latter.

The ISS can support a variety of fundamental and applied research for the United States and international partners. It provides a unique, continuously operating environment in which to test countermeasures for long-term human space travel hazards, to develop and test technologies and engineering solutions in support of exploration, and to provide ongoing practical experience living and working in space. However, with the retirement of the space shuttle fleet, there will be no U.S. government space transportation system available to carry astronauts or payloads to the ISS.

The main advantages for conducting life and physical sciences research on the ISS are the access to the microgravity environment, long-duration time periods for research, and the extended flexibility the crew and principal investigators will have to perform the experiments onboard. The ISS also provides the opportunity to repeat or modify experiments in real time when necessary. In addition, the human aspect of crew participation, as both experimenter and subject, is invaluable in human life sciences research. Finally, the ISS provides an analog environment for simulating long-term deep-space human exploration, which allows NASA the opportunity to prepare humans, machines, and organizational and mission planning for the rigors of the next chapter in human space exploration.

Science payload operations on the ISS are supported by a wide variety of programs, equipment, and laboratory modules. The following are significant payload components:

- U.S. Laboratory,
- Facility-Class Payloads,
- Attached External Payloads,
- Centrifuge Accommodation Module (this program was canceled),[‡]
- Japanese Experiment Module,
- Columbus Orbital Facility, and
- Russian Research Modules.

The U.S. Laboratory, also known as Destiny, is the major U.S. contribution of scientific capacity to the ISS. It provides equipment for research and technology development and houses all the necessary systems to support a controlled-environment laboratory. Destiny provides a year-round, shirtsleeve atmosphere for research in areas such as life sciences, physical sciences, Earth science, and space science research. This pressurized module is

[†]See http://www.nasa.gov/mission_pages/station/research/facilities_by_name.html.

[‡]Utilization of this major component for onboard experiments was to commence in 2008, and was included because it has been cited in past NRC reports as a critical capability for space life sciences research.

designed to accommodate pressurized payloads and has a capacity for 24 rack locations, of which the International Standard Payload Racks (ISPRs) will occupy 13 (see Figure 3.1).

The ISPRs inside the ISS pressurized modules provide the only means of accommodating payload experiments. Each ISPR consists of an outer shell that holds interchangeable racks and that maintains a set of standard interfaces, a support structure, and equipment for housing research hardware. It can accommodate one or several experiments. Through the ISPRs, the ISS payload experiments will be able to interface with the following ISS resources contained on Destiny:

- Electrical power,
- Thermal control,
- Command/data/video,
- Vacuum exhaust/waste gas, and
- Gaseous nitrogen.

The EXPRESS (expedite the processing of experiments to space station) Rack System was developed to provide ISS accommodations and resources such as power, data, and cooling to small, subrack payloads and is housed within ISPRs aboard the space station. An EXPRESS rack accommodates payloads originally fitted to shuttle middeck lockers and International Subrack Interface Standard drawer payloads, allowing previously flown payloads to evolve to flight on the ISS.

Figure 3.2 depicts one possible standard EXPRESS rack configuration with eight middeck size drawers and two smaller, International Subrack Interface Standard drawers. The EXPRESS Rack System provides payload accommodations that allow quick, simple integration by using standardized hardware interfaces through which ISS resources can be distributed to the experiment, experiment commands can be given, and data and video can be transmitted.

Facility-Class Payloads

A facility-class payload is a long-term or permanent ISS-resident facility that provides services and accommodations for experiments in a particular science discipline. It includes general capabilities for main areas of science research. Facility-class payloads are located on ISPRs. These facilities are designed to allow easy change-out of experiments by the crew and to accommodate varied experiments in the same area of research. There are facility-class payload racks dedicated to life sciences, material sciences, and fluids and combustion.

- *Window Observational Research Facility.* The Window Observational Research Facility (WORF)[1] provides a crew workstation window in the U.S. Destiny module to support research-quality optical Earth observations. Some of these observations include "rare and transitory Earth surface and atmospheric phenomena."[2]
- *Life Sciences Glovebox.* This glovebox, which occupies one rack location, provides a sealed workspace[3] within which biological specimens and chemical agents can be handled while remaining isolated from the ISS cabin. Its design was based on experience with other gloveboxes flown on previous Spacelab[§] missions aboard the space shuttle.
- *Microgravity Science Glovebox.* The Microgravity Science Glovebox (MSG) also provides a sealed environment and is intended to enable scientists from multiple disciplines to participate actively in the assembly and operation of experiments in space with much the same degree of involvement that they have in their own research laboratories.[4] The MSG core work volume slides out of the rack to provide additional crew access capability from the side ports.[5]
- *Cold Storage: Minus Eighty Degree Laboratory Freezer for the ISS, Glacier, and Microgravity Experiment Research Locker/Incubator.* The Minus Eighty Degree Laboratory Freezer for the ISS (MELFI)[6] is designed

[§]Spacelab was a reusable laboratory module flown in the space shuttle's cargo bay and used for microgravity experiments that were operated and/or monitored by astronauts.

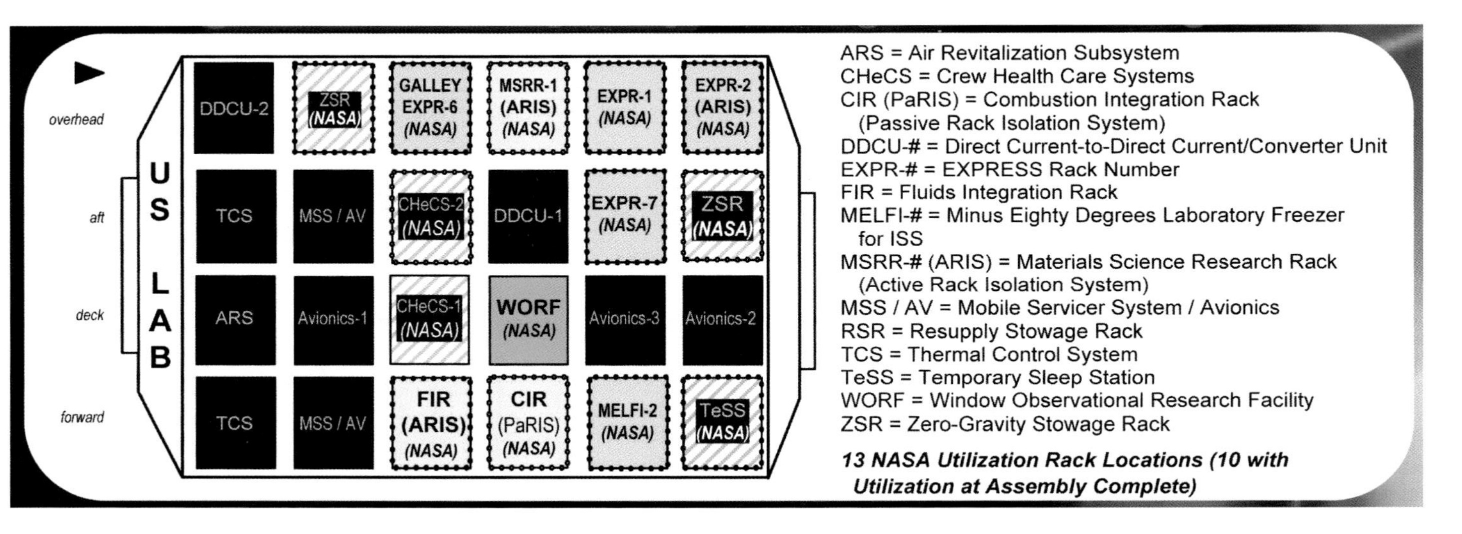

FIGURE 3.1 Destiny module research rack topology at assembly complete, Flight STS-130, Stage 19A. NOTE: 13 NASA Utilization Rack Locations (10 with Utilization at Assembly complete). ARS, Air Revitalization Subsystem; CHeCS, Crew Health Care Systems; CIR (PaRIS), Combustion Integration Rack (Passive Rack Isolation System); DDCU-#, Direct-Current-to-Direct Current/Converter Unit; EXPR-#, EXPRESS Rack Number; FIR, Fluids Integration Rack; MELFI-#, Minus Eighty Degrees Laboratory Freezer for ISS; MSRR-# (ARIS), Materials Science Research Rack (Active Rack Isolation System); MSS/AV, Mobile Servicer System/Avionics; RSR, Resupply Stowage Rack; TCS, Thermal Control System; TeSS, Temporary Sleep Station; WORF, Window Observational Research Facility; ZSR, Zero-Gravity Stowage Rack. SOURCE: J. Robinson, ISS Program Scientist, "International Space Station: Research Capabilities in Life and Physical Sciences, Early Utilization Results," presentation to the Committee for the Decadal Survey on Biological and Physical Sciences in Space, May 7, 2009.

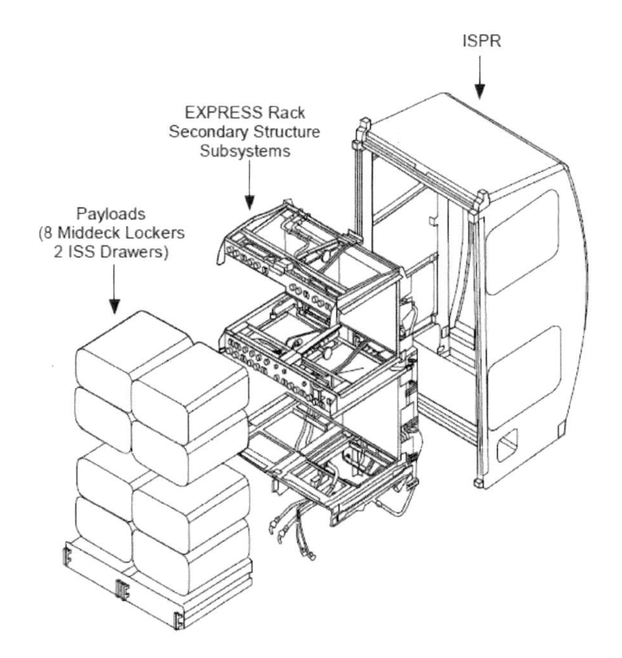

FIGURE 3.2 EXPRESS rack configuration. SOURCE: *Left:* NASA, *International Space Station Familiarization Training Manual*, ISS FAM C TM 21109, June 3, 2004. *Right:* NASA, *International Space Station Users Guide*, ISS User's Guide-Release 2.0, NASA Johnson Space Center, Houston Tex., 2000

to operate optimally at –80°C (–112°F) but has independent activation/deactivation and temperature control for each Dewar. Dewars are certified to operate at one of three set points: –95°C, –35°C, and +2°C. The MELFI can preserve samples that require all biochemical action to be stopped but do not require cryogenic temperatures. Cell culture media and bulk plant material, as well as blood, urine, and fecal samples, are among the types of items that can be stored in the MELFI.[7]

Although the MELFI is not certified for operation at temperatures other than its current three set points, the Glacier freezer aboard the ISS is capable of operating at any temperature between +4°C and –99°C, and can go as low as –130°C (which requires water cooling and more power available via the EXPRESS racks). There will be two GLACIER freezers aboard the ISS after shuttle retirement. Glacier provides powered cooling on ascent and descent, and following shuttle retirement can return to Earth with payloads via SpaceX's Dragon capsule.[8]

The Microgravity Experiment Research Locker/INcubator (MERLIN) is an on-orbit incubation and low-temperature storage facility for science experiments, as well as stowage transportation to and from orbit. It contains seven flight units owned by the University of Alabama. It can support samples from +4°C to +48.5°C in air cooling mode while stored in the space shuttle's middeck, and –20°C to +48.5°C in water cooling mode when housed in an EXPRESS rack.[9]

- *Attached External Payloads.* Attached payload sites for external experiments are located on the trusses of the ISS outside the pressurized volume. Attached payloads have access to station power, the command and data handling system, and video. The crew interfaces with attached payloads using robotics for installation and removal of the payload, with no nominal extravehicular activity operations anticipated.[10]
- *Centrifuge Accommodation Module (canceled).* The Centrifuge Accommodation Module was a research facility especially designed to study the effects of selected gravity levels (0.01 *g* to 2 *g*) on the structure and function of plants and animals, as well as to test potential countermeasures for the changes observed in microgravity.[11]

More detailed descriptions of the above facilities are given below in this chapter, in the section "Major ISS Facilities by Discipline."

U.S. Partner ISS Facilities and Modules

Meaning "dawn" in Russian, the Rassvet Mini-Research Module-1[12] is used for science research and cargo storage, as well as providing an additional docking port for Russian Soyuz and Progress vehicles at the ISS. This facility will serve as a home to biotechnology and biological science experiments, fluid physics experiments, and education research. The module contains a pressurized compartment with eight workstations, including a sealed glovebox to keep experiments separated from the cabin environment, two incubators to accommodate high- and low-temperature experiments, and a vibro-protective platform to protect payloads and experiments from onboard vibrations.[13]

The Poisk Mini-Research Module-2 is a multipurpose extension of the ISS connected to the Zvezda module. It provides an additional docking port for visiting Russian spacecraft and serves as an extra airlock for spacewalkers wearing Russian Orlan spacesuits.

ESA contributed the Columbus module and three Multi-Purpose Logistics Modules (MPLMs). One of the modules will become a permanent fixture of the ISS through an agreement with the Italian Space Agency, which built the modules. Columbus is ESA's single largest contribution to the space station, and at 23 ft long and 15 ft wide, it can accommodate 10 internal ISPRs of experiments and four external payload facilities (up to 370 kg each). Five of the ISPRs belong to NASA. Columbus houses a Biolab, Fluid Science Laboratory, European Physiology Module, and European Drawer Rack. It is equipped with two video cameras and a monitor audio system, as well as an external payload facility. The laboratory can support up to three crew members and can operate in temperatures between 16°C and 27°C and at air pressures between 959 and 1013 hPa.

The MPLMs and Columbus share a similar architecture and basic systems and have essentially the same dimensions. Approximately 21 ft long and 15 ft in diameter, the MPLMs can accommodate 16 payload racks, 5 of which can be furnished with power, data, and fluid to support a refrigerator-freezer.[14] Built by the Italian Space Agency but owned and operated by NASA, the MPLMs ferry pressurized material to the station and return waste, experiment racks, and other miscellaneous items (e.g., failed equipment) to Earth via the space shuttle. One of the MPLMs, Leonardo, was converted into the Permanent Multipurpose Module of the ISS, which was launched to the ISS on the STS-133 mission on February 24, 2011 (ISS Assembly Mission ULF5).

The Japanese Experiment Module Kibo, which is Japanese for "hope," is Japan's first human space facility and its primary contribution to the ISS. The Kibo laboratory consists of a pressurized module and an exposed facility, which have a combined focus on space medicine, biology, Earth observations, material production, biotechnology, and communications research. The laboratory consists of six components:[15]

- Two research facilities, the pressurized module and the exposed facility;
- A logistics module attached to each facility;
- Remote Manipulator System; and
- Inter-Orbit Communication System unit.

Kibo's pressurized module can hold up to 23 racks, including 10 experiment racks. Five of these racks belong to NASA. They provide access to power, communications, air conditioning, hardware cooling, water control, and experiment support functions.[16]

Kibo's exposed facility is located outside the pressurized module and is continuously exposed to the space environment. Experiments conducted on the exposed facility focus on Earth observation, communications, science, engineering, and materials science. The facility measures 18.4 ft wide, 16.4 ft high, and 13.1 ft long, and can support up to 10 experiment payloads at a time. Kibo uses Experiment Logistics Modules, which are both pressurized and exposed to the space environment and which serve as on-orbit storage areas housing materials for experiments, maintenance tools, and supplies. The pressurized section of each logistics module is a cylinder attached to the top of Kibo's pressurized module and can hold eight experiment racks. The exposed section of each logistics module is a pallet that can hold three experiment payloads, measuring 16.1 ft wide, 7.2 ft high, and 13.8 ft long.

Laboratory Support Equipment

Laboratory support equipment aboard the ISS includes automatic temperature controlled stowage, centrifuges, combustion chambers, biological culture apparatus, experiment preparation units, human restraint system, incubators (with a temperature range of −20°C to +48.5°C, interferometers, laptop computers, microscopes, multi-electrode electroencephalogram mapping module, optical and infrared cameras, optics benches, refrigerators and freezers that can store samples down to −185°C, solid-state power control module, spectrophotometers, temperature-controlled units, ultrasound, vacuum access system, and vibration isolation frames.

Major ISS Facilities by Discipline

Life Sciences

Human Research Facility 1

The Human Research Facility 1 (HRF-1)[17] provides investigators with a laboratory platform to study how long-duration spaceflight affects the human body. The HRF-1 includes a clinical ultrasound and a device for measuring mass.[18] Equipment is housed either on a rack based on the EXPRESS rack design or in stowage until needed. HRF-1 operates at ambient temperature, utilizing the ISS moderate temperature cooling loop, and has access ports for a nitrogen delivery system, vacuum system, and laptop. HRF-1's Workstation 2 is a computer system that provides operators with a platform to install and run software used in investigations. Components and equipment include the following: ultrasound drawer containing ultrasound/Doppler equipment, Workstation 2, two cooling stowage drawers that maintain a uniform temperature, Space Linear Acceleration Mass Measurement Device,[19] and the Continuous Blood Pressure Device.

Human Research Facility 2

The Human Research Facility 2 (HRF-2),[20] like HRF-1, addresses the effects of long-duration spaceflight on the human body. HRF-2 contains different equipment than HRF-1 contains, such as a refrigerated centrifuge and devices for measuring blood pressure and heart and lung functions. It is based on the EXPRESS Rack System and, like HRF-1, is able to provide power, data handling, cooling air and water, pressurized gas, and vacuum to experiments. Components and equipment include the following: refrigerated centrifuge, Workstation 2 (same as in HRF-1), two cooling stowage drawers, and the Pulmonary Function System. Included in the Pulmonary Function System are the Photoacoustic Analyzer Module, Pulmonary Function Module, Gas Analyzer System for Metabolic Analysis Physiology,[21] and the Gas Delivery System.

The 6-chamber refrigerated centrifuge can hold samples sized from 2 to 50 mL, while the 24-chamber refrigerated centrifuge can hold samples ranging from 0.5 to 2.2 mL. These refrigerated centrifuges can spin from 500 to 5,000 revolutions per minute for durations of 1 to 99 minutes, or they can be set to run continuously. The centrifuges were designed to be capable of maintaining a chamber temperature of +4°C, with selectable set points in increments of 1°C. The refrigeration capability of the centrifuges does not work currently, but they still function nominally as centrifuges.[22]

The Pulmonary Function System is a NASA-ESA collaboration that allows two different respiratory instruments to be created through the interconnection of components: the Mass spectrometer-based Analyzer System and the Photoacoustic-based Analyzer System. Combined with the Gas Analyzer System for Metabolic Physiology, Pulmonary Function Module, and Gas Delivery System, these instruments can "determine the concentration of respired gas components," take "cardiovascular and respiratory measures, including breath-by-breath lung capacity and cardiac output," and "measure the concentration of gases in inspired and expired air."[23]

Biological Experiment Laboratory

The Biological Experiment Laboratory (BioLab)[24] is an on-orbit biology laboratory located in the Columbus module that is used to study how microgravity and space radiation affect unicellular and multicellular organisms including bacteria, insects, protists (simple eukaryotic organisms), seeds, and cells. The facility is divided into two sections: one automated (that can also receive commands from ground operators) and one designed for crew interaction with experiments. Onboard instrumentation includes a large incubator, two centrifuges, a microscope, a spectrophotometer, a sample-handling mechanism (robotic arm), automatic and manual temperature-controlled stowage units, an experiment preparation unit, and a glovebox (called the BioGloveBox). BioLab's life support system can also regulate atmospheric content, including humidity. Biological samples for experimentation are transported from the ground in experiment containers or in small vials and manually inserted into the BioLab rack.

European Physiology Module

The European Physiology Module (EPM)[25,26,27] is designed to improve understanding of the effects of space flight on the human body. EPM research covers neurological, cardiovascular, and physiological studies and investigations of metabolic processes. It consists of three science modules: two active modules (Cardiolab and the Multi Electrodes Encephalogram Measurement Module) and one sample collection kit that enables collection of biological samples (blood, urine, and saliva). Up to three "active" human body experiments can be tested at one time; each subject is required to supply baseline samples and data before spaceflight for comparison to experimental data.

Muscle Atrophy Research And Exercise System

The Muscle Atrophy Research and Exercise System (MARES)[28,29,30] is used in studying human musculoskeletal, biomechanical, and neuromuscular physiology with respect to microgravity. This general-purpose instrument includes a human restraint system, a vibration isolation frame, a laptop, and a direct drive motor to provide the core mechanical stimulus. The system can be used in conjunction with other instrumentation, such as the Percutaneous Electrical Muscle Stimulator II and an electromyogram device. Although its primary function is research, it can also be used solely for exercise purposes. Components and equipment include an electromechanical box, human restraint system, linear adapter, vibration isolation frame, and laptop computer.

Saibo Experiment Rack

The Saibo Rack[31,32] is a Japanese experiment platform that includes subracks designed for living-cell biology experiments. It includes a glovebox (called the Clean Bench) and a cell biology experiment facility that includes a CO_2 gas incubator with controlled atmosphere and centrifuges that support operations from 0.1 *g* to 2.0 *g*. The Clean Bench includes a HEPA filter and high-performance optical microscope. Located in the Japanese Kibo module, Saibo is housed in an ISPR that can be divided in multiple payload segments.

Physical Sciences

Fluids and Combustion Facility

The Fluids and Combustion Facility consists of two racks, the Combustion Integrated Rack (CIR) and the Fluids Integrated Rack (FIR). The CIR[33–36] features a 100-liter combustion chamber and is used to perform combustion experiments in microgravity and consists of an optics bench, a combustion chamber, a fuel and oxidizer management system, environmental management systems, interfaces for science diagnostics and experiment-specific equipment, five cameras, gas supply package, exhaust vent system and gas chromatograph, and environmental control subsystems (including water and air thermal control, and fire detection and suppression). The CIR has been designed for use with the Passive Rack Isolation System, which connects the rack to the ISS structure and attenuates much of the U.S. Laboratory's vibration.[37] CIR experiments are conducted by remote control from the Telescience Support Center at NASA Glenn Research Center.[38]

The FIR[39,40] is a fluid physics research facility, complementary to the CIR, that is designed to support investigations in areas such as colloids, gels, bubbles, wetting and capillary action, and phase changes in microgravity such as boiling and cooling. It uses the Active Rack Isolation System and is capable of incorporating different modules that support widely varying types of experiments. Components and equipment include an optics bench and a light microscopy module, gas interface panel (providing ISS gaseous nitrogen and vacuum), Fluids Science

Avionics Package, and Mass Data Storage Unit. While most FIR experiment operations are performed from the ground by teams at NASA Glenn Research Center, the ISS crew installs and configures the necessary hardware.

Materials Science Research Rack-1

The Materials Science Research Rack-1 (MSRR-1)[41] is used for materials science experiments and research in microgravity. It provides instrumentation and thermal chambers for the study and mixing of materials, growing crystals, and quenching/solidifying metals or alloys. It occupies a single ISPR and is equipped with the Active Rack Isolation System. While MSRR-1 is a highly automated facility, it requires crew attention for maintenance and to install exchange modules. The first experiment module installed in the MSRR-1 is the Materials Science Laboratory (MSL) built by ESA. MSL takes up almost half of the MSRR-1 housing, is designed for materials processing and advanced diagnostics, and features multiple on-orbit, replaceable, module inserts also developed by ESA. The MSL can also be used for stowage and transportation back to Earth.[42,43,44] Components and equipment include a solid-state power control module, master controller, vacuum access system, and thermal and environmental control system. MSRR-1 currently hosts ESA's Material Science Laboratory.

Fluid Science Laboratory

The Fluid Science Laboratory (FSL)[45,46] was developed by ESA and designed to conduct fluid physics research in microgravity. It can be operated as a fully automatic or semi-automatic facility controlled either by ISS crew or in telescience mode from the ground. A drawer system allows for different configurations to accommodate a variety of experiments and for easy access for upgrades and maintenance of the system. FSL experiments must be installed in an FSL experiment container with a typical mass of 30 to 35 kg and dimensions of $400 \times 270 \times 280$ mm^3. Researchers may choose to activate the Canadian Space Agency-developed Microgravity Vibration Isolation Subsystem (via magnetic levitation) to isolate the experiment from space station "*g*-jitter" perturbations.[47] Components and equipment include optical and infrared cameras, multiple interferometers, illumination sources, two central experiment modules, a video management unit, storage, and a workbench.

Microgravity Science Glovebox

More than twice the size of gloveboxes flown aboard the space shuttle, MSG[48,49] (Figure 3.3) is an extendable and retractable 9-ft^3 sealed work area (at negative pressure relative to ISS cabin pressure) accessible to the crew through glove ports and to ground-based scientists through real-time data links. MSG is well suited for small and medium-sized investigations in many different kinds of microgravity research, such as fluid physics, combustion science, materials science, biotechnology, and fundamental physics. Components and equipment include three stowage drawers, a powered video drawer containing four video cameras, four recorders, two monitors (digital or 8 mm), standard and wide-angle lenses for each camera, and a laptop computer.

Mini-Research Module 1 Rassvet

As noted above, MRM1[50,51] is used for science research and cargo storage, as well as providing an additional docking port for Russian Soyuz and Progress vehicles at the ISS. Measuring 19.7 ft long and 7.7 ft in diameter, the facility has been planned to serve as a site for biotechnology and biological science experiments, fluid physics experiments, and education research. The module plans describe a pressurized compartment with eight workstations, a sealed glovebox to keep experiments separated from the cabin environment, two incubators to accommodate high- and low-temperature experiments, and a vibroprotective platform to protect payloads and experiments from onboard vibrations.

The eight workstations or arc-frames have extendable module-racks. The MRM1 glovebox can handle sterile or hazardous substances or bulk matter, as well as providing airlock, cleaning, and sterilization aids at 99.9 percent pure atmosphere. Its usable volume is 0.25 m^3. The High Temperature Universal Biotechnological Thermostat is designed to provide the temperature conditions required for handling biological objects (+2°C to +37°C). The Low Temperature Biotechnological Thermostat is similar, but it has an operating temperature of −20°C. The Universal Vibration Protection Platform protects scientific equipment up to 50 kg in mass from ambient vibration at frequencies from 0.4 to 250 Hz with a coefficient not less that 20 dB.

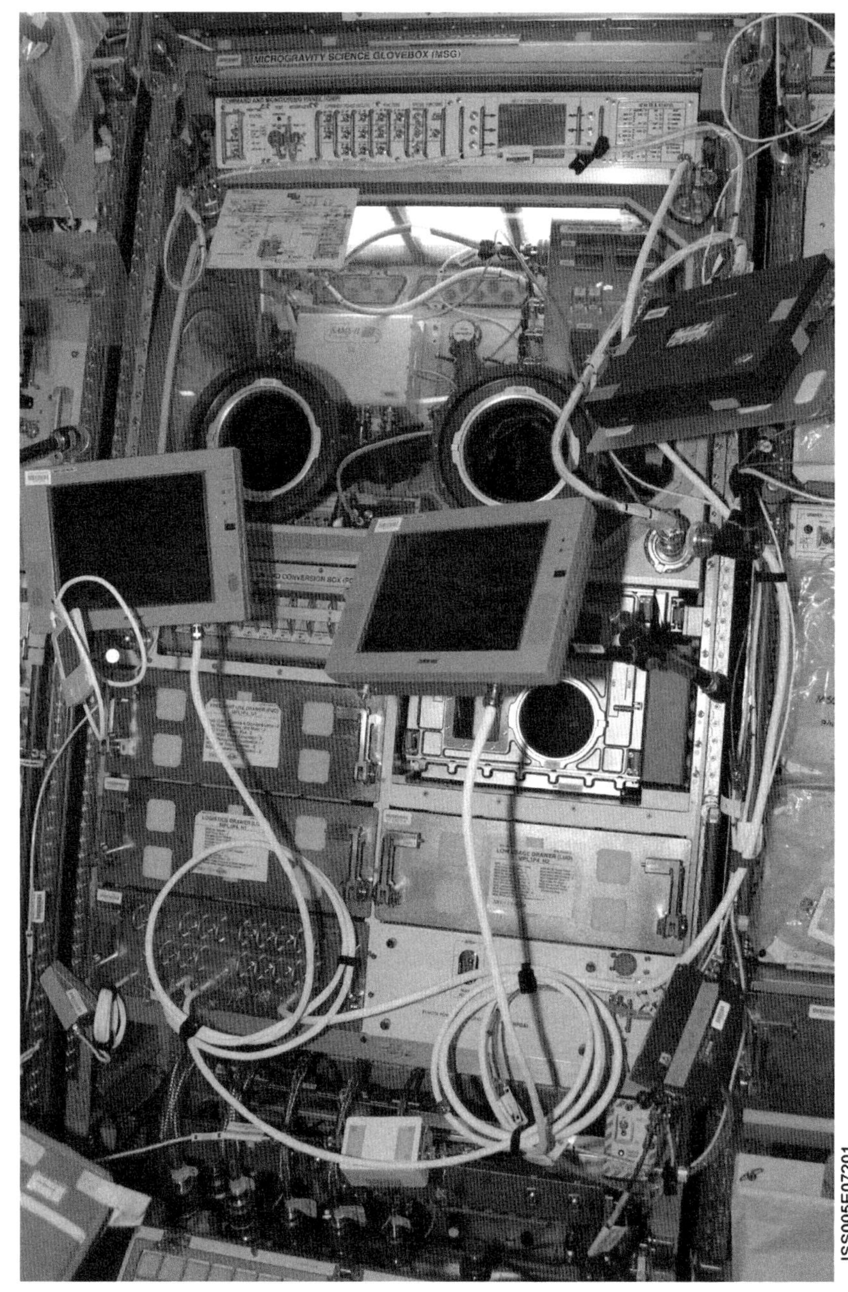

FIGURE 3.3 Microgravity Science Glovebox on the ISS. SOURCE: NASA Marshall Space Flight Center, available at http://msglovebox.msfc.nasa.gov/capabilities.html.

Ryutai Experiment Rack

Japanese for "fluid," the Ryutai Experiment Rack is housed in an ISPR in the Japanese Kibo module. Ryutai[52,53] provides standard interfaces to accommodate modular payloads in various research areas. It is designed to contain multiple JAXA subrack facilities. Components and equipment and subracks include a fluid physics experiment facility, protein crystallization research facility, solution crystallization observation facility, and image processing unit.

Space Sciences Payloads

EXPRESS Racks

The EXPRESS Rack System[54] was developed to provide the ISS with standardized accommodations for small, subrack payloads. Currently, eight EXPRESS racks are in use or scheduled for use on the ISS. The EXPRESS Rack System also includes transportation racks to transport payloads to and from the ISS and suitcase simulators to allow a payload developer to verify ISS power and data interfaces at the development site. Each EXPRESS rack can accommodate 10 "standard-sized small experiments" (the equivalent in up to 80 individual experiments), and approximately 50 percent of the EXPRESS payload housing capability is available for use as part of the U.S. National Laboratory aboard the space station. Of the eight EXPRESS racks, one rack, EXPRESS-6, is used in part for crew galley purposes. EXPRESS racks 7 and 8 were added specifically for the National Laboratory.[55] Subracks accommodated on an EXPRESS rack include the European Module Cultivation System installed on EXPRESS Rack 3A, Advanced Protein Crystal Facility on EXPRESS Rack 1, Biotechnology Specimen Temperature Controller on EXPRESS Rack 4, Commercial Generic Bioprocessing Apparatus, BioServe Culture Apparatus, General Laboratory Active Cryogenic ISS Experiment Refrigerator, Microgravity Experiment Research Locker Incubator, Biotechnology Temperature Refrigerator, ARCTIC Refrigerator/Freezer, Portable Astroculture Chamber, Advanced Space Experiment Processor, Advanced Biological Research System, and Common Refrigerator Incubator Module-Modified.

Columbus External Payload Facility

The Columbus External Payload Facility[56,57] (attached to the outside of ESA's Columbus module) has four powered external attachment points for scientific payloads: one on the nadir side, one on the zenith side, and two on the starboard sides of the Columbus module. Each attachment can provide 1.25 kW of power via two 120-Vdc redundant power feeds. Modules attached to these locations are interchangeable via extravehicular activity. The maximum mass the Columbus External Payload Facility can accommodate (including the adapter plate) is 290 kg, and payload dimensions should not exceed 86.4 × 116.8 × 124.5 cm, not including the adapter plate. The first set of experiments attached to the Columbus External Payload Facility consisted of the European Technology Exposure Facility[58] and the Sun Monitoring on the External Payload Facility of Columbus.[59] The Atomic Clock Ensemble in Space[60] will be delivered at a date yet to be determined.

European Drawer Rack

The European Drawer Rack[61,62] is a multidiscipline experiment rack housed in an ISPR with seven experiment modules, each of which has separate access to power, cooling, data communications, vacuum, nitrogen supply, and venting. Experiments are largely autonomous, but can also be controlled remotely via telescience, or in real-time by the crew through a dedicated laptop. The European Drawer Rack provides power, cooling, and communications equipment for each of its payloads and is also equipped to supply vacuum, vents, and nitrogen if necessary.

Japanese Experiment Module—Exposed Facility

This exposed facility on the Japanese Experiment Module[63] provides an external platform that can accommodate up to 10 experiments in the space environment. The first JAXA instruments installed in this facility are the Space Environment Data Acquisition Equipment-Attached Payload, and the Monitor of All-sky X-ray Image Payload. The first NASA instruments will be a hyperspectral imager and ionosphere detector.

Earth Sciences Payloads

WORF

WORF[64] is based on the ISPR and utilizes avionics and hardware adapted from the EXPRESS Rack System, providing 0.8 m^3 of payload volume. It is used in conjunction with and in support of the U.S. Laboratory Science Window for Earth observation science in the U.S. Destiny module. WORF's primary function is to control the external shutter of the window and as a mounting for imaging hardware. It is designed to minimize reflections and glare and hosts both crew-tended and automated activities. WORF can handle up to three payloads simultaneously, depending on available space and resources. WORF also includes the Agricultural Camera (AgCam), which provides images within 2 days directly to requesting farmers, ranchers, foresters, natural resource managers, and tribal officials for use in environmental and land management. It can take images in the visible and infrared light spectra.[65] WORF components include standard ISPR avionics, support systems, plus imaging systems.

Other ISS Payloads

European Transportation Carrier

The European Transportation Carrier (ETC)[66,67,68] is used for on-orbit stowage for ESA payload items and support for other European facilities. Its primary use is as a transport rack in conjunction with the MPLM to and from the ISS aboard the space shuttle or European Automated Transfer Vehicle, but it can also be used as a workbench for other experiments. The ETC can hold up to 410 kg (881 lb) of payload.

EXPRESS Logistics Carriers on External Trusses

The EXPRESS Logistics Carrier[69] is designed to support external payloads mounted to the starboard and port external trusses of the ISS with either Earth or space views. Five carriers are to be delivered by the time that the space shuttle is retired. Each can accommodate up to 12 fully integrated payloads, Orbital Replacement Units, or loads of outfitting cargo to the ISS in the space shuttle cargo bay. The truss segments can provide each carrier with two 3-kW, 120-Vdc electric feeds, and there are two types of data ports (High-Rate Data Link and Low-Rate Data Link) to connect the carriers to the ISS.

GLOBAL SPACE TRANSPORTATION SYSTEMS

Of the ISS partners, four provide launch services for crew or payload: the United States (NASA), Russia (Roscosmos), Europe (ESA), and Japan (JAXA). Only the United States, Russia, and China currently launch humans into space, and the United States has the greatest payload-to-orbit capability, but all four ISS partners are instrumental in supplying and maintaining the orbiting laboratory.

The debate surrounding the use of commercial launch services is unique to the United States, and the committee has indicated, to the best of its knowledge, commercial companies either involved in or trying to enter the launch services industry. Because this is a rapidly changing field, the committee focused on companies that either have Space Act agreements with NASA for hardware development specifically related to commercial crew and cargo or currently provide actual services.[70]

United States

Space Transportation System (NASA)

Commonly known as the "space shuttle," the U.S. Space Transportation System has been the primary means of U.S. support to the ISS. The system consists of two reusable solid rocket boosters, external liquid hydrogen and liquid oxygen fuel tanks, and a reusable orbiter that can carry large payloads of up to 24,400 kg to low Earth orbit (LEO), along with typically seven crew members.[71] It has been used to deliver entire modules and segments of the ISS and for logistics, resupply, and sample retrieval purposes, typically using the MPLM. The space shuttle is launched from Kennedy Space Center in Florida.

Currently, there is no U.S. government launch vehicle that can be counted on with certainty for U.S. crew and payload operations to the ISS beyond 2011. As of this writing, the space shuttle is slated for retirement in 2011, but the U.S. Congress may choose to extend shuttle operations beyond that date. The successor system to the space shuttle, the Ares I rocket and Orion Crew Vehicle, were in development, although recent policy decisions may end their development in favor of commercial industry-designed and industry-built launch vehicles and capsules or a combination of government and industry development.

Under the original plan devised during the administration of President George W. Bush in 2004, the Constellation Program would build two new rockets, a crewed rocket (Ares I) and a heavy-lift payload rocket (Ares V), to replace the space shuttles. Ares I was planned to go on line in 2015.

Taurus II-Cygnus (Orbital Sciences Corporation)—In Development

The Taurus II launch vehicle[72] is currently under development through NASA's Commercial Orbital Transportation Services program and has been awarded a Space Station Commercial Resupply Services contract for future resupply missions to the ISS.[73] It is slated for its first launch in 2011. The Taurus II has a payload capacity of up to 7,000 kg to LEO.[74] For ISS resupply missions, it will carry an uncrewed Cygnus spacecraft to deliver up to 2,700 kg of pressurized and unpressurized cargo.[75] Taurus II missions are initially planned to be launched from the NASA Wallops Flight Facility, but the rocket is compatible with many other U.S. launch facilities that could be used, depending on demand.[76]

Falcon 9-Dragon (Space Exploration Technologies Corporation)—In Development

Following its selection for the Commercial Orbital Transportation Services program, the Space Exploration Technologies Corporation (SpaceX) has been awarded a Space Station Commercial Resupply Services contract for ISS resupply using its Falcon 9 launch vehicle.[77,78] The Falcon 9 launch vehicle is partially reusable and is capable of lifting 10,450 kg to LEO when launched from Cape Canaveral Air Force Station. The reusable SpaceX Dragon spacecraft could be available for ISS resupply missions of up to 6,000 kg payloads to LEO or for crew transfer missions of up to seven crew members. The capsule will also be able to return up to 3,000 kg of payload to Earth. For non-ISS missions, the spacecraft operates under the "DragonLab" name as an emergent microgravity research and sample return capability.[79]

Inflatable Habitats (Bigelow Aerospace)—In Development

Bigelow Aerospace is developing inflatable space-based habitats. Two prototypes, Genesis I and Genesis II, were launched and operated in LEO in 2006 and 2007, respectively. Future versions of larger modules and concepts for modular space stations based on the Genesis concept are also under development.[80]

CST-100 (The Boeing Company and Bigelow Aerospace)—In Development

A cooperative project between Bigelow Aerospace and the Boeing Company has been awarded a Space Act Commercial Crew Development (CCDev) contract valued at $18 million for the development of the CST-100 crew capsule,[81,82] which could be launched by either the United Launch Alliance Evolved Expendable Launch Vehicle or the SpaceX Falcon 9 rocket.[83]

United Launch Alliance CCDev Project—In Development

Although not currently tasked with resupply for the ISS or actively engaged in microgravity research support missions, the United Launch Alliance has been awarded a $6.7 million CCDev contract for development of an Emergency Detection System (EDS) to be used on Delta IV, Atlas V, and other launch vehicles.[84,85] Development includes detection system definition, testing, demonstration, and crew interface design.[86]

DreamChaser CCDev Project (Sierra Nevada Corporation)—In Development

The DreamChaser[87,88] is a reusable crewed spacecraft concept based on the NASA HL-20 lifting body concept launched atop an Atlas V Evolved Expendable Launch Vehicle. It is supported by a $20 million NASA CCDev contract.[89]

Blue Origin, LLC—In Development

Blue Origin[90,91] has a Space Act Agreement with NASA to develop two technologies that will help reduce risk associated with orbital spaceflight. The first of these technologies is a "pusher escape system" that replaces the traditional emergency capsule escape tower affixed to the front of a crew capsule, with a separation system mounted on the back of the capsule. Unlike an escape tower, the pusher escape engine will not be consumed during normal launch operations and can be reused, thus lowering costs. The other technology is a "composite pressure vehicle" that will use composite panels bonded together.

Currently, Blue Origin is developing a suborbital vehicle that will carry three or more astronauts to an altitude of 350,000 ft.

Russia

Soyuz-Soyuz/Progress (Roscosmos, Starsem)

The Soyuz launch vehicle is a significant asset for transferring both crew and cargo to and from the ISS. It is utilized by Russian, U.S., and European astronauts. Soyuz and the space shuttle are the only human-rated launch vehicles currently available. For crew rotation missions, the Soyuz family of rockets carries the Soyuz spacecraft with three seats; the Soyuz-derived Progress spacecraft is used for cargo missions and has a 1,700-kg cargo capacity.[92] Soyuz capsules, which remain on the ISS for extended periods, also provide a crew rescue capacity in case of emergencies.

Proton (Roscosmos, Khrunichev)

The Proton rocket is an expendable launch vehicle capable of transferring payloads up to 20,700 kg to LEO. It is not used in a regular ISS resupply capacity, but it is used in directly delivering larger ISS modules, including the Zarya and Zevsda modules.

Europe

Ariane 5–Automated Transfer Vehicle (ESA, Centre National d'Etudes Spatiales, Arianespace, EADS Astrium)

The Ariane 5 is a heavy-lift launch vehicle operated by Arianespace for both commercial and ESA services. It is operated and launched from the Guyana Space Center in Kourou, French Guyana.[93] In support of the ISS, the Ariane 5 is used to launch the Automated Transfer Vehicle (ATV) developed by EADS Astrium.

The ATV is an autonomous (but human-rated) resupply vessel capable of ferrying a total of 7,667 kg of pressurized and unpressurized cargo, as well as transferable fuel, to the ISS. Furthermore, the ATV is capable of providing orbit-raising boosts to the ISS and can remain berthed to it for extended periods of time to provide for additional living space.[94] The pressurized cargo section of the ATV is derived from the MPLM and can accommodate up to eight standard racks.[95] The ATV is also able to remove up to 6,300 kg of waste from the ISS on its destructive atmospheric re-entry. Re-entry and return systems for ATV evolution concepts have been under study by ESA to allow for eventual options to return cargo and crew to Earth.[96]

The first ATV, named "Jules Verne," docked with the ISS on April 3, 2008, after an extensive on-orbit testing phase and re-entered the atmosphere on September 5, 2008.[97] The second ATV, "Johannes Kepler," is scheduled to launch in late 2010,[98] with the third ATV, "Edoardo Amaldi," undergoing development for the following year.[99]

Vega Intermediate Experimental Vehicle—In Development

ESA and Arianespace are developing the Vega launch vehicle to place smaller (300-2,000 kg) payloads into orbit economically.[100] While specifically designed to place scientific Earth-observation satellites into polar orbits, Vega is slated to launch the Intermediate Experimental Vehicle in 2012 as a test vehicle for a comprehensive atmospheric re-entry technology development and demonstration program.[101] This test vehicle features a lifting-body configuration and will be used to test dynamic guidance and control technologies during re-entry.[102] While the Vega is not designed to directly support microgravity research, results from the Intermediate Experimental Vehicle are expected to transfer to the development of a crew and cargo transfer capability.

Japan

The **H-IIB–H-II Transfer Vehicle (JAXA, Mitsubishi Heavy Industries)** is a Japanese launch vehicle used in ISS resupply and other missions. It has a lift capacity of 16,500 kg to a LEO with a 51.6° inclination.[103] This is the insertion orbit for the H-II Transfer Vehicle (HTV), a partially pressurized cargo spacecraft to resupply the ISS that has a resupply payload capacity of approximately 6,000 kg.[104] The Japanese have successfully launched the HTV on resupply missions to the ISS.

FREE-FLYERS

Free-flyers are satellites that can be used for automated microgravity research in both biological and physical sciences, such as growing bacteria in space or exposing materials to the space environment, among many other uses. Mission durations, satellite bus and payload sizes, and mission purposes vary widely. Free-flyers can operate either with or without human interaction, and may or may not return samples or data back to Earth autonomously. Some free-flyers will only transmit data back to Earth and are not designed for re-entry.

In many respects, the ISS is itself a very large free-flyer, albeit a permanently crewed one. Traditional free-flyers are typically not designed for or expected to interact with human operators following their launch, unless samples are returned to Earth from orbit. Although it has been proposed that the ISS act as a node for free-flyers, at which visiting vehicles can rendezvous and be refurbished with new payloads, hardware, and software,[105] there is no indication that the ISS will be used in this way.

United States

The NASA Authorization Act of 2005 established a provision that mandated the use of free-flyers as part of the 15 percent allocation of ISS research funds beginning in fiscal year (FY) 2006.[106,107]

NASA's Ames Research Center is home to NASA's Microsatellite Free Flyer program, which is part of the center's ISS Non-Exploration Projects effort focused on implementing peer-reviewed fundamental space biology investigations on a microsatellite free-flyer platform. The Microsatellite Free Flyer program was created to add additional research capacity to U.S. scientists in fundamental space biology: life at molecular and cellular levels, interactions between organisms, and life across generations. Many of these flights, like NASA's GeneSat and PharmaSat, are flown on a relatively new type of satellite platform known as a CubeSat.

The CubeSat[108] was developed in 1999 at the California Polytechnic State University with a universal standard that can be adopted and built anywhere in the world. A CubeSat consists of one, two, or three cube units (1U, 2U, and 3U, respectively) to make a single satellite. An individual cube measures 10 cm per side with a mass of up to 1.33 kg. The primary mission of the CubeSat is to provide access to space for small payloads at costs that are inexpensive compared with traditional satellite platforms. CubeSats are traditionally launched as secondary payloads, making them subject to the launch parameters of the host payload and thus imposing constraints on CubeSat experiments out of the researchers' control. They can be used for experiments in the biological, physical, materials, and Earth sciences. With the mass and size restraints of the CubeSat platform, it remains to be seen what future capabilities CubeSats will have, but already they have been used to expose materials to the space environment, grow cultures in space, and take pictures of Earth.

Europe

The Foton is an uncrewed, Russian-built retrievable capsule, providing an intermediate microgravity platform. It was first launched by the Soviet Union in 1985 and today is launched out of the Baikonur Cosmodrome in Kazakhstan. The Kazakhstan-Russia border is the general area from which capsules are retrieved. Fotons are launched into near-circular LEOs by a three-stage Soyuz-U rocket. Such flights provide researchers with gravity levels less than 10^{-5} g for missions lasting around 2 weeks. ESA's participation in the Foton program began in 1991 with a protein crystallization experiment. The Foton capsule is useful for experiments in biology, fluid and combustion physics, astrobiology/exobiology, and materials science.[109]

Payloads usually fly about 2 years after experiment approval, with the possibility of relatively late experiment installation ("late access") 48-72 h before launch. The Foton also allows for the use of interactive experiment operations (telescience).

The capsule measures 3.2 m long and 2.5 m wide, with an attitude control system that is used for spacecraft alignment in preparation for its re-entry. While in orbit, the attitude control system is not used, and the spacecraft does experience a low level of spin (around 0.1 rpm), which has little effect on the magnitude of microgravity. It can hold up to 650 kg of scientific payload, with a volume measuring 1.6 m^3 for experiment hardware. The size and mass of a single payload is not limited by any specific criteria and will be established by ESA on a case-by-case and mission-by-mission basis.

Power comes from a battery module containing lithium cells and AgZn batteries, providing an average daily power electrical budget of 800 W during a typical 2-week mission.

Capsule pressure is generally kept around 1 atm but can range from 0.454 atm to 1.5 atm; the temperature range is 19°C to 26°C. The capsule is subjected to three types of radiation sources: background radiation (0.055 rad/day), solar flares (50 rad, possible at any point during the mission), and gamma-ray sources inside the re-entry capsule with a total radiation dose of 0.104 rad/day at a distance of 500 mm from the source.

GROUND-BASED FACILITIES

To orient the reader, general types of ground-based facilities are discussed first, according to the general field of research in which they are used: physical sciences, life sciences (including biomedical research), and space radiation research. A more specific inventory of facilities relevant to microgravity research follows, starting with U.S. facilities and then describing major capabilities in Europe, Russia, Japan, and China.

General Types by Field of Research

Facilities for Physical Sciences

There are three major types of ground-based facility that can be used for microgravity experiments in the physical sciences: drop towers, parabolic flights, and sounding rockets. A drop tower is a tall vertical shaft, multiple stories high, where drop experiments can be conducted. As they free-fall down the shaft, in a casement that protects the experiment from the effects of drag, a microgravity environment will be experienced for a short time, usually a few seconds.

Similar to the idea of a drop tower, sounding rockets provide microgravity by allowing experiments to free-fall but through a much larger distance. Sounding rockets can reach altitudes of up to 700 to 800 km before releasing the experimental payload and allowing it to free-fall. The NASA program in polar ballooning (with its fruitful partnership with the National Science Foundation)[110] also offers the possibility of reaching the edge of space with potential for free-fall payloads similar to sounding rockets.

Microgravity is achieved during a parabolic flight by flying an airplane on a parabolic trajectory. At the start of the parabolic climb, a period of increased gravity is followed by approximately 20 s of microgravity before another period of increased gravity, after which the plane pulls out of the parabolic trajectory and gravity returns to a 1-*g* state.

Facilities for Life Sciences

In addition to parabolic flights and sounding rockets (drop towers have not been routinely used by life scientists), space life sciences researchers use a range of specialized ground-based facilities to understand effects of the spaceflight environment on biological systems. These facilities are often highly specialized and target a specific component of spaceflight. For example, the Controlled Environment Systems Research Facility at the University of Guelph in Ontario, Canada, allows specific research on closed environment activities related to plant growth, such as testing whether low-pressure environments could be used to reduce system mass when growing plants in space as part of a bioregenerative life support system. Similar specialized ground facilities are essential components of research to understand the effects of space radiation on microbial, plant, and animal biology in space. Facilities for ground-based research in biomedical sciences are available in the United States, Europe, and Asia.

Facilities for Space Radiation Research

Exposure to radiation in space involves predominantly exposure to galactic cosmic rays and solar particle events. Although galactic cosmic rays may be formed from most of the elements in the periodic table, about 90 percent of the particles are protons. The remaining 10 percent are helium, carbon, oxygen, magnesium, silicon, or iron ions, exposure to which is much more damaging per unit dose than similar exposures to conventional medical x-rays or gamma rays used in radiation therapy. Solar particle events involve exposures to energetic protons, which are similar in their radiobiological effects to x-rays and gamma rays. However, protons as well as most heavy ions have dose-distribution features in biological systems that are different from conventional radiotherapy qualities of radiation and may have unique biological effects on the host. Space radiation of various types has been only poorly studied in the literature and requires unique facilities available at only a few places on Earth.

Recent approaches to patient radiotherapy have found protons to be beneficial in treating some forms of cancer, and several radiation therapy groups in the United States have constructed proton irradiation devices to carry out specialized proton therapy for particular cancers. Among these are the Loma Linda University and the University of Pennsylvania facilities, both of which are permitting NASA-funded proton irradiation experiments when patients are not being treated. These facilities have been used to mimic solar particle irradiation that involves mostly exposure to protons. Nevertheless, because the needs of the space radiation community cannot be met by such limited resources, NASA developed the NSRL.[111]

U.S. Ground-Based Facilities

The United States has a multitude of facilities across the country for studying microgravity sciences and the effects of spaceflight on humans. The bulk of these facilities are operated at NASA centers, but others operate out of U.S. National Laboratories and some universities.

NASA Ames Research Center

Microgravity Test Facility (Propulsion and Navigation Testing)[112]

The MicroGravity Test Facility was originally developed for testing and validating propulsion, autonomous navigation, and control of the Personal Satellite Assistant and was designed to simulate the microgravity environment on board the ISS's U.S. Laboratory module. Six degrees of freedom (DOF) are achieved with a 3-DOF gimbal system suspended from a 3-DOF translation crane, allowing vehicle motion in all directions. Precision sensors measure vehicle thrust in the x, y, and z axes to within a fraction of an ounce. These measurements are then used to actuate external motor controllers in the suspension system, which simulate the vehicle's motion in a microgravity environment. The technology developed for this facility could be scaled to support even larger and more complex simulations such as rendezvous and docking maneuvers for two independently operated vehicles, as well as terrestrial lander studies for various microgravity environments.

Genome Research Facility (Fundamental Biology)[113]

The goal of the Genome Research Facility is to support NASA research objectives in the areas of nanotechnology, fundamental space biology, and astrobiology, specifically through the development of devices that can detect single molecules of nucleic acids, decode DNA sequence variations in the genome of any organism, and apply functional genomic assays to determine molecular information processing functions in model organisms. The Genome Research Facility also makes use of NASA Advanced Supercomputing[114] capabilities to develop bioinformatics algorithms used to support the optimization of oligonucleotide array design and molecular dynamic modeling of ion signatures in nanopores.

Life Sciences Data Archive (Fundamental Biology)[115]

The NASA Life Sciences Data Archive is intended to support NASA's Human Research program by facilitating the capture and flow of life science evidence data at a single site. These data document the effects of spaceflight and are available for use by researchers. In addition, this archive also includes space-flown biospecimens available to scientific researchers who are pursuing answers to questions relevant to the Human Research Program.

Centrifuge Facilities (Gravitational Biology)[116]

NASA Ames centrifuge facilities consist of four main centrifuges: a human-rated 20-*g* centrifuge and three nonhuman accelerator facilities. Of the latter, the 24-ft-diameter centrifuge was designed to create hypergravitational research conditions for small animal, plant, and hardware payloads, while the 8-ft-diameter centrifuge was designed specifically to accommodate habitats developed for the ISS.[117] The last of the nonhuman accelerator facilities, the Low Vibration Rotational Device, is a single-arm centrifuge with a 10-ft radius and hydrostatic oil film bearing. When this centrifuge is configured with an onboard tissue culture incubator to study the effects on cultured cells of exposure to short- or long-duration hypergravity, it is referred to as the Hypergravity Facility for Cell Culture.[118]

Bioengineering Branch (Human Research Facility)[119]

The Bioengineering Branch in the Space Biosciences Division at NASA Ames is developing advanced technologies required for future human exploration missions in space. It includes NASA's Exploration Life Support program, which is charged with developing the advanced technologies and systems that support humans in extended space exploration. The development of these technologies is focused on the need to increase mission self-sufficiency by minimizing mass, power, and volume requirements through regeneration of vital resources.

NASA Glenn Research Center

Exercise Physiology and Countermeasures Project (Human Research Facility)[120]

The Exercise Physiology and Countermeasures Project supports the lead project office at NASA JSC in developing exercise countermeasure prescriptions and exercise devices for space exploration that are effective, optimized, and validated to meet medical, vehicle, and habitat requirements. Current projects include the development of a more comfortable harness for use on the ISS treadmill; an enhanced zero-gravity locomotion simulator, which is a new ground-based simulator developed to address the negative physiological effects of spaceflight on the musculoskeletal system; and assessments of locomotion in simulated lunar gravity relating to critical mission tasks that may be required by a crew member on a lunar mission.

Digital Astronaut Project (Human Research Facility/Fundamental Biology)[121]

As described by NASA, "The Digital Astronaut Project, led out of the JSC and in partnership with Glenn Research Center and the University of Mississippi Medical Center, is an effort to create a detailed computer model of the entire functioning human physiology that can be used to predict the effects of spaceflight on each physiological system. All body systems, such as the cardiovascular and vestibular systems, will be simulated at the level of detail required to understand the effects of spaceflight. As part of this computational effort, Glenn Research Center is responsible for creating detailed modules that predict functional cardiac changes, alterations in bone remodeling physiology, and changes in muscle activation resulting from extended-duration reduced-gravity exposure. Additionally, Glenn Research Center recently completed work on a module simulating renal stone formation and transport in microgravity. This center is also responsible for leading project-wide verification and validation of the integrated model."

Exploration Medical Capability Project (Human Research Facility)[122]

The portion of the Exploration Medical Capability Project performed at the Glenn Research Center includes development of the Intravenous Fluid Generation for Exploration (IVGEN) project, the Integrated Medical Model, biosensors, and in-flight lab analysis techniques. The IVGEN project's goal is to meet the requirement for intravenous fluid during exploration missions by constructing a filtration system that will generate fluid using in situ resources. The Integrated Medical Model program develops protocols relating to planned responses for potential injuries to astronauts in space such as bone fracture, insomnia, kidney stones, head injuries, and other ailments. The current biosensors being developed can be used to measure a variety of biological data and are designed to be worn with or without spacesuits. The in-flight laboratory analysis group is developing in-flight biological test methods that do not require disposable testing components, as this waste would be cumbersome on a longer-duration flight.

2.2 Second Drop Tower[123]

The NASA Glenn 2.2 Second Drop Tower is one of two drop towers located at the NASA site in Brookpark, Ohio. As detailed by NASA, "The drop tower's 2.2-second microgravity test time is created by allowing an experiment package to free-fall a distance of 79 ft (24 m). The drop tower uses an experiment/drag shield system to minimize the aerodynamic drag on the free-falling experiment. Experiments are assembled in a rectangular aluminum frame, which is enclosed in an aerodynamically designed drag shield (which weighs 725 lb, 330 kg). This package is hoisted to the top of the tower, where it is connected to monitoring equipment (e.g., high-speed video cameras and onboard computers) before being dropped. The experiment itself falls 7.5 inches (19 cm) within the drag shield while the entire package is falling. The drop ends when the drag shield and experiment are stopped by an airbag at the bottom of the tower. The drop tower can accommodate experiments up to 350 kg."

Zero Gravity Research Facility[124]

The Zero Gravity Research Facility provides a near-weightless or microgravity environment for a duration of 5.18 s by allowing the experiment vehicle to free-fall in a vacuum for a distance of 432 ft (132 m). A five-stage vacuum pumping process is used to reduce the pressure in the chamber to 0.05 torr. Evacuating the chamber to this pressure reduces the aerodynamic drag on the freely falling experiment vehicle to less than 0.00001 *g*.[125] The Zero Gravity Research Facility can perform experiments with payloads of up to 1,000 lb and up to 66 inches in height and 38 inches in diameter.

Microgravity Emissions Laboratory[126]

The Microgravity Emissions Laboratory was developed for the support, simulation, and verification of the ISS microgravity environment. It uses a low-frequency acceleration measurement system to characterize rigid-body inertial forces generated by various operating components of the ISS. These acceleration emissions could, if too large, hinder the science performed on the ISS by disturbing the microgravity environment. Typical test components are disk drives, pumps, motors, solenoids, fans, and cameras. Other test articles have included onboard electric power systems for spacecraft, optical measurement systems, and crystal growth experiment package assemblies.

Microgravity Data Archive[127]

The Microgravity Data Archive is a database intended to hold both Experiment Data Management Plans (EDMPs) and any publications or presentations that were generated from combustion or fluids experiments (flight- and ground-based). The EDMPs contain information about flight experiments such as the project scientist, a summary of the experiment, what data were collected during the experiment, what data are available from the experiment, and what the results of the experiments were. Also included on an EDMP is a list of what publications or presentations were written about the experiment. The publications and presentations information includes author, title, where and when published, and the abstract from the paper.

Telescience Support Center[128]

The Telescience Support Center allows researchers on Earth to operate experiments onboard the ISS and the space shuttles. NASA's continuing investment in the required software, systems, and networks provides distributed ISS ground operations that enable payload developers and scientists to monitor and control their experiments via this ground-based center. The goal of the center is to enhance the quality of scientific and engineering data, while reducing the long-term operational costs of experiments by allowing principal investigators and engineering teams to operate their payloads from their home institutions.

The Cleveland Clinic Foundation Center for Space Medicine[129]

Supported by a cooperative agreement from NASA's Glenn Research Center, the Cleveland Clinic's Center for Space Medicine provides a focal point for the clinic's overall space medicine research, giving researchers access to the network of more than 2,000 physicians and scientists employed by the Cleveland Clinic Foundation. The center's creation coincided with President George W. Bush's Vision for Space Exploration and NASA's push for human lunar missions. The center conducts research in the major physiological research fields that affect astronauts in space: musculoskeletal, neurosensory, and cardiovascular systems and radiation. The center also works with the John Glenn Biomedical Engineering Consortium and receives grants from NASA Headquarters and the National Space Biomedical Research Institute.

NASA Goddard Space Flight Center

NASA Sounding Rocket Program, Wallops Flight Facility[130]

Sounding rockets carry scientific instruments into space along parabolic trajectories, providing nearly vertical traversals along their upleg and downleg, while appearing to "hover" near their apogee location. Microgravity missions are conducted on high-altitude, free-fall parabolic trajectories, which provide microgravity environments that lack the vibrations frequently encountered on human-tended platforms. Currently, Wallops Flight Facility in Wallops Island, Virginia, is the only facility in the United States that designs, manufactures/fabricates, integrates, tests, and launches sounding rockets.[131]

High Capacity Centrifuge[132]

NASA Goddard's 120-ft-diameter centrifuge can accelerate a 2.5-ton payload up to 30 *g*, well beyond the force experienced in a launch. This centrifuge is used only for equipment testing and has not been rated for human testing at high speeds.

Space Environment Simulator[133]

The Space Environment Simulator is a thermal vacuum chamber that exposes spacecraft components and other payloads to environmental conditions similar to those they will experience in space. The chamber has mechanical vacuum pumps augmented by cryopumps. These pumps work together to eliminate nearly all of the air in the chamber, achieving conditions down to about a billionth of Earth's normal atmospheric pressure. To simulate the hot and cold extremes possible in space, the thermal vacuum chamber can reach temperatures in a 600-degree range from 302°F to −310°F.[134] The cylindrical chamber is 40 ft tall and 27 ft wide.

NASA Johnson Space Center

Johnson Space Center is the primary site in the United States for astronaut operations, and as such maintains extensive resources that support both intramural and extramural investigations in microgravity science. For example, active research labs are maintained for cardiovascular, neuro-vestibular, nutrition, exercise, musculoskeletal, and behavioral health research. Each laboratory is led by a dedicated scientist with an experienced research team. These investigators both initiate their own intramural research programs and are also available to collaborate with extramural investigators on NASA-approved science experiments. A fully equipped biochemistry laboratory provides laboratory services to approved investigators. Crew quarters are present on site for behavioral research as well as support for landing day research activities. The Houston Mission Control Payload Operations Center provides continuous support for flight experiments, including, for example, real-time remote guidance for ultrasound and other clinical research activities. An extensive system of high-fidelity training sites for all experimental modules is available in dedicated buildings both for science and for operations. These are supported by a large neutral buoyancy laboratory for simulating EVA activities and ISS operations, as well as four human-rated vacuum chambers, two human-rated hypobaric chambers, and one human-rated hyperbaric chamber. A partial-gravity simulation system (POGO) is available to simulate space operational activities, including spacesuit function at reduced gravitational gradients. These research activities are supported by the Johnson Space Center Flight Medicine Clinic, which is affiliated with the University of Texas Medical Branch at Galveston (UTMB) aerospace medical residency. The UTMB Clinical Translational Sciences Center has a bed rest facility that is available to provide support for bed rest studies.

Human Test Subject Facility[135]

The Human Test Subject Facility is responsible for providing qualified test participants for ground-based research. The Flight Analogs Project Team at JSC is planning a series of studies over the next 10 years that support the scientific needs of the space program. Two studies for which participants are currently being recruited are a bed rest study[136] and the Lunar Analog Feasibility Study.[137]

Microgravity University[138]

The Reduced Gravity Student Flight Opportunities Program provides an opportunity for undergraduate students to propose, design, fabricate, fly, and evaluate a reduced-gravity experiment of their choice over the course of 4 to 6 months. The overall experience includes scientific research, hands-on experimental design, test operations, and educational/public outreach activities.[139] Each accepted submission is flown on a NASA reduced-gravity aircraft, which generally flies 30 parabolic maneuvers over the Gulf of Mexico. Student experiments must be organized, designed, and operated by student team members alone.

Reduced Gravity Research Program[140]

The NASA Reduced Gravity Research Program, operated out of JSC, provides NASA researchers with a free-fall environment via parabolic flights to simulate a microgravity environment for test and training purposes. The program uses a C-9B aircraft (a McDonnell-Douglas DC-9) to conduct the reduced-gravity parabolic flights tests, which last for 2 to 3 h and average 40 to 50 parabolas. The aircraft has a cargo test area approximately 45 ft long, 104 inches wide, and 80 inches high.

Currently, the C-9B is not NASA's primary vehicle for conducting parabolic flight tests, and the agency awarded a contract to Zero Gravity Corporation in 2008 to provide these services. The C-9B is still operational at JSC, and on occasion it conducts parabolic flight tests as well as meets other miscellaneous agency needs.¶

NASA Kennedy Space Center

Baseline Data Collection Facility[141]

The Baseline Data Collection Facility (BDCF) provides a research infrastructure and a technical workforce to support human research and testing in response to spaceflight and the conditions of a microgravity environment with potential research applications for the general population. This series of laboratories housing experiment-unique equipment is used to perform physiology testing on space shuttle crew members before, during (monitoring and/or ground controls), and after flight. Kennedy Space Center provides physicians, nurses, and specialized technicians for these activities. The BDCF is one of only two facilities in the United States capable of studying astronaut response to spaceflight immediately upon an astronaut's return to Earth. (The other facility is the Post-flight Science Support Facility, located at Dryden Flight Research Center, Edwards Air Force Base, California.) Astronauts who land in Russia can either be studied in Star City, Russia, or flown immediately back to the United States where delayed investigations can be performed at JSC <24 h after landing.

The BDCF is equipped with multiple kinds of microscopy, including transmitted-light brightfield, darkfield, differential interference contrast, epi-fluorescence, and phase contrast. The facility has microbial, sterility, clinical, and hematology testing, as well as indoor air quality investigative surveys. The facility can create specialized gas mixtures and contains autoclaving services (steam, dry heat), an ethylene oxide sterilization system, and radioisotope-rated laboratories. There are both refrigerated and nonrefrigerated centrifuges, as well as refrigerators and freezers for controlled specimen and reagent storage (+4°C, –20°C, –80°C). The facility provides calibration,

¶More information on NASA and Zero-G Corporation's relationship, and services provided by Zero-G Corp., can be found below in the section "Zero-G Corporation."

installation, and operation of specialized equipment such as magnetic resonance imaging assemblies, densitometers, cardiovascular devices, and vestibular testing equipment (rotating chair devices, treadmills, head-and-gaze systems, and obstacle courses). BDCF personnel also coordinate, schedule, and perform experiment protocol reviews and validate the integrity of research methods and relevant device systems. BDCF personnel also coordinate customer use of unique chemicals, radioisotopes, and custom-blended breathing gases.

NASA Marshall Space Flight Center

Marshall Space Flight Center is home to the ISS Payload Operations Center, linking researchers around the world with their experiments and astronauts aboard the ISS. The Payload Operations Center integrates research requirements, plans science missions, integrates crew and ground team training and research mission timelines, manages use of space station payload resources, handles science communications with the crew, and manages commanding and data transmissions to and from the ISS. The Operations Center is staffed 24 h every day by three shifts of flight controllers.

Marshall Space Flight Center also manages the MSG, which was launched to the ISS in June 2002.[142]

NASA Space Radiation Laboratory

The $34 million NSRL at BNL is the result of an agreement between the Department of Energy, which owns BNL, and NASA, which designed the NSRL and makes it available to investigators through a beamtime-request-based approach. The NSRL is dedicated to studying the effects of space radiation on biological specimens with the ultimate goal of developing effective countermeasures for deep-space human exploration.[143] For several years, research supported by NASA on the radiobiological effects of high-energy heavy ions had been conducted at the Lawrence Berkeley National Laboratory BEVALAC linear accelerator in California. That operation ended in the early 1990s, and now BNL's Alternating Gradient Synchotron (AGS) is the only accelerator in the United States capable of providing heavy-ion beams at the energies of interest for space radiobiology.[144]

NSRL became operational in 2003. Radiobiology and physics experiments are conducted three to four times per year for 6 weeks, for up to several weeks per run.[145] NASA uses the facility for many different kinds of experiments, such as studying model organisms, cell and tissue cultures, and various materials bombarded with beams of carbon, silicon, iron, and gold ions at energies generally ranging from 0.3 to 1.0 billion electron volts (GeV) per nucleon.[146] Investigators with NASA funding can apply to the NSRL for beam time for an experiment; the experiment is reviewed by a committee and either approved for beam time or rejected (and thereby sent back for modification if possible).

Facility users include NASA (JSC, NASA Specialized Center of Research and Training, and the National Space Biomedical Research Institute), national laboratories and research institutes (BNL, Lawrence Berkeley National Laboratory, Medical Research Council in England, and the National Institute of Health in Italy), and numerous universities in the United States and around the world.[147] Most of these users have funding from NASA to conduct studies on space radiation effects in biological systems (cells and animals).

Currently the NSRL has the ability to produce not only protons but also other types of high linear energy transfer (LET) radiation,** including mixed-field irradiation.†† Upgrades are planned to increase the number of potential particles that are available for a particular run. Although the NSRL was designed with biological experiments in mind (including incubators and microscopes, cell counters, other equipment for cell culture, and animal housing capabilities for rodents and some larger mammals), studies of shielding effects are also possible.

The continued availability of the NSRL is considered critical by the NASA space radiation biology community because there are few if any places in the world that can produce radiation with the characteristics appropriate for space studies. While some high LET radiation sources are available in Germany and Japan, these do not exactly

**Linear energy transfer (LET) is the amount of energy deposited per unit distance that a charged particle travels. "High LET" radiation includes the heavier-than-protons charged particle radiations found in galactic cosmic rays.

††"Mixed fields" are mixtures of protons with heavier charged particles or of a variety of heavy particles.

mimic space radiation and have limited time available for experimental studies and limited facilities to handle cell and animal research.

Zero-G Corporation

NASA first purchased parabolic flights for microgravity research from the Zero-G Corporation in 2005, and in 2008 signed a contract with the company to become the operational branch of the existing Reduced Gravity Program at JSC and replace JSC's C-9B aircraft. In addition to services provided for NASA, Zero-G also offers its services to private researchers. Private researchers can buy a portion of the parabolas in a given flight, thus sharing the aircraft with other researchers, or purchase the entire aircraft for a flight. The number of parabolas and specifications thereof are negotiable between researchers and Zero-G. The cargo area for testing is 20 ft long and 5 ft wide.

When private researchers charter Zero-G's services, they are subject to Federal Aviation Administration rules and regulations regarding flight safety and conduct. However, when NASA sponsors flight tests, it rather than the Federal Aviation Administration has authority over the aircraft. This impacts the type of experiments that can be conducted on Zero-G's aircraft.

European Ground-Based Facilities

ESA has three types of ground-based microgravity facilities, but it also supports other facilities and environments on Earth that simulate the space environment. The three facility types are drop towers, parabolic flights, and sounding rockets.[148] ESA also has two primary locations for human spaceflight research: one in Toulouse, France; the other in Cologne, Germany. Finally, ESA has multiple user support and operation centers (USOCs), which are based in national centers distributed throughout Europe. These centers are responsible for the use and implementation of European payloads onboard the ISS.

Zentrum für Angewandte Raumfahrt Microgravitation (ZARM) Drop Tower

The ZARM drop tower,[149] located in Bremen, Germany, is ESA's primary drop tower facility. At 146 m tall, the concrete shaft can provide near-weightlessness for experiments up to three times a day. Experiments dropped from the top of the tower experience 4.74 s of microgravity, and experiments catapulted up the tower from the bottom experience 9.48 s of microgravity measurements. The catapult system doubles the standard drop microgravity time by accelerating the capsule (anywhere from 300 kg up to 500 kg) up the shaft at a speed of up to 48 m/s within 0.28 s.

The cylindrical experiment containment capsule has a diameter of 800 mm and a length of 1.6 m or 2.4 m depending on the space required. The capsule can be dropped through the drop tube vacuum from a maximum height of 120 m, reaching an ultimate microgravity quality with residual accelerations less than 10^{-5} g. Nominal capsule pressure is set to 1.013 hPa, and the temperature can be adjusted to between −20°C and +60°C. The operational voltage range for experiments is from 26.4 to 35 Vdc. The ZARM drop tower has been used as a platform for fundamental physics experiments, materials science, cell biology, and fluid and combustion physics.

Airbus A-300 "Zero-G"

ESA has been using its Airbus A-300 "Zero-G"[150] since 1997 to conduct parabolic flights, based out of the Bordeaux-Mérignac airport in France. The aircraft generally executes a series of 31 parabolic maneuvers per flight. Starting from normal horizontal flight at 6,000 m and 810 km/h, the A-300 ascends for about 20 s, experiencing acceleration between 1.5 and 1.8 g. At an altitude of 7,500 m, the aircraft assumes an upward angle of 47 degrees relative to the horizontal and an airspeed of 650 km/h; the engine thrust is reduced to the minimum required to compensate for air drag. The aircraft then follows a free-fall ballistic trajectory lasting approximately 20 s, during which weightlessness is achieved, reaching the peak of the parabola at around 8,500 m, by which time the speed of the aircraft has dropped to 390 km/h. The period between the start of each parabola is 3 min, with a 1-min parabolic phase (20 s at 1.8 *g*, 20 s of weightlessness, 20 s at 1.8 *g*) followed by a 2-min period at steady level 1-*g* flight. Parabolas are executed in sets of five, after which a longer time is allowed to elapse to allow for modifications to experiments.

Cabin pressure is around 0.79 atmospheres during parabolic maneuvers, with a cabin temperature maintained between 18°C and 25°C.

Parabolic flights can be used for experiments in fundamental physics, materials science, biology, technology, fluid and combustion physics, and physiology.

ESA Sounding Rockets

ESA has been using sounding rockets for microgravity research since 1982.[151] Today, the agency has four types of sounding rockets (from smallest to largest): MiniTEXUS, TEXUS, MASER, and MAXUS. All of the rockets are launched from the Esrange launch site east of Kiruna in northern Sweden, located 200 km above the Arctic Circle. ESA's sounding rockets are useful for experiments in fundamental physics, biology, fluid and combustion physics, and materials science.

MiniTEXUS is a two-stage solid propellant short-duration sounding rocket capable of carrying one to two experiment modules for 3-4 min of microgravity ($\leq 10^{-4}$ *g*) at a spin rate of 5 Hz, reaching a peak acceleration of 21 *g* during the first stage. The rocket can handle scientific payloads up to 100 kg, with a payload diameter of 43.8 cm and a length of 1 m. MiniTEXUS reaches an apogee of 140 km.

The TEXUS two-stage solid propellant rocket, which has been in operation since 1977, is the workhorse of the ESA sounding rocket family, having launched more than 75 experiments. TEXUS is able to haul up to 260 kg of scientific hardware for approximately 6 min of microgravity ($\leq 10^{-4}$ *g*) at a spin rate of 3-4 Hz, reaching a peak acceleration of 10 *g* during the first stage. The payload capsule measures 0.43 m in diameter and can accommodate a maximum payload length of 3.4 m. TEXUS reaches an apogee of 260 km. Both TEXUS and MiniTEXUS are operated by an industrial consortium led by EADS-ST out of Bremen, Germany.

MASER (Material Science Experiment Rocket) is a Swedish two-stage solid propellant rocket that has been in operation since 1987. It is managed by the Swedish Space Corporation. MASER is capable of launching up to 260 kg of scientific hardware for approximately 6 min of microgravity ($\leq 10^{-4}$ *g*) at a spin rate of 3-4 Hz, reaching a peak acceleration of 10 *g* during the first stage. Like TEXUS, MASER reaches an apogee of 260 km.

ESA's MAXUS rocket is the most powerful of the ESA sounding rocket family. Operated by a joint venture formed by EADS-ST and the Swedish Space Corporation, the one-stage solid propellant rocket is capable of launching up to 480 kg of scientific hardware for 12-13 min of microgravity ($\leq 10^{-4}$ *g*) at a spin rate of ≤0.5 Hz, reaching a peak acceleration of 13 *g*. At apogee, the MAXUS rocket reaches 705 km, about 250 km higher than the orbit of the ISS.

Although these four rockets provide experimental environments that differ in some regards (e.g., outer structure heating during re-entry), all payloads are protected prior to launch in environmentally controlled capsules at temperatures around 18°C with a range of ±5°C. After impact of the payloads on the ground, the experiment modules are exposed to snow and cold air for a period of up to 2 h.

Institute for Space Medicine and Physiology

The Institute for Space Medicine and Physiology (MEDES)[152] was created in 1989 to develop expertise in human spaceflight medicine, prepare for future interplanetary human missions, and apply the results of space research in health care back on Earth. MEDES, which is based in Toulouse, France, has four main research areas focusing on space missions support, clinical research, health applications, and telemedicine. The institute provides medical support to the integrated team at the European Astronauts Center. The MEDES role includes selection and training of astronauts, performing aptitude tests, providing medical support during space flights, and ensuring crew rehabilitation and safety.

MEDES clinical research takes place at its 1,000 m^2 Space Clinic, located within the Toulouse Rangueil

Hospital. Primary areas of research are in physiology, pharmacology, and the evaluation of biomedical devices. Part of this research includes simulating the effects of the space environment (involving but not limited to bed rest, confinement, and circadian rhythms), both to study physiological effects and to develop preventive methods and medicine.

MEDES health applications research is focused on mitigating the hazardous effects of the space environment on astronauts, and on identifying potential benefits for Earth-based health care for diseases linked to conditions such as aging, lack of physical activity, and osteoporosis. In the area of telemedicine, MEDES is developing satellite applications for remote medical consultation, epidemiology, training (in particular through the use of interactive satellite television), and patient care.

Simulation Facility for Occupational Medicine Research

The 300 m^2 Simulation Facility for Occupational Medicine Research (AMSAN)[153] is operated by the German Aerospace Center's (DLR) Department of Flight Physiology. It investigates aviation and space medicine-related issues and conducts human factors research. The facility can hold up to eight subjects at a time, providing environmental control (e.g., artificial light, climate, and sound-proofing). AMSAN can be used by other departments of DLR and by external clients. Research is focused on:

- Examinations of the effects of nocturnal aircraft noise on sleep and performance (assessment and evaluation of criteria);
- Standardized metabolic balance studies in a controlled environment for question in space physiology (calcium, muscle, and bone metabolism; cardiovascular and volume regulation);
- Investigations under the conditions of simulated weightlessness (bed rest, 6° head-down-tilt);
- Simulating critical duty rosters of aircrews;
- Assessing changes in the circadian system through time zone flights (jet-lag), shift work, and irregular duty-hours; and
- Clinical studies for assessing the efficacy of drugs in the aerospace environment.

European User Support and Operations Centers

ESA operates nine USOCs throughout Europe, each of which is responsible for oversight and operations relating to specific ISS ESA payloads.[154] USOCs act as the link between experiment users and the ISS and are also responsible for preparation of payloads before launch, experimental procedure development, optimization and calibration of payloads, and supporting training activities for ISS crew. These centers are supported by the Columbus Control Centre, located at the German Space Operations Center of DLR in Oberpfaffenhofen, Germany.

Columbus Control Centre, Oberpfaffenhofen, Germany

Although most of the Columbus laboratory's functions are automated, the Columbus Control Centre[155] is staffed and operated 24 h/day, 7 d/wk. All ground services for Columbus, such as communications (voice, video, and data), are provided by the Columbus Control Centre for ISS users, the ATV Control Centre, the European Astronaut Centre, industrial support sites, and ESA management. The center is in close contact with mission control centers in Houston and Moscow. In addition, it coordinates operations with the ISS Payload Operations and Integration Center at NASA Marshall Space Flight Center.

Biotechnology Space Support Centre, Zurich, Switzerland

The Biotechnology Space Support Centre[156] is a program of the Space Biology group at the Swiss Federal Institute of Technology, Zurich.[157] This USOC is responsible for the operations of Swiss programs on the ISS and acts as the Facility Responsible Center for KUBIK, a transportable incubator, and as the Facility Support Center for Biolab, an experimental biology facility in Columbus. It also provides ground infrastructure and training for ESA-approved space experiments and acts as an information center and public outreach group.

Belgian User Support and Operations Centre, Brussels, Belgium

Located at the Belgian Institute for Space Aeronomy, the Belgian USOC[158] is tasked with promoting space research programs and flight opportunities for Belgian scientists in industry, academia, and federal and regional institutions. It provides scientific support in a variety of fields, including microgravity, Earth observation, space sciences, and technology. The Belgian USOC acts as the Facility Responsible Center for the external Solar Monitoring Facility on the Columbus module on the ISS and as a co-Facility Support Center for the European Drawer Rack/Protein Crystallization Diagnostics Facility.

Centre d'Aide au Développement des activités en Micro-pesanteur et des Opérations Spatiales, Toulouse, France

Operated by the French Centre National d'Etudes Spatiales, the Centre d'Aide au Développement des activités en Micro-pesanteur et des Opérations Spatiales (CADMOS)[159] is a program designed to help scientific teams prepare and develop experiments for the microgravity environment. CADMOS was founded as the office responsible for all French human flights performed on the Mir space station or shuttle spacecraft. It has directly overseen several recent missions including a joint ISS French-Russian mission in 2001, for which CADMOS assumed all responsibilities including preparation, development, and certification of payloads and direct mission operations.

Danish Medical Centre of Research, Odense, Denmark

The Danish Medical Centre of Research (Damec)[160] is a high-technology company with a focus on advanced medical instrumentation and other engineering fields pertaining to space applications. Primary ISS contributions include respiratory equipment, the Cevis ergometer for the NASA Destiny laboratory, and the Pulmonary Function System for ESA.[161] Damec is the USOC and Facility Support Center for the Pulmonary Function System. In addition, the center has participated in several ESA parabolic flight campaigns to test scientific equipment in simulated microgravity conditions.[162]

Erasmus User Support and Operations Centre, Noordwijk, The Netherlands

The Erasmus USOC is located in the Erasmus Building of ESA's European Space Research and Technology Centre in Noordwijk, the Netherlands.[163] This USOC is the Flight Responsible Center for the European Drawer Rack, and it has overall responsibility for the rack, as well as for stand-alone Columbus module payloads aboard the ISS. This USOC also acted as the Flight Responsible Center for the European Technology Exposure Facility. Via teleoperations, the Erasmus USOC operates specific payloads and experiments onboard the ISS and works with multiple Facility Support Centers and User Home Bases located at institutes and universities across Europe. This USOC prepares, plans, and coordinates payload flights and ground operations; monitors experiments and payloads around the clock; tests and validates new operations; and prepares new facilities and experiments.

Spanish User Support and Operations Centre, Madrid, Spain

The Spanish USOC[164] is a center of the Polytechnic University of Madrid and acts on behalf of ESA as the point of contact for Spanish user teams developing experiments requiring a microgravity environment. The Spanish USOC is the Facility Support Center for the Fluid Science Laboratory, an ESA experimental payload on the ISS Columbus module. It is also the Spanish point of contact for all of ESA's low-gravity platforms, including drop towers, parabolic flights, sounding rockets, and Foton retrievable capsules.

Telespazio, Naples, Italy

In 2009, Telespazio incorporated the Microgravity Advanced Research and Support Center[165] and now acts as an Italian support center for European experiments on the ISS. Telespazio works in all areas of development and experimentation, including payload integration, in-orbit operations, astronaut training, and dissemination of scientific data. In particular, Telespazio has devoted resources to the European Union's scientific and technological research project ULISSE, a program designed to assimilate data from ISS experiments with data from other microgravity missions. In addition to contributions to ISS operations, Telespazio participates in research for direct exploration missions to near-Earth objects through ESA's Space Situational Awareness program and in lunar and Mars exploration programs promoted by ESA.

Microgravity User Support Centre, Cologne, Germany

DLR's primary microgravity research facility is the Microgravity User Support Centre,[166] which supports users in materials physics, aerospace medicine, and astrophysics. Specific tasks include preliminary research on functionally identical ground models of flight hardware, on-orbit operation of payloads, experiment analysis, and data archiving. Its ISS facilities include Biolab, Expose, Matroshka, and the Materials Science Laboratory. In addition, this center is responsible for space readiness qualification of space experiments to be flown on the ISS, largely through testing on the European A300 parabolic flight test aircraft.[167]

Norwegian User Support and Operations Centre, Trondheim, Norway

The Norwegian USOC[168] is the Facility Responsible Center for the European Modular Cultivation System,[169] and the USOC has contributed to planning, development, integration, and execution of the different experiments that utilize this system onboard the ISS. The center provides system users with communication and data-processing capabilities that call for real-time data monitoring and control of experiments.

Russian Ground-Based Facilities

Atlas Aerospace

Atlas Aerospace[170] is a private company based in Russia and offering parabolic microgravity flights aboard an IL-76 aircraft. It also operates the "hydrospace" neutral buoyancy services and centrifuge tests and simulators at the Yuri Gagarin Cosmonaut Training Center in support of space research, as well as for tourism and recreational purposes.

Yuri Gagarin Cosmonaut Training Center

The Yuri Gagarin Cosmonaut Training Center[171] operates several facilities to support microgravity operations including spacecraft simulators and mock-ups, a neutral buoyancy hydrolab, centrifuges, airplanes, and pressure chambers.

Japanese Ground-Based Facilities

JAXA Tsukuba Space Center (TKSC), Tsukuba

The TKSC,[172] which sits on a 530,000 m^2 site located in Tsukuba Science City, is a consolidated operations facility with world-class equipment and testing facilities. As the center of Japan's space network, the TKSC plays an important role in Japan's spacefaring activities. For example, the Japanese Experiment Module Kibo for the ISS was developed and tested at the TKSC. Astronaut training and space medicine research are also in progress there, including simulation of the effects of weightlessness and other factors of the space environment (confinement, circadian rhythms), ground-based control experiments, testing of equipment, medical screening and check-up of astronauts, practice with various tasks in simulated weightlessness (water buoyancy), and wearing spacesuits specifically designed for this purpose. TKSC capabilities include expertise in human factors, confinement, isolation, simulated weightlessness, human physiology, and research on circadian rhythms, human performance, sleep, etc.

Parabolic Flight Center, Diamond Air Service (DAS) Incorporation, Aichi

Parabolic pattern flights by jet aircraft provide a microgravity condition for about 20 s (at less than $3 \times 10^{-2}\ g$) in the cabin per pattern. MU-300 and G-II aircraft are used for these microgravity flights. The experimental support system in the cabin was developed by JAXA.

Short Arm Human Centrifuge Center, Nihon University School of Medicine, Tokyo

This human centrifuge has a short-radius arm, which can be adjusted to lengths between 1.6 m and 2.1 m. It has a seat located in the cabin, which is freely movable, allowing its top to lean toward the center of rotation. The long axis of the subject's body is thus parallel to the resultant force vector determined by the gravitation force of Earth and the force generated by the centrifuge. This centrifuge accelerates and maintains gravity levels of one to three times Earth's gravity.

Chinese Ground-Based Facilities

China Astronaut Research and Training Center, Beijing

Facility capabilities include simulation of the effects of weightlessness by bed rest, etc.

Short Arm Human Centrifuge at the Fourth Military Medical University, Xi' an Shaanxi

Facility capabilities include the examination of human physiology under various gravitational forces by short-arm human centrifuge.

REFERENCES

1. NASA. International Space Station. Fact Sheet. Window Observational Research Facility. Available at http://www.nasa.gov/mission_pages/station/research/experiments/WORF.html.
2. NASA. 2001. *NASA Fiscal Year 2002 Congressional Budget.* Released April 9, p. HSF 1-29. Available at http://www.nasa.gov/pdf/118805main_Budget2002.pdf.
3. NASA Ames Research Center. Humans in Space: Life Sciences Glovebox. Available at http://www.nasa.gov/centers/ames/research/humaninspace/humansinspace-lifescienceglovebox.html.
4. NASA Science and Mission Systems. Microgravity Science Glovebox. Overview. Available at http://msglovebox.msfc.nasa.gov/.
5. European Space Agency. Microgravity Science Glovebox. Document EUC-ESA-FSH-023, Revision 1.2. Available at http://www.spaceflight.esa.int/users/downloads/factsheets/fs023_11_msg.pdf.
6. NASA. International Space Station. Fact Sheet. Minus Eighty-Degree Laboratory Freezer for ISS. Available at http://www.nasa.gov/mission_pages/station/research/experiments/MELFI.html.
7. Hutchison, S., and Campana, S., ISS Payloads Office, NASA. 2010. "NASA Conditioned Stowage Capability," presentation for the NASA ISS Research Academy/Pre-application Meeting, August 3-5, 2010, League City, Texas, dated August 2. Available at http://www.nasa.gov/pdf/478102main_Day2_P11m_IP_JSC_ColdStow_Hutchison.pdf.
8. Hutchison, S., and Campana, S., ISS Payloads Office, NASA. 2010. "NASA Conditioned Stowage Capability," presentation for the NASA ISS Research Academy/Pre-application Meeting, August 3-5, 2010, League City, Texas, dated August 2. Available at http://www.nasa.gov/pdf/478102main_Day2_P11m_IP_JSC_ColdStow_Hutchison.pdf.
9. Hutchison, S., and Campana, S., ISS Payloads Office, NASA. 2010. "NASA Conditioned Stowage Capability," presentation for the NASA ISS Research Academy/Pre-application Meeting, August 3-5, 2010, League City, Texas, dated August 2. Available at http://www.nasa.gov/pdf/478102main_Day2_P11m_IP_JSC_ColdStow_Hutchison.pdf.
10. NASA. Human Space Flight. Space Station Assembly. Elements: EXPRESS Pallet. Available at http://spaceflight.nasa.gov/station/assembly/elements/ep/index.html.
11. Lomax, T. 2003. "ISS Centrifuge Accommodation Module (CAM) and Contents," presentation to the Space Station Utilization Advisory Subcommittee, July 29. Available at http://spaceresearch.nasa.gov/docs/ssuas/lomax_8-2003.pdf.
12. NASA. International Space Station. A New "Dawn" in Space. Available at http://www.nasa.gov/mission_pages/station/science/10-051.html.
13. NASA. International Space Station. A New "Dawn" in Space. Available at http://www.nasa.gov/mission_pages/station/research/10-051.html.
14. NASA Marshall Space Flight Center. 2005. International Space Station: Marshall Space Flight Center's Role in Development and Operations. NASA Facts. FS-2005-05-49-MSFC. Pub 8-40398. May. Available at http://www.nasa.gov/centers/marshall/pdf/115945main_iss_fs.pdf.

15. NASA. International Space Station. Space Station Assembly. Kibo Japanese Experiment Module. Available at http://www.nasa.gov/mission_pages/station/structure/elements/jem.html.
16. NASA. International Space Station. Space Station Assembly. Kibo Japanese Experiment Module. Available at http://www.nasa.gov/mission_pages/station/structure/elements/jem.html.
17. NASA. International Space Station. Fact Sheet. Human Research Facility-1 (HRF-1). Available at http://www.nasa.gov/mission_pages/station/science/experiments/HRF-1.html.
18. NASA. International Space Station. Fact Sheet. Human Research Facility-1 (HRF-1). Available at http://www.nasa.gov/mission_pages/station/science/experiments/HRF-1.html.
19. The Space Linear Acceleration Mass Measurement Device was moved to HRF-1 from HRF-2 during Expedition 11 and swapped with GASMAP.
20. NASA. International Space Station. Fact Sheet. Human Research Facility-2 (HRF-2). Available at http://www.nasa.gov/mission_pages/station/research/experiments/HRF-2.html.
21. GASMAP was moved to HRF-2 from HRF-1 and swapped with the Space Linear Acceleration Mass Measurement Device during Expedition 11.
22. NASA. International Space Station. Fact Sheet. Refrigerated Centrifuge (RC). Available at http://www.nasa.gov/mission_pages/station/science/experiments/RC.html.
23. NASA. International Space Station. Fact Sheet. Human Research Facility-2 (HRF-2). Available at http://www.nasa.gov/mission_pages/station/research/experiments/HRF-2.html.
24. NASA. International Space Station. Fact Sheet. Biological Experiment Laboratory (BioLab). Available at http://www.nasa.gov/mission_pages/station/research/experiments/BioLab.html.
25. NASA. International Space Station. Fact Sheet. European Physiology Module (EPM). Available at http://www.nasa.gov/mission_pages/station/science/experiments/EPM.html.
26. European Space Agency. *European Physiology Module.* Document EUC-ESA-FSH-014, Revision 1.1. Available at http://www.spaceflight.esa.int/users/downloads/factsheets/fs014_11_epm.pdf.
27. European Space Agency. Human Spaceflight and Exploration. *European Physiology Modules.* Available at http://esamultimedia.esa.int/docs/hsf_research/EPM_description.pdf.
28. NASA. International Space Station. Fact Sheet. Muscle Atrophy Research and Exercise System (MARES). Available at http://www.nasa.gov/mission_pages/station/research/experiments/MARES.html.
29. European Space Agency. *MARES.* Document EUC-ESA-FSH-046, Revision 1.0. Available at http://www.spaceflight.esa.int/users/downloads/factsheets/fs046_10_mares.pdf.
30. European Space Agency. Human Spaceflight Research. *ISS User Guide. Counter Measure Devices.* Available at http://esamultimedia.esa.int/docs/hsf_research/ISS_User_Guide/33_CounterMeasures_web.pdf.
31. NASA. International Space Station. Fact Sheet. Saibo Experiment Rack (Saibo). Available at http://www.nasa.gov/mission_pages/station/research/experiments/Saibo.html.
32. Japanese Aerospace Exploration Agency. Experiment Facilities Onboard Kibo's Pressurized Module (PM). Available at http://kibo.jaxa.jp/en/experiment/pm/.
33. NASA. International Space Station. Fact Sheet. Combustion Integrated Rack (CIR). Available at http://www.nasa.gov/mission_pages/station/science/experiments/CIR.html.
34. Weiland, K.J., Gati, F.G., Hill, M.E., and O'Malley, T.F., and Zurawski, R.L. 2005. *The Fluids and Combustion Facility: Enabling the Exploration of Space.* NASA/TM-2005-213973. NASA Glenn Research Center, Cleveland, Ohio. December. Available at http://gltrs.grc.nasa.gov/reports/2005/TM-2005-213973.pdf.
35. NASA Glenn Research Center. Space Flight Systems. ISS Research Project. Combustion Integrated Rack (CIR). Available at http://issresearchproject.grc.nasa.gov/FCF/CIR/.
36. NASA Glenn Research Center. Space Flight Systems. ISS Research Project. ISS Fluids and Combustion Facility. Available at http://issresearchproject.grc.nasa.gov/FCF/.
37. NASA. International Space Station. Fact Sheet. Combustion Integrated Rack (CIR). Available at http://www.nasa.gov/mission_pages/station/research/experiments/CIR.html.
38. NASA. International Space Station. Fact Sheet. Combustion Integrated Rack (CIR). Available at http://www.nasa.gov/mission_pages/station/research/experiments/CIR.html.
39. NASA. International Space Station. Fact Sheet. Fluids Integrated Rack-Fluids and Combustion Facility (FIR). Available at http://www.nasa.gov/mission_pages/station/science/experiments/FIR.html.
40. NASA Glenn Research Center. Space Flight Systems. ISS Research Project. Fluids Integrated Rack (FIR). Available at http://issresearchproject.grc.nasa.gov/FCF/FIR/.
41. NASA. International Space Station. Fact Sheet. Materials Science Research Rack-1 (MSRR-1). Available at http://www.nasa.gov/mission_pages/station/science/experiments/MSRR-1.html.

42. NASA. International Space Station. Fact Sheet. Materials Science Research Rack-1 (MSRR-1). Available at http://www.nasa.gov/mission_pages/station/research/experiments/MSRR-1.html.
43. European Space Agency. *Materials Science Laboratory.* Document EUC-ESA-FSH-022, Revision 1.1. Available at http://www.spaceflight.esa.int/users/downloads/factsheets/fs022_11_msl.pdf.
44. European Space Agency. ESA Human Spaceflight. Users. MSL. Available at http://www.spaceflight.esa.int/users/index.cfm?act=default.page&level=11&page=782.
45. NASA. International Space Station. Fact Sheet. Fluid Science Laboratory (FSL). Available at http://www.nasa.gov/mission_pages/station/science/experiments/FSL.html.
46. European Space Agency. Fluid Science Laboratory (FSL). About FSL. Welcome. Available at http://spaceflight.esa.int/users/virtualinstitutes/fsl/index1.html.
47. European Space Agency. Fluid Science Laboratory (FSL). About FSL. Welcome. Available at http://spaceflight.esa.int/users/virtualinstitutes/fsl/index1.html.
48. NASA Science and Mission Systems. Microgravity Science Glovebox. Overview. Available at http://msglovebox.msfc.nasa.gov/.
49. NASA Science and Mission Systems. Microgravity Science Glovebox. Overview. Available at http://msglovebox.msfc.nasa.gov/.
50. S.P. Korolev Rocket and Space Corporation. News. May 18, 2010. MCC-M - S.P. Korolev RSC Energia, Korolev, Moscow Region. Available at http://www.energia.ru/en/iss/mim1/photo_05-18.html.
51. NASA. International Space Station. Fact Sheet. A New "Dawn" in Space. Available at http://www.nasa.gov/mission_pages/station/science/10-051.html.
52. NASA. International Space Station. Fact Sheet. Ryutai Experiment Rack (Ryutai). Available at http://www.nasa.gov/mission_pages/station/science/experiments/Ryutai.html.
53. Japanese Aerospace Exploration Agency. RYUTAI Rack. Available at http://iss.jaxa.jp/kibo/kibomefc/ryutai_rack.html.
54. NASA. International Space Station. Fact Sheet. Expedite the Processing of Experiments for Space Station Racks (EXPRESS_Racks). Available at http://www.nasa.gov/mission_pages/station/research/experiments/EXPRESS_Racks.html.
55. Jones, R., ISS Payloads Office, NASA. 2008. "International Space Station National Lab Status," presentation to the Biotechnology Utilization Planning for the International Space Station National Laboratory, November 13-14, 2008. Available at http://www.nasa.gov/pdf/296699main_R.%20Jones.pdf.
56. NASA. International Space Station. Fact Sheet. Columbus-External Payload Facility (Columbus-EPF). Available at http://www.nasa.gov/mission_pages/station/science/experiments/Columbus-EPF.html.
57. European Space Agency. Columbus payload accommodation. Section 7.6.3 of *European Users Guide to Low Gravity Platforms*. UIC-ESA-UM-001, Issue 2, Revision 0. Available at http://www.spaceflight.esa.int/users/downloads/userguides/colaccom.pdf.
58. European Space Agency. Columbus. European Technology Exposure Facility (EuTEF). Available at http://www.esa.int/esaMI/Columbus/SEM7ZTEMKBF_0.html.
59. NASA. International Space Station. Fact Sheet. Sun Monitoring on the External Payload Facility of Columbus (Solar). Available at http://www.nasa.gov/mission_pages/station/research/experiments/Solar.html.
60. European Space Agency. *Atomic Clock Ensemble in Space.* Document EUC-ESA-FSH-031, Revision 1.0. Available at http://www.spaceflight.esa.int/users/downloads/factsheets/fs031_10_aces.pdf.
61. NASA. International Space Station. Fact Sheet. European Drawer Rack (EDR). Available at http://www.nasa.gov/mission_pages/station/research/experiments/EDR.html.
62. European Space Agency. Columbus. European Drawer Rack (EDR). Available at http://www.esa.int/esaMI/Columbus/ESA5NM0VMOC_0.html.
63. NASA. International Space Station. Fact Sheet. Japanese Experiment Module-Exposed Facility (JEM-EF). Available at http://www.nasa.gov/mission_pages/station/science/experiments/JEM-EF.html.
64. NASA. International Space Station. Fact Sheet. Window Observational Research Facility (WORF). Available at http://www.nasa.gov/mission_pages/station/science/experiments/WORF.html.
65. NASA. International Space Station. Fact Sheet. Agricultural Camera-AgCam name used historically from 2005-2010, later version known as ISAAC (AgCam). Available at http://www.nasa.gov/mission_pages/station/science/experiments/AgCam.html.
66. NASA. International Space Station. Fact Sheet. European Transportation Carrier (ETC). Available at http://www.nasa.gov/mission_pages/station/science/experiments/ETC.html.
67. European Space Agency. Available at http://www.spaceflight.esa.int/users/downloads/factsheets/fs032_10_etc.pdf.

68. European Space Agency. *European Transport Carrier.* Document EUC-ESA-FSH-032, Revision 1.0. Available at http://www.esa.int/esaMI/Columbus/SEM39T63R8F_0.html.
69. NASA. International Space Station. Fact Sheet. EXPRESS Logistics Carrier (ELC). Available at http://www.nasa.gov/mission_pages/station/science/experiments/ELC.html.
70. NASA. Commercial Crew and Cargo Program Office. Commercial Partners. Available at http://www.nasa.gov/offices/c3po/partners/index.html.
71. Aerospace-technology.com. NASA Space Shuttle Orbiter Vehicles—Discovery, Atlantis and Endeavor, USA. Specifications. Available at http://www.aerospace-technology.com/projects/discovery/specs.html.
72. NASA. Commercial Crew and Cargo Program Office. C3PO. Orbital. Available at http://www.nasa.gov/offices/c3po/partners/orbital/index.html.
73. NASA. 2008. NASA Awards Space Station Commercial Resupply Services Contracts. Press Release. December 23. Available at http://www.nasa.gov/home/hqnews/2008/dec/HQ_C08-069_ISS_Resupply.html.
74. Orbital Sciences Corporation. 2011. *Taurus® II Medium-Class Space Launch Vehicle.* Fact Sheet. Available at http://www.orbital.com/NewsInfo/Publications/TaurusII_fact.pdf.
75. Orbital Sciences Corporation. *Cygnus™ Advanced Maneuvering Spacecraft.* Fact Sheet. Available at http://www.orbital.com/NewsInfo/Publications/Cygnus_fact.pdf.
76. Orbital Sciences Corporation. *Taurus II. Medium-Class Launch Vehicle.* Brochure. Available at http://www.orbital.com/NewsInfo/Publications/TaurusII_Bro.pdf.
77. NASA. Commercial Crew and Cargo Program Office. C3PO. Space Exploration Technologies (SpaceX). Available at http://www.nasa.gov/offices/c3po/partners/spacex/index.html.
78. NASA. 2008. NASA Awards Space Station Commercial Resupply Services Contracts. Press Release. December 23. Available at http://www.nasa.gov/home/hqnews/2008/dec/HQ_C08-069_ISS_Resupply.html.
79. SpaceX. Falcon 9 Overview. Available at http://www.spacex.com/falcon9.php.
80. Bigelow Aerospace. Orbital Complex Construction. Available at http://www.bigelowaerospace.com/orbital-complex-construction.php.
81. NASA. Commercial Crew and Cargo Program Office. C3PO. Boeing. Available at http://www.nasa.gov/offices/c3po/partners/boeing/index.html.
82. Boeing Space Exploration. 2010. NASA Selects Boeing for American Recovery and Reinvestment Act Award to Study Crew Capsule-based Design. News Release. February 2. Available at http://boeing.mediaroom.com/index.php?s=43&item=1054.
83. Boeing Space Exploration. 2010. NASA Selects Boeing for American Recovery and Reinvestment Act Award to Study Crew Capsule-based Design. News Release. February 2. Available at http://boeing.mediaroom.com/index.php?s=43&item=1054.
84. NASA. Commercial Crew and Cargo Program Office. C3PO. United Launch Alliance. Available at http://www.nasa.gov/offices/c3po/partners/unitedlaunchalliance/index.html.
85. Lindenmoyer, A., NASA Johnson Space Center. 2010. "Commercial Crew and Cargo Program," presentation to the 13th Annual FAA Commercial Space Transportation Conference, Arlington, Virginia, February 11. Available at http://www.aiaa.org/pdf/industry/presentations/Lindenmoyer_C3PO.pdf.
86. Lindenmoyer, A., NASA Johnson Space Center. 2010. "Commercial Crew and Cargo Program," presentation to the 13th Annual FAA Commercial Space Transportation Conference, Arlington, Virginia, February 11. Available at http://www.aiaa.org/pdf/industry/presentations/Lindenmoyer_C3PO.pdf.
87. Lindenmoyer, A., NASA Johnson Space Center. 2010. "Commercial Crew and Cargo Program," presentation to the 13th Annual FAA Commercial Space Transportation Conference, Arlington, Virginia, February 11. Available at http://www.aiaa.org/pdf/industry/presentations/Lindenmoyer_C3PO.pdf.
88. NASA. Commercial Crew and Cargo Program Office. C3PO. Sierra Nevada. Available at http://www.nasa.gov/offices/c3po/partners/sierranevada/index.html.
89. Lindenmoyer, A., NASA Johnson Space Center. 2010. "Commercial Crew and Cargo Program," presentation to the 13th Annual FAA Commercial Space Transportation Conference, Arlington, Virginia, February 11. Available at http://www.aiaa.org/pdf/industry/presentations/Lindenmoyer_C3PO.pdf.
90. NASA. *Blue Origin Space Act Agreement.* Space Act Agreement No. NNJ1OTAO2S. Commercial Crew and Cargo Program Office, NASA Johnson Space Center, Houston, Texas. Available at http://www.nasa.gov/centers/johnson/pdf/471971main_NNJ10TA02S_blue_origin_SAA_R.pdf.
91. NASA. Commercial Crew and Cargo Program Office. C3PO. Blue Origin. Available at http://www.nasa.gov/offices/c3po/partners/blueorigin/index.html.

92. NASA. International Space Station. Space Station Assembly. Available at http://www.nasa.gov/mission_pages/station/structure/elements/progress.html.
93. Arianespace. Launch Services. Launcher Family. Available at http://www.arianespace.com/launch-services/launch-services-overview.asp.
94. European Space Agency. *Automated Transfer Vehicle.* Document EUC-ESA-FSH-003, Revision 1.2. Available at http://www.spaceflight.esa.int/users/downloads/factsheets/fs003_12_atv.pdf.
95. European Space Agency. Automated Transfer Vehicle. Mission Concept and the Role of ATV. Available at http://www.esa.int/esaMI/ATV/SEMOP432VBF_0.html.
96. European Space Agency. Automated Transfer Vehicle. ATV Evolution: Advanced Reentry Vehicle (ARV). Available at http://www.esa.int/esaMI/ATV/SEMNFZOR4CF_0.html.
97. European Space Agency. Automated Transfer Vehicle. ATV-1: Jules Verne. Available at http://www.esa.int/esaMI/ATV/SEM92X6K56G_0.html.
98. European Space Agency. Automated Transfer Vehicle. ATV-2: Johannes Kepler. Available at http://www.esa.int/esaMI/ATV/SEM8HX6K56G_0.html.
99. European Space Agency. Automated Transfer Vehicle. ATV-3: Edoardo Amaldi. Available at http://www.esa.int/esaMI/ATV/SEMIZP6K56G_0.html.
100. European Space Agency. Launch Vehicles. Launchers. Vega. Available at http://www.esa.int/SPECIALS/Launchers_Access_to_Space/ASEKMU0TCNC_0.html.
101. European Space Agency. Launchers. Intermediate Experimental Vehicle—New Video. News. October 9, 2008. Available at http://www.esa.int/esaMI/Launchers_Home/SEMQDO4N0MF_0.html.
102. European Space Agency. 2006. IXV: The Intermediate Experimental Vehicle. *ESA Bulletin* 128. November. Available at http://www.esa.int/esapub/bulletin/bulletin128/bul128h_tumino.pdf.
103. Japanese Aerospace Exploration Agency. HTV/H-IIB Special Site. Overview of the H-IIB Launch Vehicle. Available at http://www.jaxa.jp/countdown/h2bf1/overview/h2b_e.html.
104. Japanese Aerospace Exploration Agency. HTV/H-IIB Special Site. Overview of the HTV. Available at http://www.jaxa.jp/countdown/h2bf1/overview/htv_e.html.
105. Antol, J., and Headley, D.E. 1999. *The International Space Station As a Free Flyer Servicing Node.* NASA Langley Research Center, Hampton, Va. Available at http://ntrs.nasa.gov/archive/nasa/ casi.ntrs.nasa.gov/20040086756_2004091347.pdf.
106. NASA Ames Research Center. ISS Research Project. Microsatellite Free Flyer. Available at http://microsatellitefreeflyer.arc.nasa.gov/nasa_a_a.html.
107. NASA Authorization Act of 2005, Section 204. Public Law 109-155. Available at http://legislative.nasa.gov/PL109-155.pdf.
108. CubeSat Program. *CubeSat Design Specification.* Rev. 12. California Polytechnic State University, San Luis Obispo, Calif. Available at http://cubesat.org/images/developers/cds_rev12.pdf.
109. European Space Agency. Foton retrievable capsules. Section 6 of *European Users Guide to Low Gravity Platforms.* UIC-ESA-UM-001, Issue 2, Revision 0. Available at http://www.spaceflight.esa.int/users/downloads/userguides/chapter_6_foton.pdf.
110. NASA Goddard Space Flight Center. NASA's Scientific Ballooning Program. Available at http://astrophysics.gsfc.nasa.gov/balloon/.
111. NASA/Brookhaven National Laboratory Space Radiation Program. NASA Space Radiation Laboratory at Brookhaven. Available at http://www.bnl.gov/medical/NASA/NSRL_description.asp.
112. NASA Ames Research Center. Ames Technology Capabilities and Facilities. Microgravity Test Facility. Available at http://www.nasa.gov/centers/ames/research/technology-onepagers/microgravity_test_facility.html.
113. NASA Ames Research Center. Genome Research Facility. Objectives. Available at http://phenomorph.arc.nasa.gov/.
114. NASA. High-End Computing Capability. Resources. Columbia Supercomputer. Available at http://www.nas.nasa.gov/Resources/Systems/columbia.html.
115. NASA Ames Research Center. Space Biosciences Division. Data Archive Project. Available at http://spacebiosciences.arc.nasa.gov/page/data-archive-project.
116. NASA Ames Research Center. Space Biosciences Division. 20-G Centrifuge. Available at http://spacebiosciences.arc.nasa.gov/page/20-g-centrifuge.
117. NASA Ames Research Center. Space Biosciences Division. 24-Foot Diameter Centrifuge. Available at http://spacebio sciences.arc.nasa.gov/page/24-foot-diameter-centrifuge.
118. NASA Ames Research Center. Space Biosciences Division. Low Vibration Rotational Device. Available at http://spacebiosciences.arc.nasa.gov/page/low-vibration-rotational-device.

119. NASA Ames Research Center. Greenspace. Bioengineering Branch. Available at http://www.nasa.gov/centers/ames/greenspace/bioengineering.html.
120. NASA Glenn Research Center. Space Flight Systems. Human Research Program. Exercise Physiology and Countermeasures Project (ExPC): Keeping Astronauts Healthy in Reduced Gravity. Available at http://spaceflightsystems.grc.nasa.gov/Advanced/HumanResearch/Exercise/.
121. NASA Glenn Research Center. Space Flight Systems. Human Research Program. Digital Astronaut Simulates Human Body in Space. Available at http://spaceflightsystems.grc.nasa.gov/Advanced/HumanResearch/Digital/.
122. NASA Glenn Research Center. Space Flight Systems. Human Research Program. Exploration Medical Capability. Available at http://spaceflightsystems.grc.nasa.gov/Advanced/HumanResearch/Medical/.
123. NASA Glenn Research Center. 2.2 Second Drop Tower. Facility Overview. Available at http://facilities.grc.nasa.gov/drop/.
124. NASA Glenn Research Center. Zero Gravity Research Facility. Quick Facts. Available at http://facilities.grc.nasa.gov/zerog/quick.html.
125. NASA Glenn Research Center. Zero Gravity Research Facility. Facility Overview. Available at http://facilities.grc.nasa.gov/zerog/.
126. NASA Glenn Research Center. Microgravity Emissions Laboratory. Facility Overview. Available at http://facilities.grc.nasa.gov/mel/index.html.
127. NASA Glenn Research Center. Microgravity Data Archive. Our Mission. Available at http://microgravity.grc.nasa.gov/fcarchive/.
128. NASA Glenn Research Center. Space Flight Systems. ISS Research Project. GRC Telescience Support Center. Available at http://spaceflightsystems.grc.nasa.gov/Advanced/ISSResearch/TSC/.
129. Cleveland Clinic. The Cleveland Clinic Center for Space Medicine. Available at http://www.lerner.ccf.org/csm/.
130. NASA Goddard Space Flight Center. NASA Sounding Rocket Program Science and User's Website. Available at http://rscience.gsfc.nasa.gov/index.html.
131. NASA Sounding Rocket Operations Contract (NSROC). NSROC Online. Available at http://www.nsroc.com/front/html/mmframe.html.
132. NASA Goddard Space Flight Center. About Goddard. Unique Resources at Goddard. Available at http://www.nasa.gov/centers/goddard/about/unique_resources.html.
133. NASA Goddard Space Flight Center. About Goddard. Unique Resources at Goddard. Available at http://www.nasa.gov/centers/goddard/about/unique_resources.html.
134. NASA. 2008. NASA Goddard Provides Environmental Testing for Hubble Components. March 13. Available at http://www.nasa.gov/mission_pages/hubble/servicing/series/testing_chambers.html.
135. NASA Johnson Space Center. Human Test Subject Facility. Human Test Subject Facility (HTSF). Available at https://bedreststudy.jsc.nasa.gov/.
136. NASA Johnson Space Center. Human Test Subject Facility. Bed Rest Study. Available at https://bedreststudy.jsc.nasa.gov/Bedrest.aspx.
137. NASA Johnson Space Center. Human Test Subject Facility. Lunar Analog Feasibility Study. Available at https://bedreststudy.jsc.nasa.gov/Lunar.aspx.
138. NASA Johnson Space Center. Microgravity University. Welcome. Available at http://microgravityuniversity.jsc.nasa.gov/.
139. NASA. 2009. Undergraduate Students Fly High for Weightless Science. Press Release 09-295. December 22. Available at http://www.nasa.gov/home/hqnews/2009/dec/HQ_09-295_Edu_Micro-g_Flights.html.
140. NASA Johnson Space Center. NASA Reduced Gravity Program. About the NASA Reduced Gravity Research Program. Available at http://jsc-aircraft-ops.jsc.nasa.gov/Reduced_Gravity/index.html.
141. NASA. Exploration. Human Research. Kennedy Space Center. Available at http://www.nasa.gov/exploration/humanresearch/HRP_NASA/research_at_nasa_KSC.html.
142. NASA Science and Mission Systems. Microgravity Science Glovebox. Overview. Available at http://msglovebox.msfc.nasa.gov/.
143. NASA/Brookhaven National Laboratory Space Radiation Program. NASA Space Radiation Laboratory at Brookhaven. Available at http://www.bnl.gov/medical/nasa/NSRL_description.asp.
144. NASA/Brookhaven National Laboratory Space Radiation Program. Space Radiobiology. Available at http://www.bnl.gov/medical/nasa/LTSF.asp.
145. NASA/Brookhaven National Laboratory Space Radiation Program. NASA Space Radiation Laboratory at Brookhaven. Available at http://www.bnl.gov/medical/nasa/NSRL_description.asp.
146. NASA/Brookhaven National Laboratory Space Radiation Program. Discovering Space Travelers' Exposure Risks. Available at http://www.bnl.gov/medical/NASA/CAD/Administrative/NSRL_Fact_Sheet.asp.

147. Brookhaven National Laboratory. 2003. NASA space radiobiology research takes off at new Brookhaven facility. *Discover Brookhaven*, Volume 1, No. 3, Fall. Available at http://www.bnl.gov/discover/Fall_03/NSRL_3.asp.
148. See the section "Free-Flyers" for information on the Foton capsule.
149. European Space Agency. Drop tower. Section 3 of *European Users Guide to Low Gravity Platforms*. UIC-ESA-UM-001, Issue 2, Revision 0. Available at http://www.spaceflight.esa.int/users/downloads/userguides/chapter_3_drop_tower.pdf.
150. European Space Agency. Parabolic flights. Section 4 of *European Users Guide to Low Gravity Platforms*. UIC-ESA-UM-001, Issue 2, Revision 0. Available at http://www.spaceflight.esa.int/users/downloads/userguides/chapter_4_parabolic_flights.pdf.
151. European Space Agency. Sounding rockets. Section 5 of *European Users Guide to Low Gravity Platforms*. UIC-ESA-UM-001, Issue 2, Revision 0. Available at http://www.spaceflight.esa.int/users/downloads/userguides/chapter_5_sounding_rockets.pdf.
152. See the Institute for Space Medicine and Physiology home page at http://www.medes.fr/home_en.html.
153. Institute of Aerospace Medicine. Biomedical Science Support Center. Facilities. Simulation Facility for Occupational Medicine Research (AMSAN). Available at http://www.dlr.de/me/en/desktopdefault.aspx/tabid-1961/2779_read-4525/.
154. European Space Agency. International Space Station. Human Spaceflight and Exploration. European Participation. On the Ground. Control Centres. Available at http://www.esa.int/esaHS/ESAOYJ0VMOC_iss_0.html.
155. European Space Agency. International Space Station. Human Spaceflight and Exploration. European Participation. On the Ground. Control Centres. Available at http://www.esa.int/esaHS/ESAOYJ0VMOC_iss_0.html.
156. ETH Zurich. Space Biology Group. BIOTESC (Biotechnology Space Support Center). Available at http://www.spacebiol.ethz.ch/biotesc/.
157. ETH Zurich. Space Biology Group. BIOTESC (Biotechnology Space Support Center). Available at http://www.spacebiol.ethz.ch/biotesc/index.
158. B.USOC (Belgian User Support and Operation Centre). What is the B.USOC? Available at http://www.busoc.be/en/whatisbusoc.htm.
159. See the CADMOS (Centre d'Aide au Développement des activités en Micro-pesanteur et des Opérations Spatiales) home page at http://cadmos.cnes.fr/en/index.html.
160. Damec Research Aps. Background. Available at http://www.damec.dk/damec2/02-Background/022-Background.asp.
161. Damec Research Aps. Space Experience. Space Station. Available at http://www.damec.dk/damec2/05-SpaceExperience/055-ISS.asp.
162. Damec Research Aps. Space Experience. Parabolic Flights. Available at http://www.damec.dk/damec2/05-SpaceExperience/055-ParbFlight.asp.
163. European Space Agency. Human Spaceflight Research. Erasmus Centre. Erasmus User Support and Operations Centre. Available at http://www.esa.int/SPECIALS/HSF_Research/SEMCICF280G_0.html.
164. E-USOC (Spanish User Support and Operations Centre). About Us. Available at http://www.eusoc.upm.es/en/e-usoc/introduction.html.
165. Telespazio. Scientific Programs. Available at http://www.telespazio.com/programm_S_e.html.
166. German Aerospace Center (DLR). Space Operations and Astronaut Training. Microgravity User Support Center (MUSC). Available at http://www.dlr.de/musc.
167. German Aerospace Center (DLR). Space Operations and Astronaut Training. Microgravity User Support Center (MUSC). Parabolic Flights. Available at http://www.dlr.de/rb/en/desktopdefault.aspx/tabid-4536/admin-1//7433_read-11192/.
168. NTNU Samfunnsforskning AS (Center for Interdisciplinary Research in Space). Projects. Norwegian User Support and Operations Centre. Available at http://www.n-usoc.no/.
169. NTNU Samfunnsforskning AS (Center for Interdisciplinary Research in Space). Projects. Norwegian User Support and Operations Centre. Available at http://www.n-usoc.no/.
170. Atlas Aerospace. Simulators. General Description. Available at http://www.atlasaerospace.net/eng/tren.htm.
171. Yu. A. Gagarin Cosmonaut Training Center. Facilities of the Centre. Cosmonaut Training Base. Available at http://www.gctc.ru/eng/facility/default.htm.
172. Japanese Aerospace Exploration Agency. About JAXA. Field Centers. Tsukuba Space Center (TKSC). Overview. Available at http://www.jaxa.jp/about/centers/tksc/index_e.html.

4

Plant and Microbial Biology

THE ROLE OF PLANT AND MICROBIAL RESEARCH IN EXPLORING THE EFFECTS OF MICROGRAVITY

Earth's gravity has a profound effect on biological systems. The most obvious effect is to define the weight of organisms, shaping the evolution of features such as bone and muscle mass in animals and the development of supportive tissues in plants. However, subtler gravitational effects are also integral to biological processes on Earth. For example, gravity-driven buoyancy is responsible for convection and so contributes to processes as diverse as the cooling of the surface of leaves or the distribution of signaling molecules between bacteria. Such gravity-related processes are connected to the functioning of terrestrial biology at scales from the whole organism to individual cells. Life has evolved with this constant background of Earth-normal 1 *g* and so has never had to develop the capacity to adapt to any other gravitational environment. The reduced gravity of spaceflight is therefore outside the limits that have shaped terrestrial biology. This fact provides the overarching principle for the need to understand the biological impacts of spaceflight environments. Understanding how terrestrial biology responds to microgravity and partial gravity will reduce exploration risks to crews by allowing us to understand responses and so design countermeasures to potential problems. However, spaceflight also offers the unique environment of reduced gravity in which to probe and dissect biological mechanisms. Therefore, research into how terrestrial biology responds to spaceflight will both refine our understanding of what it takes to explore space and help define how conditions on Earth have shaped biology here.

Novel environments, such as spaceflight, provide science with a tool to help define the range of adaptive processes present in biological systems and set the limits on where terrestrial biology can thrive. For the National Aeronautics and Space Administration (NASA), there is also a practical and tactical reason for this research: to best enable the entire exploration process. A deeper understanding of biological processes—all biological processes—within the spaceflight environment produces a more resilient, safer, and better informed exploration architecture. Yet there is also a fundamental question at stake: can life as we know it survive and thrive off the face of Earth, and by extension, what are the limits of our kind of life, terrestrial life, in the universe? There was a time when "life" in this context would have been limited to the human organism, and the question might have been limited to the human dimension. However, given the results of recent genome sequencing projects, we now more clearly understand the relatedness and interrelatedness of all life on Earth down to its molecular underpinnings. Over the past decade, the unraveling of genomes within the tree of life has highlighted structural and functional similarities among terrestrial life forms, while exposing the key differences that allow species and individuals to be distinct.

Therefore, there is a realization that insights into plants and microbes inform us not only about their responses but also about ourselves and our relationships within the entire tree of life as it has developed on the planet Earth.

Plants and microbes are also projected to be key elements in long-term life support efforts in extraterrestrial habitats through providing fresh food and recycling air, water, and waste products. These organisms will likely have other positive, as well as potentially negative, impacts on the success of long-duration spaceflight. For example, in numerous anecdotal reports, Russian cosmonauts and U.S. astronauts mention that cultivating plants in space is different from the usual mission operations in that it provides a link with Earth. Indeed, plants have been shown to be useful as countermeasures for the mental health-related difficulties experienced by humans living in a variety of isolated, stressful, or extreme environments.[1,2,3] Similarly, maintaining proper internal microbial ecosystems is essential to human health and survival, even though microbes can present engineering challenges due to their fouling of equipment. Indeed, the effects of spaceflight on microbial populations represent a poorly defined risk to astronauts on long-term missions. Understanding the impacts of spaceflight and reduced-gravity environments on plants and microbes thus becomes an important goal to support safe, long-term human habitation in extraterrestrial environments.

This chapter focuses on the value of fundamental biological research on plants and microbes related to the exploration process. Model systems, underlying molecular mechanisms, and the need for rigorous experimental design are highlighted as the means to best develop a mechanistic understanding of the responses of microbes and plants to the spaceflight environment. That knowledge can be applied to solve exploration needs and predict solutions to anticipated challenges. In addition, and as an added benefit, a better understanding of the underlying molecular mechanisms of gravity sensing and adaptation to microgravity environments may also lead to unanticipated advances that help human life and health on Earth. For example, an improved understanding of how to optimize plant growth in the extreme environment of spaceflight may lead to strategies to increase the efficiency of terrestrial crop production, and insights into how microbial populations change in response to the stresses of spaceflight could provide clues to how microbial populations might be managed in terrestrial settings.

RESEARCH ISSUES

Overview: The Need for Modern Analyses Applied to Model Systems

The past decade has redefined our understanding of biology in terrestrial settings at the molecular, developmental, and cellular levels. The study of spaceflight biology is poised to take advantage of this new knowledge. It is clear that the recent massive strides in genome sequencing, for example, could revolutionize the design of experiments that can be conducted in space and allow scientists to answer fundamental questions about the role of gravity in biological systems. Such knowledge can also be used to optimize plant and microbial systems for the spaceflight environment. It is also clear that our collective understanding of environmental sensing and response is beginning to illuminate underlying mechanisms at the cellular and molecular levels, largely due to the rapid progress enabled by well-characterized model systems. Therefore use of such systems and of molecular analyses that allow us to directly probe these mechanisms will be key to expanding our understanding of biological processes in space. Importantly, insights from these model systems have far-reaching, cross-kingdom applications. Presented below are key issues for space-related research relating to specific sensory and response systems such as those used to react to changes in gravity, mechanical forces, atmosphere, and radiation, and to the integration of these varied stimuli acting simultaneously into a response that is likely unique to terrestrial biology in the extraterrestrial environment.

Sensory Mechanisms I: Gravity Sensing and Response Mechanisms in Plants

All Earth organisms have evolved in an environment at a constant 1 *g*, and much of their development is entrained to this all-pervasive cue. Yet the mechanisms whereby plants sense and respond to the gravity vector remain in large part unknown. Plants have developed a sensitive system to monitor the gravity vector and to respond with directional, or gravitropic, growth.[4,5,6] Such gravitropism allows the plant to maintain the correct orientation of its organs and so helps define the architecture of both the root and shoot systems. In fact, diverse aspects of plant development can be altered by gravity; for example, the weight of a branch leads to stress within the plant and the production of additional supporting materials called reaction wood.[7] The past decade has seen significant advances in our understanding of how gravity is sensed and translated into changes in plant growth and development; it has also highlighted critical gaps in our knowledge of how these events are induced.

Perception of gravity in plants is now generally accepted to involve specialized cells containing mobile starch-filled organelles, called amyloplasts, that likely act as the gravity-responsive masses triggering cellular responses.[8] In the aerial parts of the plant, these sensors lie in the vasculature, a sheath of endodermal cells surrounding tissues specialized for water and nutrient transport, whereas in the root, they are localized to the extreme tip in the root cap. Evidence also points to an unidentified second root gravity sensory system outside of the root cap.[9,10] A range of molecular components that are linked to gravity perception have been identified,[11,12] but currently still unknown are the precise molecular identities of the receptors that translate the physical force of gravity to cellular signal(s). Similarly, the identity of the immediate signals generated by this sensory system and the associated response components that encode the directional information remain to be defined. As stated in a previous National Research Council (NRC) report on space biology[13] our ignorance of the cellular gravity perception machinery remains a fundamental gap in our understanding of how gravity can affect plant growth and development. A NASA research thrust into these fundamental control mechanisms underlying plant growth and development would provide knowledge needed to design plant-based systems as an integral component of bioregenerative life support systems for extended human spaceflight, as well as provide a better understanding of plant growth control mechanisms on Earth.

The mechanisms underlying the control of subsequent plant growth responses have received intensive study, with directional transport of the plant hormone auxin emerging as a significant regulatory element. For example, proteins of the AUX/LAX, PIN, and ABCB families are now known to represent the major transporters that direct the flow of this growth-regulating hormone.[14] However, the mechanisms linking gravity perception to the correct placement and relocalization of these transporters, and to the systems that regulate their activities, still remain to be defined.[15,16] Other hormones and signals such as cytokinins, ethylene, and reactive oxygen species have also been proposed to be integral regulators of plant gravity-responsive growth,[17] and there remains a significant open question as to the interrelationships between these regulatory systems. Advancing and integrating our knowledge of plant growth control are further critical components of research for NASA to pursue. Such analyses will contribute fundamental knowledge of the controls of plant form, with potentially widespread application on Earth, where features such as growth habit (which underpinned the green revolution[18]) and even responsiveness to gravity (crop recovery after lodging,[19] where the weather has bent a crop flat to the ground) have important impacts on crop yields and harvesting. This insight into plant development and physiological responses would also be critical to our ability to design bioregenerative life support systems that incorporate plants to provide sustained replenishment of water and air and to provide food for extended crewed missions into space.

The past decade has also seen an increasing realization that the responses of plants to gravity are inextricably intermingled with those to other stimuli. Thus, perceptions of light, touch, and water gradients have all been shown to modulate the gravity response[20] and vice versa. For example, touch stimulation of a plant causes decreased gravity response,[21] whereas gravitropism may suppress the growth response of roots toward water sources.[22,23] In the spaceflight environment, plants are exposed to many stimuli other than reduced gravity. Light levels and quality, atmospheric composition, nutrient levels, and water availability are all critical elements shaping plant growth in space. Profiling the alterations in gene expression seen upon changes in gravity has also exposed a complex response network that shares common elements with reactions to other stimuli such as touch and light.[24,25] Illuminating the integration of multiple stimuli is key to understanding the effects of gravity and spaceflight. A program focused on such mechanistic understanding would present opportunities for collaboration with programs at the National Science Foundation (NSF), the U.S. Department of Agriculture, and the National Institutes of Health (NIH).

Considering the presence of a multiplicity of stimuli and the often extreme environments of spaceflight, these interactions of the gravity perception machinery with other signaling systems may have important and likely unexpected effects on plant growth in space. Robust transcriptional profiling, coupled to proteomic and metabolic analyses of the changes elicited by extended growth in microgravity and partial-gravity environments, will be essential to characterize the responses of plants and microbes to the unique challenges of spaceflight. Release of such data to the scientific community for intensive study as rapidly as possible will maximize the science return from each experiment. Raw datasets, including unprocessed molecular data that can be subjected to multiple subsequent analyses, will broaden the community of researchers engaged in elucidating the mechanisms underlying plant responses to space environments. It is therefore imperative for NASA to develop both guidelines and tools for the rapid and efficient dissemination of such datasets, in addition to publishing research studies. This effort will require balancing the needs for assessment of the robustness of the data and the interests of the principal investigators collecting the data with the need for public dissemination.

Sensory Mechanisms II: Gravity and Mechanical Sensing in Microbes

The space environmental factors that shape plant growth (e.g., microgravity, light and gas levels and quality, and nutrient and water availability) represent key common parameters across kingdoms and have critical effects on other biological systems that will be intentionally and unintentionally taken into space. These include the microbiota present within the spaceflight environment (Figure 4.1), including the large microbial populations present within and upon the crew. Despite the strictest decontamination procedures, it will not be possible to remove microbial populations from spaceflight environments and, in fact, microbiota are essential to human health and survival. For future long-duration missions, it is thus imperative to more fully understand the responses of microorganisms to the unique environment of space at levels from the individual cell and its sensory machinery to populations and microbial communities. Our current lack of an understanding of how microbial populations change during extended spaceflight represents a significant limitation on our ability to ensure the safety of the crew on such missions. Research in this area should receive high priority. In addition, an understanding of the molecular mechanisms underlying how these systems respond to conditions in space may one day allow us to genetically engineer beneficial microbes (and plants)—for example, those involved in a bioregenerative life support system—so they can better handle the environments involved in space exploration.

There has been an ongoing interest in the responses of bacteria to the spaceflight environment,[26] and these studies indicate that when grown in a liquid environment in microgravity, bacteria behave differently than when grown in the same environment under normal gravity. As described below, the limited data available indicate that spaceflight alters gene expression patterns, the transcriptome, of at least one bacterial species. There is some evidence to suggest that bacteria can acquire an increased resistance to stresses such as acid shock after experiencing a low-shear or microgravity environment.[27] However, it is unclear if bacteria and other microbes *directly* sense gravity. An alternative possibility that has yet to be fully explored is that the microgravity environment generates secondary physical parameters that affect microbial activity. Regardless of whether microbes sense gravity directly, the study of mechanisms by which bacteria sense and respond to microgravity environments will still serve several purposes: It provides a simple and tractable model system for how an organism detects and adapts to a microgravity environment and/or mechanical forces, it may lead to protocols that help avoid infections during long-duration spaceflight, and it will provide fundamental research that may enhance the treatment of infections on Earth.

Although it has not been conclusively demonstrated that bacteria directly sense gravitational forces, it is undeniable that they do detect mechanical signals. One of the primary ways any organism detects such forces is by mechanosensitive channels. Bacteria harbor at least two classes of such channels: the MscL and MscS families.[28] A well-defined and supported function of these channels is as biological "emergency release valves" that allow bacteria to survive acute decreases in external osmotic strength.[29] Although further study is required to determine if any of them play a role in sensing microgravity environments, these bacterial mechanosensors have given researchers insight into the mechanisms by which bacteria detect osmotic forces.[30] They have served, and will continue to serve, as a paradigm for how mechanosensors, including those in plants and animals, detect and respond to mechanical forces including gravity,[31] indicating that a robust basic research program will have far-reaching impact on our understanding of how organisms in general may be affected by spaceflight. For example, although plants are exquisitely sensitive to mechanical forces, such as the touch stimulations and pervasive vibrations experienced in the spaceflight environment, a plant touch sensor has yet to be characterized at the molecular level. Similarly, the

FIGURE 4.1 Microbial contaminants growing on the interior wall of the International Space Station where crew placed their clothing after working out. The panel eventually had to be replaced after attempts at decontamination failed. SOURCE: NASA Image ISS010E11563, available at http://www.nasa.gov/mission_pages/station/science/experiments/Environmental_Monitoring.html#images.

molecular sensors that detect amyoplast settling related to plant gravity detection within roots and shoots have yet to be identified. Some of the strongest candidates for such plant mechanosenors are homologs of the MscS family of bacterial channels.[32,33,34] Thus, a comparative analysis of sensors derived from model systems has enormous potential to discover fundamental principles underlying mechano- and gravity-sensing across kingdoms.

Recent work has revealed that bacteria also contain homologs of the cytoskeletal elements actin and tubulin found in eukaryotic cells. Previous work suggested that mammalian cells use tensegrity-based cytoskeletal architecture (triangular, geodesic shapes) to translate changes in mechanical forces at the membrane to corresponding molecular responses inside the cell, including changes in cell signaling.[35,36] Consistent with these ideas, in plants the actin cytoskeleton appears to play an inhibitory role in gravity perception.[37,38,39] Thus, deformation of the cytoskeleton may itself serve as a widespread component of mechano- or gravity-sensors. However, it is presently unclear if this mechanism is used by bacteria, and the concept warrants study. Here again the observations reinforce the need to develop a research portfolio aimed at searching across kingdoms for commonalities and differences in the themes of sensing and response.

Finally, bacteria have intercellular communication systems that work through molecules known as quorum sensing signals.[40,41] We have no idea how these signals may function in spaceflight environments. It is possible that what appears to be microgravity sensing may actually be a modification of other sensing systems, perhaps because of the decrease in shear forces and convection currents associated with reduced gravity environments. Studies of the activity of these intercellular communication systems and the intracellular signaling events they elicit may help resolve effects of partial gravity on bacterial activities as being direct or indirect, and may ultimately provide a probe for effects secondary to microgravity.

The study of mechanosensitive channels and cytoskeletal elements therefore holds promise for understanding the molecular mechanisms underlying how microbes detect mechanical forces. Further study should be performed to determine how these systems, as well as quorum sensing, are affected by a microgravity environment. The study of bacterial cell signaling affords the possibility of improving our understanding of the effects of microgravity on bacterial metabolic activities and provides a model to understand how microgravity might affect cellular signaling systems in general. The broad study of all of these potential sensory systems and their potential activation of microbial virulence or other morphological factors is of mutual interest to many funding agencies including NIH, NSF, and the Department of Defense, thus providing opportunities for collaboration with these agencies. Evolutionary similarities between mechanosensitive channels and cytoskeletal elements across kingdoms indicate that microbial studies would have implications over a wide range of other organisms. Only through an integrated, cross-kingdom research portfolio will NASA be able to approach these critical questions of the degree to which biological systems share fundamental themes of sensory systems and signal processing responsive to the environmental milieu of spaceflight.

Sensory Mechanisms III: Cells

Similar questions about mechanisms of sensing and response in microbes arise when considering the cells of more complex organisms. Although it is clear that some single-celled organisms can sense and respond to gravity, such as the directional swimming response of Euglena in a gravitational field (so called gravitaxis), the answers to the most basic questions such as, How does a eukaryotic cell sense and respond to gravity? or, Do all cells sense and/or respond to gravity? remain unclear. Yet cell research is a vital tool to extend our presence in space. Single-celled organisms and cells in culture afford specific research advantages for spaceflight because, relative to more complex forms of life, they present fewer logistical demands, replicate more frequently, and are a microcosm for ways in which more complex systems respond to space environments. Understanding their response to spaceflight represents a first step toward establishing what levels of gravity are necessary to sustain normal cellular function. Alterations in gravity, space radiation, and the host of environmental factors in spacecraft are thought to impact cells in many ways, including affecting proliferation, inducing chromosomal aberrations, and affecting gene expression. To determine how and why cellular reactions in space occur, consequences of the physical environment of spaceflight and reduced gravity such as alterations in convection must be differentiated from direct biological effects through sensory systems directly monitoring and responding to gravitational force.

Because complex organisms consist of many different cells and cell systems, responses of one cell type are not necessarily predictive of other cell types, but developing models of simple cells and their methods of detecting and responding to gravity is a critical step in understanding the reaction of complex multicellular organisms and their capacity to thrive in space. Modern cell biology techniques, such as analyses of gene expression, protein expression, and structure that help identify signals and their responses arising in space environments, will enable elucidation of underlying mechanisms.

Radiation Effects on Plants and Microbes

The sessile, photosynthetic lifestyle of plants means that many have evolved to tolerate levels of solar radiation reaching the surface of Earth, such as ultraviolet wavelengths of light, that can be extremely deleterious to animal biology. Similarly, microbes represent some of the most radiation-resistant organisms known.[42] However, the response of these organisms to the radiation experienced in space remains extremely poorly defined. As described more fully in Chapter 7 and a recent report from the NRC,[43] the space radiation environment presents significant challenges to biology because it differs substantially from the radiation environment on Earth and includes high-energy protons and atomic nuclei of the heavier elements. The highly charged and energetic nuclei, referred to as HZE particles, are important components of galactic cosmic radiation. HZE particles are less abundant than protons but far more hazardous. Solar disturbances can also contribute to the space radiation environment when they include solar particle events, in which high-energy protons are often the principal form of radiation. An additional source of radiation can be found in low Earth orbit, where protons are trapped in radiation belts at certain altitudes. Protons and HZE particles interact with the material of space habitats, space suits, and biological organisms. Consequently, the intensity, energy spectral characteristics, and quality of radiation inside a spacecraft or habitat differ from that in free space. The radiation dose received by biological organisms therefore varies according to the architecture of the spacecraft/habitat and when and where that organism moves around within the spacecraft.

The implementation of radiation protection is based on current understanding of the biological effects of exposure to radiation. Techniques to mitigate radiation risks include operational and scheduling procedures and physical shielding to reduce exposure to a level "as low as reasonably achievable" (based on known risks to humans). NASA's research strategy for implementing better radiation health protection centers on the premise that space radiation can be simulated in ground-based laboratories. Much NASA-funded radiation research uses ground-based accelerators at the Brookhaven National Laboratory, where dedicated beam lines and research support facilities constitute the NASA Space Radiation Laboratory (NSRL). Both HZE particles and protons can be produced at the NSRL to perform focused, mechanistic studies on the biological consequences of exposure to the components of space radiation environments. The basic knowledge obtained from these experiments enables the prediction of effects attributable to space radiation and will reduce the associated uncertainties, thus reducing the need for large safety margins and expensive shielding. At present there are indications that microbes can show high resistance to some kinds of radiation exposure,[44] but lacking is the systematic analysis of the sensitivity and responses of plants and microbes to the expected radiation environment of extended spaceflight or possible planetary outposts. Thus, research needs in spaceflight radiation include determining the fundamental effects of radiation on plants and microbes. Understanding these effects and the underlying mechanisms will foster development and implementation of strategies for operational countermeasures.

Plant and Microbial Growth Under Altered Atmospheric Pressures

The pressure and composition of Earth's atmosphere evolved over the eons to its current levels, approximately 101 kPa (1 atmosphere) with approximately 20 percent oxygen, which support current terrestrial biology. However, providing a constant 1 atmosphere of pressure throughout the entire habitable portion of a spacecraft presents significant engineering challenges. Reducing pressures where possible has been seen as an attractive option to reduce both structural mass requirements and the levels of stored gases needed to support the crew. The lower limit of pressure for human comfort during routine activities is about 34 kPa, but only if the partial pressure of oxygen is maintained at 15 to 20 kPa. While humans could safely breathe an atmosphere almost entirely of oxygen at even lower pressures, fire hazard becomes a concern unless a quenching gas and water vapor are added such that the total atmospheric pressure is 30 to 50 kPa.[45] The Mercury, Gemini, Apollo, and Skylab programs ran at 34 kPa. The space shuttle and the ISS usually run at 101 kPa, though the shuttle pressures are reduced to approximately 75 kPa during extravehicular activities (EVAs).[46,47] The atmospheric pressures aboard future vehicles and habitats are likely to be similarly variable. The Orion Crew Exploration Vehicle, for example, is designed to run at total atmospheric pressures ranging from approximately 100 kPa during ISS docking phases to approximately 54 kPa while associated with Altair and within Altair and lunar surface systems.[48,49] The choices of atmospheric pressure, which are driven by operational and engineering limitations that involve containment and transition to EVA, are likely to have profound effects on plant and microbial development.

To survive, the microbes and plants present in any spaceflight habitat or environment must be able to adapt to the environmental pressures imposed upon them. However, understanding basic physiology at lower atmospheric pressures has been largely unaddressed by NASA science. This is due in part to the fact that microbial and plant spaceflight experiments have occurred largely in the shuttle and ISS eras in which 101 kPa has been the operational norm. In addition to considerations relative to plants and microbes within human spaceflight environments, a primary and long-term goal of sustaining life in remote space locations such as the Moon or Mars is to minimize the overall cost by employing bioregenerative life support systems. Because plants tolerate pressures much lower than those required for humans, well below 25 kPa depending on the plant and its stage of growth,[50,51] plants could be grown within low-pressure habitats, thereby saving on the resources needed to maintain high atmospheric pressures.

During the past 10 years, some fundamental strides have been made in understanding the effects of atmospheric pressure and gas composition on plants and microbes. Studies at universities and international partner institutions have revealed that plants undergo a dramatic shift in gene expression as they alter metabolism in low atmospheric pressures, but they do adapt successfully in terms of plant productivity.[52,53,54] Microbial survival has been shown to depend on atmospheric pressure, especially at extremely low pressures.[55] However, for both plant and microbial responses, these studies have been limited and leave much unexplored. Such experiments can be performed in ground-based facilities.

It is critical that fundamental biological responses of plants and microbes be examined at the pressures and gas compositions that have been chosen for future spaceflight and planetary outpost systems and EVA subsystems. Because biological life support systems could be managed at lower atmospheric pressures, biological responses over a wide range of operational pressures should be investigated. Given that some vehicles and habitats may be inactivated for extended periods, the ecology of closed systems at low pressures and inactive conditions (such as reduced temperature regulation or air flow) should also be explored.

Spaceflight Syndrome I: Response to the Integrated Spaceflight Environment

Understanding the role of spaceflight environments—as a collective set of environmental components—in metabolic and physiological processes in biology, including the biology of plants and microbes, is necessary to understanding both the fundamental impacts of spaceflight on biological systems and how those impacts will influence human life support options. Although the most obvious environmental component in space is the lack of a gravity vector, there are many additional factors influencing the spaceflight environment that must also be considered and overcome. Some of those factors are independently identifiable. Radiation, for example, certainly influences biological systems. However, the spaceflight environment is actually a complex and interrelated environmental collective that arises from a range of inputs that are either intrinsic and natural (e.g., radiation, gravity) or derived from the spacecraft habitat (atmospheric composition, pressure, volatile organic compounds, variations in light spectrum, vibrations, noise, etc.). The spacecraft environment is a highly engineered volume that brings multiple potential components into environmental interactions. The integrated response to these multiple signals is likely to yield unexpected effects, especially against the unique background of microgravity. While the past decade has seen successful plant and microbial growth experiments in space, a full understanding of the deeper impacts of spaceflight on plant and microbial growth and development remains an imperative yet to be achieved.

There remain, for example, unresolved questions of whether the microgravity environment itself is deleterious to basic physiological processes in plants.[56,57] Plants grown in orbiting vehicles display reduced or altered gravitropic orientations and growth forms.[58,59] Physical processes that depend on gravity are diverse and include such important functions as particle sedimentation, isothermal settling, and buoyancy-driven convection. These functions in turn determine cellular processes such as sedimentation of organelles, chromosome movement, macromolecular assembly, convection and intracellular transport, diffusion of molecules between cells, and cytoskeletal organization.[60] Thus, the absence of gravity may by itself have profound effects on plant morphology and cellular physiology.

That said, the past decade has seen many experiments that resulted in essentially normal plant growth and development, including demonstrations of generational seed-to-seed growth on orbit,[61,62,63] which was called out as a priority in the 1998 NRC report.[64] However, changes in plant growth relative to Earth-bound controls have been documented.[65-69] Many of the recent successful plant growth experiments, for example, took place in advanced hardware that removed engineering-imposed environmental stresses and compensated for altered physical components such as diffusion and convection.[70,71] These successes were based on fundamental studies that had preceded them. Other examples have shown less than optimal growth or altered states of physiology or metabolism.[72,73] Some alterations appear to be caused by hypoxic root zone[74,75,76] conditions due to the lack of convection and plant-atmosphere gas exchange.[77,78] The data indicate that such secondary effects as root zone hypoxia remain influential in directing spaceflight changes in metabolism.[79-82] These aeration-related problems, as well as chromosomal abnormalities, lack of gravitropism, and reduced lignification, have all contributed to difficulties in growing completely normal plants in spaceflight environments.[83,84,85]

Data from the past decade show that the practice of growing plants in space continues to improve, therefore allowing a direct approach to fundamental understanding of spaceflight-induced (rather than hardware-induced) changes in metabolism. Moderately comprehensive gene expression studies are revealing genome-wide changes in gene expression of plants grown in space, relative to the controls,[86,87] but the number of replicated experiments is very low, limiting the depth of understanding. Similarly, detailed examinations of plant physiology during spaceflight have revealed both fundamental conservation of some metabolisms on orbit[88] and differences in others.[89,90] The data suggest that much remains to be learned about plant responses to the spectrum of spaceflight and exploration environments. A high-priority research program in this area will drive both practical insight into how plants (and microbes, see below) can be grown successfully in space and advances in understanding how plants respond to their environment. The improvements in hardware for plant growth in space suggest that future experiments will need to address fewer of the syndrome components due to intrinsic factors such as hardware limitations and can instead focus on fundamental extrinsic factors such as microgravity and partial gravity or radiation exposure during both spaceflight and excursions to planetary surfaces such as the Moon and Mars. Thus, continuing to apply insights gained from basic research to improving the hardware for plant-growing systems should be integral to NASA's vision for the next decade and beyond.

Transitions between microgravity, partial gravity, hyper-gravity, and 1 *g* occur as integral components of launches such as those for flights to the ISS and will occur during excursions to planetary surfaces including the Moon or Mars. Current evidence suggests that plants and microbes undergo multiple molecular and cellular responses during such transitions, yet survive launch and microgravity without gross disruption of growth and development. However, complete understanding of those transitional responses is yet to be achieved. Moreover, the fundamental question of how organisms that have spent extended periods, or their entire life, in space respond to such gravitational transitions remains unanswered. Yet these transitions will be experienced both in research settings where samples are delivered to, or recovered from spaceflight and as part of exploration missions. Exposure to spaceflight may affect subsequent plant and microbial responses to gravity and help define the response systems these organisms have to monitoring and/or responding to gravitational signals. The ability to conduct research in transitional gravity environments, together with the ability to grow plants and microbes through multiple generations offered by facilities such as the ISS, should allow for these studies to be performed on organisms entrained at specific gravity levels and thereby elucidate the biological responses to gravity transitions.

Spaceflight Syndrome II: Microbial Ecosystems and Environments

There are more bacterial than human cells in and on the human body. This microbiota is beneficial for several physiological functions including food digestion. However, the influences of microgravity and partial-gravity environments on the human microbiota have yet to be fully examined. Disturbances in the spectrum of these microbes are known to affect human health; for example, the overgrowth of certain bacteria, such as *Clostridium difficile*, in the large intestine can lead to potentially life-threatening illness. Studies need to be performed on scraped skin and fecal samples to determine the bacterial species that thrive under flight conditions, determine the effectiveness of hygiene within spacecraft, and establish whether the composition of the microbiota differs consistently from that of normal terrestrial microbiota. Some environmental conditions may accelerate changes in microbiota. For example, antibiotics can alter the normal intestinal flora. How repopulation by microbes after removal of antibiotic selection might occur in space is not known, yet a normal intestinal microbiota is critical for human health. Thus, if astronauts are treated with antibiotics, a study should be done of whether their intestinal flora returns and if so, how rapidly, whether it reflects a normal intestinal flora, and whether some oral treatment with viable microbial cultures would be beneficial in re-establishing a microbial population conducive to health.

Species that are uncommon, or that have significantly increased or decreased in number, can be studied in a "microbial observatory" on the ISS, in ground-based facilities, or both. If these studies suggest that permanent changes have occurred within the species, approaches such as microarray analysis and whole-genome sequencing can be used to determine what modifications or mutations may have occurred to shift the microbial population dynamics. The continuing decline in the cost and increase in speed of genomic analysis should facilitate the comprehensive study of any changes in these microbial populations in space. Wide dissemination of this rich collection of raw data within the scientific community will allow a variety of scientific investigations. Recent advances in defining the human microbiome make this decade an ideal time to probe the influence of spaceflight on human microbiota.

Spaceflight Syndrome III: Changes in the Virulence of Pathogens

Although several studies report reactivation of latent viruses during spaceflight, the data so far suggest that this reactivation is due to dysregulation or decreases in the host's immune response. This area is covered in depth in Chapter 6 of this report, where recommendations for future study are proposed. One must also consider fungal and bacterial infections of immunocompromised humans as potential issues during spaceflight. Bacterial pathogens pose a danger to astronauts, particularly on long spaceflights during which medical evacuation is not an option. When challenged by certain stresses, bacteria can become more resistant to antimicrobial agents. Infections forming biofilms within the body or at a tissue/catheter interface can be extremely difficult to treat. Studies have shown that *Pseudomonas aeruginosa* and *E. coli* can form biofilms in low-shear modeled microgravity,[91] suggesting that biofilms could be a problem in space as they are on Earth. Spaceflight experiments are needed to augment these limited preliminary studies to clarify whether biofilms are more common or functionally different in microgravity conditions than they are at 1 *g*.

Other studies suggest that virulence of bacteria may increase independent of biofilm formation. The best-studied case is *Salmonella enterica serovar Typhimurium*, which consistently becomes more resistant to an acid shock and modestly more virulent to mice at 1 *g* after bacterial growth under microgravity conditions.[92] Experiments aimed at understanding the underlying mechanism for this phenomenon have not yielded a solid explanation.

The ISS provides a limited platform to understand the relationship of microgravity and other aspects of space and spaceflight environments to bacterial virulence. Currently, bacterial virulence is assayed post-flight in ground-based facilities. Ultimately, one would like to study bacterial-host interactions under microgravity conditions. However, given the difficulties of containment of the potentially virulent bacteria from astronauts in a crewed flight on the ISS and the current lack of an independent life support system for such studies on these platforms, such protocols are complicated. To pursue this question, NASA should consider the use of alternative platforms such as free-flyers, as well as simple animal model systems including invertebrates, which would not pose a risk to crew health.

Microbe-Microbe Interactions

It remains unclear how space environments influence different species of bacteria and other microbes. It is conceivable, if not highly probable, that the microgravity environment, as well as other variables such as increased radiation, changes the dynamics between species competing for specific niches. Quorum sensing is a mechanism whereby bacteria influence each other's behavior.[93] Because of the changes in fluid dynamics and convection currents in microgravity and partial-gravity environments, sensing of quorum signals could be altered in the space environment. Thus, the study of quorum sensing signals in microgravity environments could yield important insight into how bacteria compete and cooperate during spaceflight.

The NASA program called Surface, Water, and Air Biocharacterization (SWAB) carries out continued classical identification and characterization of microbes, including bacteria and molds, on surfaces and in the atmosphere of the ISS.[94] Data from SWAB, including the relative abundance of each organism, may provide clues to microbial species that are more resilient in the space environment and thus should be made available to the general scientific community as a ready-made "microbial observatory." In addition to the collection of ambient samples, direct experiments that compare mixed cultures with ground-based controls should be carried out to determine how the spaceflight environment influences microbial competition. The subsequent analyses and pursuit of the molecular mechanisms underlying any selective advantages found would be of extraordinary scientific benefit, as they would yield clues about how beneficial species (e.g., microbes for a bio-recycling center or for nitrogen fixation for plants) could be genetically altered to better compete and survive the space environment.

With current capabilities and costs for whole-genome sequencing of bacteria, it is within reach to use model bacteria to follow genomic evolution of bacteria over time in space. This will provide information about selective forces and the targets of selection during spaceflight. As reflected in the recommendations at the end of this chapter, the panel favors conducting simple genome sequencing-based experiments during the next decade on the evolution of bacteria in space.

Microbe-Plant Interactions

Plants are subject to microbial and viral pathogens. As with microbe-human interactions, studies will be required to determine any changes that may occur for these pathogen-host interactions in a low-gravity environment or generally within spaceflight or surface habitats. Currently, essentially all of the programs studying the possibility of growing viable food crops in low-gravity environments have used relatively sterile conditions and hydroponics for nutrients, yet outbreaks of plant pathogens have been seen in space-flown materials.[95,96,97] However, it is unclear what microbes are likely to infect the plants, fluids, and workings of the hydroponic system in a low-gravity environment. If such facilities are to be established, for example on a lunar station, programs should be established to determine the microbiota likely to be established within these systems. As mentioned above, a long-term goal of these studies should be to determine if beneficial microbes can be selected or genetically modified to better compete with other organisms and survive the low-gravity environment.

It will be critical to determine whether changes in microbial populations and plant pathogenicity are a response to the unique elements of the spaceflight environment, such as direct effects of microgravity on microbial or plant physiology, or are effects that relate to suboptimal growth conditions such as poor humidity control or nutrient delivery in plant growth chambers designed for use in space. Such studies may not only have long-term benefits for life support systems in low-gravity environments; an increase in our understanding of how to optimize plant growth conditions/responses in the controlled environments of spaceflight may also benefit crop production in controlled environments on Earth.

Role of Plants and Microbes in Long-Term Life Support Systems

Eventually space travel will require the ability for self-sufficiency. Once mission profiles extend beyond short trips to the lunar surface, the duration of each mission will mean it will no longer remain cost-efficient or indeed feasible to dispose of all waste and resupply oxygen, water, and food to crew members from Earth. NASA has acknowledged this reality for more than two decades with programs exploring the development of both physicochemical and bioregenerative life support systems (Figure 4.2). The program on bioregenerative capabilities arose from observations that the only truly long-term, self-sustaining life support system that has a demonstrated stability and efficacy relies upon biological systems for its function; that system is the life support afforded by Earth.

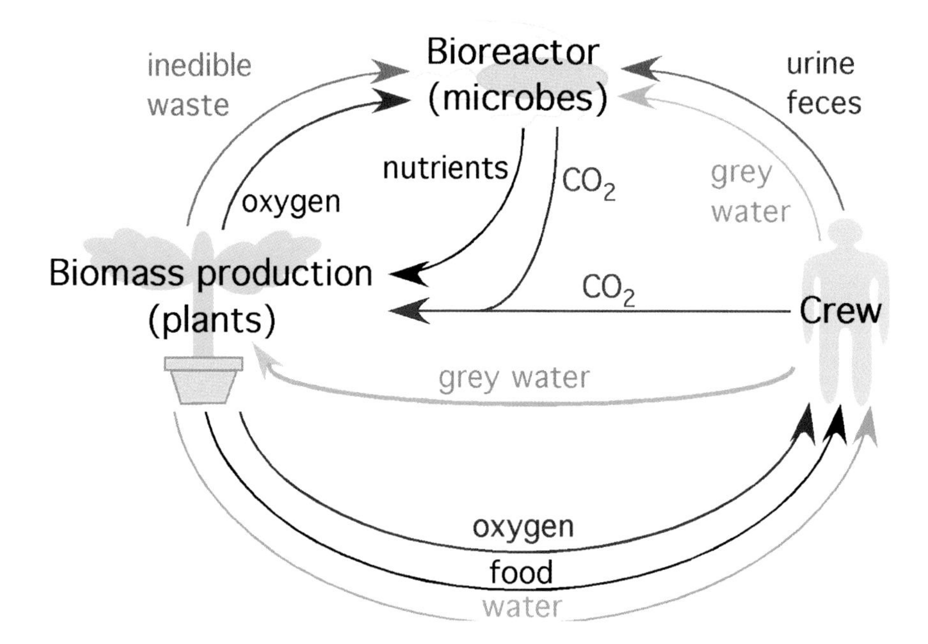

FIGURE 4.2 Bioregenerative life support systems may utilize plants and microbes to recycle waste and provide food, oxygen, and water for crews on long-duration missions. SOURCE: After D.A. Koerner, *Planting Spaces in Outer Spaces*, NASA Kennedy Space Center, Plantation, Fla., 2004; available at http://www.plantation.org/docs/landscape/planting-spaces.pdf; and P.L. Barry, Leafy Green Scientists, NASA Marshall Space Flight Center, Huntsville, Ala., 2001; available at http://science.nasa.gov/science-news/science-at-nasa/2001/ast09apr_1/.

NASA's Closed Ecological (or Controlled Environment) Life Support System (CELSS) Program of the 1980s focused primarily on food crops as key for sustained production of essential life support consumables (i.e., food, oxygen, water) to make space habitats independent of periodic resupply from Earth.[98,99,100] Initial research and development efforts maximized crop productivity and yield in controlled environments by elevating light levels, enriching atmospheric CO_2 concentrations, optimizing growth temperatures, and using well-managed hydroponic culture to prevent water and mineral nutrients from limiting crop yield. Early studies focused on wheat, potato, soybean, and lettuce as model CELSS candidate crop species.[101-104] The common outcome of these studies was yield rates exceeding world records for those crops in the field.[105,106] Calculations indicated that caloric and nutritional requirements for a space-crew diet could be met by ≤ 50 m^2 of crop-growth space per person, with air revitalization occurring by default.[107,108]

The subsequent Advanced Life-Support Program supported research on resource recovery, systems analysis, food technology, and some plant research during the 1990s.[109-114] Several multi-institutional NASA Specialized Centers of Research and Training (NSCORTs) were established in the 1990s and early 2000s to address systems integration issues for closing loops between edible-biomass production, food processing, human activities, and resource recovery in a complex life support system that would be closed with respect to mass but open with respect to energy.[115,116] Hypobaric pressure was also evaluated to reduce the structural mass required for growing plants in space.[117] Physico-chemical resource recovery was added to bioregenerative approaches. A major conclusion from this work and associated systems analysis was that growing plants for maximum productivity would cost large amounts of energy, and those requirements were well beyond the likely capabilities of an early outpost.[118] Since these energy needs have been identified as a key potential limitation to the large-scale cultivation of plants in space, it is expected that research will address the development of solar collection/photovoltaic-electrical storage to enable fiber-optic plant-growth lighting and energize light-emitting diodes (LEDs) to sustain crop production.

Investigating planetary in situ resource utilization may reduce the need for complete self-sufficiency of life support systems for habitation scenarios that are energy and/or resource limited.[119] It should therefore be an integral component of research aimed at lunar or Mars outposts. Food plants and microbes should be selected, bred, and/or engineered for functional efficacy as well as for adaptability to particular space environments. Such a selection strategy will be viable only if supported by a vigorous program in basic plant and microbial research to define how these organisms sense and respond to the varied environments presented in space.

NASA therefore should commit to a program aimed at defining the degree to which a bioregenerative life support system could enable human habitation of space that eventually becomes independent of resupply from Earth. Implementation will need to occur in stages that are affordable, incremental, and have long-term commitment of resources. Food plants will need to be a cornerstone of this effort because they alone can synthesize nutritious, edible biomass from CO_2, inorganic nutrients, and water while revitalizing the atmosphere using the energy of light. However, microbial reactors will likewise need to receive attention to provide reliable and efficient processing of the solid, liquid, and gaseous wastes of habitation. Bioregenerative systems must compete effectively with the physico-chemical/resupply strategy, so equivalent system mass needs to be reduced for crop production, for food preparation/storage, and for resource recovery. This research area may also inform the critical emerging field of sustainability science.

Because international collaborations will be essential to make rapid progress with these aims, NASA should support collaborations, where appropriate, with partners that are already pursuing these goals, such as European scientists developing component parts to the Micro-Ecological Life Support System Alternative (MELiSSA).[120]

AVAILABLE AND NEEDED PLATFORMS

To elucidate the effects of space environments on life, provide an understanding of life's foundations on Earth, and facilitate exploration, access to a variety of research platforms will be required. A portfolio approach that relies on communities of investigators using model organisms, robust technology development, and all available ground and flight platforms will greatly facilitate this endeavor. It will allow repeat investigations for critical new discoveries, decrease time from selection to flight, shorten the discovery confirmation process, and enhance the mission-driven nature of life sciences research.

Ground-Based Facilities

Ground-based research provides the basis for the design of flight research and broadens our fundamental knowledge. Space radiation in particular can be well simulated in ground-based laboratories. Accelerators at the NSRL produce both high-energy protons and the energetic nuclei of heavier elements that are important components of galactic cosmic radiation. Access to this facility will allow focused, mechanistic studies on the biological consequences of exposure of plants and microbes to the components of space radiation environments, although analysis of the effects of long-term radiation exposure for months to years simulating extended exploration missions will likely still need to be made in space.

Ground-based analysis of shared specimens and data from space experiments using new technological approaches such as transcript profiling would increase the value of research and build a broad community of researchers engaged in solving the central issues of space biology. In addition, ground-based facilities that would allow systems integration and validation for a bioregenerative life support system will need to be implemented, with the goal of testing such integration toward the end of the next 10 years, once the applicability of individual components has been rigorously verified.

Flight Platforms

Only in flight can key questions about biological responses to microgravity be addressed, providing the unique opportunity of examining terrestrial organisms without the otherwise ever-present stimulus of gravity. The ISS will provide a platform for experiments to be performed in low Earth orbit to both understand the effects of micrograv-

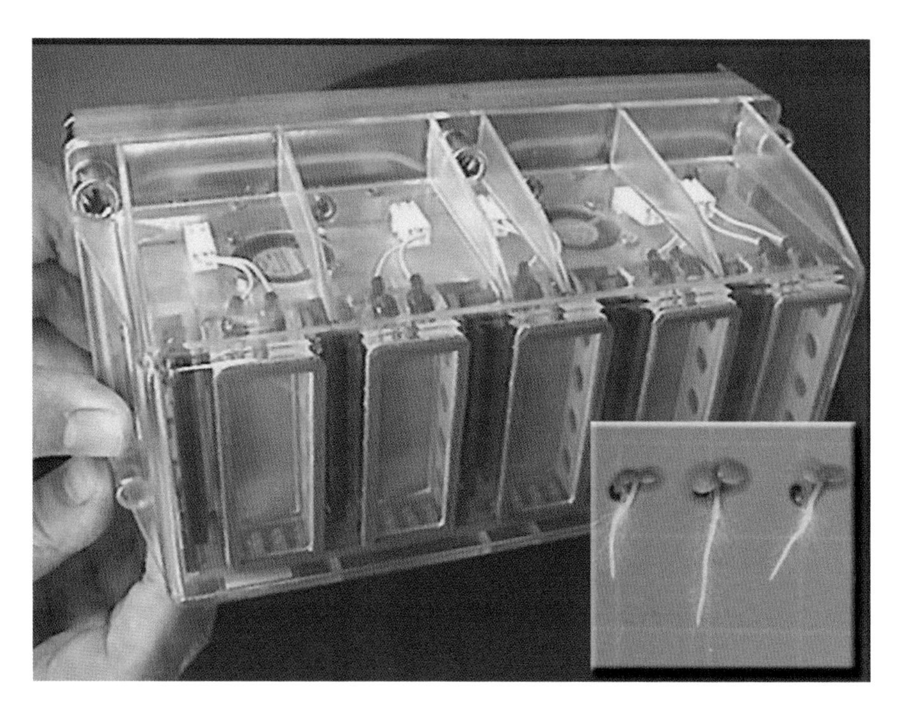

FIGURE 4.3 Current growth facilities on the ISS, such as the plant culture chambers of the European Modular Cultivation System pictured above, already provide limited access to experimental platforms to understand how the spaceflight environment affects plant growth and development. SOURCE: NASA Image 183375, available at http://www.nasa.gov/images/content/183375main_EMCS2.jpg.

ity and serve as a preparatory step to interplanetary journeys. The upcoming completion of the ISS is leading to a new era in which operations can focus on research and the development and validation of technologies to enable space exploration. In this new era, a combination of experimental design and appropriate equipment can begin to exclude the extraneous variables of the spacecraft environment to focus on critical features inherent to spaceflight such as reduced gravity and altered radiation exposures (Figure 4.3). Free-flyers would complement the research possible on the ISS. They would be especially well suited for experiments involving virulent organisms or toxic, radioactive, or otherwise dangerous materials that pose a risk to humans. Short-duration microgravity programs address the critical transition between gravity at 1 *g* and microgravity, over time frames in which many biological responses involved in adaptation are likely to occur. Suborbital platforms and parabolic flights are key platforms in the delivery of short-duration microgravity for biological studies.

The Lunar Surface as a Platform

While the currently existing and planned hardware on the ISS will allow microgravity experiments, manipulating gravitational stimulus via centrifuges to provide fractional gravity will only be possible for cells and very small organisms. We currently know that biological processes that operate properly at 1 *g* do not in microgravity, but the threshold for restoring proper function is unknown. To reduce risk and uncertainty in planning for human exploration of Mars, we need to know which biological functions will be normal in 1/3 *g*. The capability to carry out biological experimentation (especially if centrifuges are available) and test bioregenerative life support systems at lunar bases or potentially on robotic lunar landers will allow those questions to be answered. If biological functions are normal in the 1/6 *g* environment of the Moon, then they should certainly be fine on Mars. If they are still altered on the Moon, then centrifugation studies can be used to determine the impact of 1/3 *g*.

ENABLING TECHNOLOGIES

Progress in understanding the research issues outlined in this chapter will require access to space-adapted versions of the advanced tools and techniques that now support research in Earth-based laboratories, as well as to new tools developed to take advantage of the space environment. The following are some of the more important of these enabling technologies:

- Microanalytical technologies—molecular tags, liquid protein and gene arrays, reporter-based polymerase chain reaction, high-throughput sequencing;
- Miniaturized, autonomous processing and control systems for space biological research;
- In situ imaging systems to visualize changes in cell shape, configuration, and molecular tags;
- Advanced in-flight systems and modules for growth and nurturing of cells, microbes, and plants;
- In situ analysis and automated sample management/handling systems to permit remote measurements and data return;
- Advanced fixation and cryopreservation systems;
- Flight and lunar surface centrifuges;
- Noninvasive technologies to quantify radiation-induced damage to cells and tissue; and
- Ionizing radiation sources for synergistic studies on space-based platforms.

Some critical analytical approaches that can be performed on Earth in support of in-space experimentation include:

- Bioinformatics for discovery of key cellular and molecular systems necessary for biological organisms to thrive in space environments; and
- Computational models of molecular systems.

PRIORITIZED RESEARCH RECOMMENDATIONS

The research recommendations set forth below provide a delineated set of goals and approaches that encompass the science needed in support of the exploration mission and the science that is uniquely enabled by that exploration. These two closely connected concepts—the science that enables exploration and the science enabled by exploration—speak strongly to the powerful role of science within the human spaceflight endeavor. These research recommendations also depend on and define the resources needed to accomplish the delineated goals. Those resources include hardware and flight opportunities together with robust ground-based programs that place highly evolved experiments in the best position for spaceflight access.

1. NASA should establish a microbial observatory program on the ISS to conduct long-term, multigenerational studies of microbial population dynamics. The effects of the spaceflight environment on microbial population dynamics are largely unknown and represent both a significant gap in our knowledge and an important opportunity to study the evolution of microbial populations and predict health and engineering risks during long-term space exploration. As part of this effort, NASA should:

   a. Capitalize on the technological maturity, low cost, and speed of genomic analyses and the rapid generation time of microbes to monitor the evolution of microbial genomic changes in response to the selective pressures present in the spaceflight environment;

   b. Study changes in microbial populations from the skin and feces of the astronauts, plant and plant growth media, and environmental samples taken from surfaces and the atmosphere of the ISS; and

   c. Establish an experimental program targeted at understanding the influence of the spaceflight environment on defined microbial populations.

Once fully implemented, within 10 years this microbial observatory program could provide significant insight into spaceflight-induced changes in the populations of beneficial and potentially harmful microbes. The program would also provide both mechanistic understanding of these changes, for example cataloging population changes and mapping/linking these to environmental niche and genomic changes, as well as insight into practical countermeasures for mitigating risks to humans and hardware. **(P1)**

2. NASA should establish a robust spaceflight program of research analyzing plant and microbial growth in spaceflight environments and physiological responses to the multiple stimuli encountered in those environments. The effects of the complex environment of spaceflight on plant and microbial growth, physiology, and development, such as effects of altered gravity or radiation, remain poorly characterized. Understanding these responses will be invaluable for defining how biological systems respond to spaceflight. It will provide critical information required for the successful incorporation of plants and microbes into a bioregenerative life support system and have critical impact on understanding effects of importance to human exploration of space, such as possible changes in virulence of plant and animal pathogens in space. A successful research program addressing these questions would:

a. Establish a robust spaceflight program of research analyzing plant and microbial growth and physiological responses to the multiple stimuli encountered in spaceflight environments;
b. Encourage research studying the responses to individual components of spaceflight environments, such as altered gravity, radiation, and atmospheric composition, and to the integrated effects of these multiple factors; and
c. Establish as goals for this research program both characterizing the changes elicited by the components of the spaceflight environments and conducting fundamental research to understand the basic mechanisms of plant and microbial sensing and response to these stimuli.

This program should take advantage of the many recently emerged, systems-level analytical technologies such as genomics, transcriptomics, proteomics, and metabolomics. It should also apply modern cellular and molecular approaches and integrate a vigorous flight-based and ground-based research program.

By the end of 2020, a well-targeted program using model systems and looking at the spaceflight environmental factors that are currently believed to have the largest effects (such as radiation, partial gravity, and atmospheric composition) should be able to assess the individual contributions of these stimuli to the changes that have been observed in plants and microbes growing in space. This time frame also allows for significant progress toward providing a mechanistic basis for these observed changes and the development of countermeasures to mitigate changes having significant detrimental effects on the growth of plants and beneficial microbes in space **(P2)**.

3. NASA should develop a research program aimed at demonstrating the roles of microbial-plant systems in long-term life support systems. Incorporation of plants and microbes into a bioregenerative life support system represents one highly attractive avenue to sustain the crew on long-duration missions without resupply. However, the state of the technology is far from testing the feasibility of developing a robust, sustainable life support system incorporating biological systems as a major component. Such a program should:

a. Ensure careful development of each component of such a life support system in a rigorous ground-based research program with eventual validation in space; and
b. Establish the long-term goal of the program to be integration of each validated component into a life support system that can contribute to water and air purification, waste processing, and crew nutrition either alone or in conjunction with physico-chemical approaches.

A realistic time frame for this work would be construction of a ground-based test bed for systems integration toward the end of the next 10 years, after component systems have matured. Efficient systems integration could be expected in the 10-20 year time frame, but the technology may not be ready for deployment as an integrated, operational life support system on the Moon and beyond for at least 20 years. Planning should begin for ground-based life support test beds including crops, food and waste processing, resource recovery, humans or their surrogates, and systems control, but such integration efforts should not dominate the space life support agenda or budget until critical components of the system have been identified, developed, and rigorously tested for appropriate functionality. Table 4.1 gives an overview of the research timeline. **(P3)**

Table 4.1 breaks down these broad research recommendations and technology/platform requirements along a timeline based on the current status of knowledge, goals for the next decade, and long-term aims for 2020 and beyond. It also defines advances in knowledge and capabilities that can be expected if these research emphases are pursued.

PROGRAMMATIC ISSUES AND RECOMMENDATIONS

As emphasized in the 1998 NRC strategy report on space biology,[121] spaceflight opportunities should unconditionally give maximal access to peer-reviewed experiments that have a strong basis in ground tests and spaceflight performance verifications. Given the intense limitations of actual spaceflight access, no spaceflight opportunities should carry science that is not vetted by peer review within NASA or formal NASA science partnerships. NASA should coordinate agency assets, commercial payload developers, and flight systems developers in a manner that serves the best science.

Spaceflight science should also maximize repeated, multiple-sample experiment designs. All experiment profiles should include a statistical treatment that allows strong conclusions from the data. Experiments that are single runs without replication of samples and that are devoid of statistical treatment should be eliminated.

A long-term, well-supported funding base in fundamental and applied biology in space will develop a scientific community to carry out the research required to meet the prioritized science objectives. However, recent funding activities and policies have left the space biology community fragmented and less than fully committed to NASA activities. Given the time frame required for completion of the types and scales of experiments indicated in this report, typical grant funding durations should cover multiple years, with contingencies for delays in flight experiments. Stable funding of multiyear durations is essential for implementing projects that will enable a scientific community that is not only immediately responsive to short-term issues but also capable of educating the next generations of space biology scientists.

The space biology research programs will advance rapidly when supported by a robust ground research program. The ground research program will produce and refine the questions to be addressed in space. Ground research will refine the technologies to be employed in space. Ground programs also produce the range of mutant strains and other biological resources that allow spaceflight experiments to embrace and engage the most modern innovations in biological science. Ground research is critical to the longer-term biological sciences that will be engaged for life support systems. With limited orbital capacity for plant growth production, life support principles that address the scalability and applicability of plant growth in life support functions must be conducted in ground-based facilities.

Modern analytical techniques such as those employed in genomics, transcriptomics, proteomics, and metabolomics offer an immense opportunity to understand the effects of spaceflight on biological systems. Such techniques generate considerable amounts of data that can be mined and analyzed for information by multiple researchers. The creation of formalized program to promote the sharing and analysis of such data would greatly enhance the science derived from flight opportunities. Elements of such a program would include guidelines on data sharing and community access, with a focus on rapid release of these datasets while respecting the rights of the investigators conducting the experiments. A program of analysis grants, dedicated to the analysis of spaceflight-derived datasets, would provide value-added interpretation while ensuring that all data are maximally mined for information. Larger-scale multiple investigator experiments, with related science objectives, methods, and data products, would result in the production of large datasets and would emphasize analysis over implementation. Key aspects of such large-scale experiments would be replicates and statistical strength.

Biological experiments in space will benefit from a considered intermingling of automated and scientist-in-the-loop implementation. Increased automation will be required for sophisticated experiments on the ISS, free-flyers, and other platforms. Telemetric science without sample return will greatly facilitate increased sophistication in the

TABLE 4.1 Overview of Research Timeline

Research Targets	Current Status	2010-2020	2020 and Beyond	Outcomes
Understand the effects of space environments on microbial populations	SWAB program samples ISS populations, little experimentation	• Establish Microbial Observatory on the ISS to study population dynamics and genomic alterations • Generational experiments and metagenomics to investigate microbial evolution in space		Increased understanding of the fundamental mechanisms and designs of life on Earth
Determine how space environments affect organisms at critical stages of growth and development	Impacts documented, but mechanisms poorly characterized and understood Components of spaceflight syndrome not clearly separated	• Carry out controlled comprehensive analysis with model systems • Focus rigorous, highly replicated "omic"[a] analyses on mechanistic questions using a single stimulus	• Apply systems-level analyses to responses to multiple stimuli	Ability to predict the adaptation processes of cell, microorganisms, plants, and ecosystems in response to space environments
Understand gravity sensing and response systems	No consensus on gravity sensing in microbes Components identified in plants, but little systems understanding	• Resolve basis of cell and microbial responses to microgravity • Define and test molecular basis using spaceflight experiments in combination with centrifugation	• Test interactions with other stimuli • Systems analysis establishing interactions of gravity responses with other stimuli	
Mitigate and manage human infectious disease risks	Indications of altered virulence of single species of microorganisms in spaceflight conditions	• Characterize and assess critical risks by assessing effects of space environments on pathogenic and cooperative interactions among species • Advance understanding of mechanisms	• Develop and evaluate candidate countermeasures with ground analogs and spaceflight	Risks to the exploration process decreased by a mechanistic appreciation of the effects of radiation, gravity, and closed environments on plant and microbe systems, and on human health and performance
Reduce exploration costs and risks through robust, sustainable, bioregenerative life support components	ISS baseline is 90-day resupply Low-level effort in United States after five decades of research International efforts strong	• Identify critical components for rigorous ground-based programs • Emphasis on fresh food first • Trade studies for low Earth orbit, the Moon, Mars	• Integrated testing of lower equivalent system mass life support technologies and subsystems in relevant environments • Select space-optimized plants and microbes • On-orbit/lunar surface validation	
Reduce uncertainties about the risks of space radiation environments to microbes and plants	Little information about the impacts of space radiation on plants and microbes	• Expand knowledge of risk using comprehensive analyses of model systems in ground facilities, especially the NASA Space Radiation Laboratory • Validation in the combined space environment using free-flyers and external platforms on the ISS	• Establish acceptable levels of risk • Develop and test countermeasures including genetic resistance	

TABLE 4.1 Continued

Research Targets	Current Status	2010-2020	2020 and Beyond	Outcomes
Enabling technologies	Very limited on-orbit technology for molecular or cellular analysis; no technology development program	• Robust development program for on-orbit/lunar-surface technologies, esp. molecular and cellular • Coordinated, comprehensive large-scale flight experiments • Programs to facilitate community analysis of datasets	• Microanalytical technologies • Miniaturized growth, processing, and control systems for space biological research • In situ imaging systems • Centrifuges	Automated analytical capacity off-planet (ISS, free-flyers, lunar surface) and data-sharing program to maximize research return
Research platform	Ground labs including analogs; little access to shuttle, ISS	Ground laboratories including analogs; robust programs on the ISS and free-flyers	Ground laboratories including analogs; shuttle, the ISS, free-flyers, and lunar science stations	

[a]"Omic" refers to genomics, proteomics, transcriptomics, and metabolomics.

design of space biology experimentation. However, there should be a continued emphasis on keeping scientists engaged during the conduct of the experiment, to allow the experiment to be facile and responsive to the flight profile or experiment progress. That emphasis could be accomplished by designing autonomous hardware to be communicative and responsive to remote input. Future science will be enhanced by a robust technology development program that advances these principles. That emphasis can also be accomplished by keeping scientists actively involved in the conduct of the scientist-tended experiments wherever possible, such as on parabolic, suborbital, and when possible, orbital platforms.

Space biology represents a potential opportunity for coalescing disparate programmatic elements within NASA and its international partners. Biological studies discussed in this report currently have representation in multiple parts of NASA, including astrobiology, planetary protection, fundamental space biology, and exploration life sciences. A cohesive and visible voice at NASA headquarters would leverage the biological representation among programs such as planetary protection, astrobiology, and bioastronautics. Coordination with international partners, the ISS National Laboratory partners, and commercial partners would help complete the vision of space biology.

REFERENCES

1. Kanas, N., and Manzey, D. 2003. *Space Psychology and Psychiatry.* Kluwer Academic Publishers, Dordecht, The Netherlands.
2. Ulrich, R.S. 1984. View through a window may influence recovery from surgery. *Science* 224(4647):420-421.
3. Ulrich, R.S., Simons, R.F., Losito, B.D., Fiorito, E., Miles, M.A., and Zelson, M. 1991. Stress recovery during exposure to natural and urban environments. *Journal of Environmental Psychology* 11:201-230.
4. Hoson, T., and Soga, K. 2003. New aspects of gravity responses in plant cells. *International Review of Cytology* 229:209-244.
5. Morita, M.T., and Tasaka, M. 2004. Gravity sensing and signaling. *Current Opinion in Plant Biology* 7(6):712-718.
6. Perrin, R.M., Young, L.S., Murthy, U.M.N., Harrison, B.R., Wang, Y., Will, J.L., and Masson, P.H. 2005. Gravity signal transduction in primary roots. *Annals of Botany (London)* 96(5):737-743.
7. Hoson, T., and Soga, K. 2003. New aspects of gravity responses in plant cells. *International Review of Cytology* 229:209-244.
8. Perrin, R.M., Young, L.S., Murthy, U.M.N., Harrison, B.R., Wang, Y., Will, J.L., and Masson, P.H. 2005. Gravity signal transduction in primary roots. *Annals of Botany (London)* 96(5):737-743.

9. LaMotte, C.E., and Pickard, B.G. 2004. Control of gravitropic orientation. II. Dual receptor model for gravitropism. *Functional Plant Biology* 31(2):109-120.
10. Wolverton, C., Mullen, J.L., Ishikawa, H., and Evans, M.L. 2002. Root gravitropism in response to a signal originating outside of the cap. *Planta* 215(1):153-157.
11. Morita, M.T., and Tasaka, M. 2004. Gravity sensing and signaling. *Current Opinion in Plant Biology* 7(6):712-718.
12. Blancaflor, E.B., and Masson, P.H. 2003. Plant gravitropism. Unraveling the ups and downs of a complex process. *Plant Physiology* 133(4):1677-1690.
13. National Research Council. 1998. *A Strategy for Research in Space Biology and Medicine in the New Century.* National Academy Press, Washington, D.C.
14. Titapiwatanakun, B., and Murphy, A.S. 2009. Post-transcriptional regulation of auxin transport proteins: Cellular trafficking, protein phosphorylation, protein maturation, ubiquitination, and membrane composition. *Journal of Experimental Botany* 60(4):1093-1107.
15. Titapiwatanakun, B., and Murphy, A.S. 2009. Post-transcriptional regulation of auxin transport proteins: Cellular trafficking, protein phosphorylation, protein maturation, ubiquitination, and membrane composition. *Journal of Experimental Botany* 60(4):1093-1107.
16. Feraru, E., and Friml, J. 2008. PIN polar targeting. *Plant Physiology* 147(4):1553-1559.
17. Morita, M.T., and Tasaka, M. 2004. Gravity sensing and signaling. *Current Opinion in Plant Biology* 7(6):712-718.
18. Khush, G.S. 1999. Green revolution: Preparing for the 21st century. *Genome* 42(4):646-655.
19. Pinthus, M.J. 1973. Lodging in wheat, barley and oats; the phenomenon—Its causes and preventative measures. *Advances in Agronomy* 25:209-263.
20. Gilroy, S., and Masson, P.H. 2008. *Plant Troipisms.* Blackwell, Oxford, U.K.
21. Massa, G.D., and Gilroy, S. 2003. Touch modulates gravity sensing to regulate the growth of primary roots of Arabidopsis thaliana. *Plant Journal* 33(3):435-445.
22. Kobayashi, A., Takahashi, A., Kakimoto, Y., Miyazawa, Y., Fujii, N., Higashitani, A., and Takahashi, H. 2007. A gene essential for hydrotropism in roots. *Proceedings of the National Academy of Sciences U.S.A.* 104(11):4724-4729.
23. Takahashi, H., Miyazawa, Y., and Fujii, N. 2009. Hormonal interactions during root tropic growth: Hydrotropism versus gravitropism. *Plant Molecular Biology* 69(4):489-502.
24. Kimbrough, J.M., Salinas-Mondragon, R., Boss, W.F., Brown, C.S., and Sederoff, H.W. 2004. The fast and transient transcriptional network of gravity and mechanical stimulation in the Arabidopsis root apex. *Plant Physiology* 136(1):2790-2805.
25. Salinas-Mondragon, R., Brogan, A., Ward, N., Perera, I., Boss, W., Brown, C.S., and Sederoff, H.W. 2005. Gravity and light: Integrating transcriptional regulation in roots. *Gravitational and Space Biology Bulletin* 18(2):121-122.
26. Horneck, G., Mancinelli, R., and Klaus, D. 2010. Space microbiology. *Microbiology and Molecular Biology Reviews* 74(1):121-156.
27. Wilson, J.W., Ott, C.M., Höner zu Bentrup, K., Ramamurthy, R., Quick, L., Porwollik, S., Cheng, P., McClelland, M., Tsaprailis, G., Radabaugh, T., Hunt, A., et al. 2007. Space flight alters bacterial gene expression and virulence and reveals a role for global regulator Hfq. *Proceedings of the National Academy of Sciences U.S.A.* 104(41):16299-16304.
28. Pivetti, C.D., Yen, M.R., Miller, S., Busch, W., Tseng, Y.H., Booth, I.R., and Saier, M.H., Jr. 2003. Two families of mechanosensitive channel proteins. *Microbiology and Molecular Biology Reviews* 67(1):66-85.
29. Levina, N., Tötemeyer, S., Stokes, N.R., Louis, P., Jones, M.A., and Booth, I.R. 1999. Protection of Escherichia coli cells against extreme turgor by activation of MscS and MscL mechanosensitive channels: Identification of genes required for MscS activity. *The EMBO Journal* 18(7):1730-1737.
30. Blount, P., Iscla, I., and Li, Y. 2008. Mechanosensitive channels and sensing osmotic stimuli in bacteria. Pp. 25-47 in *Sensing with Ion Channels* (B. Martinac, ed.). Springer-Verlag Press, Berlin, Germany.
31. Blount, P., Li, Y., Moe, P.C., and Iscla, I. 2008. Mechanosensitive channels gated by membrane tension: Bacteria and beyond. Pp. 71-101 in *Mechanosensitive Ion Channels* (A. Kamkin and I. Kiseleva, eds.). Mechanosensitivity in Cells and Tissues, Volume 1. Springer Press, New York.
32. Haswell, E.S., and Meyerowitz, E.M. 2006. MscS-like proteins control plastid size and shape in Arabidopsis thaliana. *Current Biology* 16(1):1-11.
33. Haswell, E.S., Peyronnet, R., Barbier-Brygoo, H., Meyerowitz, E.M., and Frachisse, J.M. 2008. Two MscS homologs provide mechanosensitive channel activities in the Arabidopsis root. *Current Biology* 18(10):730-734.
34. Peyronnet, R., Haswell, E.S., Barbier-Brygoo, H., and Frachisse, J.M. 2008. AtMSL9 and AtMSL10: Sensors of plasma membrane tension in Arabidopsis roots. *Plant Signaling and Behavior* 3(9):726-729.
35. Ingber, D.E. 1997. Integrins, tensegrity, and mechanotransduction. *Gravitational and Space Biology Bulletin* 10(2):49-55.

36. Ingber, D.E. 2003. Tensegrity II. How structural networks influence cellular information processing networks. *Journal of Cell Science* 116(Pt 8):1397-1408.
37. Hou, G., Kramer, V.L., Wang, Y.S., Chen, R., Perbal, G., Gilroy, S., and Blancaflor, E.B. 2004. The promotion of gravitropism in Arabidopsis roots upon actin disruption is coupled with the extended alkalinization of the columella cytoplasm and a persistent lateral auxin gradient. *Plant Journal* 39(1):113-125.
38. Hou, G., Mohamalawari, D.R., and Blancaflor, E.B. 2003. Enhanced gravitropism of roots with a disrupted cap actin cytoskeleton. *Plant Physiology* 131(3):1360-1373.
39. Yamamoto, K., and Kiss, J.Z. 2002. Disruption of the actin cytoskeleton results in the promotion of gravitropism in inflorescence stems and hypocotyls of Arabidopsis. *Plant Physiology* 128(2):669-681.
40. Ng, W.L., and Bassler, B.L. 2009. Bacterial quorum-sensing network architectures. *Annual Review of Genetics* 43:197-222.
41. Fuqua, C., and Greenberg, E.P. 2002. Listening in on bacteria: Acyl-homoserine lactone signalling. *Nature Reviews Molecular Cell Biology* 3(9):685-695.
42. DeVeaux, L.C., Müller, J.A., Smith, J., Petrisko, J., Wells, D.P., and DasSarma, S. 2007. Extremely radiation-resistant mutants of a halophilic archaeon with increased single-stranded DNA-binding protein (RPA) gene expression. *Radiation Research* 168:507-514.
43. National Research Council. 2008. *Managing Space Radiation Risk in the New Era of Space Exploration.* The National Academies Press, Washington D.C.
44. DeVeaux, L.C., Müller, J.A., Smith, J., Petrisko, J., Wells, D.P., and DasSarma, S. 2007. Extremely radiation-resistant mutants of a halophilic archaeon with increased single-stranded DNA-binding protein (RPA) gene expression. *Radiation Research* 168:507-514.
45. Paul, A.L., and Ferl, R. 2006.The biology of low atmospheric pressure—Implications for exploration mission design and advanced life support. *Gravitational and Space Biology* 19:3-17.
46. Paul, A.L., and Ferl, R. 2006. The biology of low atmospheric pressure—Implications for exploration mission design and advanced life support. *Gravitational and Space Biology* 19:3-17.
47. NASA Exploration Atmospheres Working Group. 2006. *Recommendations for Exploration Spacecraft Internal Atmospheres.* NASA Johnson Space Center, Houston, Tex.
48. NASA Exploration Atmospheres Working Group. 2006. *Recommendations for Exploration Spacecraft Internal Atmospheres.* NASA Johnson Space Center, Houston, Tex.
49. Anderson, M., Curley, S., Stambaugh, I., and Rotter, H. 2009. Altair lander life support: Design analysis cycles 1, 2, and 3. SAE 39th International Conference on Environmental Systems. Savannah, Ga., July 12-16, 2009. SAE Technical Paper 2009-01-2477. SAE International, Warrendale, Pa.
50. Andre, M., and Richaux, C. 1986. Can plants grow in quasi-vacuum? Pp. 395-404 in *CELSS 1985 Workshop.* NASA Publication TM 88215. NASA Ames Research Center, Moffett Field, Calif.
51. McKay, C.P., and Toon, O.B. 1991. Making Mars habitable. *Nature* 352:489-496.
52. Paul, A.L., Schuerger, A.C., Popp, M.P., Richards, J.T., Manak, M.S., and Ferl, R.J. 2004. Hypobaric biology: Arabidopsis gene expression at low atmospheric pressure. *Plant Physiology* 134(1):215-223.
53. Richards, J.T., Corey, K.A., Paul, A.L., Ferl, R.J., Wheeler, R.M., and Schuerger, A.C. 2006. Exposure of Arabidopsis thaliana to hypobaric environments: implications for low-pressure bioregenerative life support systems for human exploration missions and terraforming on Mars. *Astrobiology* 6(6):851-866.
54. Corey, K.A., Barta, D.J., and Henninegr, D.L. 1999. Photosynthesis and respiration of a wheat stand at reduced atmospheric pressure and reduced oxygen. *Advances in Space Research* 20:1869-1877.
55. Schuerger, A.C., and Nicholson, W.L. 2006. Interactive effects of low pressure, low temperature, and CO_2 atmospheres inhibit the growth of Bacillus spp. under simulated martian conditions. *Icarus* 181:52-62.
56. Halstead, T.W., and Dutcher, F.R. 1984. Status and prospects. *Annals of Botany* 54(Suppl. 3):3-18.
57. Dutcher, F.R., Hess, E.L., and Halstead, T.W. 1994. Progress in plant research in space. *Advances in Space Research* 14(8):159-171.
58. Volkmann, D., Behrens, H.M., and Sievers, A. 1986. Development and gravity sensing of Cress roots under microgravity. *Naturwiss* 73:438-441.
59. Kern, V.D., Schwuchow, J.M., Reed, D.W., Nadeau, J.A., Lucas, J., Skripnikov, A., and Sack, F.D. 2005. Gravitropic moss cells default to spiral growth on the clinostat and in microgravity during spaceflight. *Planta* 221(1):149-157.
60. Todd, P. 1989. Gravity-dependent phenomena at the scale of the single cell. *American Society for Gravitational and Space Biology Bulletin* 2:95-113.
61. Musgrave, M.E., and Kuang, A. 2003. Plant reproductive development during spaceflight. *Advances in Space Biology and Medicine* 9:1-23.

62. Musgrave, M.E., and Kuang, A. 2001. Reproduction during spaceflight by plants in the family Brassicaceae. *Journal of Gravitational Physiology* 8(1):P29-P32.
63. Musgrave, M.E., Kuang, A., Xiao, Y., Stout, S.C., Bingham, G.E., Briarty, L.G., Levenskikh, M.A., Sychev, V.N., and Podolski, I.G. 2000. Gravity independence of seed-to-seed cycling in Brassica rapa. *Planta* 210(3):400-406.
64. National Research Council. 1998. *A Strategy for Research in Space Biology and Medicine in the New Century.* National Academy Press, Washington, D.C.
65. Musgrave, M.E., and Kuang, A. 2003. Plant reproductive development during spaceflight. *Advances in Space Biology and Medicine* 9:1-23.
66. Musgrave, M.E., and Kuang, A. 2001. Reproduction during spaceflight by plants in the family Brassicaceae. *Journal of Gravitational Physiology* 8(1):P29-P32.
67. Paul, A.L., Daugherty, C.J., Bihn, E.A., Chapman, D.K., Norwood, K.L., and Ferl, R.J. 2001. Transgene expression patterns indicate that spaceflight affects stress signal perception and transduction in arabidopsis. *Plant Physiology* 126(2):613-621.
68. Stutte, G.W., Monje, O., Hatfield, R.D., Paul, A.L., Ferl, R.J., and Simone, C.G. 2006. Microgravity effects on leaf morphology, cell structure, carbon metabolism and mRNA expression of dwarf wheat. *Planta* 224(5):1038-1049.
69. Stout, S.C., Porterfield, D.M., Briarty, L.G., Kuang, A., and Musgrave, M.E. 2001. Evidence of root zone hypoxia in Brassica rapa L. grown in microgravity. *International Journal of Plant Sciences* 162(2):249-255.
70. Paul, A.L., Daugherty, C.J., Bihn, E.A., Chapman, D.K., Norwood, K.L., and Ferl, R.J. 2001. Transgene expression patterns indicate that spaceflight affects stress signal perception and transduction in arabidopsis. *Plant Physiology* 126(2):613-621.
71. Stutte, G.W., Monje, O., Hatfield, R.D., Paul, A.L., Ferl, R.J., and Simone, C.G. 2006. Microgravity effects on leaf morphology, cell structure, carbon metabolism and mRNA expression of dwarf wheat. *Planta* 224(5):1038-1049.
72. Halstead, T.W., and Dutcher, F.R. 1984. Status and prospects. *Annals of Botany* 54(Suppl. 3):3-18.
73. Dutcher, F.R., Hess, E.L., and Halstead, T.W. 1994. Progress in plant research in space. *Advances in Space Research* 14(8):159-171.
74. Cowles, J.R., Scheld, H.W., Lemay, R., and Peterson, C. 1984. Growth and lignification in seedlings exposed to eight days of microgravity. *Annals of Botany* 54(Suppl. 3):3-18.
75. Krikorian, A.D., and O'Conner, S.A. 1984. Karyological observations. *Annals of Botany* 54(Suppl. 3):49-63.
76. Slocum, R.D., Gaynor, J.J., and Galston, A.W. 1984. Cytological and ultrastructural studies on root tissues. *Annals of Botany* 54(Suppl. 3):65-76.
77. Musgrave, M.E., Kuang, A., and Matthews, S.W. 1997. Plant reproduction during spaceflight: Importance of the gaseous environment. *Planta* 203(Suppl.):S177-S184.
78. Michael, G., Musgrave, M., Bard, V., Thwaites, S., Beney, J., Sumners, S., Krasner, D., and Wyndham, M. 1988. Self referral to consultants [letter]. *British Medical Journal (Clinical Research Ed.)* 296(6622):640.
79. Slocum, R.D., Gaynor, J.J., and Galston, A.W. 1984. Cytological and ultrastructural studies on root tissues. *Annals of Botany* 54(Suppl. 3):65-76.
80. Hullinger, R.L. 1993. The avian embryo responding to microgravity of space flight. *Physiologist* 36(1 Suppl):S42-S45.
81. Vartapetian, B.B. 1991. Flood sensitive plants under primary and secondary anoxia: Ultrastructure and metabolic responses. Pp. 201-216 in *Plant Life Under Oxygen Deprivation* (M.B. Jackson, D.D. Davies, and H. Lambers, eds.). SPB Academic Publishing, The Hague, Netherlands.
82. Davies, W.J., Metcalfe, J.C., Schurr, U., Taylor, G., and Zhang, J. 1987. Hormones as chemical signals involved in root to shoot communication of effects of changes in the soil environment. Pp. 206-216 in *Hormone Action in Plant Development, A Critical Appraisal* (G.V. Hoad, M.B. Jackson, J.R. Lenton, and R. Atkin, eds.). Butterworths, London, U.K.
83. Cowles, J.R., Scheld, H.W., Lemay, R., and Peterson, C. 1984. Growth and lignification in seedlings exposed to eight days of microgravity. *Annals of Botany* 54(Suppl. 3):3-18.
84. Krikorian, A.D., and O'Conner, S.A. 1984. Karyological observations. *Annals of Botany* 54(Suppl. 3):49-63.
85. Slocum, R.D., Gaynor, J.J., and Galston, A.W. 1984. Cytological and ultrastructural studies on root tissues. *Annals of Botany* 54(Suppl. 3):65-76.
86. Paul, A.L., Poppb, M.P., Gurleyc, W.B., Guyd, C., Norwoode, K.L., and Ferla, R.J. 2005. Arabidopsis gene expression patterns are altered during spaceflight. *Advances in Space Research* 36(7):1175-1181.
87. Solheim, B.G., Johnsson, A., and Iversen, T.H. 2009. Ultradian rhythms in Arabidopsis thaliana leaves in microgravity. *The New Phytologist* 183(4):1043-1052.
88. Monje, O., Stutte, G., and Chapman, D. 2005. Microgravity does not alter plant stand gas exchange of wheat at moderate light levels and saturating CO_2 concentration. *Planta* 222(2):336-345.

89. Stutte, G.W., Monje, O., Goins, G.D., and Tripathy, B.C. 2005. Microgravity effects on thylakoid, single leaf, and whole canopy photosynthesis of dwarf wheat. *Planta* 223(1):46-56.
90. Tripathy, B.C., Brown, C.S., Levine, H.G., and Krikorian, A.D. 1996. Growth and photosynthetic responses of wheat plants grown in space. *Plant Physiology* 110(3):801-806.
91. McLean, R.J., Cassanto, J.M., Barnes, M.B., and Koo, J.H. 2001. Bacterial biofilm formation under microgravity conditions. *FEMS Microbiology Letters* 195(2):115-119.
92. Wilson, J.W., Ott, C.M., Höner zu Bentrup, K., Ramamurthy, R., Quick, L., Porwollik, S., Cheng, P., McClelland, M., Tsaprailis, G., Radabaugh, T., Hunt, A., et al. 2007. Space flight alters bacterial gene expression and virulence and reveals a role for global regulator Hfq. *Proceedings of the National Academy of Sciences U.S.A.* 104(41):16299-16304.
93. Ng, W.L., and Bassler, B.L. 2009. Bacterial quorum-sensing network architectures. *Annual Review of Genetics* 43:197-222.
94. Evans, C.A., Robinson, J.A., Tate-Brown, J., Thumm, T., Crespo-Richey, J., Baumann, D., and Rhatigan, J. 2008. *International Space Station Science Research Accomplishments During the Assembly Years: An Analysis of Results from 2000-2008.* NASA/TP-2009-213146-REVISION A. NASA Center for AeroSpace Information, Hanover, Md.
95. Bishop, D.L., Levine, H.G., Kropp, B.R., and Anderson, A.J. 1997. Seedborne fungal contamination: Consequences in space-grown wheat. *Phytopathology* 87(11):1125-1133.
96. Leach, J.E., Ryba-White, M., Sun, Q., Wu, C.J., Hilaire, E., Gartner, C., Nedukha, O., Kordyum, E., Keck, M., Leung, H., and Guikema, J.A. 2001. Plants, plant pathogens, and microgravity—A deadly trio. *Gravitational and Space Biology Bulletin* 14(2):15-23.
97. Ryba-White, M., Nedukha, O., Hilaire, E., Guikema, J.A., Kordyum, E., and Leach, J.E. 2001. Growth in microgravity increases susceptibility of soybean to a fungal pathogen. *Plant Cell Physiology* 42(6):657-664.
98. Galston, A.W. 1992. Photosynthesis as a basis for life-support on Earth and in space. *Bioscience* 42(7):490-493.
99. MacElroy, R.D., and Bredt, J. 1985. *Controlled Ecological Life Support System: Life Support Systems in Space Travel.* NASA Conference Publication 2378. NASA Ames Research Center, Moffett Field, Calif.
100. NASA Office of Space Science and Applications, Life Sciences Division. 1985. *Controlled Ecological Life Support Systems (CELSS) Program Plan.* NASA, Washington, D.C.
101. Bugbee, B.G., and Salisbury, F.B. 1988. Exploring the limits of crop productivity. 1. Photosynthetic efficiency of wheat in high irradiance environments. *Plant Physiology* 88(3):869-878.
102. Knight, S.L., and Mitchell, C.A. 1983. Enhancement of lettuce yield by manipulation of light and nitrogen nutrition. *Journal of the American Society for Horticultural Science* 108(5):750-754.
103. Raper, C.D., Vessey, J.K., and Henry, L.T. 1991. Increase in nitrate uptake by soybean plants during interruption of the dark period with low intensity light. *Physiologia Plantarum* 81(2):183-189.
104. Wheeler, R.M., and Tibbitts, T.W. 1987. Utilization of potatoes for life-support-systems in space. 3. Productivity at successive harvest dates under 12-H and 24-H photoperiods. *American Potato Journal* 64(6):311-320.
105. Bugbee, B., and Monje, O. 1992. The limits of crop productivity. *Bioscience* 42(7):494-502.
106. Knight, S., and Mitchell, C.A. 1987. Stimulating productivity of hydroponic lettuce by foliar application of triacontanol. *Horticultural Science* 22:1307-1309.
107. Salisbury, F.B., and Bugbee, B. 1984. Wheat farming in a lunar base. Pp. 635-645 in *Lunar Bases and Space Activities of the 21st Century* (W.W. Mendel, ed.). Lunar and Planetary Institute, Houston, Tex.
108. Wheeler, R.M. 1996. Gas balances in a plant-based CELSS. Pp. 207-216 in *Plants in Space Biology* (H. Suge, ed.). Tohoku University, Sendai, Japan.
109. Averner, M. 1993. *NASA Advanced Life Support Program Plan.* Office of Life and Microgravity Sciences and Applications Division, NASA, Washington, D.C.
110. Hill, W.A., Loretan, P.A., Bonsi, C.K., Morris, C.E., Lu, J.Y., and Ogbuehi, C. 1989. Utilization of sweet potatoes in controlled ecological life support systems (CELSS). *Advances in Space Research* 9:1631-1635.
111. Hoff, J.E., Howe, J.M., and Mitchell, C.A. 1982. *Nutritional and Cultural Aspects of Plant Species Selection for a Controlled Ecological Life Support System.* NASA Contractor Report 166324. NASA Ames Research Center, Moffett Field, Calif.
112. Ohler, T.O., and Mitchell, C.A. 1995. Effects of carbon dioxide level and plant density on cowpea canopy productivity for a bioregenerative life support system. *Life Support and Biosphere Science* 2:3-9.
113. Tibbits, T.W., and Alford, D.K. 1982. *Controlled Ecological Life Support System Use of Higher Plants.* NASA Conf. Publ. 2231. NASA Ames Research Center, Moffett Field, Calif.
114. Volk, G.M., and Mitchell, C.A. 1995. Photoperiod shift effects on yield characteristics of rice. *Crop Science* 35(6):1631-1635.
115. Stutte, G.W., Monje, O., Goins, G.D., and Tripathy, B.C. 2005. Microgravity effects on thylakoid, single leaf, and whole canopy photosynthesis of dwarf wheat. *Planta* 223(1):46-56.

116. Tripathy, B.C., Brown, C.S., Levine, H.G., and Krikorian, A.D. 1996. Growth and photosynthetic responses of wheat plants grown in space. *Plant Physiology* 110(3):801-806.
117. Corey, K.A., Barta, D.J., and Henninegr, D.L. 1999. Photosynthesis and respiration of a wheat stand at reduced atmospheric pressure and reduced oxygen. *Advances in Space Research* 20:1869-1877.
118. Mitchell, C.A., Dougher, T.A.O., Nielsen, S.S., Belury, M.A., and Wheeler, R.M. 1996. Costs of providing edible biomass for a balanced vegetarian diet in a controlled ecological life-support system. Pp. 245-254 in *Plants in Space Biology* (H. Suge, ed.). Tohoku University, Sendai, Japan.
119. Monje, O., Stutte, G., and Chapman, D. 2005. Microgravity does not alter plant stand gas exchange of wheat at moderate light levels and saturating CO_2 concentration. *Planta* 222(2):336-345.
120. de Micco, V., Aronne, G., Joseleau, J.P., and Ruel, K. 2008. Xylem development and cell wall changes of soybean seedlings grown in space. *Annals of Botany (London)* 101(5):661-669.
121. National Research Council. 1998. *A Strategy for Research in Space Biology and Medicine in the New Century.* National Academy Press, Washington, D.C.

5

Behavior and Mental Health

Among the myriad challenges that will be faced by astronauts during long-duration space expeditions and extended stays on extraterrestrial "way stations" are three unique stressors: (1) prolonged exposure to solar and galactic radiation; (2) prolonged periods of exposure to microgravity; and (3) confinement in close, relatively austere quarters along with a small number of other crew members with whom the astronaut will need to live and work effectively for months (or even years) on end, with limited contact with family and friends and no possibility of direct outside human intervention.[1]

Accordingly, research in the next decade must include studies aimed at (1) determining the mission-specific effects of these and other relevant stressors, alone and in combination, on the astronauts' general psychological and physical well-being and (2) development of interventions to prevent, minimize, or reverse deleterious effects during extended missions. For the first of these aims, the emphasis should be on determining the extent to which such stressors impact astronauts' capacity to perform mission-related tasks, both as individuals and as members of a team—and thus, the extent to which such stressors constitute a risk to mission success.

From a behavioral standpoint, the most salient threats to astronaut effectiveness during long-duration missions include the following risks:

- Psychological and physiological well-being of individual astronauts might decline over time, impacting astronaut effectiveness and accomplishment of mission goals.
- Team cohesiveness and effectiveness might erode to the extent that team performance and mission success are jeopardized.
- Brain physiology and cognitive functioning of astronauts might be adversely affected either directly (e.g., by factors such as long-term radiation exposure) or indirectly (e.g., by chronically inadequate sleep) to the extent that mission performance and success are jeopardized.

Nevertheless, scientific studies aimed at identifying and quantifying the effects of such factors are sparse. This is largely because of understandable concerns about protecting the confidentiality of astronauts' behavioral and mental health data. For example, the 2010 National Aeronautics and Space Administration (NASA) Human Research Road Map[2] lists the following risks under the Behavioral Health and Performance category: risk of performance errors due to fatigue resulting from sleep loss, circadian desynchronization, extended wakefulness, and work overload; risk of performance decrements due to inadequate cooperation, coordination, communication, and psychosocial adaptation within a team; and risk of adverse behavioral conditions and psychiatric disorders. Yet NASA's policy has been to restrict access to psychological test material and other personal information needed to investigate how these factors affect performance, behavior, or mental health. As a result, empirical studies assessing the dynamics of these risks and their impact on mission performance and health are sparse; most of the relevant information is more anecdotal than evidence-based. Therefore, prominent among current priorities is the need to improve access to relevant data in a manner that is consistent with both the confidentiality rights of the astronauts and the scientific imperatives of NASA's mission.

COGNITIVE FUNCTIONING

To realize the full potential of crewed space missions, it is critical that appropriately selected astronauts are provided an environment that ensures not only their physical health and well-being but also their psychological health and their continued capacity for higher-order cognitive abilities such as problem solving, situational awareness, and judgment—as well as those human qualities such as inquisitiveness, courage, and determination upon which the successful exploration of space ultimately depends.

Although recent work (e.g., using functional brain imaging techniques) has begun to reveal the physiological basis of cognitive performance, there is currently no physiological marker of "general cognitive performance capacity" or "cognitive reserve."[3] At present, the only way to determine cognitive performance capacity is to administer cognitive tests designed to assess specific aspects of cognitive ability such as learning, problem-solving, emotional lability, and so on.

Because space is a hostile and unforgiving environment, even small errors in judgment or coordination can potentially have profoundly adverse consequences. Construction, repair, and exploration in space depend upon extravehicular activities, involving elaborate preparation by astronauts. In their review, Mallis and DeRoshia list a number of documented incidents involving operational performance errors by astronauts, along with evidence of irritability, impatience, declines in performance on a reasoning task, etc., during American and Soviet space missions.[4] Figure 5.1 shows a photograph of an astronaut performing an extravehicular activity with his boots on the wrong feet. Although this error resulted in no damage to astronaut, suit, or vehicle, it nevertheless serves as a prosaic reminder that lapses in attention and cognitive performance can and do occur during space missions. It is not difficult to envision serious (even catastrophic) consequences of such lapses.

Cognitive functioning considerations also pertain to ground crew. As pointed out by Mallis and DeRoshia, ground control personnel provide around-the-clock coverage of critical tasks during space missions, and they can be affected in unique ways.[5] For example, ground crews working on Mars Exploration Rover operations were required to adapt their work schedules to the 24.6-h Mars sol, which meant that work shifts were initiated 39 min later each day. Because they were exposed to Earth's 24-h light/dark cycle while continuously shifting their work schedules later each Earth day, it is likely that alertness/performance deficits accrued in these individuals due to circadian desynchrony (experienced by travelers as "jet lag"). Additional research is needed to determine the effects of such schedules on ground crew performance and to devise strategies to ensure that adequate alertness and performance in these individuals can be sustained.

Cognitive Testing

Several neuropsychological and cognitive assessment metrics and batteries currently exist; most were initially developed for the purpose of identifying deficits in neuropsychological functioning that reflect some underlying pathological state. Identification of such deficits or states during space missions is, of course, critical; accurate identification of depression, attention deficits, sleepiness, anxiety, etc., is needed to ensure application of appropriate countermeasures and to subsequently assess the efficacy of those countermeasures. However, the ultimate utility for NASA of cognitive performance testing lies in its potential to (1) inform the astronaut selection process and (2) provide meaningful data for detecting trends in the astronauts' (and the ground crew's) status during actual missions. In this context, "status" includes their capacity to perform mission-related tasks and their mood and level of psychological health/well-being at both individual and group levels. This means that cognitive tests must

FIGURE 5.1 The astronaut in this photograph has his boots on the wrong feet. It is not yet clear how often such cognitive errors occur, but astronaut error clearly constitutes a significant threat to the success of long-duration missions in space. SOURCE: NASA Image S122-E-008916 (February 15, 2008), available at http://spaceflight.nasa.gov/gallery/images/shuttle/sts-122/html/s122e008916.html.

be selected or developed that are not only optimally valid and reliable but also optimally sensitive to potentially subtle neuropsychological deficits caused by stressors unique to the astronauts' environment (e.g., microgravity, exposure to radiation, isolation, monotony, and environmental pollution) and/or the ground crew's environment (e.g., circadian desynchrony).

To the extent possible, the cognitive capacity of astronauts should be monitored using embedded measures—for instance, reaction time when working at a computer monitor or efficiency in operating a robotic arm during typical mission-related duties. The use of noninvasive performance measures could obviate the need for implementation of (at least some) dedicated assessment/testing. Any additional cognitive tests to be administered should be (1) validated against specific, mission-relevant tasks (e.g., associated with monitoring and maintenance of equipment) and/or (2) selected to reflect abilities deemed desirable for the human exploration aspects of the mission (problem solving, situational awareness, etc.).

Specific Recommendations

1. *Cognitive tests (including noninvasive embedded measures) that are relevant to or predictive of mission-related tasks should be identified or developed.* Research could be conducted mostly in laboratories and analog environments on Earth; this research enables space missions.

2. *Cognitive tests should be validated against actual mission performance.* Research should be conducted in actual work environments of interest or in high-fidelity simulators or analog environments (e.g., ground control facilities or the International Space Station [ISS]).

3. *Astronauts, ground crew, and other NASA personnel should be made aware of how cognitive and psychological research is increasingly critical for mission success and should be encouraged to participate in such research.* Psychological and cognitive research should be prominent in the astronaut training curriculum, with

an emphasis on eliminating misconceptions, negative attitudes, and fears regarding such research. This research enables space missions.

4. *NASA management should ensure that relevant data are made available for psychological and cognitive research, including previously collected (albeit de-identified) data.* This is research that enables space missions. In addition, this is research that could well be completed in the near term.

5. *Cognitive testing during the selection process should be aimed at determining not only the extent of an individual's abilities within relevant cognitive domains but also the extent to which candidates are likely to retain and effectively use cognitive capacity when experiencing mounting exposure to stressors (i.e., cognitive resilience).* Thus, measures of cognitive resilience should be identified or developed to facilitate longitudinal assessment of astronauts' capacity for sustaining performance in the face of significant stressors, particularly in the context of challenging situations such as docking, EVAs, equipment failures, etc. Efforts to measure individual characteristics such as "cognitive reserve" (which may underlie various aspects of psychological and cognitive resilience) should be developed. Test development strategies in which cognitive testing is administered under challenging conditions such as sleep loss or hypoxia might be appropriate. This research, which enables space missions, could take place in ground-based laboratories.

Although new measures may be salient, it is likely to be more productive to construct or adapt cognitive test batteries for astronauts from existing tests/resources—taking advantage of previous efforts to establish validity, reliability, and norms. Resources such as the National Institutes of Health Toolbox should be used to leverage development of optimally useful cognitive testing regimens.

6. Cognitive testing and monitoring should include ground crew, who may also be subjected to unique stressors during missions (e.g., adaptation of work schedules to the martian light/dark schedule) and upon whom mission success also depends. This is research that enables space missions.

7. Behavioral scientists should be included in the initial phases of planning for future missions involving protracted flight and for studies in which extended expeditions are simulated or undertaken.

INDIVIDUAL FUNCTIONING

The adaptation of the individual astronaut to the novel conditions of space and increasing distance from Earth must be considered in directing research activities over the next decade. Major priorities are methods and interventions to maintain optimal psychological and behavioral functioning and to prevent or treat mental disorders that might develop in the space environment.

Current research needs and priorities to support optimal individual functioning during long-duration missions in many respects mirror recommendations made in 1998 by the NRC Committee on Space Biology and Medicine on topics to be examined.[6]

Selection

The first opportunity to identify candidates who are better able to withstand the rigors of space missions is during astronaut selection. Astronauts, and later exploration crews, are chosen not only for their technical skills but also for their psychological hardiness and ability to live and work with others in a confined and extreme environment. Tests that accurately determine the optimal configuration of personality traits, coping skills, and cognitive capabilities are needed to assess functioning at the time of selection, for monitoring during space missions, and to predict the likelihood that functioning will become impaired (i.e., some type of behavioral or mental disorder will develop over time in the operational environment).

Personality Measures

Astronaut candidates have been evaluated by means of self-report personality inventories and formal psychiatric interviews to screen for subclinical and clinical disorders and family history of mental disorders, following a "select out" strategy. However, greater attention needs to be paid to the identification of less obvious subclinical and clinical personality patterns and traits that could be exacerbated during a long-duration mission and could have deleterious effects on both individual and group performance.[7] Although "select-in" strategies have been applied less frequently, they are important in identifying psychological strengths that optimize adaptive functioning in the space environment, for example, resilience in coping with significant adversity. These positive characteristics also can be assessed in different types of analog environments. Evaluation of the individual's behavior in such settings, supplemented by the use of established psychometric instruments, may serve two purposes: (1) improved screening and selection based on possession of assets or individual strengths rather than the absence of weaknesses or deficits and (2) development of interventions defined to develop resilience under adverse field situations.

Further, selection will not be entirely effective if different nations participating in exploration missions use different selection criteria. The administration of a core battery of personality measures with norms standardized for each country could improve the selection process and provide informative data on the prediction of both positive and negative outcomes involving individuals in mixed gender and international crews.

Training

Selection is, of course, only the first step toward ensuring positive psychological functioning. NASA has implemented stress management training and informal behavioral observation to ascertain how individuals deal with stress and interact with other team members. However, standardization of training programs and comprehensive research to evaluate the effectiveness of these programs need to be conducted. More effective training programs focused on dealing with family problems will help ensure the continued behavioral health of astronauts and their families in all phases of their involvement with NASA and beyond. Family members, both in the immediate nuclear family and in the extended family, have myriad concerns not just in coping with the absence of the astronaut but also about matters such as how and what to communicate with astronauts during long-duration missions.

Psychological Symptoms

Even though astronauts are psychologically healthy at the time of selection, they have reported symptoms reflecting psychological dysfunction during space missions.[8,9] To the extent that these psychological problems developed in response to the rigors of the space environment (rather than indicating inadequate screening), it is likely that such problems will be exacerbated during exploration missions involving greater isolation from Earth and separation from family and significant others over increasingly extended periods of time. Table 5.1 lists several psychological symptoms that have been reported in analog settings (polar expeditions) and space missions, respectively. Such endeavors are similar in several critical ways, and as shown, there are commonalities among the symptoms produced under both conditions.

Studies of individuals in analog environments suggest that the experience of spaceflight and the novelty of the environment may also generate positive mood and behaviors, thus enhancing positive adaptation and mitigating negative mood symptoms and possible decrements in performance.[10] These findings highlight the complexity of human adaptation to such environments and reinforce the need for additional studies utilizing analogs and high-fidelity simulations. Research on factors that might facilitate astronaut habituation to (and thereby partially offset the deleterious effects of) the harsh space capsule environment should be investigated in such analog environments. Examples of analog and simulation settings on Earth of relevance to human space missions include the following, as listed in Kanas and Manzey:[11]

- Arctic and Antarctic expeditions,
- Mountain climbing expeditions,
- Submarines and ships at sea,
- Remote sea-based oil drilling platforms,
- Underwater simulators (e.g., marine science habitats),
- Land-based simulators (e.g., hyperbaric chambers),
- Aircraft cockpit simulators, and
- Hypodynamia (bed rest) study settings.

TABLE 5.1 Psychological Symptoms Reported During Space Missions and an Analogous Setting: Polar Expeditions

Psychological Symptoms Noted During Spaceflight[a]	Psychological Symptoms During Polar Expeditions[b]
Fatigue	**Somatic symptoms**
Somatic concerns	Fatigue
Stress-related cardiac arrhythmias	Weight gain
Memory impairments	Gastrointestinal complaints
Anxiety	Rheumatic aches and pains
Depression	Headaches
Tension	**Disturbed sleep**
Interpersonal conflict	Difficulty falling asleep
Withdrawal	Difficulty staying asleep
	Loss of slow-wave sleep
	Loss of rapid eye movement sleep
	Impaired cognition
	Reduced accuracy and increased response time for cognitive tasks of memory, vigilance, attention, and reasoning
	Easily hypnotized and susceptible to suggestion
	Intellectual inertia
	Spontaneous fugue states (Antarctic stare)
	Negative affect
	Depressed mood
	Anger and irritability
	Anxiety
	Interpersonal tension and conflict
	Toward group members
	Toward people external to the group

[a]N. Kanas and D. Manzey, *Space Psychology and Psychiatry*, 2nd Edition, Microcosm Press, El Segundo, Calif., and Springer, Dordrecht, The Netherlands, 2008.

[b]L.A. Palinkas and P. Suedfeld, Psychological effects of polar expeditions, *Lancet* 371(9607):153-163, 2008.

The possibility of mental disorders developing in space is a significant concern. Both psychological and biological approaches are being pursued to identify vulnerability to psychological dysfunction. Based on extended experience with long-duration missions, the Russian space program has identified a psychophysiological condition termed asthenia, evident in approximately 60 percent of cosmonauts who have flown in space.[12] This syndrome is characterized by symptoms of fatigue and exhaustion, physical weakness, cardiovascular problems, emotional lability and irritability, concentration difficulties, and other decrements in cognitive function,[13] suggesting a disorder similar to chronic fatigue syndrome and including both psychological and physiological symptoms. Interdisciplinary research is needed to better understand and document the etiology of this condition and to inform the development of effective countermeasures. In addition, the development of strategies for managing agitation is particularly important for long-duration flight in tightly confined quarters.

Therapeutic Strategies

As missions extend beyond low Earth orbit, crews will become more autonomous, and two-way communication with Earth will be delayed (around 40 min for signals sent from Mars). Consequently, direct communication between a troubled crew member and a mental health specialist will become increasingly difficult as missions travel farther from Earth. It is therefore necessary to develop alternative, evidence-based, effective methods to prevent or alleviate mental health problems. Computer-interactive intervention and treatment programs may be more "user friendly" for astronauts than self-disclosure of personal problems by talking with a specialist. Treatment outcome and follow-up findings demonstrate that computer-based cognitive behavior therapy and other types of problem-solving intervention programs, either free-standing or with minimal additional therapist input, have proven effective.[14] Research is underway to develop multimedia computer-based, self-directed treatment programs for astronauts to autonomously manage problems with depression,[15] stress, and anxiety.[16] It is important to bring the programs to a technology readiness level (TRL) that will enable effectiveness to be systematically evaluated in comparative treatment outcome studies.

Psychoactive medications have been part of the formulary used onboard crewed space vehicles.[17] Indeed, diverse drugs acting on the central nervous system (sedative hypnotics, anxiolytics, anti-nausea agents) are the most widely used medications consumed in space. However, physiological changes due to microgravity and other effects of space may change the pharmacokinetic characteristics of medications, influencing both dosage and route of administration. These possibilities need to be studied more fully.

Also, recent medical research shows that efficacy of interventions and therapies can sometimes be enhanced by taking into account individual differences among patients. For example, increased efficacy can sometimes be achieved by customizing medications based on the individual patient's genetic profile.[18] It is likely that similar benefits will accrue with development of empirically validated, individually tailored countermeasures to prevent or mitigate psychological dysfunctions during space missions.

Integrating behavioral health evaluations into the annual flight physical examination could help identify, and perhaps even prevent, behavioral and mental health problems and would support the astronaut throughout his/her career.[19] Routine psychological evaluations may also enhance participation in behavioral research by virtue of familiarizing astronaut personnel with the nature and importance of these issues and concomitantly reducing the stigma associated with them. The post-flight period can be particularly stressful; post-flight personality changes have been reported, although in both positive and negative directions. In addition, it is important to provide psychosocial support to the astronaut's family. More comprehensive data are needed to assess the long-term psychological health of the astronaut and family.

Specific Recommendations

1. *Mission performance measures should be developed and empirically evaluated to provide criteria for assessing optimal functioning on exploration missions.* The required research could be completed in the near term if personality data already collected from the NASA space program are used in addition to current data. Such research enables space missions.
2. *To improve the selection process, a standardized core battery of robust psychological measures with established reliability and validity should be developed with adequate norms for each nation participating in exploration missions.* The test battery should assess both personality and social skill characteristics and pay greater attention to the assessment of personality patterns. This research enables space missions.
3. *Mission control personnel should also complete the personality test battery.* Their test results will contribute to a scientific understanding of individual characteristics that influence the interactions between space crews and ground personnel and the consequent effects on mission performance, particularly under conditions of high autonomy. This research enables space missions.
4. *Comprehensive training programs should be developed and standardized to provide and empirically assess the effectiveness of stress management and other training procedures for astronauts, mission control personnel, and family members across all phases of the mission, including the post-mission period.* Such research enables space missions.
5. *Interdisciplinary research should be undertaken to better understand asthenia and to develop and assess the effectiveness of countermeasures to prevent or alleviate this condition.* This research both enables and is enabled by space missions.
6. *Additional research should be performed to study the biological and psychological factors that mediate adaptation and psychological resilience to the space environment.* This research both enables and is enabled by space missions.

7. *The influence of microgravity on the psychopharmacology, efficacy, and side effect profile of psychoactive medications should be studied.* Recent findings suggest a genetic basis for individual differences in the pharmacokinetic profile of certain drugs. For example, the apolipoprotein E epsilon 4 variant, which is a known marker for susceptibility to Alzheimer's disease, is also associated with relatively exaggerated cognitive impairment following administration of benzodiazepines—a class of medications that is frequently used in space.[20] This research both enables and is enabled by space missions.

8. *Research should be conducted to identify and prevent or mitigate decrements in individual functioning. The effectiveness of verbal content analysis and other telemedicine-based techniques to identify stress and the development of mental disorders and to treat them through onboard counseling or computer-interactive programs should be studied. A variety of computer-interactive intervention programs should be developed and compared in head-to-head trials to identify the most effective and astronaut-compatible programs for use in space.* Systematic research needs to be conducted during space missions and in analog environments to evaluate the effectiveness of the currently available prevention and intervention programs, with development of new, astronaut-specific versions/programs as needed. This research both enables and is enabled by space missions.

9. In addition to astronauts, future crewed space activities may include space tourists or scientific payload specialists. *Careful attention should be given to selection characteristics of such individuals as well as to the extent of their training for the space environment.* While some guidelines are currently available for short-term space tourists, it is likely that more stringent selection criteria and training will be necessary for non-astronaut crew who are participating in longer space missions.

GROUP FUNCTIONING

In the 2001 Institute of Medicine report *Safe Passage: Astronaut Care for Exploration Missions*,[21] the following was noted: "Perhaps the matter of highest priority in the performance and general living conditions domain is the development of an evidence-based approach to the management of harmonious and productive, small, multinational groups who must live and work together in isolated, confined, and hazardous environments" (p. 145). Factors that mediate group functioning are depicted in Figure 5.2.

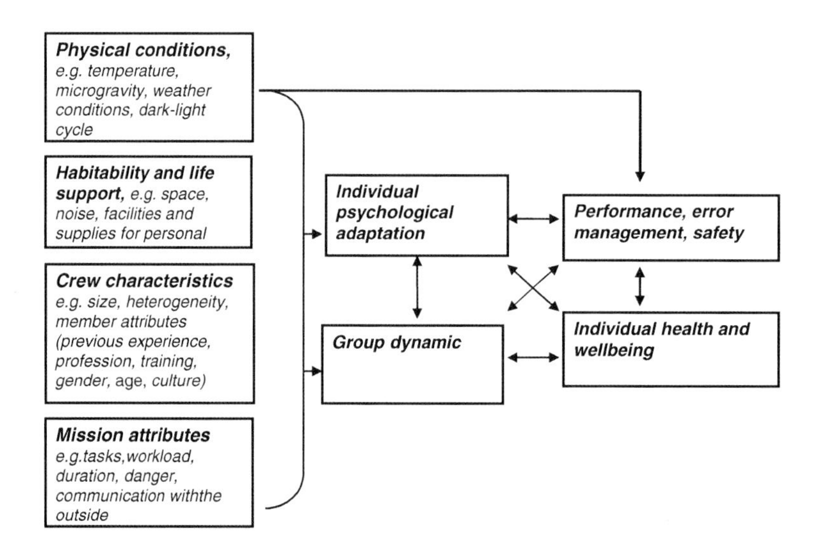

FIGURE 5.2 Interactions among factors that mediate group functioning. SOURCE: With kind permission from Springer Science+Business Media: *Reviews in Environmental Science and Biotechnology*, Human challenges in polar and space environments, Volume 5, 2006, pp. 281-296, G.M Sandal, G.R. Leon, and L. Palinkas, Figure 1. © Springer Science+Business Media B.V. 2006.

Despite the progress made in the past decade, continued research is required to identify individual, interpersonal, cultural, and environmental determinants of crew cohesion, crew performance, and ground-crew interaction and to develop and evaluate evidence-based programs and interventions that promote optimal group functioning and prevent threats to crew cohesion and productivity.

Individual Factors

Analog studies and surveys of astronaut personnel have identified individual characteristics that are predictors of social compatibility. These positive characteristics include low extraversion and high introversion,[22] high positive instrumentality (goal-oriented, active, self-confident), high expressiveness (kind, aware of others' feelings), low negative instrumentality (not arrogant, hostile, boastful, egotistical), and low communion (self-subordinating, subservient, unassertive).[23,24] However, more research is required to determine the influence of individual personality and performance on group functioning and vice versa.[25] In this regard, there is considerable overlap between research to be conducted on group functioning and research on individual functioning. Such research would include observational and experimental studies of the phenomena of "approach and avoidance"—i.e., when do individual crew members need to be a part of the larger group, and when should they be apart from the group? A better understanding of strategies for dealing with crew-imposed ostracism and marginality or self-imposed isolation and withdrawal is also required, since such understanding will be the key to development of effective countermeasures that promote crew cohesion and prevent conflict.

Interpersonal Factors

A number of important social characteristics that affect group functioning in space have been identified. One factor relates to the homogeneity and heterogeneity of crews with respect to social (e.g., age, gender, cultural background), psychological (need for achievement, aggressiveness, autonomy), and other (e.g., interest in leisure activities) characteristics that have been shown to predict crew cohesion and conflict, team decision making, and response to crises. For example, as noted by Kanas and colleagues,[26] in a situation where astronauts from one country are treated as guests in a facility or craft that is nominally under control of another (host) country, such individuals may feel subordinate and marginalized, leading to increased tension, reduced communication, and increased risk of failing to appropriately coordinate activities when performing routine assignments or responding to crises.[27]

Crews of future long-duration missions (whether for exploration or colonization) will most likely be composed of a heterogeneous mix of national, organizational, and professional cultures. Studies of cultural issues that may enhance or degrade performance among such blended groups are scarce to nonexistent. Because of large cultural differences in preferred leadership styles among likely mission participants, the question of how to optimize leadership for such a group in isolation is critical.

Culture, often described as the software of the mind, is a major determinant of performance and adjustment behavior in endeavors that require individuals to work harmoniously in teams, especially under conditions of isolation and danger.[28] Three kinds of cultures have been identified: national, organizational, and professional. National cultures are the internalized attitudes and normative beliefs associated with being a citizen of a particular country. Organizational cultures consist of norms and practices associated with membership in an organization such as a space agency or research group. Finally, professional cultures are those norms and practices internalized when achieving membership in an occupation such as physician, scientist, engineer, or astronaut.

National culture is likely to be influential in long-duration spaceflight or colonization with crews of diverse origin. Differences in organizational and professional cultures are also likely to mediate team performance. Helmreich and Merritt[29] examined cultural dimensions in organization groups such as pilots and physicians. Their results confirmed the ubiquity of cultural dimensions revealing, for example, a professional culture in medicine that stresses the need for perfection and a strong denial of the effects of stress and fatigue on performance. NASA and other space agencies have their own organizational cultures and, of course, contain diverse professional cultures, including those of astronauts and ground personnel.

Autonomy and Other Environmentally Mediated Factors

Some characteristics likely to influence group functioning are environmental in nature. Such features may include the physical environment (e.g., a lunar or martian base, spacecraft) and the organizational environment (e.g., military versus civilian control, multinational administration versus administration by predominately one organization or nation). One salient aspect of the organizational environment noted in the *Safe Passage* report is crew autonomy: the extent to which crews will be required to, or allowed to, act as autonomous entities.[30] The degree of crew autonomy will vary as a function of mission parameters (e.g., distance from Earth, length of time) and the ability of ground control systems to provide and offer timely support to the crew. However, other factors—such as the individual need for control, expression of crew hostility and stress displaced to mission control personnel, and the benefits of autonomy for team-building—must also be considered in both planning for autonomy and developing training programs to promote autonomy.

Leadership

Leadership is a critical predictor of team functioning in isolated and confined environments. Leadership issues are intimately connected with issues of crew autonomy,[31] and leadership style is significantly associated with measures of team cohesion and conflict.[32,33,34] However, additional research is required to determine whether cultural differences in the exercise of leadership are likely to produce crew tension when leaders and followers belong to different cultural groups. Additional research is also required to determine whether distributed leadership in the form of team decision making is facilitated by homogeneous crews that foster cohesion and reduce conflict or by heterogeneous crews that avoid groupthink and are therefore more likely to produce innovative solutions in emergencies.

Tools and Facilities

Research is needed to develop additional, robust measures of group processes that provide specificity and sensitivity to subtleties of group interactions and their effects on team performance. A balance of research in naturally occurring analog settings (e.g., polar and undersea research facilities) and rigorously designed experimental simulations (e.g., long-duration chamber studies) that faithfully mirror actual mission parameters (e.g., isolation, confinement, workload, long and uncertain duration, communication delays, disruption of diurnal sleep-wake cycles) will be critical.

Interventions

Development of evidence-based countermeasures designed to facilitate group functioning and prevent decrements in group functioning due to inadequate communication and coordination or due to increased tension and conflict will remain a high priority for the next decade. The issue of crew member selection also remains a primary focus, as there are currently no evidence-based guidelines for selection of individuals who will be expected to live and work together for extended periods of time. Presumably, many of the requirements for the development of such guidelines will come from studies in which optimally and non-optimally functioning crews are compared. Characteristics that distinguish crew members from one another (age, gender, cultural background, personality, interest in leisure activities) and characteristics that make crew members similar (mission identity, shared sense of purpose, socioeconomic status) are likely to be important. Identification of individual characteristics associated with the assumption of statuses and roles that facilitate or promote group cohesion (i.e., a mediator or culture broker) or exacerbate group conflict (i.e., by an instigator or nonconformist) will also be necessary.

Training

Evidence-based training programs are needed to promote behaviors that facilitate crew cohesion, conflict management, and utilization of prosocial skills under conditions of prolonged isolation and confinement. Some training may involve adaptation and evaluation of existing evidence-based practices like social-skills training under a specific set of circumstances (long-duration missions characterized by prolonged isolation and confinement).

For instance, social-skills training may include modules that enable astronauts to identify when to approach other crew members to address individual or group decrements in performance and when to avoid such interactions.

Although a number of multinational, multicultural organizations have instituted training programs to increase cultural awareness and help personnel deal with potential cultural conflicts, the efficacy of these programs has not been well validated.

Finally, more research on noninvasive techniques for monitoring group functioning and administration of support to ensure optimal functioning during extended-duration missions is recommended. Strategies for assessing and monitoring the status and performance of individual crew members should be complemented by strategies for assessing and monitoring team performance.

Specific Recommendations

To achieve an understanding of fundamental mechanisms underlying group functioning that will be critical to the success of long-duration missions and to facilitate development of countermeasures that promote optimal group functioning and prevent or defuse group tension and conflict that might lead to a decline in performance, the panel offers the following recommendations:

1. *NASA should develop new or adapt existing evidence-based tools for crew selection and for training astronauts to assume roles that facilitate crew cohesion, conflict management, and utilization of social skills under conditions of prolonged isolation and confinement.* This research both enables and is enabled by space missions.
2. *Research should be conducted to identify individual, interpersonal, cultural, and environmental determinants of crew cohesion, crew performance, and ground-crew interaction.* Such determinants include but are not limited to the reciprocal influence of individual personality and performance on group functioning; the phenomena of approach and avoidance and strategies for dealing with crew-imposed ostracism and marginality or self-imposed isolation and withdrawal; heterogeneous and homogeneous social, personality, and cultural characteristics of crew members associated with optimally functioning crews; the costs and benefits to group functioning of crew heterogeneity, formation of dyads or subgroups on the basis of shared interests, and increased crew autonomy; the impact of cultural differences in the exercise of leadership on crew cohesion and tension; and crew size dynamics for both smaller ($n = 3$) and larger crews. *The efficacy of cultural training programs should be investigated using multicultural teams as subjects in relevant environments.* This research both enables and is enabled by space missions.
3. *Development should be undertaken of additional, robust measures of group processes that could provide greater sensitivity and specificity in the measurement of the subtleties of group interactions and their relationships to team performance.* Such measures could be employed to identify potential problems early and facilitate the timing of interventions to prevent frictions from escalating into overt problems. This line of research both enables and is enabled by space missions.
4. *NASA should develop a balanced research portfolio in analog settings (e.g., the ISS, polar and undersea research facilities) and rigorously designed experimental simulations (e.g., long-duration chamber studies) that faithfully mirror actual space mission parameters (e.g., isolation, confinement, workload, long and uncertain duration, communication delays, disruption of diurnal sleep-wake cycles).* This research enables space missions. This research in analog environments can be accomplished in the intermediate term, that is, the next 3 to 5 years. The ISS also offers several advantages that allow it to function as an analog setting for future expeditionary missions to the Moon or Mars, including small crews of similarly trained and employed individuals and prolonged exposure to microgravity and radiation.
5. *Development and validation should be conducted for noninvasive techniques for monitoring group functioning and support to ensure optimal functioning during extended-duration missions.* This research enables space missions.

SLEEP AND SPACE

Historically, NASA has recognized the importance of sleep and circadian rhythms for the sustainment of cognitive functioning and has funded and conducted seminal studies in the relevant areas of sleep loss, circadian rhythms, alertness, and performance. Studies including objective measures of sleep (i.e., polysomnography and/or wrist actigraphy) have generally revealed that sleep is altered during space missions, with reduced sleep duration[35–38] and evidence of circadian desynchrony[39,40,41] that may become more severe as the duration of the space mission is extended (beyond 90 days).[42] Consistent with these findings, sleep architecture has been found to be affected during spaceflight, with some studies showing a shift in the timing of slow wave (deep) sleep toward the latter third of the sleep period[43,44] and others finding reduced slow wave sleep during the last third of the sleep period.[45] Rapid eye movement (REM) sleep is also affected, with reduced REM onset latency[46,47] and an increased amount of stage REM sleep upon return to Earth (i.e., evidence of possible REM rebound, which typically occurs in response to REM deprivation).[48] Subjective ratings of fatigue have been found to increase during some spaceflights,[49,50] but subjective fatigue is not always evident during space missions.[51] This discrepancy may reflect variations in the intensity and timing of operational demands across missions (factors that can mediate sleep duration, timing, and quality in operational environments on Earth as well), suggesting that at least some sleep and circadian rhythm-related disturbances during space missions may be amenable to work scheduling-related interventions.[52]

A variety of neurobehavioral performance deficits have been found during spaceflight, and those deficits have clear implications for astronaut safety and mission success (for a review, see Mallis and DeRoshia[53]). Although it is difficult to specify the extent to which sleep loss and fatigue have contributed to actual errors and accidents during missions in space, sleep loss and fatigue have been recognized as likely contributory factors in incidents such as the Mir-Progress collision on June 25, 1997.[54]

NASA's long-term focus on sleep research has been, and remains, appropriate since it is clear that adequate sleep, by means of physiological processes that are not yet understood, constitutes a necessary precondition for normal cognitive functioning—and thus for safe and successful space exploration.

Effects of Acute Sleep Loss

Prior studies have shown that performance on a variety of cognitive tasks ranging from simple reaction time to situation awareness and problem solving varies as a function of sleep duration, time since awakening, and phase of the circadian rhythm of alertness.[55] The biochemical processes that underlie the alertness and cognitive performance deficits that accrue with acute sleep loss—and the complementary biochemical processes by which alertness and mental agility are restored during subsequent sleep—are as yet unknown.[56] However, functional brain imaging studies have shown that sleepiness is characterized by deactivation of regional brain activity, with the greatest deactivations occurring in the thalamus (which mediates alertness and attention); the inferior parietal/superior temporal cortex (a heteromodal association area that mediates arithmetic problem solving); and, most notably, the prefrontal cortex, which mediates virtually all higher-order mental abilities including judgment, planning, and problem solving (see Figure 5.3).[57] It is clear that such regional deactivations underlie, or at least reflect, specific performance deficits, since it is possible to actually predict specific sleep-loss-induced performance deficits based on this pattern of regional brain deactivation.[58]

Effects of Chronic Sleep Restriction

Many sleep-loss studies have been conducted over the past century. Typically, the behavioral and physiological effects of one to three nights without any sleep (total sleep deprivation) were measured in these studies. However, only recently have the potential effects of chronic (weeks, months, or longer) sleep restriction become a focus of scientific research—despite the fact that chronic sleep restriction is far more prevalent, and therefore potentially a bigger problem, than total sleep deprivation in both Earth- and space-based operational environments.

Although relatively few in number, sleep restriction studies have revealed two important consequences: First, subjects do not realize that their performance is affected—i.e., *subjective* habituation to chronically inadequate sleep can occur—although no *objective* adaptation occurs and their cognitive performance and alertness remain decremented for the duration of the sleep restriction period.[59] Second, long-term changes occur in brain physiology and hormonal profiles, and these changes can have potentially detrimental effects on health. For example,

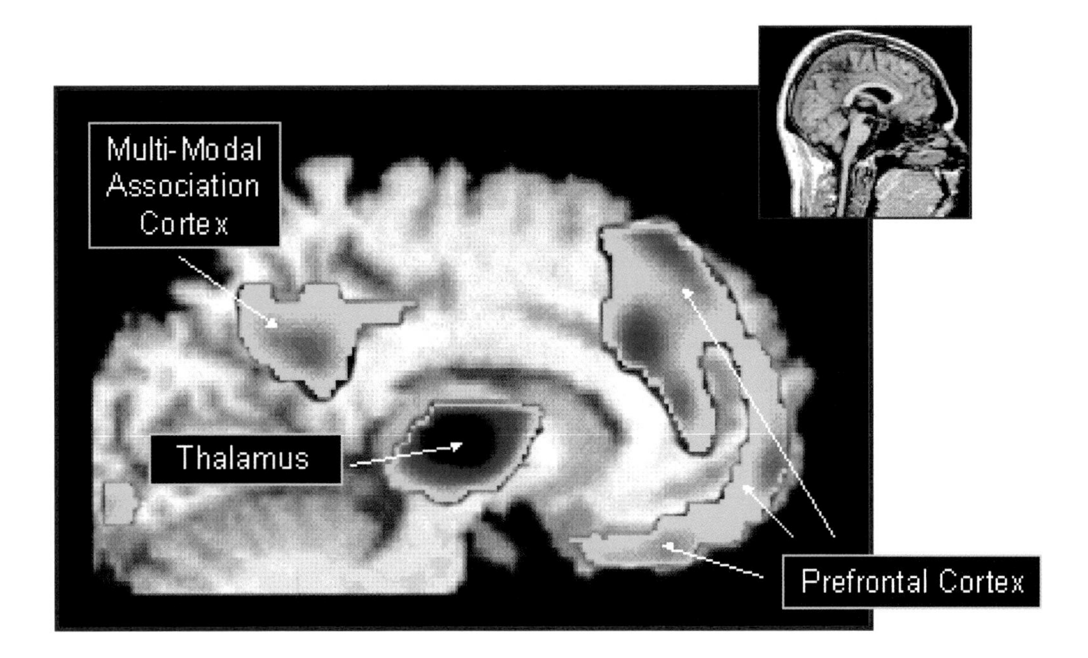

FIGURE 5.3 Prominent among those brain regions showing partial deactivation (as indicated by reduced fluorodeoxyglucose uptake) over 72 h of sleep loss are the prefrontal cortex (which mediates higher-order cognitive functions like planning and problem solving); inferior parietal/superior temporal cortex; and thalamus. SOURCE: Based on data from M. Thomas, H. Sing, G. Belenky, H. Holcomb, H. Mayberg, R. Dannals, H. Wagner, D. Thorne, K. Popp, L. Rowland, A. Welsh, S. Balwinski, and D. Redmond, Neural basis of alertness and cognitive performance impairments during sleepiness. I. Effects of 24 h of sleep deprivation on waking human regional brain activity, *Journal of Sleep Research* 9(4):335-352, 2000.

evidence is accruing that individuals who habitually get less sleep tend to gain weight faster, which is thought to be the result of sleep-restriction-induced changes in leptin and ghrelin levels and a slowed metabolic rate.[60]

Clearly, adequate sleep is a basic physiological need that is necessary for sustaining optimal mental functioning. During extended space missions, adequate sleep will therefore be part of the foundation that must be maintained to ensure that crew members are at least nominally prepared to deal with those challenges requiring cognitive agility and astuteness.

Sleep and Resilience

The potential effects of chronic sleep loss on those aspects of cognitive functioning that mediate individual and group functioning (cohesiveness and cooperation) are less clear than the effects on alertness and simple cognitive processes discussed above. Because group cohesion and cooperation will be far more challenging to sustain during extended-duration space missions, research that explores these effects will be particularly important during the next decade.

Sleep disturbance is a common complaint in (and perhaps an integral component of) disorders such as depression and post-traumatic stress disorder (PTSD). In fact, sleep disturbance is among the most common complaints of PTSD patients.[61] However, it is not yet known whether chronic sleep disturbance prior to stressor exposure sensitizes individuals (thus making them more susceptible to developing PTSD) and/or whether PTSD-associated sleep disturbance exacerbates symptoms and interferes with recovery.

Of particular interest is the possibility that chronic sleep difficulties in space could lower psychological resilience and increase the incidence of stressor-induced psychophysiological symptoms/problems such as asthenia. Among the arguments in favor of this hypothesis are that (1) emotional lability is a prominent symptom of asthenia (and a known effect of sleep loss)[62] and (2) increased ambient light intensity (which could strengthen the circadian rhythm of alertness) is thought to help reverse symptoms of asthenia.

Specific Recommendations

1. *NASA should continue to support studies to determine and quantify the extent to which sleep plays a role in maintaining mental, physical, and cognitive resilience during space missions and the extent to which sleep-enhancing interventions reverse stress-related symptoms and restore or sustain mental resilience.* This research both enables and is enabled by space missions. In addition, this research can be accomplished in the short term.
2. *Given the importance of sleep for sustaining cognitive performance, NASA should support research to determine whether stressors unique to the space environment impact restorative sleep processes, and if so, to what extent.* Stressors of interest include, but are not limited to, microgravity, exposure to radiation, environmental pollution, and lack of privacy/personal space. This research both enables and is enabled by space missions.
3. *NASA should support research to determine the efficacy of interventions (both pharmacological and nonpharmacological) for improving and sustaining adequate sleep in the space environment.* This research both enables and is enabled by space missions.
4. *NASA should sponsor research to determine the utility of sleep monitoring during extended missions in space.* Sleep not only impacts mental state and cognitive abilities but also can be a sensitive barometer that reflects emotional health status. This research is important to ensure that adequate sleep is obtained and also to gauge how well astronauts are coping with stressors during missions of extended duration. This research both enables and is enabled by space missions.
5. *Comprehensive, multidisciplinary, extended-duration simulation studies in the ISS and/or in high-fidelity analog settings should be conducted to determine and model the potentially complex, bidirectional interactions among sleep, cognitive functioning, individual functioning, and group dynamics.* These variables will ultimately determine the chances for success in extended-duration space missions. This research enables space missions.
6. *Sleep-related research programs should continue to explore the physiological mechanisms by which exposure to light resets and maintains the circadian rhythm of alertness and sleep/wake cycles.* Alignment with current and future NASA missions and goals should be considered. This research enables space missions.

HIGHEST-PRIORITY RECOMMENDATIONS

Research imperatives for the next decade should include studies aimed at (1) determining the potential effects of stressors likely to be encountered during extended space missions, alone and in combination, on astronauts' general psychological and physical well-being—with particular emphasis on sustaining astronauts' capacity to perform mission-related tasks, both as individuals and as members of a team; and (2) developing interventions to prevent, minimize, or reverse deleterious effects of such stressors.

The panel selected and prioritized the following recommendations on the basis of (1) a logical progression from assessment to monitoring to intervention and (2) the level of existing knowledge and potential for important discoveries. In particular, research is needed to look at the interactive effects of all space exploration-relevant stressors on the psychological status and performance of astronauts. Although such studies are logistically difficult, they are critical for understanding and predicting stress in astronauts and for devising and testing interventions. Recommendations 1, 2, and 3 enable space missions. Recommendation 4 both enables and is enabled by space missions.

1. *Development of sensitive, meaningful, and valid measures of mission-relevant performance for both astronauts and mission control personnel.* A combination of embedded performance measures (e.g., indicators of performance capacity that are maximally unobtrusive by virtue of being obtained/monitored during completion of actual mission-related tasks) and sensitive, well-validated (against actual mission tasks) cognitive tests is a basic requirement. This major gap in assessment must be filled in order to achieve both the research-related

goals of understanding and quantifying the effects of stressors on mission-critical performance capacity and the mission-related need for sensitive tools to monitor cognitive performance, detect problems, and facilitate timely and effective interventions. It is estimated that significant progress toward accomplishment of this research goal could be achieved in 2 to 5 years. **(B1)**

2. *Integrated translational research in which long-duration missions are simulated (in analogs or on the ISS) specifically for the purpose of studying interrelationships among individual functioning, cognitive performance, sleep, and group dynamics.* The ultimate research aim would be to build predictive models of astronaut performance and well-being during extended missions in space that can be used to plan and manage such missions. It is estimated that significant progress toward accomplishment of this research goal could be achieved in 5 to 8 years. **(B2)**

3. *Research to determine genetic, physiological (e.g., sleep-related), and psychological underpinnings of individual differences in resilience to stressors during extended missions in space, with development of an "individualized medicine-like" approach to sustaining astronauts during such missions.* It is estimated that significant progress toward accomplishment of this research goal could be achieved in 5 to 10 years. **(B3)**

4. *Research to enhance the cohesiveness, team performance, and effectiveness of multinational crews, especially under conditions of extreme isolation and autonomy.* It is estimated that significant progress toward accomplishment of this research goal could be achieved in 2 to 5 years. **(B4)**

The preceding estimates of time required to achieve significant scientific progress are based on the presumption that most astronauts will agree to participate in the relevant studies. More detailed research priorities within the areas of cognitive performance, sleep, individual functioning, and group functioning are described elsewhere in this chapter.

REFERENCES

1. U.S. Human Spaceflight Plans Committee. 2009. *Seeking a Human Spaceflight Program Worthy of a Great Nation.* NASA, Washington, D.C. Available at http://www.nasa.gov/pdf/396117main_HSF_Cmte_FinalReport.pdf.
2. NASA. 2010. *NASA Human Research Road Map.* Available at http://humanresearchroadmap.nasa.gov/.
3. Stern, Y. 2009. Cognitive reserve. *Neuropsychologia* 47(10):2015-2028.
4. Mallis, M.M., and DeRoshia, C.W. 2005. Circadian rhythms, sleep, and performance in space. *Aviation, Space, and Environmental Medicine* 76(6 Suppl.):B94-B107.
5. Mallis, M.M., and DeRoshia, C.W. 2005. Circadian rhythms, sleep, and performance in space. *Aviation, Space, and Environmental Medicine* 76(6 Suppl.):B94-B107.
6. National Research Council. 1998. *A Strategy for Research in Space Biology and Medicine into the Next Century.* National Academy Press, Washington, D.C.
7. Leon, G.R. 2008. Strategies to optimize individual and team performance. *Proceedings of the 3rd International Association for the Advancement of Space Safety Conference (IAASS).* October 22, 2008, Rome, Italy. International Association for the Advancement of Space Safety, Noordwijk, The Netherlands.
8. Kanas, N., Sandal, G.M., Boyd, J.E., Gushin, V.I., Manzey, D., North, R., Leon, G.R., Suedfeld, P., Bishop, S., Fiedler, E.R., Inoue, N., et al. 2009. Psychology and culture during long-duration space missions. *Acta Astronautica* 64:659-677.
9. Institute of Medicine. 2001. *Safe Passage: Astronaut Care for Exploration Missions.* National Academy Press, Washington D.C.
10. Palinkas, L.A., and Suedfeld, P. 2008. Psychological effects of polar expeditions. *Lancet* 371(9607):153-163.
11. Kanas, N., and Manzey, D. 2008. *Space Psychology and Psychiatry.* 2nd Edition. Microcosm Press, El Segundo, Calif., and Springer, Dordrecht, The Netherlands.
12. Gushin, V.I., Institute for Biomedical Problems (Russia), communication to the National Research Council's Human Behavior and Mental Health Panel via teleconference, December 15, 2009.
13. Myasnikov, V.I., and Zamaletdinov, I.S. 1996. Psychological states and group interactions of crew members in flight. Pp. 419-432 in *Space Biology and Medicine III: Humans in Spaceflight* (C.S. Leach Huntoon, A.E. Nicogossian, V.V. Antipov, and A.E. Grigoriev, eds.). American Institute of Aeronautics and Astronautics, Reston, Va.
14. Gega, L., Marks, I., and Mataix-Cols, E. 2004. Computer-aided CBT self-help for anxiety and depressive disorders: Experience of a London clinic and future directions. *Journal of Consulting and Clinical Psychology* 60:147-157.

15. Carter, J.A., Buckey, J.C., Greenhalgh, L., Holland, A.W., and Hegel, M.T. 2005. An interactive media program for managing psychosocial problems spaceflights. *Aviation, Space, and Environmental Medicine* 76(6 Suppl.):B213-B223.
16. Craske, M.G., Rose, R.D., Lang, A., Welch, S.S., Campbell-Sills, L., Sullivan, G., Sherbourne, C., Bystritsky, A., Stein, M.B., and Roy-Byrne, P.P. 2009. Computer-assisted delivery of cognitive behavioral therapy for anxiety disorders in primary-care settings. *Depression and Anxiety* 26:235-242.
17. Saivin, S., Pavy-Le Traon, A., Soulez-LaRiviere, C., Guell, A., and Houin, G. 1997. Pharmacology in space: Pharmacokinetics. In *Advances in Space Biology and Medicine* (S.L. Bonting, ed.). Volume 6. JAI Press, Greenwich, Conn.
18. Langreth, R., and Waldholz, M. 1999. New era of personalized medicine: Targeting drugs for each unique genetic profile. *Oncologist* 4(5):426-427.
19. NASA. 2007. *NASA Astronaut Health Care System Review Committee Report to the Administrator.* February-June. Available at http://www.nasa.gov/pdf/183113main_NASAhealthcareReport_0725FINAL.pdf.
20. Pomara, N., Willoughby, L., Wesnes, K., Greenblatt, D.J., and Sidtis, J.J. 2005. Apolipoprotein E epsilon4 allele and lorazepam effects on memory in high-functioning older adults. *Archives of General Psychiatry* 62(2):209-216.
21. Institute of Medicine. 2001. *Safe Passage: Astronaut Care for Exploration Missions.* National Academy Press, Washington D.C.
22. Palinkas, L.A., and Suedfeld, P. 2008. Psychological effects of polar expeditions. *Lancet* 371(9607):153-163.
23. Rose, R.M., Fogg, L.F., Helmreich, R.L., and McFadden, T. 1994. Psychological predictors of astronaut effectiveness. *Aviation, Space, and Environmental Medicine* 65:910-915.
24. McFadden, T.J., Helmreich, R.L., Rose, R.M., and Fogg, L.F. 1994. Predicting astronauts' effectiveness: A multivariate approach. *Aviation, Space, and Environmental Medicine* 65:904-909.
25. Palinkas, L.A., Gunderson, E.K.E., Holland, A.W., Miller, C., and Johnson, J.C. 2000. Predictors of behavior and performance in extreme environments: The Antarctic Space Analogue Program. *Aviation, Space, and Environmental Medicine* 71:619-625.
26. Kanas, N., Sandal, G.M., Boyd, J.E., Gushin, V.I., Manzey, D., North, R., Leon, G.R., Suedfeld, P., Bishop, S., Fiedler, E.R., Inoue, N., et al. 2009. Psychology and culture during long-duration space missions. *Acta Astronautica* 64:659-677.
27. Gushin, V.I., Efimov, V.A., Smirnova, T.M., Vinokhodova, A.G., and Kanas, N. 1998. Subject's perception of the crew interaction dynamics under prolonged isolation. *Aviation, Space, and Environmental Medicine* 69:556-561.
28. Kanas, N. 2009. *Psychology and Culture During Long-Duration Space Missions.* International Academy of Astronautics, Paris, France.
29. Helmreich, R.L., and Merritt, A. 1999. *Culture at Work in Aviation and Medicine.* Ashgate, New York.
30. Institute of Medicine. 2001. *Safe Passage: Astronaut Care for Exploration Missions.* National Academy Press, Washington D.C., pp. 146-147.
31. Kanas, N., Sandal, G.M., Boyd, J.E., Gushin, V.I., Manzey, D., North, R., Leon, G.R., Suedfeld, P., Bishop, S., Fiedler, E.R., Inoue, N., et al. 2009. Psychology and culture during long-duration space missions. *Acta Astronautica* 64:659-677.
32. Blair, S.M. 1992. *Community Organization Under Differing South Pole Leaders.* AIAA 92-1528. American Institute of Aeronautics and Astronautics, Washington, D.C.
33. Johnson, J.C., Palinkas, L.A., and Bosterm, J.S. 2003. Informal social roles and the evolution and stability of social networks. Pp. 121-132 in *Dynamic Social Network Modeling and Analysis: Workshop Summary and Papers* (R. Breiger, K. Corley, and P. Pattison, eds.). The National Academies Press, Washington, D.C.
34. Kanas, N.A., Salnitskiy, V.P., Boyd, J.E., Gushin, V.I., Weiss, D.S., Saylor, S.A., Kozerenko, O.P., and Marmar, C.M. 2007. Crewmember and mission control personnel interactions during International Space Station missions. *Aviation, Space, and Environmental Medicine* 78:601-607.
35. Dijk, D.J., Neri, D.F., Wyatt, J.K., Ronda, J.M., Riel, E., Ritz-De Cecco, A., Hughes, R.J., Elliott, A.R., Prisk, G.K., West, J.B., and Czeisler, C.A. 2001. Sleep, performance, circadian rhythms, and light-dark cycles during two space shuttle flights. *American Journal of Physiology. Regulatory, Integrative and Comparative Physiology* 281(5):R16747-R16764.
36. Monk, T.H., Buysse, D.J., Billy, B.D., Kennedy, K.S., and Willrich, L.M. 1998. Sleep and circadian rhythms in four orbiting astronauts. *Journal of Biological Rhythms* 13(3):188-201.
37. Gundel, A., Nalashiti, V., Reucher, E., and Zulley, J. 1993. Sleep and circadian rhythm during a short space mission. *Clinical Investigation* 71(9):718-724.
38. Gundel, A., Polyakov,V.V., and Zulley, J. 1997. The alterations of human sleep and circadian rhythms during spaceflight. *Journal of Sleep Research* 6(1):1-8.
39. Dijk, D.J., Neri, D.F., Wyatt, J.K., Ronda, J.M., Riel, E., Ritz-De Cecco, A., Hughes, R.J., Elliott, A.R., Prisk, G.K., West, J.B., and Czeisler, C.A. 2001. Sleep, performance, circadian rhythms, and light-dark cycles during two space shuttle flights. *American Journal of Physiology. Regulatory, Integrative and Comparative Physiology* 281(5):R16747- R16764.

40. Monk, T.H., Kennedy, K.S., Rose, L.R., and Linenger, J.M. 2001. Decreased human circadian pacemaker influence after 100 days in space: A case study. *Psychosomatic Medicine* 63(6):881-885.
41. Gundel, A., Polyakov, V.V., and Zulley, J. 1997. The alterations of human sleep and circadian rhythms during spaceflight. *Journal of Sleep Research* 6(1):1-8.
42. Monk, T.H., Kennedy, K.S., Rose, L.R., and Linenger, J.M. 2001. Decreased human circadian pacemaker influence after 100 days in space: A case study. *Psychosomatic Medicine* 63(6):881-885.
43. Gundel, A., Nalashiti, V., Reucher, E., and Zulley, J. 1993. Sleep and circadian rhythm during a short space mission. *Clinical Investigation* 71(9):718-724.
44. Gundel, A., Polyakov, V.V., and Zulley, J. 1997. The alterations of human sleep and circadian rhythms during spaceflight. *Journal of Sleep Research* 6(1):1-8.
45. Dijk, D.J., Neri, D.F., Wyatt, J.K., Ronda, J.M., Riel, E., Ritz-De Cecco, A., Hughes, R.J., Elliott, A.R., Prisk, G.K., West, J.B., and Czeisler, C.A. 2001. Sleep, performance, circadian rhythms, and light-dark cycles during two space shuttle flights. *American Journal of Physiology. Regulatory, Integrative and Comparative Physiology* 281(5):R16747- R16764.
46. Gundel, A., Nalashiti, V., Reucher, E., and Zulley, J. 1993. Sleep and circadian rhythm during a short space mission. *Clinical Investigation* 71(9):718-724.
47. Gundel, A., Polyakov, V.V., and Zulley, J. 1997. The alterations of human sleep and circadian rhythms during spaceflight. *Journal of Sleep Research* 6(1):1-8.
48. Dijk, D.J., Neri, D.F., Wyatt, J.K., Ronda, J.M., Riel, E., Ritz-De Cecco, A., Hughes, R.J., Elliott, A.R., Prisk, G.K., West, J.B., and Czeisler, C.A. 2001. Sleep, performance, circadian rhythms, and light-dark cycles during two space shuttle flights. *American Journal of Physiology. Regulatory, Integrative and Comparative Physiology* 281(5):R16747- R16764.
49. Dijk, D.J., Neri, D.F., Wyatt, J.K., Ronda, J.M., Riel, E., Ritz-De Cecco, A., Hughes, R.J., Elliott, A.R., Prisk, G.K., West, J.B., and Czeisler, C.A. 2001. Sleep, performance, circadian rhythms, and light-dark cycles during two space shuttle flights. *American Journal of Physiology. Regulatory, Integrative and Comparative Physiology* 281(5):R16747- R16764.
50. Kelly, T.H., Heinz, R.D., Zarcone, T.J., Wurster, R.M., and Brady, J.V. 2005. Crewmember performance before, during, and after spaceflight. *Journal of the Experimental Analysis of Behavior* 84(2):227-241.
51. Monk, T.H., Buysse, D.J., Billy, B.D., Kennedy, K.S., and Willrich, L.M. 1998. Sleep and circadian rhythms in four orbiting astronauts. *Journal of Biological Rhythms* 13(3):188-201.
52. Stampi, C. 1994. Sleep and circadian rhythms in space. *Journal of Clinical Pharmacology* 34(5):518-534.
53. Mallis, M.M., and DeRoshia, C.W. 2005. Circadian rhythms, sleep, and performance in space. *Aviation, Space, and Environmental Medicine* 76(6 Suppl.):B94-B107.
54. Ellis, S.R. 2000. Collision in space. *Ergonomics in Design: The Magazine of Human Factors Applications* 8(1):4-9.
55. See, for example, T.J. Balkin, T. Rupp, D. Picchioni, and N.J. Wesensten, Sleep loss and sleepiness: Current issues, *Chest* 134(3):653-660, 2008.
56. Siegel, J.M. 2005. Clues to the functions of mammalian sleep. *Nature* 37(7063):1264-1271.
57. Thomas, M., Sing, H., Belenky, G., Holcomb, H., Mayberg, H., Dannals, R., Wagner, H., Thorne, D., Popp, K., Rowland, L., Welsh, A., Balwinski, S., and Redmond, D. 2000. Neural basis of alertness and cognitive performance impairments during sleepiness. I. Effects of 24 h of sleep deprivation on waking human regional brain activity. *Journal of Sleep Research* 9(4):335-352.
58. See, for example, W.D. Killgore, T.J. Balkin, and N.J. Wesensten, Impaired decision making following 49 h of sleep deprivation, *Journal of Sleep Research* 15:7-13, 2006; and W.D. Killgore and S.A. McBride, Odor identification accuracy declines following 24 h of sleep deprivation, *Journal of Sleep Research* 15:111-116, 2006.
59. Belenky, G.L., Wesensten, N.J., Thorne, D., Thomas, M., Sing, H., Redmond, D.P., Russo, M.B., and Balkin, T.J. 2003. Patterns of performance degradation and restoration during sleep restriction and subsequent recovery: A sleep dose-response study. *Journal of Sleep Research* 12:1-12.
60. Leproult, R., and Van Cauter, E. 2010. Role of sleep and sleep loss in hormonal release and metabolism. *Endocrine Development* 17:11-21.
61. Neylan, T.C., Marmar, C.R., Metzler, T.J., Weiss, D.S., Zatzick, D.F., Delucchi, K.L., Wu, R.M., and Schoenfeld, F.B. 1998. Sleep disturbances in the Vietnam generation: Findings from a nationally representative sample of male Vietnam veterans. *The American Journal of Psychiatry* 155:929-933.
62. Meerlo, P., Sgoifo, A., and Suchecki, D. 2008. Restricted and disrupted sleep: Effects on autonomic function, neuroendocrine stress systems and stress responsivity. *Sleep Medicine Reviews* 12(3):197-210.

6

Animal and Human Biology

Over millions of years, the structure and function of organisms have evolved under the influence of a constant gravity stimulus, which consists of the natural force of attraction exerted by celestial bodies such as Earth. To fully understand this influence of gravity, living systems must be studied by essentially eliminating the gravity variable. This task can be daunting because organisms must live for a sufficient time outside the effects of Earth's gravity, in a state of free-fall. In the United States, the agency designated by Congress to develop a space research program involving the life and physical sciences is the National Aeronautics and Space Administration (NASA). During its 50 years of existence, NASA has continued to evolve such a program, which, at the present time, is centered primarily on operational medicine objectives being pursued on the National Space Laboratory, a key component of the International Space Station (ISS).

It is now recognized that habitation of the microgravity environment poses potential deleterious consequences for essentially all the organ systems of the body, even though it is routine for human astronauts and cosmonauts to spend 180 days or longer living and performing a number of challenging tasks on the ISS. Given the typical lifespan of humans, 180 days in space may seem trivial. However, in the case of rodents, the animal model most scientists have used to study fundamental biological processes in space, such a time frame represents approximately one-fourth to one-third of the species' adult life. Thus, studies on these rodents in space have the potential to extrapolate important implications for humans living in space well beyond 6 months.

As part of the decadal survey process, the Committee for the Decadal Survey on Biological and Physical Sciences in Space formed the advisory Animal and Human Biology (AHB) Panel and tasked it to address the research needed to (1) enable humans to carry out long-term space exploration and (2) ascertain opportunities provided by the space environment that enable a greater understanding of how gravity shapes fundamental biological processes of various organisms.

To meet its objectives, the AHB Panel focused on the following topics: (1) what is known about the risk and deleterious effects of spaceflight (and ground-based analogs) on the structure and function of the musculoskeletal (bone and muscle), sensory-motor, cardiovascular, pulmonary, endocrine, and immune systems, as well as how animals develop in the absence of gravity; (2) the effectiveness of the countermeasures currently used to maintain organ system homeostasis in the face of microgravity; (3) the knowledge gaps in understanding of the above topics that need to be addressed; (4) the research platforms needed to undertake new research initiatives in the next decade; (5) the overarching issues that have to be addressed in fostering cutting-edge, integrative research in humans and animals, and spanning multiple physiological systems, to generate future countermeasure strategies; and (6) the specific high-priority research initiatives that are needed to sharpen and advance the science knowledge necessary for progress in the next decade. (It is important to note that certain topics that have a major impact across multiple physiological systems, such as nutrition, are in most cases not covered in this chapter, but rather in Chapter 7, which focuses on crosscutting issues.) Finally, in examining programmatic activities relevant to this chapter, and as discussed in the committee's interim report to NASA,[1] the AHB Panel was deeply concerned that NASA had severely reduced research initiatives in the life and physical sciences in the latter half of the past decade. In the panel's view, this action has effectively paralyzed research initiatives previously recommended by National Research Council (NRC) study committees (as reflected by the relative paucity of publications since 2005 in recommended subject areas such as bone) and poses a daunting challenge to future administrations attempting to reverse the neglect and to accomplish the life and physical sciences research initiatives recommended in this report.

RESEARCH ISSUES

Risks for Bone Loss During Long-Duration Space Missions

The skeletal (bone) system provides the solid framework for humans and mammals to oppose gravity, and its fidelity in accomplishing this fundamental process has evolved over millions of years. Given this evolutionary role, it is not surprising that bone loss occurs in astronauts at a rate that is both substantial and progressive with time spent in microgravity.[2-5] Accordingly, without appropriate countermeasures, spaceflight of 2 years or longer will present serious risks due to progressive bone fragility. Therefore, there is a need to adopt effective countermeasures that have been appropriately tested in relevant human and animal models. The 1998 NRC report *A Strategy for Research in Space Biology and Medicine in the New Century*[6] recommended several experiments to address the problem of bone loss during spaceflight. At present, several key issues raised in the 1998 NRC report have not been addressed. For instance, the report recommended that genetically altered mice be used in flight experiments to investigate the molecular mechanisms of bone loss, yet these experiments have not been completed. The report also recommended that in-flight animal facilities should house 30 adult rats or mice, but the ISS can currently house only 6 mice (in the Mice Drawer System on the Italian Space Agency investigation). These recommendations should be implemented, and additional steps should be taken to advance research into bone loss in microgravity for the development of effective countermeasures.

Effects of Spaceflight Environment on the Structure and Function of Bone

Bone loss during spaceflight appears to be due primarily to increased resorption in load-bearing regions of the skeleton.[7,8,9] There is also some evidence of a decrease in bone formation. The rate of bone loss in microgravity is roughly 10 times greater than the bone mineral density (BMD) loss per month that occurs in postmenopausal women on Earth who are not on estrogen therapy.[10-13] Results from Skylab,[14] Mir,[15,16] the space shuttle,[17] and the ISS[18,19,20] missions have shown substantial areal and volumetric bone loss in critical regions such as the proximal femur and spine. The most accurate data, derived from quantitative computed tomography, have shown that spinal volumetric BMD was lost at a rate of 0.9 percent per month and total hip volumetric BMD was lost at rate of 1.4 percent per month; there was, however, considerable variability between individuals.[21] Changes in bone strength (expressed as percentage loss) were much greater than changes in BMD.[22]

BMD lost in 6-month missions appears to be mostly reversible by 1,000 days after return to normal gravity (1 *g*).[23,24] However, changes in bone structure are not reversible and seem to mimic changes in the elderly.[25]

An important question that remains unanswered is whether any loading that is performed by simply living and working in partial gravity—such as the 1/6 *g* of the Moon or the 1/3 *g* of Mars—will provide any protection from the bone loss that occurs in microgravity. Expert opinion as presented in a recent symposia is that it will not,[26] although data from a partial-gravity mouse model is just becoming available.[27]

Animal Studies

Rodents have been flown on the Cosmos biosatellite[28-33] and on space shuttle missions[34-46] to measure bone loss. The most consistent finding was the striking decrease in bone formation with spaceflight, which stopped completely at some skeletal sites.[47] In contrast to what occurs in humans, bone resorption in rats was not substantially changed in one flight experiment.[48] whereas it was increased in another.[49] Most rodents flown in space have been immature and undergoing rapid bone growth.[50] In these studies, bone formation dominates during growth, and so it is not surprising that greater changes were observed in bone formation than in bone mineral resorption. It is difficult to extrapolate from these data to the expected changes in the mature skeleton during long-term spaceflight. A few studies have used adult animals, and in these studies bone formation was suppressed at the periosteal surface.[51,52,53] In contrast, longitudinal growth was minimally affected by either spaceflight or hindlimb unloading (HU).[54] One study has shown bone loss in spaceflight to be greater than that in ground-based models, such as HU,[55] although it is important to distinguish between bone loss, the removal of existing bone, and failure to gain bone in growing animals. In this model, traction applied to the tail of rodents elevates the lower extremities and eliminates generation of ground reaction forces.

In Vitro Studies

Bone cell culture experiments have been performed on space shuttle missions,[56,57] on Skylab,[58] and on the free-flyer Foton-M.[59] These studies demonstrated differences in gene expression and growth factor production by osteoblasts.[60] Osteoclasts were also affected by spaceflight.[61] Interpretation of these experiments is a challenge, because cells in culture behave much differently than cells embedded in bone. Furthermore, the vibration during launch can confound cell culture experiments, particularly for short-term (1 to 2 weeks) experiments.[62] The experimental complications during spaceflight and the difficulty with interpretation make bone cell experiments of lower priority compared with animal or human spaceflight studies. There is, however, a potential use of cultured bone cells in biotechnology applications.

Status of Countermeasures

Exercise Countermeasures

To date, NASA and the Russian space program have relied primarily on exercise countermeasures to attenuate bone loss,[63,64,65] but no exercise has yet proven to be uniformly effective for maintaining bone mass[66,67] during flight or bed rest. Similarly, low-magnitude, low-frequency mechanical signals were not effective in prolonged best rest.[68] There is evidence that the external loading on previous exercise devices used in space has been insufficient to provide the required stimulus to bone.[69,70] Recent (fall 2009) additions to the exercise devices on the ISS now offer the possibility for greater loading and for definitive research to examine the efficacy of exercise countermeasures. The capacity to measure loads was also added to these new devices, but interaction with NASA personnel indicates that further refinements are required to produce accurate load estimates. The ISS provides an excellent research platform for studies that are relevant to missions outside low Earth orbit because the microgravity environment on the ISS presents a greater challenge to the musculoskeletal system than does a partial-gravity environment.[71]

Ground-based research using bed rest (head-down for microgravity or head-up for lunar simulation) provides another important research platform for exercise countermeasures.[72-81] For example, bed rest studies suggest that bone may be somewhat protected with sufficient loads and exercise time. Results from Vernikos et al.[82] showed that intermittent upright posture and exercise reduce the increased blood calcium levels observed in bed rest, and data from Smith et al.,[83,84] and Zwart et al.[85] showed positive benefits of supine treadmill running within lower-body negative pressure during 30 and 60 days of bed rest. There are no studies indicating positive effects on bone in passive intermittent rotational artificial gravity.[86] It is possible that lower-body negative pressure or centrifugation coupled with exercise may be more effective than exercise alone, perhaps through modulation of some other necessary physiological factor (e.g., improved blood flow, fluid shifts, or circulating hormones).

Pharmaceutical and Nutritional Countermeasures

Vitamin D supplementation has been used by NASA during spaceflight,[87,88,89] but one report[90] showed that serum 25-hydroxycholecalciferol was decreased after flight despite supplementation with vitamin D.

Short-term calcium supplementation has not been effective in reducing bone loss during spaceflight or head-down bed rest.[91,92]

Over the past 15 years, several drugs have been developed to prevent bone loss associated with osteoporosis (e.g., bisphosphonates, selective estrogen receptor modulators, parathyroid hormone). Some of these drugs may be useful for preventing bone loss in astronauts during long-term spaceflight. As with exercise countermeasures, the ISS is ideal for testing the effectiveness of drugs. However, at the time of the writing of this report only one experiment involving oral bisphosphonates had been flown on the ISS, and only two astronauts had participated. Bisphosphonates substantially reduce bone resorption in Earth-bound patients and can be given by mouth daily or weekly, or yearly by infusion. A new potent antiresorptive bisphosphonate, zolendronic acid, effectively preserves bone mass in osteoporotic patients when given by infusion once per year.[93]

Bisphosphonates have shown promise in ground-based studies. As an example, a single injection of pamidronate maintained a slightly increased BMD in the spine and hip in a 90-day bed rest study.[94] These drugs have been shown to attenuate bone loss in hindlimb unloaded rodents.[95] In addition, excess urinary calcium excretion was reversed by pamidronate. This dual action, blocking bone loss and reducing urinary calcium excess, suggests there is potential utility for bisphosphonates in space.

A theoretical concern with use of bisphosphonates is that suppressing of resorption will also suppress bone formation, but data from spaceflight are lacking to address this issue. Should a fracture occur in space, use of bisphosphonates might slow healing. Consequently, it will be important to study fracture healing in space with antiresorptives to provide assurance that fractures will heal. Further, the long-acting nature of some bisphosphonates means that the suppression of bone turnover could persist upon return to normal gravity. The suppression of bone turnover in people who maintain vigorous levels of activity could have deleterious effects on bone quality. Other potential issues with the use of bisphosphonates include osteonecrosis of the jaw[96] and atypical sub-trochanteric fractures after prolonged bisphosphonate use.[97]

Another drug recently approved for use in postmenopausal women with osteoporosis, Denosumab, blocks an important orthoclase-stimulating peptide called RANKL and stops bone loss in osteoporotic patients when given by injection every 6 months.[98] A bone anabolic drug, teriparatide, is approved by the Food and Drug Administration for the treatment of patients at high risk for fracture,[99] but there are ethical concerns in the treatment of a person with a normal skeleton with this drug because of the possibility of undesirable side effects.

Challenges for the use of drugs in space include storage and packaging that prevent degradation in the space environment. Additionally, it is not known whether drugs will have the same bioactivity when taken in a weightless environment.[100,101] Bone-acting drugs have not yet been tested to determine the length of time they remain active in space, where, for example, they are exposed to higher radiation levels than on the ground. If bioactivity is compromised, long-acting drugs might be given pre-flight to avoid the need for in-flight dosing, although currently the longest interval between dosing of any appropriate drug is 12 months. Treatment of astronauts upon return to Earth with therapeutic drugs also needs to be explored.

An animal experiment using a myostatin inhibitor was tested on the space shuttle (STS-118). Myostatin is an antigrowth factor protein that blocks muscle growth, and so researchers expected the use of myostatin inhibitors to prevent muscle loss. In addition to preventing muscle loss during the 13-day mission, there are initial indications that the drug also preserved bone mass and strength.[102,103] In another important recent study, sclerostin knockout mice did not lose bone during ground-based disuse.[104,105] In addition, a sclerostin antibody improved bone mass in an animal model of colitis-induced bone loss.[106]

Gaps in Knowledge

Mechanism of Bone Loss

An animal research program offers many vitally needed tools to better understand bone loss during weightlessness—both to better define risks to the skeletal health of humans in prolonged spaceflight and to provide models to rigorously test recently developed pharmacologic strategies to control bone loss. However, past animal studies in spaceflight have been confusing. As discussed previously, many studies of rodents found that bone loss was due primarily to cessation of bone formation. Studies of astronauts using biomarkers have identified increased bone breakdown by resorption as the predominant mechanism, although the role of suppressed bone formation should be further investigated. These findings are complicated by the fact that most rodents studied in space were immature and growing, whereas all astronauts are mature and have stopped growing. Further experiments with mature animals will help determine the relative contributions of decreased bone formation and increased bone mineral resorption in bone loss in microgravity. This will be critical in identifying effective pharmacologic countermeasures, which target either bone mineral resorption or bone formation.

Molecular and Cellular Mechanisms

There are many unanswered questions about bone loss during spaceflight. For instance, the molecular mechanisms by which bone senses gravitational forces remain unknown. Currently, it is thought that osteocytes within the bone sense mechanical perturbations and signal to osteoblasts and osteoclasts, but the biochemical nature of these signals among cells is poorly understood. Bone cell factors that have been identified as therapeutic targets include sclerostin (an inhibitor of bone formation), RANKL (a stimulator of bone resorption), and osteoprotegerin (an inhibitor of bone resorption); others will likely emerge in the coming decade.[107] These targets are ideal for continued studies using animal models.

Fracture Repair

Without an effective countermeasure during spaceflight, bone loss occurs at an alarming rate and fracture risk is increased. Any fracture sustained by an astronaut during a long-duration mission must heal in a microgravity or partial-gravity environment. Fracture healing was studied on shuttle flight STS-29, and deficiencies in angiogenesis were noted.[108] However, the shuttle flight was too short in duration (5 days) to fully evaluate fracture healing. A subsequent 5-week ground-based study showed that fracture healing was impaired in rats subjected to HU.[109]

Effects of Radiation

A major obstacle to long-term spaceflight is the effects of space radiation on astronauts. The effects of radiation on bone during spaceflight are currently unknown. However, ground-based studies demonstrate that space-like radiation causes bone loss.[110] Further studies of the combined effects of radiation and unloading on bone structure are urgently needed.

Hormonal Issues

Endocrine hormones, such as parathyroid hormone, calcitonin, glucocorticoids, and insulin-like growth factor, influence bone homeostasis on Earth. Estrogen deficiency plays a major role in the pathogenesis of bone loss and fracture in both women and men on Earth.[111] It has also recently been shown that the leptin receptor plays a key role in mechano-signal transduction.[112] There is no direct evidence that these endocrine hormones play a major role in bone loss in space, as evidenced by the site-specific, rather than systemic, nature of bone loss in the skeleton due to microgravity. However, experiments with the rodent model have shown that estrogen status alters the skeletal response to spaceflight and HU.[113,114] Recent evidence suggests that local (autocrine or paracrine) effects are more important in regulating bone lost as a result of disuse,[115,116] and local expression of autocrine and paracrine factors have been investigated in the rat.[117,118,119] However, traditional systemic endocrine factors (e.g., estrogen, cortisol) can also act in a paracrine manner because bone cells contain the enzymatic machinery to produce these factors locally.

Fluid Shifts

Bone loss in human volunteers subjected to bed rest is greatest in the lower extremities, particularly the calcaneus. In contrast, the upper extremities do not lose bone, and there is a net gain in bone mass in the cranial bones.[120] This pattern of bone loss matches the expected changes in fluid pressures caused by change in body position during bed rest, relative to the gravitational vector. A similar pattern in bone loss occurs in rats subjected to HU.[121] Because fluid shifts are an important physiological adaptation to spaceflight, further studies of animals and humans during spaceflight are warranted to allow a better understanding of the mechanisms of bone loss and their possible associations with fluid shifts.

McCarthy[122] postulated that the shear forces created by interstitial fluid flow influence bone loss in microgravity. He has created a pneumatic venous tourniquet that can modulate fluid flow within tissues when placed around the ankle of an astronaut or a rat. More recently Yokota and colleagues[123] have shown that lateral loading of joints increases both fluid flow within bone and bone formation. These findings demonstrate the link between bone fluid flow and the maintenance of bone mass. Further research using tourniquets or other means will help to determine to what extent fluid shifts are responsible for bone loss.

Calcium Metabolism and Kidney Stones

Risk factors for renal stone formation (high urinary calcium excretion, low urinary volume) are exaggerated in spaceflight.[124,125] Urinary calcium excretion was highly variable both before and during flight, but impressively elevated calcium excretions were noted in several individuals,[126,127] and the source was likely increased bone resorption.[128] Calcium oxalate and calcium phosphate concentrations during flight increased to levels that favored crystallization.[129] Countermeasures should include increased fluid intake to increase urine volumes, but such efforts are complicated by fluid shifts in microgravity, reduced plasma volume, fluid retention from hormonal adjustments, and practical constraints encountered in spaceflight. Studies on agents to prevent reductions in urinary pH and citrate in flight[130] and during bed rest[131] have been explored, but their effectiveness is uncertain. The key logical countermeasure would be to reduce bone breakdown and thereby achieve a dual benefit. For example, pamidronate both blocked bone resorption and reduced excessive urinary calcium excretion in a bed rest study.[132] Nutrition plays a role in calcium metabolism and bone health during long-duration spaceflight that needs further investigation. Astronauts have been reported to consume only 80 percent of the recommended energy intake during long-duration spaceflight, and food restriction has been shown to exaggerate the effects on bone of unloading in animals and humans.[133,134,135] Nutrition, a crosscutting issue affecting multiple systems, is discussed in detail in Chapter 7.

Research Models and Platforms

Animal Experiments

An active animal research program is critical both to better understand the adaptive response of bone to weightlessness and to better define risks to the skeletal health of humans in prolonged spaceflight. In addition, animal experiments are necessary to rigorously test pharmacological strategies to control bone loss. Studies of genetically modified mice, such as the sclerostin knockout mouse, provide a means of isolating the importance of specific signaling factors in bone.[136,137] Further studies of genetically altered mice subjected to weightlessness, both as ground-based models and on the ISS, are urgently needed.

The sclerostin knockout mouse is an example of the wide variety of highly informative and newly available[138] genetically modified animals. In fact, there are several hundred genetically modified mouse strains that selectively delete (gene knockout or replacement) or overexpress (transgenic animals) specific genes and gene products that are important in bone biology.[139-144] The rapid pace of advances in molecular biology pertinent to bone metabolism provides the basis for breakthroughs in space biology, thus emphasizing the importance of reinvigorating basic research on bone biology in altered gravity. In ground-based studies, rodent HU is a proven model for disuse and fluid shift caused by spaceflight,[145] and this model should continue to be exploited and supported. The ISS would be an excellent platform for spaceflight studies of rodents, if adequate rodent housing facilities were added. Given the breadth of this field, a rigorous selection process will be needed to prioritize the use of particular genetically modified mice, including conditional knockouts, which are best suited to answer the research questions. The limitations in comparing effects in the mouse to those in humans should be carefully considered, and the recent breakthroughs in gene technology in rats may provide alternatives to mice for genetic studies.[146,147]

Human Experiments

The six-degree head-down bed rest model has been successfully exploited in a number of U.S. and European studies.[148,149,150] In human bed rest studies, the rate of bone resorption increased two-fold and bone formation was significantly reduced.[151] However, changes in bone formation markers did not reflect the histologic evidence of reduced bone formation.[152] This finding calls attention to the relative insensitivity of bone formation markers to detect reduced bone formation. More research on bone cell turnover is needed.

Research Recommendations

Need for More Basic Research

The severe drop in NASA funding between 2005 and 2009 has had a chilling effect on basic research to explain the changes seen in physiological systems during spaceflight. In the ISS era (2000 to the present), the emphasis of NASA research has been on exercise countermeasures to bone loss in humans (see relevant NASA Research Announcements).[153]

Basic research into the mechanisms of bone adaptation in altered gravity using animal and human models needs to be reinvigorated. The agreement between, and the joint solicitations from, the National Institutes of Health and NASA could be exploited to expand the research focus from primarily operational issues to include fundamental science that will inform future missions. Maximizing the possibilities offered by the ISS National Laboratory over the next decade will also be critical to the future success and safety of long-duration missions.

Recommended Experiments

Human Studies. All experiments listed below can be completed within the next decade.

1. Ongoing human research on the efficacy of exercise to preserve bone during ISS missions should continue. There is a need for studies with adequate statistical power of all the available ISS exercise devices (including the latest devices: the advanced resistive exercise device and T2 treadmill), with accurate quantification of external loads and compliance with prescribed exercise in order to determine whether exercise is an effective countermeasure for bone loss.
2. The efficacy of bisphosphonates and other anti-osteoporosis drugs should be tested on the ISS during approximately 6-month missions in an adequate population of astronauts. The interaction of pharmacological and exercise countermeasures also needs to be studied, since prior use on Earth has invariably involved weight bearing, and only a single study has examined the use of bisphosphonates in the bed rest analog, and the subjects in this study did not exercise.[154]
3. Use of any drug as a countermeasure during long-term spaceflight will require that the drug can be stored in space without losing its effectiveness. Whether bone-active drugs can be stored in space is currently not known, and this question should be studied further using the ISS.
4. The possibility exists that neither exercise nor currently available pharmaceutical countermeasures will stop bone loss completely during spaceflight. In this case, studies that include exercise and existing or new pharmacological therapies should be undertaken to test the synergy between these countermeasures. These studies should include the evaluation of changes in bone structure and strength.[155,156,157]
5. The bone health of women during long-duration spaceflight requires further study. The practice of inhibiting menstrual periods during flights of up to 6 months is likely to contribute to marked, and possibly irreversible, bone loss in longer missions. These issues are further addressed in the section of Chapter 7 titled "Biological Sex/Gender Considerations."
6. Interventions to accelerate skeletal recovery following long-duration spaceflight should be conducted.
7. More studies are required on the lifetime bone health of astronauts who have flown on long-duration missions. In particular, the risks of fracture and renal stones need to be examined and, if found to be elevated, addressed.
8. Future studies should address issues of bone quality and not just bone mineral density, because the former is more relevant to performance and fracture resistance.

Animal Studies. All ground-based studies listed below can be completed within the next decade, but flight studies will be limited by the inadequate animal housing on the ISS.

1. Animal experiments should be conducted on rodents that are skeletally mature for relevance to adult organisms.

2. Studies of genetically altered mice exposed to weightlessness in space and to the newly developed partial-gravity analogs on Earth are strongly recommended.
3. The efficacy of existing and new osteoporosis drugs under clinical development should be tested in animal models of weightlessness (both ground-based and in spaceflight).
4. Fracture healing, methods to improve fracture healing, and effects of antiresorptive drugs on fracture healing should be further evaluated in animal models of weightlessness (both ground-based and in spaceflight).
5. The combined effects on bone of space radiation and altered-gravity should be evaluated in ground-based animal models.
6. The precise cellular signaling mechanisms responsible for initiating increased bone resorption and reduced bone formation in weightlessness should be studied further in animal models of weightlessness.

Consideration should be given to using the rat model because of its successful track record in predicting the actions of pharmacological interventions on human bone and because of new technology to genetically manipulate rats.

Obstacles to Progress

Progress in identifying countermeasures to bone loss during long-duration spaceflight has been hindered by a number of factors that need to be addressed in the next decade.

Historically, the exercise devices that have been flown on the ISS (with the exception of the advanced resistive exercise device[158]) have undergone limited pre-flight testing to establish their efficacy. This is evidenced by the lack of published studies in the literature. In addition, conversations and direct interaction with NASA personnel indicate that the devices have not had the longevity required to survive programmed use by crew members without large investments of crew time in maintenance. For example, large maintenance and redesign costs are known to have been incurred for repairs of the interim resistive exercise device (iRED) and TVIS. A further issue is whether or not the stimuli provided by the exercise devices are sufficient to generate the required responses to preserve musculoskeletal homeostasis. This has not been the case in the past.[159,160]

NASA should develop a larger bed rest facility that will allow more rapid evaluation of ground-based simulations of countermeasures with adequate statistical power. As indicated a number of times elsewhere in this chapter, having appropriate facilities on the ISS for conducting animal studies is also an important need.

Pharmaceutical countermeasures include bisphosphonates, but two rare potential problems with this class of drugs have received much negative public attention and may have prevented their more widespread use. Atypical subtrochanteric femoral fractures have been reported after long-term use.[161] Osteonecrosis of the jaw has also been observed, but the frequency from bisphosphonate use is generally agreed to be quite low, estimated to be 0.7 per 100,000 person-years of exposure.[162,163] The American Dental Association has published guidelines that propose careful examination of patients for underlying dental conditions. Despite this recommendation, community dental care has been denied to individuals (including astronauts) who have used bisphosphonates. Thus the negative perception of some of the rare side effects of bisphosphonate use has prevented more widespread use of this class of drugs among astronauts. NASA should help allay concerns by assisting with careful selection of dentists who do needed dental work for astronauts in advance of bisphosphonate dosing and agree to take on dental care later if required. Further exploration of the effects of artificial gravity on bone is also warranted. A specific discussion of the use of artificial gravity as an integrated countermeasure for a wide range of systems can be found in Chapter 7.

Risks for Skeletal Muscle During Long-Duration Spaceflight

While the skeletal (bone) system evolved to provide a solid foundation in animals and humans in opposing the force of gravity during weight bearing, the skeletal muscle system, which is the largest organ system of the body, also evolved in response to gravity. The skeletal system developed the capacity for generating high-force contraction processes not only to synergize with bone in opposing gravity but also to enable individuals to perform a wide range of activity patterns under normal gravity loading conditions such as running, jumping, lifting, and moving heavy objects. Hence, over millions of years mammalian skeletal muscle fibers evolved into two general functional types, referred to as motor units. A motor unit consists of a group of fibers of relatively similar structural and functional properties that is innervated by a common neuron. Motor units can be classified as either slow contracting (type I) or fast contracting (type II).[164] These contrasting fibers types, under the influence of the nervous system, account for the great diversity in activity pattern that humans and animals can achieve in transitioning from the physiological state of inactivity to activity of varying intensity.

Similar to the bone system, both the slow and fast muscle fiber types are negatively affected by reduced gravitational loading, which occurs during spaceflight, as well as along the long axis of the skeleton in ground-based analogs such as prolonged bed rest. Therefore, the goal of this section is to summarize what is known and not known about the risks of spaceflight for the skeletal muscle system.[165] The concern is that when human and animal subjects are exposed to microgravity, their lower limb and core trunk muscles atrophy and lose strength and stamina, thereby reducing the fidelity of movement spanning a wide range of activities.[166,167,168] This, in turn, can negatively affect the overall fitness of the astronaut when functioning in gravity environments, whether on Earth or other celestial bodies.[169] These alterations in muscle structure and function were clearly identified by the 1998 NRC report *A Strategy for Research in Space Biology and Medicine in the New Century*[170] and are elaborated further in this section.

Effects of the Spaceflight Environment on the Structure and Function of the Skeletal Muscle System

Muscle Mass

Rodent Studies. Exposure to microgravity during the Russian Cosmos Program and NASA Space Lab missions showed that skeletal muscle fibers rapidly atrophy. This alteration occurs principally in the soleus (ankle plantar flexor), the vastus intermedius (deep quadriceps knee extensor), and the adductor longus (femur adductor) muscles, all of which predominantly express slow type I fibers.[171,172] The muscle atrophy of these slow muscles is greater than that of their fast type synergists such as the gastrocnemius and vastus lateralis muscles.[173,174] As much as a 40 to 45 percent loss in muscle fiber mass/size can occur in the soleus muscle, depending on the duration of the unloading state.[175] As a result, both slow and fast muscle fibers shrink in size.[176] The ground-based analog for spaceflight involving rodents is the HU model.[177-180] This analog is described in the previous section. Interestingly, this model mimics the muscle loss seen in response to spaceflight, suggesting that HU is a good model to undertake studies on the rodent skeletal muscle system, given the current lack of opportunities to study animal subjects in space.

Human Studies. A similar response of muscle wasting has been reported in humans for muscles such as the soleus and the vastus, thereby resulting in loss of muscle volume and muscle fiber size.[181-184] However, in humans the reduction in fast fiber cross-sectional area can equal or even exceed the loss in slow fiber size.[185] Since the fast fibers are larger than the slow fibers in humans, it appears that the larger fibers may be more susceptible to the unloading stimulus. In bed rest studies, which are the primary analog to mimic spaceflight microgravity conditions in humans, losses in muscle mass (volume) and reductions in the size of the individual fibers closely resemble the responses seen in both short-duration spaceflight on the space shuttle and long-duration missions on the ISS.[186] These losses in muscle fiber mass are the signature alteration affecting muscle fiber homeostasis.

Alterations in Protein Balance, Expression, and Contractile Phenotype

Rodent Studies. The Cosmos and shuttle spaceflight animal studies have provided insight on alterations in the subcellular muscle protein milieu.[187-190] The myofibril fraction, which accounts for more than 50 percent of a muscle's total protein pool, is the primary target for degradation, especially of the key proteins such as myosin heavy chain (referred to as MHC) and actin,[191] which govern force development and hence the strength of the contraction.[192,193,194] Also, there are shifts in MHC isoform gene expression,[195-198] showing that the slow MHCs become repressed while the fast type II MHCs are turned on, which indicates a significant shift from a slow anti-gravity to a faster contractile phenotype. Moreover, such studies in spaceflight conditions were corroborated by studies using the rodent HU model.[199] Taken together, it is apparent that the HU model is an important analog to the spaceflight environment in terms of altering muscle mass, strength, and contractile phenotype.[200-203] Unfortunately after the completion of the NASA flight program in 1998 involving animals, there has been little further progress in ascertaining the effects of long-duration spaceflight on the homeostasis not only of the skeletal muscle system but also of other important systems such as bone, cardiovascular, pulmonary, sensory-motor, and immune systems. This has left a tremendous void in understanding the biological processes governing muscle atrophy and phenotype plasticity in response to long-duration spaceflight missions.

Human Studies. In the early 1990s, Edgerton and colleagues were the first to obtain biopsy samples from astronauts before and after short-duration shuttle spaceflight missions (5 and 11 days).[204] Their findings suggest that shifts in slow to fast MHC gene expression also occur in humans. Additional studies obtained from missions of longer duration revealed that individual fibers demonstrated lower force per cross-sectional area as well as shifts to fast type IIa and IIx MHC expression.[205,206] Such losses in muscle mass and shifts in contractile phenotype have important functional consequences as presented below.

Functional Alterations in Skeletal Muscle

Rodent Studies. While only a few studies have been performed to examine the functional properties of rodent muscle immediately following spaceflight, these studies clearly show that there are alterations in the contractile processes as delineated by force-velocity tests involving the antigravity soleus muscle.[207,208] These alterations involve (1) a reduction in force output for any given velocity of contraction, (2) a reduction in power output, and (3) a decrease in the resistance to fatigue in response to repetitive contraction output. These observations of reduced function are consistent with the atrophy process and the transformation from a slow to faster contractile phenotype as discussed above. In additional studies, there is strong evidence that the slow muscle fibers show evidence of susceptibility to injury as a result of initially readapting to the normal gravity environment.[209,210] Collectively, these observations suggest that the performance of individuals undertaking physical activity in a gravity environment could be compromised and that the muscle could be prone to further injury in performing tasks demanding high functional output. Such deficits are illustrated by marked changes in rodent posture (low center of gravity), as well as the extensive use of the tail for support. Also, there is an inability to move quickly while pushing off from the balls of the feet for locomotion.[211] Thus, these observations point to deficits in the sensory-motor system of rodents following spaceflight that warrant further investigation, especially in response to long-term spaceflight.

Human Studies. Studies on humans following both spaceflight and ground-based bed rest exposure demonstrate alterations similar to those reported in rodents. The signature response involves a reduction in absolute strength of the target muscle group and decrements in the torque-velocity relationship.[212-215] These functional alterations appear to be greater than the deficits in muscle mass, especially early on in the time course of spaceflight.[216] The differential responses in muscle strength could be due, in part, to sensory-motor alterations, which impair the nervous system's ability to recruit motor units in response to high loading stimuli. Individual-fiber analyses further suggest that the loss in force capability could also be due to deficits in the intrinsic properties of the myofibers.[217,218] Also, there appears to be a wide range of response in such muscle function deficits among human subjects.[219,220] Whether such diversity is due to the responsiveness of astronaut subjects to the unloading state or to differences in countermeasure strategies that are being employed among the astronaut subjects remains to be determined. (Note: astronauts do not perform a prescribed exercise routine.) In humans, little information is available as to whether skeletal muscle is prone to injury during early recovery from spaceflight. However, significant soreness has been reported anecdotally by astronauts; such soreness could impact high-intensity emergency egress capability during the early recovery period.[221] In a previous report on astronauts and cosmonauts following spaceflight of varying duration, evidence based on magnetic resonance T2 analyses suggested that muscle injury was probably occurring in some of the subjects during the early stages (days) of recovery.[222]

Key Synergies with Other Systems

Bone. It is well recognized that both skeletal muscle and bone homeostasis are negatively affected by prolonged exposure to spaceflight as well as to ground-based simulations of spaceflight.[223,224] As noted in the bone section, pre- and post-flight quantitative computed tomography analysis has shown that long-duration spaceflight missions induced average volumetric BMD losses of about 0.9 percent per month and 1.4 percent per month in the spine and hip, respectively.[225,226] Findings on skeletal muscles similarly suggest a range of atrophy averaging about 6 percent to 8 percent per month, also with greater losses in the lower extremities compared with the upper extremities.[227] These respective deficit profiles for bone and muscle actually exceed what is observed during the aging-induced disorders of osteoporosis of bone and muscle sarcopenia. Hence, the question arises as to whether the structural and functional integrity of the two systems are physiologically linked.

Recent bed rest study findings provide evidence that the mechanical stress strategically imposed on skeletal muscles by physical exercise during spaceflight or ground-based analogs can have a positive impact on the homeostasis of bone. However, pharmacological strategies specifically targeting bone homeostasis do not synergistically affect skeletal muscle.[228] These findings suggest that while both resistance exercise (RE) and bisphosphonate treatment (an inhibitor of bone resorption) have a positive effect on bone homeostasis, only the RE treatment has a positive impact on both skeletal muscle and bone, particularly in those regions where the mechanical stress on the muscle system is enhanced.

Similar findings were provided by Shackelford et al.,[229] who compared bed rest plus RE with bed rest alone. The RE consisted of a vigorous loading program targeting multiple muscle groups for a period of 17 weeks. Volitional strength increased significantly compared to pretraining values in the RE group, whereas it declined in the bed rest control group. Losses in muscle mass across the muscle groups were significantly less than that which occurred for the control group, indicating that muscle atrophy was markedly retarded by the RE program. Interestingly, losses in BMD were significantly less in the RE group than in the controls. In fact, in the calcaneal region the BMD was actually increased somewhat over the pre-exposure values, indicating that RE can have a powerful impact on bone even under unloading states. These studies point to the potential positive value of RE programs, when carried out under appropriate training conditions, in reducing the deleterious effects of chronic unloading on muscle strength and muscle and bone mass.

Sensory-Motor. As noted above, during the early stages of unloading (as seen in microgravity and bed rest), muscle group strength is compromised before significant muscle atrophy occurs, providing evidence that the ability to recruit motor units likely is compromised during the early stages of unloading states. In animal studies, locomotor patterns are compromised, as reported above. These observations indicate that the combined skeletal muscle and sensory-motor systems are highly integrated; dysfunction in either system has deleterious consequences when the systems are challenged following spaceflight. In the future, the two systems should be studied as a functional entity. Such research not only should examine muscle structure and performance but also should examine function originating from different areas of the cortex, the activation of muscle motor units, and the properties of the neuromuscular junction, in order to dissect the complete pathways in the control of movement.

Status of Countermeasures

Animal Studies. Studies on animals have used a variety of manipulations to counteract atrophy responses induced by HU. The most physiologically relevant to the human resistance exercise program involves two different resistance loading paradigms: one employed a paradigm of repetitive isometric contractions;[230] the second used a sequenced combination of isometric, concentric, and eccentric muscular actions during each contraction cycle.[231] Both studies used an experimental strategy of studying the effectiveness of the training paradigm during the rapid state of atrophy, which occurs during the first 7 to 10 days of HU, during which the rodent gastrocnemius muscles atrophy by approximately 25 percent. The isometric-only paradigm was not fully successful in maintaining muscle mass. This result was attributed to imposing an insufficient amount of loading stimuli on the muscle. This interpretation was supported by the inability of the muscle to maintain sufficient signaling pathway stimuli to optimize protein synthesis capability.[232] However, in the second study involving the integrated contraction mode paradigm,[233] muscle mass and contractile protein concentration and content were maintained. This response was consistent with the muscle maintaining a normal protein translational signaling cascade. These studies point to the potential for using RE loading that targets different muscle groups in order to mitigate muscle atrophy.

Human Studies. During spaceflight on the ISS typically spanning 6 months, NASA astronauts are instructed to perform an activity regimen of their own choosing, typically consisting of combinations of treadmill exercise with a loading support system, cycle ergometry, and RE. As noted by Trappe et al., there is no specific prescription that each subject follows.[234] Also, no control group is used as a reference. Thus, it is difficult to compare responses among the subjects as to which exercise combinations are most effective. Given these caveats, the general consensus indicates that the current paradigms being used are not fully successful in preventing muscle strength loss, muscle fiber atrophy, and the contractile phenotype shift from slow to fast properties, e.g., the muscle system alterations that are well documented.[235,236] However, there is evidence that a combination of endurance and resistance loading does have a significant positive effect on some of the subjects.[237,238]

Bed Rest Studies. Recent bed rest studies have provided important information concerning the effects of RE on skeletal muscle homeostasis. The importance of these studies is that (1) bed rest results in muscle alterations that are similar in scope to that seen in spaceflight;[239-243] (2) control bed rest groups were used routinely to serve as a reference to any countermeasure paradigm imposed; (3) the studies mimicked the time frame that has been examined on the ISS; and (4) novel RE equipment and paradigms were used, including a flywheel device that imposes high concentric/eccentric forces on the target muscle groups, e.g., imposed loads that are greater than those apparently attained with the equipment currently used on the ISS. For example, on the ISS an interim resistance exercise device (iRED) is used for loading skeletal muscle. Its loading unit consists of two canisters capable of producing loads up to about 68 kg per canister (a loading amount insufficient for loading large muscle groups). A known limitation of the iRED is the inability to precisely set the load and quantify the workloads.[244] Overall, the results show that the quadriceps muscles can be maintained at normal volume (size) with the training paradigms imposed.[245] However, with regard to calf muscle, especially the slow soleus muscle, the findings indicate that this muscle cannot be maintained at both normal mass and functional capacity.[246] This result suggests that a more robust training program is needed to protect the calf muscles. On the other hand, there is encouraging evidence that current paradigms are effective in maintaining the slow contractile phenotype in soleus muscle.[247]

Experiments Using Alternative Loading Countermeasures

A new type of countermeasure, explored recently by NASA in a pilot study, employs the principle of artificial gravity (AG) or hypergravity. This pilot study used a short-arm centrifuge to expose bed rest subjects for 1 hour per day to a dose of acceleration force vector that was directed from the head to the feet (2 *g* at the foot).[248] A separate control group of bed rest subjects was given a simulated posture without the applied 2 *g* force. While this study focused on several systems including bone, cardiovascular, sensory-motor/vestibular, and skeletal muscle, this section discusses only the analyses that focused on the skeletal muscle system. Post/pre-torque-velocity determinations revealed greater decrements in knee extensor performance in the control group than in the bed rest plus AG group. Also, muscle strength in the AG group was preserved, especially in the calf muscle groups, because the plantar flexors of the AG group actually produced a net gain in torque-velocity properties, whereas the control group showed a significant decrement. While these findings provide evidence that AG has potential as a countermeasure, it should be noted that other devices also may have beneficial effects on skeletal muscle and other organ systems (see the section "Interaction of Muscle and Other Systems"). While AG suggests promising opportunities for sustaining skeletal muscle homeostasis, more research is needed regarding the potential impact of AG on other organ systems such as bone, immune, and cardiovascular systems. Further, it remains to be seen how AG paradigms applied to animal subjects affect both developmental processes and other fundamental processes across the organ systems.

Gaps in Knowledge

High-Priority Gaps in Fundamental Knowledge: Animal Studies

Contractile Protein Turnover. It is well known that maintenance of muscle mass depends on the balance between processes regulating the synthesis of contractile proteins and processes governing their degradation. This is normally a continuous balancing process, referred to as "protein turnover." Although great strides have been made in understanding protein turnover, more research is critical to unravel the cascade from the perspective of identifying (1) the mechanoreceptor system(s) that are sensitive to gravity stimuli and (2) the downstream signaling pathways and transcription factors that control the expression of the contractile protein genes comprising the contractile muscle synthesis apparatus. Recent findings on atrophy of rodent muscle in response to unloading suggest that transcriptional regulation of both myosin and actin (the two most abundant proteins expressed in muscle) may be the pivotal link, because these genes are rapidly turned off within several hours of initiating unloading stimuli. Figure 6.1 shows the rapidity with which the two most abundantly expressed genes in slow antigravity muscle fibers, that is, slow type I myosin and actin, are markedly inhibited within 24 hours of unloading. These two proteins are the drivers of the muscle contraction process. This loss in expression causes a reduction in the ability of the muscle fibers to maintain muscle size, which is also shown in Figure 6.1. These findings thus provide new insight concerning target areas that must be regulated in order to maintain skeletal muscle homeostasis.

To degrade the contractile proteins, each protein must be separated from its intact cohesive myofibril structure and targeted and labeled for degradation,[249] but very little information is known about these latter critical steps of the degradation process. Thus, additional studies on protein turnover are pivotal to understanding the homeostasis of muscle in response to altered loading states.

Substrate Energy Turnover. Studies on rodent skeletal muscle metabolic pathways in the context of spaceflight and other states of unloading have revealed a variety of responses with no clear-cut adaptive responses in the oxidative enzyme systems.[250,251,252] For example, Baldwin et al.[253] observed that in rodent skeletal muscle immediately following spaceflight, no reduction occurred in the capacity of skeletal muscle mitochondria to metabolize pyruvate, a derivative substrate of carbohydrate. However, they found a reduction in the capacity of different muscle types to oxidize a long-chain fatty acid, palmitate. This latter finding is in agreement with an observed increase in the accumulation of lipid in the fibers of skeletal muscles exposed to spaceflight.[254] Also, the metabolic pathway

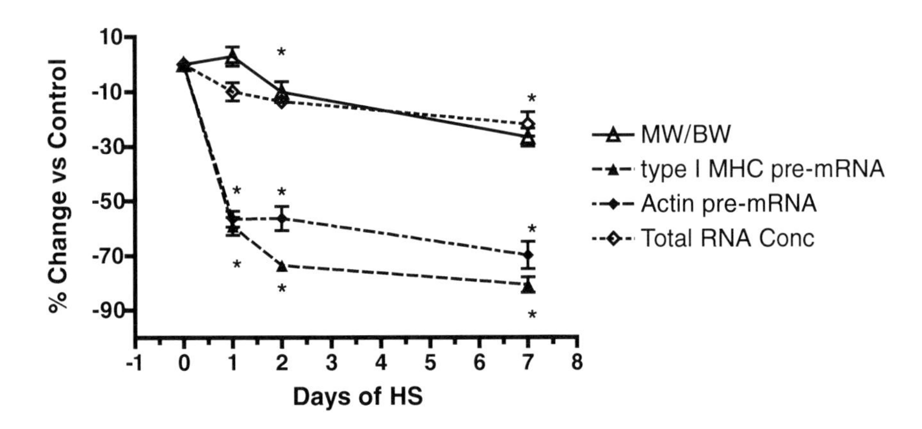

FIGURE 6.1 Time course of the change in the soleus after 1, 2, and 7 days of hindlimb unloading (HU) in rodents. Percentage change versus control for soleus muscle weight (relative to body weight), total RNA concentration, type I MHC pre-mRNA, and skeletal α actin pre-mRNA expression. Asterisk (*) is for $P < 0.05$ versus control. SOURCE: Data from J.M. Giger, P.W. Bodell, M. Zeng, K.M. Baldwin, and F. Haddad, Rapid muscle atrophy response to unloading: Pretranslational processes involving MHC and actin, *Journal of Applied Physiology* 107:1204-1212, 2009.

for glucose uptake is increased in muscles exposed to reduced gravity.[255] Although data on enzyme activity are equivocal, it is possible that as a result of a shift in substrate preference in response to reduced gravity, carbohydrates are used preferentially to provide energy to support muscle contraction. Further studies are needed that focus on organismal metabolic processes in humans during long-duration spaceflight and/or bed rest to ascertain (1) if there are intrinsic lesions in the mitochondrial system limiting metabolism of fatty acids and (2) if alterations in the substrate preference for sources of energy production during exercise in spaceflight under varying loading conditions affect the utilization of carbohydrates during prolonged exercise, such as during extravehicular activity on different planets with reduced-gravity conditions.

The Role of Reactive Oxygen Species, Satellite Cells, and Growth Factors in Protein Balance Regulation. Recent findings suggest that treatment with antioxidants can slow the atrophy of unloaded respiratory muscles involved in breathing.[256] Further, any stimuli to improve protein balance appear to be predicated on expressing Insulin-Like Growth Factor-1, which functions as an autocrine-paracrine regulator within skeletal muscle and plays a regulatory role in muscle growth, even in the absence of the pituitary-growth hormone axis.[257,258] This growth factor is thought to regulate satellite cell proliferation and differentiation, which play a role in maintaining protein balance in muscle tissue during both hypertrophy and atrophy.[259,260] These observations suggest that this factor, along with others such as reactive oxygen species and other growth regulators such as myostatin, may play an integrative role in buffering muscle atrophy stimuli if such target genes are fully activated. Studies on this topic in animal models using the HU model could be of value, as they would focus on limb muscles that are sensitive to gravity.

Maintaining the Slow Contractile Phenotype. One of the key questions in muscle biology concerns the mechanism of contractile phenotype switching that occurs in response to muscle unloading. New findings on this topic in the evolving area of epigenetics may provide keys to understanding such mechanisms of gene switching.[261] Epigenetic gene regulation phenomena constitute an evolving field in the study of transcription regulation that involves altering the function of DNA without changing its nucleotide sequence. Instead, alterations are generated in the chromatin, histone, and DNA structural properties by a variety of processes that respond to environmental stimuli. These alterations in turn result in alterations in the transcriptional activity of the target gene. Thus, the field of epigenetics may provide insights into understanding contractile protein expression. Studies on this topic could be easily accomplished using the HU model.

High-Priority Gaps in Applied Knowledge: Human Studies

Highly loaded resistance exercise is a promising countermeasure for ameliorating loss in muscle mass and strength, but additional research is needed to develop (1) the appropriate training devices and (2) the optimal prescriptions to maintain muscle homeostasis and sensory-motor function. These latter properties are pivotal for complex movement paradigms such as performing a variety of extravehicular activities in space, as well as performing emergency egress from the spacecraft on landing. Further, it is most probable that those exercise paradigms that benefit the skeletal muscle system will also affect the functional capacity of other systems in the context of integrated organ system function. It is imperative to design modalities and exercise protocols that can condition the body for cardiovascular/aerobic fitness, skeletal muscle strength and endurance, and sensory-motor fidelity, as well as bone and connective tissue homeostasis. Such broadly effective exercise modalities would further the strategy of designing optimal exercise prescriptions for improving the widest range of physiological systems. To NASA's credit, since 2007 NASA/NSBRI has indicated, in its Research Focus Announcements[262] soliciting ground-based research studies for human health in space, a keen interest concerning applications related to new exercise devices that can affect multiple physiological systems. This is an important direction for advancing knowledge of how to enhance countermeasures mitigating the deleterious effects of long-duration spaceflights on astronaut health and performance.

Research Models and Platforms

Animal Research

Ground-based research on animal models such as rats and mice have played a major role in generating fundamental knowledge concerning the effects of microgravity on muscle alterations and in developing countermeasures to microgravity-induced alterations in muscle mass, phenotype, and function. This fundamental research can be continued using the HU model.[263,264,265] As discussed above, new avenues of animal research are unfolding in the fields of epigenetics of gene expression and protein turnover in response to unloading stimuli.

Equally important is the capacity to expose adult mammals such as rodents to spaceflight for long durations, because this is the only model in which to study long-term effects on the physiology of the organism without any interference from the countermeasures that are obligatory when human subjects are tested. Furthermore, normal adult mice, as well as mice with either knockout or overexpressed genes, can provide fundamental information about skeletal muscle function in microgravity (see discussion of this topic in the section above, "Risks for Bone Loss During Long-Duration Space Missions"). For example, one can examine the time course of the alterations and compare such deleterious effects with effects on ground-based control animals of the same age and sex. Also, given a capability to bring test animals back to Earth following long-term exposure to microgravity, their recovery process across several physiological systems can be studied to ascertain whether there are changes in response to long-term microgravity that are not reversible with re-exposure to a normal 1-*g* environment. Currently, the lack of an animal facility for rodents on the ISS suitable for long-duration studies on adult animals is a major research impediment that will hamper the ability to obtain information important for maintaining astronaut health and fitness for duty. Furthermore, research on animal models will be constrained without the ability to manipulate the gravity variable as a factor modulating the fundamental processes underlying organ system homeostasis.

Human Research

Currently, appropriate flight opportunities on the ISS, as well as ground-based analogs (bed rest and other models of unloading) are available to pursue both basic biological science and translational science initiatives. The key is to take greater advantage of the ISS as a platform for conducting long-duration studies on the effects of countermeasures and interactions of different countermeasures in terms of human health and performance. For example, exercise equipment with a greater potential to improve the functional capacity of more than one physiological system needs to be tested. The key is to design facilities that can accommodate a variety of exercise modalities and to perform integrative studies that simultaneously monitor the functions of multiple organ systems. Mechanistic studies on humans are essential to ascertain how the results of mechanistic research in animal models translate to human organ systems.

High-Priority Research Recommendations

Animal Studies

1. Studies should be conducted to identify the underlying mechanism(s) regulating net protein balance and protein turnover in skeletal muscle during states of unloading and recovery. These studies are essential to understanding the process of muscle wasting. Such studies should examine the roles of growth factors, hormones, signaling pathways, protease and myostatin inhibitors, possible pharmacological interventions such as antioxidants, and nitric oxide signaling. This research could potentially be concluded in a 10-year time frame and is not dependent on access to space if the HU model is used.

2. Studies should be conducted to ascertain the regulatory processes controlling expression of the contractile protein genes in response to states of unloading. The focus should be activity-induced transcriptional mechanisms with a thrust toward the evolving field of epigenetics. These studies are needed because the slow contractile phenotype is repressed during exposure to both spaceflight and chronic bed rest, thereby jeopardizing muscle function and movement fidelity. Such research, using the HU model, could be conducted on rodents within an 8-year time frame.

3. Research should be pursued on small animal systems that can delve into the developmental biology of

the undifferentiated skeletal muscle system, as well as other organ systems.[266,267] Early studies during the Neurolab project clearly demonstrated that skeletal muscles fail to grow and develop when infant rodents are flown in space.[268-272] The same impediment is seen for other systems such as bone and sensory-motor development. A major gap in developmental biology studies on microgravity/spaceflight is that there are no suitable mammalian models to fill in the evolutionary gap currently linking very small living creatures (e.g., worms) to the human species. This is a major flaw that needs correction.

Human Studies

Studies should be undertaken to (1) develop and test new prototype exercise devices; (2) optimize physical activity paradigms/prescriptions targeting multisystem countermeasures, preferably with the same training device; and (3) assess substrate energy turnover during exercises of different intensities as a function of prolonged spaceflight. Also, NASA should continue to explore the possibility of alternative devices (e.g., flywheel, lower-body negative-pressure/running, rowing, and artificial gravity) that have the potential to affect multiple organ systems, in addition to more recent conventional exercise equipment (such as the advanced resistive exercise device (ARED) and combined operation load bearing external resistance treadmill (COLBERT)). As discussed in the preceding sections, these studies are important because the current exercise devices and corresponding physical activity countermeasure prescriptions are reported to be insufficient for optimally maintaining physical fitness and organ system homeostasis. Depending on how rapidly such new devices come onboard, this research could be completed within 15 years.

Risks for Sensory-Motor and Vestibular Deficits During Long-Duration Spaceflight

While this section is titled "sensory-motor and vestibular function," the sensory-motor functions covered encompass all sensory systems, e.g., vision, proprioception, pain, and even odor and taste. The central nervous system continuously monitors environmental elements that provide the responses needed to survive, such as gathering food and eating, reproducing, and so on. Traveling safely for extended times in different gravitational environments depends on the successful integration of multiple sensory and motor systems so that movement can be controlled effectively and safely. Precision in movement is critical in that even small errors in control of movement during flight maneuvers, extravehicular activities, controlling robotic devices, and traversing the surface of the Moon while in a spacesuit can have serious consequences. Thus from a safety perspective preservation of well-controlled movements within these variable environments is of utmost importance.[273] For example, neural control of movement in the spaceflight environment must take into account not only the vestibular sensory system but also information related to the dynamics of the interaction of head/neck, trunk, and limb muscle positions, as well as control of eye position.[274,275] Full awareness of the integrated function of the sensory-motor system is likely to provide a clearer perception into the problems and solutions related to the performance of humans in space and to their postflight recovery. A well-quantified illustration of this integrative function in postural control is reported by Speers et al.,[276] who showed that the instabilities of posture and gait following spaceflight resulted at least in part from highly interactive adaptations of each of the sensory inputs associated with vestibular, proprioceptive, and visual systems.

Effects of the Spaceflight Environment on the Function of the Sensory-Motor System

Mismatch of Functional Integration of Sensory and Motor Circuits

In travels from 1 *g* to microgravity and back to 1 *g*, disturbances in occulomotor control, vestibular function, pain sensitivity, muscle stretch sensitivity, joint position sense, and cutaneous sensitivity to vibration have been observed.[277-287] All of these sensory systems must be integrated effectively to successfully control movement and position.[288] The numerous and consistently observed changes to the responses in gravitational fields altered from 1 *g* demonstrate that the sensory-motor system is essentially calibrated for a 1-*g* environment. Illusions of movement of self relative to the environment reflect a mismatch between the expected and real patterns of sensory information.[289,290,291]

Upon withdrawal of a vertically oriented gravitational force on the whole body, it has been observed on Mir, space shuttle missions, and the ISS that a general state of flexion mechanically and neurophysiologically is assumed, with the exception of the ankle.[292,293] This change in posture undoubtedly involves a new state of activity of muscle spindles, Golgi tendon organs, and other proprioception afferents that would occur in conjunction with altered vestibular control in the absence of the gravitational loading that normally occurs at 1 *g*. While the ankle position tends to become more plantarflexed mechanically in the 1-*g* environment, the neural response is one of active dorsiflexion; that is, upon removal of weight bearing, the tibialis anterior, the primary dorsiflexor of the ankle, becomes much more active than when the individual is in a standing position in a 1-*g* environment. Thus, the flexors, including the ankle—i.e., dorsiflexors, not plantar flexors—become hyperactive, and remain so for prolonged periods during spaceflight.[294] The most logical explanation for this relatively plantar-flexed mechanical position in the ankle in spite of the hyperactivity of the dorsiflexors involves the balance of force generated passively by dorsiflexors versus plantar flexors of the ankle.

Numerous experiments have demonstrated disturbances in balance capability for several days or more following return to 1 *g*. Postural responses before and after spaceflight reflect changes in vestibular function and its interactions with vision when traveling from 1 *g* to microgravity and back to 1 *g*.[295,296,297] The variability in the magnitude of these disturbances in different crew members has been substantial, and the differences are not readily attributable to the length of duration of the spaceflight.[298] Perhaps such variability in response to gravity changes is related to the consistency of adhering to a given type of countermeasure.[299]

Extensive attention has been paid to the role of vestibular dysfunction in microgravity as having a key role in causing motion sickness.[300,301] It has been hypothesized that this sensory mismatch compared to 1 *g* is the reason for motion sickness. Although this is an appealing concept, it has never been tested stringently. Many unanswered questions remain about the role of the vestibular system in the process of adaptation to variable gravity, but future priorities related to neuroscience must be considered from a broader perspective: not only how the vestibular system responds to the challenges of space missions but also how understanding this system could provide insights into fundamental issues regarding other organ systems.

Functional Recalibration of Sensory-Motor Circuits Controlling Flexion Versus Extension

Modulation of monosynaptic and polysynaptic reflexes has been observed both during and following flight.[302,303] There have been shifts in the relative activation of motor unit pools associated with flexion versus extension and with fast versus predominantly slow motor pools when tested at 1 *g* after returning from a microgravity environment lasting for days to weeks.[304,305] For example, when a rhesus monkey walked on a treadmill after only 12 days of spaceflight, the relative recruitment of fast muscles was significantly elevated compared to preflight, while that of slow muscles was depressed (Figure 6.2). There was a marked decrease in the maximum force that could be generated voluntarily, and most of this decrease appeared to be neural in nature rather than muscular.[306,307] Also, in rats, monkeys, and humans, loss of extensor muscle function significantly exceeds that of the flexors. As observed in crew members, the susceptibility to the activation of sensory-motor circuits becomes more biased toward flexion compared to extension and more biased to fast versus slow muscles and motor units within a period of only 2 weeks of exposure to a microgravity environment. This means that the neural circuitry for controlling locomotion has adapted by selectively changing the translation of sensory input to motor output and therefore must require some combination of rapid re-adaptation and compensatory modulations that can correct for the acquired changes that occurred in microgravity.

Gaps in Knowledge

The chronic patterns and levels of activity of sensory-motor circuits define to a large degree the efficacy of neuronal pathways and the interlinking of these different circuits. These chronic patterns shape the ability to perform any routine or specialized motor task with some well-defined level of predictability of success.[308] These patterns of activity are also known to modulate nerve growth factors and endocrine responses, both of which play supportive roles in neural reorganization of circuits and the homeostasis of multiple tissues.[309] Clearly, the pattern and amount of activity play an important role in the homeostasis of the metabolic and contractile properties of skel-

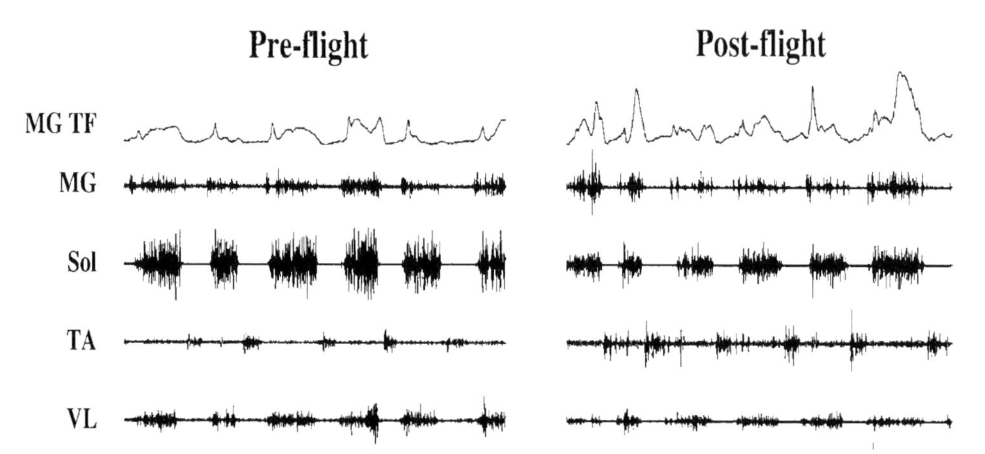

FIGURE 6.2 Medial gastrocnemius tendon force (TF) and raw electromyographic signals of the medial gastrocnemius (MG), soleus (Sol), tibialis anterior (TA), and vastus lateralis (VL) muscles from a Rhesus monkey are shown during pre-flight and post-flight stepping after 12 days of spaceflight. SOURCE: M.R. Recktenwald, J.A. Hodgson, R.R. Roy, S. Riazanski, G.E. McCall, I. Kozlovskaya, D.A. Washburn, J.W. Fanton, and V.R. Edgerton, Effects of spaceflight on rhesus quadrupedal locomotion after return to 1G, *Journal of Neurophysiology* 81:2451-2463, 1999, reproduced with permission.

etal muscles.[310] These neuromuscular activity patterns in turn modulate the homeostasis of the connective tissues that transmit stresses and strains associated with muscle function, which, in turn, affect intra- and extra-muscular connective tissues such as aponeuroses and tendons, as well as bone. Therefore, some "dose" of neuromuscular activity-exercise of the neural pathways affected in spaceflight is one of multiple potential solutions for maintaining homeostasis and function within and among multiple gravitational environments.

The critical information needed to formulate an activity-based strategy for spaceflight is to know the *dose-response* features of alterations of these tissues to unloading stimuli. Important progress has been made in some ground-based models, particularly for muscles. But there has been little or no opportunity to systematically define this dose-response relationship for different tissues in the space environment, primarily because of the lack of programmatic synergies involving bone, muscle, and sensory-motor integrative research, as noted previously.

Careful documentation of changes in the activity patterns of specific muscle groups is important for any countermeasure. For example, the levels of muscular activity of a major flexor, the tibialis anterior (TA), and an extensor, the soleus, throughout a 17-day spaceflight were markedly elevated compared with either pre- or post-flight (Figure 6.3). In a given 24-hour period the soleus muscle was more than twice as active during flight compared to pre-or post-flight, and the activity of the tibialis anterior was as much as 50 times greater. These results illustrate why one cannot assume or even expect that spaceflight is associated simply with a reduction in activity level. They suggest that the mechanisms underlying the common changes in sensory-motor performance are unlikely to be attributable simply to low levels of activity of the neural control circuits when exposed to microgravity or other features of the space environment. From a general perspective, one can reasonably claim that an integrative approach has been the mode of operation throughout spaceflight endeavors. However, when evaluating the details, critical failures at multiple levels have persisted, and these shortcomings have precluded a reasonable rate of progress toward solving the problems at hand.

Thus, a critical need is better understanding of the amount and patterns of activity that can maintain reasonably normal properties of the sensory-motor circuits, muscle, connective tissue, and hormonal and growth factor components known to play a role in the homeostasis of these tissues.[311,312] Research to answer these questions should incorporate experiments ranging from the molecular control of specific proteins, transcription factors, etc., to in vivo experiments on mammals in the space environment. Significant progress toward this goal can be made most rapidly with a systematic, sustained, and well-planned effort.

All of the modulations of sensory-motor function that have been documented demonstrate that the neuromus-

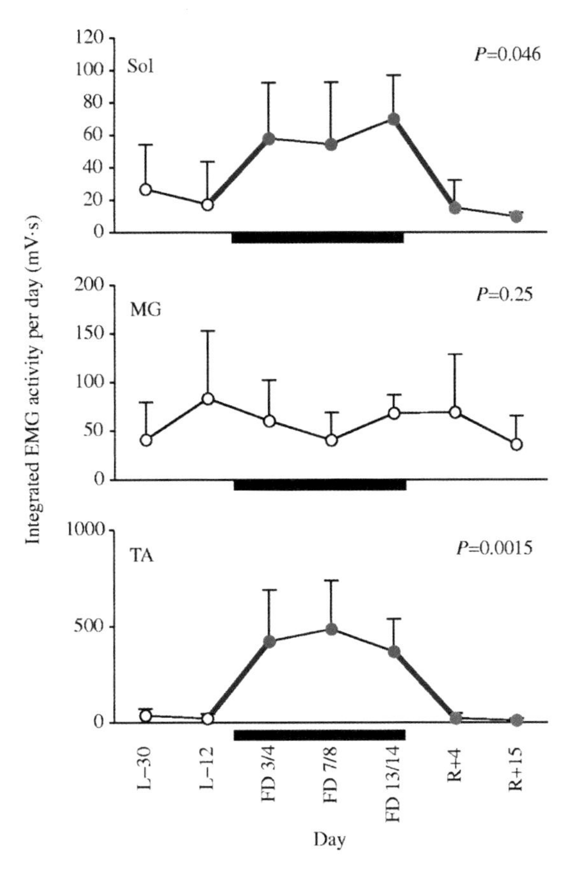

FIGURE 6.3 Mean integrated electromyographic activity (millivolt-seconds, mV·s) observed over the entire recording day for the soleus (Sol), medial gastrocnemius (MG), and tibialis anterior (TA) muscles for test days pre-flight (L-30 and L-12), in-flight (FD 3/4, FD 7/8, and FD 13/14), and post-flight (R+4 and R+15). The filled bar on the *x* axis indicates flight days. Values are mean values and ± standard error of the mean. SOURCE: V.R. Edgerton, G.E. McCall, J.A. Hodgson, J. Gotto, C. Goulet, K. Fleischmann, and R.R. Roy, Sensorimotor adaptations to microgravity in humans, *Journal of Experimental Biology* 204:3217-3224, 2001; available at http://jeb.biologists.org/cgi/content/full/204/18/3217, reproduced with permission.

cular system can adapt rapidly to varying gravitational environments. Nevertheless, it is necessary to determine the magnitude and pattern of adaptations that undoubtedly reflect the duration of exposure to a given gravitational environment and the activities that are performed while in a novel gravitational environment.[313,314,315] Similarly the rate and magnitude of these modulations will vary in duration following return to a 1-*g* environment.

Based on the experience to date on the ISS, it appears that all of the adaptations in the sensory-motor system to microgravity and back to 1 *g* can be accommodated safely, if appropriate accommodations are made in the transition to different gravitational environments, at least for the duration of the flights experienced thus far.[316] Whether these adaptations become more critical in more prolonged flights is completely unknown.

Consequently, for very long-duration flights, as would occur on a trip to and from Mars, it seems prudent to accept the ability of the sensory-motor system to adapt to the new gravitational environments encountered, but to develop countermeasure strategies to adapt to these new environments. To do this, one needs to know the time course and the specific changes that occur, as well as ascertaining the countermeasure strategies needed to modulate the specific sensory-motor functions that are so responsive to changes in the gravitational environment. It seems unlikely that a single or even a few countermeasure interventions can address each of the adaptations that must occur among the different organ systems.

Status of Countermeasures

While the technical capability and scientific expertise exist to understand important mechanisms of sensory-motor function under different gravity states, there has been little integration between the study of the skeletal muscle and sensory-motor systems to unravel how each system impacts the other. As a result of this lack of integrated research, minimum progress has been made toward developing countermeasures to avoid dysfunctional control of movement.[317,318] A combination of different models of spaceflight as well as data from spaceflight using activity-based interventions appears to have shown some effectiveness at least in preserving muscle mass and some physiological properties of muscle. While crew members on the Mir space station exercised for up to 3 hours per day,[319] it seems unlikely that this volume of exercise is necessary to maintain skeletal muscle volume and function. The more critical variable is likely to be more related to the type of activity, for example, activity in which high forces are generated.[320] More than 3 percent of the astronaut's time in space has been reported to be involved in exercise. Astonishingly, there has been no comprehensive reporting of activity patterns prior to this recent account of the activities of space patient crew members. There remain no prescribed exercise programs, and this has limited the potential for identifying a scientifically based program of activity with known consequences for different physiological systems, including those involved in the neural control of movement.[321] A number of studies have demonstrated pre- and post-flight changes in movement performance, but there has been little attention to countermeasures that might prevent a loss in gross and fine motor skills when astronauts return to Earth.

Some of these programmatic barriers are as follows:

1. There have been inadequate opportunities to form the multidisciplinary teams needed to conduct well-designed and comprehensive studies. Instead, scientists have worked as single-discipline teams. For instance, there has been minimal interaction between experts in neurophysiology and muscle physiology.
2. There have been inadequate opportunities to address any one problem in a consistent, systematic manner.[322] A researcher might have an opportunity to conduct an experiment on one flight, but rarely is there an opportunity for follow-up experiments, thereby precluding systematic efforts to resolve a given issue.
3. While the technology is theoretically available for answering many of the questions at hand, resources are needed to develop these technologies so that they can be used effectively in the space environment. This will require significant support for cross-discipline teams drawn from both the biological and the engineering communities.
4. After effective multidisciplinary teams have been developed, a concerted effort is needed to redefine the operational challenges related to safety, health, and productivity while performing routine and challenging motor tasks during and after arriving at different gravitational environments. For example, the focus within the muscle is frequently expressed as a problem related to strength and endurance (fatigability) and as problems related to maintaining muscle mass.

Are these really the key problems? Or are the key problems related more to how to maximize accurate and safe performance of a specific or even routine motor task, most of which demand neither maximum strength nor maximum endurance? How do the problems of accuracy and precision of relatively fine movements relate to changes in vestibular function, ocular motor function, and the interaction of these control systems with the spinal neural control mechanisms? Is the aerobic capacity of a crew member a critical factor for performance, health, and safety? Is it important to maintain levels of cardiopulmonary function similar to that at Earth's gravity so as to maintain general body metabolic homeostasis and to avoid, for example, the reduction in insulin sensitivity associated with chronic deconditioning? These issues illustrate the importance of cross-disciplinary solutions, particularly when considering the potential role of exercise countermeasures. The issues raised in this section are relevant as well to the areas covered in Chapter 7, which discusses integration and translational strategies.

Research Models and Platforms

Animals

Multiple models and platforms should be employed to address the highest-priority questions. While one of the most popular ground-based models has been hindlimb unloading of rats and mice, critical answers can be derived for specific questions using a wide range of models. While research can proceed most rapidly by taking advantage of multiple platforms on Earth, the ISS is an essential platform. Further, its value would be greatly enhanced with a centrifuge facility. As in the past, access to biosatellites will also provide a very effective platform.

Humans

There should be a greater focus on performing comprehensive in vivo in-flight experiments using animals, but such in vivo experiments are even more important for studies of humans. With proper biotechnical development, major improvements can be made in measuring a vast number of interrelated variables, which are likely to provide more insight than has been possible to date into the critical mechanisms that control the homeostasis of multiple tissues.

Research Recommendations

Problems related to sensory-motor function in the spaceflight environment require a highly multidisciplinary approach from concept to solution. Although some significant experiments, both ground-based and during spaceflight, have provided novel insights and have given a better understanding of the possibility of successful long-duration flights and what the potential problems are likely to be,[323] the sparseness of sensory-motor knowledge precludes confidently establishing a comprehensive strategy that will maximize the probability of successful mobility in space environments and after the ultimate return to Earth. To close these gaps, there will need to be fundamental changes in programmatic policies that determine distribution of resources and means of decision making.

More specifically, the following prototype studies are needed:

1. Determine the daily levels and pattern of recruitment of flexor and extensor muscles of the neck, trunk, arms, and legs at 1 *g* and after being in a novel gravitational environment with accommodating life support systems for up to 180 days. These changed patterns of neuromuscular activity over different time periods need to be carefully examined with respect to changes in the accuracy of movements and the type and severity of functional disruption to other tissues, particularly muscle, connective tissue, and cardiovascular and hormonal systems.
2. Identify the neuromuscular mechanisms that underlie the loss of accuracy in controlling movement with respect to (1) changes in neural control at the cortical, subcortical, and spinal levels; (2) changes in muscle properties; and (3) changes in visual, vestibular, and/or proprioceptive perception.
3. Determine whether changes in the accuracy of movement of the head, trunk, and limbs due to changing gravitational environments can be prevented or corrected with an exercise countermeasure designed to preserve aerobic function using a treadmill or stationary bike or other exercise device that maintains muscle strength and increases resistance to fatigue.

Effects of the Spaceflight Environment on Fluid Shifts

Fluid shifts due to the elimination of gravitational gradients are a fundamental consequence of the entrance into and existence in microgravity. Such shifts are particularly prominent in humans because of their predominantly upright posture. Although the 1998 NRC report did not recommend fluid shifts for future study,[324] the uncertainty of the effects of cephalic fluid shifts during extended flights in microgravity indicates that this subject does need to be reviewed. For decades, the loss in plasma volume with fluid shifts was cited as the reason for the decrease in body mass experienced during the early stages of spaceflight; however, careful research by Drummer and associates demonstrated that this explanation was incomplete in that the decrease in body mass was due, in part, to inadequate hydration and salt intake coupled with a reduced caloric intake.[325] Furthermore, Leach et al. reported that with short spaceflights not every astronaut who exhibited a decline in plasma volume had a reduction in total body water.[326] However, the astronauts did demonstrate sodium retention.[327]

Cardiovascular system performance is closely tied to fluid shifts because the latter can result in decreased plasma, cardiac, and stroke volume in microgravity. Fluids shifts can be associated with the risks of (1) orthostatic hypotension,[328] (2) inadequate physical performance for egress or for vigorous activities, and (3) a loss in visual acuity with flights of extended durations.[329]

To understand the acute and chronic effects of fluid shifts in microgravity, it is essential to specify the reference point to which comparisons are being made. Thus, when compared to the fully adapted upright standing position in a 1-*g* environment, the transition to microgravity is dramatic and predictable, with hydrostatic changes becoming the dominating stimulus. Prior to entering space, astronauts often dehydrate themselves to avoid urination during launch, then spend many hours on their backs with their legs elevated in a prelaunch position, and frequently experience the nausea and vomiting of "space sickness"[330] during the early hours in space. These circumstances compromise the initial fluid measurements made in microgravity and raise questions concerning their relevance.

Observed Effects of the Spaceflight Environment

Human Studies

Entry into space is associated with diminution of hydrostatic forces[331] and a fluid volume translocation of approximately 2 liters[332] from the lower body to the thorax and cephalic regions. The first invasive measures of central venous pressures (CVP), referenced to the atmosphere, were quite surprising in that CVP decreased to near 0 mm Hg in the acute transition to space.[333,334] However, left ventricular end-diastolic volume and stroke volume clearly increased at the same time,[335,336] which argued strongly for an increase in transmural pressure (intracardiac pressure minus pericardial pressure).[337] This hypothesis was supported by studies during parabolic flight, which demonstrated that esophageal pressure decreased more than CVP referenced to the atmosphere.[338] Thus spaceflight altered the external constraining forces of the chest and thorax, leading to increased cardiac distending pressure despite an apparently reduced CVP.

These fluid shifts were associated not only with an increased cardiac transmural pressure[339,340,341] but also with reduced intravascular volumes and pressures in the legs[342] and elevated transcapillary fluid filtrations into regions of the upper body.[343,344] Ultrasound measurements of the thickness of facial and tibial tissues of a cosmonaut early in flight demonstrated that fluids had shifted into and remained within facial tissues for 3 days (+7 percent), while leaving the tissues located over the tibia (−17 percent).[345] After landing, the fluids shifted from the head region but exhibited no changes at the tibia level.

Intraocular pressure was measured on the D-2 mission (N = 4). After an initial increase of 114 percent, the pressures rapidly returned to within pre-flight values (10 mm Hg) where they remained during the 10-day mission.[346] Concerns have recently been raised by an expert panel on papilledema about the clinical possibility of visual acuity being affected not only by elevated intraocular pressures during long-duration missions but also by optic disc edema, choroidal folds, cotton wool spots, and intracranial hyptertension.[347]

A ground-based head-down analog study was conducted by Hargens and associates, who used wick catheters inserted into the lower legs of subjects. They found significant decreases in interstitial pressures located in muscle (4.6 mm Hg to −2.8 mm Hg) and subcutaneous (8.5 mm Hg to 6.7 mm Hg) tissues with no significant changes in capillary and in interstitial colloid osmotic pressures.[348] On the other hand, measurement of Starling forces in the head and neck regions in a similar analog by Parazynski et al. revealed an increased capillary hydrostatic pressure of approximately 7 mm Hg in the lower lip with a decreased colloid osmotic pressure of 3 mm Hg.[349] Plasma colloid osmotic pressure and plasma volume were measured in subjects participating in a head-down bed rest study, and after 16 days, colloid osmotic pressure had significantly increased (29 percent), while plasma volume was significantly reduced, by 18 percent. Although no measurements were made, the authors attributed these changes to a reduction in extracellular volume.[350] From the results of the SLS-1 and SLS-2 space shuttle missions, Leach et al. hypothesized that an increased permeability of the capillary membrane was an important mechanism for the reduction of plasma volume.[351] However, there were no measurements that directly supported this hypothesis.

Fluid shifts can also alter intracranial pressures (ICPs), which can influence cerebrospinal fluids, intraocular pressures, the activation of the space adaptation syndrome, and possibly, visual acuity.[352] Using a non-invasive tympanic membrane displacement technique based on a sigmoidal relationship between membrane displacement and ICP, Murthy and colleagues reported that 6° head-down tilt was associated with a significant displacement that corresponded to an estimated ICP value of 17 mm Hg, an estimate of some clinical importance.[353]

Animal Studies

In Cosmos 1229, ICP was evaluated in monkeys using the pulse waveform and was found to be significantly increased on the first day but to have decreased to within normal values at the end of flight.[354] Histological assessment of brains from rats suspended for 93 days exhibited changes that indicated to the investigators that the animals had experienced increased ICP and brain edema.[355] Kawai concluded from his direct measurements of ICP in rats and rabbits that larger animals had higher ICP values, that it was unlikely that brain edema would occur in 24 hours but likely after 84 hours, and that the initial increase in ICP was related to the shift of cerebrospinal fluid to the head region and to the increased volume from intracranial veins. For decreases in ICP to occur with head-down tilt, increased absorption of cerebrospinal fluid by the arachnoidal villi was expected.[356]

Specific Effects of Fluid Shifts on Plasma, Cardiac, and Stroke Volumes of Humans

Associated with the intravascular fluid shifts into the interstitial and intracellular spaces[357,358] have been the following volume changes.

Plasma Volume. As summarized in the first row of Table 6.1, for both spaceflight and head-down bed rest analogs, fluid and protein shifts from the plasma were predominately responsible for the acute response, which represented a 17 percent reduction in plasma volume.[359-362] Although Leach et al. proposed that the set point for plasma volume is re-established by microgravity, this concept has yet to be verified. Among the numerous factors interacting to regulate plasma volume in microgravity, which is consistently reduced by 10 to 15 percent, select hormones[363] must be considered.

Cardiac Volume. As summarized in the second row of Table 6.1, after 48 hours of head-down bed rest, presumably because of a loss in plasma volume, left ventricle end diastolic volume decreased when compared to the supine position.[364-370] However, the bed rest value was higher than recorded for the standing position.[371] Entrance into microgravity significantly influenced diastolic filling volumes so that the acute stage represented an approximate increase of 20 percent.[372] With continued existence in microgravity or bed rest, and in the absence of countermeasures, the left ventricle atrophied, resulting in a decrease in left ventricle end diastolic volume (even at the same filling pressure)[373] at a rate that approximated 1 percent per week for both men and women.[374,375] Ultimately, the combination of a reduced plasma volume and progressive atrophy will result in a reduced left ventricle end diastolic volume (compared to the supine position) being available for perfusion purposes.

Stroke Volume. As summarized in the third row of Table 6.1, the changes with both analogs and spaceflight followed the pattern exhibited by cardiac volumes, resulting in a significantly lower stroke volume being available for situations that require maximal performance in an upright posture (e.g., astronaut egress) or for returning to Earth and facing conditions associated with orthostatic hypotension.[376-380] Because spaceflight can result in cardiac atrophy and hypovolemia, the supine stroke volume value measured after landing is approximately halfway between the standing and supine values recorded prior to flight.

Sodium Retention. Sodium balance data from space shuttle flights SLS-1 and SLS-2, Mir 97, and EuroMIR 94 missions have effectively demonstrated that sodium retention occurs with an existence in microgravity presumably because of a dissociation between sodium and water.[381,382] However, it is not known at what stage (early or chronic) in microgravity an equilibrium is established.

TABLE 6.1 Fluid Shifts and Select Volume Changes

	Analog Results			Spaceflight Results			
Measurement	Acute Responses	Early Adaptive Response	Chronic Adaptive Response	Parabolic Response	Acute Responses	Early Adaptive Response	Chronic Adaptive Response
Plasma volume	▼	▼[a]	▼[a]	Not measured	▼[a]	▼	▼
Cardiac volume	▲	▼[a]	▼[a]	▲[a]	▲[a]	▼[a]	▼[a]
Stroke volume	▲[a]	▼[a]	▼[a]	▲	▲[a]	▼[a]	▼[a]

NOTE: Acute, responses in 24 hours or less; early, responses in 2 to 18 days; chronic, responses in 84 to 180 days; cardiac volume, left ventricle end diastolic volume, reference body position was supine. Either decreased or increased volumes are indicated.

[a]Response was statistically significant.

SOURCE: *Plasma volume:* Data from C.S. Leach, C.P. Alfrey, W.N. Suki, J.L. Leonard, P.C. Rambaut, L.D. Inners, S.M. Smith, H.W. Lane, and J.M. Krauhs, Regulation of body fluid compartments during short-term space flight, *Journal of Applied Physiology* 81:105-116, 1996; M.A. Perhonen, F. Franco, L.D. Lane, J.C. Buckey, G.C. Blomqvist, R.M. Zerwekh, R.M. Peshock, P.T. Weatherall, and B.D. Levine, Cardiac atrophy after bed rest and space flight, *Journal of Applied Physiology* 91:645-653, 2001; S.M. Smith, J.M. Krauh, and C.S. Leach, Regulation of body fluid volume and electrolyte concentration in spaceflight, *Advances in Space Biology and Medicine* 6:123-165, 1997; S.H. Platts, D.S. Martin, M.B. Stenger, and S.A. Perez, Cardiovascular adaptations to long-duration head-down bed rest, *Aviation, Space, and Environmental Medicine* 80(5 Suppl.):A29-A36., 2009.

Cardiac volume: Data from P. Arbeille, G. Fomina, J. Roumy, I. Alferova, N. Tobal, and S. Herault, Adaptation of the left heart, cerebral arteries and jugular and femoral veins during short-and long-term head-down tilt and space flights, *European Journal of Applied Physiology* 86:157-168, 2001; B.D. Levine, J.H. Zuckerman, and J.A. Pawelczyk, Cardiac atrophy after bed-rest deconditioning: A nonneural mechanism for orthostatic intolerance, *Circulation* 96:517-525, 1997; M.A. Perhonen, J.H. Zuckerman, and B.D. Levine, Deterioration of left ventricular performance after bed rest, *Circulation* 103:1851-1857, 2001; M.A. Perhonen, F. Franco, L.D. Lane, J.C. Buckey, G.C. Blomqvist, R.M. Zerwekh, R.M. Peshock, P.T. Weatherall, and B.D. Levine, Cardiac atrophy after bed rest and space flight, *Journal of Applied Physiology* 91:645-653, 2001; E.G. Caini, L. Weinert, R.M. Lang, and P. Vaida, The role of echocardiology in the assessment of cardiac function in weightlessness—Our experience during parabolic flights, *Respiratory Physiology and Neurobiology* 169(Suppl. 1):S6-S9, 2009; J.B. Charles and C.M. Lathers, Cardiovascular adaptation to spaceflight, *Journal of Clinical Pharmacology* 31:1010-1023, 1991; D.S. Martin, D.A. South, M.I. Wood, M.W. Bungo, and J.V. Meck, Comparison of echocardiographic changes after short-and long-duration spaceflight, *Aviation, Space, and Environmental Medicine* 73:532-536, 2002.

Stroke volume: Data from P. Arbeille, G. Fomina, J. Roumy, I. Alferova, N. Tobal, and S. Herault, Adaptation of the left heart, cerebral arteries and jugular and femoral veins during short-and long-term head-down tilt and space flights, *European Journal of Applied Physiology* 86:157-168, 2001; B.D. Levine, J.H. Zuckerman, and J.A. Pawelczyk, Cardiac atrophy after bed-rest deconditioning: A nonneural mechanism for orthostatic intolerance, *Circulation* 96:517-525, 1997; E.G. Caini, L. Weinert, R.M. Lang, and P. Vaida, The role of echocardiology in the assessment of cardiac function in weightlessness—Our experience during parabolic flights, *Respiratory Physiology and Neurobiology* 169(Suppl. 1):S6-S9, 2009; J.B. Charles and C.M. Lathers, Cardiovascular adaptation to spaceflight, *Journal of Clinical Pharmacology* 31:1010-1023, 1991; D.S. Martin, D.A. South, M.I. Wood, M.W. Bungo, and J.V. Meck, Comparison of echocardiographic changes after short-and long-duration spaceflight, *Aviation, Space, and Environmental Medicine* 73:532-536, 2002.

Fluid Shifts and Hormonal Influences in a Spaceflight Environment

The inability to secure numerous in-flight hormonal measurements from subjects is a condition that needs to be corrected in future experiments. This is especially true for animals because most, if not all, measurements have been obtained after landing. The following hormones are of interest because of their role in restoring fluid volumes.

Vasopressin. In the SLS-1 and SLS-2 flights, urinary concentrations of arginine vasopressin (AVP) were found to have increased when compared to pre-flight results.[383] In the important metabolic studies conducted on Mir, extensive blood measurements from one cosmonaut demonstrated that AVP was significantly elevated.[384] Mean post-flight blood AVP levels from 27 cosmonauts returning from missions lasting from 120 to 366 days were increased by 135 percent compared to pre-flight values.[385] In contrast, two cosmonauts on days 216-219 in space had AVP blood levels that were reduced by 21 percent and 40 percent, respectively; these changes were attributed to a possible change in sensitivity of the kidneys.[386] Vasopressin or ADH was measured in nine subjects after a 90-day head-down bed rest study, and no significant changes were reported. The authors suggested that the decrease in plasma volume of approximately 5 percent was insufficient to alter plasma osmolality.[387] Among available animal studies, measurements of vasopressin concentration per unit of protein in posterior pituitary tissues from rats on Cosmos 2044 found significantly lower concentrations than in ground-based controls.[388]

Plasma Renin Activity (PRA). During the early stages of SLS-1 and SLS-2 missions, PRA values were variable but, near the end of the flights, they remained elevated over pre-flight values.[389] A summary of PRA data from Spacelab D-2, Mir, and EuroMIR 97 missions reported increased values that ranged from 69 percent to 472 percent when compared to pre-flight supine conditions.[390] The percentage changes were markedly lower when crew members were in a supine position as opposed to seated, demonstrating the importance of position when securing a pre-flight PRA measure. Serial PRA results from the MIR 97 metabolic mission also showed increased levels from pre-flight measurements.[391]

Aldosterone. Summarized plasma and urinary results from short- and long-term flights showed both increased and decreased levels of aldosterone, as well as combinations of increases and decreases.[392] Values from SLS-1 and SLS-2 showed that aldosterone decreased,[393] while plasma results from the Mir metabolic study found that aldosterone levels were significantly increased.[394] However, data from both flights demonstrated positive sodium balances. Post-flight results from 23 cosmonauts who had been in microgravity between 120 to 366 days demonstrated a 61 percent average increase.[395] As with PRA levels, prolonged bed rest was associated with a marked increase (87 percent) and extensive variability.[396]

Atrial Natriuretic Peptide (ANP). In-flight plasma ANP levels were measured in two astronauts during an 8-day flight, and marked reductions were noted.[397] Similar plasma results were obtained with one astronaut after 26 days in microgravity,[398] whereas no changes were noted for a single cosmonaut after 20 days.[399] Hinghofer-Szalkay et al. reported that an astronaut in microgravity for 438 days had post-flight plasma ANP values that were lower (12 percent) than baseline values; additionally, when an astronaut was subjected to conditions of lower body negative pressure (LBNP) throughout the flight, ANP levels were consistently reduced.[400] After a 90-day head-down bed rest study, ANP levels were associated with a 33 percent reduction in ANP, which was attributed to a reduction in plasma volume.[401]

Norepinephrine (NE). In the 16-day Neurolab mission, NE spillover and clearance values were elevated over pre-flight values, as was muscle sympathetic nerve activity (MSNA).[402] Compared to pre-flight values, plasma NE levels from three cosmonauts measured after 217 to 219 days in microgravity were markedly increased but were regarded to be within normal limits. Their urinary NE levels were more variable and revealed no observable trend.[403] Goldstein and coworkers reported that 14 days of head-down bed rest was associated with a nonsignificant 27 percent decrease in plasma NE and a significant 32 percent decrease in urinary NE levels.[404] In a 42-day bed rest study, plasma NE levels decreased by 14 percent after 21 days but increased by 33 percent at the end.[405] Christensen et al. utilized platelet NE concentrations to assess sympathetic activity in cosmonauts and in ground-based subjects in a head-down bed rest study; they found increased platelet NE values in the cosmonauts and significant decreases in the bed rest subjects.[406] Shoemaker and colleagues investigated the changes in MSNA in subjects participating in 14-day head-down bed rest studies and reported that a significant decrease occurred in baseline MSNA.[407] Furthermore, in subjects prone to hypotension an inadequate increase in sympathetic discharge had occurred.[408] However, subjects who were not prone to hypotension experienced an increase in supine baseline MSNA.[409] A similar result was reported by Pawelczyk et al. for subjects participating in an 18-day head-down bed rest study.[410] MSNA measurements throughout a long-duration head-down bed rest study would help clarify the uncertainties associated with the previous findings.

Commentary

Entry into microgravity is associated with a marked central fluid shift, leading to cardiac distension, compared with the standing upright position on Earth. If all other factors were unchanged, this central hypervolemia would likely lead to a neurohumoral environment that would result in a volume loss, with most astronauts exhibiting reductions in total body water. However, this response is variable and seldom achieves statistical significance.

Because of confounding variables and limited access to crew members, this acute neurohumoral response has been difficult to characterize in microgravity, making it hard to determine what afferent or efferent pathways have been altered by a central fluid shift. However, after the early stage of spaceflight (see Table 6.1), it appears that cardiac volumes, fluid hemodynamics, and the neural and hormonal adaptive responses reach an equilibrium that is approximately halfway between the values associated with the supine and upright positions on Earth and close to the values seen in the seated upright position.[411,412]

Status of Countermeasures for Fluid Shifts

A previous NASA task force recommended that astronauts consume a liter of isotonic saline solution 2 hours before re-entry to enhance plasma volume.[413] To prevent blood pooling in the legs and to minimize the loss of plasma volume, LBNP at 30 mm Hg has been advocated.[414] Russian authorities promote the utilization of the "Penguin suit," an elastic loading suit that enhances venous return during the activity of extensor muscles.[415] In addition, thigh occlusion cuffs known as "Brasselyets" have been combined with exercise and are regarded as an effective countermeasure. Prior to descent, the use of an antigravity suit is recommended to protect astronauts or cosmonauts from negative hypergravity effects of landing.[416] Although systematic centrifugation in a small-arm centrifuge has potential as an effective countermeasure for use on the ISS or possibly in a lunar module, its implementation remains uncertain. Animals that were centrifuged in Cosmos 936 exhibited fewer deleterious effects of microgravity than did their flight controls.[417] With the increasing number of female astronauts, it should be noted that ground-based research has indicated that estrogen and progesterone supplementation has promise as a countermeasure against loss of plasma volume that occurs in microgravity (see Table 6.1).[418,419] However, because these hormone supplements have many other systemic effects, their utility as a fluid countermeasure is uncertain.

Gaps in Fundamental and Applied Knowledge Concerning Fluid Shifts

The longitudinal effects of the influence of microgravity on the Starling forces in human and animals warrant investigation, since, to date, there is incomplete information pertaining to humans and none from animals on this fundamental topic. Inherent in this recommendation is the need to develop and maintain a database that includes fluid shift data that incorporate the information on volume change listed in Table 6.1.

Gaps in applied knowledge pertain to whether the loss of tissue weight combined with changes in blood pressure shift the Starling-Landis equation toward greater filtration into tissues,[420] whether lymphatic function is compromised by simulated and actual gravitational conditions,[421] and whether fluid shifts are a primary or a secondary reason for the decrease in visual acuity associated with long-duration flights. Systematic investigations into the efficacy of mechanical countermeasures (e.g., Brasselyets, LBNP, centrifugation) that restore Earth-normal head-to-foot hydrostatic gradients are important because inhibiting this fundamental response to microgravity may provide no meaningful benefit during prolonged existence in space. Tests of new quantitative strategies to restore plasma volume prior to landing, such as augmentation of venous return, increased salt intake several days earlier, or hormonal manipulation with or without salt consumption and with careful attention to the timing of interventions, should be undertaken before long-duration flights are initiated.

Research Models (Analogs) and Platforms

For research in the next decade, the head-down bed rest model has recently been selected by NASA Johnson Space Center in conjunction with the University of Texas Medical Branch to become the analog for testing human countermeasures for lunar and Mars missions.[422] For rats and mice, the HU model is endorsed by the AHB Panel.[423] The adoption of the bed rest model incorporates the essence of the 6° head-down position of Kakurin et al.,[424] which has been an analog for simulated microgravity experiments for more than three decades because of its effectiveness in producing cephalic fluid shifts. An equally effective analog in producing the cephalic fluid shifts of microgravity has been the immersion model advocated by Epstein.[425] During the past decade, these two models have been challenged because they have been unable to duplicate the results that have been predicted by the Henry-Gauer reflex,[426,427] namely the increased central venous pressure,[428,429] diuresis,[430] and natriuresis[431] that occur during the acute and early stages of spaceflight. In a white paper submitted to the study committee, Norsk and Christensen argue that a limitation of prevailing ground-based analogs has been their inability to simultaneously reproduce the increased functional residual capacity (compared to the supine position in 1 *g*) of thoracic expansion and the cephalic shifts.[432] Elsewhere, these authorities on the subject have suggested perfecting an analog with thoracic compression for assessing entry into space or a dry immersion analog for oral water loads.[433]

Numerous scientists believe that the Henry-Gauer fluid shift model for humans fails because it is too simplistic in defining the afferent and efferent pathways that mediate the responses to microgravity. Moreover, it is unable to account for changes in capillary permeability, increased upper-body capillary pressures, hydration, and elevated feet posture or for the changes associated with motion sickness. For the reasons noted previously, it is necessary to secure meaningful measurements of fluid shifts (as shown in Table 6.1) during the acute and early stages of spaceflight; thus modifications of existing analogs are in order. However, for flights longer than the early stages, and especially for long-duration flights on the ISS and beyond, the long-duration head-down bed rest analog adopted by NASA[434] should be utilized.

Besides ground-based platforms, it is essential that investigators have access to the facilities of the ISS for basic animal and human studies. Access to a free-flyer platform is not recommended because the fluid equilibrium in the subject would be disrupted after landing.

Recommendations for Fundamental and Applied Research Studies

1. Investigations are needed to determine the basic mechanisms, adaptations, and clinical significance of changes in regional vascular/interstitial pressures (Starling forces) during long-duration space missions. Fundamental studies should be done with both animals and humans that characterize the effect of microgravity on Starling forces and test the hypothesis that microgravity will increase capillary permeability. Investigations should be undertaken to determine whether a set point for plasma volume exists and to determine the mechanisms responsible for its existence. Cellular and molecular studies should be conducted on the suppressional influences of simulated and actual microgravity on the secretion of vasopressin by cells in the posterior pituitary. Related investigations should be undertaken to determine whether microgravity will alter the sensitivity of kidney tubules to the presence of vasopressin.
2. Applied studies are recommended that use invasive procedures with animals and non-invasive techniques to determine the early and late adaptive changes in response to microgravity on intracranial pressure, intraocular pressure, and cerebral edema. Systematic investigations should be conducted on the efficacy of mechanical countermeasures (e.g., Brasselyets, LBNP, centrifugation) that restore Earth-normal head-to-foot hydrostatic gradients. Such studies are important because inhibiting this fundamental response to microgravity may provide no meaningful benefit during prolonged existence in space. Tests of new quantitative strategies to restore plasma volume in the period before landing, such as augmentation of venous return, increased salt intake several days earlier, or hormonal manipulation with or without salt consumption, should be undertaken before long-duration flights are initiated.

Interactions should begin between vision researchers and flight surgeons to conduct long-term simulated and real microgravity research on the influences of fluid shifts on visual acuity and relevant cardiovascular parameters.

Risks for the Cardiovascular System During Long-Duration Spaceflight

The cardiovascular system is particularly affected by gravity and weightlessness because of its dynamic nature and dependence on hydrostatic pressure for the delivery of blood flow to all organs at an optimal perfusion pressure. The adaptation and adjustments that characterize the responses to postural changes and to the metabolic demands of activities on Earth are altered significantly under microgravity conditions and the associated reductions in mechanical forces during physical activities. Chronic reductions in metabolic demands and oxygen uptake reduce the demands on cardiac output and tissue perfusion. The magnitude of the neurohumoral responses to activity and posture and of cardiac and vascular structural changes will likely be a function of the duration of spaceflights.

Such changes may result in significant loss of essential enabling cardiovascular functions and thus present a clinically significant risk. Specifically, (1) the capacity for physical stress (oxygen transport and aerobic power) and thermal regulation in space or during landing may be severely limited at times of greatest need for increased systemic perfusion; (2) the "adaptive" structural and functional changes in space over prolonged periods may result in debilitating orthostatic intolerance upon return to partial or full gravity, with potential failure to reverse structural changes that could become pathological; and (3) inadequate screening for subclinical cardiovascular disease, or acceleration of the atherosclerotic process by prolonged (or high-energy bursts of) total body irradiation, may lead to potentially catastrophic coronary events in space that could be life- or mission-threatening. In the new paradigm of very prolonged exposures to space (i.e., 6 months or more), cardiovascular pathobiology will need to be better defined not just at the integrated systems level but also at the molecular, cellular, and genetic levels in order to adopt definitive countermeasures or therapeutic strategies.

Comprehensive reviews of this topic include one by Blomqvist and Stone in the *Handbook of Physiology* in 1983 titled "Cardiovascular Adjustments to Gravitational Stress"[435] and more recently one by Aubert et al. in *Acta Cardiologica* in 2005 titled "Cardiovascular Function and Basics of Physiology in Microgravity."[436] An important workshop on cardiovascular research in space was held at the International Space University in Strasbourg, France, in 2008. Its proceedings, published in *Respiratory Physiology and Neurobiology* (2009),[437] cover the spectrum of cardiovascular adjustments in microgravity and provide an excellent baseline for future direction of research using the resources of the ISS.[438-443]

Effects of the Spaceflight Environment on the Structure and Function of the Cardiovascular System

Increased Cardiopulmonary Volume

Entry into microgravity elicits acute physiological responses, some of which become chronic adaptations. The most obvious response, which is a direct effect of altered gravitational and hydrostatic gradients, is the central and cephalic fluid shift from the lower extremities and abdomen, estimated to be approximately 2 liters.

Because of the central role of fluid shifts in the multisystem adjustments to microgravity, a separate, previous section of this report is devoted to fluid shifts. Acute consequences have included facial puffiness, headaches, and "bird legs," with increased capillary hydrostatic pressure and transcapillary fluid shifts in tissues of the head and neck[444] as a result of an increase in venular pressure.[445] It is not known whether the cephalic fluid shift contributes to a chronic increase in intracranial pressure[446,447] or to the observed increase in cephalic bone mass.[448]

A direct consequence of the redistribution of blood volume with a central shift is an increase in cardiopulmonary volume compared to upright standing on Earth. With that increase in central volume, stroke volume increases as well as cardiac output, since the decline in heart rate is minimal.

Despite the increase in cardiopulmonary volume, central venous and atrial pressures are decreased when referenced to atmospheric pressure, as indicated by direct invasive measurements in space.[449,450] This reduction is not due to reduced cardiac filling, since left ventricular end diastolic volume and stroke volume also increase[451] (see Table 6.1). Thus the transmural ventricular diastolic filling pressure must actually increase.[452,453] This conclusion is supported by measurements during parabolic flight of esophageal pressure (an estimate of intrathoracic pressure and an index of the directional change in pericardial pressure), which decreased more than central venous pressure.[454] It is not likely that myocardial compliance changes within a few seconds of exposure to microgravity; however, extracardiac constraining forces, generated predominantly by compression from the lungs and thorax,[455] also likely decrease due to expansion of the rib cage in space, thereby decreasing extracardiac pressure and allowing expansion of the cardiac chambers. Arterial systolic pressure and pulse pressure at heart level both increase.

Decreased Total Blood Volume

These early adjustments to the increase in central blood volume last for a few hours or days. Subsequently, stroke volume and cardiac output return to levels that are less than those associated with a supine position, greater than those in a standing position, and equivalent to a head-up tilt of about 30° on Earth.[456] This gradual reduction in stroke volume is associated with a clear loss of plasma volume[457] without any evidence that systolic cardiac function deteriorates. These early rapid hemodynamic changes do not seem to present a problem during flight and for the most part are within the range of changes in the cardiovascular system observed during postural adjustments in daily life.

Compounding the acute reduction in plasma volume are reductions in red cell mass[458-464] and hence in total blood volume.[465-475] These reductions take place more slowly over time and certainly contribute to post-flight orthostatic intolerance by further impairing cardiac filling when returning astronauts stand upright on Earth.[476-486] While there is no simple explanation for the decrease in blood volume, the fact remains that in microgravity total body water may not change consistently in all astronauts, whereas in-flight measurements of plasma volume consistently show reductions of 10 to 17 percent. Neurohumoral and hormonal factors regulate fluid intake, diuresis, natriuresis, pre- and post-capillary resistances, capillary filtration, and permeability, thereby causing this reduction in plasma volume. These factors include sympathetic nerve activity, renin/angiotensin/aldosterone, vasopressin, and atrial natriuretic factors. A reduction in erythropoietin may be a factor in the reduction of red cell mass.[487-489] (See details in the "Fluid Shifts" section above.)

Decreased Cardiac Work and Cardiac Atrophy

There is evidence of cardiac atrophy and reduced cardiac distensibility (reduced volume at the same transmural distending pressure), which further impair cardiac filling in the upright position.[490-497] Magnetic resonance imaging has shown a 7 to 10 percent decrease in cardiac muscle mass following 10 days of spaceflight, although only four astronauts were examined in these studies.[498] Diastolic function was further compromised by reduced left ventricular untwisting, leading to diminished diastolic filling, which compounded the structural changes and together with them resulted in a substantially reduced upright stroke volume on return to Earth.[499] Numerous studies involving 2 to 12 weeks of bed rest, including both men and women as subjects, suggest that without countermeasures the heart atrophies at a rate of about 1 percent per week.[500-503] The mechanism of this atrophy is unknown, but it is likely due to reductions in cardiac work from a combination of reduced gravitational loading and decreased physical activity and metabolic demands.[504-508]

Cardiac mass is highly regulated in response to changes in loading conditions. When unloading is marked, cardiac atrophy is rapid (within 7 days) and dramatic (40 percent decrease in heart weight and myocyte volume).[509] These adaptations appear to be localized exclusively to cardiac myocytes.[510] According to McGowan et al., "As a function of decreased myocyte volume, the relative concentration of collagen is increased, which contributes to increased chamber stiffness."[511] As reported recently by Lisy et al., a dog model of cardiac atrophy (matching quite closely the human bed rest deconditioning model) confirmed that chronic reduction in left ventricular volume (from inferior venocaval banding) causes a significant reduction in left ventricular mass and myocyte volume "despite marked neurohumoral stimulation by the growth promoters endothelin and antiotensin II."[512] The molecular mechanisms underlying cardiac atrophy and remodeling, which have been reviewed recently, suggest novel targets for this process such as autophagy.[513]

Cardiac atrophy can be prevented when bed rest—and presumably spaceflight—are accompanied by exercise training and significant increases in cardiac work.[514,515] Intriguingly, protein supplementation during bed rest, even without exercise training, also appears to minimize cardiac atrophy, though the mechanism underlying this observation is unknown.[516] When combined with skeletal muscle atrophy, the reduction in cardiac mass and blood volume limits maximal cardiac output, decreases aerobic exercise capacity, and may limit performance. If cardiac atrophy during bed rest is prevented (by exercise) and hypovolemia is corrected (by volume infusion), orthostatic intolerance can be eliminated,[517] highlighting the importance of these factors in mediating orthostatic intolerance after bed rest.

At present, there are limited data from long-duration spaceflight lasting 6 months or longer, which are the durations of a typical ISS sojourn. It is likely that, with prolonged flight, cardiac morphological changes may be accentuated and could even limit in-flight performance, although such an in-flight limitation has not been identified to date.

Cardiac Arrhythmias

Cardiac rhythm irregularities including premature ventricular contractions, atrial fibrillation, and a short run of ventricular tachycardia have been reported, especially during long-duration flights,[518,519] and could have been, at least theoretically, associated with the known changes in cardiac structure. However the arrhythmias reported do not appear to have been life threatening or even clearly induced by spaceflight.[520-523] There have been reports of minor prolongations in the corrected QT interval (longest reported, 420 ms),[524] which never reached durations in the clinically worrisome range (i.e., >500 ms),[525,526] and there are no reports of torsade de pointes, the arrhythmia that would likely be induced by clinically significant QT prolongation. It is not known whether cardiac atrophy is a substrate for electrical remodeling in space. Although changes in autonomic control have been implicated as potential contributing factors to arrhythmias, the sympathetic activation associated with spaceflight is quite modest[527,528] and well within the normal adjustments that occur during daily life. Nonetheless, rigorous quantification of arrhythmia frequency and its variability pre-flight as well as in-flight, along with non-invasive assessments of cardiac electrophysiological properties, will be necessary to determine the magnitude and significance of these observations. Recovery of left ventricular mass and the time needed for it are uncertain, and this aspect has been poorly studied, especially after long-duration flight.

Cardiovascular Control Systems

The regulation of the cardiovascular system is defined by intrinsic properties of the heart and blood vessels as well as extrinsic neurohumoral influences.

- *Intrinsic properties* are related to cardiac or vascular muscle behavior and include the force of contraction, relaxation, vascular tone, resistance, volume distribution, protein expression, myosin heavy or light chains, and cytoskeletal structures.
- *Extrinsic regulation* refers specifically to autonomic neurohumoral and hormonal influences, including efferent innervation, renal and adrenal hormones, central neuronal nuclei, sensory receptors, and both autocrine and paracrine factors.

The intrinsic properties of the heart and blood vessels, including their contractile properties, are closely linked to the neurohumoral regulatory process by the presence of a rich afferent sensory network within the heart itself[529] and by the distortion of arterial baroreceptors that are influenced by both flow[530] and pulsatility.[531,532]

Data from spaceflights suggest that there are modest reductions in the vagally mediated regulation of heart rate that explain the mild tachycardia seen in space relative to the supine position on Earth.[533,534] The sympathetic nervous system seems to be slightly upregulated.[535-539] Direct measurements of sympathetic nerve activity indicate appropriate increases in response to reductions in central blood volume with lower-body negative pressure[540] and with upright tilt,[541] to the fall in arterial pressure with the Valsalva maneuver,[542] and to other autonomic stimuli such as nociception or static exercise.[543] After 2 weeks of spaceflight, the magnitude of sympathetic activation associated with upright posture appears to be precisely calibrated to the degree of reduction in stroke volume.[544,545]

Most ground-based bed rest studies are similar in reporting mild increases in muscle sympathetic nerve activity after 2 to 3 weeks of 6° head-down rest; these increases also appear appropriate for the reduction in cardiac filling with either lower-body negative pressure[546] or head-up tilt.[547] Of interest is the observation that the increase in sympathetic nerve activity with mental stress is augmented following the period of head-down bed rest.[548] Despite the increases in sympathetic nerve activity, orthostatic intolerance has been reported in some individuals following 14 days of head-down bed rest and in some astronauts after spaceflights lasting 16 days.[549,550] In the former, increases in nerve activity were significantly less during 10 to 15 minutes of 60° head-up tilt in subjects with orthostatic intolerance, compared with those who were tolerant to the tilt.[551] Similarly, the astronauts who were intolerant to 10 minutes of standing without assistance on landing day had significantly lower plasma NE levels than those who were tolerant,[552] although their NE release in response to tyramine was normal.[553] Thus, the extensive studies on cardiovascular adjustments primarily to bed rest but also to microgravity, which have been reviewed recently,[554-561] indicate that the reflex responses of the sympathetic nervous system to changes in cardiac filling, stroke volume, and arterial pressure are preserved,[562-569] while vagal responses are reduced.[570,571] However, in individuals who may be susceptible to orthostatic intolerance,[572,573] the magnitude of the sympathetic response and the subsequent cardiovascular adjustment may be insufficient to offset the effects of reductions in blood volume and stroke volume. Susceptible individuals may therefore be unable to maintain an adequate increase in peripheral vascular resistance[574-578] to support cerebral perfusion in the face of this prominent hemodynamic compromise.[579]

Animal Studies. There is an extensive literature investigating the mechanisms of animal cardiovascular adaptation to adjustments in hydrostatic gradients, predominantly using the hindlimb unloading (HU) rat model.[580] Because quadrupeds are fundamentally different from bipedal upright humans,[581] the relevance of this model for human orthostatic tolerance can be questioned, and indeed HU rats do not develop cardiac atrophy in short-duration studies.[582] However, they do if the hindlimb unloading is continued for at least 28 days.[583] Key findings have included differential atrophy/hypertrophy and vascular responsiveness of hindlimb/mesenteric and cerebral vasculature, to a large extent because of directionally opposite changes in local vascular perfusion pressure when normal hydrostatic gradients are reversed by hindlimb unloading.[584,585] Additional observations have included altered mesenteric vasoconstriction through changes in ryanodine receptor function,[586] reduced aortic contractile function[587,588] and increased aortic stiffness,[589,590] and altered central nervous system processing of baroreceptor inputs at the rostral ventrolateral medulla (RVLM).[591] High-resolution studies in the HU model have suggested that tonic GABA-mediated inhibition of the RVLM is enhanced after hindlimb unloading, which could restrain sympathetic activation.[592]

Risks of Cardiovascular Events

During a prolonged space mission to Mars, astronauts will not have access to comprehensive health care services for periods of 2 to 3 years or more. Since the majority of experienced astronauts are middle aged (the average age of the current astronaut corps is 46 years, with a range 33 to 58 years), they are at risk for developing serious cardiovascular events such as a myocardial infarction or sudden cardiac death, especially during high-intensity exertion. Such events are of course life-threatening for the astronaut and mission-threatening for NASA.

One of the greatest challenges facing the cardiovascular community is termed the "sudden death paradox": although the patients at highest relative risk for sudden cardiac death can be clearly identified, based on known risk factors, the greatest number of sudden deaths on an absolute basis occur in patients not previously determined to be at high risk.[593] Thus, the ability to identify at-risk individuals who are currently asymptomatic is a topic of intense research within the cardiovascular community that is relevant for both NASA and public health.

Although astronauts are now carefully screened prior to selection, they often must wait a decade or longer to fly select missions. NASA invests considerable resources in training astronauts, and so screening and monitoring strategies should be implemented to follow astronauts from selection to flight, identify individuals whose short-term (i.e., 2- to 3-year) risk for a cardiovascular event may have increased, and develop risk mitigation strategies, either pharmacological (such as administration of statins)[594] or physiological (training for high levels of fitness),[595] that will keep the risk sufficiently low.

Post-Flight Maladjustments

Optimal performance of astronauts upon return to partial or full gravity has to be a primary goal of research in the next decade.

Orthostatic Intolerance. Orthostatic intolerance even after a few hours of spaceflight has been a significant problem[596-602] for NASA since the early days of crewed spaceflight. During post-flight orthostasis, the decreased blood volume, even in the absence of excessive pooling, compounded by cardiac atrophy causes a marked fall in cardiac filling pressure and in stroke volume. Indeed, the sine qua non of the cardiovascular maladjustment to spaceflight is a reduced stroke volume in the upright position. A reflex tachycardia and a reflex increase in sympathetic nerve activity do occur.[603] However, the reflex increase in vascular resistance known to occur in splanchnic and limb circulation during venous pooling in humans[604,605] is insufficient to prevent a fall in blood pressure in the most susceptible astronauts.[606-609] This uncoupling of sympathetic activity from the vasoconstrictor response and the reduced vasoconstrictor reserve in the face of a reduced stroke volume are what ultimately result in orthostatic hypotension and hemodynamically mediated collapse.[610,611] In some individuals, a true neurally mediated syncopal reaction may also occur. This may be due to a sudden sympathetic withdrawal, with or without bradycardia,[612] or to a reflex neurogenic vasodilation.[613-616] The frequency of these different types of responses has not been studied systematically. To provoke syncope or orthostatic intolerance, the fall in pressure must be sufficient to decrease cerebral blood flow, although it has been reported that cerebrovascular autoregulation is preserved after spaceflight.[617] There may, however, be a shift in the cerebral autoregulatory curve to a higher pressure, which may be related to a reduction in cerebral vessel wall shear stress with a reduced pulsatile blood flow[618] or to a vascular adaptation to microgravity[619] or to endothelial damage.[620]

Orthostatic intolerance has been reported to be more frequent in female, compared to male, astronauts.[621] Careful analysis of gender-based differences in blood pressure control points primarily to a smaller heart and blood volume associated with a greater reduction in stroke volume as the primary underlying mechanism for this difference,[622,623,624] although women may also have a lower blood pressure and altered sympathetic signaling in the upright posture, especially during the early follicular phase of the menstrual cycle.[625]

Reduced Exercise Capacity. Exercise capacity is decreased significantly after spaceflight partly because the ability to increase stroke volume during exercise is reduced by more than 30 percent.[626-632] Other contributing factors include reduced red cell mass and oxygen-carrying capacity, cardiac muscle atrophy, reduced reflex vasoconstriction in non-exercising vascular beds to appropriately redistribute blood flow, and skeletal muscle atrophy.[633,634,635] Microgravity exercise studies conducted after short-duration shuttle missions have shown convincingly that maximal oxygen uptake (VO_2 max) during post-flight exercise is reduced by roughly 20 to 25 percent, because of insufficient stroke volume and cardiac output responses.[636] Similar observations were made after the longer-duration Skylab missions,[637] but VO_2 max has not been reported after long-duration flight on the ISS.

The reduction in upright stroke volume and cardiac output is of special concern because of expectations that heavy launch and entry suits will be a future requirement that will present an added physical and heat stress. Improvement in exercise capacity may be rapid after spaceflights of less than 1-month duration, but 1 week is required for full aerobic capacity to recover, even after short-duration missions.[638] Longer recovery time would be expected with longer flights, although this has not been proven and may depend on the extent of in-flight exercise training.[639] Recovery of autonomic control and blood volume as well as red cell mass may be faster than recovery from cardiac and skeletal muscle atrophy. Both of the latter contribute to the reduction in exercise capacity. There are no data on recovery from these or other unexplored complications after long flights.

Status of Countermeasures

Countermeasures used by the United States and Russia have focused on (1) restoring plasma volume immediately prior to re-entry (fluid loading), (2) reducing peripheral pooling and enhancing central blood volume during the first few hours after landing (compression devices and skin surface cooling), (3) augmenting peripheral vasoconstriction pharmacologically, (4) utilizing LBNP in the Chibis garment, and (5) using a variety of exercise training strategies to limit cardiovascular atrophy and deconditioning during a mission.

Restoring Plasma Volume Immediately Prior to Re-entry

Because of the well-described loss of plasma volume during spaceflight, one of the earliest and most persistent countermeasures used by astronauts has been fluid loading. In 1985, Bungo and Charles described a combination of salt tablets and water drinking that decreased standing heart rate by about 30 percent and ameliorated to some degree the orthostatic fall in blood pressure.[640] Since then, fluid loading has been employed by nearly all crew members despite causing nausea and vomiting in some. More recently, this type of countermeasure has been confirmed in bed rest studies to be somewhat effective for restoring plasma volume and improving upright hemodynamics.[641] Intriguingly, a recent study showed that the intravenous infusion of dextran sufficient to normalize both plasma volume and cardiac filling pressure did not by itself prevent orthostatic intolerance;[642] only when cardiac morphology was preserved with exercise training (supine cycling), in addition to restoration of plasma volume, was orthostatic intolerance completely prevented. Preliminary work by this same group[643] has employed 48 hours of a high-salt diet along with fluorocortisone to increase renal salt retention with similar efficacy. Earlier work had shown that both glucocorticoids and mineralocorticoids increase cardiovascular responses to NE.[644,645] An alternative to fluid loading has been a single dose of maximal exercise, which has also been shown to acutely restore plasma volume and improve orthostatic tolerance after bed rest.[646] The utility of fluid loading to improve orthostatic tolerance in situations where it is compromised was also confirmed recently by Keller et al., who administered dextran intravenously during heat stress and completely prevented orthostatic intolerance in this setting.[647] Although heat activation of thermal receptors suppresses the baroreceptor reflex,[648] the primary determinant of orthostatic intolerance in whole-body heat stress is marked expansion of peripheral vascular capacity. This results in reduction of the effective central blood volume and cardiac filling pressure, hence the effectiveness of fluid loading.[649]

Non-Fluid-Loading Strategies to Restore Central Blood Volume After Landing

Anti-shock garments have been used by astronauts similar to their use by high-performance jet pilots and are part of the standard approach to re-entry. These devices have been shown to increase cardiac filling pressure and stroke volume by redistributing blood from the peripheral veins centrally,[650] but they have not been studied systematically during spaceflight. Skin cooling is also used and causes both skin vasoconstriction and a shift in the operating point of the baroreflex control of sympathetic nerve activity, which improves tilt tolerance.[651,652] Again, this approach has not been studied systematically and may place a crew at risk for decompensation should the cooling garment fail at a critical time.

Restoring Vasoconstrictor Tone

Despite the increase in sympathetic activity that has been measured in the upright position post-flight, there is a reduced vasoconstrictor reserve that prevents the redistribution of blood flow from the splanchnic and peripheral circulation to the brain during orthostasis. Promising preliminary data exist for adrenergic vasoconstrictor agonists, e.g., midodrine as a countermeasure.[653,654] Octreotides are also promising as supplements to compression garments.[655]

Exercise Training During Bed Rest to Preserve Upright Exercise Capacity

Exercise training has been employed in many different ways and at different levels with the goal of preventing cardiovascular deconditioning.[656] Three recent strategies have been particularly effective. The combination of treadmill exercise with lower-body negative pressure to simulate head-to-foot gravitational loads and normal 1-*g* ground reaction forces has been studied extensively by Hargens and colleagues.[657] This strategy is quite effective at preventing cardiac atrophy[658] while preserving upright exercise capacity[659] and muscle strength, but it is only partially protective against orthostatic intolerance.[660] The study by Shibata[661] described above prevented cardiac atrophy, preserved upright exercise capacity, and prevented orthostatic intolerance but required 90 minutes of exercise per day and invasive volume restoration. Equally promising is a combination of rowing and strength training.[662] A self-powered centrifuge has also been proposed. The fact that at least three different modes, frequencies, and intensities of exercise training have been effective at preventing cardiac atrophy and preserving exercise tolerance is strongly suggestive that cardiovascular loading is the key variable. However, to date none of these interventions have been tested against each other directly or in space. Ultimately, a multidisciplinary approach to exercise countermeasures will be required, as astronauts will likely not do separate exercises for each physiological system. Rather, a single exercise prescription, involving a well-periodized combination of endurance and strength training for different durations and intensities, while allowing for astronaut preference, will likely be most effective. Periodic in-flight testing with feedback to the crew and modification of exercise practices based on clearly defined changes would also improve the practical implementation and efficacy of exercise countermeasures.

Gaps in Knowledge

Is Orthostatic Intolerance Always Due to Orthostatic Hypotension?

Orthostatic intolerance after spaceflight has generally been presumed to be due to hypotension and cerebral hypoperfusion. However, at least some astronauts have been reported to be unable to stand even with a normal blood pressure.[663] To date, virtually all of the studies examining orthostatic tolerance have been conducted under controlled but artificial conditions of either quiet standing or a tilt test. The latter is known to cause false positive results, even in patients who have never had syncope.[664] Moreover, simply moving the legs during a tilt study may dramatically improve orthostatic tolerance, as shown after exercise in soldiers more than half a century ago.[665] It is also not clear how much other factors such as muscle weakness, neurovestibular abnormalities, a change in cerebrovascular autoregulation, or sympatho-vascular uncoupling with decreased vasoconstriction reserve contribute to symptoms of intolerance. Understanding these factors would be particularly relevant after long-duration spaceflight, when tolerance may be especially low.[666] Although orthostatic symptoms can be easily managed under controlled conditions in the supine position with a combination of intravenous fluids and flight surgeon support, it is unclear how severe these symptoms will be under conditions of fractional gravitational loads (i.e., 3/8 *g* on Mars) when such support will be unavailable or under conditions of emergency egress when the muscle pump is activated, optimal physical performance is needed, and tolerance would be especially critical.

What Is the Best Strategy or Combination of Strategies to Prevent or Manage Orthostatic Intolerance and Preserve Exercise Capacity After Spaceflight?

As mentioned above, different exercise strategies have been employed. Exercise within a lower-body negative-pressure chamber is effective at preventing cardiac atrophy, preserving upright exercise capacity, and improving—but not eliminating—orthostatic intolerance. Supine cycling combined with acute intravenous volume loading with dextran preserved both exercise capacity and orthostatic tolerance,[667] although this exercise model is time-consuming and impractical. Rowing ergometry combined with strength training plus oral volume loading, with a combination of 48 hours of salt loading and fludrocortisone, is also promising.[668] It is not clear which of these approaches is most effective.

Augmenting peripheral vasoconstriction pharmacologically with drugs such as midodrine has been beneficial after bed rest[669] and may be helpful for reducing orthostatic intolerance after spaceflight.[670] This intervention has not been tested rigorously in flight because of possible interactions with other commonly used in-flight drugs.

It is important to know the amount and patterns of activity that can maintain reasonably normal properties of the sensory-motor circuits, muscle, connective tissue, and hormonal and growth factors that are known to play a role in the homeostasis of these tissues. Also needed is information on the reason for the uncoupling of sympathetic activity from the vasoconstrictor response, the changes in vascular function and structure that can influence volume shifts and blood flow redistribution, and the means of enhancing sympathetic nerve activity. The approach to answering these questions should incorporate experiments ranging from the molecular control of specific proteins, transcription factors, etc., to in vivo experiments in the space environment.

Space Radiation and Progression of Atherosclerosis

The optimal strategy to screen and monitor middle-aged astronauts to prevent flying a crew member at increased risk for a cardiovascular event is unknown. Such an objective may be particularly challenging within a population of relatively low-risk individuals such as the astronaut corps. Moreover, although the specific effect of space radiation on the vasculature is unknown, exposure to high-dose radiation (x-ray therapy for cancer, atomic bomb survivors, radiated stents) is well known to accelerate atherosclerosis on Earth[671,672] and therefore is considered a potentially serious risk during long-duration spaceflight such as a mission to Mars. Radiation risks are covered in more detail in the radiation biology discussion in Chapter 7.

Integrated Cardiovascular Response to Microgravity and Implications for Astronaut Health

Hundreds of studies have been conducted on animals and humans, on Earth and in space, by hundreds of investigators from many different laboratories in different countries. Much has been learned about the acute and short-term adjustments of the cardiovascular system to microgravity. Most of these studies, however, have been observational and not hypothesis-driven. Some results have led to effective countermeasures, but many more have been controversial, incomplete, and inconclusive.

Of particular concern are the structural changes that would develop after months or years of existence in microgravity. Such changes include reduced cardiac mass; cardiac and skeletal muscle atrophy; reduced bone density; cardiac and vascular, as well as autonomic and sensory motor, neuronal remodeling; chronic reduction in red cell mass and plasma volume; and changes in vascular myogenic tone, endothelial shear stress and function, and capillary hydrostatic pressure and permeability. The irreversibility of these changes; their potential for accelerating latent pathological processes such as hypertension, atherosclerosis, coronary diseases, or diabetes; and the reduced capacity to respond to demands for increased physical activity and greater cardiovascular stresses are significant risks. Information on the fundamental cellular and molecular mechanisms (such as mechanosensors and metaboreceptors, or protein synthesis and catabolism) responsible for these chronic changes is essential for their prevention or reversibility.

A strategic, coordinated, integrated, research plan and an interdisciplinary approach by international teams of scientists pursuing research programs with specific goals, aims, and hypotheses will be necessary to define and overcome the operational challenges related to the safety, health, and productivity of astronauts while they perform both routine and challenging motor tasks during and after arriving at different gravitational environments.

Is the Aerobic Capacity of a Crew Member a Critical Factor for Performance?

It is important to maintain safe levels of cardiopulmonary function and general body metabolic homeostasis, especially during long-duration flights. NASA must define the physical demands of astronaut tasks and their tolerance levels. For example, many tasks may be relatively unimpeded by a 20 to 25 percent reduction in VO_2 max, depending on the fitness of an individual crew member. Such issues again emphasize the importance of cross-disciplinary solutions, particularly when addressing the potential role of exercise countermeasures.

Research Models and Platforms

As discussed at length above in the "Fluid Shifts" section of this chapter, multiple models have been used to simulate the confinement and loss of hydrostatic gradients associated with spaceflight. From the perspective of effects on the cardiovascular system, the highest-resolution model available for humans is the 6° head-down tilt bed rest model.[673] The key elements of the physical effects of microgravity—namely, increased cardiac transmural filling pressure and acute increase in left ventricle volume—are reproduced faithfully by the head-down tilt model. Moreover, the ultimate outcome in terms of changes in supine and upright hemodynamics has been demonstrated to be quite similar between head-down tilt and spaceflight.[674] There are concerns, however, that no ground-based model can exactly reproduce the changes in external cardiac mechanical constraint experienced in spaceflight. Similarly, the reduced work load and metabolic demands during physical activity in microgravity cannot be reproduced. These may be partly responsible for the observations of cardiac atrophy and the uncoupling of sympathetic nerve activity from vasoconstrictor reserve.

For high-resolution basic science studies involving experimental animals, the HU rodent model is a suitable model for the hormonal factors and for the component of the fluid shifts that deals with peripheral and cephalic shifts. However, concern remains regarding differences between quadrupeds and upright bipedal humans and the lack of cardiac atrophy in the HU model, at least with the most common, short-duration exposures.

The other effect of long-duration spaceflight that cannot be easily reproduced in ground-based models is chronic radiation exposure. The effects of long-term, low-level "spacelike" radiation, accompanied by intermittent high-dose bursts of radiation equivalent to that delivered by solar flares, can at the present time be experienced only in space (although a dedicated effort to reproduce this type of exposure at the NASA Space Radiation Laboratory at Brookhaven National Laboratory is theoretically possible). Therefore only a true space-based platform with access to animal models for periods ranging from months to years—such as a lunar base, for example—can definitively examine the effects of space radiation on the cardiovascular system, as well as explore the mechanisms of observed structural and functional maladjustments and the likelihood of their reversal. (Note that because solar flares are rare and unpredictable events, use of an irradiator on the space platform would allow easier study of the effects of high-dose bursts of radiation.)

Required Platforms

The ground-based platform will be the laboratory of the investigator and is expected to be suitable for both human and animal research. Ground-based laboratories can be (but do not necessarily have to be) supported by integrated head-down tilt bed rest facilities such as exist at the University of Texas Medical Branch in Galveston, which are coordinated by staff and investigators from Johnson Space Center. For microgravity platforms, the foremost environment will be the National Laboratory on the ISS, which will require both animal and human research facilities. Free-flyer platforms that can house and fly animals for designated periods are better suited for investigations of muscle and bone but could also be useful for cardiovascular investigations.

Research Recommendations

A comprehensive research program in cardiovascular sciences should be carried out at two levels: One, at the system level, would be used to describe and define changes in integrative cardiovascular functions. The second, at the cellular, molecular, or even genetic level, would be used to define basic mechanisms that lead to dysfunction. It will also be essential to determine the integrative effect of these cardiovascular adjustments on other systems, e.g., musculoskeletal, hematopoietic, and renal-endocrine systems.

Human Studies: Enabling Cardiovascular Research Recommendations Targeted to Specific Risks

The following three enabling areas of cardiovascular research are considered essential in the next decade.

Research Area 1. Maintaining an optimal capacity for the level of physical activity required to complete the tasks demanded of astronauts, and for thermal regulation during extravehicular activity (EVA) or unexpectedly high-risk egress from the vehicle.

a. Investigate the effects of prolonged periods of microgravity and partial gravity (3/8 and 1/6 *g*) on the determinants of task-specific, enabling levels of work capacity. Specifically, are there changes in energy availability, oxygen consumption, tissue perfusion, stroke volume, blood volume, blood flow redistribution to active muscle, or the capacity for heat dissipation that make it impossible to safely meet the excessive demands for physical activity?

- Determine whether microgravity induces a redistribution of blood flow as hydrostatic pressure is eliminated. Changes in tissue perfusion and intravascular pressures may cause changes in vascular structure and vasomotor tone (vascular proliferation or atrophy, secretion of endothelial-derived vasoactive substances, and microcirculatory autoregulatory mechanisms) and changes in organ function (pulmonary gas exchange, renal clearance mechanisms, blood-brain barrier and cerebral pressure including the retina, etc.).
- Determine the effects on the microcirculation and on capillary filtration and permeability as intravascular pressures change in microgravity. Determine the effects of such changes on tissue pressure, edema, intravascular volume loss, and shifts between interstitial, extracellular, and intracellular fluid.
- Determine the changes in oxygen consumption and oxidative and glycolytic metabolism as energy needs are reduced by confinement and loss of gravitational force. Determine consequential changes in cardiac output, blood flow distribution, and perfusion of exchange capillaries.

b. Conduct fundamental studies that are both spaceflight-enabling and enabled by spaceflight on myocardial mass and contractility, capillary filtration, hormonal changes, signaling pathways, and transcriptional regulation of cardiac myosin and actin. Such studies will provide insight into mechanisms that contribute to the decrease in stroke volume and into their future prevention by more specific countermeasures (discussed in more detail below).

Research Area 2. Orthostatic intolerance after restoration of gravitational gradients (both 1 *g* and 3/8 *g*): Determining its severity as a function of prolonged microgravity, especially under "real life" task demands such as emergency egress or Mars-like tasks, and the likelihood of functional recovery as well as the time necessary for it.

a. Determine the integrative mechanisms of orthostatic intolerance after restoration of gravitational gradients. Specifically, determine the relative contribution of hypotension and cerebral hypoperfusion, compared with neurovestibular, kinesthetic, or muscular weakness, to orthostatic intolerance following spaceflight.

b. If hypotension is the primary factor in orthostatic intolerance, consider whether reductions in cerebral blood flow, changes in neurovascular control, or fluid shifts and severe loss of central blood volume or cardiac atrophy (or their interaction) cause this hemodynamic compromise.

- Investigate the specific mechanisms underlying the inadequate increase in total peripheral resistance during orthostatic stress observed post-flight, and develop effective countermeasures. Determine whether there is a functional sympatholysis with decreased vascular responses to sympathetic nerve activity. Attempt to identify individuals who may have more limited capacity for orthostatic tolerance, and use the understanding gained about differences in tolerance to help target countermeasures more effectively.
- Determine if the baroreflex-mediated increase in sympathetic nerve activity is blunted or the pattern of sympathetic neural firing is altered, causing inefficient vasoconstriction.

c. Confirm and examine more systematically the efficiency of promising post-flight countermeasures such as use of the pharmacologic agent α_1-agonist midodrine, and the use of post-flight gravitational countermeasures such as thigh cuffs, support stockings, or inflation of G-suits (Penguin suit); also determine the efficacy of promising in-flight countermeasures such as exercise of different modes, durations, and intensities; centrifuge-simulated artificial gravity; and novel fluid loading strategies that will more effectively restore blood and plasma volume.

Research Area 3. Alterations in coronary vascular disease with prolonged irradiation and exposure to microgravity. Early undetected coronary pathology may become manifest or be accelerated, with potential catastrophic consequences.

a. Refine and develop more sensitive techniques for detecting early coronary artery disease and its surrogate indicators: e.g., coronary calcification, cardiac CT angiography, intima-media carotid index, and pulse wave analysis and velocity in humans. Because of the small size of the astronaut corps, such studies would have to be undertaken in partnership with the cardiovascular research community.

b. Investigate the influences of total body irradiation in addition to exposure to microgravity on the acceleration of the atherosclerotic process, and determine whether specific medical therapy or other biologic factors (genetics, levels of fitness, etc.) can mitigate this risk over a period (3 years) equivalent to a mission to Mars.

Animal Studies: Basic Research in Microgravity Science Enabled by Simulated Microgravity and Space Studies

In addition to the enabling research areas identified above, which are considered the key areas of research for the next decade necessary to safely support a human presence in space, basic research into critical cardiovascular processes could be enabled by a robust research effort in space involving experimental animals. Areas of interest could include:

Research Area 1. Autonomic neurobiology.

a. Determine alterations in sensory signaling, for example;

- Mechanosensory adaptation, baroreceptor nerve activity, mechanosensitivity, compliance of carotid sinus, neuronal adaptation, resetting, ion channels expression, structural changes, fiber characteristics, and paracrine and autocrine changes; and
- Chemosensory adaptation, pH, pO_2, pCO_2 responsiveness, carotid body size and structure, glomus cell sensitivity, ion channels, and neuronal changes in petrosal ganglia.

b. Determine alterations in central neuronal nuclei involved in autonomic regulation and circulatory control. These include nucleus tractus solitarius, RVLM, caudal ventrolateral medulla neuronal structures, nucleus ambiguous, dorsal motor vagal nucleus, hypothalamic neuron nuclei, paraventricular nucleus, and others. Assess changes in activation, excitability, neuronal density, and autocrine regulators and receptors.

c. Characterize changes in sympathetic and parasympathetic innervation: density, NE release and turnover, innervation of vascular and cardiac muscle, receptor density, vascular reactivity, and autoregulation (cerebral).

Research Area 2. Cardiac and vascular muscle biology.

a. Heart: areas of interest could include regulators of cardiac mass, apoptosis, hypertrophy, gap junction, compliance, stiffness, contractility, ion channels, pacemaker function, and developmental changes. Identify changes in cardiac proteins, cytoskeleton, matrix, signaling pathways, reactive O_2, and gene arrays.

b. Vascular: myogenic tone, innervation, endothelial function, muscle atrophy, proliferation, hypertrophy, protein structure, permeability, and capillary density. Identify cellular and molecular changes as mentioned above.

Research Area 3. Other specific areas relevant to cardiovascular interactions with other systems affected by spaceflight.

a. Circadian rhythm and cardiovascular function (related to sleep disturbances):

- Molecular clocks and cardiometabolic syndrome,
- Circadian proteins and genotoxic stress, and
- Circadian clocks and vascular function.

b. Lipid oxidation and cardiovascular disease (related to immune response alterations):

- Oxidative stress, phospholipid oxidation, and the innate immune system in atherosclerosis; and
- Isoprostanes as biomarkers and effectors in cardiovascular disease.

c. Pathobiology of calcific vasculopathy (related to bone/calcium metabolism):

- Osteogenic Wnt* signaling in valvular and vascular sclerosis,
- Fetuin regulation of calcified matrix metabolism, and
- Molecular imaging of vascular mineral metabolism.

d. Wnt proteins in cardiovascular development (related to developmental biology); Wnt signaling in cardiac hypertrophy/remodeling.

Benefits from the combination of enabling and basic space studies will translate not only to healthier astronauts during space exploration but also to a better understanding of the aging process and the pathophysiology and treatment of patients with syncope, heart failure, and atherosclerosis.

Risks for Pulmonary Function During Long-Duration Space Missions

Effects of the Spaceflight Environment on the Structure and Function of the Pulmonary System

The effects of microgravity on the lungs of crew members were studied extensively on several Space Lab missions in the 1990s, and a few additional measurements have been made on the ISS.[675,676] Overall gas exchange

*Wnt signaling represents a large family of proteins that are involved in the differentiation, proliferation, and maturation of multiple cell types, including bone and cardiac myocytes; such proteins also are active in mediating vascular calcification, a key component of atherosclerosis.

as measured from oxygen uptake and carbon dioxide output during rest and exercise was essentially unchanged compared with the 1-*g* environment, and the same was true of alveolar pO_2 and pCO_2 values. The only exception was when the environmental pCO_2 was allowed to rise during one of the Space Lab missions. The distributions of ventilation and blood flow became more uniform as expected, although some inequality remained, which is understandable considering the complexity of the airway and blood vessel systems. The diffusing capacity of the lung for carbon monoxide was increased, both because of an increase in the diffusing capacity of the blood-gas barrier and the increased volume of blood in the pulmonary capillaries. This could be explained by the increased thoracic blood volume resulting from the lack of gravity-induced pooling of blood in dependent regions of the body.

They were also minor changes in lung volumes. Functional residual capacity during microgravity was intermediate between that seen in the upright and supine postures at 1 *g*. This was explained by the influence of the weight of the abdominal contents on the diaphragm and was in fact predicted prior to any spaceflights. An unexpected finding was a reduction in residual volume. This was probably caused by the lung parenchyma being equally expanded in all regions in microgravity, which is not the case at 1 *g*. All these changes remained during the Space Lab missions, each of which was 9 to 16 days in duration. Total pulmonary blood flow (cardiac output) and cardiac stroke volume increased early in Space Lab flights but then decreased over the duration of the mission. An unexpected and still unexplained finding was an alteration in the intrapulmonary distribution of two gases of very different molecular weights, helium and sulfur hexafluoride, following a single inhalation. The provisional explanation is that there was a change in the topography of the acinar region of the lung, possibly as a result of interstitial edema, but this needs further investigation.

Additional studies were made on the ISS, although these were less sophisticated than those on Space Lab because of the limited equipment and crew time. However, an important finding was that pulmonary function in general returned to the pre-flight state within 2 or 3 days of returning from a 6-month mission.[677] The net result of all these measurements suggests that the basic functions of the lung are unlikely to cause a health problem during a 3-year mission, for example to Mars.

Few animal studies of pulmonary function have been made in microgravity partly because much of the important information can be obtained from human investigations. The animal studies that have been made are generally consistent with the human studies. An exception is some invasive studies involving the injection of small microspheres and subsequent cutting of the lung into small pieces. These procedures clearly cannot be carried out in humans, but they show substantial inequality of blood flow at the acinar level. No gender differences in pulmonary function responses to microgravity have been reported.

In the event that pulmonary disease develops during spaceflight—for example, bronchitis or pneumonia—the consequences could be more severe than in the normal 1-*g* environment if there is an impairment of the immune response. This topic is discussed in the section below on the immune response in microgravity.

Status of Countermeasures

Countermeasures are of limited value in the context of pulmonary function. There is no feasible way of altering the breathing pattern of astronauts to modify the amount and site of aerosol deposition in the lung. The solution to this problem is to prevent exposure to lunar or Mars dust, presumably by appropriate filtering equipment in the breathing circuit. For the denitrogenation problem, it would be valuable to improve understanding of the nitrogen washout rate from the tissues in microgravity or low-gravity fields so that decompression sickness can be avoided. However, in the absence of this information, decompression sickness can be prevented by long nitrogen washout times using oxygen inhalation. The penalty here is loss of working time for the crew member. The occurrence of the presumed subclinical interstitial pulmonary edema resulting from the increase in thoracic blood volume appears to have no measurable effects on pulmonary function for the durations of microgravity encountered so far but conceivably will become an issue for longer flights. The change in chemosensitivity that results in a reduced ventilatory response to hypoxia seems to be of minor importance, at least in the present flight durations.

Gaps in Knowledge

The amount and the site of the deposition of aerosols of different sizes in microgravity and reduced gravitational fields have been investigated to some extent. For example, a few studies made in aircraft flying in parabolic profile patterns have found that, in both microgravity and lunar levels of gravity, dust of 1-micron diameter is deposited in more peripheral parts of the lung compared with 1-*g* conditions.[678] However, much additional research is required. The same is true of denitrogenation rates in microgravity and reduced-gravity fields, for which more information is needed. Note that these rates will depend on the partial pressure of nitrogen to which a crew member has been exposed prior to denitrogenation, and this may be altered in exploration of the Moon and Mars. For example, an astronaut who lives in a reduced-barometric-pressure environment but breathes oxygen at an increased concentration to prevent hypoxia will have a very different denitrogenation rate than someone living in the normal atmospheric pressure of the Earth environment.

Research Models and Platforms

Human studies on the ISS are necessary to obtain more information on the amount and site of inhaled aerosols of different sizes. This can be done using single breath washouts as previously conducted on parabolic profile flights. In addition, animal studies in microgravity and reduced-gravity fields will be valuable for obtaining more information about aerosol behavior in the lung. Measurements in microgravity should be made on the ISS, and appropriate animal holding facilities will be required for this. In addition if a centrifuge for rodents was available on that facility, the measurements could be made at various partial gravitational levels.

Animal studies of the rate of denitrogenation are less applicable to the human situation but nevertheless could elucidate some of the basic biology, such as the rate of nitrogen elimination from organs having different amounts of lipid.

Some information about aerosol deposition in the lung can be obtained from parabolic flights, but the short duration of microgravity or reduced gravity there is a serious limitation. Future suborbital flights may provide additional opportunities.

Research Recommendations

1. Determine the amount and site of deposition of aerosols of different sizes in the lungs of humans and animals in microgravity. The rationale is that deposition is different in microgravity compared with normal gravity, and there is evidence that lunar dust and Mars dust are potentially toxic. The research would be carried out on the ISS and could potentially be concluded in a 10-year time frame.

2. Determine the rate of washout of nitrogen from the body for humans and animals in microgravity—for example, on the ISS. The measurements in humans should be made by recording the nitrogen concentration in expired gas. Measurements in animals could elucidate the washout rate of nitrogen from different organs of the body.

Risks for Specific Endocrine Influences During Long-Duration Space Missions

The endocrine system is essential for normal organism homeostasis on Earth, and its importance in maintaining and in re-establishing homeostasis in response to spaceflight and recovery was effectively documented in the 1998 NRC report *A Strategy for Research in Space Biology and Medicine in the New Century*.[679] That rationale is not repeated in this report. However, the prospect and uncertainties of long-duration exploration and research flights combined with non-endorsed or non-implemented recommendations provided in 1998 indicate that select hormonal influences warrant further review. Thus, the goal of this section is to discuss those hormones whose influences could be enabling within a microgravity environment and those that have the potential to be better understood by study in the same environment. A complete discussion is difficult because existing in-flight data on hormonal influences and reports on humans are limited, and virtually all animal results have been obtained after landing. As hormones obviously play a key role in many systems, relevant endocrine topics are included in a number of other sections in this chapter.

Alterations and Effects of the Spaceflight Environment on the Endocrine System

Growth Hormone

Human Studies. Growth hormone (GH) is important for bone metabolism, maintaining muscle mass and strength, stimulating afferent neurons, and promotion of growth and development of young animals and humans.[680,681,682] However, measurement of GH is complicated by the heterogeneous nature of the hormone and by the presence and heterogeneity of GH-binding protein in the plasma.[683] Further, methodological problems have not been resolved. Extensive human GH research has been conducted in Russia, where subjects participating in a long-term bed rest study had 16 percent lower GH levels and altered circadian rhythm profiles after 75 days, compared with the baseline. These changes were attributed to a decrease in motor neuron activity.[684] Measurements obtained from 53 cosmonauts participating in Salyut and Mir flights for 114 to 438 days had post-flight GH levels that remained elevated for 10 weeks. The results were complex, given that 32 cosmonauts had values significantly higher than their pre-flight means, whereas 21 had significantly lower results. Additional subgroup investigations ($n = 33$) indicated an association between resting human GH levels and performing scheduled exercises during flight. Specifically, cosmonauts who did not adhere to prescribed exercise schedules had lower mean values than those with higher compliance records.[685] McCall et al.[686] reported that the release of bio-assayable GH was suppressed in microgravity, whereas the opposite results were obtained by Macho and associates, who used a similar experimental design but measured isoforms.[687] Thus, although results have been mixed, on balance the findings suggest a tendency for reduced GH titers during a prolonged spaceflight, especially in subjects not performing routine physical exercise (see below).

Animal Studies. Hymer and associates reported that the secretory capacity of GH cells from rats flown on three different flights was significantly lower than in control rats, specifically when biologic rather than immunologic assays were used. They also reported similar trends occurring with HU rats.[688] Bigbee and associates reported that basal plasma bio-assayable GH levels of HU rats were significantly reduced (51 percent lower after 1 week and 55 percent lower after 8 weeks), compared with those of ambulatory control rats.[689] The pituitary cells flown on Cosmos 2044 had reduced secretion of growth hormone, which was attributed to the GH releasing factor peptide.[690] Thus, in animals it appears that unloading conditions induced a reduction in GH levels; these results complement those of investigations on the hypothalamic-pituitary axis discussed in the 1998 NRC report.[691]

Thyroid Hormones

Human Studies. Thyroid hormones are important for metabolism and for the growth of young animals and humans. Slight increases in thyroid stimulating hormone were found for astronauts in the Apollo mission; significant elevations were reported for the astronauts in Skylab[692] and for the four astronauts on the D-2 flight.[693] In-flight thyroxine (T4) values of Skylab astronauts were significantly increased, whereas triodothyronine (T3) levels were decreased. These observations were thought to result from an inhibitory effect of cortisol on the conversion of T3 to T4.[694] Reductions in in-flight urinary T3 and T4 levels were found on SLS flights 1 and 2. In cosmonauts on missions ranging from 115 to 438 days, post-flight plasma levels of T4 were elevated for 7 days before returning to near pre-flight means, while T3 levels were lower than their pre-flight means.[695] Thus, it appears that T3, the biologically active thyroid hormone, is slightly reduced during and following spaceflight.

Animal Studies. Post-flight plasma levels of thyroid stimulating hormone were elevated after an 18.5-day flight and decreased after a 7-day mission.[696] Summarized results from three Cosmos flights lasting from 14.0 to 19.5 days indicated that the post-flight T3 and T4 concentrations were significantly lower than those in controls.[697] In the same report, thyroid gland concentrations of both hormones were significantly lower than those in controls.

Rats suspended for 12 days had significantly lower T4 values after 6 days.[698] Surprisingly, there is a paucity of published literature on changes in the pituitary thyroid axis. Adams and colleagues appear to be the first to include thyroid-deficient rats in flight experiments to determine the interactions between thyroid deficiency and MHC gene expression.[699] Using neonatal pups, they found that microgravity significantly reduced the expression of the fast type I MHC gene, whereas thyroid deficiency prevented the normal expression of the fast MHC phenotype and prevented the gravity-dependent expression of the embryonic/neonatal isoforms. They concluded that a normal expression of the fast IIb MHC required an intact thyroid state that appeared to be independent of normal weight-bearing activities.

Cortisol and Cortisone

Human Studies. Cortisol changes have been assessed in flight by measurements of blood, urine, and saliva. Stimulated by the secretion of adrenocorticotrophic hormone (ACTH) and known as the "stress hormone," cortisol becomes an important consideration with long-duration missions because it:

1. Is a major cause of secondary osteoporosis in humans;[700]
2. Exerts proteolytic actions on both smooth and fast twitch (type II) skeletal muscles;[701]
3. Stimulates energy expenditure and fat metabolism;[702] and
4. Has negative effects on the immune system.

Elevated cortisol levels during flight were characterized in the 1998 NRC report *A Strategy for Research in Space Biology and Medicine in the New Century*[703]as a "frequent but not invariable finding." Data pertaining to short flights (2 to 14 days, N = 182) have indicated that post-flight ACTH and cortisone were significantly decreased, whereas after longer flights (120 to 366 days, N = 21), cortisone was significantly increased, by 171 percent.[704] Measurement of plasma cortisol during the Skylab flights revealed an elevation by 20 to 80 percent throughout the flight.[705] According to Grigoriev et al.,[706] flights of 12-month duration resulted in a 10-fold increase in ACTH levels with lesser changes in cortisol concentrations, which strongly suggested an impact on the hypothalamic-pituitary-adrenal axis. Similar findings were reported for the SLS-1 and SLS-2 flights. Results from flights before 1998 demonstrated that fast twitch (type IIa) muscle fibers had experienced significant atrophy with reduced function.[707]

Metabolic studies using head-down bed rest conditions for 42 days showed significantly elevated urinary cortisol levels during the study.[708] Further, a positive correlation between the change in tibia epiphyseal bone mineral content and elevations in serum cortisol levels was reported in a 90-day bed rest study to evaluate resistive exercise and pamidronate as a countermeasure.[709] However, cortisol concentrations did not change in a 7-day bed rest study.[710] A longer, 42-day bed rest study of the interactions between body composition, energy expenditure, energy, and water metabolism, along with catabolic and anabolic hormone expression, found marked reductions in body weight, lean body mass, total energy expenditure, energy intake, and the energy available for physical activity.[711] The authors suggested that the hormonal profile, including a significant increase in cortisol, interfered with nutrient oxidation, and they proposed nutritional countermeasures. Collectively, these findings suggest that during the stress of spaceflight and bed rest, elevated cortisol can exert a potential confounding effect on protein balance.

Animal Studies. Overall, post-flight plasma corticosteroid results for rats from six missions of durations between 7 and 20 days demonstrated that the flight animals had higher values than did the controls, and in four of the missions, their mean post-flight plasma levels were significantly elevated (approximately four-fold) over the pre-flight mean values.[712] Rats tail-suspended (HU model) for 12 days had significant increases in corticosterone at 6 days, which subsequently returned to near the baseline at the end of the study.[713]

Insulin and Diabetogenic Trends

Human Studies. The prospect of space research and explorations in the future involving long-duration flights merits a re-evaluation of the impact on insulin of prolonged exposure to microgravity. Notably, the 1998 NRC report underscored the difficulty in interpreting insulin data from flights in isolation without parallel ground-based data.[714] That report also noted the presence of conflicting results.

Using a 7-day head-down bed rest analog design that included men and women, Blanc et al. in 2002 demonstrated increased fasting insulin levels and the presence of insulin resistance, which the authors attributed to the magnitude of the skeletal muscle tissue in males and to both skeletal muscles and liver tissue in females.[715] Decreased in-flight concentrations of insulin have been reported for crew members of Skylab and of select space shuttle flights,[716] whereas nonsignificant elevations of insulin and C-peptide were reported for four astronauts in the SLS-1 mission[717] and for the 10-day D-2 flight.[718] Post-flight increases in insulin levels have been reported for nearly all Russian and U.S. flights regardless of the duration. It is not clear to what extent an increased dietary intake might be contributory, as suggested by the 1998 NRC report.[719] Notably, limited glucose tolerance test results have been reported for astronauts, although flight results show higher insulin and C-peptide levels compared to data collected after landing or during recovery.[720,721] Anecdotal findings in two cosmonauts, tested at days 321 and 322, suggest a response of "delayed utilization."[722] Markin et al. observed higher blood glucose concentrations for individuals who had been in microgravity for 6 months or longer.[723] These latter findings emphasize the importance of regular, moderate-to-heavy exercise during long-duration flights to minimize increases in insulin resistance and the possibility of type II diabetes-like symptoms. Clearly, more research is needed on this important topic.

Animal Studies. The HU rat model is associated with the presence of insulin resistance.[724] A decline in oral glucose tolerance and a decreased insulin-stimulated transport in isolated soleus muscle tissue, associated with an activation of the p38 mitogen-activated protein kinase, was found after 24 hours of head-down tail suspension in juvenile female rats.[725] However, the unloading of the soleus muscle did not alter insulin-stimulated activities of various signaling factors. Tobin[726,727] perfected a cell rotating culture system for simulating microgravity (high aspect rotation vessel or HARV) for use with islets of Langerhans. Using lipopolysaccharide to stimulate tumor necrosis factor α activity, Tobin and colleagues reported a significant decrease in insulin secretion and an increase in glucose levels. Overall, these findings on rodents support the conclusions derived above for humans.

Testosterone and Gonadal Function

Human Studies. Because the 1998 NRC report was based on information generated between 1977 and 1992, an additional focus on testosterone during long-duration flights is warranted.[728] Testosterone influences growth, development, and reproduction, and its secretion is regulated by luteinizing hormone. When measured in the plasma, urine, and saliva during the 10-day D-2 mission (n = 4 astronauts), levels of luteinizing hormone were significantly increased, while levels of urinary and salivary testosterone and testicular androgen 3-α-androstanediol were significantly decreased.[729,730]

To date, there have been no studies in microgravity involving humans that have pertained to spermatogenesis. In a 120-day Russian bed rest study, Nichiporuk et al. reported that sperm collected after 50 to 60 days and 100 days exhibited a reduction in live spermatozoa with active mobility and an increase in the percentage of morphologically/structurally altered spermatozoa.[731]

Animal Studies. After 14 days in microgravity, post-flight rat testosterone concentrations were significantly reduced and 60 percent lower than in synchronous ground-based animals.[732] A lower basal testosterone secretion rate was seen in rats after an 18.5-day flight.[733] Suspended rats, included in the Cosmos 2044 study, exhibited markedly lower testosterone concentrations when compared to synchronous control. Both flight and suspended animals had significantly lighter testes than did controls.[734] Because of the possibility of cryptorchism, Deaver et al. ligated the inguinal canal and found significantly lower testosterone levels in serum and intestinal fluid and lower testes mass in HU rats.[735] Further, mice suspended for 12 days had significantly lower concentrations of testosterone and less testes mass.[736] However, rats centrifuged at 2 *g* had elevated urinary testosterone levels, suggesting centrifugation as an appropriate countermeasure to restore homeostasis.[737]

Spermatogenesis research in microgravity began decades ago with two dogs that were in space for 22 days. The main finding was an increase in atypical spermatozoa characterized by tail curling or the absence of a tail.[738]

However, an extensive spermatogenesis study of rats returning from a 14-day flight reported essentially normal conditions, although the authors appropriately recommended additional studies in longer-duration flights.[739] The negative results with the 7- and 14-day Cosmos 1667 and 2044 flights add credence to their statements.[740] In 2002, Tash et al. conducted a 6-week HU study in which cryptorchism was prevented and stated "spermatogenesis is severely inhibited by long term HLS [hindlimb suspension]." However, the testosterone concentration was not depressed. Additionally, they reported that spermatogenic cells beyond the round spermatid were not present.[741] Motababagni investigated the influence of HU on spermatogenesis in groups of rats that had inguinal canal ligation and observed a severely affected spermatogenesis after 6 weeks, because no late elongated spermatids and spermatozoa were present in the testes.[742]

Similar findings of impaired testes cell function have also been reported on isolated cell lines studied under simulated microgravity conditions.[743] Collectively, both the human and animal studies suggest that gonadal function and spermatogenesis are impaired in response to chronic unloading conditions.[744,745] Thus, future research is needed especially in humans to expand on these interesting observations during prolonged spaceflight.

Estrogen and Gonadal Function

In the approximately 5,600 publications listed in MEDLINE since 1948 under the term "weightlessness," none indicated that estrogen and gonadal function have ever been measured in microgravity or in the recovery period. Moreover, this void is barely improved by simulated microgravity studies, as fewer than five investigations have been reported. Therefore, studies should be undertaken to address female endocrine function during exposure to microgravity and during analog studies such as bed rest.

Status of Countermeasures

The single most important countermeasure for any given hormonal alteration is direct hormonal replacement. Exercise is a likely next countermeasure for both humans and animals, given that exercise intensity affects hormonal responses.[746] However, the choice of regimens and implementation of exercise need additional attention to go beyond what is currently advocated by NASA for astronauts scheduled for a 6-month tour on the ISS. Notably, a 2009 assessment of nine astronauts concluded: "Future long-duration space missions should modify the current ISS exercise prescription and/or hardware to better preserve muscle mass and function, thereby reducing the risk imposed to crew members."[747]

Astronauts must be cognizant of glucocorticoid-induced type IIa fiber atrophy, which has been demonstrated in human spaceflight[748] but can be prevented in fast twitch fibers (in animals) by moderate to heavy dynamic exercise.[749] They must also be cognizant of the finding that the inactivity of bed rest will suppress the release of testosterone.[750] While the release of insulin is reduced with progressive intensities of dynamic exercise, exercise will facilitate glucose uptake into cells by its action on glucose transporters.[751] As to the specifics of the exercise prescription, countermeasures need to be designed in conjunction with the advice of experts in skeletal muscle biology.

In addition to its use as an exercise prescription, centrifugation has the potential to be an effective countermeasure for the endocrine reproductive issues associated with microgravity.[752] For human use, a feasible system could be a short-arm centrifuge with the capability of achieving 2-g force vector, similar to one utilized by the University of California, Irvine.[753]

Gaps in Fundamental Knowledge of High Priority

The Influence of Microgravity on Human Spermatogenesis and Gonadal Function

Although humans have been exposed to space for nearly half a century, there is no published information on the effects of microgravity on human spermatogenesis and gonadal function. In addition, the paucity of simulation studies is striking. Minimal insights can be drawn from animal post-flight data because of the brevity of flights, the limited number of animals, and the inability to eliminate the possibility of testes abnormalities existing before flight. Consequently, current knowledge of the subject depends on ground-based animal studies with conflicting results. The seriousness of this knowledge gap is emphasized by Tash et al.,[754] who wrote concerning their results (spermatogenic cells beyond spermatid were not present), "if these findings hold true in μG [microgravity], it implies that male astronauts may become infertile after long-term exposure to μG." This knowledge gap is well documented and needs to be addressed to ensure the safety of astronauts participating in future long-duration space missions.

The Influence of Microgravity on Smooth Muscle Atrophy with an Emphasis on the Role of Glucocorticoids

While much remains unknown concerning the effects of microgravity on the mechanisms of atrophy in both skeletal and cardiac muscle, the need for additional research is well recognized. For smooth muscle, assumptions from ground-based models and uterine myometrial results[755] associated with parturition form the foundations of current knowledge. Whether smooth muscle not associated with the events of parturition actually exhibits atrophy processes in response to microgravity is unknown. Beyond the fact that glucocorticoids can enhance proteolysis in smooth muscle and are elevated in microgravity, their long-term influences remain to be assessed.

Microgravity and Its Suppressive Influence on Growth Hormone Secretion or Release

As emphasized in the animal studies pertaining to GH (see previous discussion) and when compared to control conditions, the secretory capacities of GH cells or pituitary cells flown in microgravity, as measured by their biological activity, were marked reduced. A similar trend was observed with HU rats.[756–759] Since reports indicate that systemic insulin-like growth factor 1 levels were suppressed, mechanistic studies are needed to explain these observations and to determine their influence on the growth and development of neonatal rats or mice in microgravity. A coordinated ground- and flight-based research program is recommended to acquire this needed information. However, the survival, growth, and development of neonatal rats in microgravity will require both the availability and the modification of the existing animal habitat.

Gaps in Applied Knowledge of High Priority

Role of Hormone Changes as Mediators of Other Biological Responses to the Space Environment

It is unclear whether disruptions in multiple hormonal systems (e.g., thyroid, cortisol, GH) have a significant impact on physiologic changes during spaceflight (loss of bone, muscle, etc.), independent of the effects of microgravity. It is also not known whether changes in hormonal systems during spaceflight reduce the effectiveness of countermeasures (e.g., reduced effectiveness of exercise in a low-androgen state to preserve bone). Research in these areas is needed to determine whether hormonal homeostasis represents an important countermeasure target.

Quantifying the Diabetogenic Potential of Microgravity

Beginning in 1977 and continuing today, inferences concerning diabetic tendencies in astronauts continue to be made based on in-flight and/or post-flight hormonal measurements, especially those associated with C-peptide and insulin concentrations. Because analogs for both humans and animals follow profiles similar to those exhibited by flight animals, the significance and risk associated with this gap can be minimized by combining a diabetogenic awareness program with glucose tolerance tests during prolonged durations in simulated and actual microgravity.

Research Models and Platforms

NASA's recent acceptance of the head-down bed rest model for the testing of potential countermeasures[760] will end the debate as to the merits of the head-down bed rest position versus the horizontal bed rest position in maximizing the benefits of fluid shifts, and will standardize the analog for the research needed in the next decade. However, it is essential that the ISS become a platform for securing serial in-flight hormonal measurements to establish an endocrine database. Such a database will serve as a foundation for long-duration investigations pertaining to gap research, countermeasure effectiveness, propensity for diseases, and hormonal interactions with other systems. Access to partial-gravity platforms is recommended for the same reason. Because the free-flyer platform does not allow sufficient time for an equilibrium to be established with regard to fluid shifts, its use is not recommended here for these types of studies.

Results from the HU model for rats and mice[761] compare favorably with human data, but extrapolation to flight findings is tenuous because the overwhelming majority of these results were obtained after landing. Thus, it is essential to ensure access to the ISS platform for studies on animal models in order to establish an in-flight hormonal database, initiate long-duration investigations, examine pituitary cell secretion profiles, perform gap-related research, interact with other systems, and become involved with neonatal growth and development studies.

Research Recommendations

Fundamental Science

This section has recommendations for research on specific hormones, given that only the small number of hormones considered relevant to spaceflight are discussed. Because of the absence of in-flight results when post-flight data on hormones have been reported, future investigations (clinical or experimental) should not be endorsed or undertaken unless in-flight measurements are obtained and functional objectives have been defined. In the next decade, endocrine studies should be initiated with females and a database established. Needed areas of research include the following:

- *Growth hormone*

—Suppression of GH secretion in microgravity. The early reports that GH secretion from anterior pituitary cells was suppressed by microgravity[762,763] should be pursued with animals to investigate responsible mechanisms.

—Relationship between GH levels and motor neuron activity. Researchers should explore the relationships between the observations of Russian investigators who reported that reduced basal human GH levels reflected a decreased motor neuron activity[764] and reports from Edgerton's laboratory that indicated the release of the bioassayable form of GH was triggered by muscle spindle afferents.[765]

—GH influences on growth and development in microgravity require investigation with special attention devoted to neonatal animals.

- *Thyroid hormones*

—Adaptation of thyroid hormones to microgravity. The variability or absence of free T4 measurements from flight missions indicate that serial thyroid hormones profiles need to be conducted in microgravity to determine the acute, early adaptive, and chronic adaptive stages.

—Animal studies pertaining to the hypothalamic-pituitary-thyroid axis should be conducted in simulated low gravity as well as in microgravity to determine if there is a cellular basis for the reports of reduced secretion in microgravity.

—Neonatal studies with thyroid-deficient rats and the expression of MHC isoforms represent an excellent example of a direction for future research.[766]

- *Cortisol*

—Impact of microgravity on smooth muscle. Cortisol is associated with bone loss and with the atrophy of smooth muscle and type IIa skeletal muscle fibers.[767] To date, the atrophy of smooth muscle in microgravity has not been systematically investigated or acknowledged even though its occurrence has been known since before the 1998 NRC report was released.[768]

—Cortisol and bone loss. With the prospect of long-duration flights, research needs to determine whether cortisol levels are a contributing factor to bone loss.

—Impact of microgravity on hypothalamic-pituitary-adrenal axis. It is important to determine whether ACTH and cortisol will become dissociated during long-duration missions.

- *Insulin.* Long-duration missions and insulin resistance. In the next decade, it should be determined whether long-duration spaceflight exerts a diabetogenic potential on human and animal subjects.[769] Assessment by glucose tolerance tests is recommended. Unfortunately, there is limited relevant scientific data from microgravity research on this topic, with no existing databases and no established guidelines for implementation. Therefore, the recommendation for insulin research in the next decade should be aggressively developed and implemented via a plan that will decrease the possibility of insulin resistance becoming a major risk factor on long-duration spaceflights.
- *Testosterone.* Impact of microgravity on testosterone production and on spermatogenesis. Research in this area is warranted because of the continuing uncertainty and controversy regarding whether microgravity (1) alters testosterone production or (2) affects the process of spermatogenesis and alters testes mass in humans. Because there is a lack of human data from bed rest or microgravity studies, current concepts on testosterone and spermatogenesis have essentially evolved from animal analog data. This emphasis needs to be refocused on human male gonadal function to address the remote possibility that male astronauts may become infertile after long-term exposure to microgravity.

Effects of the Spaceflight Environment on the Immune System

Human Studies

Most of the studies in this area performed over the past 30 years have used a similar experimental design. Samples are taken from astronauts prior to and upon return from flight. Effects are determined by assessment of changes within an individual over time or compared to ground-based individuals. Investigations of the immune system have addressed whether (1) there was a change in the number and/or percentage of immune cells, (2) the function of the cells was altered, and (3) the changes were reversible or permanent.

Determination of changes in the human immune system has been limited to assessment of peripheral blood. As reviewed in the 1998 NRC report,[770] the data reflect a wide range of results; some of the variation can be explained by differences in study parameters (duration of flight, assay conditions, sample treatment). Considering both previously reviewed data and recent space shuttle and ISS missions, the most consistent findings have been an increase in the number of white blood cells, predominately reflected in an increased number of neutrophils and a decreased number of lymphocytes[771-774] and natural killer (NK) cells.[775,776] Although some studies suggested a shift in CD4 (helper)/CD8 (cytotoxic) T cell ratios, this change has been small in many cases.

Recent examination of the function of the innate component of the immune response continues to show a pattern similar to that reported in studies included in the 1998 NRC report:[777] (1) the number of studies is limited; (2) although decreases in function are observed, the magnitude of the change is not great and/or the pattern of change across various functions is not consistent.[778,779] Most of the studies have focused on the function of lymphocytes. No significant changes have been observed in immunoglobulin levels.[780] Although this may suggest no change in B cell function, studies addressing the production of antibodies after immunization during flight have not been performed. Using whole blood or cells isolated from blood, lymphocytes have been stimulated with mitogens (e.g., phytohemagglutinin) or antibodies (anti-CD3 and anti-CD28). Most studies report a decrease in the proliferative responses of the T cells to these stimuli.[781] In addition, cytokine production has been consistently altered, although which cytokines are significantly changed varies among studies. Although interleukin 2 (IL-2) is decreased in most studies, leading some investigators to conclude that there is a shift from Th1 (IL-2, IFN-γ) to Th2 (IL-4, IL-10) cytokines, the effect of spaceflight on IFN-γ production has not been as consistent.[782,783] More recent studies have addressed the induction of activation markers on the surface of T cells after mitogenic stimulation. Early activation markers CD69 and IL-2 receptor CD25 appear to be elevated after shorter (9 to 10 day) space shuttle missions, while CD69 appears unchanged but CD25 is decreased upon longer ISS missions. These data suggest that differences in dysregulation of the immune system may depend on flight duration. Interestingly, a similar pattern of decreased proliferation, IL-2 receptor activation, and cytokine dysregulation, including the results with IFN-γ, have been observed in the elderly.[784] The major difference between the changes observed in the immune response with aging and with spaceflight is that the effects of spaceflight are reversed days to weeks after flight.

The prevailing study design compares parameters of interest pre-flight with those obtained immediately after return to Earth. Therefore, the differences observed may reflect the forces and stressors of re-entry rather than the effect of flight. Two approaches have been used in humans to study changes in immune response during flight. In the first approach, delayed type hypersensitivity (DTH) responses to the same multi-antigen panel were compared when the antigen challenge was administered 2 days prior to landing and the response evaluated upon landing, and when both antigen challenge and subsequent 48-hour evaluation were performed 2 months prior to flight. In the two studies performed, decreased responses were observed after in-flight antigen administration.[785,786] The second approach, which has been used fairly extensively in recent studies, assesses the reactivation of latent herpes viruses.[787-790] Because an immune response to the virus is important in maintaining latency, expression of latent herpes viruses has been considered a surrogate for a decrease in virus-specific immune responses. Studies assessing viral DNA in saliva samples every several days have demonstrated expression of varicella zoster virus (VZV, the latent virus causing shingles) during and after flight in 6 of 8 astronauts[791] and an increase in both the percentage and the level of Epstein-Barr virus (EBV) DNA in 32 astronauts during and after flight, compared to ground-based controls.[792] Both the decreased DTH responses and the increased expression of latent herpes virus (e.g., VZV) are similar to responses observed in the elderly.

Although a significantly decreased immune response will render an individual susceptible to infections and a severely compromised immune system will make normally innocuous organisms life-threatening to individuals, the question is whether the level of change in the immune response that occurs with spaceflight is sufficient to result in increased susceptibilities. The 1998 NRC report concluded that, although initial studies suggested an increase in infection, the number of infections during flight decreased once multiple procedures were implemented to minimize exposure of astronauts to infectious agents prior to flight. Epidemiological review of recent flights indicates 29 incidents of probable infectious diseases during 106 shuttle flights with 742 crew members.[793] While the changes in the immune response due to flight may have contributed to the development of infections, other conditions, i.e., confined living spaces or difficulties in maintaining good hygiene, may also have increased crew members' exposure to potential pathogens, resulting in infection. In addition, because it is difficult to eliminate exposure to agents for which prevention is understood, it seems highly unlikely that it will be possible to develop measures to prevent exposure of astronauts to microbial agents that may develop enhanced potential for disease during flight (see Chapter 4) or that are introduced from the new environment. In these conditions, an immune system that is functioning at an optimum level will be necessary to ensure the health of extended-mission astronauts. Even if no negative event occurs during flight, the question remains whether dysregulation of the immune response during extended missions can have long-term effects on astronauts.

Animal Studies

Studies of animals that were flown had similar design limitations. Most of the studies were performed in rats before 1998, and examinations for effects were performed after the animals returned from flight. The results are similar to those for humans: the most consistent changes occurred in T cell responses.[794] However, due to the availability of multiple tissues for examination, it was possible to examine immune responses of various lymphoid organs in both rat and mouse models. Interestingly, although decreases were detected in proliferative responses in lymphocytes isolated from lymph nodes, no changes were observed in the spleen.[795,796] In another study, decreases were observed in both spleen and lymph node.[797,798] In the latter study, changes in the percentages of white blood cells in both lymph node and spleen paralleled those observed in humans: neutrophils were increased and lymphocytes decreased. The animal studies provide the unique ability, therefore, to determine if spaceflight affects distribution and recirculation of lymphocytes. For example, if Moon dust affects pulmonary function and immune cells cannot appropriately migrate to the lung, there may be increases in respiratory infections.

Countermeasures

If the change in the immune system in response to microgravity is found to be biologically significant, it is unlikely that there will be a simple countermeasure. However, because the immune system is affected by a number of parameters that are currently being explored as countermeasures for other adverse effects of spaceflight, any or all of those approaches may alter the immune response. For example, exercise is being proposed to counter the loss of bone and muscle. However, intense exercise can decrease the level of immune response.[799] Adequate nutrition is a continuing concern during flight. Although long-term caloric restriction can delay the decrease in immune response that occurs with age,[800] deficiencies in specific vitamins or minerals or malnutrition can result in significant decreases in immune response, which are reversible upon correction of the defect.[801] Because nutritional status has not been correlated to the level of immune dysregulation that has been reported, it is possible that focused nutritional supplementation could minimize some of the decreases observed in immune response. Similarly, alterations in sleep patterns affecting circadian rhythms have also been demonstrated to decrease immune responses.[802] These observations, however, must be interpreted in the appropriate context: although both exercise and nutrition have been repeatedly reported to alter immune responses, the effects are often statistically significant but not necessarily biologically important. However, by utilizing multiple approaches (exercise, nutrition, sleep), with each influencing the immune response in small increments, the magnitude of the dysregulation of the immune response may be minimized, thus allowing the immune system to protect the individual adequately.

Gaps in Knowledge

Several major gaps in knowledge identified in the 1998 NRC report[803] still have not been addressed. First, most of the studies since then have assessed the changes in immune response upon re-entry to Earth's atmosphere. Due to the stresses of re-entry, it is unclear to what extent the observed changes were due to re-entry or to exposure to microgravity. Because consistent decreases in lymphocyte function have been observed, it is essential that more studies be done during spaceflight to determine the point of stress. Second, while the reported changes in the immune response are statistically significant and reproducible, it has not been possible to establish if the changes in the immune system would result in increased susceptibility to infection. As indicated in the 1998 NRC report, studies can be performed in humans to address this question: astronauts can be immunized during flight, cells and sera can be collected both prior to and immediately upon re-entry, and results can be compared. T cell and antibody responses specific to the immunogen, as well as the general status of the immune response as defined by number, phenotype, and response to mitogens, not only will provide information on the specific immunization protocol but also will allow better interpretation of the previous data and assessment of potential increased susceptibility to infections. Unfortunately, these studies have not yet been performed.

However, response to immunization remains only a surrogate for the immune response to infection. The ideal situation is immunization and challenge with an infectious agent. That experimental design will never be safe for implementation in humans. The experimental model of choice for comprehensive examination of response to immunization in flight would be animal analogs, and preferably mice, due to the extensive reagents available for characterization of the immune response in mice, the ability to house more animals in a small space, and the high relevance to the human system. Immunization and infectious challenge of mice during flight might not be possible due to concerns about infection spreading to the spacecraft crew. However, as previously recommended in the 1998 NRC report,[804] immunization during flight with a collection of immune samples prior to and upon re-entry, with challenge upon re-entry, needs to be performed to definitively establish whether or not the changes observed represent an important risk for astronauts.

In addition to these animal studies, further investigation is needed of the data that are currently available from astronauts. For example, although there is increased activation of latent herpes viruses (e.g., EBV and VZV) during flight as determined by increases in the number of copies of viral DNA, the relationship of these changes to symptoms is unclear. Symptoms related to reactivation of EBV are difficult to detect, but reactivation of VZV as shingles can be assessed. It was noted that one astronaut developed shingles upon return, but the expression of VZV DNA by this individual was not mentioned.[805] In addition, the immunity parameters evaluated before and after flight should be expanded to include more markers reflective of aging. The current profile of changes in immune response after spaceflight reflect those of aging. However, aging is associated with additional changes (e.g., shortening of telomeres, CD28 and Treg expression) that are not currently evaluated in astronauts. If the changes induced by spaceflight reflect the additional parameters that are unique to aging, spaceflight or spaceflight analogs may be useful in exploring the biological underpinnings of the initial stages of aging—a study impossible on Earth because the stressors that induce the aging phenomenon on Earth are not sufficiently defined to allow assessment at initiation of the process.

One of the concerns regarding long-duration spaceflights that has not been addressed is the potential effect of radiation on the immune system. It is well established that radiation associated with x-rays and gamma rays (low-LET radiation; described in Chapter 7) can significantly decrease the immune response: radiation can prevent a primary immune response, with higher doses inhibiting both primary and memory immune responses. While there are fewer studies that have examined immune system effects of the radiation present in space, such as protons and high-LET radiation, a similar decrease in the ability of lymphocytes to proliferate was observed after irradiation with gamma rays or protons.[806] More recently, exposure to heavy ions (high LET) resulted in significant loss of telomeres of lymphocytes,[807] a situation seen frequently with aging and associated with decreased proliferation. Although these data suggest that the radiation encountered in long-duration spaceflight would decrease immune response, the limited ability to study forms of high-LET radiation and particularly low-dose but long-term exposure to high-LET radiation, makes it difficult to determine the magnitude of the effect radiation would have on immune cells during flight. Studies that involve microgravity in conjunction with radiation would have to be performed in order to definitively assess the impact of these combined stressors on the generation of immune response and the development of abnormalities within the immune system, such as the development of lymphoma and leukemia.

A number of ground-based systems have been developed using both human and animal subjects to allow further exploration of the changes induced by flight.[808-813] Although each system may simulate several of the aspects of flight (e.g., stress, alterations in circadian rhythm), none of the systems has the ability to mimic all of the stressors. Until definitive information is developed regarding the parameters that are consistently altered in humans during spaceflight, it is impossible to determine which of the ground-based models should be explored further. In humans, the Antarctic studies appear to have many similarities with flight, including decreased proliferative responses, altered cytokines, and reactivation of latent herpes viruses.[814] However, these studies also have significant limitations in the range of experiments that can be performed.

Recommendations

Although no clinically deleterious effects of spaceflight on the immune system have been observed to date, any consistent changes in the immune response due to flight need to be pursued because of the potential for major negative consequences during longer flights or altered external conditions. The panel therefore makes the following recommendations:

1. Existing data from both published studies and longitudinal assessments of astronaut health should be carefully evaluated together to determine whether observed changes in immune response upon re-entry are associated with symptoms related to reactivation of herpes viruses, infections, and autoimmune processes.
2. Multiple parameters of T cell activation should be obtained from astronauts before and after re-entry to establish which parameters are altered during flight. Because many studies were conducted 20 to 30 years ago, the parameters examined need to reflect the current state of the technology for immune assessments.
3. It is essential that studies with mice be performed on the ISS to establish the biological relevance of the changes observed in the immune system. These studies should include immunization and challenge, with samples of lymphocytes and sera acquired both prior to and immediately upon re-entry. The parameters examined need to be aligned with those human immune response components influenced by spaceflight.

4. The changes observed in the immune system may reflect the impact of multiple stressors. To both address the mechanism(s) of the changes in the immune system and develop measures to limit the changes, data from multiple organ/system-based studies need to be integrated. For example, data from cortisol assessments (stress studies), nutritional diaries, weight changes, exercise volume and intensity, and level of activity (e.g., EVA) need to be considered when designing and interpreting immunity data from individuals.

5. The reversibility of the changes that occur during and after flight has to be examined more carefully. Incorporation of parameters used to identify aging (e.g., shortening of telomeres, CD28 and Treg expression) in the expanded evaluation of immune responses in spaceflight may provide insight regarding either how the two systems are different or how the changes in flight might be utilized to identify early aging.

Reproduction and Development

Developmental space biology is the fundamental research discipline concerned with the influence of gravity variations and other factors inherent to the space environment on reproduction, genetic integrity, differentiation, growth development, lifespan, senescence, and subsequent generations.[815] Previous NRC study committees[816,817] and other advisory groups[818,819] have consistently underscored the importance of research on developmental biology for advancing understanding of how gravity shapes life on Earth. These reports have clearly articulated priorities for developmental space biology research, namely: (1) utilizing the space environment to enhance understanding of the nervous system, and (2) understanding effects of the space environment on the complete life cycles of mammals and other vertebrates. The topics discussed in this section focus exclusively on basic research *enabled by* the unique environment of space.

Prior to 2003, there was a growing emphasis on spaceflight studies of the reproduction and development of insects, fish, amphibians, birds, rats, and mice.[820,821] U.S. animal experiments in space have since been dramatically curtailed due to funding shortfalls, retirement of the Spacelab program in 1998, impending loss of the space shuttle, and inadequate animal research capabilities.[822] While past recommendations remain important, new vistas have significantly broadened the range of discovery opportunities.

Importance of Developmental Biology Research in Space

Developmental biology is concerned with early determinants of fertilization and embryogenesis and with the subsequent development and maturation of organisms throughout the lifespan and across generations. It is a core discipline within the contemporary life sciences that crosses the boundaries of neurobiology, systems biology, molecular biology, genomics, and epigenetics. Developmental analysis focuses on mechanisms of development, differentiation, growth, and maturation at the molecular, cellular, genetic, and organismal levels, as well as the evolution of ontogenetic mechanisms. Because different organisms solve the problems of development differently, multiple model systems are used to uncover guiding principles and identify molecular mechanisms that are conserved across phylogeny.

In the past decade, major conceptual and technological advances in genetics, molecular conservation, and genome sequencing have considerably expanded the scope and depth of developmental biology in ways that will significantly enhance the ability to uncover fundamental biological principles governing how bodies and brains organize, develop, maintain, and adapt under the constant force of gravity.

- *Epigenetics* ("above the genome") refers to gene-environment interactions that establish an individual's phenotype beginning early in life.[823] Epigenetic analysis provides the first mechanistic representation of how environmental factors modulate genetic expression without alterations in DNA sequences. This new approach is increasing basic knowledge of how observable molecular changes (e.g., chromatin remodeling, DNA methylation, and histone modifications of gene products) result in stable, heritable changes in gene expression and later-life phenotypes. Epigenetic modifications can persist across generations, raising important questions regarding the ability of offspring of multigenerational subjects of studies in space to adjust to normal gravity.

- *"Omics" systems biology* is based on high-throughput gene array technologies that allow for simultaneous examination of multiple, complex changes in DNA, messenger RNA, proteins, metabolites, and methylation state.[824,825] These exciting new approaches are rapidly advancing genomics and related fields, thereby significantly advancing perspectives on and understanding of biological subsystems and their functional integration. Next-generation deep sequencing machines promise to yield whole-genome and organ-specific expression profiles at unprecedented speed, thereby enabling impressive scientific achievements and novel biological applications.

In the era of the ISS, developmental biology research is poised to enable new and fundamentally important discovery domains. Developmental studies using model organisms that allow for detailed comparative analyses at cellular, genetic, and molecular levels will provide a wide range of new opportunities for space biology research that were not present in the past decade.

Effects of the Spaceflight Environment on Reproduction and Development

For invertebrates and some vertebrates, fertilization and development can proceed in space. Because of short flight durations, high costs, and limited opportunities for research in space, only brief reproductive phases have been studied in vertebrates. The most promising empirical approaches in developmental biology rely on studies of multiple model systems, comprising key organisms spanning invertebrates to mammals, for which genomic information is available.

Invertebrates

The fruit fly (*Drosophila melanogaster*) and nematode (*Caenorhabditis elegans*) are highly organized, multicellular organisms well suited to elucidating certain molecular, genetic, cellular, and physiological responses to the space environment within and across multiple generations.[826] Most importantly for mutational analyses, these and other invertebrates are excellent models for studies of DNA damage/repair and programmed cell death (apoptosis)[827] and for identifying additive influences of the space environment, particularly microgravity and space radiation[828] (important for enabling space habitation rather than fundamental biology). Invertebrates have short life cycles, producing hundreds of offspring within weeks, thereby contributing to the branch of developmental biology concerned with evolution and development—in this case, the adaptive effects of multigenerational exposure to microgravity. Such adaptations could be enhanced using flies with or without gravity sensation. Comparative gene microarray analyses of spaceflight-exposed flies and nematodes can be used to identify gravity-induced changes in expression of orthologous genes.[829] No overt structural or functional effects of spaceflight exposure on invertebrate development have been reported, although genes related to embryonic and larval development, gametogenesis, and reproduction are up-regulated in *C. elegans*.[830]

Vertebrates

Amphibian embryonic development is adaptive and characterized by a high degree of plasticity. Studies of Medaka fish, Urodele amphibians, and frogs in space have demonstrated successful mating, external fertilization, and hatching.[831] Structural abnormalities during early post-fertilization events have not been consistently observed across species, and percentages of fertilization and survival are typical. Space-born fish possessed the normal complement of germ cells and subsequently produced offspring at 1 *g*.

Mammalian reproduction comprises an intricate and complex series of events, including internal fertilization, implantation, placentation, organogenesis, fetal development, birth, lactation, parental care, and postnatal maturation. Until the first mammal undergoes an entire life cycle in space, it is difficult to specify whether precise developmental phases are gravity-dependent. The minimal existing data, derived from rats, suggest that early reproductive events (i.e., fertilization, implantation, placentation, organogenesis), the transition from prenatal to postnatal life, and maternal-offspring interactions are high-priority research areas. No mammal has given birth in space. It is unclear whether normal vaginal birth of rats or mice can occur in the space environment.[832] Cesarean deliveries may be required because of decreased connexin 43, the major gap junction protein in the myometrium that synchronizes and coordinates labor contractions.[833]

Growth and development of prenatal rats exposed to spaceflight for 4.5 to 11 days and then born on Earth either were normal or the offspring weighed significantly less than controls.[834] One-week- but not 2-week-old rat pups flown in a habitat that was suboptimal for supporting nursing litters weighed 25 to 50 percent less than controls.[835] The retarded growth of the young postnatal rats was likely due to disrupted nursing and mother-offspring interactions, rather than a direct effect of the space environment on growth and development. These observations attest to the importance of developing space habitats adequate for supporting reproducing and developing rodents.

Vestibular Development. As the primary vertebrate sensory system that detects changes in gravitational force, the vestibular system is responsive to linear accelerations (gravitational and translational) acting on the head. Rotation, or angular acceleration, is detected by the semicircular canals within the inner ear, but also affects the gravistatic receptors through centrifugal force. Information is relayed to the central nervous system via vestibular ganglion cells from the vestibular receptors to vestibular neurons in the brainstem to motor neurons. The vestibular system has widespread interactions throughout the neuraxis, including motoric, autonomic,[836] homeostatic,[837] circadian,[838] affective,[839] and spatial navigation systems.[840] Studies in the space environment provide the most useful and meaningful approaches to advancing knowledge of how these many systems organize, develop, and become integrated.

Experience transduced into neural activity programs the developing nervous system and determines its ultimate form and function. Early sensory experience plays active, formative roles in shaping both the neural architecture and neurobehavioral function,[841] leading to the fundamental biological tenet that specific age-dependent experiences are required for proper development of the central nervous system. The concepts of *critical* and *sensitive* periods refer to particular times in development when there is heightened sensitivity to certain environmental stimuli that determine subsequent function. For example, delayed cochlear implants will not restore hearing in deaf-born children when they are past the age to learn to decode sound. In alternating amblyopia, stereovision is lost because of a misalignment of the sensory input (foveation) during the child's developing years, often because of strabismus. Thus, there is perfect vision in both eyes, but the visual world is viewed by only one eye at a time.

These concepts are broadly applicable to our understanding of vestibular system formation, development, and function. Yet knowledge of vestibular development has progressed far more slowly than knowledge of other sensory systems, in part because of the historic difficulty of depriving young organisms of the Earth-constant stimulus of gravity. Studies of rats that underwent development in space[842,843] (Neurolab) provided early indications of the existence and length of the critical period, but additional studies are needed. Mutants without gravity perception can provide some insights but cannot substitute for exposure to microgravity, because such mutants lack all sensation and cannot easily regain it unless vestibular implants are fully functional.

Studies of the effects of different gravitational environments can be an invaluable tool for advancing fundamental understanding of the organizing principles of the vestibular and other neuromotor systems. Gravity deprivation during a critical period in development and maturation is likely to alter each of these systems.[844] Progress has been made in specifying molecular and cellular events involved in normal vestibular development,[845,846] yet relatively little work has focused on consequences of gravity deprivation. The limited spaceflight data suggest that vestibular system development occurs differently in the absence of gravity. Morphological changes in gravistatic receptors and in central connectivity and/or function have been reported in a variety of species, including tadpoles,[847] fish,[848] chicks,[849] and rats[850,851] after varying durations of embryonic or post-hatching/postnatal development during spaceflight. Central vestibular and behavioral changes were noted in perinatal rats that underwent the second half of gestation in space. Studies have also reported changes following development in hypergravity,[852] with some effects opposite those observed in microgravity.[853] These studies support the importance of dose-response comparisons between gravity load and biological response, using hypergravity to predict changes in microgravity.

Recently, a number of genetically altered mouse strains that fail to develop various elements of the inner ear vestibular apparatus have become available.[854] Mice that fail to develop otoconia, thereby lacking gravistatic receptor function,[855] and other mutants (e.g., members of the Hmx homeobox gene family) exhibiting depletions of vestibulosensory cells can provide fundamental new information on molecular and cellular bases of vestibular development and evolution. These models should be incorporated into research on space biology platforms, particularly using conditional mutants with defects that allow direct comparison of normal and mutated littermates with little or no other systemic defects and good viability.[856] Importantly, while vestibular responses to linear acceleration are absent in some of these genetically altered mice, they appear to compensate for much of this loss.[857] The microgravity of space provides the gold standard for advancing fundamental biological principles of vestibular system development and function.

Motor Development and Sensory-Motor Integration. Neuromuscular development begins in utero and continues throughout the first year of life in humans and throughout the first postnatal month in rats.[858] The neuromuscular system requires gravitational loading during early skeletal muscle development. Space-flown neonatal rats exhibited profound changes in development of the weight-bearing soleus muscle, including decreased muscle fiber growth, increased sensitivity to muscle reloading injury, reduced growth of motor neuron terminals, and lowered ability of the muscles to utilize oxygen.[859] Changes in the expression of myosin isoforms were observed in neonatal rats flown on the same mission.[860] Studies involving hindlimb unloading have recapitulated some of these effects,[861] but the definitive tests are studies in space.

The presence of gravity may be as fundamental to the development of posture and movement as images are to the development of the visual system.[862] Whereas exposure to spaceflight of prenatal rats had no effect on the development of somatic motor skills, young postnatal rats exposed to spaceflight for either 9 or 16 days displayed a repertoire of motor behaviors different from those of 1-*g* controls.[863]

Both size and complexity of the dendritic architecture of the medial spinal motor neurons that receive vestibular input and are involved in postural control were reduced in 1-week-old rats exposed to spaceflight.[864] Existing data suggest that these enduring experience-related changes are due to altered numbers and locations of synapses, rather than altered potency of pre-existing stable synapses. Neurons subserving motor function may thus undergo activity-dependent maturation in early postnatal life in a manner analogous to the sensory systems.

Cortical Changes. Studies of prenatal rats flown in space provide preliminary evidence for an aberrant or decelerated schedule of neuronal births in cortical layers, neuronal degeneration, and fewer glial cells and capillaries, compared to controls, indicating possible developmental retardation. Young postnatal rats flown on the space shuttle showed laminar-specific changes in the number and morphology of cortical synapses,[865] suggesting that gravity may be an important environmental parameter for normal cortical synaptogenesis.

Cognitive Mapping. A single cognitive mapping study has been done, barely scratching the surface of this important area of inquiry. Young postnatal rats exposed to spaceflight for 16 days showed remarkably normal performance on Morris water maze and radial arm maze tasks,[866] suggesting minimal long-term impact of exposure to microgravity, at this particular developmental age and flight duration, on spatial learning and memory.

Gaps in Knowledge

Lifespan and Multigenerational Studies

Lifespan studies of mammals to enable identification of gravity-dependent processes in the organization and development of the central nervous system and other organ systems, including their integration, function and maintenance, and transmission across generations, have not been accomplished. The small amount of existing data is derived from brief (16 days or less) spaceflight studies with post-flight analyses at 1 *g*. High-priority studies include lifespan and multigeneration studies in the space environment, incorporating developmental programming, epigenetics, and omics systems biology approaches during key reproductive and developmental phases in different model systems, especially mammalian models.

Developmental biology research is needed to fill gaps in knowledge of the effects of the space environment on (1) gene expression changes over time in model organisms; (2) DNA replication and repair processes, and their long-term consequences; (3) changes in single-nucleotide polymorphisms in large populations across multiple generations; (4) replication and reproduction across multiple generations; (5) intracellular molecular changes; and (6) changes in tissues, organs, physiological systems, and whole organisms throughout the lifespan and across generations in space. Throughout these studies, it will be important to determine effects of gravity in relation to those of other space-related factors (e.g., radiation) and their interactions.

Vestibular Development

Gaps exist in knowledge of gravity's influence on (1) vestibular system development from peripheral end organs to central nervous system targets, including ground-based studies focused on genetically altered models and species in which peripheral organs can be selectively ablated during development (e.g., chick); (2) vestibular interactions with other neural systems (i.e., homeostatic and circadian systems, autonomic system, formation and maintenance of neural maps, affective system); (3) influence of efferent systems on the development and maintenance of peripheral vestibular apparatus; (4) development of postural, gait, and locomotive control systems; (5) mechanisms for vestibular compensation; and (6) role of vestibular experience in complex navigational behavior. There is a major gap in understanding of the organization and development of structural and functional linkages among these diverse systems. Basic research enabled by the unique environment of space and ground-based studies are needed to address this deficiency.

Motor Development and Sensory-Motor Integration

A major gap exists in understanding of how exposure to the space environment from pre-conception throughout adulthood affects the formation, development, function, and maintenance of the motor system. It is important to ascertain whether these effects reflect reorganization of a motor system developing in ways that are appropriate to the environment within which it has developed and the extent to which post-development adaptation to a different environment can occur. It may be that gravitational loading is required for proper development and integration of vestibulomotor function. Understanding the mechanisms underlying these changes will shed important new light on motor development and its controls.

Neuroplasticity

Knowledge is lacking of how experience with gravity influences neural development. *Neuroplasticity* refers to changes, through new experiences,[867] in neurons, neural networks, and their function that occur in target neurons, pathways, or entire sensory systems through apoptosis (programmed cell death) and cell atrophy. Neuroplasticity is not limited to development but persists in adulthood as an inherent feature of everyday brain function; it is critical for learning and memory and the adaptability of primary sensory maps.[868] During early life, neuroplasticity is shaped by sensory input during critical periods of development.

Neurotrophins (e.g., nerve growth factor, brain-derived neurotrophic factor, and their Trk receptors) play major roles in central nervous system development, plasticity, and experience, including learning.[869] During development, neurotrophins control neuronal survival, target innervation, and synaptogenesis. In the vestibular system, TrkB activation and, to a lesser extent, TrkC activation are important for ganglion neuron survival, innervation, and synaptogenesis.[870] Studies in the space environment and ground-based platforms can address this major gap in fundamental understanding of how gravity shapes neural development and plasticity.

Research Models and Platforms

A robust and meaningful research effort in developmental space biology will require access to spaceflight as well as extensive use of ground-based platforms and approaches, including centrifugation, HU, and genetically altered animals. Invertebrate, non-mammalian vertebrate, and mammalian models should be utilized.

- *Free-flyers.* Free-flyers should be used to conduct shorter-duration missions, optimally with an animal centrifuge to provide proper 1-*g* controls for mammalian studies, an important experimental control that was eliminated from the ISS.
- *Celestial bodies.* Beyond 2020, celestial bodies such as Earth's Moon are important potential research platforms for ongoing studies in partial gravity. They could eventually provide platforms for sustainable research laboratories for developmental biological research across generations.
- *Centrifugation.* Ground-based centrifugation should be used to apply fractional increments in *g*-loads exceeding 1 *g* along a "dose-response" continuum to facilitate understanding of responses in decreased gravity and to help develop a program of research in artificial gravity.

The ISS is a necessary platform for long-duration studies (60+ days). To date, NASA has not added Advanced Animal Habitats (AHHs) to the ISS, and so its current capacity is 6 mice. However, with the addition of the four AHHs recommended in this report (capable of housing 6 rats or 8 to 10 mice) it would be possible to bisect some or all of the AHHs so that long-term developmental studies could be undertaken. Four habitats could be bisected to provide eight compartments for housing a single breeding pair or nursing litter to maintain statistically meaningful numbers of mice for lifespan and multigenerational studies. This plan utilizes a staged approach for mating and for birth and weaning of eight litters within a 30-day time frame, with a second generation produced within 90 days. Offspring can be removed at specific ages for analysis at key developmental time points. This approach serves to limit the overall number of mice to required levels and, once established, is particularly amenable to investigations of genetically altered mouse models.

Research Recommendations

1. Studies should be conducted on transmission across generations of structural and functional changes induced by exposure to space during development. Such research will provide vital fundamental information about how genetic and epigenetic factors interact with the environment to shape gravity-dependent processes and about the penetrating influence of those factors across subsequent generations. Spaceflight experiments offer unique insights into the role of forces omnipresent on Earth (but absent in orbital flight) that can actively shape genomes in ways that are heritable. Such spaceflight experiments would place gravitational biology at the leading edge of modern developmental and evolutionary science.
2. Spaceflight and ground-based (e.g., centrifugation, hindlimb unloading) studies should be conducted to ascertain the role(s) of gravity in the organization, development, and maintenance of the sensory and motor systems in mammalian systems and their functional integration. These studies will yield important new fundamental biological knowledge regarding the development and regulation of gravity-dependent brain, physiological, and behavioral systems. High-priority research should examine structural and functional changes in peripheral and central elements of the sensory and motor systems and their underlying mechanisms, critical periods of development, sensory-motor integration, neural plasticity and modulation of nerve growth factors, and adaptation. Important research in this area could be completed within a 10-year time frame.
3. Model systems offer increasingly valuable insights into basic biology. There should be a coordinated emphasis on utilization of invertebrates, non-mammalian vertebrates, and mammals to identify functional and evolutionary commonalities. Genetically altered models including mice should be used for analysis of key genes, gene products, and signaling pathways. This research should include an organized effort to identify orthologous genes, common changes in gene expression, epigenetic analysis, and key model systems for spaceflight. Experimentation with rats should also be conducted to analyze gravity-based changes in certain systems (e.g., muscle) and to retain continuity with existing developmental space biology research.

Merging of Disciplines to Study Gravity-Dependent Adaptations

The previous sections of this chapter focus on gravity-related alterations occurring in specific organ systems: bone, muscle, sensory-motor, cardiovascular, etc. Each of these systems is affected by alterations in gravity stimuli, especially microgravity. Also, it is apparent that certain systems are structurally and functionally aligned with one another, with muscle and bone and muscle or sensory-motor function as primary examples of such alignment. One of the deficiencies in the evolution of the NASA life sciences research mission concerns the lack of programs designed to foster such integration in terms of addressing specific problems or themes on a more global scale. This deficiency needs to be corrected well beyond the approach of tissue-sharing programs; effective integration can be achieved by sponsoring research initiatives that require the integrated study of multiple systems in the experimental design of individual projects. There should be a strategy to take advantage of research platforms (e.g., hindlimb unloading, centrifugation, bed rest, spaceflight modules/ISS) affecting both animal and human research objectives, whether the focus is on basic science or on more translational research directions. Several examples are given below to amplify the issue of integration.

Integrated Animal Studies on Bone, Microcirculation, and Skeletal Muscle Structure and Function

Rodent studies indicate that musculoskeletal and cardiovascular structure and function depend importantly on weight bearing, as well as gravitational blood pressures and flows within the body.[871,872,873] Moreover, Stevens et al.[874] have documented that microvascular to interstitial-fluid pressure gradients are important in regulating bone formation in the murine femur during hindlimb unloading. In addition, preservation of neuron-motor control and sensory feedback during exposure to microgravity for normal ambulation on the Moon and Mars may require walking and running against artificial gravity while in transit to these destinations. Such observations are important in understanding why exercise countermeasures used to date have not fully protected the bone, muscle, orthostatic tolerance, and fitness of individuals exposed to prolonged microgravity.

Hargens and Richardson point out that adult humans "spend about two-thirds of their existence in upright, sitting, and standing postures. During upright posture on Earth, blood pressures are greater in the feet than at heart or head levels due to gravity's effects on columns of blood in the body (hydrostatic or gravitational pressures). For example, mean arterial pressure at heart level is normally about 100 mm Hg, whereas that in the head is slightly lower (e.g., 70 mm Hg) and that in the feet is much greater (e.g., 200 mm Hg)."[875] During exposure to microgravity and probably on the Moon as well, arterial, venous, and microcirculatory blood pressure gradients are more or less absent, so that blood immediately shifts to chest and head tissues. According to earlier work, loss during hindlimb unloading of transmural stresses associated with local blood pressures and with maintenance of the structure and function of blood vessels affects musculoskeletal, brain, and cardiovascular structure and function.[876,877,878]

In ground-based projects involving the rodent HU model, the experimental design can be integrated to examine the time course of simultaneous alterations in bone, connective tissue, microcirculation, and muscle structure and function in normal animals as well as in transgenic models, while targeting specific regulatory pathways that are likely to be involved in microcirculatory, bone, and muscle maladaptations in the lower and upper body. These ground-based approaches are ideal for studies addressing pharmacological interventions such as the use of bisphosphonates, transgenic manipulation, and altered growth factor manipulations such as with myostatin and insulin-like growth factor-1. Based on these results, ground-based approaches can be transformed into similar projects that use the same animal models on the ISS but that can examine long-duration (out to 180 days) time course alterations in space. Additional studies also could be conducted to determine the ability of animals to recover back on Earth from maladaptations to microgravity. Such studies would provide a foundation for subsequent research with humans either during prolonged bed rest or in space.

Integrated Muscle, Bone, Sensory-Motor, Cardiovascular, and Organismic Metabolic Studies: Countermeasure Strategies in Response to Chronic Bed Rest

It is generally accepted within the NASA science community that a variety of exercise paradigms have the potential for maintaining homeostasis of essentially all of the organ systems covered in this report. Thus, it seems reasonable that in formulating an exercise countermeasure in the context of bed rest and/or spaceflight intervention, the primary objective should be to integrate research objectives concerning specific topics or research themes (e.g., control of movement, tissue atrophy/deterioration, regulation of organ system fitness, circulatory/microcirculation alterations, or sensory-motor and skeletal muscle function), across qualified research teams to ensure such integration. These integrative projects could be conducted not only in NASA facilities but also in national or international research centers and within universities or in clinical facilities. For example, it is interesting to note that such cooperation has been achieved by NASA and the National Institutes of Health in areas of common interest on a national level and by NASA, the European Space Agency (ESA), and the French Centre National d'Etudes Spatiales (WISE-2005) on an international level. In this way the knowledge base and effectiveness of such integrative research likely will be much greater than what is learned from the individual projects.

As Hargens and Richardson have pointed out,[879]

> Exercise countermeasures for astronauts in space are still unresolved, although recent calculations suggest that all exercise in space to date has lacked sufficient loads to maintain preflight bone mass.[880] Although Russian cosmonauts walk and run on a treadmill for 2 to 3 hours per day in an attempt to prevent bone loss, their bungee-cord loading

apparatus is uncomfortable at loads over 70 percent body weight. Furthermore, blood pressure stimuli at their feet are abnormally low because gravitational blood pressures are absent with their treadmill hardware.

In the context of these integrated studies targeting any given combination of organ systems, it is important to note that certain modality prototypes have begun to evolve for delivering a broader range of exercise/loading stimuli capable of simultaneously affecting the homeostasis of a number of these physiological systems. These prototypes include (1) modification of the flywheel device used in several European bed rest studies summarized above, such that both high resistance-loading stimuli and high aerobic-metabolic stimuli can be achieved with the same instrument, (2) human-powered artificial gravity, which can also provide both total body aerobic-metabolic challenges and high muscle group loading (2 to 3 *g* body weight), (3) aerobic treadmill exercise within an LBNP chamber combined with resistive flywheel exercise during WISE-2005, and (4) other activity devices, such as rowing machines, designed to challenge the musculoskeletal and cardiovascular systems. These devices, when used in prolonged bed rest studies, could provide a broad range of physiological alterations across the spectrum of organ systems.

Further, as these alternative devices become better characterized, it will be important to compare modality prototypes in order to select the most viable device for inducing alterations that affect the homeostasis of the broadest range of physiological systems being targeted.

Consequently, it is envisioned that integrated studies could simultaneously address skeletal muscle alterations (muscle size, phenotype, and endurance properties) and changes in sensory-motor function (cortical functional magnetic resonance imaging, electromyography, and accelerometer parameters).

In another paradigm, cardiovascular/metabolic alterations (cardiac output regulation, muscle blood flow regulation), as well as microcirculatory function in muscle and skin, could be examined. Further, studies are needed to address the kinetics of oxygen utilization and substrate metabolism turnover (anaerobic threshold, substrate utilization crossover points) as a function of the duration of exposure to chronic bed rest. These alterations would be compared to those in subjects experiencing both bed rest and specific countermeasure stimuli. Obviously, other combinations of studies could be conducted such as studies of bone (quantitative computed tomography, finite element analyses), muscle fitness, and orthostatic and neurosensory interactions. The key is that multiple systems should be studied in the same subject at the same time, particularly organ system combinations that may yield unforeseen outcomes that could lead to new insights—for example, are bone alterations impacted by alterations in the shear stress of blood flow in different anatomical regions?

Thus, in all integrated studies, a multidisciplinary approach is recommended to develop and evaluate integrated exercises that simulate normal daily loads on the cardiovascular, microcirculatory, neuron-motor control, and musculoskeletal systems. In validating the success of the countermeasure program objectives, NASA should define the limits of exercise stress that astronauts are likely to encounter in performing their duties, so that test criteria can be established that address whether the various countermeasures are achieving physiological end points appropriate to the level of fitness needed. Additional information on this important topic can be found in Chapter 7, "Crosscutting Issues for Humans in the Space Environment."

RESEARCH PRIORITIES AND PLATFORMS

Based on deliberations concerning each of the preceding discipline-specific sections in this chapter, the AHB Panel rank-ordered the following disciplines as high priority, starting with the highest priority: (1) bone and connective tissue, (2) skeletal muscle and sensory-motor performance, and (3) heart and cardiovascular system homeostasis. The high priority given these areas is based on consideration of factors such as the magnitude of the spaceflight-induced changes that have been observed, the potential mission risk posed by spaceflight-induced changes, and the short- and long-term risks to the health of astronauts. Structural and functional deficits in these systems have the potential to severely affect the homeostasis of the organism, as well as cause a significant negative impact on the "fitness for duty" of astronauts, thereby compromising spaceflight missions lasting longer than 6 months. The key high-priority research recommendation for these systems and other ancillary high-priority recommendations are presented below. Very general estimates are given regarding the time that might be needed to fulfill the individual research recommendations, assuming robust programmatic support and reasonable access to flight opportunities, but the actual length of time will depend on a variety of factors, including the nature of the studies, the complexity of the experimental design, the number of subjects needed for statistical power, and the validity of the findings.

Bone

Human Studies

All experiments under this heading can be completed within the next decade.

1. Bisphosphonates have been used to successfully treat bone loss associated with osteoporosis, and bed rest studies suggest that bisphosphonates can reduce bone loss associated with weightlessness. In addition, these drugs may reduce the incidence of kidney stones. The efficacy of bisphosphonates should be tested on the ISS during ~6-month missions in an adequate population of astronauts. This testing should include pre-flight dosing with long-acting bisphosphonates. **(AH1)**
2. It is known that bone mass decreases during spaceflight, but recent data demonstrate that the structure and strength of bone also deteriorate. The preservation/reversibility of bone structure/strength should be evaluated when assessing countermeasures.[881,882,883] These measurements can be made during the countermeasure assessments recommended above. **(AH2)**

Animal Studies

All ground-based studies under this heading can be completed within the next decade, but flight studies will be limited by the inadequate animal housing on the ISS.

1. More studies of genetically altered mice exposed to weightlessness are strongly recommended. These studies were recommended in the 1998 NRC report[884] but not implemented. In particular, further ground-based and spaceflight studies of mice with altered gene expression are warranted. These studies are essential to allow a better understanding of the molecular mechanisms of bone loss and also may allow the development of more effective countermeasures. **(AH3)**
2. New osteoporosis drugs under clinical development (for example, sclerostin antibody and Denosumab as a candidate) should be tested in animal models of weightlessness (both ground-based and in spaceflight). These studies are essential to identify new pharmacological countermeasures. **(AH4)**

Skeletal Muscle

Animal Studies

Studies should be conducted to identify the underlying mechanism(s) regulating net protein balance and protein turnover in skeletal muscle during states of unloading and recovery. These studies are essential to understanding the process of muscle wasting. Such studies should examine the roles of growth factors, hormones, signaling pathways, protease and myostatin inhibitors, possible pharmacological interventions such as use of antioxidants, and nitric oxide signaling. This research could potentially be concluded in a 10-year time frame and would not depend on access to space if the hindlimb unloading model is utilized. **(AH5)**

Human Studies

Studies are recommended to (1) develop and test new prototype exercise devices, and (2) optimize physical activity paradigms/prescriptions targeting multisystem countermeasures, in addition to skeletal muscle, preferably with the same training device(s). Also, consideration should be given to new developed devices, in addition to the conventional exercise equipment onboard the ISS. These studies are needed because the current exercise devices and corresponding physical activity countermeasure prescriptions are ineffective for optimally maintaining physical fitness and organ system homeostasis. Depending on how rapidly such newly improved devices come onboard, this research could be completed within 15 years. **(AH6)**

Sensory-Motor Function

How patterns of neuromuscular activity change in the spaceflight environment and upon return to 1 *g* is not known. Therefore, it is critical to determine the daily levels and pattern of recruitment of flexor and extensor muscles of the neck, trunk, arms, and legs at 1 *g* and after exposure to a novel gravitational environment for up to 180 days. These altered patterns of neuromuscular activity over different time periods need to be carefully examined with respect to changes in the accuracy of movements and the type and severity of their impact on other tissues, particularly muscle, connective tissue, and cardiovascular and hormonal systems. Studies are needed to:

1. Identify the neuromuscular mechanisms that underlie the loss of accurate control of movement with respect to (1) changes in neural control at the cortical, subcortical, and spinal levels; (2) changes in muscle properties; (3) changes in visual or proprioceptive perception; and (4) vestibular function.
2. Determine whether changes in accuracy of movement due to changing gravitational environments can be prevented or corrected with an exercise countermeasure designed to preserve aerobic function, such as use of a treadmill, stationary bike, or other exercise device designed to maintain muscle strength. **(AH7)**

Fluid Shifts

A focus on fluid shifts represents a crosscutting research priority that addresses virtually all systems and is particularly relevant for humans because of the large gravitational gradients involved. Studies are needed to determine the basic mechanisms, adaptations, and clinical significance of changes in regional vascular/interstitial pressures (Starling forces) during long-duration space missions. These changes may be especially important in the brain, where chronic and sustained increases in intracranial, intraocular, and retinal pressures may induce clinically important structural and functional abnormalities that alter visual acuity. Studies are recommended of mechanical and/or pharmacologic countermeasures both as interventions to test mechanistic hypotheses regarding the pathophysiology of prolonged, sustained fluid shifts and simultaneously to develop evidence-based clinical treatments. Strategies to restore Earth-like plasma volume and fluid distribution in the period before landing, such as augmentation of venous return, increased salt intake several days earlier, or hormonal manipulation with or without salt consumption, should be studied quantitatively to determine their efficacy in minimizing orthostatic hypotension with restoration of gravitational gradients. **(AH8)**

Cardiovascular Function

Three enabling areas of cardiovascular research are essential in the next decade.

1. Investigate the effect of prolonged periods of microgravity and partial gravity (3/8 or 1/6 *g*) on the determinants of task-specific enabling levels of work capacity. Specifically, are there changes in energy availability, oxygen consumption, tissue perfusion, stroke volume, blood volume, blood flow redistribution to active muscle, or the capacity for heat dissipation that make it impossible to meet either routine or emergency demands for physical activity? Fundamental studies enabled by this research priority, including basic studies of myocardial mass and contractility, capillary filtration, hormonal changes, signaling pathways, and transcriptional regulation of cardiac structural proteins, will provide insight into mechanisms that contribute to the decrease in stroke volume and their future amelioration by more specific countermeasures. **(AH9)**

2. Determine the integrative mechanisms of orthostatic intolerance after restoration of gravitational gradients (both 1 *g* and 3/8 *g*); examine its severity as a function of prolonged microgravity, especially under real-life conditions of spaceflight such as emergency egress or Mars-like tasks, and the likelihood of as well as the time necessary for functional recovery. Specifically, what must be determined is the relative importance of hypotension and cerebral hypoperfusion, compared with neurovestibular, kinesthetic, or muscular weakness, to the orthostatic intolerance following long-duration spaceflight.

If it is confirmed by systematic examination that orthostatic intolerance after long-duration spaceflight is due to hypotension, determine whether reductions in cerebral blood flow, or neurovascular control, loss of blood volume, or cardiac atrophy (or their interaction) cause this hemodynamic compromise. Use the results from these investigations of mechanism to confirm and examine more systematically the efficiency of promising post-flight countermeasures, such as use of the pharmacologic agent α_1-agonist midodrine and the use of post-flight gravitational countermeasures such as thigh cuffs, support stockings, or inflation of G-suits; also determine the efficacy of promising in-flight countermeasures such as exercise of different modes, durations, and intensities, as well as centrifuge-simulated artificial gravity, along with novel fluid loading strategies that will more effectively restore blood and plasma volume. Studying sufficient numbers of astronauts of both genders over a wide age range will help to define the mechanisms underlying individual variability in these responses. **(AH10)**

3. The primary emergent, mission- and life-threatening medical event (besides trauma) relevant to middle-aged astronauts is likely to be an acute coronary syndrome. Collaborative studies among flight medicine and cardiovascular epidemiologists are recommended to determine the best screening strategies (such as vascular imaging with or without cardiac biomarkers) to avoid flying astronauts with subclinical coronary heart disease that could become manifest during a long-duration exploration-class mission (3 years). It is also recommended that collaborative studies with radiation biologists be conducted at a basic level to determine whether atherosclerotic vascular disease could be accelerated by the chronic exposure to radiation in deep space. **(AH11)**

Pulmonary Function

Determine the amount and site of the deposition of aerosols of different sizes in the lungs of humans and animals in microgravity. The rationale is that deposition is different in microgravity compared with normal gravity, and there is evidence that lunar dust and Mars dust are potentially toxic. The research can be carried out on the ISS and could potentially be concluded in a 10-year time frame. **(AH12)**

Immunology

Although no clinically deleterious effects of spaceflight on the immune system have been observed to date, studies of the acute and chronic effects of the space environment on the immune system have been limited. Further studies should be pursued because of the potential for adverse events during longer flights or in response to altered external conditions. The high-priority recommendations below can be completed in 5 to 10 years, depending on the availability of flight time:

1. Multiple parameters of T cell activation should be obtained from astronauts before and after re-entry to establish which parameters are altered during flight. The parameters examined should represent the current state of technology in assessments of immunity, including those that reflect aging-related changes in the immune system (e.g., shortening of telomeres, expression of CD28 and Treg). **(AH13)**

2. Changes observed in the immune system may reflect the impact of multiple stressors. To both address the mechanism(s) of changes in the immune system and develop measures to limit the changes, data from multiple organ/system-based studies have to be integrated. For example, data from assessments of cortisol levels (stress studies), nutritional diaries, and logs of weight changes, exercise volume and intensity, and level of activity (e.g., EVA) need to be considered when interpreting data on immune responses from individuals. **(AH14)**

3. It is essential that studies of mice be performed on the ISS to establish the biological relevance of the changes observed in the immune system. These studies should include immunization and challenge, with samples

acquired both prior to and immediately upon re-entry. The parameters examined need to be aligned with those human immune response components influenced by spaceflight. **(AH15)**

Reproduction and Development

Studies should be conducted on transmission across generations of structural and functional changes induced by exposure to space during development. Such research will provide vital fundamental information about how genetic and epigenetic factors interact with the environment to shape gravity-dependent processes and about the penetrating influence of these factors across subsequent generations. Spaceflight experiments offer unique insights into the role of forces omnipresent on Earth (but absent in orbital flight) that can actively shape genomes in ways that are heritable. Such spaceflight experiments would place gravitational biology at the leading edge of modern developmental and evolutionary science. Ground-based studies should be conducted to develop specialized habitats to support reproducing and developing rodents in space. This research could be accomplished within 10 years. **(AH16)**

Research Platforms

To address the above recommendations, the following platforms are needed to carry out integrated research projects involving both animal and human subjects.

Animal Platforms

From a practical perspective, the AHB Panel envisions an animal research program centered on rodents (mice and/or rats). The reason is that rodents, especially rats, have been used more extensively than any other animal model for spaceflight studies. Also, the genomic material of rodents is closely aligned to that of humans. Studies utilizing both ground-based analogs and spaceflight are recommended. For ground-based studies, the panel recommends continuing the research foundation generated by the HU model and centrifugation, which have produced extensive data spanning hundreds of studies over the past three decades. As noted in the section above titled "Merging of Disciplines to Study Gravity-Dependent Adaptations," the HU model can be used to develop time course databases for alterations in the musculoskeletal system, along with generating potential countermeasures (endurance and loading exercise, hormonal and growth factor modulations) through a variety of models in addressing most of the recommendations pertinent to basic science research.

In order to bolster research in more fundamental biology, the panel strongly recommends establishing a flight-based research platform on the ISS. The ISS platform is preferred over other types of free-flyer opportunities because long-duration studies can be carried out on the ISS for approximately 180 days, thus spanning a significant component of an adult rodent's lifespan. Also, animal research on the ISS provides the only means for establishing the fundamental alterations unfolding during long-duration exposure to microgravity without any countermeasure intervention, which cannot be studied in the humans living on the ISS. This is important because it is only through the physiological responses of animal models that one can ascertain the degree to which microgravity affects the vital systems of the organism. Many of the alterations observed in human physiological homeostasis in response to spaceflight have also been noted in animal models, as delineated in the prior sections of this chapter. Further, the AHB Panel envisions exposing animals to spaceflight for long durations and then returning them to Earth, where their ability to correct the structural and functional deficits that occurred in space can be studied. The findings of the animal flight research program have far-reaching consequences not only for understanding the nuances of spaceflight but also for gaining insight into how humans adapt on Earth in response to lifestyles of chronic inactivity. See Chapter 7 for additional information on this topic.

Human Platforms

Ground-based research opportunities for performing integrated research programs should use the bed rest model extensively, and in certain cases other analogs such as unilateral limb suspension, for studying unloading-induced deficits in systems such as muscle and bone. For example, it is now routine to study subjects for 90 days or longer during bed rest under controlled experimental conditions. NASA should adopt the model used by ESA, in which such studies use a multicenter approach. Again, the focus should be on integrated approaches for targeting multiple systems with coordinated analyses, as outlined in the section "Integrated Muscle, Bone, Sensory-Motor, Cardiovascular, and Organismic Metabolic Studies: Countermeasure Strategies in Response to Chronic Bed Rest." These studies should be linked to a variety of countermeasure paradigms centered on exercise approaches challenging bone, skeletal muscle, pulmonary/cardiovascular fitness, and immune system homeostasis. These countermeasures include resistance and aerobic exercise devices (ergometers, treadmill, and rowing modalities), as well as artificial gravity and endurance exercise with lower-body negative pressure, which can also target multiple systems with high loading stimuli. As noted above, it may be necessary to compare different countermeasure modalities for their respective effectiveness in modulating multiple organ systems.

The ISS Platform

Such integrated studies should be extended to the ISS research platform, to validate the science outcomes seen in ground-based analogs. Such studies should heavily involve the science community so that research objectives for understanding underlying mechanisms can be achieved, rather than merely focusing on outcomes from an operational medical perspective. The participants in this decadal survey were aware of situations in which extramural scientists whose projects had been approved were unable to access, or were able to access only with great difficulty, operational medicine data needed for the project. This suggests that mechanisms for engaging the scientific community and for publishing the valuable human research data generated from limited spaceflight opportunities need to be improved. The panel also strongly recommends that any research performed on the ISS should involve systematic studies in which all the subjects perform identical protocols in order to enhance the fidelity of the data that are obtained. This approach is more likely to yield clear insights as to which countermeasure paradigms are optimally successful in maintaining organism homeostasis. Only by establishing a confirmed evidence base can NASA evolve a validated countermeasure program that is truly optimal for astronauts exposed to long-duration space travel.

OVERARCHING AND PROGRAMMATIC ISSUES

As the AHB Panel compared the recommendations and findings from its discipline subgroups, certain overarching issues emerged that have program-wide relevance. These issues are summarized below.

The Need for Animal Research on the ISS and Other Space Platforms

For the past decade, biological research in space has emphasized human countermeasures, as conducted by the operational medicine program. While such research is appropriate, given the dependence of the U.S. space program on a human presence in space, this focus has limited the ability to answer fundamental questions about the response of biological systems to altered gravity. The panel is unanimous in its recommendation that an animal habitat should be incorporated as soon as possible into the ISS.

Flight experiments on animals began in 1782 when a duck, a sheep, and a rooster were sent aloft in a hot air balloon by the Montogolfier Brothers of Paris.[885] Moreover, animal models have enabled humans to initiate previous exploration and discovery missions by serving as subjects to determine the effects of radiation, establishing biological and safety limits, perfecting life support systems, evaluating countermeasures, and providing insights into anatomical and physiological responses and the mechanism responsible for them.[886-889]

The human space age began with Gagarin's epic orbital flight of April 12, 1961, and Shephard's suborbital flight less than a month later;[890] however, each was enabled by Sputnik flights containing dogs, guinea pigs, and mice[891] and by the U.S. flight of the chimpanzee Ham.[892] The importance of and need for animals in space were officially recognized in the Goldberg report of 1987,[893] which listed as the third of its four goals "to understand the role gravity plays in the biological processes of both plants and animals." This goal, which has yet to be achieved, has become more important because the ongoing human tasks of exploration, habitation, and discovery require the results of animal research to address the critical human risk factors for human exploration.

From the beginning of the human space age and continuing up to January 16, 2003, the United States and Russia flew 34 missions in which mammalian experiments were conducted on the biological systems mentioned in this report.[894,895] However, the average duration of those flights was 11 ± 4 days,[896] with the longest being 23 days.[897,898] In the overwhelming majority of these flights, the biological data were collected at or after recovery, which means that they are important but are of limited use in determining and interpreting in-flight effects. Meaningful in-flight results have been obtained with monkeys and rodents from the Russian Bion or Cosmos flights[899-904] and from monkeys and rodents on Spacelab Life Sciences 1 and 2[905,906,907] and from Neurolab or STS-90.[908] While these results were obtained from flights lasting only from 9 to 16 days, in-flight measurements enabled investigators to secure molecular, cellular, and systems data encompassing growth and development; motor control; and select structure and functional relationships for the nervous, cardiovascular, musculoskeletal, immune, temperature regulatory, and circadian rhythm systems. Moreover, in-flight studies provided the opportunity to separate gravitational influences from landing effects, secure repeat measurements, secure video recordings of animals performing manual tasks, and demonstrate the biotelemetric effectiveness of surgically implanted electrodes and sensors in animals as a means of acquiring meaningful physiological data on multiple biological systems.[909]

Recommendation Concerning the National Laboratory

To maximize the contribution of animal experimentation to reducing human risks during space exploration and discovery while enhancing the acquisition of fundamental knowledge, it is essential that single and repeated long-duration in-flight animal experiments be conducted. This can best be accomplished by having NASA make available as soon as possible, within the National Laboratory on the ISS, four Advanced Animal Habitats[910] with the capability of accommodating 8 to 10 mice or 4 to 5 rats per habitat for animal investigations devoted to the priorities recommended in this report. It is assumed that animals would be transferred to a clean habitat after each 30-day period. To accommodate longitudinal investigations, including studies of growth, development, and reproduction, while being in accord with current procedures for astronauts, the study time frame would be 6 months.[911] The caging infrastructure can be easily modified to create birthing dens to facilitate studies of reproduction. Such an arrangement is consistent with Article 3 of Section 305 in the National Aeronautics and Space Administration Authorization Act of 2005, which directs the NASA administrator to conduct animal research for which it is essential that the United States have an operational laboratory for space-directed animal and human research.[912] Finally, despite its awareness that the large centrifuge program has little likelihood of being restarted, the AHB Panel would be remiss if it did not strongly recommend an animal centrifuge capable of accommodating rats/mice at variable gravity levels. This capability would enhance the research potential of the ISS laboratory by creating both partial gravity and hypergravity stimuli to allow analysis of the role of the gravity vector in modulating functional capacity across many organ systems, especially the bone, muscle, neurosensory, and cardiovascular systems.

Research access to the National Laboratory will strengthen existing relationships with U.S. international partners in Russia, Europe, and Japan and promote possible new relationships with China and India. Besides improving existing relationships, research access enhances the outsourcing of animals to international partners for shorter-term landing and recovery experiments. Access will also augment possibilities for commercial transportation to reduce costs and increase durations of parabolic flights and to replace the loss of the space shuttle system for access to the National Laboratory.

Inherent in the recommendation for animal and human experimentation in the National Laboratory is the understanding that the implementation of this decadal survey's recommended research objectives will have sufficient governmental priority between 2010 and 2020 to supersede existing NASA operational plans.

Animal Research on Other Platforms

Uncrewed flight opportunities on free-flyers should be used to conduct shorter-duration missions than can be accomplished on the ISS, optimally with the availability of an animal centrifuge to provide proper 1-*g* controls for animal specimens. Interaction with, and utilization of facilities provided by, international partners should be exploited in this regard.

Improved Access to Biological Samples and Data from Astronauts

The medical and scientific communities interested in human health during long-duration spaceflight have been consistent in their requests for greater access to biological samples and other data collected from astronauts during spaceflight missions. The astronauts' rights to privacy have, at times, appeared to conflict with the need for access to valuable in-flight data in order to benefit future space travelers. In 2001, the Institute of Medicine published *Safe Passage: Astronaut Care for Exploration Missions*,[913] which included suggestions for resolving this conflict. Among the recommendations were (1) that NASA should establish a comprehensive health care system for astronauts for the purpose of collecting and analyzing data, and (2) that NASA should develop a strategic health care research plan designed to increase the knowledge base about the risks to astronaut health. While some of these goals have been met, much remains to be done to provide more widespread access to data on astronaut health. For example, there is still little information in the public domain about the bone health of women astronauts who have flown on long-duration space missions. Current policies on access to data are limiting progress toward effective countermeasures. NASA is urged to study and take action on this issue.

Limitations of Ground-Based Facilities

Experiments conducted in space are complex and expensive. It is recommended that NASA make an enhanced commitment to ground-based analogs of spaceflight for purposes of human subject research. One example is the bed rest analog.[914,915] At the time of writing, NASA has only one facility in the United States for bed rest study, the Flight Analogs Research Project, located at the University of Texas Medical Branch in Galveston, Texas. This unit has a maximum occupancy of 10 subjects and is located in an area that is significantly at risk for hurricane damage. The small number of beds means that progress on the many proposed experiments is slow, and the geographical location has, in the past, led to evacuation of subjects and the premature end of bed rest campaigns. It is thus strongly recommended that NASA consider establishing a bed rest facility allowing study of at least 20 patients simultaneously and that this facility be located in a region of the country that is not at high risk for severe weather that could interrupt experiments. Alternatively, contract arrangements could be established with one or more of the European bed rest facilities to conduct studies for NASA.

Limitations on Sample Delivery Back to Earth

Many biological samples need to be stored under tightly controlled environmental conditions, such as in a −80°C freezer. With the end of the space shuttle era in 2011, an important mechanism for delivery back to Earth of biological samples collected on the ISS will no longer be available. Because much of the up-mass to and down-mass from the ISS will likely be delivered by commercial contractors after 2011, it is essential that the contract specifications for such vehicles include adequate capacity for transporting biological samples. Conditioned down-mass is of particular importance in this regard, because there are facilities on the ISS for storage of samples. Unless suitable down-mass transportation is made available, only the relatively simple analyses that can be conducted onboard the ISS will be feasible.

Space Platform for Research Beyond 2020 Will Be Needed

While most of the recommendations in this report deal with the decade 2010-2019, the AHB Panel recognizes the long time constant inherent in the implementation of some recommendations and thus the importance of planning for the period 2020-2029 when, under current plans, the ISS will have been de-orbited.

The effort in the third decade of this century must begin by extending the findings of research that was conducted in 2010-2019 and filling in the gaps remaining from that work. While these gaps are hard to predict, it seems reasonable to assume that not everything that will be started in the decade to 2019 will reach maturity by 2020.

A major new factor in the third decade will be the presence of new transport vehicles capable of carrying astronauts well beyond low Earth orbit. This will open up at least three new areas of study: (1) Countermeasures will be needed during long voyages in microgravity, in environments that will be drastically smaller than the ISS. This will require small-footprint, low-power yet highly effective devices that will provide countermeasures for changes in multiple systems. (2) It is likely the astronauts will spend time in locations where there is a fraction of Earth's gravity. The role of partial gravity in preventing deterioration in important physiological systems will need to be clearly understood, and countermeasures to supplement these effects, if necessary, will need to be developed. (3) Current plans appear to call for the design of a new generation of surface rovers, which may not allow for normal conditioning stimuli during EVA. There will, therefore, be a need for exercise facilities for both recreation and countermeasures at a lunar or planetary base.

It is highly likely that, by the end of the present decade, much will remain to be learned about the fundamental molecular mechanisms of the responses to microgravity and altered gravity. NASA should therefore consider a flexible infrastructure of experimental facilities that can be easily upgraded when going from the ISS era to systems used in the next step in human exploration. Currently, genetically modified animals, particularly mice, are the most productive research model on Earth, and it is vital that maximum use of these models occur in space research.

The establishment of a lunar outpost could provide the opportunity for an important research platform for ongoing studies in partial gravity. Among other important uses, this outpost could eventually provide a means for a sustainable research laboratory for developmental biological research on model systems across generations. In such a laboratory, growth, development, maturation, and longevity of animals could be carefully examined. NASA might consider naming such a facility a "National Laboratory" to highlight its importance as a key national scientific resource.

A number of responses of the cardiovascular system to altered gravity will need to be studied in the third decade of this century. Among the important issues left to be answered are likely to be the effects of prolonged weightlessness and irradiation on the pathobiology of the aging process, atherosclerosis, and coronary vascular disease; the regulation of molecular clocks, circadian proteins, and genetic signaling of cardiac and vascular muscle apoptosis and hypertrophy; the autonomic neurobiology of the circulation and cardiac pacemaker function and automaticity; and the genetic profile that determines the risk of accelerated cardiovascular disease and stroke.

If the recommendation to restore NASA's ground-based and in-flight fundamental biology and non-human life science programs is implemented in the next decade, it is likely to pay dividends in the third decade by opening new vistas for human and animal research.

Relevance of the Report to NASA Fundamental Space Biology Strategic Planning

In June 2010, NASA released the Fundamental Space Biology (FSB) Science Plan for the next decade.[916] The FSB Science Plan lists three primary goals: (1) to effectively use microgravity (and an altered-gravity continuum) to enhance understanding of fundamental biological processes; (2) to develop scientific and technological foundations for a safe, productive human presence in space for extended periods in the context of space exploration; and (3) to apply the knowledge gained in science and technology to improve national competitiveness, education, and quality of life on Earth.

The central elements of the plan's research focus would involve (1) cell, molecular, and microbial biology; (2) organelle, organ, and organismal function; and (3) developmental biology. These scientific inquiries and approaches will continue to utilize ground-based models and facilities as well as space platforms such as the ISS and a variety of free-flyer vehicles.

The driving force for these initiatives includes past reports from the NRC with recommendations presented in 1989 and 1998[917] that to date have not been fully implemented.

One of the important thrusts in the FSB Science Plan involves the proposed program's commitment to enhance research on the ISS that can accommodate animal research opportunities. If pursued, such a thrust could lead to the rejuvenation of NASA's commitment to animal research for both fundamental science and translational research.

The committee notes that the overall thrust of the FSB Science Plan is aligned with the subject matter, the scientific issues, and the research recommendations presented in this chapter. The key issue that will define the success of the FSB program in the next decade is whether meaningful animal science can unfold on the ISS.

REFERENCES

1. National Research Council. 2010. *Life and Physical Sciences Research for a New Era of Space Exploration: An Interim Report.* The National Academies Press, Washington, D.C.
2. Milstead, J.R., Simske, S.J., and Bateman, T.A. 2004. Spaceflight and hindlimb suspension disuse models in mice. *Biomedical Sciences Instrumentation* 40:105-110.
3. Frost, H.M., and Jee, W.S. 1992. On the rat model of human osteopenias and osteoporoses. *Bone and Mineral* 18:227-236.
4. Keyak, J.H., Koyama, A.K., LeBlanc, A., Lu, Y., and Lang, T.F. 2009. Reduction in proximal femoral strength due to long-duration spaceflight. *Bone* 44:449-453.
5. Smith, S.M., Wastney, M.E., O'Brien, K.O., Morukov, B.V., Larina, I.M., Abrams, S.A., Davis-Street, J.E., Oganov, V., and Shackelford, L.C. 2005. Bone markers, calcium metabolism, and calcium kinetics during extended-duration space flight on the Mir space station. *Journal of Bone and Mineral Research* 20:208-218.
6. National Research Council. 1998. *A Strategy for Research in Space Biology and Medicine in the New Century.* National Academy Press, Washington, D.C.
7. Smith, S.M., Wastney, M.E., O'Brien, K.O., Morukov, B.V., Larina, I.M., Abrams, S.A., Davis-Street, J.E., Oganov, V., and Shackelford, L.C. 2005. Bone markers, calcium metabolism, and calcium kinetics during extended-duration space flight on the Mir space station. *Journal of Bone and Mineral Research* 20:208-218.
8. Smith, S.M., Wastney, M.E., Morukov, B.V., Larina, I.M., Nyquist, L.E., Abrams, S.A., Taran, E.N., Shih, C.Y., Nillen, J.L., Davis-Street, J.E., Rice, B.L., and Lane, H.W. 1999. Calcium metabolism before, during, and after a 3-mo spaceflight: Kinetic and biochemical changes. *American Journal of Physiology* 277:R1-R10.
9. Zerwekh, J.E., Ruml, L.A., Gottschalk, F., and Pak, C.Y. 1998. The effects of twelve weeks of bed rest on bone histology, biochemical markers of bone turnover, and calcium homeostasis in eleven normal subjects. *Journal of Bone and Mineral Research* 13:1594-1601.
10. Lang, T.F., LeBlanc, A.D., Evans, H.J., Lu, Y., Genant, H., and Yu, A. 2004. Cortical and trabecular bone mineral loss from the spine and hip in long-duration spaceflight. *Journal of Bone and Mineral Research* 19:1006-1012.
11. LeBlanc, A., Lin, C., Shackelford, L., Sinitsyn, V., Evans, H., Belichenko, O., Schenkman, B., Kozlovskaya, I., Oganov, V., Bakulin, A., Hedrick, T., and Feeback, D. 2000. Muscle volume, MRI relaxation times (T2), and body composition after spaceflight. *Journal of Applied Physiology* 89:2158-2164.
12. Iki, M., Kajita, E., Dohi, Y., Nishino, H., Kusaka, Y., Tsuchida, C., Yamamoto, K., and Ishii, Y. 1996. Age, menopause, bone turnover markers and lumbar bone loss in healthy Japanese women. *Maturitas* 25:59-67.
13. Sirola, J., Kröger, H., Honkanen, R., Jurvelin, J.S., Sandini, L., Tuppurainen, M.T., and Saarikoski, S. 2003. OSTPRE Study Group Factors affecting bone loss around menopause in women without HRT: A prospective study. *Maturitas* 45:159-167.
14. Whedon, G.D., Lutwak, L., Rambaut, P., Whittle, M., Leach, C., Reid, J., and Smith, M. 1976. Effect of weightlessness on mineral metabolism, metabolic studies on Skylab orbital space flights. *Calcified Tissue Research* 21(Suppl.):423-430.
15. Smith, S.M., Wastney, M.E., O'Brien, K.O., Morukov, B.V., Larina, I.M., Abrams, S.A., Davis-Street, J.E., Oganov, V., and Shackelford, L.C. 2005. Bone markers, calcium metabolism, and calcium kinetics during extended-duration space flight on the Mir space station. *Journal of Bone and Mineral Research* 20:208-218.
16. LeBlanc, A., Schneider, V., Shackelford, L., West, S., Oganov, V., Bakulin, A., and Voronin, L. 2000. Bone mineral and lean tissue loss after long duration space flight. *Journal of Musculoskeletal and Neuronal Interactions* 1:157-160.
17. LeBlanc, A.D., Spector, E.R., Evans, H.J., and Sibonga, J.D. 2007. Skeletal responses to space flight and the bed rest analog: A review. *Journal of Musculoskeletal and Neuronal Interactions* 7:33-47.
18. Keyak, J.H., Koyama, A.K., LeBlanc, A., Lu, Y., and Lang, T.F. 2009. Reduction in proximal femoral strength due to long-duration spaceflight. *Bone* 44:449-453.
19. Lang, T.F., LeBlanc, A.D., Evans, H.J., Lu, Y., Genant, H., and Yu, A. 2004. Cortical and trabecular bone mineral loss from the spine and hip in long-duration spaceflight. *Journal of Bone and Mineral Research* 19:1006-1012.
20. Sibonga, J.D., Evans, H.J., Sung, H.G., Spector, E.R., Lang, T.F., Oganov, V.S., Bakulin, A.V., Shackelford, L.C., and LeBlanc, A.D. 2007. Recovery of spaceflight-induced bone loss: Bone mineral density after long-duration missions as fitted with an exponential function. *Bone* 41:973-978.
21. Vico, L., Collet, P., Guignandon, A., Lafage-Proust, M.H., Thomas, T., Rehaillia, M., and Alexandre, C. 2000. Effects of long-term microgravity exposure on cancellous and cortical weight-bearing bones of cosmonauts. *Lancet* 355:1607-1611.
22. Keyak, J.H., Koyama, A.K., LeBlanc, A., Lu, Y., and Lang, T.F. 2009. Reduction in proximal femoral strength due to long-duration spaceflight. *Bone* 44:449-453.

23. Sibonga, J.D., Evans, H.J., Sung, H.G., Spector, E.R., Lang, T.F., Oganov, V.S., Bakulin, A.V., Shackelford, L.C., and LeBlanc, A.D. 2007. Recovery of spaceflight-induced bone loss: Bone mineral density after long-duration missions as fitted with an exponential function. *Bone* 41:973-978.
24. Lang, T.F., Leblanc, A.D., Evans, H.J., and Lu, Y. 2006. Adaptation of the proximal femur to skeletal reloading after long-duration spaceflight. *Journal of Bone and Mineral Research* 21:1224-1230.
25. Lang, T.F., Leblanc, A.D., Evans, H.J., and Lu, Y. 2006. Adaptation of the proximal femur to skeletal reloading after long-duration spaceflight. *Journal of Bone and Mineral Research* 21:1224-1230.
26. Cavanagh, P., and Rice, A., eds., 2007. *Bone Loss During Spaceflight: Etiology, Countermeasures, and Implications for Bone Health on Earth*, Proceedings of the Symposium on Bone Loss During Spaceflight, June 23-24, 2005. Cleveland Clinic Press, Cleveland, Ohio.
27. Wagner, E.B. 2007. Musculoskeletal Adaptation to Partial Weight Suspension: Studies of Lunar and Mars Loading. Ph.D., Massachusetts Institute of Technology.
28. Morey, E.R., and Baylink, D.J. 1978. Inhibition of bone formation during space flight. *Science* 201:1138-1141.
29. Wronski, T.J., Morey-Holton, E., and Jee, W.S. 1981. Skeletal alterations in rats during space flight. *Advances in Space Research* 1:135-140.
30. Wronski, T.J., and Morey, E.R. 1983. Effect of spaceflight on periosteal bone formation in rats. *American Journal of Physiology* 244:R305-R309.
31. Jee, W.S., Wronski, T.J., Morey, E.R., and Kimmel, D.B. 1983. Effects of spaceflight on trabecular bone in rats. *American Journal of Physiology* 244:R310-R314.
32. Cann, C.E., and Adachi, R.R. 1983. Bone resorption and mineral excretion in rats during spaceflight. *American Journal of Physiology* 244:R327-R331.
33. Vico, L., Novikov, V.E., Very, J.M., and Alexandre, C. 1991. Bone histomorphometric comparison of rat tibial metaphysis after 7-day tail suspension vs. 7-day spaceflight. *Aviation, Space, and Environmental Medicine* 62:26-31.
34. Milstead, J.R., Simske, S.J., and Bateman, T.A. 2004. Spaceflight and hindlimb suspension disuse models in mice. *Biomedical Sciences Instrumentation* 40:105-110.
35. Smith, S.M., Wastney, M.E., Morukov, B.V., Larina, I.M., Nyquist, L.E., Abrams, S.A., Taran, E.N., Shih, C.Y., Nillen, J.L., Davis-Street, J.E., Rice, B.L., and Lane, H.W. 1999. Calcium metabolism before, during, and after a 3-mo spaceflight: Kinetic and biochemical changes. *American Journal of Physiology* 277:R1-R10.
36. Morey, E.R., and Baylink, D.J. 1978. Inhibition of bone formation during space flight. *Science* 201:1138-1141.
37. Vico, L., Novikov, V.E., Very, J.M., and Alexandre, C. 1991. Bone histomorphometric comparison of rat tibial metaphysis after 7-day tail suspension vs. 7-day spaceflight. *Aviation, Space, and Environmental Medicine* 62:26-31.
38. Ortega, M.T., Pecaut, M.J., Gridley, D.S., Stodieck, L.S., Ferguson, V., and Chapes, S.K. 2009. Shifts in bone marrow cell phenotypes caused by spaceflight. *Journal of Applied Physiology* 106:548-555.
39. Shaw, S.R., Vailas, A.C., Grindeland, R.E., and Zernicke, R.F. 1988. Effects of a 1-wk spaceflight on morphological and mechanical properties of growing bone. *American Journal of Physiology* 254:R78-R83.
40. Simmons, D.J., Russell, J.E., and Grynpas, M.D. 1986. Bone maturation and quality of bone material in rats flown on the space shuttle 'Spacelab-3 Mission.' *Bone and Mineral* 1:485-493.
41. Bateman, T.A., Zimmerman, R.J., Ayers, R.A., Ferguson, V.L., Chapes, S.K., and Simske, S.J. 1998. Histomorphometric, physical, and mechanical effects of spaceflight and insulin-like growth factor-I on rat long bones. *Bone* 23:527-535.
42. Wronski, T.J., Morey-Holton, E.R., Doty, S.B., Maese, A.C., and Walsh, C.C. 1987. Histomorphometric analysis of rat skeleton following spaceflight. *American Journal of Physiology* 252:R252-R255.
43. Kaplanskĭ, A.S., Durnova, G.N., Sakharova, Z.F., and Il'ina-Kakueva, E.I. 1987. Histomorphometric analysis of the bones of rats on board the Kosmos 1667 biosatellite. *Kosmicheskaia Biologiia I Aviakosmicheskaia Meditsina* 21:25-31.
44. Vico, L., Chappard, D., Alexandre, C., Palle, S., Minaire, P., Riffat, G., Novikov, V.E., and Bakulin, A.V. 1987. Effects of weightlessness on osseous tissue of the rat after a space flight of 5 days (Cosmos 1514). *Journal of Physiology (Paris)* 82:1-11.
45. Platts, S.H., Martin, D.S., Stenger, M.B., Perez, S.A., Ribeiro, L.C., Summers, R., and Meck, J.V. 2009. Cardiovascular adaptations to long-duration head-down bed rest. *Aviation, Space, and Environmental Medicine* 80:A29-A36.
46. Reschke, M.F., Bloomberg, J.J., Paloski, W.H., Mulavara, A.P., Feiveson, A.H., and Harm, D.L. 2009. Postural reflexes, balance control, and functional mobility with long-duration head-down bed rest. *Aviation, Space, and Environmental Medicine* 80:A45-A54.
47. Wronski, T.J., and Morey, E.R. 1983. Effect of spaceflight on periosteal bone formation in rats. *American Journal of Physiology* 244:R305-R309.

48. Cann, C.E., and Adachi, R.R. 1983. Bone resorption and mineral excretion in rats during spaceflight. *American Journal of Physiology* 244, R327-R331.
49. Cavolina, J.M., Evans, G.L., Harris, S.A., Zhang, M., Westerlind, K.C., and Turner, R.T. 1997. The effects of orbital spaceflight on bone histomorphometry and messenger ribonucleic acid levels for bone matrix proteins and skeletal signaling peptides in ovariectomized growing rats. *Endocrinology* 138(4):1567-1576.
50. Turner, R.T. 2000. Invited review: What do we know about the effects of spaceflight on bone? *Journal of Applied Physiology* 89:840-847.
51. Cavanagh, P., Rice, A. Cavanagh, P., and Rice, A., eds. 2007. *Bone Loss During Spaceflight: Etiology, Countermeasures, and Implications for Bone Health on Earth,* 1st ed. Cleveland Clinic Press, Cleveland, Ohio.
52. Wronski, T.J., and Morey, E.R. 1983. Effect of spaceflight on periosteal bone formation in rats. *American Journal of Physiology* 244:R305-R309.
53. Wronski, T.J., Morey-Holton, E.R., Doty, S.B., Maese, A.C., and Walsh, C.C. 1987. Histomorphometric analysis of rat skeleton following spaceflight. *American Journal of Physiology* 252:R252-R255.
54. Sibonga, J.D., Zhang, M., Evans, G.L., Westerlind, K.C., Cavolina, J.M., Morey-Holton, E., and Turner, R.T. 2000. Effects of spaceflight and simulated weightlessness on longitudinal bone growth. *Bone* 27(4):535-540.
55. Vico, L., Novikov, V.E., Very, J.M., and Alexandre, C. 1991. Bone histomorphometric comparison of rat tibial metaphysis after 7-day tail suspension vs. 7-day spaceflight. *Aviation, Space, and Environmental Medicine* 62:26-31.
56. Hughes-Fulford, M., Tjandrawinata, R., Fitzgerald, J., Gasuad, K., and Gilbertson, V. 1998. Effects of microgravity on osteoblast growth. *Gravitational and Space Biology Bulletin* 11:51-60.
57. Grindeland, R., Hymer, W.C., Framington, M., Fast, T., Hayes, C., Mottler, K., Patil, L., and Vasqyes, M. 1987. Changes in pituitary growth hormone cells prepared from rats flown on Spacelab 3. *American Journal of Physiology* 252:R209-R215.
58. Hughes-Fulford, M., and Scheld, H.W. 1989. Thin film bioreactors in space. *Advances in Space Research* 9(11):111-117.
59. Tamma, R., Colaianni, G., Camerino, C., Di Benedetto, A., Greco, G., Strippoli, M., Vergari, R., Grano, A., Mancini, L., Mori, G., Colucci, S., Grano, M., and Zallone, A. 2009. Microgravity during spaceflight directly affects in vitro osteoclastogenesis and bone resorption. *FASEB Journal* 23:2549-2554.
60. Hughes-Fulford, M., Tjandrawinata, R., Fitzgerald, J., Gasuad, K., and Gilbertson, V. 1998. Effects of microgravity on osteoblast growth. *Gravitational and Space Biology Bulletin* 11:51-60.
61. Tamma, R., Colaianni, G., Camerino, C., Di Benedetto, A., Greco, G., Strippoli, M., Vergari, R., Grano, A., Mancini, L., Mori, G., Colucci, S., Grano, M., and Zallone, A. 2009. Microgravity during spaceflight directly affects in vitro osteoclastogenesis and bone resorption. *The FASEB Journal* 23:2549-2554.
62. Hughes-Fulford, M. 2004. Lessons learned about spaceflight and cell biology experiments. *Journal of Gravitational Physiology* 11:105-109.
63. Cavanagh, P.R., Licata, A.A., and Rice, A.J. 2005. Exercise and pharmacological countermeasures for bone loss during long-duration space flight. *Gravitational and Space Biology Bulletin* 18:39-58.
64. Grigoriev, A.I., Kozlovskaya, I.B., and Potapov, A.N. 2002. Goals of biomedical support of a mission to Mars and possible approaches to achieving them. *Aviation Space and Environmental Medicine* 73(4):379-384.
65. Kozlovskaya, I.B., and Egorov, A.D. 2003. Some approaches to medical support for martian expedition. *Acta Astronautica* 53(4-10):269-275.
66. Lang, T.F., LeBlanc, A.D., Evans, H.J., Lu, Y., Genant, H., and Yu, A. 2004. Cortical and trabecular bone mineral loss from the spine and hip in long-duration spaceflight. *Journal of Bone and Mineral Research* 19:1006-1012.
67. Cavanagh, P.R., Licata, A.A., and Rice, A.J. 2005. Exercise and pharmacological countermeasures for bone loss during long-duration space flight. *Gravitational and Space Biology Bulletin* 18:39-58.
68. Rubin, C., Adler, B., Qin, Y., and Judex, S. 2007. Mechanical signals may provide an effective countermeasure to bone and muscle loss during long-term spaceflight. Pp. 175-187 in *Bone Loss During Spaceflight: Etiology, Countermeasures, and Implications for Bone Health on Earth* (P. Cavanagh and A. Rice, eds.). Cleveland Clinic Press, Cleveland, Ohio.
69. Gopalakrishnan, R., Genc, K.O., Rice, A.J., Lee, S.M.C., Evans, H.J., Maender, C.C., Ilaslan, H., and Cavanagh, P.R. 2010. Muscle volume, strength, endurance, and exercise loads during 6-month missions in space. *Aviation, Space, and Environmental Medicine* 81(2):91-102.
70. Trappe, S., Costill, D., Gallagher, P., Creer, A., Peters, J.R., Evans, H., Riley, D.A., and Fitts, R.H. 2009. Exercise in space: Human skeletal muscle after 6 months aboard the International Space Station. *Journal of Applied Physiology* 106:1159-1168.
71. Cavanagh, P.R., Gopalakrishnan, R., Rice, A.J., Genc, K.O., Maender, C.C., Nystrom, P.G., Johnson, M.J., Kuklis, M.M., and Humphreys, B.T. 2009. An ambulatory biomechanical data collection system for use in space: Design and validation. *Aviation, Space, and Environmental Medicine* 80:870-881.

72. LeBlanc, A.D., Spector, E.R., Evans, H.J., and Sibonga, J.D. 2007. Skeletal responses to space flight and the bed rest analog: A review. *Journal of Musculoskeletal and Neuronal Interactions* 7:33-47.
73. Crucian, B.E., Stowe, R.P., Mehta, S.K., Yetman, D.L., Leal, M.J., Quiriarte, H.D., Pierson, D.L., and Sams, C.F. 2009. Immune status, latent viral reactivation, and stress during long-duration head-down bed rest. *Aviation, Space, and Environmental Medicine* 80:A37-A44.
74. Inniss, A.M., Rice, B.L., and Smith, S.M. 2009. Dietary support of long-duration head-down bed rest. *Aviation, Space, and Environmental Medicine* 80:A9-A14.
75. Meck, J.V., Dreyer, S.A., and Warren, L.E. 2009. Long-duration head-down bed rest: Project overview, vital signs, and fluid balance. *Aviation, Space, and Environmental Medicine* 80:A1-A8.
76. Platts, S.H., Martin, D.S., Stenger, M.B., Perez, S.A., Ribeiro, L.C., Summers, R., and Meck, J.V. 2009. Cardiovascular adaptations to long-duration head-down bed rest. *Aviation, Space, and Environmental Medicine* 80:A29-A36.
77. Reschke, M.F., Bloomberg, J.J., Paloski, W.H., Mulavara, A.P., Feiveson, A.H., and Harm, D.L. 2009. Postural reflexes, balance control, and functional mobility with long-duration head-down bed rest. *Aviation, Space, and Environmental Medicine* 80:A45-A54.
78. Seaton, K.A., Bowie, K.E., and Sipes, W.A. 2009. Behavioral and psychological issues in long-duration head-down bed rest. *Aviation, Space, and Environmental Medicine* 80:A55-A61.
79. Seaton, K.A., Slack, K.J., Sipes, W.A., and Bowie, K.E. 2009. Cognitive functioning in long-duration head-down bed rest. *Aviation, Space, and Environmental Medicine* 80:A62-A65.
80. Spector, E.R., Smith, S.M., and Sibonga, J.D. 2009. Skeletal effects of long-duration head-down bed rest. *Aviation, Space, and Environmental Medicine* 80:A23-A28.
81. Zwart, S.R., Oliver, S.A.M., Fesperman, J.V., Kala, G., Krauhs, J., Ericson, K., and Smith, S.M. 2009. Nutritional status assessment before, during, and after long-duration head-down bed rest. *Aviation, Space, and Environmental Medicine* 80:A15-A22.
82. Vernikos, J., Ludwig, D.A., Ertil, A., Wade, C.E., Keil, L., and O' Hara, D.B. 1996. Effect of standing or walking on physiological changes induced by head down bed rest: Implications for space flight. *Aviation, Space, and Environmental Medicine* 67(11):1069-1079.
83. Smith, S.M., Davis-Street, J.E., Fesperman, J.V., Calkins, D.S., Bawa, M., Macias, B.R., Meyer, R.S., and Hargens, A.R. 2003. Evaluation of treadmill exercise in a lower body negative pressure chamber as a countermeasure for weightlessness-induced bone loss: A bed rest study with identical twins. *Journal of Bone and Mineral Research* 18:2223-2230.
84. Smith, S.M., Zwart, S.R., Heer, M., Lee, S.M., Baecker, N., Meuche, S., Macias, B.R., Shackelford, L.C., Schneider, S., and Hargens, A.R. 2008. WISE-2005: Supine treadmill exercise within lower body negative pressure and flywheel resistive exercise as a countermeasure to bed rest-induced bone loss in women during 60-day simulated microgravity. *Bone* 42:572-581.
85. Zwart, S.R., Hargens, A.R., Lee, S.M., Macias, B.R., Watenpaugh, D.E., Tse, K., and Smith, S.M. 2007. Lower body negative pressure treadmill exercise as a countermeasure for bed rest-induced bone loss in female identical twins. *Bone* 40:529-537.
86. Smith, S.M., Zwart, S.R., Heer, M.A., Baecker, N., Evan, H.J., Feiveson, L., Shackelford, C., and LeBlanc, A.D. 2009. Effects of artificial gravity during bed rest and bone metabolism in humans. *Journal of Applied Physiology* 107:47-53.
87. Smith, S.M., Zwart, S.R., Block, G., Rice, B.L., and Davis-Street, J.E. 2005. The nutritional status of astronauts is altered after long-term space flight aboard the International Space Station. *Journal of Nutrition* 135(3):437-443.
88. Zwart, S.R., Kloeris, V.L., Perchonok, M.H., Braby, L., and Smith, S.M. 2009. Assessment of nutrient stability in foods from the space food system after long-duration spaceflight on the ISS. *Journal of Food Science* 74(7):H209-H217.
89. Heer, M. 2002. Nutritional interventions related to bone turnover in European space missions and simulation models. *Nutrition* 18(10):853-856
90. Smith, S.M., Zwart, S.R., Block, G., Rice, B.L., and Davis-Street, J.E. 2005. The nutritional status of astronauts is altered after long-term space flight aboard the International Space Station. *Journal of Nutrition* 135(3):437-443.
91. Heer, M. 2002. Nutritional interventions related to bone turnover in European space missions and simulation models. *Nutrition* 18(10):853-856.
92. Baecker, N., Frings-Meuthen, P., Smith, S.M., and Heer, M. 2010. Short-term high dietary calcium intake during bedrest has no effect on markers of bone turnover in healthy men. *Nutrition* 26(5):522-527.
93. Black, D.M., Delmas, P.D., Eastell, R., Reid, I.R., Boonen, S., Cauley, J.A., Cosman, F., Lakatos, P., Leung, P.C., Man, Z., Mautalen, C., et al. 2007. HORIZON Pivotal Fracture Trial Once-yearly zoledronic acid for treatment of postmenopausal osteoporosis. *New England Journal of Medicine* 356:1809-1822.

94. Watanabe, Y., Oshina, H., Mizuno, K., Sekiguchi, C., Fukunaga, M., Kohri, K., Rittweger, J., Felsenberg, D., Matsumoto, T., and Nakamura, T. 2004. Intravenous pamidronate prevents femoral bone loss and renal stone during 90 day bed rest. *Journal of Bone and Mineral Research* 19:1771-1778.
95. Turner, R.T., Evans, G.L., Lotinun, S., Lapke, P.D., Iwaniec, U.T., and Morey-Holton, E. 2007. Dose-response effects of intermittent PTH on cancellous bone in hindlimb unloaded rats. *Journal of Bone and Mineral Research* 22(1):64-71.
96. American Dental Association Council on Scientific Affairs. 2006. Dental management of patients receiving oral bisphosphonate therapy: Expert panel recommendations. *Journal of the American Dental Association* 137:1144-1150.
97. Sellmeyer, D.E. 2010. Atypical fractures as a potential complication of long-term bisphosphonate therapy. *Journal of the American Medical Association* 304(13):1480-1484.
98. Cummings, S.R., San Martin, J., McClung, M.R., Siris, E.S., Eastell, R., Reid, I.R., Delmas, P., Zoog, H.B., Austin, M., Wang, A., Kutilek, S., Adami, S., Zanchetta, J., Libanati, C., Siddhanti, S., and Christiansen, C. 2009. FREEDOM Trial Denosumab for prevention of fractures in postmenopausal women with osteoporosis. *New England Journal of Medicine* 361:756-765.
99. File, E., and Deal, C. 2009. Clinical update on teriparatide. *Current Rheumatology Reports* 11(3):169-176.
100. Aronsohn, A., Brazeau, G., and Hughes, J. 1999. The impact of space travel on dosage form design and use. *Journal of Gravitational Physiology* 6:P169-P172.
101. Czarnik, T.R., and Vernikos, J. 1999. Physiological changes in spaceflight that may affect drug action. *Journal of Gravitational Physiology* 6:P161-P164.
102. Ferguson, V., Paietta, R., Stodieck, L., Hanson, A., Young, M., Bateman, T., Lemus, M., Kostenuik, P., Jiao, E., Zhou, X., Lu, J., Simonet, W., Lacey, D., and Han, H. 2009. Inhibiting myostatin prevents microgravity associated bone loss in mice. American Society for Bone and Mineral Research 31st Annual Meeting, Denver, Colo., September 15. Abstract A09003077. Available at http://www.asbmr.org/Meetings/AnnualMeeting/AbstractDetail.aspx?aid=9901d87a-8a25-4d83-a01e-f891cff99030.
103. Elkasrawy, M.N., and Hamrick, M.W. 2010. Myostatin (GDF-8) as a key factor linking muscle mass and bone structure. *Journal of Musculoskeletal and Neuronal Interactions* 10(1):56-63.
104. Li, X., Ominsky, M.S., Niu, Q.T., Sun, N., Daugherty, B., D'Agostin, D., Kurahara, C., Gao, Y., Cao, J., Gong, J., Asuncion, F., et al. 2008. Targeted deletion of the sclerostin gene in mice results in increased bone formation and bone strength. *Journal of Bone and Mineral Research* 23:860-869.
105. Lin, C., Jiang, X., Dai, Z., Guo, X., Weng, T., Wang, J., Li, Y., Feng, G., Gao, X., and He, L. 2009. Sclerostin mediates bone response to mechanical unloading through antagonizing Wnt/beta-catenin signaling. *Journal of Bone and Mineral Research* 24:1651-1661.
106. Eddleston, A., Marenzana, M., Moore, A.R., Stephens, P., Muzylak, M., Marshall, D., and Robinson, M.K. 2009. A short treatment with an antibody to sclerostin can inhibit bone loss in an ongoing model of colitis. *Journal of Bone and Mineral Research* 24:1662-1671.
107. Robling, A.G., and Turner, C.H. 2009. Mechanical signaling for bone modeling and remodeling. *Critical Reviews in Eukaryotic Gene Expression* 19:319-338.
108. Kirchen, M.E., O'Connor, K.M., Gruber, H.E., Sweeney, J.R., Fras, I.A., Stover, S.J., Sarmiento, A., and Marshall, G.J. 1995. Effects of microgravity on bone healing in a rat fibular osteotomy model. *Clinical Orthopaedics and Related Research* Sep(318):231-242.
109. Midura, R.J., Su, X., and Androjna, C. 2006. A simulated weightlessness state diminishes cortical bone healing responses. *Journal of Musculoskeletal and Neuronal Interactions* 6:327-328.
110. Bandstra, E.R., Thompson, R.W., Nelson, G.A., Willey, J.S., Judex, S., Cairns, M.A., Benton, E.R., Vazquez, M.E., Carson, J.A., and Bateman, T.A. 2009. Musculoskeletal changes in mice from 20-50 cGy of simulated galactic cosmic rays. *Radiation Research* 172:21-29.
111. Compston, J. 2009. Clinical and therapeutic aspects of osteoporosis. *European Journal of Radiology* 71(3):388-391.
112. Kapur, S., Amoui, M., Kesavan, C., Wang, X., Mohan, S., Baylink, D.J., and Lau, K.H. 2010. Leptin receptor (LEPR) is a negative modulator of bone mechanosensitivity and genetic variations in LEPR may contribute to the differential osteogenic response to mechanical stimulation in the C57Bl/6j and C3H/HeJ pair of mouse strains. *Journal of Biological Chemistry* 285(48):37607-37618.
113. Turner, R.T., Evans, G.L., Cavolina, J.M., Halloran, B., and Morey-Holton, E. 1998. Programmed administration of parathyroid hormone increases bone formation and reduces bone loss in hindlimb-unloaded ovariectomized rats. *Endocrinology* 139(10):4086-4091.
114. Hefferan, T.E., Evans, G.L., Lotinun, S., Zhang, M., Morey-Holton, E., and Turner, R.T. 2003. Effect of gender on bone turnover in adult rats during simulated weightlessness. *Journal of Applied Physiology* 95(5):1775-1780.

115. Lerner, U.H., and Persson, E. 2008. Osteotropic effects by the neuropeptides calcitonin gene-related peptide, substance P and vasoactive intestinal peptide. *Journal of Musculoskeletal and Neuronal Interactions* 8(2):154-165.
116. Datta, H.K., Ng, W.F., Walker, J.A., Tuck, S.P., and Varanasi, S.S. 2008. The cell biology of bone metabolism. *Journal of Clinical Pathology* 61(5):577-587.
117. Zhang, M., and Turner, R.T. 1998. The effects of spaceflight on mRNA levels for cytokines in proximal tibia of ovariectomized rats. *Aviation, Space, and Environmental Medicine* 69(7):626-629.
118. Cavolina, J.M., Evans, G.L., Harris, S.A., Zhang, M., Westerlind, K.C., and Turner, R.T. 1997. The effects of orbital spaceflight on bone histomorphometry and messenger ribonucleic acid levels for bone matrix proteins and skeletal signaling peptides in ovariectomized growing rats. *Endocrinology* 138(4):1567-1576.
119. Westerlind, K.C., and Turner, R.T. 1995. The skeletal effects of spaceflight in growing rats: Tissue-specific alterations in mRNA levels for TGF-beta. *Journal of Bone and Mineral Research* 10(6):843-848.
120. LeBlanc, A., Lin, C., Shackelford, L., Sinitsyn, V., Evans, H., Belichenko, O., Schenkman, B., Kozlovskaya, I., Oganov, V., Bakulin, A., Hedrick, T., and Feeback, D. 2000. Muscle volume, MRI relaxation times (T2), and body composition after spaceflight. *Journal of Applied Physiology* 89:2158-2164.
121. Hefferan, T.E., Evans, G.L., Lotinun, S., Zhang, M., Morey-Holton, E., and Turner, R.T. 2003. Effect of gender on bone turnover in adult rats during simulated weightlessness. *Journal of Applied Physiology* 95:1775-1780.
122. McCarthy, I.D. 2005. Fluid shifts due to microgravity and their effects on bone: A review of current knowledge. *Annals of Biomedical Engineering* 33:95-103.
123. Zhang, P., Malacinski, G.M., and Yokota, H. 2008. Joint loading modality: Its application to bone formation and fracture healing. *British Journal of Sports Medicine* 42:556-560.
124. Whitson, P.A., Pietrzyk, R.A., and Pak, C.Y. 1997. Renal stone risk assessment during space shuttle flights. *Journal of Urology* 158:2305-2310.
125. Zerwekh, J.E. 2002. Nutrition and renal stone disease in space. *Nutrition* 18(10):857-863.
126. Smith, S.M., Wastney, M.E., O'Brien, K.O., Morukov, B.V., Larina, I.M., Abrams, S.A., Davis-Street, J.E., Oganov, V., and Shackelford, L.C. 2005. Bone markers, calcium metabolism, and calcium kinetics during extended-duration space flight on the Mir space station. *Journal of Bone and Mineral Research* 20:208-218.
127. Whitson, P.A., Pietrzyk, R.A., and Pak, C.Y. 1997. Renal stone risk assessment during space shuttle flights. *Journal of Urology* 158:2305-2310.
128. Smith, S.M., Wastney, M.E., O'Brien, K.O., Morukov, B.V., Larina, I.M., Abrams, S.A., Davis-Street, J.E., Oganov, V., and Shackelford, L.C. 2005. Bone markers, calcium metabolism, and calcium kinetics during extended-duration space flight on the Mir space station. *Journal of Bone and Mineral Research* 20:208-218.
129. Whitson, P.A., Pietrzyk, R.A., and Pak, C.Y. 1997. Renal stone risk assessment during space shuttle flights. *Journal of Urology* 158:2305-2310.
130. Whitson, P.A., Pietrzyk, R.A., and Pak, C.Y. 1997. Renal stone risk assessment during space shuttle flights. *Journal of Urology* 158:2305-2310.
131. Zerwekh, J.E., Ruml, L.A., Gottschalk, F., and Pak, C.Y. 1998. The effects of twelve weeks of bed rest on bone histology, biochemical markers of bone turnover, and calcium homeostasis in eleven normal subjects. *Journal of Bone and Mineral Research* 13:1594-1601.
132. Watanabe, Y., Ohshima, H., Mizuno, K., Sekiguchi, C., Fukunaga, M., Kohri, K., Rittweger, J., Felsenberg, D., Matsumoto, T., and Nakamura, T. 2004. Intravenous pamidronate prevents femoral bone loss and renal stone formation during 90-day bed rest. *Journal of Bone and Mineral Research* 19:1771-1778.
133. Smith, S.M., Zwart, S.R., Block, G., Rice, B.L., and Davis-Street, J.E. 2005. The nutritional status of astronauts is altered after long-term space flight aboard the International Space Station. *Journal of Nutrition* 135(3):437-443.
134. Baek, K., Barlow, A.A., Allen, M.R., and Bloomfield, S.A. 2008. Food restriction and simulated microgravity: Effects on bone and serum leptin. *Journal of Applied Physiology* 104(4):1086-1093.
135. Biolo, G., Ciocchi, B., Stulle, M., Bosutti, A., Barazzoni, R., Zanetti, M., Antonione, R., Lebenstedt, M., Platen, P., Heer, M., and Guarnieri, G. 2007. Calorie restriction accelerates the catabolism of lean body mass during 2 wk of bed rest. *American Journal of Clinical Nutrition* 86(2):366-372.
136. Li, X., Ominsky, M.S., Niu, Q.T., Sun, N., Daugherty, B., D'Agostin, D., Kurahara, C., Gao, Y., Cao, J., Gong, J., Asuncion, F., Barrero, M., Warmington, K., Dwyer, D., Stolina, M., Morony, S., Sarosi, I., Kostenuik, P.J., Lacey, D.L., Simonet, W.S., Ke, H.Z., and Paszty, C. 2008. Targeted deletion of the sclerostin gene in mice results in increased bone formation and bone strength. *Journal of Bone and Mineral Research* 23:860-869.
137. Suva, L. 2009. Commentary: Sclerostin and the unloading of bone. *Journal of Bone and Mineral Research* 24:1649-1650.

138. National Human Genome Research Institute. Newsroom. 2005. NIH news researchers to gain wider access to knockout mice. October 25. Available at http://www.genome.gov/17015131.
139. Li, X., Ominsky, M.S., Niu, Q.T., Sun, N., Daugherty, B., D'Agostin, D., Kurahara, C., Gao, Y., Cao, J., Gong, J., Asuncion, F., et al. 2008. Targeted deletion of the sclerostin gene in mice results in increased bone formation and bone strength. *Journal of Bone and Mineral Research* 23:860-869.
140. Suva, L. 2009. Commentary: Sclerostin and the unloading of bone. *Journal of Bone and Mineral Research* 24:1649-1650.
141. National Human Genome Research Institute. Newsroom. 2005. NIH news researchers to gain wider access to knockout mice. October 25. Available at http://www.genome.gov/17015131.
142. Han, H.Q., Stodieck, L., Ferguson, V., Zhou, X., Lu, J., Hanson, A., Young, M., Jiao, E., Kwak, K., Rosenfeld, R., Boone, T., et al. 2008. Pharmacological myostatin antagonism effectively mitigates spaceflight-induced muscle atrophy in mice. 24th Annual American Society for Gravitational and Space Biology Joint Life in Space for Life on Earth, June 25. Session 3: Muscle and Metabolism Physiology, Abstract. Available at http://www.congrex.nl/08a09/Sessions/25-06%20Session%203a.htm.
143. Kearns, A.E., Khosla, S., and Kostenuik, P.J. 2008. Receptor activator of nuclear factor kappaB ligand and osteoprotegerin regulation of bone remodeling in health and disease. *Endocrine Reviews* 29:155-192.
144. Ominsky, M.S., Stolina, M., Li, X., Corbin, T.J., Asuncion, F.J., Barrero, M., Niu, Q.T., Dwyer, D., Adamu, S., Warmington, K.S., Grisanti, M., et al. 2009. One year of transgenic overexpression of osteoprotegerin in rats suppressed bone resorption and increased vertebral bone volume, density, and strength. *Journal of Bone and Mineral Research* 24:1234-1246.
145. Morey-Holton, E., Globus, R.K., Kaplansky, A., and Durnova, G. 2005. The hindlimb unloading rat model: Literature overview, technique update and comparison with space flight data. *Advances in Space Biology and Medicine* 10:7-40.
146. Iwaniec, U.T., Yuan, D., Power, R.A., and Wronski, T.J. 2006. Strain-dependent variations in the response of cancellous bone to ovariectomy in mice. *Journal of Bone and Mineral Research* 21(7):1068-1074.
147. Jacob, H.J., Lazar, J., Dwinell, M.R., Moreno, C., and Geurts, A.M. 2010. Gene targeting in the rat: Advances and opportunities. *Trends in Genetics* September 23.
148. Spector, E.R., Smith, S.M., and Sibonga, J.D. 2009. Skeletal effects of long-duration head-down bed rest. *Aviation, Space, and Environmental Medicine* 80:A23-A28.
149. Pavy-Le Traon, A., Heer, M., Narici, M.V., Rittweger, J., and Vernikos, J. 2007. From space to Earth: Advances in human physiology from 20 years of bed rest studies (1986-2006). *European Journal of Applied Physiology* 101:143-194.
150. Thomsen, J.S., Morukov, B.V., Vico, L., Alexandre, C., Saparin, P.I., and Gowin, W. 2005. Cancellous bone structure of iliac crest biopsies following 370 days of head-down bed rest. *Aviation, Space, and Environmental Medicine* 76:915-922.
151. Zerwekh, J.E., Ruml, L.A., Gottschalk, F., and Pak, C.Y. 1998. The effects of twelve weeks of bed rest on bone histology, biochemical markers of bone turnover, and calcium homeostasis in eleven normal subjects. *Journal of Bone and Mineral Research* 13:1594-1601.
152. Zerwekh, J.E., Ruml, L.A., Gottschalk, F., and Pak, C.Y. 1998. The effects of twelve weeks of bed rest on bone histology, biochemical markers of bone turnover, and calcium homeostasis in eleven normal subjects. *Journal of Bone and Mineral Research* 13:1594-1601.
153. See NASA Research Opportunities at the NASA Solicitation and Proposal Integrated Review and Evaluation System website, available at http://nspires.nasaprs.com/external/.
154. LeBlanc, A.D., Driscol, T.B., Shackelford, L.C., Evans, H.J., Rianon, N.J., Smith, S.M., Feeback, D.L., and Lai, D. 2002. Alendronate as an effective countermeasure to disuse induced bone loss. *Journal of Musculoskeletal and Neuronal Interactions* 2(4):335-343.
155. Sibonga, J.D., Evans, H.J., Sung, H.G., Spector, E.R., Lang, T.F., Oganov, V.S., Bakulin, A.V., Shackelford, L.C., and LeBlanc, A.D. 2007. Recovery of spaceflight-induced bone loss: Bone mineral density after long-duration missions as fitted with an exponential function. *Bone* 41:973-978.
156. Lang, T.F., Leblanc, A.D., Evans, H.J., and Lu, Y. 2006. Adaptation of the proximal femur to skeletal reloading after long-duration spaceflight. *Journal of Bone and Mineral Research* 21:1224-1230.
157. Sibonga, J.D., Cavanagh, P.R., Lang, T.F., LeBlanc, A.D., Schneider, V., Shackelford, L.C., Smith, S.M., and Vico, L. 2007. Adaptation of the skeletal system during long-duration spaceflight. *Clinical Reviews in Bone and Mineral Metabolism* 5(4):249-261.
158. Loehr, J.A., Lee, S.M., English, K.L., Sibonga, J., Smith, S.M., Spiering, B.A., and Hagan, R.D. 2010. Musculoskeletal adaptations to training with the advanced resistive exercise device. *Medicine and Science in Sports and Exercise* May 13.

159. Genc, K.O., Gopalakrishnan, R., Kuklis, M.M., Maender, C.C., Rice, A.J., Bowersox, K.D., and Cavanagh, P.R. 2010. Foot forces during exercise on the International Space Station. *Journal of Biomechanics* 43(15):3020-3027.
160. Cavanagh, P.R., Genc, K.O., Gopalakrishnan, R., Kuklis, M.M., Maender, C.C., and Rice, A.J. 2010. Foot forces during typical days on the international space station. *Journal of Biomechanics* 43(11):2182-2188.
161. Sellmeyer, D.E. 2010. Atypical fractures as a potential complication of long-term bisphosphonate therapy. *Journal of the American Medical Association* 304(13):1480-1484.
162. American Dental Association Council on Scientific Affairs. 2006. Dental management of patients receiving oral bisphosphonate therapy: Expert panel recommendations. *Journal of the American Dental Association* 137:1144-1150.
163. Delmas, P.D., Munoz, F., Black, D.M., Cosman, F., Boonen, S., Watts, N.B., Kendler, D., Eriksen, E.F., Mesenbrink, P.G., and Eastell, R. 2009. HORIZON-PFT Research Group Effects of yearly zoledronic acid 5 mg on bone turnover markers and relation of PINP with fracture reduction in postmenopausal women with osteoporosis. *Journal of Bone and Mineral Research* 24:1544-1551.
164. Adams, G.R., Haddad, F., and Baldwin, K.M. 2003. Gravity plays an important role in muscle development and the differentiation of contractile protein phenotype. Pp. 111-122 in *The Neurolab Mission: Neuroscience Research in Space* (J.C. Buckley and J.L. Homack, eds.). NASA SP 2003-535. NASA, Washington, D.C.
165. NASA. 2010. Critical Path Roadmap Document. January. Available at http://bioastroroadmap.nasa.gov/index.jsp.
166. Adams, G.R., Caiozzo, V.J., and Baldwin, K.M. 2003. Skeletal muscle unweighting: Spaceflight and ground-based models. *Journal of Applied Physiology* 95:2185-2201.
167. Fitts, R.H., Riley, D.A., and Widrick, J. 2000. Physiology of a microgravity environment. Invited review: Microgravity and skeletal muscle. *Journal of Applied Physiology* 89:823-839.
168. Trappe, A., Costill, D., Gallagher, P., Creer, A., Peters, J.R., Evans, H., Riley, D.A., and Fitts, R.H. 2009. Exercise in space: Human skeletal muscle after 6 months aboard the International Space Station. *Journal of Applied Physiology* 106:1159-1168.
169. Trappe, A., Costill, D., Gallagher, P., Creer, A., Peters, J.R., Evans, H., Riley, D.A., and Fitts, R.H. 2009. Exercise in space: Human skeletal muscle after 6 months aboard the International Space Station. *Journal of Applied Physiology* 106:1159-1168.
170. National Research Council. 1998. *A Strategy for Research in Space Biology and Medicine in the New Century.* National Academy Press, Washington, D.C.
171. Roy, R.R., Baldwin, K.M, and Edgerton, V.R. 1996. Response of the neuromuscular unit to space flight: What have we learned from the rat model. *Exercise and Sport Science Reviews* 24:399-420.
172. Thomason, D.B., Herrick, R.E., Surdyka, D., and Baldwin, K.M. 1987. Time course of soleus muscle myosin expression during hindlimb suspension and subsequent recovery. *Journal of Applied Physiology* 63:130-137.
173. Fitts, R.H., Riley, D.A., and Widrick, J. 2000. Physiology of a microgravity environment. Invited Review: Microgravity and skeletal muscle. *Journal of Applied Physiology* 89:823-839.
174. Shackelford, L.C., LeBlanc, A.D., and Driscoll, T.B. 2004. Resistance exercise as a countermeasure to disuse-induced bone loss. *Journal of Applied Physiology* 97:119-127.
175. Thomason, D.B., Herrick, R.E., Surdyka, D., and Baldwin, K.M. 1987. Time course of soleus muscle myosin expression during hindlimb suspension and subsequent recovery. *Journal of Applied Physiology* 63:130-137.
176. Roy, R.R., Baldwin, K.M., and Edgerton, V.R. 1996. Response of the neuromuscular unit to space flight: What have we learned from the rat model. *Exercise and Sport Science Reviews* 24:399-420.
177. Fitts, R.H., Riley, D.A., and Widrick, J. 2000. Physiology of a microgravity environment. Invited review: Microgravity and skeletal muscle. *Journal of Applied Physiology* 89:823-839.
178. Giger, J.M., Bodell, P.W., Zeng, M., Baldwin, K.M., and Haddad, F. 2009. Rapid muscle atrophy response to unloading: Pretranslational processes involving MHC and actin. *Journal of Applied Physiology* 107:1204-1212.
179. Riley, D.A., Ellis, S., Slocum, G.R., Sedlak, F.R., Bain, J.L.W., Krippendorf, B.B., Lehman, C.T., Macias, M.Y., Thompson, J.L., Vijayan, K., and DeBruin, J.A. 1996. In-flight and post flight changes in skeletal muscles of SLS-1 and SLS-2 spaceflown rats. *Journal of Applied Physiology* 81:133-144.
180. Roy, R.R., Baldwin, K.M., and Edgerton, V.R. 1996. Response of the neuromuscular unit to space flight: What have we learned from the rat model. *Exercise and Sport Science Reviews* 24:399-420.
181. Adams, G.R., Caiozzo, V.J., and Baldwin, K.M. 2003. Skeletal muscle unweighting: Spaceflight and ground-based models. *Journal of Applied Physiology* 95:2185-2201.
182. Fitts, R.H., Riley, D.A., and Widrick, J. 2000. Physiology of a microgravity environment. Invited review: Microgravity and skeletal muscle. *Journal of Applied Physiology* 89:823-839.

183. LeBlanc, A., Schneider, V., and Shackelford, L. 2000. Bone mineral and lean tissue loss after long duration spaceflight. *Journal of Musculoskeletal and Neuronal Interactions* 1:157-160.
184. Trappe, A., Costill, D., Gallagher, P., Creer, A., Peters, J.R., Evans, H., Riley, D.A., and Fitts, R.H. 2009. Exercise in space: Human skeletal muscle after 6 months aboard the International Space Station. *Journal of Applied Physiology* 106:1159-1168.
185. Fitts, R.H., Riley, D.A., and Widrick, J. 2000. Physiology of a microgravity environment. Invited review: Microgravity and skeletal muscle. *Journal of Applied Physiology* 89:823-839.
186. Trappe, A., Costill, D., Gallagher, P., Creer, A., Peters, J.R., Evans, H., Riley, D.A., and Fitts, R.H. 2009. Exercise in space: Human skeletal muscle after 6 months aboard the International Space Station. *Journal of Applied Physiology* 106:1159-1168.
187. Baldwin, K.M., Herrick, R.E., Ilyina-Kakeuva, E., and Oganov, V.S. 1990. Effects of zero gravity on myofibril content and isomyosin distribution in rodent skeletal muscle. *The FASEB Journal* 4:79-83.
188. Caiozzo, V.J., Baker, M.J., Herrick, R.E., Tao, M., and Baldwin, K.M. 1994. Effect of spaceflight on skeletal muscle: Mechanical properties and myosin isoform content of a slow antigravity muscle. *Journal of Applied Physiology* 76:1764-1773.
189. Caiozzo, V.J., Haddad, F., Baker, M.J., Herrick, R.E., Prietto, N., and Baldwin, K.M. 1996. Microgravity induced transformations of myosin isoforms and contractile properties of skeletal muscle. *Journal of Applied Physiology* 81:123-132.
190. Haddad, F., Herrick, R.E., Adams, G.R., and Baldwin, K.M. 1993. Myosin heavy chain expression in rodent skeletal muscle: Effects of exposure to zero gravity. *Journal of Applied Physiology* 75:2471-2477.
191. Thomason, D.B., Morrison, P.R., Organov, V., Ilyina-Kakueva, E., Booth, F.W., and Baldwin, K.M. 1992. Altered actin and myosin expression in muscle during exposure to microgravity. *Journal of Applied Physiology* 73(Suppl.):90S-93S.
192. Bottinelli, R., Schiaffino, S., and Reggiani, C. 1991. Force velocity relations and myosin heavy chain isoform compositions of skinned-fibres from rat skeletal muscle. *Journal of Physiology* 437:655-672.
193. Caiozzo, V.J., Baker, M.J., Herrick, R.E., Tao, M., and Baldwin, K.M. 1994. Effect of spaceflight on skeletal muscle: Mechanical properties and myosin isoform content of a slow antigravity muscle. *Journal of Applied Physiology* 76:1764-1773.
194. Caiozzo, V.J., Haddad, F., Baker, M.J., Herrick, R.E., Prietto, N., and Baldwin, K.M. 1996. Microgravity induced transformations of myosin isoforms and contractile properties of skeletal muscle. *Journal of Applied Physiology* 81:123-132.
195. Caiozzo, V.J., Baker, M.J., Herrick, R.E., Tao, M., and Baldwin, K.M. 1994. Effect of spaceflight on skeletal muscle: Mechanical properties and myosin isoform content of a slow antigravity muscle. *Journal of Applied Physiology* 76:1764-1773.
196. Caiozzo, V.J., Haddad, F., Baker, M.J., Herrick, R.E., Prietto, N., and Baldwin, K.M. 1996. Microgravity induced transformations of myosin isoforms and contractile properties of skeletal muscle. *Journal of Applied Physiology* 81:123-132.
197. Haddad, F., Adams, G.R., Bodell, P.W., and Baldwin, K.M. 2006. Isometric resistance exercise fails to counteract skeletal atrophy processes during the early states of unloading. *Journal of Applied Physiology* 100:433-441.
198. Roy, R.R., Baldwin, K.M., and Edgerton, V.R. 1996. Response of the neuromuscular unit to space flight: What have we learned from the rat model. *Exercise and Sport Science Reviews* 24:399-420.
199. Roy, R.R., Baldwin, K.M., and Edgerton, V.R. 1996. Response of the neuromuscular unit to space flight: What have we learned from the rat model. *Exercise and Sport Science Reviews* 24:399-420.
200. Caiozzo, V.J., Baker, M.J., Herrick, R.E., Tao, M., and Baldwin, K.M. 1994. Effect of spaceflight on skeletal muscle: Mechanical properties and myosin isoform content of a slow antigravity muscle. *Journal of Applied Physiology* 76:1764-1773.
201. Caiozzo, V.J., Haddad, F., Baker, M.J., Herrick, R.E., Prietto, N., and Baldwin, K.M. 1996. Microgravity induced transformations of myosin isoforms and contractile properties of skeletal muscle. *Journal of Applied Physiology* 81:123-132.
202. Haddad, F., Herrick, R.E., Adams, G.R., and Baldwin, K.M. 1993. Myosin heavy chain expression in rodent skeletal muscle: Effects of exposure to zero gravity. *Journal of Applied Physiology* 75:2471-2477.
203. Shackelford, L.C., LeBlanc, A.D., and Driscoll, T.B. 2004. Resistance exercise as a countermeasure to disuse-induced bone loss. *Journal of Applied Physiology* 97:119-127.
204. Edgerton, V.R., Shou, M.-Y., Klitgaard, H., Jiang, B., Bell, G., Harris, B., Saltin, B., Gollnick, P.D., Roy, R.R., Day, M.K., and Greenisen, M. 1995. Human fiber size and enzymatic properties after 5 and 11 days of spaceflight. *Journal of Applied Physiology* 78:1733-1739.
205. Fitts, R.H., Riley, D.A., and Widrick, J. 2000. Physiology of a microgravity environment. Invited review: Microgravity and skeletal muscle. *Journal of Applied Physiology* 89:823-839.

206. Gallagher, P., Trappe, S., Harber, M., Creer, A., Mazzatti, S., Trappe, T., Alkner, B., and Tesch, P. 2005. Effects of 84-days of bed rest and resistance exercise training on single muscle fibre myosin heavy chain distribution in human vastus lateralis and soleus muscles. *Acta Physiologica Scandinavica* 185:61-69.
207. Caiozzo, V.J., Baker, M.J., Herrick, R.E., Tao, M., and Baldwin, K.M. 1994. Effect of spaceflight on skeletal muscle: Mechanical properties and myosin isoform content of a slow antigravity muscle. *Journal of Applied Physiology* 76:1764-1773.
208. Caiozzo, V.J., Haddad, F., Baker, M.J., Herrick, R.E., Prietto, N., and Baldwin, K.M. 1996. Microgravity induced transformations of myosin isoforms and contractile properties of skeletal muscle. *Journal of Applied Physiology* 81:123-132.
209. Riley, D.A., Ellis, S., Slocum, G.R., Sedlak, F.R., Bain, J.L.W., Krippendorf, B.B., Lehman, C.T., Macias, M.Y., Thompson, J.L., Vijayan, K., and DeBruin, J.A. 1996. In-flight and post flight changes in skeletal muscles of SLS-1 and SLS-2 spaceflown rats. *Journal of Applied Physiology* 81:133-144.
210. Vijayan, K., Thompson, J.L., and Riley, D.A. 1998. Sarcomere lesion damage mainly in slow fibers of reloaded rat adductus longus muscles. *Journal of Applied Physiology* 85:1017-1023.
211. Riley, D.A., Ellis, S., Slocum, G.R., Sedlak, F.R., Bain, J.L.W., Krippendorf, B.B., Lehman, C.T., Macias, M.Y., Thompson, J.L., Vijayan, K., and DeBruin, J.A. 1996. In-flight and post flight changes in skeletal muscles of SLS-1 and SLS-2 spaceflown rats. *Journal of Applied Physiology* 81:133-144.
212. Adams, G.R., Caiozzo, V.J., and Baldwin, K.M. 2003. Skeletal muscle unweighting: Spaceflight and ground-based models. *Journal of Applied Physiology* 95:2185-2201.
213. Fitts, R.H., Riley, D.A., and Widrick, J. 2000. Physiology of a microgravity environment. Invited review: Microgravity and skeletal muscle. *Journal of Applied Physiology* 89:823-839.
214. Trappe, S., Creer, A., Minchev, K., Slivka, D., Louis, E., Luden, T., and Trappe, T. 2007. Human soleus single muscle fiber function with exercise or nutritional countermeasures during 60 days of bed rest. *American Journal of Physiology. Regulatory, Integrative and Comparative Physiology* 294:R939-R347.
215. Trappe, A., Costill, D., Gallagher, P., Creer, A., Peters, J.R., Evans, H., Riley, D.A., and Fitts, R.H. 2009. Exercise in space: Human skeletal muscle after 6 months aboard the International Space Station. *Journal of Applied Physiology* 106:1159-1168.
216. Adams, G.R., Caiozzo, V.J., and Baldwin, K.M. 2003. Skeletal muscle unweighting: Spaceflight and ground-based models. *Journal of Applied Physiology* 95:2185-2201.
217. Adams, G.R., Caiozzo, V.J., and Baldwin, K.M. 2003. Skeletal muscle unweighting: Spaceflight and ground-based models. *Journal of Applied Physiology* 95:2185-2201.
218. Fitts, R.H., Riley, D.A., and Widrick, J. 2000. Physiology of a microgravity environment. Invited review: Microgravity and skeletal muscle. *Journal of Applied Physiology* 89:823-839.
219. Trappe, S., Creer, A., Minchev, K., Slivka, D., Louis, E., Luden, T., and Trappe, T. 2007. Human soleus single muscle fiber function with exercise or nutritional countermeasures during 60 days of bed rest. *American Journal of Physiology. Regulatory, Integrative and Comparative Physiology* 294:R939-R347.
220. Trappe, A., Costill, D., Gallagher, P., Creer, A., Peters, J.R., Evans, H., Riley, D.A., and Fitts, R.H. 2009. Exercise in space: Human skeletal muscle after 6 months aboard the International Space Station. *Journal of Applied Physiology* 106:1159-1168.
221. Trappe, A., Costill, D., Gallagher, P., Creer, A., Peters, J.R., Evans, H., Riley, D.A., and Fitts, R.H. 2009. Exercise in space: Human skeletal muscle after 6 months aboard the International Space Station. *Journal of Applied Physiology* 106:1159-1168.
222. LeBlanc, A., Lin, C., Shackelford, L., Sinitsyn, V., Evans, H., Belichenko, O., Schenkman, B., Kozlovskaya, I., Oganov, V., Bakulin, A., Hedrick, T., and Feeback, D. 2000. Muscle volume, MRI relaxation times (T2), and body composition after spaceflight. *Journal of Applied Physiology* 89:2158-2164.
223. LeBlanc, A., Schneider, V., and Shackelford, L. 2000. Bone mineral and lean tissue loss after long duration spaceflight. *Journal of Musculoskeletal and Neuronal Interactions* 1:157-160.
224. Organov, V.S., Grigoriev, A.I., and Veronin, L.I. 1992. Bone mineral density in cosmonauts after flights lasting 4.5-6 months on the Mir orbital station. *Aviakosm Ekolog Med.* 26:20-24.
225. LeBlanc, A., Schneider, V., and Shackelford, L. 2000. Bone mineral and lean tissue loss after long duration spaceflight. *Journal of Musculoskeletal and Neuronal Interactions* 1:157-160.
226. Organov, V.S., Grigoriev, A.I., and Veronin, L.I. 1992. Bone mineral density in cosmonauts after flights lasting 4.5-6 months on the Mir orbital station. *Aviakosm Ekolog Med.* 26:20-24.
227. LeBlanc, A., Schneider, V., and Shackelford, L. 2000. Bone mineral and lean tissue loss after long duration spaceflight. *Journal of Musculoskeletal and Neuronal Interactions* 1:157-160.

228. Rittweger, J., Frost, H.M., Schiessl, H., Oshima, H., Alkner, B., Tesch, P., and Felsenberg, D. 2005. Muscle atrophy and bone loss after 90 days' bed rest and the effects of flywheel resistive exercise and palmidronate: Results from the LTBR study. *Bone* 36:1019-1029.
229. Shackelford, L.C., LeBlanc, A.D., and Driscoll, T.B. 2004. Resistance exercise as a countermeasure to disuse-induced bone loss. *Journal of Applied Physiology* 97:119-127.
230. Haddad, F., Adams, G.R., Bodell, P.W., and Baldwin, K.M. 2006. Isometric resistance exercise fails to counteract skeletal atrophy processes during the early states of unloading. *Journal of Applied Physiology* 100:433-441.
231. Adams, G.R., Haddad, F., Bodell, P.W., Tran, P.D., and Baldwin, K.M. 2007. Combined isometric, concentric, and eccentric resistance exercise prevents unloading-induced muscle atrophy in rats. *Journal of Applied Physiology* 103:1644-1654.
232. Haddad, F., Adams, G.R., Bodell, P.W., and Baldwin, K.M. 2006. Isometric resistance exercise fails to counteract skeletal atrophy processes during the early states of unloading. *Journal of Applied Physiology* 100:433-441.
233. Adams, G.R., Haddad, F., Bodell, P.W., Tran, P.D., and Baldwin, K.M. 2007. Combined isometric, concentric, and eccentric resistance exercise prevents unloading-induced muscle atrophy in rats. *Journal of Applied Physiology* 103:1644-1654.
234. Trappe, A., Costill, D., Gallagher, P., Creer, A., Peters, J.R., Evans, H., Riley, D.A., and Fitts, R.H. 2009. Exercise in space: Human skeletal muscle after 6 months aboard the International Space Station. *Journal of Applied Physiology* 106:1159-1168.
235. Fitts, R.H., Riley, D.A., and Widrick, J. 2000. Physiology of a microgravity environment. Invited review: Microgravity and skeletal muscle. *Journal of Applied Physiology* 89:823-839.
236. Trappe, A., Costill, D., Gallagher, P., Creer, A., Peters, J.R., Evans, H., Riley, D.A., and Fitts, R.H. 2009. Exercise in space: Human skeletal muscle after 6 months aboard the International Space Station. *Journal of Applied Physiology* 106:1159-1168.
237. Trappe, A., Costill, D., Gallagher, P., Creer, A., Peters, J.R., Evans, H., Riley, D.A., and Fitts, R.H. 2009. Exercise in space: Human skeletal muscle after 6 months aboard the International Space Station. *Journal of Applied Physiology* 106:1159-1168.
238. Fitts, R.H., Trappe, S.W., and Costill, D.L., Gallagher, P.M., Creer, A.C., Colloton, P.A., Peters, J.R., Romantowski, J.G., Bain, J.L., and Riley, D.A. 2010. Prolonged space flight-induced alterations in the structure and function of human skeletal muscle fiber. *Journal of Physiology* 588:3567-3592.
239. Adams, G.R., Caiozzo, V.J., and Baldwin, K.M. 2003. Skeletal muscle unweighting: Spaceflight and ground-based models. *Journal of Applied Physiology* 95:2185-2201.
240. Fitts, R.H., Riley, D.A., and Widrick, J. 2000. Physiology of a microgravity environment. Invited review: Microgravity and skeletal muscle. *Journal of Applied Physiology* 89:823-839.
241. Gallagher, P., Trappe, S., Harber, M., Creer, A., Mazzatti, S., Trappe, T., Alkner, B., and Tesch, P. 2005. Effects of 84-days of bed rest and resistance exercise training on single muscle fibre myosin heavy chain distribution in human vastus lateralis and soleus muscles. *Acta Physiologica Scandinavica* 185:61-69.
242. Lemoine, J.K., Jaus, J.M., and Trappe, S. 2009. Muscle proteins during 60-day bedrest in woman: Impact of exercise or nutrition. *Muscle and Nerve* 39:463-471.
243. Trappe, S., Creer, A., Minchev, K., Slivka, D., Louis, E., Luden, T., and Trappe, T. 2007. Human soleus single muscle fiber function with exercise or nutritional countermeasures during 60 days of bed rest. *American Journal of Physiology. Regulatory, Integrative and Comparative Physiology* 294:R939-R347.
244. Trappe, S., Costill, D., Gallagher, P., Creer, A., Peters, J.R., Evans, H., Riley, D.A., and Fitts, R.H. 2009. Exercise in space: Human skeletal muscle after 6 months aboard the International Space Station. *Journal of Applied Physiology* 106(4):1159-1168.
245. Gallagher, P., Trappe, S., Harber, M., Creer, A., Mazzatti, S., Trappe, T., Alkner, B., and Tesch, P. 2005. Effects of 84-days of bed rest and resistance exercise training on single muscle fibre myosin heavy chain distribution in human vastus lateralis and soleus muscles. *Acta Physiologica Scandinavica* 185:61-69.
246. Trappe, S., Creer, A., Minchev, K., Slivka, D., Louis, E., Luden, T., and Trappe, T. 2007. Human soleus single muscle fiber function with exercise or nutritional countermeasures during 60 days of bed rest. *American Journal of Physiology. Regulatory, Integrative and Comparative Physiology* 294:R939-R347.
247. Gallagher, P., Trappe, S., Harber, M., Creer, A., Mazzatti, S., Trappe, T., Alkner, B., and Tesch, P. 2005. Effects of 84-days of bed rest and resistance exercise training on single muscle fibre myosin heavy chain distribution in human vastus lateralis and soleus muscles. *Acta Physiologica Scandinavica* 185:61-69.
248. Caiozzo, V.J., Haddad, F., Lee, S., Baker, M., Paloski, W., and Baldwin, K.M. 2009. Artificial gravity as a countermeasure to microgravity: A pilot study examining the effects on knee extensor and plantar flexor muscle groups. *Journal of Applied Physiology* 107:39-46.

249. Nakao, R., Hirasaka, K., Goto, J., Ishidoh, K., Yamada, C., Ohno, A., Okumura, Y., Nonaka, I., Yasutomo, K., Baldwin, K.M., Kominami, E., et al. 2009. Ubiquitin ligase Cbl-b is a negative regulator for insulin-like growth factor 1 signaling during muscle atrophy caused by unloading. *Molecular and Cell Biology* 29:4798-4811.
250. Baldwin, K.M., Herrick, R.E., and McCue, S.A. 1993. Substrate oxidation capacity in rodent skeletal muscle: Effects of exposure to zero gravity. *Journal of Applied Physiology* 75:2666-2470.
251. Fitts, R.H., Riley, D.A., and Widrick, J. 2000. Physiology of a microgravity environment. Invited review: Microgravity and skeletal muscle. *Journal of Applied Physiology* 89:823-839.
252. Jiang, B.Y., Ohira Y., Roy R.R., Nguyen, Q., Ilyina-Kakeuva, E.I., Oganov, V., and Edgerton, V.R. 1992. Adaptation of fibers in fast twitch muscles of rats to spaceflight and hindlimb suspension. *Journal of Applied Physiology* (Suppl.):58S-65S.
253. Baldwin, K.M., Herrick, R.E., and McCue, S.A. 1993. Substrate oxidation capacity in rodent skeletal muscle: Effects of exposure to zero gravity. *Journal of Applied Physiology* 75:2666-2470.
254. Musacchia, X.J., Steffen, J.M., Fell, R.D., Dombrowski, M.J., Oganov, V.W., Ilyina-and Kakeuva, E.I. 1992. Skeletal muscle atrophy response to 14 days of weightlessness: Vastus intermedius. *Journal of Applied Physiology* (Suppl.):44S-50S.
255. Henriksen, E.J., Tischler, M.E., and Johnson, D.G. 1986. Increased response to insulin of glucose metabolism in six day unloaded rat soleus. *Journal of Biological Chemistry* 261:10708-10712.
256. McClung, J.M., Whidden, M.A., Kavazis, A.N., Falk, D.J., Deruisseau, K.C., and Powers, S.K. 2008. Redox regulation of diaphragm proteolysis during mechanical ventilation. *American Journal of Physiology. Regulatory, Integrative and Comparative Physiology* 294:R1608-R1617.
257. Adams, G.R., and Haddad, F. 1996. The relationship among IGF-1, DNA content and protein accumulation during skeletal muscle hypertrophy. *Journal of Applied Physiology* 81:2509-2516.
258. Adams, G.R., Haddad, F., and Baldwin, K.M. 1999. Time course of changes in markers of myogenesis in overloaded rat skeletal muscles. *Journal of Applied Physiology* 87:1705-1712.
259. Adams, G.R., and Haddad, F. 1996. The relationship among IGF-1, DNA content and protein accumulation during skeletal muscle hypertrophy. *Journal of Applied Physiology* 81:2509-2516.
260. Adams, G.R., Haddad, F., and Baldwin, K.M. 1999. Time course of changes in markers of myogenesis in overloaded rat skeletal muscles. *Journal of Applied Physiology* 87:1705-1712.
261. Pandorf, C.E., Haddad, F., Wright, C., Bodell, P.W., and Baldwin, K.M. 2009. Differential epigenetic modifications of histones at the myosin heavy chain genes in fast and slow skeletal muscle fibers and in response to muscle unloading. *American Journal of Physiology. Cell Physiology* 297(1):C6-C16.
262. See Funding Opportunities at the National Space Biomedical Research Insitute website, available at http://www.nsbri.org/funding-opportunities.
263. Giger, J.M., Bodell, P.W., Zeng, M., Baldwin, K.M., and Haddad, F. 2009. Rapid muscle atrophy response to unloading: Pretranslational processes involving MHC and actin. *Journal of Applied Physiology* 107:1204-1212.
264. Jiang, B.Y., Ohira Y., Roy R.R., Nguyen, Q., Ilyina-Kakeuva, E.I., Oganov, V., and Edgerton, V.R. 1992. Adaptation of fibers in fast twitch muscles of rats to spaceflight and hindlimb suspension. *Journal of Applied Physiology* (Suppl.):58S-65S.
265. Roy, R.R., Baldwin, K.M., and Edgerton, V.R. 1996. Response of the neuromuscular unit to space flight: What have we learned from the rat model. *Exercise and Sport Science Reviews* 24:399-420.
266. Adams, G.R., McCue, S.A., Bodell, P.W., Zeng, M., and Baldwin, K.M. 2000. Effects of spaceflight and thyroid deficiency on hindlimb development I: Muscle mass and IGF-I expression. *Journal of Applied Physiology* 88:894-903.
267. Adams, G.R., Haddad, F., McCue, S.A., Bodell, P.W., Zeng, M., Qin, A.X., and Baldwin, K.M. 2000. Effects of spaceflight and thyroid deficiency on rat hindlimb development II: Expression of MHC isoforms. *Journal of Applied Physiology* 88:904-916.
268. Adams, G.R., McCue, S.A., Bodell, P.W., Zeng, M., and Baldwin, K.M. 2000. Effects of spaceflight and thyroid deficiency on hindlimb development I: Muscle mass and IGF-I expression. *Journal of Applied Physiology* 88:894-903.
269. Adams, G.R., Haddad, F., McCue, S.A., Bodell, P.W., Zeng, M., Qin, A.X., and Baldwin, K.M. 2000. Effects of spaceflight and thyroid deficiency on rat hindlimb development II: Expression of MHC isoforms. *Journal of Applied Physiology* 88:904-916.
270. Ikemoto, M., Nikawa, T., Takeda, S., Watanabe, C., Kitano, T., Baldwin, K., Izumi, R., Nonaka, I., Towatari, T., Teshima, S., Rokutan, K., and Kishi K. 2001. Space shuttle flight (STS 90) enhances degradation of rat myosin heavy chain in association with activation of ubiquitin-proteasome pathway. *FASEB Journal* 15:1279-1282.
271. Nakao, R., Hirasaka, K., Goto, J., Ishidoh, K., Yamada, C., Ohno, A., Okumura, Y., Nonaka, I., Yasutomo, K., Baldwin, K.M., Kominami, E., et al. 2009. Ubiquitin ligase Cbl-b is a negative regulator for insulin-like growth factor 1 signaling during muscle atrophy caused by unloading. *Molecular and Cellular Biology* 29:4798-4811.

272. Nikawa, T., Ishidoh, K., Hirasaka, K., Ishihara, I., Ikemoto, M., Kano, M., Kominami, E., Nonaka, I., Ogawa, T., Adams, G.R., Baldwin, K.M., et al. 2004. Skeletal muscle gene expression in space-flown rats. *The FASEB Journal* 18:522-524.
273. Paloski, W.H., Oman, C.M., Bloomberg, J.J., Reschke, M.F., Wood, S.J., Harm, D.L., Peters, B.T., Mularvara, A.P., Locke, J.P., and Stone, L.S. 2008. Risk of sensory-motor performance failures affecting vehicle control during space missions: A review of the evidence. *Journal of Gravitational Physiology* 15(2):1-29.
274. Young, L.R., Yajima, K., and Paloski, W.H., eds. 2009. *Artificial Gravity Research to Enable Human Space Exploration.* International Academy of Astronautics, Paris, France.
275. Clement, G., and Bukley, A. 2007. *Artificial Gravity.* Springer, New York, N.Y.
276. Speers, R.A., Paloski, W.H., and Kuo, A.D. 1998. Multivariate changes in coordination of postural control following spaceflight. *Journal of Biomechanics* 31(10):883-889.
277. Young, L.R., and Shelhamer, M. 1990. Microgravity enhances the relative contribution of visually induced motion sensation. *Aviation, Space, and Environmental Medicine* 61:525-530.
278. Lackner, J.R. 1992. Spatial orientation in weightless environments. *Perception* 21:803-812.
279. Lackner, J.R., and DiZio, P. 1992. Gravitoinertial force level affects the appreciation of limb position during muscle vibration. *Brain Research* 592:175-180.
280. Lackner, J.R., DiZio, P., and Fisk, J. 1992. Tonic vibration reflexes and background force level. *Acta Astronautica* 26:133-136.
281. Fisk, J., Lackner, J.R., and DiZio, P. 1993. Gravitoinertial force level influences arm movement control. *Journal of Neurophysiology* 69:504-511.
282. Clement, G., Gurfinkel, V.S., Lestienne, F., Lipshits, M.I., and Popov, K.E. 1984. Adaptation of postural control to weightlessness. *Experimental Brain Research* 57:61-72.
283. Paloski, W.H., Reschke, M.F., Black, F.O., Doxey, D.D., and Harm, D.L. 1992. Recovery of postural equilibrium control following spaceflight. *Annals of the New York Academy of Sciences* 656:747-754.
284. Bloomberg, J.J., Peters, B.T., Smith, S.L., Huebner, W.P., and Reschke, M.F. 1997. Locomotor head-trunk coordination strategies following space flight. *Journal of Vestibular Research* 7:161-177.
285. Jaekl, P., Zikovitz, D.C., Jenkin, M.R., Jenkin, H.L., Zacher, J.E., and Harris, L.R. 2005. Gravity and perceptual stability during translational head movement on Earth and in microgravity. *Acta Astronautica* 56:1033-1040.
286. Lackner, J.R., and DiZio, P. 2009. Angular displacement perception modulated by force background. *Experimental Brain Research* 195:335-343.
287. Lackner, J.R., and DiZio, P. 1996. Motor function in microgravity: Movement in weightlessness. *Current Opinion in Neurobiology* 6:744-750.
288. Speers, R.A., Paloski, W.H., and Kuo, A.D. 1998. Multivariate changes in coordination of postural control following spaceflight. *Journal of Biomechanics* 31(10):883-889.
289. Paloski, W.H., Reschke, M.F., Doxey, D.D., and Black, F.O. 1992. Neurosensory adaptation associated with postural ataxia following space flight. Pp. 311-315 in *Posture and Gait: Control Mechanisms* (M. Woolacott and F. Horak, eds.). University of Oregon Press, Eugene, Ore.
290. Reschke, M.F., Bloomberg, J.J., Paloski, W.H., Mulavara, A.P., Feiveson, A.H., and Harm, D.L. 2009. Postural reflexes, balance control, and functional mobility with long-duration head-down bed rest. *Aviation, Space, and Environmental Medicine* 80(5 Suppl.):A45-A54.
291. Jain, V., Wood, S.J., Feiveson, A.H., Black, F.O., and Paloski, W.H. 2010. Diagnostic accuracy of dynamic posturography testing after short-duration spaceflight. *Aviation, Space, and Environmental Medicine* 81(7):625-631.
292. Clement, G., and Lesienne, F. 1988. Adaptive modifications of postural attitude in condition of weightlessness. *Experimental Brain Research* 72:381-389.
293. Thornton, W.E. 1978. Anthropometric changes in weightlessness. In *Anthropometry for Designers.* Anthropometric Source Book, Volume 1 (Anthropology Research Staff, eds.). NASA RP-1024. NASA Johnson Space Center, Houston, Tex.
294. Edgerton, V.R., McCall, G.E., Hodgson, J.A., Grotto, J., Goulet, C., Fleischmann, K., and Roy, R.R. 2001. Sensorimotor adaptations to microgravity in humans. *Journal of Experimental Biology* 204:3217-3224.
295. Clement, G., and Andre-Deshays, C. 1987. Motor activity and visually induced postural reactions during two-g and zero-g phases of parabolic flight. *Neuroscience Letters* 79:113-116.
296. Reschke, M.F., Anderson, D.J., and Homick, J.L. 1984. Vestibulospinal reflexes as a function of microgravity. *Science* 225:212-214.
297. Watt, D.G., K.E. Money, and Tomi, L.M. 1986. M.I.T./Canadian vestibular experiments on the Spacelab-1 mission: 3. Effects of prolonged weightlessness on a human otolith-spinal reflex. *Experimental Brain Research* 64:308-315.

298. Mulavara, A.P., Cohen, H.S., and Bloomberg, J.J. 2009. Critical features of training that facilitate adaptive generalization of over ground locomotion. *Gait and Posture* 29(2):242-248.
299. Edgerton, V.R., and Roy, R.R. 1996. Neuromuscular adaptations to actual and simulated spaceflight. Pp. 721-763 in *Handbook of Physiology: Environmental Physiology* (M.J. Fregley and C.M. Blatteis, eds.). Volume 1. Oxford University Press, New York, N.Y.
300. Oman, C.M. 1998. Sensory conflict theory and space sickness: our changing perspective. *Journal of Vestibular Research* 8:51-56.
301. Lackner, J.R., and Dizio, P. 2006. Space motion sickness. *Experimental Brain Research* 175:377-399.
302. Kozlovskaya, I.B., Dmitrieval, I.F., Grigorieva, A., Kirenskaya, A., and Kreidich, Y. 1988. Gravitational mechanisms in the motor systems: Studies in real and simulated weightlessness. Pp. 37-48 in *Stance and Motion: Fact and Concepts* (V.S. Gurfinkel et al., eds.). Plenum, New York, N.Y.
303. Kozlovskaya, I.B., Kreidich, Yu.V., Oganov, V.S., and Koserenko, O.P. 1981. Pathophysiology of motor functions in prolonged manned space flights. *Acta Astronautica* 8:1059-1072.
304. Roy, R.R., Hodgson, J.A., Aragon, J., Day, M.K., Kozlovskaya, I., and Edgerton, V.R. 1996. Recruitment of the Rhesus soleus and medial gastrocnemius before, during and after spaceflight. *Journal of Gravitational Physiology* 3:11-15.
305. Recktenwald, M.R., Hodgson, J.A., Roy, R.R., Riazanski, S., McCall, G.E., Kozlovskaya, I., Washburn, D.A., Fanton, J.W., and Edgerton, V.R. 1999. Effects of spaceflight on rhesus quadrupedal locomotion after return to 1 G. *Journal of Neurophysiology* 81:2451-2463.
306. Edgerton, V.R., and Roy, R.R. 1996. Neuromuscular adaptations to actual and simulated spaceflight. Pp. 721-763 in *Handbook of Physiology: Environmental Physiology* (M.J. Fregley and C.M. Blatteis, eds.). Volume 1. Oxford University Press, New York, N.Y.
307. Kozlovskaya, I.B., Kreidich, Yu.V., Oganov, V.S., and Koserenko, O.P. 1981. Pathophysiology of motor functions in prolonged space flights. *Acta Astronautica* 8:1059-1072.
308. Fong, A.J., Roy, R.R., Ichiyama, R.M., Lavrov, I., Courtine, G., Gerasimenko, Y., Tai, Y.C., Burdick, J., and Edgerton, V.R. 2009. Recovery of control of posture and locomotion after a spinal cord injury: Solutions staring us in the face. *Progress in Brain Research* 175:393-418.
309. Neeper, S.A., Gómez-Pinilla, F., Choi, J., and Cotman, C. 1995. Exercise and brain neurotrophins. *Nature* 373:109.
310. Edgerton, V.R., and Roy, R.R. 1996. Neuromuscular adaptations to actual and simulated spaceflight. Pp. 721-763 in *Handbook of Physiology: Environmental Physiology* (M.J. Fregley and C.M. Blatteis, eds.). Volume 1. Oxford University Press, New York, N.Y.
311. Alford, E.K., Roy, R.R., Hodgson, J.A., and Edgerton, V.R. 1987. Electromyography of rat soleus, medial gastrocnemius, and tibialis anterior during hind limb suspension. *Experimental Neurology* 96:635-649.
312. Edgerton, V.R., and Roy, R.R. 1996. Neuromuscular adaptations to actual and simulated spaceflight. Pp. 721-763 in *Handbook of Physiology: Environmental Physiology* (M.J. Fregley and C.M. Blatteis, eds.). Volume 1. Oxford University Press, New York, N.Y.
313. Edgerton, V.R., and Roy, R.R. 1996. Neuromuscular adaptations to actual and simulated spaceflight. Pp. 721-763 in *Handbook of Physiology: Environmental Physiology* (M.J. Fregley and C.M. Blatteis, eds.). Volume 1. Oxford University Press, New York, N.Y.
314. Grigoriev, A.I., Bugrov, S.A., Bogomolov, V.V., Yegorov, A.D., Kozlovskaya, I.B., Pestov, I.D., and Tarasov, N.K. 1990. Review of the major medical results of the 1-year flight on space station "Mir." *Kosmicheskaia Biologiia I Aviakosmicheskaia Meditsina* 24(5)5:3-10.
315. Layne, C.S., McDonald, P.V., and Bloomberg, J.J. 1997. Neuromuscular activation patterns during treadmill walking after space flight. *Experimental Brain Research* 113(1):104-116.
316. Grigoriev, A.I., Bugrov, S.A., Bogomolov, V.V., Yegorov, A.D., Kozlovskaya, I.B., Pestov, I.D., and Tarasov, N.K. 1990. Review of the major medical results of the 1-year flight on space station "Mir." *Kosmicheskaia Biologiia I Aviakosmicheskaia Meditsina* 24(5)5:3-10.
317. Edgerton, V.R., and Roy, R.R. 1996. Neuromuscular adaptations to actual and simulated spaceflight. Pp. 721-763 in *Handbook of Physiology: Environmental Physiology* (M.J. Fregley and C.M. Blatteis, eds.). Volume 1. Oxford University Press, New York, N.Y.
318. Lackner, J.R., and DiZio, P. 2000. Artificial gravity as a countermeasure in long-duration space flight. *Journal of Neuroscience Research* 62:169-176.
319. Edgerton, V.R., and Roy, R.R. 1996. Neuromuscular adaptations to actual and simulated spaceflight. Pp. 721-763 in *Handbook of Physiology: Environmental Physiology* (M.J. Fregley and C.M. Blatteis, eds.). Volume 1. Oxford University Press, New York, N.Y.

320. Trappe, S., Costill, D., Gallagher, P., Creer, A., Peters, J.R., Evans, H., Riley, D.A., and Fitts, R.H. 2009. Exercise in space: Human skeletal muscle after 6 months aboard the International Space Station. *Journal of Applied Physiology* 106(4):1159-1168.
321. Trappe, S., Costill, D., Gallagher, P., Creer, A., Peters, J.R., Evans, H., Riley, D.A., and Fitts, R.H. 2009. Exercise in space: Human skeletal muscle after 6 months aboard the International Space Station. *Journal of Applied Physiology* 106(4):1159-1168.
322. Cohen, H.S. 2003. Update on the status of rehabilitative countermeasures to ameliorate the effects of long-duration exposure to microgravity on vestibular and sensorimotor function. *Journal of Vestibular Research* 13:405-409.
323. Grigoriev, A.I., Bugrov, S.A., Bogomolov, V.V., Yegorov, A.D., Kozlovskaya, I.B., Pestov, I.D., and Tarasov, N.K. 1990. Review of the major medical results of the 1-year flight on space station "Mir." *Kosmicheskaia Biologiia I Aviakosmicheskaia Meditsina* 24(5)5:3-10.
324. National Research Council. 1998. *A Strategy for Research in Space Biology and Medicine in the New Century.* National Academy Press, Washington, D.C., p. 139.
325. Drummer, C., Hesse, C., Baisch, F., Norsk, P., Elmann-Larsen, B., Gerzer, R., and Heer, M. 2000. Water and sodium balances and their relation to body mass changes in microgravity. *European Journal of Clinical Investigation* 30:1066-1075.
326. Leach, C.S., Alfrey, C.P., Suki, W.N., Leonard, J.L., Rambaut, P.C., Inners, L.D., Smith, S.M., Lane, H.W., and Krauhs, J.M. 1996. Regulation of body fluid compartments during short-term flight. *Journal of Applied Physiology* 81:105-116.
327. Drummer, C., Norsk, P., and Heer, M. 2001. Water and sodium balance in space. *American Journal of Kidney Diseases* 38:684-690.
328. Blomqvist, C.G., Buckey, J.C., Gaffney, F.A., Lane, L.D., Levine, B.D., and Watenpaugh, D.E. 1994. Mechanisms of post-flight orthostatic intolerance. *Journal of Gravitational Physiology* 1:122-124.
329. Polk, J.D. 2009. "Flight Surgeon Perspective: Gaps in Human Health, Performance, and Safety," presentation to Committee of the Decadal Survey on Biological and Physical Sciences in Space, August. National Research Council, Washington, D.C.
330. Heer, M., and Paloski, W.H. 2009. Space motion sickness: Incidence, etiology and countermeasures. *Autonomic Neuroscience: Basic and Clinical* 129:77-79.
331. Smith, S.M., Krauh, J.M., and Leach, C.S. 1997. Regulation of body fluid volume and electrolyte concentration in spaceflight. *Advances in Space Biology and Medicine* 6:123-165.
332. Moore, T.P., and Thornton, W.E. 1987. Space shuttle inflight and postflight fluid shifts measured by leg volume changes. *Aviation, Space, and Environmental Medicine* 58(9 Suppl.):A91-A96.
333. Buckey, J.C., Gaffney, F.A., Lane, L.D., Levine, B.D., Watenpaugh, D.E., Wright, S.J., Yancy, C.W., Jr., Meyer, D.M., and Blomqvist, C.G. 1995. Central venous pressure in space. *Journal of Applied Physiology* 81:19-25.
334. Foldager, N.T., Anderson, A.E., Jensen, F.B., Ellegaard, P., Stadeager, C., Videbaek, R., and Norsk, P. 1996. Central venous pressure in humans during microgravity. *Journal of Applied Physiology* 81:408-412.
335. Buckey, J.C., Gaffney, F.A., Lane, L.D., Levine, B.D., Watenpaugh, D.E., Wright, S.J., Yancy, C.W., Jr., Meyer, D.M., and Blomqvist, C.G. 1995. Central venous pressure in space. *Journal of Applied Physiology* 81:19-25.
336. Prisk, G.M., Guy, H.G.B., Elliott, A.R., Deutschman III, R.A., and West, J.B. 1993. Pulmonary diffusing capacity, capillary blood volume, and cardiac output during sustained microgravity. *Journal of Applied Physiology* 75:15-26.
337. White, R.G., and Blomqvist, C.G. 1998. Central venous pressure and cardiac function during spaceflight. *Journal of Applied Physiology* 85:738-746.
338. Videbaek, R., and Norsk, P. 1997. Atrial distension in humans during microgravity induced by parabolic flights. *Journal of Applied Physiology* 83:1862-1966.
339. Tyberb, J.V., and Hamilton, D.R. 1996. Orthostatic hypotension and the role of changes in venous capacitance. *Medicine and Science in Sports and Exercise* 28(Suppl.):S29-S31.
340. White, R.G., and Blomqvist, C.G. 1998. Central venous pressure and cardiac function during spaceflight. *Journal of Applied Physiology* 85:738-746.
341. Videbaek, R., and Norsk, P. 1997. Atrial distension in humans during microgravity induced by parabolic flights. *Journal of Applied Physiology* 83:1862-1966.
342. Watenpaugh, D.E. 2001. Fluid volume control during short-term space flight and implications for human performance. *Journal of Experimental Biology* 204:3209-3215.
343. Diedrich, A.S., Paranjape, Y., and Robertson, D. 2007. Plasma and blood volume in space. *American Journal of the Medical Sciences* 334:80-85.
344. Watenpaugh, D.E. 2001. Fluid volume control during short-term space flight and implications for human performance. *Journal of Experimental Biology* 204:3209-3215.

345. Kirsch, K.A., Baartz, F.-J., Gunga, H.C., and Rocker, L.C. 1993. Fluid shifts in and out of superficial tissues under microgravity and terrestrial conditions. *Clinical Investigation* 71:687-689.
346. Kuipers, A. 1996. First results from experiments performed with the ESA Anthrorack rack during the D-2 Spacelab mission. *Acta Astronautica* 38:865-875.
347. Wakins, S.D., and Barr, Y.R. 2010. *Papilledema Summit: Summary Report.* NASA/TM-2010-2I6114. March. NASA, Washington, D.C., pp. 1-12.
348. Hargens, A.R., Tipton, C.M., Gollnik, P.D., Murbarak, S.J., et al. 1983. Fluid shifts and muscle function in humans during acute simulated weightlessness. *Journal of Applied Physiology* 54:1003-1009.
349. Parazynski, S.E., Hargens, A.R., Tucker, B., Aratow, M.F., and Crenshaw, A. 1991. Transcapillary fluid shifts in tissues of the head and neck and after simulated microgravity. *Journal of Applied Physiology* 71:2469-2475.
350. Shi-Tong, T., Ballard, R.E., Murthy, G.E., Hargens, A.R., Convertino, V.A. 1998. Plasma colloid osmotic pressure in humans during simulated microgravity. *Aviation, Space, and Environmental Medicine* 69:23-26.
351. Leach, C.S., Alfrey, C.P., Suki, W.N., Leonard, J.L., Rambaut, P.C., Inners, L.D., Smith, S.M., Lane, H.W., and Krauhs, J.M. 1996. Regulation of body fluid compartments during short-term space flight. *Journal of Applied Physiology* 81:105-116.
352. Polk, J.D. 2009. "Flight Surgeon Perspective: Gaps in Human Health, Performance, and Safety," presentation to the Committee on Decadal Survey on Biological and Physical Sciences in Space, August. National Research Council, Washington, D.C.
353. Murthy, G., Marchbanks, R.J., Watenpaugh, D.E., Meyer, J.-U., Eliashberg, N., and Hargens, A.R. 1992. Increased intracranial pressure in humans during simulated microgravity. *The Physiologist* 35(Suppl.):S184-S185.
354. Krotov, V.P., Trambovetski, E.V., and Korolkov, V.I. 1994. [Intracranial pressure in monkeys under microgravity.] [Russian]. *Fziol. Zh. Im. I. M. Sechenova* 80:1-8.
355. Krasnov, I.B., Gulevskaia, T.S., and Morgunov, V.A. 2005. [Morphology of vessels and vascular plexus in the rat's brain following 93-day simulation of the effects of microgravity.] [Russian]. *Aviakosmicheskaia i Ekologicheskaia Meditsina* 39:32-36.
356. Kawai, Y., Doi, M., and Rimoyama, R. 1998. Intracranial pressure in rabbits exposed to head-down tilt. *Journal of Gravitational Physiology* 5:P23-P24.
357. Leach, C.S., Alfrey, C.P., Suki, W.N., Leonard, J.L., Rambaut, P.C., Inners, L.D., Smith, S.M., Lane, H.W., and Krauhs, J.M. 1996. Regulation of body fluid compartments during short-term space flight. *Journal of Applied Physiology* 81:105-116.
358. Drummer, C., Norsk, P., and Heer, M. 2001. Water and sodium balance in space. *American Journal of Kidney Diseases* 38:684-690.
359. Leach, C.S., Alfrey, C.P., Suki, W.N., Leonard, J.L., Rambaut, P.C., Inners, L.D., Smith, S.M., Lane, H.W., and Krauhs, J.M. 1996. Regulation of body fluid compartments during short-term space flight. *Journal of Applied Physiology* 81:105-116.
360. Perhonen, M.A., Franco, F., Lane, L.D., Buckey, J.C., Blomqvist, G.C., Zerwekh, R.M., Peshock, R.M., Weatherall, P.T., and Levine, B.D. 2001. Cardiac atrophy after bed rest and space flight. *Journal of Applied Physiology* 91:645-653.
361. Smith, S.M., Krauh, J.M., and Leach, C.S. 1997. Regulation of body fluid volume and electrolyte concentration in spaceflight. *Advances in Space Biology and Medicine* 6:123-165.
362. Platts, S.H., Martin, D.S., Stenger, M.B., and Perez, S.A. 2009. Cardiovascular adaptations to long-duration head-down bed rest. *Aviation, Space, and Environmental Medicine* 80(5 Suppl.):A29-A36.
363. Diedrich, A., Paranjape, S.Y., and Robertson, D. 2007. Plasma and blood volume in space. *American Journal of the Medical Sciences* 334:80-85.
364. Arbeille, P., Fomina, G., Roumy, J., Alferova, I., Tobal, N., and Herault, S. 2001. Adaptation of the left heart, cerebral arteries and jugular and femoral veins during short-and long-term head-down tilt and space flights. *European Journal of Applied Physiology* 86:157-168.
365. Levine, B.D., Zuckerman, J.H., and Pawelczyk, J.A. 1997. Cardiac atrophy after bed-rest deconditioning: A nonneural mechanism for orthostatic intolerance. *Circulation* 96:517-525.
366. Perhonen, M.A., Zuckerman, J.H., and Levine, B.D. 2001. Deterioration of left ventricular performance after bed rest. *Circulation* 103:1851-1857.
367. Perhonen, M.A., Franco, F., Lane, L.D., Buckey, J.C., Blomqvist, G.C., Zerwekh, R.M., Peshock, R.M., Weatherall, P.T., and Levine, B.D. 2001. Cardiac atrophy after bed rest and space flight. *Journal of Applied Physiology* 91:645-653.
368. Caini, E.G., Weinert, L., Lang, R.M., and Vaida, P. 2009. The role of echocardiology in the assessment of cardiac function in weightlessness-our experience during parabolic flights. *Respiratory Physiology and Neurobiology* 169(Suppl. 1):S6-S9.

369. Charles, J.B., and Lathers, C.M. 1991. Cardiovascular adaptation to spaceflight. *Journal of Clinical Pharmacology* 31:1010-1023.
370. Martin, D.S., South, D.A., Wood, M.I., Bungo, M.W., and Meck, J.V. 2002. Comparison of echocardiographic changes after short-and long-duration spaceflight. *Aviation, Space, and Environmental Medicine* 73:532-536.
371. Levine, B.D., Zuckerman, J.H., and Pawelczyk, J.A. 1997. Cardiac atrophy after bed-rest deconditioning: A nonneural mechanism for orthostatic intolerance. *Circulation* 96:517-525.
372. Caini, E.G., Weinert, L., Lang, R.M., and Vaida, P. 2009. The role of echocardiology in the assessment of cardiac function in weightlessness—Our experience during parabolic flights. *Respiratory Physiology and Neurobiology* 169(Suppl. 1):S6-S9.
373. Levine, B.D., Zuckerman, J.H., and Pawelczyk, J.A. 1997. Cardiac atrophy after bed-rest deconditioning: A nonneural mechanism for orthostatic intolerance. *Circulation* 96:517-525.
374. Perhonen, M.A., Franco, F., Lane, L.D., Buckey, J.C., Blomqvist, G.C., Zerwekh, R.M., Peshock, R.M., Weatherall, P.T., and Levine, B.D. 2001. Cardiac atrophy after bed rest and space flight. *Journal of Applied Physiology* 91:645-653.
375. Dorfman, T.A., Levine, B.D., Tillery, T., Peshock, R.M., Hastings, J.L., Schneider, S.M., Macias, B.R., Biolo, G., and Hargens, A.R. 2007. Cardiac atrophy in women following bed rest. *Journal of Applied Physiology* 103:8-16.
376. Arbeille, P., Fomina, G., Roumy, J., Alferova, I., Tobal, N., and Herault, S. 2001. Adaptation of the left heart, cerebral arteries and jugular and femoral veins during short-and long-term head-down tilt and space flights. *European Journal of Applied Physiology* 86:157-168.
377. Levine, B.D., Zuckerman, J.H., and Pawelczyk, J.A. 1997. Cardiac atrophy after bed-rest deconditioning: A nonneural mechanism for orthostatic intolerance. *Circulation* 96:517-525.
378. Caini, E.G., Weinert, L., Lang, R.M., and Vaida, P. 2009. The role of echocardiology in the assessment of cardiac function in weightlessness—Our experience during parabolic flights. *Respiratory Physiology and Neurobiology* 169(Suppl. 1):S6-S9.
379. Charles, J.B., and Lathers, C.M. 1991. Cardiovascular adaptation to spaceflight. *Journal of Clinical Pharmacology* 31:1010-1023.
380. Martin, D.S., South, D.A., Wood, M.I., Bungo, M.W., and Meck, J.V. 2002. Comparison of echocardiographic changes after short-and long-duration spaceflight. *Aviation, Space, and Environmental Medicine* 73:532-536.
381. Drummer, C., Hesse, C., Baisch, F., Norsk, P., Elmann-Larsen, B., Gerzer, R., and Heer, M. 2000. Water and sodium balances and their relation to body mass changes in microgravity. *European Journal of Clinical Investigation* 30:1066-1075.
382. Drummer, C., Norsk, P., and Heer, M. 2001. Water and sodium balance in space. *American Journal of Kidney Diseases* 38:684-690.
383. Leach, C.S., Alfrey, C.P., Suki, W.N., Leonard, J.L., Rambaut, P.C., Inners, L.D., Smith, S.M., Lane, H.W., and Krauhs, J.M. 1996. Regulation of body fluid compartments during short-term space flight. *Journal of Applied Physiology* 81:105-116.
384. Drummer, C., Hesse, C., Baisch, F., Norsk, P., Elmann-Larsen, B., Gerzer, R., and Heer, M. 2000. Water and sodium balances and their relation to body mass changes in microgravity. *European Journal of Clinical Investigation* 30:1066-1075.
385. Grigoriev, A.I., Huntoon, C., Natochin, Yu.V. 1995. On the correlation between individual biochemical parameters of human blood serum following space flight and their basal values. *Acta Astronautica* 36:639-648.
386. Grigoriev, A.I., Morukov, B.V., and Vorobiev, D.V. 1994. Water and electrolyte studies during long-term missions on board the space stations SALYUT and MIR. *Clinical Investigation* 72:169-189.
387. Custaud, M.-A., de Chantemele, E.B., Blanc, S., Gauquelin-Koch, G., and Gharib, G. 2005. Regulation de la volemie au cours d'une stimulation d'impesanteur de longue duree. *Canadian Journal of Physiology and Pharmacology* 83:1145-1153.
388. Keil, L., Evans, J., Grindeland, R., and Krasnov, I. 1992. Pituitary oxytocin and vasopressin content of rats flown on COSMOS 2044. *Journal of Applied Physiology* 73(Suppl.):166S-168S.
389. Leach, C.S., Alfrey, C.P., Suki, W.N., Leonard, J.L., Rambaut, P.C., Inners, L.D., Smith, S.M., Lane, H.W., and Krauhs, J.M. 1996. Regulation of body fluid compartments during short-term space flight. *Journal of Applied Physiology* 81:105-116.
390. Drummer, C., Norsk, P., and Heer, M. 2001. Water and sodium balance in space. *American Journal of Kidney Diseases* 38:684-690.
391. Drummer, C., Norsk, P., and Heer, M. 2001. Water and sodium balance in space. *American Journal of Kidney Diseases* 38:684-690.
392. Huntoon, C.S., Clintron, N.W., and Whitson, P.A. 1994. Endocrine and biochemical functions. Pp. 334-350 in *Space Physiology and Medicine* (A.E. Nicogossian, C.L. Huntoon, and S.L. Pool, eds.). Lea and Febiger, Philadelphia, Pa.

393. Leach, C.S., Alfrey, C.P., Suki, W.N., Leonard, J.L., Rambaut, P.C., Inners, L.D., Smith, S.M., Lane, H.W., and Krauhs, J.M. 1996. Regulation of body fluid compartments during short-term space flight. *Journal of Applied Physiology* 81:105-116.
394. Drummer, C., Hesse, C., Baisch, F., Norsk, P., Elmann-Larsen, B., Gerzer, R., and Heer, M. 2000. Water and sodium balances and their relation to body mass changes in microgravity. *European Journal of Clinical Investigation* 30:1066-1075.
395. Grigoriev, A.I., Huntoon, C., Natochin, Yu.V. 1995. On the correlation between individual biochemical parameters of human blood serum following space flight and their basal values. *Acta Astronautica* 36:639-648.
396. Custaud, M.-A., de Chantemele, E.B., Blanc, S., Gauquelin-Koch, G., and Gharib, G. 2005. Regulation de la volemie au cours d'une stimulation d'impesanteur de longue duree. *Canadian Journal of Clinical Pharmacology* 83:1145-1153.
397. Grigoriev, A.I., Huntoon, C.L., Morukov, B.V., and Lane, H.W. 1996. Endocrine, renal and circulatory influences on fluid and electrolyte homeostasis during weightlessness: A joint Russian-U.S. project. *Journal of Gravitational Physiology* 3:83-86.
398. Drummer, C., Gerzer, R., Baisch, F., and Heer, M. 2000. Body fluid regulation in μ-gravity differs from that on Earth. *Pflügers Archiv: European Journal of Physiology* 441(Suppl):R66-R72.
399. Gauqelin, G., Geelen, E., Gharib, C., Grigoriev, A.I., Guell, A., Kvetnansky, R. Macho, L., Noskov, V., Pasi, P., Soukanov, J., and Patricot, M., et al. 1990. Volume regulating hormones, fluid and electrolyte modification during the Aragatz mission (Mir Station). Pp. 603-608 in *Proceedings of the Fourth European Symposium on Life Science Research in Space*, Trieste, Italy, May 28-June 1, 1990. ESA SP-30 7. European Space Agency, Paris, France.
400. Hinghofer-Szalkay, H.G., Noskov, V.K., Rossler, A., Ggrigoriev, A.I., Kvetnansky, R., and Polyakow, V.V. 1999. Endocrine status and LBNP-induced hormone changes during a 438-day spaceflight: a case study. *Aviation, Space, and Environmental Medicine* 70:1-5.
401. Custaud, M.-A., de Chantemele, E.B., Blanc, S., Gauquelin-Koch, G., and Gharib, C. 2005. Regulation de la volemie au cours d'une stimulation d'impesanteur de longue duree. *Canadian Journal of Clinical Pharmacology* 83:1145-1153.
402. Ertl, A.C., Diedrich, A., Biaggioni, I., Levine, B.D., Robertson, R.M., Cox, J.F., Zuckerman, J.H., Pawelczyk, J.A., Ray, C.A., Buckey, J.C., Jr., Lane, L.D., et al. 2002. Human muscle sympathetic nerve activity and plasma noradrenaline kinetics in space. *Journal of Physiology* 538(Pt 1):321-329.
403. Kvetnansky, R., Ddavydova, N.A., Noskov, V.B., Vigas, M., Popova, I.A., Usakov, A.C., Macho, L., and Grigoriev, A.I. 1988. Plasma and urine catecholamine levels in cosmonauts during long-term stay on Space Station Salyut-7. *Acta Astronautica* 17:181-186.
404. Goldstein, D.S., Vernikos, J., Holmes, C., and Convertion, V.A. 1995. Catecholaminergic effects of prolonged head-down bed rest. *Journal of Applied Physiology* 78:1023-1029.
405. Sigaudo, D., Fortrat, J.-O., Allevard, A.-M., Maillet, A., Cottet-Emard, J.M., Vouillarmet, A., Hughson, R.L., Gauquelin-Koch, G., and Gharib, C. 1998. Changes is the sympathetic nervous system induced by 42 days of head-down bed rest. *American Journal of Physiology* 274 (6 Pt 2):H1875-H1884.
406. Christensen, N.J., Heer, M., Ivanova, K., and Norsk, P. 2005. Sympathetic nervous activity decreases during head-down bed rest but not during microgravity. *Journal of Applied Physiology* 99:1552-1557.
407. Shoemaker, J.K., Hogeman, C.S., Leuenberger, U.A., and Herr, M.D., Gray, K., Silber, D.H., and Sinoway, L.I. 1998. Sympathetic discharge and vascular resistance after bed rest. *Journal of Applied Physiology* 84:612-617.
408. Shoemaker, J.K., Hogeman, C.S., and Sinoway, L.T. 1999. Contributions of MSNA and stroke volume to orthostatic intolerance following bed rest. *American Journal of Physiology* 277(4 Pt 2):R1084-R1090.
409. Shoemaker, J.K., Hogeman, C.S., and Sinoway, L.T. 1999. Contributions of MSNA and stroke volume to orthostatic intolerance following bed rest. *American Journal of Physiology* 277(4 Pt 2):R1084-R1090.
410. Pawelczyk, J.A., Zuckerman, J.H., Blomqvist, C.G., and Levine, B.D. 2001. Regulation of muscle sympathetic nerve activity after bed rest deconditioning. *American Journal of Physiology. Heart and Circulatory Physiology* 280:H2230-H2239.
411. Pancheva, M.V., Panchev, V.S., and Suvandjieva, A.V. 2006. Lower body negative pressure vs. lower body positive pressure to prevent cardiac atrophy after bed rest and spaceflight. What caused the controversy? *Journal of Applied Physiology* 100:1090-1093.
412. Norsk, P., Drummer, C., Rocker, L., Strollo, F., Christensen, N.J., Warberg, J., Bie, P., Stadeager, C., Johansen, L.B., Heer, M., Gunga, H.C., and Gerzer, R. 1995. Renal and endocrine responses in humans to isotonic saline infusion during microgravity. *Journal of Applied Physiology* 78:2253-2259.
413. Baldwin, K.M. 1997. *Task Force on Countermeasures.* Report to NASA Headquarters, Washington, D.C., pp. 12-15.
414. Traon, A.P., Sigaudo, D., Vasseur, P., Maillet, A., Fortrat, J.O., Hughson, R.L., Gauquelin-Koch, G., and Gharib, C. 1998. Cardiovascular responses to orthostatic tests after a 42-day head-down bed rest. 1998. *European Journal of Applied Physiology and Occupational Physiology* 77:50-59.

415. Kozlovskaya, I.B., and Grigoriev, A.I. 2004. Russian system of countermeasures on board of the International Space Station (ISS): The first results. *Acta Astronautica* 55:2233-2237.
416. Kozlovskaya, I.B., and Grigoriev, A.I. 2004. Russian system of countermeasures on board of the International Space Station (ISS): The first results. *Acta Astronautica* 55:2233-2237.
417. Gurovsky, N.N., Gazenko, O.G., Adamovich, B.A., Ilyin, E.A., Genin, A.M., Korolkov, V.I., Shipov, A.A., Kotovskaya, A.R., Kondratyeva, V.A., Serova, L.V., and Kondratyev, Yu. I. 1980. Study of physiological effects of weightlessness and artificial gravity in the light of the biosattelite Cosmos-936. *Acta Astronautica* 7:113-121.
418. Fortney, S.M., Beckett, W.S., Carpenter, A.J., Davis, J., Drew, H., LaFrance, N.D., Rock, J.A., Tankersley, C.G., and Vroman, N.B. 1988. Changes in plasma volume during bed rest, effects of menstrual cycle and estrogen administration. *Journal of Applied Physiology* 65:525-533.
419. Stachenfeld, N.S., and Taylor, H.S. 2005. Progesterone increases plasma volume independent of estradiol. *Journal of Applied Physiology* 98:1991-1997.
420. Hargens, A.R., Richardson, S. 2009. Cardiovascular adaptations, fluid shifts, and countermeasures related to space flight. *Respiratory Physiology and Neurobiology* 169S:S30-S33.
421. Hargens, A.R., and Richardson, S. 2009. Cardiovascular adaptations, fluid shifts, and countermeasures related to space flight. *Respiratory Physiology and Neurobiology* 169S:S30-S33.
422. Meck, J.V., Dreyer, S.A., Warren, L.E. 2009. Long-duration head-down bed rest: Project overview, vital signs, and fluid balance. *Aviation, Space, and Environmental Medicine* 80(Suppl.):A1-A8.
423. Morey-Holton, E., and Globus, K. 2002. Hindlimb unloading rodent model. Technical aspects. *Journal of Applied Physiology* 92:1367-1377.
424. Kakurin, L.I., Lobachik, V.I., Mikhailov, V.M., and Senkevich, Y.A. 1976. Antiorthostatic hypokinesia as a method of weightlessness simulation. *Aviation, Space, and Environmental Medicine* 47:1083-1086.
425. Epstein, M. 1992. Renal effects of head-out water immersion in humans, a 15-year update. *Physiological Reviews* 81:563-621.
426. Drummer, C., Gerzer, R., Baisch, F., and Heer, M. 2000. Body fluid regulation in μ-gravity differs from that on Earth: An overview. *Pflügers Archiv: European Journal of Physiology* 441:R66-R72.
427. DeSanto, N.G., Christensen, N.J., Drummer, C., Kramer, H.J., Regnard, J., Heer, M., Cirillo, M., and Norsk, P. 2001. Fluid balance and kidney function in space: Introduction. *American Journal of Kidney Diseases* 38(3):664-667.
428. Buckey, J.C., Gaffney, F.A., Lane, L.D., Levine, B.D., Watenpaugh, D.E., Wright, S.J., Yancy, C.W., Jr., Meyer, D.M., and Blomqvist, C.G. 1995. Central venous pressure in space. *Journal of Applied Physiology* 81:19-25.
429. Foldager, N.T., Anderson, A., Jensen, F.B., Ellegaard, P., Stadeager, C., Videbaek, R., and Norsk, P. 1996. Central venous pressure in humans during microgravity. *Journal of Applied Physiology* 82:408-412.
430. Norsk, P., Christensen, N.J, Gabrielsen, A., Heer, M., and Drummer, C. 2000. Unexpected renal responses in space. *Lancet* 357:146.
431. Drummer, C., Heer, M., Dressendorfer, R.A., Strasburger, C.J., and Gerzer, R. 1993. Reduced natriuresis during weightlessness. *Clinical Investigation* 71:678-686.
432. Norsk, P., and Christensen, N.J. 2009. "Adaptation of the human cardiovascular system to prolonged spaceflight: Health aspects and questions to be addressed," white paper submitted to the Committee of the Decadal Survey on Biological and Physical Sciences in Space, National Research Council, Washington, D.C.
433. Norsk, P., Christensen, N.J., Gabrielsen, A., Heer, M., and Drummer, C. 2000. Unexpected renal responses in space. *Lancet* 357:146.
434. Meck, J.V., Dreyer, S.A., and Warren, E. 2009. Long-duration head-down bed rest: Project overview, vital signs, and fluid balance. *Aviation, Space, and Environmental Medicine* 80(Suppl.):A1-A8.
435. Blomqvist, C.G., and Stone, H.L. 1983. Cardiovascular adjustments to gravitational stress. Pp. 1025-1063 in *Handbook of Physiology. The Cardiovascular System. Peripheral Circulation and Organ Blood Flow* (J.T. Shepherd, F.M. Abboud, and S.R. Geiger, eds.). American Physiological Society, Bethesda, Md.
436. Aubert, A.E., Beckers, F., and Verheyden, B. 2005. Cardiovascular function and basics of physiology in microgravity. *Acta Cardiology* 60:129-151.
437. Norsk, P., and Linnarsson, D., eds. 2009. Cardio-respiratory physiology in space. *Respiratory Physiology and Neurobiology* 169 (Suppl.). Elsevier, Paris, France.
438. Norsk, P. 2009. Cardiovascular research in space. *Respiratory Physiology and Neurobiology* 169(S1):S2-S3.
439. Hoffmann, U. 2009. Results from recent spaceflight experiments. *Respiratory Physiology and Neurobiology* 169(S1):S4-S5.
440. Navasiolava, N., and Custaud, M.A. 2009. What are the future top priority questions in cardiovascular research and what new hardware needs to be developed? *Respiratory Physiology and Neurobiology* 169(S1):S73-S74.

441. Coupé, M., Fortrat, J.O., Larina, I., Gauquelin-Koch, G., Gharib, C., and Custaud, M.A. 2009. Cardiovascular deconditioning: From autonomic nervous system to microvascular dysfunctions. *Respiratory Physiology and Neurobiology* 169(S1):S10-S12.
442. Furlan, R., Barbic, F., Casella, F., Severgnini, G., Zenoni, L., Mercieri, A., Mangili, R., Costantino, G., and Porta, A. 2009. Neural autonomic control in orthostatic intolerance. *Respiratory Physiology and Neurobiology* 169(S1):S17-S20.
443. Verheyden, B., Liu, J., Beckers, F., and Aubert, A.E. 2009. Adaptation of heart rate and blood pressure to short and long duration space missions. *Respiratory Physiology and Neurobiology* 169(S1):S13-S16.
444. Parazynski, S.E., Hargens, A.R., Tucker, B., Aratow, M.F., and Crenshaw, A. 1991. Transcapillary fluid shifts in tissues of the head and neck and after simulated microgravity. *Journal of Applied Physiology* 71:2469-2475.
445. Leach, C.S., Alfrey, C.P., Suki, W.N., Leonard, J.L., Rambaut, P.C., Inners, L.D., Smith, S.M., Lane, H.W., and Krauhs, J.M. 1996. Regulation of body fluid compartments during short-term spaceflight. *Journal of Applied Physiology* 81:105-116.
446. Krotov, V.P., Trambovetskii, E.V., and Korokov, V.I. 1994. [Intracranial pressure in monkeys under microgravity.] [Russian]. *Fziolog Zhurnal Imeni M Sechenova* 80(10):1-8.
447. Steinbach, G.C., Macias, B.R., Tanaka, K., Yost, W.T., and Hargens, A.R. 2005. Intracranial pressure dynamics assessed by noninvasive ultrasound during 30 days of bed rest. *Aviation, Space, and Environmental Medicine* 76(2):85-90.
448. LeBlanc, A., Lin, C., Shackelford, L., Sinitsyn, V., Evans, H., Belichenko, O., Schenkman, B., Kozlovskaya, I., Oganov, V., Bakulin, A., Hedrick, T., and Feeback, D. 2000. Muscle volume, MRI relaxation times (T2), and body composition after spaceflight. *Journal of Applied Physiology* 89:2158-2164.
449. Buckey, J.C., Gaffney, F.A., Lane, L.D., Levine, B.D., Watenpaugh, D.E., Wright, S.J., Yancy, C.W., Jr., Meyer, D.M., and Blomqvist, C.G. 1995. Central venous pressure in space. *Journal of Applied Physiology* 81:19-25.
450. Foldager, N., Andersen, T.A., Jessen, F.B., Ellegaard, P., Stadeager, C., Videbaek, R., and Norsk, P. 1996. Central venous pressure in humans during microgravity. *Journal of Applied Physiology* 81:408-412.
451. Buckey, J.C., Gaffney, F.A., Lane, L.D., Levine, B.D., Watenpaugh, D.E., Wright, S.J., Yancy, C.W., Jr., Meyer, D.M., and Blomqvist, C.G. 1995. Central venous pressure in space. *Journal of Applied Physiology* 81:19-25.
452. Tyberg, J.V., Belenkie, I., Manyari, D.E., and Smith, E.R. 1996. Ventricular interaction and venous capacitance modulate left ventricular preload. *Canadian Journal of Cardiology* 2(10):1058-1064.
453. White, R.J., and Blomqvist, C.G. 1998. Central venous pressure and cardiac function during spaceflight. *Journal of Applied Physiology* 85(2):738-746.
454. Videbaek, R., and Norsk, P. 1997. Atrial distension in humans during microgravity induced by parabolic flights. *Journal of Applied Physiology* 83:1862-1966.
455. Hsia, C.C., Herazo, L.F., and Johnson, R.L., Jr. 1992. Cardiopulmonary adaptations to pneumonectomy in dogs. I. Maximal exercise performance. *Journal of Applied Physiology* 73(1):362-367.
456. Pancheva, M.V., Panchev, V.S., and Suvandjieva, A.V. 2006. Lower body negative pressure vs. lower body positive pressure to prevent cardiac atrophy after bed rest and spaceflight. What caused the controversy? *Journal of Applied Physiology* 100:1090-1093.
457. Diedrich, A., Paranjape, S.Y., and Robertson, D. 2007. Plasma and blood volume in space. *American Journal of Medical Science* 334:80-85.
458. Alfrey, C.P., Udden, M.M., Leach-Huntoon, C., Driscoll, T., and Pickett, M.H. 1996. Control of red cell mass in spaceflight. *Journal of Applied Physiology* 81:98-104.
459. Smith, S.M. 2002. Red cell and iron metabolism during space flight. *Nutrition* 18:864-866.
460. Rice, L., and Alfrey, C.P. 2000. Modulation of red cell mass by neocytolysis in space and on Earth. *Pflügers Archiv: European Journal of Physiology* 441(Suppl.):R91-R94.
461. Allebban, Z., Gibson, L.A., Lange, R.D., Jago, T.L., Strickland, K.M., Johnson, D.M., and Ichki, A.T. 1996. Effect of spaceflight on rat erythroid parameters. *Journal of Applied Physiology* 81:117-122.
462. Leach, C.S., and Johnson, P.C. 1984. Influence of space flight on erythrokinetics in man. *Science* 225:216-218.
463. Gunga, H.-C., Kirsch, K., Baartz, F., Maillet, A., Gharib, C., Nalishiti, W., Rich, I., and Röcker, L. 1996. Erythropoietin under real and simulated conditions in humans. *Journal of Applied Physiology* 81:761-773.
464. De Santo, N.G., Cirillo, M., Kirsch, K.A., Correale, G., Drummer, C., Frassl, W., Perna, A.F., Di Stazio, E., Bellini, L., and Gunga, H.-C. 2005. Anemia and erythropoietin in space flights. *Seminars in Nephrology* 25:379-387.
465. Parazynski, S.E., Hargens, A.R., Tucker, B., Aratow, M.F., and Crenshaw, A. 1991. Transcapillary fluid shifts in tissues of the head and neck and after simulated microgravity. *Journal of Applied Physiology* 71:2469-2475.
466. Parazynski, S.E., Hargens, A.R., Tucker, B., Aratow, M.F., and Crenshaw, A. 1991. Transcapillary fluid shifts in tissues of the head and neck and after simulated microgravity. *Journal of Applied Physiology* 71:2469-2475.

467. Leach, C.S., Alfrey, C.P., Suki, W.N., Leonard, J.I., Rambaut, P.C., Inners, L.D., Smith, S.M., Lane, H.W., and Krauhs, J.M. 1996. Regulation of body fluid compartments during short-term spaceflight. *Journal of Applied Physiology* 81:105-116.
468. Diedrich, A., Paranjape, S.Y., and Robertson, D. 2007. Plasma and blood volume in space. *American Journal of the Medical Sciences* 334:80-85.
469. Convertino, V.A. 2002. Mechanisms of microgravity induced orthostatic tolerance: Implications for effective countermeasures. *Journal of Gravitational Physiology* 9:1-14.
470. Drummer, C., Valenti, G., Cirillo, M., Perna, A., Bellini, L., Nenov, V., De Santo, N.G. 2002. Vasopressin, hypercalciuria and aquaporin-the key elements for impaired renal water handling in astronauts? *Nephron* 92:503-514.
471. Grigoriev, A.I., Huntoon, C.L., Morukov, B.V., and Lane, H.W. 1996. Endocrine, renal and circulatory influences on fluid and electrolyte homeostasis during weightlessness: A joint Russian-U.S. project. *Journal of Gravitational Physiology* 3:83-86.
472. Norsk, P., Christensen, N.J., Bie, P., Gabrielsen, A., Heer, M., and Drummer, C. 2000. Unexpected renal responses in space. *Lancet* 356:1577-1578.
473. Drummer, C., Heer, M., Dressendorfer, R.A., Strasburger, C.J., and Gerzer, R. 1993. Reduced natriuresis during weightlessness. *Clinical Investigation* 71:678-686.
474. Drummer, C., Norsk, P., and Heer, M. 2001. Water and sodium balance in space. *American Journal of Kidney Diseases* 38:684-690.
475. Lenz, W., Herten, M., Gerzer, R., and Drummer, C. 1999. Regulation of natriuretic peptide (urodilatin) release in a human kidney line. *Kidney International* 55:91-99.
476. Convertino, V.A. 2002. Mechanisms of microgravity induced orthostatic tolerance: Implications for effective countermeasures. *Journal of Gravitational Physiology* 9:1-14.
477. Meck, J.V., Reyes, C.J., Perez, S.A., Goldberger, A.L., and Ziegler, M.G. 2001. Marked exacerbation of orthostatic intolerance after long- vs. short-duration spaceflight in veteran astronauts. *Psychosomatic Medicine* 63(6):865-873.
478. Fritsch-Yelle, J.M., Whitson, P.A., Bondar, R.L., and Brown, T.E. 1996. Subnormal norepinephrine release relates to presyncope in astronauts after spaceflight. *Journal of Applied Physiology* 81:2134-2141.
479. Waters, W.W., Ziegler, M.G., and Meck, J.V. 2002. Postspaceflight orthostatic hypotension occurs mostly in women and is predicted by low vascular resistance. *Journal of Applied Physiology* 92:586-594.
480. Meck, J.V., Waters, W.W., Ziegler, M.G., deBlock, H.F., Mills, P.J., Robertson, D., and Huang, P.L. 2004. Mechanisms of postspaceflight orthostatic hypotension: Low alpha1-adrenergic receptor responses before flight and central autonomic dysregulation postflight. *American Journal of Physiology. Heart and Circulatory Physiology* 286(4):H1486-H1495.
481. Goldstein, D.S., Pechnik, S., Holmes, C., Eldadah, B., and Sharabi, Y. 2003. Association between supine hypertension and orthostatic hypotension in autonomic failure. *Hypertension* 42(2)136-142.
482. Traon, A.P., Vasseur, P., Sigaudo, D., Maillet, A., Fortrat, J.O., Hughson, R.L., Gauquelin-Koch, G., and Gharib, C. 1998. Cardiovascular responses to orthostatic tests after a 42-day head-down bed rest. *European Journal of Applied Physiology and Occupational Physiology* 77:50-59.
483. Bleeker, M.W., De Groot, P.C., Pawelczyk, J.A., Hopman, M.T.E., and Levine, B.D. 2004. Effects of 18 days of bed rest on leg and arm venous properties. *Journal of Applied Physiology* 96:840-847.
484. Fu, Q., Witkowski, S., and Levine, B.D. 2004. Vasoconstrictor reserve and sympathetic neural control of orthostasis. *Circulation* 110:2931-2937.
485. Moore, A.D., Jr., Lee, S.M., Charles, J.B., Greenisen, M.C., and Schneider, S.M. 2001. Maximal exercise as a countermeasure to orthostatic intolerance after spaceflight. *Medicine and Science in Sports and Exercise* 33:75-80.
486. Cui, J., Durand, S., Levine, B.D., and Crandall, C.G. 2005. Effect of skin surface cooling on central venous pressure during orthostatic challenge. *American Journal of Physiology. Heart and Circulatory Physiology* 289:2429-2433.
487. Leach, C.S., and Johnson, P.C. 1984. Influence of space flight on erythrokinetics in man. *Science* 225:216-218.
488. Gunga, H.-C., Kirsch, K., Baartz, F., Maillet, A., Gharib, C., Nalishiti, W., Rich, I., and Röcker, L. 1996. Erythropoietin under real and simulated conditions in humans. *Journal of Applied Physiology* 81:761-773.
489. De Santo, N.G., Cirillo, M., Kirsch, K.A., Correale, G., Drummer, C., Frassl, W., Perna, A.F., Di Stazio, E., Bellini, L., and Gunga, H.-C. 2005. Anemia and erythropoietin in space flights. *Seminars in Nephrology* 25:379-387.
490. Rokhlenko, K.D., and Mul'diiarov, P. 1981. [Myocardial ultrastructure of rats exposed aboard "Cosmos-936."] [Russian]. *Kosmich Biol Aviak Meditsa* 15:77-82.
491. Bungo, M.W., Goldwater, D.J., Popp, R.L., and Sandler, H. 1987. Echocardiographic evaluation of space shuttle crewmembers. *Journal of Applied Physiology* 62:278-283.
492. Goldstein, M.A., Edwards, R.J., and Schroeter, J.P. 1992. Cardiac morphology after conditions of microgravity during Cosmos 2044. *Journal of Applied Physiology* 73(Suppl.):94S-100S.

493. Perhonen, M.A., Franco, F., Lane, L.D., Buckey, J.C., Blomqvist, G.C., Zerwekh, R.M., Peshock, R.M., Weatherall, P.T., and Levine, B.D. 2001. Cardiac atrophy after bed rest and space flight. *Journal of Applied Physiology* 91:645-653.
494. Levine, B.D., Zuckerman, J.H., and Pawelczyk, J.A. 1997. Cardiac atrophy after bed-rest deconditioning: A nonneural mechanism for orthostatic intolerance. *Circulation* 96:517-525.
495. Perhonen, M.A., Zuckerman, J.H., and Levine, B.D. 2001. Deterioration of left ventricular chamber performance after bed rest: "Cardiovascular deconditioning" or hypovolemia? *Circulation* 103:1851-1857.
496. Pancheva, M.V., Panchev, V.S., and Suvandjieva, A.V. 2006. Lower body negative pressure vs. lower body positive pressure to prevent cardiac atrophy after bed rest and spaceflight. What caused the controversy? *Journal of Applied Physiology* 100:1090-1093.
497. Dorfman, T.A., Levine, B.D., Tillery, T., Peshock, R.M., Hastings, J.L., Schneider, S.M., Macias, B.R., Biolo, G., and Hargens, A.R. 2007. Cardiac atrophy in women following bed rest. *Journal of Applied Physiology* 103:8-16.
498. Perhonen, M.A., Franco, F., Lane, L.D., Buckey, J.C., Blomqvist, G.C., Zerwekh, R.M., Peshock, R.M., Weatherall, P.T., and Levine, B.D. 2001. Cardiac atrophy after bed rest and space flight. *Journal of Applied Physiology* 91:645-653.
499. Dorfman, T.A., Rosen, B.D., Perhonen, M.A., Tillery, T., McColl, R., Peshock, R.M., and Levine B.D. 2008. Diastolic suction is impaired by bed rest: MRI tagging studies of diastolic untwisting. *Journal of Applied Physiology* 104:1037-1044.
500. Perhonen, M.A., Franco, F., Lane, L.D., Buckey, J.C., Blomqvist, G.C., Zerwekh, R.M., Peshock, R.M., Weatherall, P.T., and Levine, B.D. 2001. Cardiac atrophy after bed rest and space flight. *Journal of Applied Physiology* 91:645-653.
501. Dorfman, T.A., Levine, B.D., Tillery, T., Peshock, R.M., Hastings, J.L., Schneider, S.M., Macias, B.R., Biolo, G., and Hargens, A.R. 2007. Cardiac atrophy in women following bed rest. *Journal of Applied Physiology* 103:8-16.
502. Hastings, J., Pacini, E., Krainski, F., Shibata, S., Jain, M., Snell, P.G., Fu, Q., Palmer, M.D., Creson, D., and Levine, B.D. 2008. The multi-system effect of exercise training during prolonged bed rest deconditioning: An integrated approach to countermeasure development for the heart, lungs, muscle and bones. *Circulation* 118:S682. Abstract.
503. Shibata, S., Perhonen, M., and Levine, B.D. 2010. Supine cycling plus volume loading prevent cardiovascular deconditioning during bed rest. *Journal of Applied Physiology* 108(5):1177-1186.
504. Marcus, M.L., Eckberg, D.L., Braxmeier, J.L., and Abboud, F.M. 1977. Effects of intermittent pressure loading on the development of ventricular hypertrophy in the cat. *Circulation Research* 40(5):484-488.
505. Cooper, G.I.V., Kent, R.L., and Mann, D.L. 1989. Load induction of cardiac hypertrophy. *Journal of Molecular and Cellular Cardiology* 21(Suppl. 5):11-30.
506. Zile, M.R., Tomita, M., Ishihara, K., Nakano, K., Lindroth, J., Spinale, F., Swindle, M., and Carabello, B.A. 1993. Changes in diastolic function during development and correction of chronic LV volume overload produced by mitral regurgitation. *Circulation* 87:1378-1388.
507. Lisy, O., Redfield, M.M., Jovanovic, S., Jougasaki, M., Jovanovic, A., Leskinen, H., Terzic, A., and Burnett, J.C., Jr. 2000. Mechanical unloading versus neurohumoral stimulation on myocardial structure and endocrine function in vivo. *Circulation* 102(3):338-343.
508. Perhonen, M.A., Franco, F., Lane, L.D., Buckey, J.C., Blomqvist, G.C., Zerwekh, R.M., Peshock, R.M., Weatherall, P.T., and Levine, B.D. 2001. Cardiac atrophy after bed rest and space flight. *Journal of Applied Physiology* 91:645-653.
509. Frenzel, H., Schartzkopff, B., Holtermann, W., Schnurch, H.G., Novi, A., and Hort, W. 1988. Regression of cardiac hypertrophy: Morphometric and biochemical studies in rat heart after swimming training. *Journal of Molecular and Cellular Cardiology* 20:737-751.
510. Urabe, Y., Mann, D.L., Kent, R.L., Nakano, K., Tomanek, R.J., Carabello, B.A., and Cooper, G.I.V. 1992. Cellular and ventricular contractile dysfunction in experimental canine mitral regurgitation. *Circulation Research* 70:131-147.
511. McGowan, B.S., Scott, C.B., Mu, A., McCormick, R.J., Thomas, D.P., and Margulies, K.B. 2003. Unloading-induced remodeling in the normal and hypertrophic left ventricle. *American Journal of Physiology. Heart and Circulatory Physiology* 284:H2061-H2068.
512. Lisy, O., Redfield, M.M., Jovanovic, S., Jougasaki, M., Jovanovic, A., Leskinen, H., Terzic, A., and Burnett, J.C., Jr. 2000. Mechanical unloading versus neurohumoral stimulation on myocardial structure and endocrine function in vivo. *Circulation* 102(3):338-343.
513. Hill, J.A., and Olson, E.N. 2008. Cardiac plasticity. *New England Journal of Medicine* 358(13):1370-1380.
514. Shibata, S., Perhonen, M., and Levine, B.D. 2010. Supine cycling plus volume loading prevent cardiovascular deconditioning during bed rest. *Journal of Applied Physiology* 108(5):1177-1186.
515. Hastings, J., Pacini, E., Krainski, F., Shibata, S., Jain, M., Snell, P.G., Fu, Q., Palmer, M.D., Creson, D., and Levine, B.D. 2008. The multi-system effect of exercise training during prolonged bed rest deconditioning: An integrated approach to countermeasure development for the heart, lungs, muscle and bones. *Circulation* 118:S682. Abstract.
516. Dorfman, T.A., Levine, B.D., Tillery, T., Peshock, R.M., Hastings, J.L., Schneider, S.M., Macias, B.R., Biolo, G., and Hargens, A.R. 2007. Cardiac atrophy in women following bed rest. *Journal of Applied Physiology* 103:8-16.

517. Shibata, S., Perhonen, M., and Levine, B.D. 2010. Supine cycling plus volume loading prevent cardiovascular deconditioning during bed rest. *Journal of Applied Physiology* 108(5):1177-1186.
518. Fritsch-Yelle, J. M., Leuenberger, U.A., D'Aunno, D.S., Rossum, A.C., Brown, T.E., Wood, M.T., Josephson, M.E., and Goldberger, A.L. 1998. An episode of ventricular tachycardia during long-duration spaceflight. *American Journal of Cardiology* 81:1391-1392.
519. D'Aunno, D.C., Dougherty, A.H., deBlock, H.F., and Meck, J.V. 2003. Effect of short- and long-duration spaceflight on QTc intervals in healthy astronauts. *American Journal of Cardiology* 91(4):494-497.
520. Fritsch-Yelle, J.M., Charles, J.B., Jones, M.M., and Wood, M.L. 1996. Microgravity decreases heart rate and arterial pressure in humans. *Journal of Applied Physiology* 80(3):910-914.
521. Rossum, A.C., Wood, M.L., Bishop, S.L., Deblock, H., Charles, J.B. 1997. Evaluation of cardiac rhythm disturbances during extravehicular activity. *American Journal of Cardiology* 79:1153-1155.
522. Goldberger, A.L., Bungo, M.W., Baevsky, R.M., Bennett, B.S., Rigney, D.R., Mietus, J.E., Nikulina, G.A., and Charles, J.B. 1994. Heart rate dynamics during long-term spaceflight: Report on Mir cosmonauts. *American Heart Journal* 128(1):202-204.
523. Berry, C.A., and Catterson, A.D. 1967. Pre-Gemini medical predictions versus Gemini flight results. Pp. 197-199 in *Gemini Summary Conference*. NASA SP-138. NASA, Washington, D.C.
524. D'Aunno, D.C., Dougherty, A.H., deBlock, H.F., and Meck, J.V. 2003. Effect of short- and long-duration spaceflight on QTc intervals in healthy astronauts. *American Journal of Cardiology* 91(4):494-497.
525. Hobbs, J.B., Peterson, D.R., Moss, A.J., McNitt, S., Zareba, W., Goldenberg, I., Qi, M., Robinson, J.L., Sauer, A.J., Ackerman, M.J., Benhorin, J., Kaufman, E.S., Locati, E.H., Napolitano, C., Priori, S.G., Towbin, J.A., Vincent, G.M., and Zhang, L. 2006. Risk of aborted cardiac arrest or sudden cardiac death during adolescence in the long-QT syndrome. *Journal of the American Medical Association* 296:1249-1254.
526. Basavarajaiah, S., Wilson, M., Whyte, G., Shah, A., Behr, E., and Sharma, S. Prevalence and significance of an isolated long QT interval in elite athletes. 2007. *European Heart Journal* 28:2944-2949.
527. Ertl, A.C., Diedrich, A., Biaggioni, I., Levine, B.D., Robertson, R.M., Cox, J.F., Zuckerman, J.H., Pawelczyk, J.A., Ray, C.A., Buckey, J.C., Jr., Lane, L.D., et al. 2002. Human muscle sympathetic nerve activity and plasma noradrenaline kinetics in space. *Journal of Physiology* 538(Pt 1):321-329.
528. Cox, J.F., Tahvanainen, K.U.O., Kuusela, T.A., Levine, B.D., Cooke, W.H., Mano, T., Iwase, S., Saito, M., Sugiyama, Y., Ertl, A.C., Biaggioni, I., et al. 2002. Influence of microgravity on astronauts' sympathetic and vagal responses to Valsalva's manoeuvre. *Journal of Physiology* 538(Pt 1):309-320.
529. Abboud, F.M. 1989. Ventricular syncope: Is the heart a sensory organ? (Editorial). *New England Journal of Medicine* 320:390-392.
530. Hajduczok, G., Chapleau, M.W., and Abboud, F.M. 1988. Rheoreceptors in the carotid sinus of dog. *Proceedings of the National Academy of Sciences U.S.A.* 85:7399-7403.
531. Chapleau, M.W., and Abboud, F.M. 1987. Contrasting effects of static and pulsatile pressure on carotid baroreceptor activity in dogs. *Circulation Research* 61(5):648-658.
532. Abboud, F.M., and Chapleau, M.W. 1988. Effects of pulse frequency on single unit baroreceptor activity during sine-wave and natural pulses in dogs. *Journal of Physiology* 401:295-308.
533. Cox, J.F., Tahvanainen, K.U.O., Kuusela, T.A., Levine, B.D., Cooke, W.H., Mano, T., Iwase, S., Saito, M., Sugiyama, Y., Ertl, A.C., Biaggioni, I., et al. 2002. Influence of microgravity on astronauts' sympathetic and vagal responses to Valsalva's manoeuvre. *Journal of Physiology* 538(Pt 1):309-320.
534. Eckberg, D.L., Halliwill, J.R., Beightol, L.A., Brown, T.E., Taylor, J.A., and Goble, R. 2010. Human vagal baroreflex mechanisms in space. *Journal of Physiology* 588(7):1129-1138.
535. Ertl, A.C., Diedrich, A., Biaggioni, I., Levine, B.D., Robertson, R.M., Cox, J.F., Zuckerman, J.H., Pawelczyk, J.A., Ray, C.A., Buckey, J.C., Jr., Lane, L.D., et al. 2002. Human muscle sympathetic nerve activity and plasma noradrenaline kinetics in space. *Journal of Physiology* 538(Pt 1):321-329.
536. Cox, J.F., Tahvanainen, K.U.O., Kuusela, T.A., Levine, B.D., Cooke, W.H., Mano, T., Iwase, S., Saito, M., Sugiyama, Y., Ertl, A.C., Biaggioni, I., et al. 2002. Influence of microgravity on astronauts' sympathetic and vagal responses to Valsalva's manoeuvre. *Journal of Physiology* 538(Pt 1):309-320.
537. Levine, B.D., Pawelczyk, J.A., Ertl, A.C., Cox, J.F., Zuckerman, J.H., Diedrich, A., Biaggioni, I., Ray, C.A., Smith, M.L., Iwase, S., Saito, M., et al. 2002. Human muscle sympathetic neural and haemodynamic responses to tilt following spaceflight. *Journal of Physiology* 538(Pt 1):331-340.
538. Norsk, P., and Christensen, N.J. 2009. The paradox of systemic vasodilation and sympathetic nervous stimulation in space. *Respiratory Physiology and Neurobiology* 169(Suppl. 1):S26-S29.

539. Christensen, N.J., Heer, M., Ivanova, K., and Norsk, P. 2005. Sympathetic nervous activity decreases during head-down bed rest but not during microgravity. *Journal of Applied Physiology* 99:1552-1557.
540. Ertl, A.C., Diedrich, A., Biaggioni, I., Levine, B.D., Robertson, R.M., Cox, J.F., Zuckerman, J.H., Pawelczyk, J.A., Ray, C.A., Buckey, J.C., Jr., Lane, L.D., et al. 2002. Human muscle sympathetic nerve activity and plasma noradrenaline kinetics in space. *Journal of Physiology* 538(Pt 1):321-329.
541. Levine, B.D., Pawelczyk, J.A., Ertl, A.C., Cox, J.F., Zuckerman, J.H., Diedrich, A., Biaggioni, I., Ray, C.A., Smith, M.L., Iwase, S., Saito, M., et al. 2002. Human muscle sympathetic neural and haemodynamic responses to tilt following spaceflight. *Journal of Physiology* 538(Pt 1):331-340.
542. Cox, J.F., Tahvanainen, K.U.O., Kuusela, T.A., Levine, B.D., Cooke, W.H., Mano, T., Iwase, S., Saito, M., Sugiyama, Y., Ertl, A.C., Biaggioni, I., et al. 2002. Influence of microgravity on astronauts' sympathetic and vagal responses to Valsalva's manoeuvre. *Journal of Physiology* 538(Pt 1):309-320.
543. Fu, Q., Levine, B.D., Pawelczyk, J.A., Ertl, A.C., Diedrich, A., Cox, J.F., Zuckerman, J.H., Raya, C.A., Smith, M.L., Iwase, S., Saito, M., et al. 2002. Cardiovascular and sympathetic neural responses to handgrip and cold pressor stimuli in humans before, during and after spaceflight. *Journal of Physiology* 544(Pt 2):653-664.
544. Levine, B.D., Pawelczyk, J.A., Ertl, A.C., Cox, J.F., Zuckerman, J.H., Diedrich, A., Biaggioni, I., Ray, C.A., Smith, M.L., Iwase, S., Saito, M., et al. 2002. Human muscle sympathetic neural and haemodynamic responses to tilt following spaceflight. *Journal of Physiology* 538(Pt 1):331-340.
545. Levine, B.D., Pawelczyk, J.A., Zuckerman, J.H., Zhang, R., Iwasaki, K., and Blomqvist, C.G. 2003. Neural control of the cardiovascular system in space. Pp. 175-185 in *The Neurolab Spacelab Mission: Neuroscience Research in Space: Results from the STS-90, Neurolab Spacelab Mission* (J.C. Buckey and J.L. Homick, eds.). NASA 1-11. NASA Johnson Space Center, Houston, Tex. January.
546. Pawelczyk, J.A., Zuckerman, J.H., Blomqvist, C.G., and Levine, B.D. 2001. Regulation of muscle sympathetic nerve activity (MSNA) after bed rest deconditioning. *American Journal of Physiology. Heart and Circulatory Physiology* 280:H2230-H2239.
547. Shoemaker, J.K., Hogeman, C.S., and Sinoway, L.T. 1999. Contributions of MSNA and stroke volume to orthostatic intolerance following bed rest. *American Journal of Physiology* 277(4 Pt 2):R1084-R1090.
548. Kamiya, A., Iwase, S., Michikami, D., Fu, Q., and Mano, T. 2000. Head-down bed rest alters sympathetic and cardiovascular responses to mental stress. *American Journal of Physiology. Regulatory, Integrative and Comparative Physiology* 279:R440-R447.
549. Shoemaker, J.K., Hogeman, C.S., and Sinoway, L.T. 1999. Contributions of MSNA and stroke volume to orthostatic intolerance following bed rest. *American Journal of Physiology* 277(4 Pt 2):R1084-R1090.
550. Fritsch-Yelle, J.M., Whitson, P.A., Bondar, R.L., and Brown, T.E. 1996. Subnormal norepinephrine release relates to presyncope in astronauts after spaceflight. *Journal of Applied Physiology* 81:2134-2141.
551. Shoemaker, J.K., Hogeman, C.S., and Sinoway, L.T. 1999. Contributions of MSNA and stroke volume to orthostatic intolerance following bed rest. *American Journal of Physiology* 277(4 Pt 2):R1084-R1090.
552. Fritsch-Yelle, J.M., Whitson, P.A., Bondar, R.L., and Brown, T.E. 1996. Subnormal norepinephrine release relates to presyncope in astronauts after spaceflight. *Journal of Applied Physiology* 81:2134-2141.
553. Meck, J.V., Waters, W.W., Ziegler, M.G., deBlock, H.F., Mills, P.J., Robertson, D., and Huang, P.L. 2004. Mechanisms or postspaceflight orthostatic hypotension: Low alpha 1-adrenergic receptor responses before flight and central autonomic dysregulation postflight. *American Journal of Physiology. Heart and Circulatory Physiology* 286:H1486-H1495.
554. Blomqvist, C.G., and Stone, H.L. 1983. Cardiovascular adjustments to gravitational stress. Pp. 1025-1063 in *Handbook of Physiology. The Cardiovascular System. Peripheral Circulation and Organ Blood Flow* (J.T. Shepherd, F.M. Abboud, and S.R. Geiger, eds.). American Physiological Society, Bethesda, Md.
555. Aubert, A.E., Beckers, F., and Verheyden, B. 2005. Cardiovascular function and basics of physiology in microgravity. *Acta Cardiologica* 60:129-151.
556. Norsk, P. 2009. Cardiovascular research in space. *Respiratory Physiology and Neurobiology* 169(S1):S2-S3.
557. Hoffmann, U. 2009. Results from recent spaceflight experiments. *Respiratory Physiology and Neurobiology* 169(S1) S4-S5.
558. Navasiolava, N., and Custaud, M.A. 2009. What are the future top priority questions in cardiovascular research and what new hardware needs to be developed? *Respiratory Physiology and Neurobiology* 169(S1):S73-S74.
559. Coupé, M., Fortrat, J.O., Larina, I., Gauquelin-Koch, G., Gharib, C., and Custaud, M.A. 2009. Cardiovascular deconditioning: From autonomic nervous system to microvascular dysfunctions. *Respiratory Physiology and Neurobiology* 169(S1):S10-S12.
560. Furlan, R., Barbic, F., Casella, F., Severgnini, G., Zenoni, L., Mercieri, A., Mangili, R., Costantino, G., and Porta, A. 2009. Neural autonomic control in orthostatic intolerance. *Respiratory Physiology and Neurobiology* 169(S1):S17-S20.

561. Verheyden, B., Liu, J., Beckers, F., and Aubert, A.E. 2009. Adaptation of heart rate and blood pressure to short and long duration space missions. *Respiratory Physiology and Neurobiology* 169(S1):S13-S16.
562. Ertl, A.C., Diedrich, A., Biaggioni, I., Levine, B.D., Robertson, R.M., Cox, J.F., Zuckerman, J.H., Pawelczyk, J.A., Ray, C.A., Buckey, J.C., Jr., Lane, L.D., et al. 2002. Human muscle sympathetic nerve activity and plasma noradrenaline kinetics in space. *Journal of Physiology* 538(Pt 1):321-329.
563. Cox, J.F., Tahvanainen, K.U.O., Kuusela, T.A., Levine, B.D., Cooke, W.H., Mano, T., Iwase, S., Saito, M., Sugiyama, Y., Ertl, A.C., Biaggioni, I., et al. 2002. Influence of microgravity on astronauts' sympathetic and vagal responses to Valsalva's manoeuvre. *Journal of Physiology* 538(Pt 1):309-320.
564. Levine, B.D., Pawelczyk, J.A., Ertl, A.C., Cox, J.F., Zuckerman, J.H., Diedrich, A., Biaggioni, I., Ray, C.A., Smith, M.L., Iwase, S., Saito, M., et al. 2002. Human muscle sympathetic neural and haemodynamic responses to tilt following spaceflight. *Journal of Physiology* 538(Pt 1):331-340.
565. Norsk, P., and Christensen, N.J. 2009. The paradox of systemic vasodilation and sympathetic nervous stimulation in space. *Respiratory Physiology and Neurobiology* 169(Suppl. 1):S26-S29.
566. Levine, B.D., Pawelczyk, J.A., Zuckerman, J.H., Zhang, R., Iwasaki, K., and Blomqvist, C.G. 2003. Neural control of the cardiovascular system in space. Pp. 175-185 in *The Neurolab Spacelab Mission: Neuroscience Research in Space: Results from the STS-90, Neurolab Spacelab Mission* (J.C. Buckey and J.L. Homick, eds.). NASA 1-11. NASA Johnson Space Center, Houston, Tex. January.
567. Pawelczyk, J.A., Zuckerman, J.H., Blomqvist, C.G., and Levine, B.D. 2001. Regulation of muscle sympathetic nerve activity after bed rest deconditioning. *American Journal of Physiology. Heart and Circulatory Physiology* 280:H2230-H2239.
568. Shoemaker, J.K., Hogeman, C.S., and Sinoway, L.T. 1999. Contributions of MSNA and stroke volume to orthostatic intolerance following bed rest. *American Journal of Physiology* 277(4 Pt 2):R1084-R1090.
569. Fritsch-Yelle, J.M., Whitson, P.A., Bondar, R.L., and Brown, T.E. 1996. Subnormal norepinephrine release relates to presyncope in astronauts after spaceflight. *Journal of Applied Physiology* 81:2134-2141.
570. Cox, J.F., Tahvanainen, K.U.O., Kuusela, T.A., Levine, B.D., Cooke, W.H., Mano, T., Iwase, S., Saito, M., Sugiyama, Y., Ertl, A.C., Biaggioni, I., et al. 2002. Influence of microgravity on astronauts' sympathetic and vagal responses to Valsalva's manoeuvre. *Journal of Physiology* 538(Pt 1):309-320.
571. Eckberg, D.L., Halliwill, J.R., Beightol, L.A., Brown, T.E., Taylor, J.A., and Goble, R. 2010. Human vagal baroreflex mechanisms in space. *Journal of Physiology* 588(7):1129-1138.
572. Shoemaker, J.K., Hogeman, C.S., and Sinoway, L.T. 1999. Contributions of MSNA and stroke volume to orthostatic intolerance following bed rest. *American Journal of Physiology* 277(4 Pt 2):R1084-R1090.
573. Fritsch-Yelle, J.M., Whitson, P.A., Bondar, R.L., and Brown, T.E. 1996. Subnormal norepinephrine release relates to presyncope in astronauts after spaceflight. *Journal of Applied Physiology* 81:2134-2141.
574. Norsk, P., and Christensen, N.J. 2009. The paradox of systemic vasodilation and sympathetic nervous stimulation in space. *Respiratory Physiology and Neurobiology* 169(Suppl. 1):S26-S29.
575. Fritsch-Yelle, J.M., Whitson, P.A., Bondar, R.L., and Brown, T.E. 1996. Subnormal norepinephrine release relates to presyncope in astronauts after spaceflight. *Journal of Applied Physiology* 81:2134-2141.
576. Shoemaker, J.K., Hogeman, C.S., Silber, D.H., Gray, K., Herr, M., and Sinoway, L.I. 1998. Head-down-tilt bed rest alters forearm vasodilator and vasoconstrictor responses. *Journal of Applied Physiology* 84:1756-1762.
577. Meck, J.V., Waters, W.W., Ziegler, M.G., deBlock, H.F., Mills, P.J., Robertson, D., and Huang, P.L. 2004. Mechanisms or postspaceflight orthostatic hypotension: Low alpha 1-adrenergic receptor responses before flight and central autonomic dysregulation postflight. *American Journal of Physiology. Heart and Circulatory Physiology* 286:H1486-H1495.
578. Fu, Q., Witkowski, S., and Levine, B.D. 2004. Vasoconstrictor reserve and sympathetic neural control of orthostasis. *Circulation* 110:2931-2937.
579. Kamiya, A., Iwase, S., Michikami, D., Fu, Q., and Mano, T. 2000. Head-down bed rest alters sympathetic and cardiovascular responses to mental stress. *American Journal of Physiology. Regulatory, Integrative and Comparative Physiology* 279:R440-R447.
580. Hasser, E.M., and Moffitt, J.A. 2001. Regulation of sympathetic nervous system function after cardiovascular deconditioning. *Annals of the New York Academy of Sciences* 940:454-468.
581. Rowell, L.B. 1993. *Human Cardiovascular Control.* Oxford University Press, New York, N.Y.
582. Ray, C.A., Vasques, M., Miller, T.A., Wilkerson, M.K., and Delp, M.D. 2001. Effect of short-term microgravity and long-term hindlimb unloading on rat cardiac mass and function. *Journal of Applied Physiology* 91(3):1207-1213.
583. Zhang, L.F., Yu, Z.B., Ma, J., and Mao, Q.W. 2001. Peripheral effector mechanism hypothesis of postflight cardiovascular dysfunction. *Aviation, Space, and Environmental Medicine* 72:567-575.
584. Delp, M.D. 1999. Myogenic and vasoconstrictor responsiveness of skeletal muscle arterioles is diminished by hindlimb unloading. *Journal of Applied Physiology* 86(4):1178-1184.

585. Lin, L.J., Gao, F., Bai, Y.G., Bao, J.X., Huang, X.F., Ma, J., and Zhang, L.F. 2009. Contrasting effects of simulated microgravity with and without daily -Gx gravitation on structure and function of cerebral and mesenteric small arteries in rats. *Journal of Applied Physiology* 107:1710-1721.
586. Colleran P.N., Behnke, B.J., Wilkerson, M.K., Donato, A.J., and Delp, M.D. 2008. Simulated microgravity alters rat mesenteric artery vasoconstrictor dynamics through an intracellular Ca(2+) release mechanism. *American Journal of Physiology. Regulatory, Integrative and Comparative Physiology* 294:R1577-R1585.
587. Purdy, R.E., Duckles, S.P., Krause, D.N., Rubera, K.M., and Sara, D. 1998. Effect of simulated microgravity on vascular contractility. *Journal of Applied Physiology* 85(4):1307-1315.
588. Summers, S.M., Hayashi, Y., Nguyen, S.V., Nguyen, T.M., and Purdy, R.E. 2009. Hindlimb unweighting induces changes in the p38MAPK contractile pathway of the rat abdominal aorta. *Journal of Applied Physiology* 107(1):121-127.
589. Tuday, E.C., Meck, J.V., Nyhan, D., Shoukas, A.A., and Berkowitz, D.E. 2007. Microgravity-induced changes in aortic stiffness and their role in orthostatic intolerance. *Journal of Applied Physiology* 102(3):853-858.
590. Tuday, E.C., Nyhan, D., Shoukas, A.A., and Berkowitz, D.E. 2009. Simulated microgravity-induced aortic remodeling. *Journal of Applied Physiology* 106(6):2002-2008.
591. Hasser, E.M., and Moffitt, J.A. 2001. Regulation of sympathetic nervous system function after cardiovascular deconditioning. *Annals of the New York Academy of Sciences* 940:454-468.
592. Hasser, E.M., and Moffitt, J.A. 2001. Regulation of sympathetic nervous system function after cardiovascular deconditioning. *Annals of the New York Academy of Sciences* 940:454-468.
593. Myerburg, R.J., and Castellanos, A. 2006. Emerging paradigms of the epidemiology and demographics of sudden cardiac arrest. *Heart Rhythm* 3:235-239.
594. Ridker, P.M., Danielson, E., Fonseca, F.A., Genest, J., Gotto, A.M., Jr., Kastelein, J.J., Koenig, W., Libby, P., Lorenzatti, A.J., MacFadyen, J.G., Nordestgaard, B.G., et al. 2008. Rosuvastatin to prevent vascular events in men and women with elevated C-reactive protein. *New England Journal of Medicine* 359(21):2195-2207.
595. LaMonte, M.J., Fitzgerald, S.J., Levine, B.D., Church, T.S., Kampert, J.B., Nichaman, M.Z., Gibbons, L.W., and Blair, S.N. 2006. Coronary artery calcium, exercise tolerance, and CHD events in asymptomatic men. *Atherosclerosis* 189(1):157-162.
596. Convertino, V.A. 2002. Mechanisms of microgravity induced orthostatic tolerance: Implications for effective countermeasures. *Journal of Gravitational Physiology* 9:1-14.
597. Meck, J.V., Reyes, C.J., Perez, S.A., Goldberger, A.L., and Ziegler, M.G. 2001. Marked exacerbation of orthostatic intolerance after long- vs. short-duration spaceflight in veteran astronauts. *Psychosomatic Medicine* 63(6):865-873.
598. Fritsch-Yelle, J.M., Whitson, P.A., Bondar, R.L., and Brown, T.E. 1996. Subnormal norepinephrine release relates to presyncope in astronauts after spaceflight. *Journal of Applied Physiology* 81:2134-2141.
599. Waters, W.W., Ziegler, M.G., and Meck, J.V. 2002. Postspaceflight orthostatic hypotension occurs mostly in women and is predicted by low vascular resistance. *Journal of Applied Physiology* 92:586-594.
600. Meck, J.V., Waters, W.W., Ziegler, M.G., deBlock, H.F., Mills, P.J., Robertson, D., and Huang, P.L. 2004. Mechanisms or postspaceflight orthostatic hypotension: Low alpha 1-adrenergic receptor responses before flight and central autonomic dysregulation postflight. *American Journal of Physiology. Heart and Circulatory Physiology* 286:H1486-H1495.
601. Moore, A.D., Jr., Lee, S.M., Charles, J.B., Greenisen, M.C., and Schneider, S.M. 2001. Maximal exercise as a countermeasure to orthostatic intolerance after spaceflight. *Medicine and Science in Sports and Exercise* 33:75-80.
602. Buckey, J.C., Jr., Lane, L.D., Levine, B.D., Watenpaugh, D.E., Wright, S.J., Moore, W.E., Gaffney, F.A., and Blomqvist, C.G. 1996. Orthostatic intolerance after spaceflight. *Journal of Applied Physiology* 81(1):7-18.
603. Levine, B.D., Pawelczyk, J.A., Ertl, A.C., Cox, J.F., Zuckerman, J.H., Diedrich, A., Biaggioni, I., Ray, C.A., Smith, M.L., Iwase, S., Saito, M., et al. 2002. Human muscle sympathetic neural and haemodynamic responses to tilt following spaceflight. *Journal of Physiology* 538(Pt 1):331-340.
604. Abboud, F.M., Eckberg, D.L., Johannsen, U.J., and Mark, A.L. 1979. Carotid and cardiopulmonary baroreceptor control of splanchnic and forearm vascular resistance during venous pooling in man. *Journal of Physiology* 286:173-184.
605. Pawelczyk, J.A., and Levine, B.D. 2002. Heterogeneous responses of human limbs to infused adrenergic agonists: A gravitational effect? *Journal of Applied Physiology* 92:2105-2113.
606. Meck, J.V., Waters, W.W., Ziegler, M.G., deBlock, H.F., Mills, P.J., Robertson, D., and Huang, P.L. 2004. Mechanisms or postspaceflight orthostatic hypotension: Low alpha 1-adrenergic receptor responses before flight and central autonomic dysregulation postflight. *American Journal of Physiology. Heart and Circulatory Physiology* 286:H1486-H1495.
607. Fu, Q., Witkowski, S., and Levine, B.D. 2004. Vasoconstrictor reserve and sympathetic neural control of orthostasis. *Circulation* 110:2931-2937.
608. Buckey, J.C., Jr., Lane, L.D., Levine, B.D., Watenpaugh, D.E., Wright, S.J., Moore, W.E., Gaffney, F.A., and Blomqvist, C.G. 1996. Orthostatic intolerance after spaceflight. *Journal of Applied Physiology* 81(1):7-18.

609. Fritsch-Yelle, J.M., Whitson, P.A., Bondar, R.L., and Brown, T.E. 1996. Subnormal norepinephrine release relates to presyncope in astronauts after spaceflight. *Journal of Applied Physiology* 81:2134-2141.
610. Waters, W.W., Ziegler, M.G., and Meck, J.V. 2002. Postspaceflight orthostatic hypotension occurs mostly in women and is predicted by low vascular resistance. *Journal of Applied Physiology* 92:586-594.
611. Fu, Q., Witkowski, S., and Levine, B.D. 2004. Vasoconstrictor reserve and sympathetic neural control of orthostasis. *Circulation* 110:2931-2937.
612. Kamiya, A., Michikami, D., Fu, Q., Iwase, S., Hayano, J., Kawada, T., Mano, T., and Sunagawa, K. 2003. Pathophysiology of orthostatic hypotension after bed rest: Paradoxical sympathetic withdrawal. *American Journal of Physiology. Heart and Circulatory Physiology* 285(3):H1158-H1167.
613. Abboud, F.M., and Eckstein, J.W. 1966. Reflex vasoconstrictor and vasodilator responses in man. *Circulation Research* 18/19(Suppl. 1):96-103.
614. Abboud, F.M., and Eckstein, J.W. 1966. Active reflex vasodilatation in man. *Federation Proceedings* 25(6):1611-1617.
615. Mark, A.L., Kioschos, J.M., Abboud, F.M., Heistad, D.D., and Schmid, P.G. 1973. Abnormal vascular responses to exercise in patients with aortic stenosis. *Journal of Clinical Investigation* 52(5):1138-1146.
616. Mark, A.L., Abboud, F.M., Schmid, P.G., and Heistad, D.D. 1973. Reflex vascular responses to left ventricular outflow obstruction and activation of ventricular baroreceptors in dogs. *Journal of Clinical Investigation* 52(5):1147-1153.
617. Iwasaki, K., Levine, B.D., Zhang, R., Zuckerman, J.H., Pawelczyk, J.A., Diedrich, A., Ertl, A.C., Cox, J.F., Cooke, W.H., Giller, C.A., Ray, C.A., et al. 2007. Human cerebral autoregulation before, during and after spaceflight. *Journal of Physiology* 579 (Pt 3):799-810.
618. Zhang, R., and Levine, B.D. 2007. Autonomic ganglionic blockade does not prevent reduction in cerebral blood flow velocity during orthostasis in humans. *Stroke* 38:1238.
619. Zhang, L.F. 2001. Vascular adaptation to microgravity: What have we learned? *Journal of Applied Physiology* 91:2415-2430.
620. Purdy, R.E., and Meck, J.V. 2007. Endothelium in space. Pp. 50-526 in *Endothelial Biomedicine* (W.C. Aird, ed.). Cambridge University Press, New York, N.Y.
621. Waters, W.W., Ziegler, M.G., and Meck, J.V. 2002. Postspaceflight orthostatic hypotension occurs mostly in women and is predicted by low vascular resistance. *Journal of Applied Physiology* 92:586-594.
622. Fu, Q., Arbab-Zadeh, A., Perhonen, M.A., Zhang, R., Zuckerman, J.H., and Levine, B.D. 2004. Hemodynamics of orthostatic intolerance: Implications for gender differences. *American Journal of Physiology. Heart and Circulatory Physiology* 286(1):H449-H457.
623. Fu, Q., Witkowski, S., Okazaki, K., and Levine, B.D. 2005. Effects of gender and hypovolemia on sympathetic neural responses to orthostatic stress. *American Journal of Physiology. Regulatory, Integrative and Comparative Physiology* 289(1):R109-R116.
624. Fu, Q., Vangundy, T.B., Galbreath, M.M., Shibata, S., Jain, M., Hastings, J.L., Bhella, P.S., and Levine, B.D. 2010. Cardiac origins of the postural orthostatic tachycardia syndrome. *Journal of the American College of Cardiology* 55(25):2858-2868.
625. Fu, Q., Okazaki, K., Shibata, S., Shook, R.P., VanGunday, T.B., Galbreath, M.M., Reelick, M.F., and Levine, B.D. 2009. Menstrual cycle effects on sympathetic neural responses to upright tilt. *Journal of Physiology* 587(Pt 9):2019-2031.
626. Levine, B.D., Lande, L.D., Watenpaugh, D.E., Gaffney, F.A., Buckey, J.C., and Blomqvist, C.Q. 1996. Maximum exercise performance after adaptation to microgravity. *Journal of Applied Physiology* 81:686-694.
627. Kozlovskaya, I.B., and Grigoriev, A.I. 2004. Russian system of countermeasures on board of the International Space Station (ISS): The first results. *Acta Astronautica* 55:2233-2237.
628. Macias, B.R., Groppo, E.R., Eastlack, R.K., Watenpaugh, D.E., Lee, S.M., Schneider, S.M., Boda, W.L., Smith, S.M., Cutuk, A., Pedowitz, R.A., Meyer, R.S., and Hargens, A.R. 2005. Space exercise and Earth benefits. *Current Pharmaceutical Biotechnology* 6:305-317.
629. Convertino, V.A. 1996. Exercise and adaptation to microgravity environments. Pp. 815-843 in *Handbook of Physiology, Environmental Physiology*. American Physiological Society, Bethesda, Md.
630. Fortney, S.M., Schneider, V.S., and Greenleaf, J.E. 1996. The physiology of bed rest. Pp. 889-939 in *Handbook of Physiology, Environmental Physiology*. American Physiological Society, Bethesda, Md.
631. Michel, E.L., Rummel, J.A., Sawin, C.F., Buderer, M.C., and Lem, J.D. 1977. Results of Skylab medical experiment 171: Metabolic activity. Pp. 372-387 in *Biomedical Results from Skylab* (R.S. Johnson and L.F. Dietlein, eds.). NASA, Washington, D.C.
632. Moore, A.D., S.M. Lee, M.S. Laughlin, and R.D. Hagan. 2003. Aerobic deconditioning and recovery following long duration flight onboard the International Space Station. *Medicine and Science in Sports and Exercise* 35:263.

633. Berg, H.E., Dudley, G.A., Hather, B., and Tesch, P.A. 1993. Work capacity and metabolic and morphologic characteristics of the human quadriceps muscle in response to unloading. *Clinical Physiology* 13:337-347.
634. Hikida, R.S., Gollnick, P.D., Dudley, G.A., Convertino, V.A., and Buchanan, P. 1989. Structural and metabolic characteristics of human skeletal muscle following 30 days of simulated microgravity. *Aviation, Space, and Environmental Medicine* 60:664-670.
635. Il'ina-Kakueva, E.I., and Portugalov, V.V. 1979. [Effect of artificial gravitation on the skeletal musculature of rats during space flight.] [Russian] *Arkhiv Anat Gistologii Embriologii* 76(3):22-27.
636. Levine, B.D., Lande, L.D., Watenpaugh, D.E., Gaffney, F.A., Buckey, J.C., and Blomqvist, C.Q. 1996. Maximum exercise performance after adaptation to microgravity. *Journal of Applied Physiology* 81:686-694.
637. Michel, E.L., Rummel, J.A., Sawin, C.F., Buderer, M.C., and Lem, J.D. 1977. Results of Skylab medical experiment 171: Metabolic activity. Pp. 372-387 in *Biomedical Results from Skylab* (R.S. Johnson and L.F. Dietlein, eds.). NASA, Washington, D.C.
638. Levine, B.D., Lande, L.D., Watenpaugh, D.E., Gaffney, F.A., Buckey, J.C., and Blomqvist, C.Q. 1996. Maximum exercise performance after adaptation to microgravity. *Journal of Applied Physiology* 81:686-694.
639. Moore, A.D., Lee, S.M., Laughlin, M.S., and Hagan, R.D. 2003. Aerobic deconditioning and recovery following long duration flight onboard the International Space Station. *Medicine and Science in Sports and Exercise* 35:263.
640. Bungo, M.W., Charles, J.B., and Johnson, P.C., Jr. 1985. Cardiovascular deconditioning during space flight and the use of saline as a countermeasure to orthostatic intolerance. *Aviation, Space, and Environmental Medicine* 56(10):985-990.
641. Waters, W.W., Platts, S.H., Mitchell, B.M., Whitson, P.A., and Meck, J.V. 2005. Plasma volume restoration with salt tablets and water after bed rest prevents orthostatic hypotension and changes in supine hemodynamic and endocrine variables. *American Journal of Physiology. Heart and Circulatory Physiology* 288:H839-H847.
642. Shibata, S., Perhonen, M., and Levine, B.D. 2010. Supine cycling plus volume loading prevent cardiovascular deconditioning during bed rest. *Journal of Applied Physiology* 108(5):1177-1186.
643. Hastings, J., Pacini, E., Krainski, F., Shibata, S., Jain, M., Snell, P.G., Fu, Q., Palmer, M.D., Creson, D., and Levine, B.D. 2008. The multi-system effect of exercise training during prolonged bed rest deconditioning: An integrated approach to countermeasure development for the heart, lungs, muscle and bones. *Circulation* 118:S682. Abstract.
644. Schmid, P.G., Eckstein, J.W., and Abboud, F.M. 1966. Effect of 9-α-fluorohydrocortisone on forearm vascular responses to norepinephrine. *Circulation* 34:620-626.
645. Schmid, P.G., Eckstein, J.W., and Abboud, F.M. 1967. Comparison of effects of deoxycorticosterone and dexamethasone on cardiovascular responses to norepinephrine. *Journal of Clinical Investigation* 46(4):590-598.
646. Convertino, V.A. 1996. Exercise and adaptation to microgravity environments. Pp. 815-843 in *Handbook of Physiology, Environmental Physiology*. American Physiological Society, Bethesda, Md.
647. Keller, D.M., Low, D.A., Wingo, J.E., Brothers, R.M., Hastings, J., Davis, S.L., and Crandall, C.G. 2009. Acute volume expansion preserves orthostatic tolerance during whole-body heat stress in humans. *Journal of Physiology* 587(Pt 5):1131-1139.
648. Heistad, D.D., Abboud, F.M., Mark, A.L., and Schmid, P.G. 1973. Interaction of thermal and baroreceptor reflexes in man. *Journal of Applied Physiology* 35(5):581-586.
649. Keller, D.M., Low, D.A., Wingo, J.E., Brothers, R.M., Hastings, J., Davis, S.L., and Crandall, C.G. 2009. Acute volume expansion preserves orthostatic tolerance during whole-body heat stress in humans. *Journal of Physiology* 587(Pt 5):1131-1139.
650. Gaffney, F.A., Thal, E.R., Taylor, W.F., Bastian, B.C., Weigelt, J.A., Atkins, J.M., and Blomqvist, C.G. 1981. Hemodynamic effects of Medical Anti-Shock Trousers (MAST garment). *Journal of Trauma* 21(11):931-937.
651. Durand, S., Cui, J.J., Williams, K.D., and Crandall, C.G. 2004. Skin surface cooling improves orthostatic tolerance in normothermic individuals. *American Journal of Physiology. Regulatory, Integrative and Comparative Physiology* 286:R199-R205.
652. Cui, J., Durand, S., and Crandall, C.G. 2007. Baroreflex control of muscle sympathetic nerve activity during skin surface cooling. *Journal of Applied Physiology* 103:1284-1289.
653. Ramsdell, C.D., Mullen, T.J., Sundby, G.H., Rostoft, S., Sheynberg, N., Aljuri, N., Maa, M., Mukkamala, R., Sherman, D., Toska, K., Yelle, J., et al. 2001. Midodrine prevents orthostatic intolerance associated with simulated spaceflight. *Applied Physiology* 90:2245-2248.
654. Platts, S.H., Ziegler, M.G., Waters, W.W., Mitchell, B.M., and Meck, J.V. 2004. Midodrine prescribed to improve recurrent post-spaceflight orthostatic hypotension. *Aviation, Space, and Environmental Medicine* 75:554-556.
655. Jarvis, S.S., and Pawelczyk, J.A. 2010. The location of the human volume indifferent point predicts orthostatic tolerance. *European Journal of Applied Physiology* 109(2):331-341.

656. Greenleaf, J.E., Vernikos, J., Wade, C.E., and Barnes, P.R. 1992. Effect of leg exercise training on vascular volumes during 30 days of 6 degrees head-down bed rest. *Journal of Applied Physiology* 72:1887-1894.
657. Hargens, A.R., and Richardson, S. 2009. Cardiovascular adaptations, fluid shifts, and countermeasures related to space flight. *Respiratory, Physiology, and Neurobiology* 169(Suppl 1):S30-S33.
658. Dorfman, T.A., Levine, B.D., Tillery, T., Peshock, R.M., Hastings, J.L., Schneider, S.M., Macias, B.R., Biolo, G., and Hargens, A.R. 2007. Cardiac atrophy in women following bed rest. *Journal of Applied Physiology* 103:8-16.
659. Watenpaugh, D.E., and Hargens, A.R. 1996. The cardiovascular system in microgravity. Pp. 631-674 in *Handbook of Physiology, Environmental Physiology.* Oxford University Press, New York, N.Y.
660. Watenpaugh, D.E., O'Leary, D.D., Schneider, S.M., Lee, S.M., Macias, B.R., Tanaka, K., Hughson, R.L., and Hargens, A.R. 2007. Lower body negative pressure exercise plus brief postexercise lower body negative pressure improve post-bed rest orthostatic tolerance. *Journal of Applied Physiology* 103(6):1964-1972.
661. Shibata, S., Perhonen, M., and Levine, B.D. 2010. Supine cycling plus volume loading prevent cardiovascular deconditioning during bed rest. *Journal of Applied Physiology* 108(5):1177-1186.
662. Hastings, J., Pacini, E., Krainski, F., Shibata, S., Jain, M., Snell, P.G., Fu, Q., Palmer, M.D., Creson, D., and Levine, B.D. 2008. The multi-system effect of exercise training during prolonged bed rest deconditioning: An integrated approach to countermeasure development for the heart, lungs, muscle and bones. *Circulation* 118:S682. Abstract.
663. Buckey, J.C., Jr., Lane, L.D., Levine, B.D., Watenpaugh, D.E., Wright, S.J., Moore, W.E., Gaffney, F.A., and Blomqvist, C.G. 1996. Orthostatic intolerance after spaceflight. *Journal of Applied Physiology* 81(1):7-18.
664. Abboud, F.M. 1993. Neurocardiogenic syncope (Editorial Comment). *New England Journal of Medicine* 328:1117-1120.
665. Eichna, L.W., Horvath, S.M., and Bean, W.B. 1947. Post-exertional orthostatic hypotension. *American Journal of the Medical Sciences* 213:641-654.
666. Meck, J.V., Reyes, C.J., Perez, S.A., Goldberger, A.L., and Ziegler, M.G. 2001. Marked exacerbation of orthostatic intolerance after long- vs. short-duration spaceflight in veteran astronauts. *Psychosomatic Medicine* 63(6):865-873.
667. Shibata, S., Perhonen, M., and Levine, B.D. 2010. Supine cycling plus volume loading prevent cardiovascular deconditioning during bed rest. *Journal of Applied Physiology* 108(5):1177-1186.
668. Hastings, J., Pacini, E., Krainski, F., Shibata, S., Jain, M., Snell, P.G., Fu, Q., Palmer, M.D., Creson, D., and Levine, B.D. 2008. The multi-system effect of exercise training during prolonged bed rest deconditioning: An integrated approach to countermeasure development for the heart, lungs, muscle and bones. *Circulation* 118:S682. Abstract.
669. Ramsdell, C.D., Mullen, T.J., Sundby, G.H., Rostoft, S., Sheynberg, N., Aljuri, N., Maa, M., Mukkamala, R., Sherman, D., Toska, K., Yelle, J., Bloomfield, D., Williams, G.H., and Cohen, R.J. 2001. Midodrine prevents orthostatic intolerance associated with simulated spaceflight. *Journal of Applied Physiology* 90(6):2245-2248.
670. Platts, S.H., Ziegler, M.G., Waters, W.W., Mitchell, B.M., and Meck, J.V. 2004. Midodrine prescribed to improve recurrent post-spaceflight orthostatic hypotension. *Aviation, Space, and Environmental Medicine* 75:554-556.
671. Lee, P.H., and Malik, R. 2005. Cardiovascular effects of radiation therapy: Practical approach to radiation therapy-induced heart disease. *Cardiology in Review* 13:80-86.
672. Andersen, R., Wethal, T., Gunther, A., Fossa, A., Edvardsen, T., Fossa, S.D., and Kjekshus, J. 2010. Relation of coronary artery calcium score to premature coronary artery disease in survivors >15 years of Hodgkin's lymphoma. *American Journal of Cardiolology* 105:149-152.
673. Meck, J.V., Dreyer, S.A., and Warren, E. 2009. Long-duration head-down bed rest: Project overview, vital signs, and fluid balance. *Aviation, Space, and Environmental Medicine* 80(Suppl):A1-A8.
674. Pancheva, M.V., Panchev, V.S., and Suvandjieva, A.V. 2006. Lower body negative pressure vs. lower body positive pressure to prevent cardiac atrophy after bed rest and spaceflight. What caused the controversy? *Journal of Applied Physiology* 100:1090-1093.
675. Prisk, G.K., Paiva, M., and West, J.B., eds. 2001. *Gravity and the Lung: Lessons from Microgravity.* Dekker, New York, N.Y.
676. West, J.B., Elliott, A.R., Guy, H.J.B., and Prisk, G.K. 1997. Pulmonary function in space. *Journal of the American Medical Association* 277:1957-1961.
677. Prisk, G.K., Fine, J.M., Cooper, T.K., and West, J.B. 2008. Lung function is unchanged in the 1 G environment following 6-months exposure to microgravity. *European Journal of Applied Physiology* 103:617-623.
678. Darquenne, C., and Prisk, G.K. 2008. Deposition of inhaled particles in the human lung is more peripheral in lunar than in normal gravity. *European Journal of Applied Physiology* 103:687-695.
679. National Research Council. 1998. *A Strategy for Research in Space Biology and Medicine in the New Century.* National Academy Press, Washington, D.C., pp. 132-155.
680. Ohlsson, C., Bengtsson, B.A., Isaksson, O.G., Andreassen, T.T., and Slootweg, M.C. 1998. Growth hormone and bone. *Endocrine Reviews* 19:55-79.

681. Welle, S., Thornton, S., Stalt, M., and McHenry, B. 1996. Growth hormone increases muscle mass but does not rejuvenate myofibrillar protein synthesis in healthy subjects over 60 years. *Journal of Clinical Endocrinology and Metabolism* 81:3239-3243.
682. Gooselink, K.L., Grindeland, R.E., Roy, R.R., Zhong, H., Bigbee, A.J., Grossman, E.J., and Edgerton, V.R. 1998. Skeletal muscle regulatin of bioassayable growth hormone in the rat pituitary. *Journal of Applied Physiology* 84(4):1425-1430.
683. Popil, V., and Baumann, G. 2004. Laboratory measurement of growth hormone. *Clinica Chimica Acta* 350(1-2):1-16.
684. Grigoriev, A.I., and Larina, I.M. 1999. Translated 2009. Growth hormone and other regulators of muscle metabolism in blood of humans exposed to prolonged space missions and hypokinesia. *Fiziol. Cheloveka.* 25:89-96.
685. Grigoriev, A.I., and Larina, I.M. 1999. Translated 2009. Growth hormone and other regulators of muscle metabolism in blood of humans exposed to prolonged space missions and hypokinesia. *Fiziol. Cheloveka.* 25:89-96.
686. McCall, G.E., Goulet, C., Roy, R.R., Grindeland, R.E., Boorman, G.I., Bigbee, A.J., Hodgson, J.A., Greenisen, M.C., and Edgerton, V.R. 1999. Spaceflight suppresses exercise-induced release of bioassayable growth hormone. *Journal of Applied Physiology* 87:1207-1212.
687. Macho, L., Koska, J., Ksinantova, L., Vigas, M., Noskov, V.B., Grigoriev, A.I., and Kvetnansky, R. 2001. Plasma hormone levels in human subjects during stress loads in microgravity and at readaptation to Earth's gravity. *Journal of Gravitational Physiology* 8:P131-P132.
688. Hymer, W.C., Grindeland, R.E., Kransov, I., Victorov, K., Motter, K., Mukherjee, P., Shellenberger, K., and Vasques, M. 1992. Effect of space flight on rat pituitary cell function. *Journal of Applied Physiology* 73(2 Suppl.):151S-157S.
689. Bigbee, A.J., Grindeland, R.E., Roy, R.R., Zhong, H., Gosselink, K.L., Arnaud, S., and Edgerton, V.R. 2006. Basal and evoked levels of bioassayable growth hormone are altered by hindlimb unloading. *Journal of Applied Physiology* 100:1037-1042.
690. Sawachenko, P.E., Arias, C., Kransnov, I., Grindeland, R.E. and Vale, W. 1992. Effects of spaceflight on hypothalamic peptide systems controlling growth hormone dynamics. *Journal of Applied Physiology* 73(2 Suppl.):158S-165S.
691. National Research Council. 1998. *A Strategy for Research in Space Biology and Medicine in the New Century.* National Academy Press, Washington, D.C.
692. Leach, C.L., and Rambaut, P.C. 1977. Biochemical responses of the Skylab crewman: An overview. Pp. 204-216 in *Biomedical Results from Skylab* (R.S. Johnson and L.E. Dietlein, eds.). NASA SP-377. NASA, Washington, D.C.
693. Kuipers, A. 1996. First results from experiments performed with the ESA Anthrorack during the D-2 Spacelab mission. *Acta Astronautica* 38:865-875.
694. Leach, C.S., Johnson, P.C. and Driscoll, T.B. 1977. Prolong weightlessness effect on post flight plasma hormones. *Aviation, Space, and Environmental Medicine* 48(7):595-597.
695. Grigoriev, A.I., and Larina, I.M. 1999. Translated 2009. Growth hormone and other regulators of muscle metabolism in blood of humans exposed to prolonged space missions and hypokinesia. *Fiziol. Cheloveka.* 25:89-96.
696. Macho, L., Kvetnansky, R., Fickova, M., Kolena, J., Knopp, J., Tigranian, R.A., Popova, I.A., and Grogoriev, A.I. 2001. Endocrine responses to space flight. *Journal of Gravitational Physiology* 8:P117-P120.
697. Macho, L., Kvetnansky, R., Fickova, M., Kolena, J., Knopp, J., Tigranian, R.A., Popova, I.A., and Grogoriev, A.I. 2001. Endocrine responses to space flight. *Journal of Gravitational Physiology* 8:P117-P120.
698. Wimalawansa, S.M., and Wimalawansa, S.J. 1999. Simulated weightlessness-induced attenuation of testosterone production may be responsible for bone loss.1999. *Endocrine* 10:253-260.
699. Adams, G.R., Haddad, F., McCue, S.A., Bodell, P.W., Zeng, M., Qin, A.X., and Baldwin, K.M. 2000. Effects of spaceflight and thyroid deficiency on rat hindlimb development II: Expression of MHC isoforms. *Journal of Applied Physiology* 88:904-916.
700. Mazziotti, G., Angeli, A., Bilezikian, J.P., Canalis, E. and Guistina, A. 2006. Glucocorticoid-induced osteoporosis: An update. *Trends in Endocrinology and Metabolism* 17:144-149.
701. Kelley, F.J., and Goldspink, D.F. 1982. The differing response of four muscle types to desamethasone treatment in the rat. *Biochemical Journal* 208:147-151.
702. Blanc, S., Normand, S., Pachiaudi, C., Duvareille, C., and Gharib, C. 2000. Leptin responses to physical inactivity induced by simulated weightlessness. *American Journal of Physiology. Regulatory, Integrative and Comparative Physiology* 279:R891-R898.
703. National Research Council. 1998. *A Strategy for Research in Space Biology and Medicine in the New Century.* National Academy Press, Washington, D.C.
704. Grigoriev, A.I., Huntoon, C., and Natochin, Yu.V. 1995. On the correlation between individual biochemical parameters of human blood serum following space flight and their basal values. *Acta Astronautica* 36:639-648.
705. Leach, C.S., Atchuler, S.I., and Clintron-Trevino, N.M. 1983. The endocrine and metabolic responses to space flight. *Medicine and Science in Sports and Exercise* 15:432-440.

706. Grigoriev, A.I., Bugrov, A.A., Bogomolov, V.V., Egorad, A.D., Polyakov, V.V., Tarasov, I.K., and Shulzhenko, E.B. 1993. Main medical results of extended flights on Space Station Mir in 1986-1990. *Acta Astronautica* 29:581-585.
707. Fitts, R.H., Riley, D.R., and Widrik, J.J. 2001. Functional and structural adaptations of skeletal muscle to microgravity. *Journal of Experimental Biology* 204:3201-3208.
708. Blanc, S., Normand, S., Rotz, P., Pachiaudi C., Vico, L., Gharib, C., and Gauquelin-Koch, G. 1998. Energy and water metabolism, body composition, and hormonal changes induced by 42 day of enforced inactivity and simulated weightlessness. *Journal of Clinical Endocrinology and Metabolism* 83:4289-4297.
709. Rittweger, J.H., Frost, M., Schiessl, H., Ohshima, H., et al. 2005. Muscle atrophy and bone loss after 90 days' bed rest and effects of flywheel resistive exercise and pamidronate: Results from the LTBR study. *Bone* 36:1019-1029.
710. Blanc, S., Normand, S., Pachiaudi, C., Duvareille, C. and Gharib, C. 2000. Leptin responses to physical inactivity induced by simulated weightlessness. *American Journal of Physiology. Regulatory, Integrative and Comparative Physiology* 279:R891-R890.
711. Blanc, S., Normand, S., Rotz, P., Pachiaudi C., et. al. 1998. Energy and water metabolism, body composition, and hormonal changes induced by 42 day of enforced inactivity and simulated weightlessness. *Journal of Clinical Endocrinology and Metabolism* 83:4289-4297.
712. Macho, L., Kvetnansky, R., Fickova, M., Kolena, J., et al. 2001. Endocrine responses to space flight. *Journal of Gravitational Physiology* 8:P117-P120.
713. Wimalawansa, S.M., and Wimalawansa, S.J. 1999. Simulated weightlessness-induced attenuation of testosterone production may be responsible for bone loss.1999. *Endocrine* 10:253-260.
714. National Research Council. 1998. *A Strategy for Research in Space Biology and Medicine in the New Century.* National Academy Press, Washington, D.C.
715. Blanc, S., Normand, S., Pachiaudi, C., Fortrat, J.-O., et al. 2000. Fuel homeostasis during physical inactivity induced by bed rest. *Journal of Clinical Endocrinology and Metabolism* 85:2223-2233.
716. Leach, C.S., Atchuler, S.I., and Clintron-Trevino, N.M. 1983. The endocrine and metabolic responses to space flight. *Medicine and Science in Sports and Exercise* 15:432-440.
717. Stein, T.P., Schluter, M.D., and Boden, B. 1994. Development of insulin resistance by astronauts during spaceflight. *Aviation, Space, and Environmental Medicine* 65:1091-1096.
718. Kuipers, A. First results from experiments performed with the ESA Anthrorack during the D-2 Spacelab mission. 1996. *Acta Astronautica* 38:865-875.
719. National Research Council. 1998. *A Strategy for Research in Space Biology and Medicine in the New Century.* National Academy Press, Washington, D.C., p. 142.
720. Kuipers, A. 1996. First results from experiments performed with the ESA Anthrorack during the D-2 Spacelab mission. *Acta Astronautica* 38:865-875.
721. Leach, C.L., and Rambaut, P.C. 1977. Biochemical responses of the Skylab crewman: An overview. Pp. 204-216 in *Biomedical Results from Skylab* (R.S. Johnson and L.E. Dietlein, eds.). NASA SP-377. NASA, Washington, D.C.
722. Grigoriev, A.I., Bugrov, A.A., Bogomolov, V.V., Egorad, A.D. et al. 1993. Main medical result of extended flights on Space Station Mir in 1986-1990. *Acta Astronautica* 29:581-585.
723. Markin, A., Balashov, O., Polyakov, V., and Tigner, T. 1998. The dynamics of blood biochemical parameters in cosmonauts during long-term space flights. *Acta Astronautica* 42:247-253.
724. Mondon, C.E., Rodnick, K.J., Dolkas, C.B., Azhar, S., and Reaven, G.M. 1992. Alterations in glucose and protein metabolism in animals subjected to simulated microgravity. *Advances in Space* Research 12:169-177.
725. O'Keefe, M.P., Perez, F.R., Kinnick, T.R., Tischler, M.E., and Hendriksen, E.J. 2004. Development of whole-body and skeletal muscle insulin resistance after one day of hindlimb suspension. *Metabolism* 53:1215-1522.
726. Tobin, B.W., Tobin, W.W., Uchakin, P.N., and Leeper-Woodford, S.K. 2002. Insulin secretion and sensitivity in space flight: Diabetogenic effects. *Nutrition* 18:842-848.
727. Tobin, B.W., Leeper-Woodfford, S.K., Hashemi, B.B., Smith, S.M., et al. 2001. Altered TNF-α, glucose, insulin, and amino acids in islets of Langerhans cultered in a microgravity model system. *American Journal of Physiology. Endocrinology and Metabolism* 280:E92-E102.
728. National Research Council. 1998. *A Strategy for Research in Space Biology and Medicine in the New Century.* National Academy Press, Washington, D.C., p. 137.
729. Strollo F., Riondino, G., Harris, B., Strollo, G., et al. 1998. The effect of microgravity on testicular androgen secretion. *Aviation, Space, and Environmental Medicine* 69:133-136.
730. Strollo, F., Strollo, G., More, M., Ferretti, C., et al. 1994. Changes in human adrenal and gonadal function onboard Spacelab. *Journal of Gravitational Physiology* 4:P103-P104.

731. Nichiporuk, I.A., Evdokimov, V.V., Erasova, V.I., Smirov, O.A., et. al. 1998. Male reproductive system in conditions of bed-rest in a head down tilt. *Journal of Gravitational Physiology* 5:P101-P102.
732. Merrill, A.H., Jr., Wang, E., Mullins, R.E., Grindeland, R.E., et al. 1992. Analysis of plasma for metabolic and hormonal changes in rats flown aboard COSMOS 2044. *Journal of Applied Physiology* 73(Suppl.):132S-135S.
733. Macho, L., Kvetnansky, R., Fickova, M., Kolena, J., et al. 2001. Endocrine responses to space flight. *Journal of Gravitational Physiology* 8:P117-P120.
734. Merrill, A.H., Jr., Wang, E., Mullins, R.E., Grindeland, R.E., et al. 1992. Analysis of plasma for metabolic and hormonal changes in rats flown aboard COSMOS 2044. *Journal of Applied Physiology* 73(Suppl.):132S-135S.
735. Deaver, D.R., Amann, R.P., Hammerstedt, H., Ball, K., et al. 1992. Effects of caudal elevation on testicular function in rats, separation of effects on spermatogenesis and steroidogenesis. *Journal of Andrology* 13:224-231.
736. Sharma, C.S., Sarka, S., Periyakaruppan, A., Ravichandran, P., et al. 2008. Simulated microgravity activates apoptosis and NF-kappaB in mice testis. *Molecular and Cellular Biochemistry.* 313:71-78.
737. Oritz, R.M., Wade, C.E., and Morey-Holton, E. 2000. Urinary excretion of LH and testosterone from male rats during exposure to increased gravity: Post-spaceflight and centrifugation. *Proceedings of the Society for Experimental Biology and Medicine* 225:98-102.
738. Fedotova, N.L. 1967. [Spermatogenesis of the dogs Ugolyok and Veterok after their flight on board the Satellite Kosmos 110]. [Russian]. *Kosmicheskaya Biol. Med.* 1:28.
739. Amann, R.P., Deaver, D.R, Zirkin, B.R., Grills, S., et al.1992. Effects of microgravity or simulated launch on testicular function in rats. *Journal of Applied Physiology* 73(2 Suppl.):S174-S185.
740. Serova, L.V., Denisova, L.A., and Baikova, C. V. 1989. The effect of microgravity on the reproductive function of male-rats. *The Physiologist* 32(Suppl. 1):S29-S30.
741. Tash, J.S., Johnson, D.C., and Enders, G.C. 2001.Long term (6-wk) hindlimb suspension inhibits spermatogenesis in adult male rats. *Journal of Applied Physiology* 92:1191-1198.
742. Motabagani, M.A.H. 2007. Morphological and morphometric study on the effect of simulated microgravity on rat testis. *Chinese Journal of Physiology* 50:199-209.
743. Strollo, F., Masini, M.A., Pastorino, M., Ricci, F., et al. 2004. Microgravity induced alterations in cultured testicular cells. *Journal of Gravitational Physiology* 11:P187.
744. Strollo, F., Masini, M.A., Pastorino, M., Ricci, F., et al. 2004. Microgravity induced alterations in cultured testicular cells. *Journal of Gravitational Physiology* 11:P187.
745. Uva, B.M., Strollo, F., Ricci, F., Pastorino, M., et al. 2005. Morpho-functional alterations in testicular and nervous cells submitted to modeled microgravity. *Journal of Endocrinological Investigation* 28(11 Suppl. Proceedings):84-91.
746. Borer, K. 2003. *Exercise Endocrinology.* Human Kinetics Champaign, Ill., pp. 1-259.
747. Trappe, S., Costill, D., Gallagher, P., Creer, A., Peters, J.R., Evans, H., Riley, D.A., and Fitts, R.H. 2009. Exercise in space: Human skeletal muscle after 6 months aboard the International Space Station. *Journal of Applied Physiology* 106(4):1159-1168.
748. Fitts, R.H., Riley, D.R., and Widrik, J.J. 2001. Functional and structural adaptations of skeletal muscle to microgravity. *Journal of Experimental Biolology* 204:3201-3208.
749. Falduto, M.T., Czerwinski, S.M., and Hickson, R.C. 1990. Glucocorticoid-induced muscle atrophy prevention by exercise in fast-twitch fibers. *Journal of Applied Physiology* 69:1058-1062.
750. Wade, C.E., Stanford, K.I., Stein, T.P., and Greenleaf, J.G. 2005. Intensive exercise training suppresses testosterone during bed rest. *Journal of Applied Physiology* 99:59-63.
751. Borer, K.T. 2003. *Exercise Endocrinology.* Human Kinetics, Champaign, Ill., pp. 106-108.
752. Tou, J., Ronca, A., Grindeland, R., and Wade, C. 2002. Models to study gravitational biology of mammalian reproduction. *Biology of Reproduction* 67:1681-1687.
753. Caiozzo, V.J., Haddad, F., Lee, S., Baker, M., et al. 2009. Artificial gravity as a countermeasure to microgravity: A pilot study examining the effects on knee extensor and plantar flexor muscle groups. *Journal of Applied Physiology* 107:39-46.
754. Tash, J.S., Johnson, D.C., and Enders, G.C. 2001. Long term (6-wk) hindlimb suspension inhibits spermatogenesis in adult male rats. *Journal of Applied Physiology* 92:1191-1198.
755. Burden, H.W., Poole, M.C., Zary, J., Jeansonne, B., et al. 1998. The effects of space flight during gestation on rat uterine, *Journal of Gravitational Physiology* 5:23-29.
756. Hymer, W.C., Grindeland, R.E., Kransov, I., Victorov, K., et al. 1992. Effect of space flight on rat pituitary function. *Journal of Applied Physiology* 73(Suppl.):151S-157S.
757. Bigbee, A.J., Grindeland, R.E., Roy, R.R., Zhong, H., et al. 2006. Basal and evoked levels of bioassayable growth hormone are altered by hindlimb unloading. *Journal of Applied Physiology* 100:1037-1042.

758. Sawachenko, P.E., Arias, C., Kransnov, I., Grindeland, R.E., and Vale, W. 1992. Effects of spaceflight on hypothalamic peptide systems controlling growth hormone dynamics. *Journal of Applied Physiology* 73(Suppl.):158S-165S.
759. National Research Council. 1998. *A Strategy for Research in Space Biology and Medicine in the New Century.* National Academy Press, Washington, D.C.
760. Meck, J.V., Dreyer, S.A., and Warren, E. 2009. Long-duration head-down bed rest: Project overview, vital signs, and fluid balance. *Aviation, Space, and Environmental Medicine* 80(Suppl.):A1-A8.
761. Morey-Holton, E., and Globus, K. 2002. Hindlimb unloading rodent model. Technical aspects. *Journal of Applied Physiology* 92:1367-1377.
762. Hymer, W.C., Grindeland, R.E., Kransov, I., Victorov, K., et al. 1992. Effect of space flight on rat pituitary function. *Journal of Applied Physiology* 73(Suppl.):151S-157S.
763. Hymer, W.C., Grindeland, R.E., Salada, T., Nye, P., et al. 1996. Experimental modification of rat pituitary growth hormone cell function after space flight. *Journal of Applied Physiology* 80:955-970.
764. Grigoriev, A.I., Huntoon, C., and Larina, I.M. 1999. Translated 2009. Growth hormone and other regulators of metabolism in blood of humans exposed to prolonged space missions and hypokinesia. *Fizol. Cheloveka.* 25:89-96.
765. Gooselink, K.L., Grindeland, R.E., Roy, R.R., Zhong, H., Bigbee, A.J., Grossman, E.J., and Edgerton, V.R. 1998. Skeletal muscle regulatin of bioassayable growth hormone in the rat pituitary. *Journal of Applied Physiology* 84(4):1425-1430.
766. Adams, G.R., Haddad, F., McCue, S.A., Bodell, P.W., Zeng, M., Qin, A.X., and Baldwin, K.M. 2000. Effects of spaceflight and thyroid deficiency on rat hindlimb development II: Expression of MHC isoforms. *Journal of Applied Physiology* 88:904-916.
767. Kelley, F.J., and Goldspink, D.F. 1982. The differing response of four muscle types to dexamethasone treatment in the rat. *Biochemical Journal* 208:147-151.
768. National Research Council. 1998. *A Strategy for Research in Space Biology and Medicine in the New Century.* National Academy Press, Washington, D.C.
769. Tobin, B.W., Leeper-Woodfford, S.K., Hashemi, B.B., Smith, S.M., et al. 2001. Altered TNF-α, glucose, insulin, and amino acids in islets of Langerhans cultered in a microgravity model system. *American Journal of Physiology. Endocrinology and Metabolism* 280:E92-E102.
770. National Research Council. 1998. *A Strategy for Research in Space Biology and Medicine in the New Century.* National Academy Press, Washington, D.C., pp. 155-168.
771. Taylor, G.R., Neale, L.S., and Dardano, J.R. 1986. Immunological analyses of U.S. space shuttle crewmembers. *Aviation, Space, and Environmental Medicine* 57(3):213-217.
772. Meehan, R.T., Neale, L.S., Kraus, E.T., et al. 1992. Alteration in human mononuclear leucocytes following space flight. *Immunology* 76:491.
773. Stowe, R.P., Sams, C.F., Mehta, S.K., et al. 1999. Leukocyte subsets and neutrophil function after short-term spaceflight. *Journal of Leukocyte Biology* 65:179.
774. Crucian, B.E., Stowe, R.P., Pierson, D.L., and Sams, C.F. 2008. Immune system dysregulation following short- vs long-duration spaceflight. *Aviation, Space, and Environmental Medicine* 79(9):835-843.
775. Mehta, S.K., Kaur, I., Grimm, E.A., Smid, C., Feeback, D.L., and Pierson, D.L. 2001. Decreased non-MHC-restricted ($CD56^+$) killer cell cytotoxicity after spaceflight. *Journal of Applied Physiology* 91:1814-1818.
776. Meshkov, D., and Rykova, M. 1995. The natural cytotoxicity in cosmonauts on board space stations. *Acta Astronautica* 36(8-12):719-726.
777. National Research Council. 1998. *A Strategy for Research in Space Biology and Medicine in the New Century.* National Academy Press, Washington, D.C.
778. Kaur, I., Simons, E.R., Castro, V.A., Ott, C.M., and Pierson, D.L. 2004. Changes in neutrophil functions in astronauts. *Brain, Behavior, and Immunology* 18:443-450.
779. Kaur, I., Simons, E.R., Castro, V.A., Ott, C.M., and Pierson, D.L. 2005. Changes in monocyte functions of astronauts. *Brain, Behavior, and Immunology* 19:547-554.
780. Voss, E.W. 1984. Prolonged weightlessness and humoral immunity. *Science* 225:214-215.
781. Konstantinova, I.V., Sonnenfeld, G., Lesnyak, A.T., Shaffar, L., Mandel, A., Rykova, M.P., Antropova, E.N., and Ferrua, B. 1991. Cellular immunity and lymphokine production during spaceflights. *Physiologist* 34:S52-S56.
782. Manié, S., Konstantinova, I., Breittmayer, J.P., Ferrua, B., and Schaffar, L. 1991. Effects of a long duration spaceflight on human T lymphocyte and monocyte activity. *Aviation, Space, and Environmental Medicine* 62:1153-1158.
783. Crucian, B.E., Cubbage, M.L., and Sams, C.F. 2000. Altered cytokine production by specific human peripheral blood cell subsets immediately following space flight. *Journal of Interferon and Cytokine Research* 20:547.
784. Gardner, E.M., and Murasko D.M. 2002. Age-related changes in type 1 and type 2 cytokine production in humans. *Biogerontology* 3:271-290.

785. Taylor, G.R., and Janney, R.P. 1992. In vivo testing confirms a blunting of the human cell-mediated immune mechanism during space flight. *Journal of Leukocyte Biology* 51:129-132.
786. Gmünder, F.K., Konstantinove, I., Cogoli, A., et al. 1994. Cellular immunity in cosmonauts during long duration spaceflight on board the orbital MIR station. *Aviation, Space, and Environmental Medicine* 65:419.
787. Mehta, S.K., Cohrs, R.J., Forghani, B., Zerbe, G., Gilden, D.H., and Pierson, D.L. 2004. Stress-induced subclinical reactivation of Varicella Zoster Virus in astronauts. *Journal of Medical Virology* 72:174-179.
788. Pierson, D.L., Stowe, R.P., Phillips, T.M., Lugg, D.J., and Mehta, S.K. 2005. Epstein-Barr virus shedding by astronauts during space flight. *Brain, Behavior, and Immunology* 19:235-242.
789. Mehta, S.K., Stowe, R.P., Leiveson, A.H., Tyring, S.K., and Pierson, D.L. 2000. Reactivation and shedding of cytomegalovirus in astronauts. *Journal of Infection* 182:1761-1764.
790. Stowe, R.P., Mehta, S.K., Ferrando, A.A., Feeback, D.L., and Pierson, D.L. 2001. Immune responses and latent herpevirus reactivation in spaceflight. *Aviation, Space, and Environmental Medicine* 72:884.
791. Mehta, S.K., Cohrs, R.J., Forghani, B., Zerbe, G., Gilden, D.H., and Pierson, D.L. 2004. Stress-induced subclinical reactivation of Varicella Zoster Virus in astronauts. *Journal of Medical Virology* 72:174-179.
792. Pierson, D.L., Stowe, R.P., Phillips, T.M., Lugg, D.J., and Mehta, S.K. 2005. Epstein-Barr virus shedding by astronauts during space flight. *Brain, Behavior, and Immunology* 19:235-242.
793. NASA. 2009. *Risk of Crew Adverse Health Event Due to Altered Immune Response*. HRP-47060. Human Health Countermeasures Element Evidence Book. Human Research Program, NASA Johnson Space Center, Houston, Tex., June, p. 13-8.
794. Sonnenfeld, G. 2005. Use of animal models for space flight physiology studies, with special focus on the immune system. *Gravitational and Space Biology Bulletin* 18(2):31-35.
795. Nash, P., and Mastro, A. 1992. Variable lymphocyte responses in rats after space flight. *Experimental Cell Research* 202:125-131.
796. Grove, D., Pishak, S., and Mastro, A. 1995. The effect of a 10-day space flight on the function, phenotype, and adhesion molecule expression of splenocytes and lymph node lymphocytes. *Experimental Cell Research* 219:102-109.
797. Baqai, F.P., Gridley, D.S., Slater, J.M., Luo-Owen, X., Stodieck, L.S., Ferguson, V., Chapes, S.K., and Pecaut, M.J. 2009. Effects of spaceflight on innate immune function and antioxidant gene expression. *Journal of Applied Physiology* 106:1935-1942.
798. Gridley, D.S., Slater, J.M., Luo-Owen, X., Rizvi, A., Chapes, S.K., Stodieck, L.S., Ferguson, V.L., and Pecaut, M.J. 2008. Spaceflight effects on T lymphocyte distribution, function and gene expression. *Journal of Applied Physiology* 106:194-202.
799. Pahlavani, M.A. 2004. Influences of caloric restriction on aging immune system. *Journal of Nutrition, Health, and Aging* 8:38-47.
800. Gleesen, M., Bishop, N.C. 2005. The T cell and NK cell immune response to exercise. *Annals of Transplantation* 10(4):43-48.
801. Field, C.J., Johnson, I.R., and Schley, P.D. 2002. Nutrients and their role in host resistance to infection. *Journal of Leukocyte Biology* 71:16.
802. Dinges, D.F., Douglas, S.D., Zaugg, L., Campbell, D.E., McMann, J.M., Whitehouse, W.G., Orne, E.C., Kapoor, S.C., Icaza, E., and Orne, M.T. 1994. Leukocytosis and natural killer cell function parallel neurobehavioral fatigue induced by 64 hours of sleep deprivation. *Journal of Clinical Investigation* 93:1930-1939.
803. National Research Council. 1998. *A Strategy for Research in Space Biology and Medicine in the New Century.* National Academy Press, Washington, D.C., pp. 55-168.
804. National Research Council. 1998. *A Strategy for Research in Space Biology and Medicine in the New Century.* National Academy Press, Washington, D.C.
805. Pierson, D.L., Stowe, R.P., Phillips, T.M., Lugg, D.J., and Mehta, S.K. 2005. Epstein-Barr virus shedding by astronauts during space flight. *Brain, Behavior, and Immunity* 19:235-242.
806. Pecaut, M.J., Gridley, D.S., Smith, A.L., and Nelson, G.A. 2002. Dose and dose rate effects of whole-body proton-irradiation on lymphocyte blastogenesis and hematological variables: Part II. *Immunology Letters* 80(1):67-73.
807. Durante, M., George, K., and Cucinotta, F.A. 2006. Chromosomes lacking telomeres are present in the progeny of human lymphocytes exposed to heavy ions. *Radiation Research* 165:51-58.
808. Sonnenfeld, G. 2005. Use of animal models for space flight physiology studies, with special focus on the immune system. *Gravitational and Space Biology Bulletin* 18(2):31-35.
809. Wei, L.X., Zhou, J.N., Roberts, R.I., and Shi, Y.F. 2003. Lymphocyte reduction induced by hindlimb unloading: Distinct mechanisms in the spleen and thymus. *Cell Research* 13(6):465-471.

810. Nash, P.V., Bour, B.A., and Mastro, A.M. 1991. Effect of hindlimb suspension simulation of microgravity on in vitro immunological responses. *Experimental Cell Research* 195:353-360.
811. Crucian, B.E., Stowe, R.P., Mehta, S.K., Yetman, D.L., Leal, M.J., Quiriarte, H.D., Pierson, D.L., and Sams C.F. 2009. Immune status, latent viral reactivation, and stress during long-duration head-down bed rest. *Aviation, Space, and Environmental Medicine* 80(5):A37-A44.
812. Crucian, B., Lee, P., Stowe, R., Jones, J., Effenhauser, R., Widen, R., and Sams, C. 2007. Immune system changes during simulated planetary exploration on Devon Island, high arctic. *BMC Immunology* 8(7):1471-2172.
813. Tingate, T., Lugg, D.J., Muller, H.K., Stowe, R.P., and Pierson, D.L. 1997. Antarctic isolation: Immune and viral studies. *Immunology and Cell Biology* 75:275-283.
814. Tingate, T., Lugg, D.J., Muller, H.K., Stowe, R.P., and Pierson, D.L. 1997. Antarctic isolation: Immune and viral studies. *Immunology and Cell Biology* 75:275-283.
815. International Space Life Sciences Working Group (ISLSWG). 1999. Developmental Biology Workshop, Marine Biological Laboratory, Woods Hole, Mass., September 27-30.
816. National Research Council. 1987. *A Strategy for Space Biology and Medical Science for the 1980 s and 1990s.* National Academy Press, Washington, D.C.
817. National Research Council. 1998. *A Strategy for Research in Space Biology and Medicine in the New Century.* National Academy Press, Washington, D.C.
818. Moody, S.A., and Golden, C. 2000. Developmental biology research in space: Issues and directions in the era of the International Space Station. *Developmental Biology* 228:1-5.
819. NASA Developmental Biology Review Panel Report. 1999. NASA Ames Research Center, August 10-12, 1998.
820. Buckey, J.C., and Homick, J.L. 2000. *The Neurolab Spacelab Mission: Neuroscience Research in Space.* NASA Lyndon B. Johnson Space Center, Houston, Tex.
821. Ronca, A.E. 2003. Mammalian development in space. In *Development in Space* (H.J. Marthy, ed.). *Advances in Space Biology and Medicine*, Volume 9. Elsevier, The Netherlands.
822. Morey-Holton, E.R., Hill, E.L., and Souza, K.A. 2007. Animals and spaceflight: From survival to understanding. *Journal of Musculoskeletal and Neuronal Interactions* 7:17-25.
823. Goldberg, A.D., Allis, C.D., and Bernstein, E. 2007. Epigenetics: A landscape takes shape. *Cell* 128:635-638.
824. Meaney, M.J., Szyf, M., and Seckl, J.R. 2007. Epigenetic mechanisms of perinatal programming of hypothalamic-pituitary-adrenal function and health. *Trends in Molecular Medicine* 13(7):269-277.
825. Kandpal, R., Saviola, B., and Felton, J. 2009. The era of 'omics unlimited. *Biotechniques* 46:351-352, 354-355.
826. Marthy, H.J., ed. 2003. *Development in Space. Advances in Space Biology and Medicine*, Volume 9. Elsevier, The Netherlands.
827. Ikenaga, M., Yoshikawa, I., Kojo, M., Ayaki, T., Ryo, H., Ishizaki, K., Kato, T., Yamamoto, H., and Hara, R. 1997. Mutations induced in Drosophila during space flight. *Biological Sciences in Space* 11:346-350.
828. Reitz, G., Bucker, H., Facius, R., Hornck, G., Graul, E.H., Berger, H., Ruther, W., Heinrich, W., Beaujean, R., Enge, W., Alpatov, A.M., Ushakov, I.A., Zachvatkin, Yu.A., and Mesland, D.A.. 1989. Influence of cosmic radiation and/or microgravity on development of *Carausius morosus*. *Advances in Space Research* 9:161-173.
829. Leandro, L.J., Szewczyk, N.J., Benguría, A., Herranz, R., Laván, D., Medina, F.J., Gasset, G., Loon, J.V., Conley, C.A., and Marco, R. 2007. Comparative analysis of *Drosophila melanogaster* and *Caenorhabditis elegans* gene expression experiments in the European Soyuz flights to the International Space Station. *Advances in Space Research* 40:506-512.
830. Higashibata, A., Higashitani, A., Adachi, R., Kagawa, H., Honda, S., Honda, Y., Higashitani, N., Sasagawa, Miyazawa, Y., Szewczyk, N.J., Conley, C.A., Fujimoto, N., Fukui, K., Shimazu, T., Kuriyama, K., and Ishioka, N. 2007. Biochemical and molecular biological analyses of space-flown nematodes in Japan, the First International *Caenorhabditis elegans* Experiment (ICEFirst). *Microgravity Science and Technology* 19:159-163.
831. Marthy, H.J., ed. 2003. *Development in Space. Advances in Space Biology and Medicine*, Volume 9. Elsevier, The Netherlands.
832. Ronca, A.E., and Alberts, J.R. 2000. Physiology of a microgravity environment selected contribution: Effects of spaceflight during pregnancy on labor and birth at 1 G. *Journal of Applied Physiology* 89:849-854.
833. Burden, H.W., Zary, J., and Alberts, J.R. 1999. Effects of spaceflight on the immunohistochemical demonstration of connexin 26 and 43 in the postpartum uterus in rats. *Journal of Reproduction and Fertility* 116(2):229-234.
834. Ronca, A.E. 2003. Mammalian development in space. In *Development in Space* (H.J. Marthy, ed.). *Advances in Space Biology and Medicine*, Volume 9. Elsevier, The Netherlands.
835. Buckey, J.C., and Homick, J.L. 2000. *The Neurolab Spacelab Mission: Neuroscience Research in Space.* NASA Lyndon B. Johnson Space Center, Houston, Tex.

836. Kerman, I.A., McAllen, R.M., and Yates, B.J. 2000. Patterning of sympathetic nerve activity in response to vestibular stimulation. *Brain Research Bulletin* 53:11-16.
837. Fuller, P.M., Jones, T.A., Jones, S.M., and Fuller, C.A. 2004. Evidence for macular gravity receptor modulation of hypothalamic, limbic and autonomic nuclei. *Neuroscience* 129:461-471.
838. Fuller, P.M., Jones, T.A., Jones, S.M., and Fuller, C.A. 2002. Neurovestibular modulation of circadian and homeostatic regulation: Vestibulohypothalamic connection? *Proceedings of the National Academy of Sciences U.S.A.* 99:15723-15728.
839. Balaban, C.D. 2002. Neural substrates linking balance control and anxiety. *Physiology and Behavior* 77:469-475.
840. Knierim, J.J., McNaughton, B.L., and Poe, G.R. 2000. Three-dimensional spatial selectivity of hippocampal neurons during space flight. *Nature Neuroscience* 3:209-210.
841. Hubel, D.H., and Wiesel, T.N. 1982. Ferrier lecture: Functional architecture of the macaque monkey visual cortex. *Proceedings of the Royal Society London (Biology)* 198:1-59.
842. Ronca, A.E., Fritzsch, B., Bruce, L.L., and Alberts, J.R. 2008. Orbital spaceflight during pregnancy shapes function of mammalian vestibular system. *Behavioral Neuroscience* 122(1):224-232.
843. Walton, K.D., Harding, S., Anschel, D., Harris, Y.T., and Llinás, R. 2005. The effects of microgravity on the development of surface righting in rats. *Journal of Physiology* 565(Pt 2):593-608.
844. Walton, K.D., Harding, S., Anschel, D., Harris, Y.T., and Llinás, R. 2005. The effects of microgravity on the development of surface righting in rats. *Journal of Physiology* 565(Pt 2):593-608.
845. Beisel, K.W., Wang-Lundberg, Y., Maklad, A., and Fritzsch, B. 2005. Development and evolution of the vestibular sensory apparatus of the mammalian ear. *Journal of Vestibular Research* 15:225-241.
846. Fritzsch, B., Beisel, K.W., and Hansen, L.A. 2006. The molecular basis of neurosensory cell formation in ear development: A blueprint for hair cell and sensory neuron regeneration? *Bioessays* 28(12):1181-1193.
847. Böser, S., Dournon, C., Gualandris-Parisot, L., and Horn, E. 2008. Altered gravity affects ventral root activity during fictive swimming and the static vestibuloocular reflex in young tadpoles (Xenopus laevis). *Archives Italiennes de Biologie* 146(1):1-20.
848. Wiederhold, M.L., Harrison, J.L., and Gao, W. 2003. A critical period for gravitational effects on otolith formation. *Journal of Vestibular Research* 13(4-6):205-214.
849. Jones, T.A., Fermin, C., Hester, P.Y., Vellinger, J., Kenyon, R.V., Kerschmann, R., Sgarioto, R., Jun, S., and Vellinger, J. 1993. Effects of microgravity on vestibular ontogeny: Direct physiological and anatomical measurements following space flight (STS-29). *Acta Veterinaria Brno* 62(6 Suppl.):S35-S42.
850. Raymond, J., Demêmes, D., Blanc, E., Sans, N., Ventéo, S., and Dechesne, C.J. 2000. In *The Neurolab Spacelab Mission: Neuroscience Research in Space: Results from the STS-90, Neurolab Spacelab Mission* (J.C. Buckey and J.L. Homick, eds.). NASA 1-11. NASA Johnson Space Center, Houston, Tex. January.
851. Ronca, A.E., Fritzsch, B., Bruce, L.L., and Alberts, J.R. 2008. Orbital spaceflight during pregnancy shapes function of mammalian vestibular system. *Behavioral Neuroscience* 122(1):224-232.
852. Bouët, V., Wubbels, R.J., de Jong, H.A., and Gramsbergen, A. 2004. Behavioural consequences of hypergravity in developing rats. *Brain Research. Developmental Brain Research* 153(1):69-78.
853. Wade, C.E. 2005. Responses across the gravity continuum: hypergravity to microgravity. *Advances in Space Biology and Medicine* 10:225-245.
854. Fritzsch, B., Pauley, S., Matei, V., Katz, D.M., Xiang, M., and Tessarollo, L. 2005. Mutant mice reveal the molecular and cellular basis for specific sensory connections to inner ear epithelia and primary nuclei of the brain. *Hearing Research* 206(1-2):52-63.
855. Bergstrom, R.A., You, Y., Erway, L.C., Lyon, M.F., and Schimenti, J.C. 1998. Deletion mapping of the head tilt (het) gene in mice: A vestibular mutation causing specific absence of otoliths. *Genetics* 150(2):815-822.
856. Soukup, G.A., Fritzsch, B., Pierce, M.L., Weston, M.D., Jahan, I., McManus, M.T., and Harfe, B.D. 2009. Residual microRNA expression dictates the extent of inner ear development in conditional Dicer knockout mice. *Developmental Biology* 328(2):328-341.
857. de Caprona, M.D., Beisel, K.W., Nichols, D.H., and Fritzsch, B. 2004. Partial behavioral compensation is revealed in balance tasked mutant mice lacking otoconia. *Brain Research Bulletin* 64:289-301.
858. Riley, D., and Wong-Riley, M.T.T. 2000. Neuromuscular development is altered by spaceflight. In *The Neurolab Spacelab Mission: Neuroscience Research in Space: Results from the STS-90, Neurolab Spacelab Mission* (J.C. Buckey and J.L. Homick, eds.). NASA 1-11. NASA Johnson Space Center, Houston, Tex. January.
859. Riley, D., and Wong-Riley, M.T.T. 2000. Neuromuscular development is altered by spaceflight. In *The Neurolab Spacelab Mission: Neuroscience Research in Space: Results from the STS-90, Neurolab Spacelab Mission* (J.C. Buckey and J.L. Homick, eds.). NASA 1-11. NASA Johnson Space Center, Houston, Tex. January.

860. Adams, G.R., McCue, S.A., Zeng, M., and Baldwin, K.M. 1999. Time course of myosin heavy chain transitions in neonatal rats: Importance of innervation and thyroid state. *American Journal of Physiology* 276(4 Pt 2):R954-R961.
861. Huckstorf, B.L., Slocum, G.R., Bain, J.L., Reiser, P.M., Sedlak, F.R., Wong-Riley, M.T., and Riley, D.A. 2000. Effects of hindlimb unloading on neuromuscular development of neonatal rats. *Brain Research. Developmental Brain Research* 119(2):169-178.
862. Walton, K.D., Harding, S., Anschel, D., Harris, Y.T., and Llinás, R. 2005. The effects of microgravity on the development of surface righting in rats. *Journal of Physiology* 565(Pt 2):593-608.
863. Walton, K.D., Benavides, L., Singh, N., and Hatoum, N. 2005. Long-term effects of microgravity on the swimming behaviour of young rats. *Journal of Physiology* 565(Pt 2):609-626.
864. Inglis, F.M., Zuckerman, K.E., and Kalb, R.G. 2000. Experience-dependent development of spinal motor neurons. *Neuron* 26(2):299-305.
865. DeFelipe, J., Arellano, J.I., Merchán-Pérez, A., González-Albo, M.C., Walton, K., and Llinás, R. 2002. Spaceflight induces changes in the synaptic circuitry of the postnatal developing neocortex. *Cerebral Cortex* 12(8):883-891.
866. Temple, M.D., Kosik, K.S., and Stewart, O. 2002. Spatial learning and memory is preserved in rats after early development in a microgravity environment. *Neurobiology of Learning and Memory* 78:199-216.
867. Kandel, E.R., Kupferson, I., and Iverson, S. 2000. Learning and memory. Pp. 1227-1246 in *Principals of Neural Science* (E.R. Kandel, J.H. Schwartz, and T.M. Jessell, eds.). 4th Edition. McGraw Hill.
868. Sanes, J.R., and Jessell, T.M. 2000. Pp. 1087-1113. in *Principals of Neural Science* (E.R. Kandel, J.H. Schwartz, and T.M. Jessell, eds.). 4th Edition. McGraw Hill.
869. Minichiello, L. 2009. TrkB signalling pathways in LTP and learning. Review. *Nature Reviews Neuroscience* 10(12):850-860.
870. Agerman, K., Hjerling-Leffler, J., Blanchard, M.P., Scarfone, E., Canlon, B., Nosrat, C., and Ernfors, P. 2003. BDNF gene replacement reveals multiple mechanisms for establishing neurotrophin specificity during sensory nervous system development. *Development* 130(8):1479-1491.
871. Hargens, A.R., and Watenpaugh, D.E. 1996. Cardiovascular adaptations to spaceflight. *Medicine and Science in Sports and Exercise* 28:877-882.
872. Colleran, P.N., Wilkerson, M.K., Bloomfield, S.A., Suva, R.T., and Delp, M.D. 2000. Alterations in skeletal perfusion with simulated microgravity: A possible mechanism for bone remodeling. *Journal of Applied Physiology* 89:1046-1054.
873. Hwang, S., Shelkovinkov, S.A., and Purdy, R.E. 2007. Simulated microgravity effects on the rat carotid and femoral arteries: Role of contractile protein expression and mechanical properties of the vessel wall. *Journal of Applied Physiology* 102:1595-15603.
874. Stevens, H.Y., Meays, D.R., and Frangos, J.A. 2006. Pressure gradients and transport in the murine femur upon hindlimb suspension. *Bone* 39:565-572.
875. Hargens, A.R., and Richardson, S. 2009. Cardiovascular adaptations, fluid shifts, and countermeasures related to spaceflight. *Respiratory Physiology and Neurobiology* 169(Suppl. 1):S30-S33.
876. Hargens, A.R., and Watenpaugh, D.E. 1996. Cardiovascular adaptations to spaceflight. *Medicine and Science in Sports and Exercise* 28:877-882.
877. Colleran, P.N., Wilkerson, M.K., Bloomfield, S.A., Suva, R.T., and Delp, M.D. 2000. Alterations in skeletal perfusion with simulated microgravity: A possible mechanism for bone remodeling. *Journal of Applied Physiology* 89:1046-1054.
878. Hargens, A.R., and Richardson, S. 2009. Cardiovascular adaptations, fluid shifts, and countermeasures related to space flight. *Respiratory Physiology and Neurobiology* 169(Suppl. 1):S30-S33.
879. Hargens, A.R., and Richardson, S. 2009. Cardiovascular adaptations, fluid shifts, and countermeasures related to spaceflight. *Respiratory Physiology and Neurobiology* 169(Suppl. 1):S30-S33.
880. McCrory, J.L., Derr, J., and Cavanagh, P.R. 2004. Locomotion in simulated zero gravity: Ground reaction forces. *Aviation, Space, and Environmental Medicine* 75:203-210.
881. Sibonga, J.D., Evans, H.J., Sung, H.G., Spector, E.R., Lang, T.F., Oganov, V.S., Bakulin, A.V., Shackelford, L.C., and LeBlanc, A.D. 2007. Recovery of spaceflight-induced bone loss: Bone mineral density after long-duration missions as fitted with an exponential function. *Bone* 41:973-978.
882. Lang, T.F., Leblanc, A.D., Evans, H.J., Lu, Y. 2006. Adaptation of the proximal femur to skeletal reloading after long-duration spaceflight. *Journal of Bone and Mineral Research* 21:1224-1230.
883. Sibonga, J.D., Cavanagh, P.R., Lang, T.F., LeBlanc, A.D., Schneider, V., Shackelford, L.C., Smith, S.M., Vico, L. 2007. Adaptation of the skeletal system during long-duration spaceflight. *Clinical Reviews in Bone and Mineral Metabolism* 5(4):249-261.

884. National Research Council. 1998. *A Strategy for Research in Space Biology and Medicine in the New Century.* National Academy Press, Washington, D.C.
885. Gillispie, C.C. 1983. *The Montgolfier Brothers and the Invention of Aviation, 1783-1784.* Princeton University Press, New Jersey, p. 15.
886. Sonnenfeld, G. 2005. Overview. Pp. 1-5 in *Experimentation with Animals in Space* (G. Sonnenfeld, ed.). *Advances in Biology and Medicine*, Volume 10. Elsevier, Amsterdam.
887. Il'in, YeA. 1989. The Cosmos satellites: Some conclusions and prospects. *USSR Space Life Sciences Digest* 22:109-114. NASA-CR-3922(26). Available at http://ntrs.nasa.gov/archive/ nasa/casi.ntrs.nasa.gov/19900002837_1990002837.pdf.
888. Pishcik, V., and Faybishenko, Yu. 1986. The hard road to the stars. in commemoration of the 25th Anniversary of the first manned space flight. *USSR Life Science Digest* 7:91-98. NASA-CR-3922(08). Available at http://ntrs.nasa.gov/archive/ nasa/casi.ntrs.nasa.gov/19860022611_ 1986022611.pdf.
889. Morey-Holton, E.R., Hill, E.L., and Souza, K.A. 2007. Animals and spaceflight: From survival to understanding. *Journal of Musculoskeletal and Neuronal Interactions* 7:17-25.
890. Nicogossian, A.E., Pool, S.L., and Uri, J.J. 1994. Historical perspectives. Pp. 3-29 in *Space Physiology and Medicine* (A.E. Nicogossian, C.L. Leach, and S.L. Pool, eds.). 3rd Edition. Lea and Febiger, Philadelphia.
891. Pishcik, V., and Faybishenko, Yu. 1986. The hard road to the stars. in commemoration of the 25th Anniversary of the first manned space flight. *USSR Life Science Digest* 7:91-98. NASA-CR-3922(08). Available at http://ntrs.nasa.gov/archive/ nasa/casi.ntrs.nasa.gov/19860022611_ 1986022611.pdf.
892. Pace, N., Rahlmann, D.F., Kodama, A.M., Mains, R.C., and Grunbaun, B.W. 1977. Physiological studies in space with nonhuman primates using the monkey pod. Pp 23-33 in *The Use of Nonhuman Primates in Space* (R.C. Simmonds and G.H. Bourne, eds.). NASA Conference Publication 005. NASA National Technical Office, Springfield, Va.
893. National Research Council. 1987. *A Strategy for Space Biology and Medical Sciences for the 1980s and 1990s.* National Academy Press, Washington, D.C., p. 4.
894. Grindeland, R.E., Ilyn, E.A., Holley, D.C., and Skidmore, M.G. 2005. International collaboration on Russian spacecraft and the case for free flyer biosatellites. Pp. 41-80 in *Experimentation with Animal Models in Space* (G. Sonnenfeld, ed.). *Advances in Space Biology and Medicine*, Volume 10. Elsevier, Amsterdam.
895. Tipton, C.M. 1996. Animal model and their importance to human physiological responses in microgravity. *Medicine and Science in Sports and Exercise* 28(10 Suppl.):S94-S100.
896. Committee calculation on November 5, 2009, of the data listed in reference 9.
897. Il'in, YeA. 1989. The Cosmos satellites: Some conclusions and prospects. *USSR Space Life Sciences Digest* 22:109-114. NASA-CR-3922(26). Available at http://ntrs.nasa.gov/archive/ nasa/casi.ntrs.nasa.gov/19900002837_1990002837.pdf.
898. Grindeland, R.E., Ilyn, E.A., Holley, D.C., and Skidmore, M.G. 2005. International collaboration on Russian spacecraft and the case for free flyer biosatellites. Pp.41-80 in *Experimentation with Animal Models in Space* (G. Sonnenfeld, ed.). *Advances in Space Biology and Medicine*, Volume 10. Elsevier, Amsterdam.
899. Gazenko, O.G., Genin, A.M., Ilyin, E.A., Oganov, V.S., and Serova, L.V. 1980. Adapation to weightlessness and it physiological mechanisms (Results of animal experiments aboard biosatellites). *The Physiologist* 23(Suppl. 6):S11-S15.
900. Cherniack, N.S., ed. 1992. Cosmos 2044 mission. *Journal of Applied Physiology* 73:1S-200S.
901. Korolkov, V., Helwig, D., Viso, M., and Connolly, J. 1996. Cosmos 2229 mission. Overview. *Journal of Applied Physiology* 81(1):186-208.
902. Kozlovskaya, I.B., Grindeland, R.E., Visco, M., and Korolkov, V.I. 2000. Bion 11 science objectives and results. *Journal of Gravitational Physiology* 7:S19-S26.
903. Tipton, C.M. 2003. Animals and biosatellites in space. *Journal of Gravitational Physiology* 10:1-3.
904. Cohen, B., Yakushin, S.B., Holestin, G.R., Dai, M., Tomko, D.L., Badakva, A.M., and Kozlovskaya, I.B. 2005. Vestibular experiments in space. Pp. 41-80 in *Experimentation with Animal Models in Space* (G. Sonnenfeld, ed.). *Advances in Space Biology and Medicine*, Volume 10. Elsevier, Amsterdam.
905. Riley, D.A, Ellis, S., Slocum, G.R., Sedlak, F.R., Bain, J.L.W., Krippendorf, B.B., Lehman, C.T., Macias, M.Y., Thompson, J.L., Vijayan, K., and DeBruin, J.A. 1996. In-flight and post flight changes in skeletal muscles of SLS-1 and SLS-2 spaceflown rats. *Journal of Applied Physiology* 81:133-144.
906. Durvova, G., Kaplansky, A., and Morey-Holton, E. 1996. Histomorphometric study of tibia of rats exposed aboard American Spacelab Life Sciences 2 Shuttle Mission. *Journal of Gravitational Physiology* 3:80-81.
907. National Aeronautics and Space Administration. 1993. Mission Information, Space Flight Mission STS-58. Life Sciences Data Archive. NASA Johnson Space Center, Houston, Tex. Available at http://Lsda.jsc.nasa.gov/scripts/mission/miss.cfm?mis_index=7.

908. Buckey, J.C., and Homick, J.L. 2000. *The Neurolab Spacelab Mission: Neuroscience Research in Space.* NASA Lyndon B. Johnson Space Center, Houston, Tex.
909. Golov, V.K., Magedov, V.S., Skidmore, M.G., Hines, J.W., Kozlovskaya, I.B., and Korololkov, V.I. 2000. Bion 11 mission hardware. *Journal of Gravitational Physiology* 7:S27-S36.
910. Tomko, D., and Souza, K. 2009. Advanced Animal Habitat (AAH). P. 2 in *Flight and Specialized Ground Equipment for Animal, Cell and Microbial Experiments: Status and Availability*. October 26.
911. Ground, D. 2009. Six month ISS missions have greatly expanded our knowledge of human adaptation. In *Human Research Program ISS Research Opportunities*. October 26.
912. National Aeronautics and Space Administration Authorization Act of 2005. P. L. 109-15 119 Stat. 2895, December 30, 2005.
913. Institute of Medicine. 2001. *Safe Passage: Astronaut Care for Exploration Missions* (J.R. Ball and C.H. Evans, Jr., eds.). National Academy Press, Washington, D.C.
914. LeBlanc, A.D., Spector, E.R., Evans, H.J., and Sibonga, J.D. 2007. Skeletal responses to space flight and the bed rest analog: A review. *Journal of Musculoskeletal and Neuronal Interactions* 7(1):33-47.
915. Pavy-Le Traon, A., Heer, M., Narici, M.V., Rittweger, J., and Vernikos, J. 2007. From space to Earth: Advances in human physiology from 20 years of bed rest studies (1986-2006). *European Journal of Applied Physiology* 101(2):143-194.
916. National Aeronautics and Space Administration. 2010. *The NASA Fundamental Space Biology Science Plan 2010-2020. Advanced Capabilities Division.* NASA Headquarters, Washington, D.C. June.
917. National Research Council. 1998. *A Strategy for Research in Space Biology and Medicine in the New Century.* National Academy Press, Washington, D.C.

7

Crosscutting Issues for Humans in the Space Environment

Translating knowledge from basic laboratory discoveries to human spaceflight is a challenging, two-fold task: *Horizontal integration* requires multidisciplinary and transdisciplinary approaches to complex problems; *vertical translation* requires meaningful interactions among basic, preclinical, and clinical scientists to translate fundamental discoveries into improvements in the health and well-being of crew members in space and in their re-adaptation to gravity. This chapter focuses on both aspects as a framework to expand human presence in space.

There are significant scientific gaps in a number of horizontal or crosscutting areas that span multiple physiological systems that can affect human health, safety, and performance. Specific examples discussed in this chapter include:

1. The physical stress of spaceflight and how it manifests physiological strain;
2. Optimization of nutrition, energy balance, and physical activity;
3. Enabling appropriate thermoregulation;
4. Issues related to radiation biology; and
5. Sex-specific effects.

Many of these interactions also involve interactions with behavioral factors, such as effects on cognitive function. These factors are discussed extensively in Chapter 5.

Although each area could be viewed as a relatively straightforward extension of human biology, in reality they represent crosscutting, thematic areas that require the collaboration of teams of experts representing different areas of physiological and engineering expertise. Accordingly, this chapter presents a new framework that builds on revolutionary approaches to health care research that have gained prominence in the past decade. The primary goal of these approaches is to mobilize fundamental and applied scientific knowledge and resources in a way that optimizes their utility for human spaceflight crews.

A research plan for vertical translation within individual physiological systems is well summarized in Chapter 6. So how does addressing the concepts of horizontal integration and vertical translation simultaneously complement the existing research enterprise? Consider the effects of reduced mechanical loading on bone and muscle loss in space. Whereas the effects of reduced loading per se on the individual systems are discussed thoroughly in Chapter 6 using a vertical translation strategy, detrimental changes in musculoskeletal mass and function in the space environment may be further influenced by horizontal integration of multiple factors that include, but are not limited to, low energy availability, micronutrient deficiencies, reduced physical activity, increased levels of glucocorticoids and other stress hormones, and reductions in gonadal hormones. Understanding the independent and combined effects of multiple stressors on multiple physiological systems, and how resultant changes influence health, safety, and performance, should be viewed as an obligatory step for the development of effective countermeasures. For example, intensive exercise may effectively mitigate microgravity-related effects of reduced loading on muscle and bone loss but exacerbate the adverse effects of low energy availability and gonadal hormone suppression on the musculoskeletal system. Because of the small number of potential study subjects, it will probably not be possible in the foreseeable future to carry out human studies in space that effectively evaluate the independent and combined effects of multiple stressors. However, a comprehensive characterization of both the stressors and the responses of multiple biologic systems will generate important preliminary data for hypothesis-driven research on multiple stressors (e.g., ground-based human studies).

Finally, this chapter emphasizes the need for effective coupling of biological and engineering problem-solving strategies. It is difficult, if not impossible, to replicate the multiple stressors of the space environment in a ground-based analog. However, creation of an effective exercise countermeasure for the preservation of bone, muscle, and cardiovascular function in a ground-based experiment is productive only if the countermeasure can be implemented in the space environment. This important translational step is heavily dependent on approaching the problem from a multidisciplinary perspective.

SOLVING INTEGRATIVE BIOMEDICAL PROBLEMS THROUGH TRANSLATIONAL RESEARCH

Stress—Physical and Physiological Considerations

There are numerous known biomedical stressors associated with spaceflight and re-adaptation to gravity that may benefit from a translational approach.

Decompression Sickness

Human spaceflights in low Earth orbit or on exploration missions to the Moon, asteroids, or Mars all require many thousands of hours of extravehicular activity (EVA) and inherently carry a higher risk of decompression sickness.[1] Suits are necessarily designed to operate at as low a pressure as practicable (e.g., Apollo, 3.8 psi; space shuttle/International Space Station (ISS), 4.3 psi; Russian Orlan, 5.7 psi) so as to maximize suit joint and glove flexibility while maintaining physiologically adequate oxygen and CO_2 partial pressures. During the airlock decompression from spacecraft to hypobaric suit pressure, nitrogen dissolved in blood and tissues comes out of solution, creating tiny bubbles (evolved nitrogen), which potentially can cause symptoms of decompression sickness (DCS), colloquially called "the bends." DCS at 5 psia was not a problem on Apollo, because the spacecraft operated with a 100 percent O_2 environment. Primarily for scientific reasons, the space shuttle and the ISS use an Earth-like 14.7 psia, 20.8 percent O_2 atmosphere, although the space shuttle can be depressurized to 10.2 psia to shorten EVA prebreathing of pure oxygen. DCS during EVAs is avoided by prebreathing 100 percent O_2 to reduce evolved nitrogen. Exercise during prebreathing greatly accelerates denitrogenation. Staged decompression (e.g., to 10.2 psi) followed by a shorter prebreathe interval is sometimes used. Astronauts are also routinely exposed to hyperbaric environments during underwater "neutral buoyancy" EVA training. DCS upon ascent is avoided by limiting the depth and duration of underwater training and by the use of oxygen-enriched breathing gas.

Known Effects

No cases of hypobaric DCS have been reported on Apollo flights, the space shuttle, or the ISS. Decompression using conventional staged protocols normally creates venous gas emboli (VGE or "silent bubbles").[2,3] Small numbers of bubbles produce no symptoms, but in 1-*g* testing, large numbers of VGE are usually correlated with joint and muscle pain (Type I DCS). Exercise increases the number of venous bubble micronuclei and incidence of symptoms. Most VGE are believed trapped in the lungs. However intrapulmonary (pulmonary arterial to venous) connections exist[4] that shunt blood to the systemic arterial side, particularly during exercise. VGE can be easily detected using cardiac Doppler ultrasound, so the number of bubbles is used as a correlate of decompression stress during hypobaric chamber tests of astronaut oxygen prebreathe protocols. If a significant number of bubbles reach or form on the arterial side, more serious symptoms (Type II DCS) occur, ranging from headache, shortness of breath, impaired vision, fatigue, memory loss, and vomiting to seizures, paralysis, confusion, unconsciousness, and even death. Arterial gas emboli (AGE) can be detected using transcranial Doppler techniques, but so far this is practical only in laboratory settings. Fortunately, as compared to diving, there are few documented adverse medical events of space EVA decompressive stress (e.g., neurologic signs or bone necrosis).[5] Severe hypobaric Type II DCS events are rare.[6] Almost all symptoms encountered in 1-*g* hypobaric tests are Type I, and they frequently resolve with simple 100 percent O_2 treatment. For the treatment of DCS in space, the ISS suit can deliver a total of 23 psi, 14.7 psi from station atmosphere, 4.3 psi from the suit pressure, and an additional 4 psi by overriding the positive pressure relief valve with the Bends Treatment Adapter.

Effects of Countermeasures

Astronaut candidates with a history of Type II DCS or uncorrected atrial septal defects are thought to be potentially at greater risk of AGE and are disqualified. Prebreathe protocols are designed using mathematical decompression stress models (e.g., two-phase tissue bubble dynamics models), prospective hypobaric simulations in altitude chambers, statistical analysis of data from ground and flight experience, and expert judgment. Test chamber DCS incidence must always be less than 15 percent, with less than 20 percent large VGEs and no serious DCS symptoms. Ground testing[7] has shown that 10 min of exercise (75 percent of peak oxygen uptake) during a 1-h prebreathe is equivalent to a 4-h resting prebreathe. Among divers, administration of nitroglycerin[8] or intense predive exercise[9] 24 h previous reduces bubble formation. Modeling[10] and animal[11] studies indicate that recompression after a decompression step also reduces bubble formation. The latter findings suggest that multiple short sorties from a pressurized enclosure (e.g., a habitat or rover in a reduced-gravity environment) may reduce decompression stress relative to an equivalent single time exposure.

Gaps in Fundamental and Applied Knowledge

Operational incidence of Type I DCS in orbit has been far less than that anticipated based on ground tests.[12] It is not clear whether this is due to overreporting in ground studies or underreporting in orbit. Alternatively, astronauts may have fewer bubble micronuclei, VGE, and AGE in orbit, perhaps due to exertion or fluid shift effects on bubble nucleation or shunting. Further research on the cellular and molecular mechanisms of bubble formation and tissue interaction and on intrapulmonary shunting is needed. Determining whether gravity has an important effect on DCS mechanisms is obviously important. Should prebreathe protocols be designed more stringently for lunar and Mars partial gravity than for microgravity? VGE prevalence in microgravity could be experimentally assessed via cardiac Doppler ultrasound monitoring, but this important experiment has not yet been conducted during spaceflight. Currently, astronaut candidates are not evaluated for intrapulmonary shunting.[13] There may be individual differences in AGE formation due to exercise-induced intrapulmonary bubble shunting, which can only be detected via chamber testing using transcranial Doppler.

Knowing the lowest suit pressure and compatible habitat atmospheres that would allow EVA without prebreathe is important for space exploration missions. Crews could live in an 8 psi, 32 percent O_2 habitat and perform EVAs in a conventional 4.2 psi, 100 percent O_2 suit without prebreathe. It has been suggested[14,15] that, to promote suit joint and glove flexibility, crews could adapt to a slightly hypoxic 2.7 psia, 100 percent O_2 suit, and operate without prebreathe from a 6.2 psia, 36 percent O_2 habitat. This suit pressure is equivalent to that at an elevation of 40,000 feet on Earth, and the pO_2 is equivalent to breathing air at 11,000 feet. Maintaining a relatively low habitat oxygen partial pressure has fire safety advantages, but the resulting hypoxic, hypobaric environment may impair cognition and pose challenges for verbal communication and heat transfer. Higher operating pressures can mitigate the risk of DCS, but new risks for excessive fatigue and trauma may be created. Thus, an important technology challenge is to improve the design of EVA suit joints.[16] Further research is also needed on operationally acceptable low suit pressures and hypobaric hypoxia levels. Novel perfluorocarbon therapeutic oxygen carriers, which are in human testing for stroke and brain injury,[17] also enhance nitrogen offgassing. Use of such blood additives could eventually enable novel approaches to DCS and hypoxia management.

Research Models and Platforms Needed

Research models and platforms encompass animal and human research in altitude chambers and experimental ultrasound studies of incidence of VGE formation in microgravity, using operational denitrogenation protocols aboard the ISS.

Findings and Recommendations

After four decades of EVA experience, NASA views DCS as primarily an operational problem and in the past 20 years has supported little basic research on DCS mechanisms or countermeasures. However, the amount of surface EVA activity on exploration missions potentially far exceeds that on the space shuttle and the ISS, so the statistical likelihood of a DCS episode is higher. The effect of gravitational level on DCS incidence is not well understood. Previous National Research Council (NRC) reviews[18,19] have recommended increased emphasis on advanced EVA research, with increased industry and university involvement. It is critical to understand whether gravity has a direct or indirect effect on microbubble nucleation. More research is also needed to understand VGE and AGE detection and tissue interaction; intrapulmonary right-to-left shunting; effects of exercise on DCS; effects of periodic recompression on DCS; the effects of hypobaric and hypoxic environments on DCS, cognition, vocalization and heat transfer; and alternatives for hypoxia prevention and denitrogenation, such as perfluorocarbon therapeutic blood additives.

Rotation/Artificial Gravity

Artificial gravity (AG)[20] achieved by rotation was first proposed as a multisystem physiologic countermeasure by K. Tsiolkovsky more than 100 years ago. Small centrifuges have been flown on the space shuttle, the ISS, and Bion to provide 1-*g* controls for gravitational biology experiments. NASA's planners have assumed that AG will not be required for 6-month lunar missions, although longer-term effects of living at lunar gravity remain unknown. Rotation of an entire exploration mission spacecraft to provide continuous AG for astronauts significantly complicates engineering design. Nonetheless, a 1-*g*, 4-rpm, 50-m-radius rotating truss design[21] is under consideration for Mars missions.

An alternative to rotating the entire spacecraft is to provide a facility within the spacecraft for intermittent, rather than continuous, AG exposure. Encouraging recent results from bone, muscle, and cardiovascular research on humans and animal models (see Chapter 6 of this report) suggest that intermittent rather than continuous exposure to hypogravic gravitational force levels may provide an adequate countermeasure. Intermittent AG could be provided using a short-radius centrifuge (2 to 5 m) mounted inside the vehicle and perhaps even human-powered. However, short-radius centrifuges require significant rotation rates and require that humans and animals adapt to initially unpleasant vestibular and biomechanical Coriolis effects.

An alternative strategy to induce AG in humans is to use lower-body negative pressure to provide footward loading while walking or running on a treadmill.[22] This approach has been tested extensively in bed rest studies involving both men and women, and it has shown significant beneficial effects for a multiplicity of systems.[23-32]

Known Effects

Graybiel and colleagues[33] recognized that, because of the fluid mechanics of the vestibular semicircular canals, any head movement made out of the plane of rotation in a rotating system produces a strong tumbling illusion that quickly provokes motion sickness in a 1-*g* environment because of vestibular sensory cue conflict. Symptoms included fatigue, nausea, vomiting, and sensory after-effects. Most vertebrate animals exhibit similar symptoms. Rotating chair experiments on Skylab[34] and in parabolic flight[35] showed that provocativeness depends on both ambient gravitational force level and rotation rate. However, as ambient gravitational force approaches zero, subjects could make unlimited numbers of head movements in a rotating chair with their heads near the axis of rotation, even at 30 rpm, without becoming ill. In ground-based rotating rooms and centrifuges, where total gravitational force levels are inherently greater than 1, the sickness threshold is approximately 3 rpm.[36] However, when lying supine on short-radius centrifuges, many individuals can successfully adapt to significantly higher rotation rates, and the adaptation is partially retained.[37,38] Susceptibility is expected to be lower if gravitational force levels at the head are less than 1 *g*, but the dose-response relationship cannot be determined in ground-based laboratory centrifuges. Biomechanical Coriolis forces also cause reaching errors and other limb movement control problems, although humans rapidly adapt if body orientation remains constant.

Gaps in Fundamental and Applied Knowledge

There are three basic reasons to incorporate AG in orbital studies:[39]

1. To explore the feasibility and the parameter space (e.g., gravity level, exposure duration, exposure frequency, and whether concurrent exercise is used) for this potentially universal countermeasure;
2. To allow study of the biological and physical effects of partial-gravity environments at less than 1 *g* that are not otherwise available through scientific investigations of reduced-gravity environments on Earth; and
3. To serve as a 1-*g* control for microgravity experiments on human and nonhuman subjects.

Science-based decisions on whether to use AG on very long duration missions and whether to use short-radius, intermittent rotation (e.g., via an onboard centrifuge) or long-radius, continuous rotation (e.g., by rotating all or part of the spacecraft) can be made when it becomes clear whether AG is required for medical or habitability reasons and after the necessary gravitational force level and the frequency and duration of exposure, rotation rate, and gravity gradient limits are determined from dose-response studies.

Research Models and Platforms Needed

Only plant and animal studies can be accomplished using centrifuges aboard free-flying satellites (e.g., Bion). The Japanese Aerospace Exploration Agency, JAXA, developed a 2.5-m animal and plant centrifuge for the ISS and initiated ground studies in humans using head-down bed rest and intermittent short-radius centrifugation. Ultimately, flight experiments on humans and animals will be required, especially to resolve questions about partial gravity that cannot be addressed on the ground. However, both NASA's ground-based and flight AG research was suspended in 2005 when the U.S. program was refocused on lunar missions. The 2.5-m centrifuge was dropped from the ISS. Fortunately, AG research is continuing in ground laboratories elsewhere (France, Germany, Japan, China, and Belgium). Several small centrifuges remain aboard the ISS to support blood sample separation and control experiments on very small animals, plants, and other biologicals, but no facilities or logistical support exist for AG experiments on rats or small primates. The next generation of crew taxi vehicles lack sufficient interior space for a human short-radius centrifuge. Conceivably a larger habitat, tether, or truss systems could be developed to provide AG for missions otherwise involving very long duration exposures to microgravity.

Recommendations

Resolution of the gaps in fundamental knowledge of partial gravity on human physiology may require a decade or more of research in ground laboratories and eventually aboard the ISS. In collaboration with international partners, NASA should reinitiate a vigorous program of ground research, develop small animal AG experimental facilities for the ISS, and develop a simple short-radius human centrifuge for eventual countermeasure evaluation experiments aboard the ISS.

Space Motion Sickness

Of all the physiological difficulties astronauts encounter during their first days in orbit, space motion sickness has been the most overt and prevalent. Symptoms are often triggered by head and body movements or visual disorientation in the microgravity environment.[40,41] There were no reports of space sickness in the Mercury or Gemini program, probably because the cabins were so small that the crew had limited ability to move around. More than 70 percent of space shuttle crew members report some symptoms, with moderate to severe symptoms in 30 percent of cases. These symptoms usually involve multiple vomiting episodes, but as crews adapt to microgravity, symptoms usually abate within 2 to 4 days.[42,43] Crew members who have flown previously typically have milder symptoms. Nonetheless the operational impact is initially high on space shuttle flights: crews limit their movements and avoid all EVA, since suits have no vomitus containment capability. As crews adapt, symptoms abate within several days, mitigating the overall operational impact on long-duration flights. Apollo crews reported no symptoms in 1/6 *g* on the Moon. However, when microgravity-adapted crews return to Earth, they almost universally experience head-movement-contingent vertigo for at least several hours, which causes mild ataxia and frequently triggers recurrence of motion sickness symptoms. Long-duration space station crews typically experience more severe after-effects, which limit their vehicle egress ability, and these after-effects can last several days. Returning Apollo crews landed in the ocean, and many experienced symptoms. Existing anti-motion sickness drugs are sedating and only partially effective, so they are used primarily for treatment in orbit, rather than prophylaxis. Space sickness will doubtless affect passengers and crews on commercial orbital and suborbital flights and during the early microgravity and post-return days of future transportation system options. Mars exploration crews may experience symptoms in 3/8 *g* and again upon Earth return.

Known Effects

Symptoms resemble those of familiar forms of chronic motion sickness, superimposed on those caused by fluid shift, and include drowsiness, yawning, and sometimes headache. Some individuals experience sudden vomiting, but many have prodromal nausea, at least during the first several attacks. Once nausea develops, the causal link with head and body movements and disorientation becomes unambiguous. Nausea causes impaired concentration, loss of initiative, gastric stasis, and anorexia and triggers hypothalamic-pituitary-adrenal stress hormone release. Repeated vomiting leads to loss of fluid, glucose, and electrolytes. Unless those are replaced, individuals become ketotic and prostrate. The smell or sight of vomitus can trigger nausea in others.

Effects of Countermeasures

Crews routinely restrict head and body movements and remain visually upright. Some choose to take single doses of prophylactic oral promethazine-dexedrine before launch or entry. Treatment of severe cases with intramuscular injection of 25 to 50 mg promethazine followed by sleep has been judged operationally effective, but the injections are painful and drug cognitive side effects remain a concern. Potential interaction with promethazine has prevented use of midodrine as an orthostatic countermeasure. Countermeasure effectiveness has proven difficult to determine, since crew often deny prodromal symptoms; head and body movements cannot be standardized, so formal susceptibility testing is operationally impractical. Even on Earth, individual susceptibility to motion sickness stimuli can vary considerably from day to day. For all these reasons, it has not been possible to predict an individual's susceptibility to space sickness under operational conditions on the basis of a single susceptibility test on the ground.

Gaps in Fundamental and Applied Knowledge

Although NASA has made a significant investment in neurovestibular research over the past three decades, most work has focused on eye movements, perception, and posture control, rather than fundamental emetic physiology or pharmacology. It seems clear that space sickness is a form of motion sickness, with symptoms that superimpose on discomforts associated with fluid shift (e.g., headache, head fullness, visceral elevation). Sensory conflict theories for motion sickness and space sickness[44,45,46] have provided a useful conceptual etiologic framework. However, conflict neurons and the neural or chemical linkage between neurovestibular and emetic centers posited by sensory conflict theory have not yet been identified. Much has been learned about emetic physiology in the context of managing nausea after anesthesia, cancer chemotherapy, and radiation treatment.[47] However, drugs that are effective against these stimuli have proven relatively ineffective against motion sickness.[48] Lacking knowledge of which receptors to target, pharmacologic approaches to motion sickness prevention remain empirical. Some drugs (e.g., baclofen, lorazepam, diazepam) are effective but are sedating or have other side effects that prevent operational use. Drugs believed most effective against motion sickness generally have antihistaminic (e.g., promethazine) and/or antimuscarinic (e.g., scopolamine) actions. Furthermore, promethazine is sedating, while scopolamine causes detectable short-term memory loss and therefore may interfere with neurovestibular adaptation as well as cognition.

Research Models and Platforms Needed

Ground research is required on physiological mechanisms associated with sensory conflict and the emetic linkage in motion sickness. Rodents do not vomit, so primate, ferret, dog, and cat models are preferred. ISS studies are needed of cognitive side effects of anti-motion sickness drugs under operational conditions.

Recommendations

More effective and operationally acceptable motion sickness countermeasures will soon be needed if commercial suborbital and orbital flights are to succeed. There is a strong need to re-educate the new generation of astronauts on this almost universal problem, so as to take advantage of what is already known and to improve in-flight reporting. Flight surgeons should resume systematic and complete collection of data on symptoms and drug use. Tests of the cognitive side effects of existing pharmacologic therapies should be conducted on the ISS, in parallel with similar research on sleep-aid side effects. New EVA suits must be designed for vomitus containment, so early mission EVAs can be safely undertaken and suits remain reusable. A new program of basic research on the physiology of the emetic linkage should be initiated, soliciting pharmaceutical industry collaboration for development of targeted drugs.

Performance Decrements Due to Launch/Entry Acceleration and Vibration

Human performance during Earth launch accelerations and vibrations and the ability of deconditioned crews to tolerate entry and landing accelerations and touchdown shocks, particularly in off-nominal conditions, remain major constraints on mission design.[49] Performance depends on a variety of cardiovascular, pulmonary, vestibular, neuromuscular, and biomechanical factors. Apollo crews normally experienced 4.5 g_x,* launch accelerations, but 6.5 to 10 g_x was possible in some abort scenarios. Space shuttle and Soyuz launch accelerations (3 to 4 g_x) are slightly more benign. However, oscillations produced by resonant burning in solid rocket motor stages could produce 12 Hz oscillations up to 3.8 g_x, exceeding the 0.1 to 0.2 g oscillation levels specified for space shuttle and Apollo launches. The effects of combined acceleration and vibration on crew visual acuity and performance in cockpits with electronic displays are uncertain and also depend on seat and suit design.[50] Lunar orbit and landing maneuvers involve relatively low acceleration levels; the Apollo crews landed and launched standing up. However, acceleration levels that exploration crews will encounter during atmospheric entry to Mars and on return to Earth are a concern, because of the relatively high spacecraft entry velocity, as compared to an entry from low Earth orbit. Mars mission architecture studies[51] suggest that deconditioned crews could experience Mars aerocapture decelerations at up to 5 g_x and will therefore need to be supine. When Apollo crews returned from the Moon, 5.5 to 7.2 peak g_x was typical, and returning Orion lunar mission crews will likely encounter comparable levels. Seated space shuttle crews normally experience several minutes of +1.5 g_z ("eyeballs down") acceleration during approach and landing. ISS crews returning on the space shuttle lie supine. Soyuz crews lie supine on couches also designed to dissipate touchdown shock. Immediately after landing, crews typically experience head movement contingent vertigo, oscillopsia, locomotor ataxia, and motion sickness symptoms.

Known Effects

Humans can tolerate higher sustained linear accelerations for longer durations along the positive transverse ("eyeballs in") g_x axis than in other directions, since the hydrostatic gradient reducing brain blood flow is minimized and the position is more comfortable.[52,53] Physiological symptoms can include respiration difficulty, dimming and

*Subscripts are used with the unit of force due to gravity and/or acceleration to indicate the direction of the force vector relative to the body orientation of an astronaut during the flight maneuver under discussion. The positive x-axis direction is with the force vector transverse to the head and trunk of an astronaut and pointing forward, as when astronauts are essentially lying on their backs during launch. This "g_x" orientation is also called "transverse," or more colloquially, "eyeballs in." The positive z-axis direction, g_z, is with the astronaut's head-trunk axis oriented parallel to the force vector, with the vector pointing in the direction from feet to head, as in the acceleration force felt when riding in an ascending elevator. Colloquially, this orientation is called "eyeballs down."

tunneling of vision, changes in smooth-pursuit eye movements, pain, loss of consciousness, convulsions, post-acceleration, and confusion and disorientation. Tolerance is a function of magnitude, duration, and external factors such as previous training/exposure and deconditioning. Impact on human performance also depends on interface design (required reach, arm support, etc.). Little is known about the effects of multi-axis gravitational force loads, which would be experienced primarily during launch, abort, aerocapture, or entry. Less is known about human tolerance to rotation. Yaw is generally tolerated at higher rates than pitch. Tolerance to rotational velocities is reduced when translational accelerations are superimposed. Tumbling through multiple axes leads to vestibular disorientation and performance degradations and interferes with a crew's ability to take corrective actions. These accelerations could occur in abort scenarios and off-nominal situations. Landing impact acceleration (duration <0.5 s) exposure limits are based primarily on aircraft ejection data and do not consider the possible effect of bone loss due to prolonged exposure to microgravity. Exposures to whole-body vibration for up to several minutes may cause discomfort (headache and pain) and manual, visual, and cognitive performance decrements, but usually do not cause injury.[54] Acceptable limits depend on performance criteria, seat orientation, and biomechanics of the human-equipment interface and must be investigated on a case-by-case basis. Minimal tolerance usually occurs between 4 and 8 Hz, the range in which the body exhibits abdominal mechanical resonance.

Countermeasures

Primary countermeasures have included limiting launch, abort, and entry accelerations and orienting the seats so crew members are supine. However, this is not always possible in piloted vehicles where direct forward view is required. Anti-gravity compression garments covering the abdomen and legs are used on space shuttle and other high-performance aircraft and improve "eyeballs down" linear acceleration tolerance of seated pilots by 1 to 1.5 g. Astronauts and pilots undergo centrifuge training to learn the anti-gravity straining maneuver (tensing arm, leg, and abdominal muscles and exhaling against the glottis), which temporarily provides up to 3 g_z of additional tolerance. Returning astronauts are typically volume depleted, and so fluid/salt loading prior to entry has proven somewhat effective. Supine and seated crews must eventually stand up after landing, and even with fluid loading the prevalence of orthostatic hypotension on the day of landing has been 20 to 30 percent after 1- to 2-week space shuttle missions and at least 80 percent after long-duration flights. The origin of individual differences in postflight orthostatic tolerance and neurovestibular vertigo and ataxia is not well understood. The ability of long-duration crews to egress quickly after Orion landings on the ocean remains a concern. Pharmacologic agents (midodrine, octreotide) have been evaluated to improve orthostatic tolerance, but concerns about drug interactions (e.g., with promethazine administered to prevent motion sickness) have so far limited their use.[55] AG centrifuges and lower-body negative-pressure devices have also been considered.

Gaps in Fundamental and Applied Knowledge

There is effectively no knowledge of the effects of elevated g_x forces and high vibration on crew performance in a cockpit with electronic displays, where electronic crew-vehicle interfaces require novel methods of crew-vehicle interaction not encountered in earlier vehicle designs. Virtually all existing acceleration and vibration tolerance data were obtained from Earth-normal-conditioned subjects. Although sustained gravitational force-load limits for abort and nominal exposures for conditioned and deconditioned crews for durations up to 1,000 s are specified in NASA Standard 3000 and in the new NASA Human Interface Design Handbook, no vibration limits have yet been included for normal or deconditioned subjects. Similarly there are few data on the interaction between vibration and prolonged transverse acceleration or the interaction between rotational and translational accelerations. There are no data on performance effects in deconditioned subjects in simulated launch, abort, entry, or ocean emergency egress situations. Relatively little is known about the physiology of the head movement contingent vertigos, which complicate emergency egress and locomotion, or the cause of individual differences.

Research Models and Platforms Needed

- Human head-down bed rest studies are needed of deconditioning effects on acceleration and vibration tolerance and performance, using cockpit-equipped centrifuge simulators.
- Early post-flight studies are needed on head-movement-contingent vertigo and the ability of returning ISS crews to perform emergency egress. Also needed are studies of tolerance to Earth and Mars entry accelerations.
- Mathematical models of cardiovascular system response[56,57] will continue to provide important insights.

Recommendations

1. For exploration missions, since fundamental data are lacking, NASA must perform the appropriate ISS, bed rest, and centrifuge studies to determine appropriate acceleration and vibration tolerance, task performance, and emergency egress capability of deconditioned astronauts, with and without anticipated countermeasures. Vehicle launch, abort, Earth/Mars entry acceleration, and vibration profiles must be designed appropriately.

2. The prevalence of orthostatic hypotension immediately following long-duration spaceflight has never been determined in an operational context (i.e., not in a clinical tilt-table or stand test). It is essential to identify the mechanism underlying the practical incapacitation of crews after long-duration spaceflight prior to developing pharmacological countermeasures. Research should be designed to elucidate multifactorial mechanisms, which are hypothesized to include some combination of cardiovascular deconditioning, neurovestibular vertigo and ataxia, muscle weakness, and kinesthetic abnormalities.

Food, Nutrition, and Energy Balance

As described more fully below, inadequate nutrition affects all aspects of long-duration spaceflight, from accelerating bone loss and reducing muscle function to diminishing cognitive ability and depressing immune response. To protect against inadequate nutrition, it is important to:

1. Fully understand nutrient needs under space conditions compared to terrestrial conditions;
2. Develop appropriate diets for space use and ensure their palatability; and
3. Ensure that the nutrients within a food remain bioactive for the duration of the mission.

NASA has done an excellent job in parsing out specific nutrient needs for spaceflight and a fair job in establishing a climate where 100 percent consumption of supplied foods is encouraged, but gaps remain in tracking the bioactivity of essential nutrients in foods as a function of food processing, storage, exposure to radiation, etc. Because nutrient deficiencies are rare, and overweight and obesity have reached epidemic levels in the general population, issues related to nutrient stability over time or food consumption patterns among astronauts have not been a focus. However, whenever an individual is totally dependent on a limited number of foods to meet 100 percent of his/her nutrient needs (e.g., infants fed formula, patients being fed intravenously, or astronauts in space), it is critical that the diet provided be nutritionally complete. For the success of long-duration space missions, it is also essential that the diet remains nutritionally complete for the prescribed amount of time and that steps are taken to ensure that individuals consume appropriate amounts of the diet's elements to meet their nutrient and energy needs.

Impact of Spaceflight on Nutritional Status, Nutrient and Energy Intake, and Nutrient Stability in the Food Supply

Nutritional Status in Space

A substantial amount of information has been generated about nutrient status in space, although data are lacking on the mechanisms by which nutrient status in space differs from that on Earth. A thorough and continuously updated review is provided in the *Human Research Program Evidence Book*,[58] which highlights risk factors of inadequate nutrition and of an inadequate food system. Each of these risks is categorized as being a risk with *substantial evidence* rather than one which is *not yet fully supported or refuted*.[59] A complete reiteration of the current understanding of nutrient status in space is not possible in this section, but a synopsis can be given of the most important differences between Earth and space. In particular, several deficiencies or insufficiencies are consistently reported, including inadequate energy intake (see below), depressed vitamin D and K and folate status, and diminished antioxidant capacity.

Data from individual Skylab missions show that length of mission is a factor in vitamin D status; the longer the mission, the more depressed the vitamin D status. For example, a study of 11 astronauts showed an average decrease of ~25 percent ($P < 0.01$) in serum 25-hydroxycholecalciferol (25-OH vitamin D3) after flight (128-195 d).[60] Similarly, a decrease of 32 to 36 percent in 25-OH vitamin D3 concentrations in serum was found in Mir crew members during and after 3- to 4-month missions.[61,62] Bone resorption was increased after flight, as indicated by several markers. Vitamin D is required for calcium absorption, and bone loss is a clearly documented negative consequence of spaceflight.[63] There is considerable debate at this time as to whether the Dietary Reference Intake for vitamin D is adequate. Recent data suggest that an optimal serum level of 25-OH vitamin D3 is ~80 nmol/L.[64] Using these criteria, 80 to 90 percent of the subjects in the ISS study[65] had suboptimal vitamin D status. An important difference between vitamin D status on Earth and in space is that vitamin D can be synthesized from a precursor under the skin that is activated by ultraviolet (UV) light excitation. However, because astronauts are not exposed to UV light in flight, they require a vitamin D supplement (the only nutrient that is routinely supplemented in spaceflight).

Folate status is also compromised in space. In addition to a reduction in 25-OH vitamin D3, red blood cell folate concentrations were approximately 20 percent lower ($P < 0.01$) after landing.[66] The number of multivitamins consumed per week was positively correlated ($r = 0.62$, $P < 0.05$) with red blood cell folate levels, suggesting that intake from the prescribed diet was insufficient to maintain blood folate levels. Folate deficiency, with neurological and hematological manifestations, can occur in a matter of 2 to 3 months on inadequate diets. Cell populations with rapid turnover rates (such as absorptive cells lining the small intestine) require adequate folate to replicate, and folate deficiency may result in diminished absorptive capacity. Clearly, a diminished absorptive capacity would decrease availability of all nutrients to the body.

Vitamin K status has been shown to be decreased after spaceflight,[67] which again has negative implications for bone health.[68] Vitamin K is required for the formation of gamma-carboxyglutamate in specific bone proteins such as osteocalcin and matrix Gla-protein, which are necessary for appropriate calcium binding. During the 6-month spaceflight of the EUROMIR-95 mission, both bone resorption markers and urinary calcium excretion increased about two-fold immediately after launch.

A major negative factor of long-duration spaceflight is exposure of crews to ionizing radiation. Solar radiation and galactic cosmic radiation can produce reactive oxygen species, which can cause oxidative damage to DNA measurable by urinary levels of 8-hydroxy-2'-deoxyguanosine. Levels of this biomarker have increased after long-duration flight, suggesting an inadequate response to an oxidative challenge.[69]

Nutrient and Energy Intake

Nutrient intake is closely linked to an adequate energy intake. Crew members on the ISS have been reported to consume a mean of 80 percent of their recommended energy intake,[70] whereas reports from other missions show intakes around 60 percent of recommendations.[71] Energy intake is typically 30 to 40 percent below the World Health Organization recommendation, but energy expenditure is typically unchanged or even increased.[72,73] The functional consequences of inadequate energy intake are not fully appreciated, because in the current Earth environment, many see weight loss as a desirable end point. Inadequate energy intake in space results in a catabolic state. As blood glucose levels decline, muscle protein will be catabolized to supply amino acids for gluconeogenesis and dietary amino acids will be used for energy, rather than protein synthesis. Loss of muscle mass during spaceflight[74] and during bed rest[75] is well documented. A bed rest study has documented that an essential amino acid supplement, which protects against loss of muscle mass when subjects have sufficient energy intake, is no longer protective when energy intake is intentionally reduced to 80 percent of the requirement.[76] Another consequence of a catabolic state is formation of ketone bodies from lipolysis, promoting metabolic acidosis, a known consequence of an astronaut diet of high quantities of animal protein (which supplies acidic sulfur amino acids) combined with low amounts of fruits and vegetables (and thus a low intake of potassium, which helps to alkalinize pH). Acidotic conditions promote bone loss and kidney stone formation because of release of calcium as part of a compensatory mechanism.[77]

Nutrient Stability in the Astronaut Food Supply

Under space conditions, nutrients are obtained from food, with the single exception being supplementation for vitamin D. Beyond *nutrient stability* in the food supply, other issues that relate to providing food to astronauts include food safety; food preparation and storage; packaging; and mass, volume, and waste. These issues are described elsewhere.[78] Inadequate intake of any one essential nutrient over time has critical functional consequences including blindness (vitamin A), dementia (niacin), neurological dysfunction (folate), hemorrhaging (vitamin K), compromised immune function (protein), and ultimately death. In fact, most of the essential nutrients were discovered as a result of morbidity or mortality of individuals who were lacking in a specific nutrient—e.g., sailors who developed scurvy from lack of citrus fruit, prisoners who were switched from brown rice to white rice and developed beriberi, hospitalized patients who developed essential fatty acid deficiency during parenteral nutrition, and infants fed formula lacking essential nutrients. These examples underscore the critical need to assess the stability of essential nutrients in the astronaut food supply over time.

NASA Johnson Space Center has developed a wide portfolio of space foods that have been analyzed for their nutrient content based on published data. However, although it is well known that processing and long-term storage of food on Earth can lead to loss or even destruction of nutrients, there is currently insufficient knowledge about whether food processing in space affects nutrient activity, or whether nutrient potency is affected by in-flight radiation or storage under space conditions (e.g., temperature, humidity, and effects of radiation). Two ongoing projects address nutrient stability. In one study, ground-based shelf life testing is performed on five thermostabilized food items placed in storage for up to 3 years at three different temperatures. The other study is designed to determine stability of certain nutrients in five space foods and two supplements by testing content before and after spaceflight on the ISS.[79] A set of space foods and supplements that remain on Earth are matched with the space-flown foods for humidity, time, and temperature. A recent report from the latter study suggests that storage conditions are more important than exposure to microgravity, because both ground controls and space-flown foods decreased in nutrient potency over time.[80] However, the ISS may provide limited information related to exposure to space radiation that would occur on the Moon or Mars. Thus, to date NASA has had to rely on published data from external shelf life studies as guidance in designing diets and optimizing processing conditions for long-duration spaceflight.

Known Effects and Knowledge Gaps

Countermeasures Against Negative Effects on Nutritional Status and Knowledge Gaps

A decrease in nutritional status as a function of spaceflight can be due to one or more of several factors:

1. A lack of understanding of nutrient needs in space as compared to those on Earth;
2. An inappropriate diet design that does not provide the required intake of nutrients;
3. A lack of nutrient intake; or
4. A lack of nutrient stability.

As noted above, supplementation with vitamin D is necessary. Similarly, supplementation with folic acid and vitamin K improved their status.[81] This appears to indicate that the food supplied lacked adequate amounts of these nutrients, there was insufficient consumption of supplied foods containing these nutrients, or the potency of these nutrients in the supplied diet had declined. It is important to understand the reason for observed negative changes in nutritional status during and after flight in order to properly correct them. There is a considerable database documenting the association of spaceflight with depressed nutritional status, but information is insufficient as to the specific causes of depressed nutritional status. Considerable attention has been given, recently, to the development of an antioxidant "cocktail" to protect against increased production of reactive oxygen species during spaceflight. However, there are knowledge gaps about the efficacy and overall design of such a supplement. A potential countermeasure against loss of antioxidant capacity of foods needs to be based on experimental data. Processing can actually enhance antioxidant capacity in some situations and depress it in others, and antioxidants can even become pro-oxidants.[82] Before any recommendations can be made, it is critical to understand how astronaut food antioxidant capacity changes as a function of processing and "space conditions."

Countermeasures Against Lack of Energy Intake and Knowledge Gaps

Lack of sufficient energy intake is an ongoing problem with multifactorial causes, including nausea, lack of time for food preparation, lack of a social environment for eating, and boredom with foods over time. Countermeasures that have proven useful are increasing emphasis on communal dining (cosmonauts) and one-on-one counseling with astronauts as to the importance of consuming their food allotment. If a significant proportion of individuals in space continue to consume below their recommended energy requirement, strong consideration should be given to use of nutritional supplements and adjusting diets to meet nutrient requirements based on ~80 percent of energy requirements (i.e., a more nutrient dense diet). The current philosophy of basing diets on 100 percent of energy intake without supplements, commendable for Earth-based recommendations, may not be appropriate for space. An optimal nutritional status will require appropriate food intake. Further, a better estimation of the energy cost of exercise and EVAs is required to assess energy status and to design diets to meet energy requirements. Thus, activities requiring energy expenditure need to be coordinated with nutrition and food planning (see below the section "Physical Inactivity").

Countermeasures Against Loss of Nutrient Potency Due to Storage, Exposure to Microgravity, or Radiation in Space

Loss of nutrition potency is the most critical knowledge gap with respect to nutrition and the food supply for long-duration spaceflight. Lack of this knowledge for a semi-contained environment during a long-duration flight without the ability to replenish the food supply could have serious consequences, in particular as supplements will also be subject to decay over time. New methods of food processing and storage are therefore required. It is difficult to attract food scientists to focus on preserving nutrient stability in foods for 3 to 5 years when this is not an issue on Earth. In sum, the extent to which current processing techniques and subsequent space conditions affect nutrient potency needs to be ascertained before informed decisions for food selection, possible fortification, or reformulation can be made and before changes can be made to diet recommendations or new processing, packaging, and storage systems developed. Although optimal nutrition is always important for optimal health and function, it is critical for long-duration spaceflight. For example, at this time there is no evidence-based proposed system in place for how a 3-year trip to Mars and back will be able to sustain human life without replenishment of the food supply.

Research Models and Platforms Needed for New Research Directions

Excellent platforms for nutrition and food science research exist and are being exploited by the nutrition and food science scientists at NASA Johnson Space Center. These include use of bed rest facilities, which are now operated in a standardized manner; isolation venues such as NASA Extreme Environment Mission Operations; and outstanding facilities at Brookhaven National Laboratory to measure the effects of galactic cosmic radiation on animals, cells, and even foods. However, an integrative approach across all human studies in which nutrition and energy intake could be controlled or documented is missing. Valuable data are lost when, for example, human intervention studies are done on countermeasures to protect against muscle wasting, bone loss, cardiovascular effects, and immune response without any documentation as to what the subjects were eating or if they were meeting their energy requirements. As shown above, very different outcomes can occur depending on the nutrient and energy status of the subjects, and it is important to include such documentation in future studies. Use of isolation facilities is important with respect to food intake to evaluate food palatability, the effect of isolation on food intake, and how much variety, etc., is needed in the foods over time. Even if the overall goal of an isolation study is not primarily nutritional research, important information could be gained by controlling or documenting the foods eaten and the energy intake.

In basic food science, a new emphasis is needed at the molecular level to evaluate model foods and interactions of metals, oxidizing agents, lipids, etc., over time. The ability to test every astronaut food item under every condition of processing, exposure to radiation and microgravity, long-term storage, and every interaction of each essential nutrient clearly goes beyond the realm of the feasible. A better approach may be a translational approach to design food systems that can be based on experimentally validated model systems. This will require a new direction for NASA, as the Food Science and Nutrition areas have different reporting lines and until relatively recently have not worked together as a team.

Specific Recommendations for Fundamental and Applied Science Programs

1. The panel strongly recommends that food intake or at least energy intake be an outcome variable for human intervention trials supported by NASA or the National Space Biomedical Research Institute.
2. NASA should develop a strong basic food science program (most probably with external expertise) to evaluate the effects of a variety of conditions at the molecular level. This program should address nutrient-nutrient interactions and the effects of heat, storage, radiation, etc., over time. It should result in a set of principles that NASA can use to design food systems for the future.
3. NASA should consider designing diets that are more nutrient-dense to account for less than adequate energy intake.
4. NASA should develop a consistent and concerted effort across all stakeholders to emphasize the importance of maintaining energy balance during missions.
5. NASA should coordinate the design of diets and estimates of energy requirements with energy output (EVAs and exercise).

Summary and Concluding Comments

Optimal nutrition status affects every physiological system and impacts performance. At this time there is solid information on what nutrient needs are in space plus a good but not optimal understanding of energy needs. However, more effort needs to go into convincing astronauts and flight doctors of the importance of staying in energy balance. Also, there is minimal information on nutrient stability over time and as a function of food processing. Thus, it is critical that NASA:

- Address nutrient stability over time as part of planning for long-duration exploratory missions and
- Develop a food system that can support such a mission.

In addition, since nutrition is a crosscutting issue that affects all systems, it would be beneficial to have knowledge of the food and energy intake for human trials with primary outcome variables addressing different systems (bone, muscle, cardiovascular, immune response), since in most cases different outcomes can result as a function of diet or energy intake.

Radiation Biology

Space Radiation Biology Today

The area of space radiation biology has been studied in detail in recent years, with the NRC issuing several reports on the current status of the field, including strengths and weaknesses in the existing programs sponsored by NASA.[83,84,85] Further, the most recent NRC report, *Managing Space Radiation Risk in the New Era of Space Exploration*,[86] reviewed the current knowledge of the radiation environments likely to be experienced by astronauts; the effects of radiation on biological systems, electronics, and missions; and NASA's current protection plans. The findings and recommendations from that report covered a number of issues relevant to the current report, including (1) the need for research on the biological effects of and responses to space radiation; (2) the need for continued attention to radiation protection strategies, such as use of surface habitat and spacecraft structure and components, provisions for emergency radiation shelters, implementation of active and passive dosimetry, scheduling of EVA operations, and proper consideration of the As Low As Reasonably Achievable principle; and (3) the importance of making experimental radiation data available to the scientific and engineering communities. Because there have been recent comprehensive reports on space radiation biology, the panel has not attempted to carry out a duplicative review. Readers should see the recommendations in the above referenced reports. The present report contains a brief overview of important gaps in knowledge and research needs in space radiation biology and presents related recommendations.

NASA's current goals are focused predominantly on understanding the effects of space radiation on human beings in space and developing strategies to mitigate adverse effects. While there is a huge body of existing literature on the effects of low linear energy transfer (LET) radiation such as gamma rays and x-rays on biological samples, including data from long-term animal studies, clinical studies, and others, the information on radiation of the quality encountered in space (e.g., protons and high-LET radiation such as heavy charged ions) is much less detailed.[87,88] NASA has worked to develop a cadre of scientists and facilities that can be used to study the effects on biological systems of these radiation types with unique qualities. Because the radiation types and effects are distinct from those found in exposure to more conventional radiation sources, it has been necessary to train biologists in the physics, dose-distribution features, and other unique properties of space radiation. The means of doing so have included a number of annual investigators' meetings, special training workshops, and a course sponsored by NASA at the NASA Space Radiation Laboratory (NSRL) at Brookhaven National Laboratory.[89]

In the general radiation biology field, there are currently some studies being done on the biological consequences of proton effects because protons have been shown to have medical applications. There are currently at least six proton facilities functioning in the United States, and several more are under construction.[90] Protons have been shown to have depth-dose characteristics that make them a treatment option for prostate cancer, choroidal melanoma, chordoma, some brain cancers, and several other cancers. NASA's radiation biology community has been tapped to aid and provide resources to the medical community in understanding normal tissue toxicities as a result of exposure to protons in cancer treatment. Nevertheless, the medical community is more focused on high-dose and partial-body exposures, while NASA is of necessity interested in risk issues associated with low-dose and total-body exposures. Some of these cancer-based proton facilities (particularly those at Loma Linda University and the University of Pennsylvania) are also being used for NASA-based studies of space radiation effects on cells and animals.

While protons are one quality of radiation encountered in space, galactic cosmic rays include most of the elements in the periodic table; about 90 percent of the nuclei are protons (hydrogen nuclei), but the remaining 10 percent include helium, carbon, oxygen, magnesium, silicon, and iron. The proton facilities noted above can be used to study the effects of protons, but not the other heavy ions. Protons have biological effects similar to those of low-LET x-rays and gamma rays, whereas the radiation properties and biological consequences of heavier ions are distinctly different. In particular, heavy ions have a much greater relative biological effectiveness per unit dose than do x-rays.[91] Experiments on the biological effects of heavy charged ions are therefore essential to understanding the consequences of prolonged exposure to space radiation. High-LET radiobiology is an area that is not well studied in the general radiation community because it has lacked relevance to medical therapies (although this may be changing). Facilities for performing high-LET exposures for clinical applications exist only in Germany and Japan. In the United States, only the NSRL is capable of providing high-LET radiation for biological experiments.

Ground-Based Space Radiation Studies

As noted in Chapter 3, the NSRL is the result of cooperation between the Department of Energy (DOE), which runs Brookhaven National Laboratory, and NASA, which uses the facility to generate protons and heavy charged ions that can be used for irradiation.[92] While there is one proton facility used by NASA at Loma Linda University, the NSRL is the only facility in the United States that can generate appropriate heavy ions and mixed field beams for NASA's studies requiring simulated space radiation. Thus, this facility—and as a consequence the partnership with DOE—is essential for NASA's space radiation biology endeavors not only now but also well into the future; assurance of this continued cooperation is therefore important. The current contract between NASA and DOE was due to expire in June 2011, but it has already been renewed for an additional 3 years, until June 2014. There is some concern that NASA is entirely dependent on this one facility for space-radiation-relevant biological studies. DOE and NASA also cosponsor several research activities related to low-dose effects. DOE manages a low-dose program with a focus on understanding consequences of low-dose exposure to low-LET radiation, and NASA co-funds some proposals that overlap with its interests in this program, permitting a leveraging of funding.

NASA's focus on radiation risks to be encountered in space has also limited most experiments to ground-based studies. The doses currently encountered by astronauts on the ISS or on any space shuttle mission are so low that it is difficult if not impossible with today's science to detect any biological consequence using end points that are available in the research laboratory; this does not mean, however, that no consequences exist, since it has been shown clearly that low-dose radiation induces cell signal transduction changes, gene expression changes, and other end points in cell systems, and some studies in cells and animals have suggested that exposure to low-dose radiation may result in cancer.[93-97] The usual approach used by radiation biologists is to examine effects at multiple doses, starting with very high doses and then moving to lower doses, so that low-dose effects that would normally be considered marginal can be verified by more significant effects at higher doses. In addition, doses that would be encountered in compromising situations such as exposure to a solar particle event on the Moon or to constant radiation bombardment on a trip to Mars will not be mimicked by a short trip in the space shuttle or on the ISS.

The ground-based experiments have been confined largely to studies of cells or whole animals (mostly mice) exposed to radiation in the absence of other stresses such as microgravity, lack of exercise, etc., which astronauts are likely to encounter in space.[98] The NSRL is able to provide charged particles over a wide range of doses (fluences) and dose rates, including approaching those of galactic cosmic radiation in space, and with simulated solar particle events, also at relevant dose rates. Furthermore, the NSRL is positioned to allow expansion of space-like radiation research into combinations with other stressors (e.g., microgravity) and into radiation countermeasures.

Space Radiation Biology in the Next Decade

The heart of NASA's radiation biology program during the coming decade will be the development of a better understanding of radiation risks associated with spaceflight. Some of these risks are being explored now, including biological consequences of high-LET radiation, extrapolation of low-dose effects from high-dose effects, and biological consequences for noncancer end points. One unknown currently of considerable importance to the space radiation community is the definition of the dose and dose rate effectiveness factor (DDREF), a term that defines the risks associated with low-dose and low-dose-rate exposures. This factor is based on the idea that, as the dose rate decreases, repair occurs and reduces the negative consequences of an exposure to radiation.[99] Most current experimental work is based on high-dose-rate exposures, but a trip to Mars will involve a constant low-dose-rate exposure. Most exposures to radiation on the Moon are at low dose rates, not high, just as in interplanetary space. Only in the infrequent event of a solar particle event would there be higher-dose irradiation, and that would occur both on the Moon and if astronauts are in interplanetary space. Travel to Mars and other interplanetary long-term trips will therefore include primarily low-dose-rate exposures, which the DDREF will influence. There is little work being done in this area now, but during the coming decade additional low-dose-rate work will be necessary. If radiation doses in space are an uncertainty, then all approaches toward understanding effects and eventual mitigation are equally uncertain. While much work has been done on dosimetry in space, there are still unknowns. A precise understanding of the radiation qualities and doses to be encountered in the atmosphere of Mars, or on its surface, for example, is not clearly in hand. While some work has been done with dosimetry in space and also on the Moon, radiation quality questions remain.

At the current funding level, NASA programs may require significant time to assess risks from radiation exposure in space. Much work in radiation biology during the past several decades has focused on radiation as an inducer of cancer, and NASA work has not been an exception to that focus. In addition to cancer, other late tissue toxicities that have been examined include neurologic dysfunction, cataract induction, heart disease, and others. Induction of cataracts is currently being studied in the astronaut population and in human lens epithelial cells,[100] as well as in rodent models.[101,102] These results suggest that high-LET space radiation may induce cataracts at lower doses than was previously reported. In addition, in rodents cataract induction has been shown to have sex-specific differences, a finding that is worthy of continued investigation regarding mechanisms. Controversy over radiation-induced cardiac and blood vessel toxicities has also been found in the literature. Previous studies have supported the idea that long-term consequences of exposure of Hodgkin's disease patients to high-dose gamma radiation can lead to a variety of cardiac toxicities, including damage to the vasculature.[103,104] These are high doses not likely to be encountered in space, and gamma rays are low LET, unlike the high-LET radiation encountered in space. Investigations by Virmani et al.[105] suggest a relationship between exposure to radiation (via stents or external beam) and the development of atherosclerotic plaques. In long-term studies of atomic bomb survivors

Shumizu et al. demonstrated that long-term survivors of the atomic bomb radiation may be at higher risk than non-exposed populations for developing heart disease,[106,107] even when the doses are as low as 1 Gy. Again, the quality of radiation from the atomic bombs dropped on Japan is not high-LET space radiation, and the doses may be higher than those predicted from space radiation exposure. In addition, the risks associated with these exposures are long-term risks and should not affect astronauts during spaceflight. Nevertheless, concerns about the effects of underlying disease in astronauts and the uncertainties in what is known suggest that this area is a high priority for continued research.

More recently, additional programs have focused not only on the examination of high-LET radiation effects on cancer and late-tissue toxicities as end points but also on acute radiation consequences, such as neurologic effects, skin toxicities, and nausea, that could hamper functioning in space.[108] Because this work is at the cutting edge of the field, it is not always clear even which end points should be examined, and so the projects are focused less on mechanisms and directed more at a descriptive approach. While a large number of cellular studies lead into this work, most of these recent experiments require animals for assessing biological consequences in appropriate model systems. Animal work necessarily takes much longer to complete than cellular studies, and the experimental conditions often require long-duration experiments to assess precise consequences. These studies will no doubt continue through the present decade into the next and will probably not be completed during the 10 years following release of this report.

NASA is doing little work on countermeasures and mitigators at this point. Work done on shielding has been severely reduced in recent years. NASA will need to augment its countermeasures research in order to meet its goal of reducing biological consequences of exposure to radiation. A partnership with the Countermeasures program of the National Institute of Allergy and Infectious Diseases (NIAID) within the National Institutes of Health (NIH) could be very useful and productive since the goal of the NIAID program is to develop countermeasures in the same dose range that is of interest to NASA. Ongoing countermeasure work at the Department of Defense is also of interest. It will be important to assess whether those mitigators are capable of working in association with high-LET radiation. Additional work on radioprotectors (taken before radiation exposure) may also be of benefit, since astronauts on the Moon are likely to have some warning before a solar particle event occurs.

The radiation biology program is likely to be highly affected by informatics approaches to biology that have revolutionized the biomedical community. There are several investigators in the NASA community who are generating and integrating informatics approaches for studies of gene and protein expression, assessment of genetic contributions to radiation toxicities, and other purposes.[109,110,111] This work not only will enhance a basic understanding of radiation consequences but also will facilitate work toward understanding individual variation in response to radiation effects and inform modeling approaches for development of paradigms to explain radiation injury at the molecular, cellular, and organismal levels. Throughout all of biology, systems biology approaches are being used to understand the interactions of different factors on a particular biological response—multiple tissues interacting together, cytokines and other factors released in response to a stimulus, three-dimensional organization of the tissue, etc. This is likely to impact space radiation biology significantly in the coming decade. Continued interaction of NASA with other agencies such as DOE will facilitate this transition within NASA, since DOE has a well-developed systems biology group for studying radiation effects.

NASA will also need to consider examining more integrative work examining the effects not only of mixed fields of radiation but also multiple stressors (microgravity, etc.). While some work has begun on radiation effects on hindlimb loading/unloading systems in rodents, the effects of radiation on bone metabolism and other biochemical effects of microgravity have not been studied. This is especially difficult since many of the stressors (such as microgravity) are not well mimicked in a ground-based situation, and NASA's space radiation studies are designed to be almost exclusively ground-based at this point. In fact, it would be difficult to imagine a space-based study that could be guaranteed to achieve the doses needed to observe biological consequences and not at the same time be a threat to the astronauts themselves. This difficulty is likely to pose a considerable challenge for the future.

Major cuts in funding of radiobiology and radiation sciences in most federal agencies have led to a reduction in the number of programs providing radiation training and thus also a reduction in trained radiation scientists, especially radiation biologists. A plan for long-term assurance of research funding is needed for projects that require multiple years for completion, for example, carcinogenesis studies in animals, and to encourage investigators to feel invested in the NASA program, by fostering reasonable expectations of continued funding availability. As seasoned radiation scientists enter retirement, there are likely to be difficulties in finding sufficient qualified scientists in the field to meet NASA's radiation needs. Increased development of training programs in the radiation sciences generally, and space radiation specifically, are needed for junior investigators, as well as for investigators from other fields who are moving into space radiation studies.

The inclusion of training programs for graduate students, postdoctoral fellows, and junior faculty members will be essential for ensuring a proper cadre of personnel trained in radiation biology and radiation sciences and able to address questions relevant to NASA's space radiation needs. The relationship with NIAID (counterterrorism and countermeasures) will be extremely important to this educational and training mission for NASA. NIAID is already expending large resources in the radiation countermeasures area. NASA cannot afford to, and should not, "re-invent the wheel." It should capitalize on other funded projects ongoing in the area.

Recommendations

1. Continued research on the effects of space-quality radiation is essential to explaining the consequences for human health of exposure to radiation in space. Emphasis should be placed on increasing information comparing biological effects at various LET values and at low doses (low particle fluences) and low dose rates (fluxes). To enhance an understanding of whole-body and whole-organism effects, studies in cellular systems need to continue to progress to animal studies.
2. Radiation studies should include the development of an understanding not only of short-term effects that might hamper a mission (acute radiation toxicities) but also of long-term consequences (carcinogenesis, heart disease, neurologic dysfunction, and others) that might affect astronauts after they return.
3. A clearer understanding is needed of the effects of mixed fields composed of radiation of varying LETs, mixed effects of radiation and other stressors, and possible use of countermeasures.
4. NASA should enhance its relationship with other governmental programs (at NIH, DOE) to facilitate incorporation of radiation knowledge gained from other programs into the NASA information base.
5. NASA should ensure the continued availability of space radiation exposure facilities to its investigators through either a continuing relationship with Brookhaven National Laboratory or the availability of new facilities.

Physical Inactivity

During the past decade, there has been a remarkable evolution of our understanding of the role of physical activity in human health. Much recent research has shown that the human genome is predisposed to health when physical activity is maintained. Physical inactivity of the kind that might be encountered during prolonged periods of exposure to microgravity facilitates the expression of a different genotype more commonly associated with chronic diseases such as coronary artery disease, neoplasms, diabetes, and osteoporosis.[112] Thus, a threshold of physical activity is required to normalize gene expression that suppresses disease.[113] Because spaceflight represents an abrupt change in physical activity level, a systematic evaluation of genomic changes during spaceflight will advance the understanding of gene-environment interactions that influence chronic disease risk.

Gaps in the Knowledge Base

The concept of gene-environment interaction has rarely been applied to the spaceflight environment from the standpoint of physical activity. Although microgravity alters gene expression quite dramatically[114] and, at least in skeletal muscle, produces a pattern of gene expression similar to hindlimb unloading,[115] the expression of genes associated with disease has not been fully determined.

Recommended Research

An important focus of the next decade's work should include a systematic examination of mammalian gene expression in the microgravity environment for prolonged periods, focusing on genes associated with predisposition to disease. A second issue is to determine how these genes can be suppressed. At least three types of interventions are possible: genetic, pharmacologic, and addition of activity. While much of this work may be associated with ground-based research, verification from long-duration flight experiments will be necessary. Finally, a limitation of this particular route of study needs to be recognized: most chronic diseases are polygenic; changing a single gene may not produce a clinical phenotype.

Biological Sex/Gender Considerations

Approximately 500 people have orbited Earth since Gagarin's first orbital flight in April 1961.[116] Of this number, only about 12 percent (60 people) have been women.[117] Of the 92 astronauts currently classified as active in the NASA astronaut corps, 25 percent are women.[118] Only 8 women have spent 6 months or longer in microgravity, and of the 334 passengers on space shuttle flights, only 44 have been women.[119]

In 1998, the NRC report *A Strategy for Research in Space Biology and Medicine into the Next Century*[120] included a one-paragraph section mentioning that gender-specific space physiology has been neglected. That study committee's recommendation was that "NASA should continue to examine data from in-flight and ground-based model experiments, for gender differences in the response to microgravity" (p. 146). In the 11 years since that report was published, 38 flight slots have been filled by women, 8 of them conducting long-duration stays on the ISS. During this time, three reviews have addressed gender-specific health issues in spaceflight[121,122,123] and a number of investigations specifically employing women subjects in spaceflight analogs have been conducted.[124-132] These publications complement a relatively modest earlier literature of reviews and scientific studies of the physiology of women in space.[133,134,135]

Although men and women differ physically, there is considerable overlap between the sexes, which makes prediction of differences on the basis of biological sex inaccurate, particularly within a small elite subgroup (i.e., astronauts). For example, as identified in the next subsection, "Effects of Spaceflight," men and women experience the same general decrements in function, such as orthostatic intolerance, loss of fitness, musculoskeletal weakness, etc. However, studies of large population groups have found gender differences in anthropometry, exercise capacities, and sensory function. Women tend to have a higher percentage of body fat, less muscle mass, more flexibility, and lower blood pressure.[136] Women may have a more aggressive immune response to an infectious challenge, but are thought to be more susceptible to autoimmune diseases.[137] Women tend to be more sensitive to pain[138] and have better hearing and olfactory abilities.[139] Importantly, because astronauts represent such a small fraction of the population and because there is considerable overlap between women and men in all gender-related characteristics, it is not clear what differences attributable to gender have existed within the astronaut corps.

Effects of Spaceflight

For the most part, the physiological responses of men and women during spaceflight have been similar, but there may be potential differences. The following list should be considered preliminary and in need of further study because there is a void of well-controlled gender-difference observations.

- Cardiovascular deconditioning (lower orthostatic tolerance in women, especially following short-duration spaceflights [5-16 days] where there was a four times higher incidence of orthostatic intolerance [28 percent versus 7 percent] during a 10-min head-up tilt test than in males);[140]
- Exercise capacity (lower aerobic capacity and strength in women than men, with a possible greater initial decline in strength in microgravity);[141,142]
- Risk from radiation exposure for breast and endometrial cancer in women[143] and prostate cancer in men, and possible increased risk of radiation-induced cataracts in women;[144]
- Bone mineral density[145] (lower bone mineral content and bone mineral density in women but gender-specific response to microgravity still unknown);[146]
- Reproductive health (increased risk of endometriosis in microgravity);[147]
- Skeletal muscle atrophy (on average, smaller lean body mass in women than men, with a similar[148] or faster[149] loss of muscle mass in women in microgravity);
- Pharmacokinetics (women may be more susceptible to some toxins);[150]

- Renal stone formation (although there have been no marked gender differences in renal stone formation, men produce more calcium-containing stones while women produce more struvite stones [composed of magnesium ammonium phosphate]);[151]
- Nutrition (additional supplements may be needed, such as iron for menstruating women);
- Immune response (women have a more aggressive immune response but are more susceptible to autoimmune diseases);[152]
- Neurovestibular health (no obvious gender differences in susceptibility to space adaptation sickness);[153]
- Fetal developmental issues (see Chapter 6), conception in microgravity (abnormal results from animal studies),[154,155,156] and radiation exposure (also animal studies); and
- Decompression risk (increased risk in women requires confirmation).[157]

Current Status of the Knowledge Base

Mechanisms underlying gender differences in orthostatic tolerance are not yet completely understood, but may involve possible sex-specific hormone effects on neurovascular regulation of arterial pressure,[158] splanchnic vasoconstriction,[159] or physical differences such as a smaller and less distensible heart.[160] The study of the causes and prevention of post-flight orthostatic tolerance may lead to an improved diagnosis and treatment of ground-based cases of clinical orthostatic hypotension.

Spacesuit design for lunar and martian surface activity has received some recent attention. Many preliminary suit designs have been modifications from the Apollo-era EVA suit, which was designed for men. During EVA work, handgrip strength is essential to maintain stability, and the average woman's handgrip strength is only about 60 percent of a man's.[161] However, the importance of this difference is diminished in routine EVA practice by the extensive use of foot restraints at worksites and of robotic arms for many mobile tasks. However, the use of a fixed-mass life support system sized for maximal metabolic demands is likely to be a major gender difference working against generally smaller female space explorers. The fixed-mass life support system dimensions and interfaces place constraints on spacesuit design that may compromise achievable mobility for small astronauts for some joints and motions. In terms of aerobic capacity, it is estimated that physical effort in the U.S. EVA suit requires approximately 28 percent aerobic capacity for a male and 44 percent aerobic capacity for a female.[162] On the Moon and Mars with partial gravities, the metabolic cost of EVA work will be even greater.[163] Thus, an average woman working at a higher relative aerobic capacity and upper body strength will fatigue sooner. New suit-design efforts should meet individual needs for crew members[164] of both sexes. Objective requirements for strength, aerobic capacity, and flexibility are needed to select individuals optimally for an EVA-intensive mission. With the current EVA suit design and pressure, more women than men are likely to be excluded. An effort should be made to address this issue, to be certain that physics and not legacy determines who can participate in EVA.

From limited ground-based comparisons, exercise countermeasures are similarly effective in men and women in terms of maintaining aerobic capacity[165] and muscle strength.[166] A critical area of research is to develop countermeasures for bone loss during spaceflight, because low bone mass (more common in women) may increase susceptibility to fractures. Preliminary bed rest data suggest that exercise may be less effective in preventing bone loss in women than in men.[167] Investigations are needed of pharmacological countermeasures against bone loss (e.g., bisphosphonates) during spaceflight and to determine whether there are gender differences in efficacy.

NASA has focused on radiation risk assessment rather than on radiation countermeasures. Countermeasures to radiation exposure should be developed to protect against prolonged low levels of exposure or higher levels in the event of a significant accidental exposure. In addition, information from NIAID's Radiation Countermeasures program[168] may facilitate NASA's work in this direction. While it is likely that countermeasures that are useful in one sex will be applicable in the other, there are some concerns that mechanisms responsible for gender-specific differences (e.g., the role of hormone levels in cataract induction or breast cancer induction following radiation exposure) may be important in developing countermeasures. Studies of mechanisms of gender-based differences and their possible role in countermeasure development/utilization will be important factors for protection against radiation in the space environment.

Gaps in Knowledge

A gap in knowledge exists regarding the effect of spaceflight on normal hypothalamic, pituitary, and gonadal axis function. Also, flight studies are needed to measure hormones at intervals during long-duration flight to correlate with energy balance (nutrition and exercise data), reproductive function, and bone-loss and bone-gain markers.

One critical issue that needs to be understood for men and women who perform long-duration spaceflights involves bone loss. At this time it is unknown whether women are at greater risk of bone loss and fracture in space than men. Women generally have a smaller bone mineral content than men and, after menopause, have an accelerated bone loss that can reach 1 to 5 percent per year, which translates into an increased prevalence of osteoporosis. In bed rest studies, women with normal reproductive function exposed to simulated microgravity frequently become oligomenorrheic.[169] Such menstrual cycle disruptions accelerate bone loss and increase the risk of fracture. Whether the effects of microgravity and disruptions in reproductive function have additive effects on bone loss in astronauts is unknown. A common practice of women astronauts is to take hormones to regulate in-flight menstrual cycle function. Although the controlled estrogen and progesterone levels during such treatment may attenuate bone loss during spaceflight, there is no evidence to demonstrate the efficacy of this treatment. Moreover, hormone therapy may not be an option for astronauts at risk for certain cancers[170] or blood clots. In ground-based studies, some hormone therapies, such as selective estrogen-receptor modulators, protect against bone loss with less cancer risk,[171,172] but they have not been studied in a microgravity or simulated microgravity environment. The use of bisphosphonates in post-menopausal women to reduce bone loss has led to the implementation of an experiment currently underway on the ISS to test this drug's effectiveness in astronauts.[173] However, potential side effects of this treatment in patients, such as osteonecrosis of the jaw and over-suppression of bone turnover resulting in skeletal fragility, albeit rare, require caution with long-term use.[174] Atrial fibrillation has been reported in women with low blood calcium levels who take bisphosphonates,[175] although these findings have not been consistent.[176,177]

From both studies of Japanese atomic bomb casualties and survivors and radiotherapy studies of patients exposed to low-LET radiation, gender differences in tumor incidence[178] have been reported in the years following exposure to radiation. For example, there may be gender differences in development of cancer following exposure to radiation, and the types of cancers may be different. There are insufficient numbers of humans exposed to space-quality radiation to make an assessment of gender differences, and so NASA has relied on animal studies for this work. There are a few ongoing NASA projects examining cancer risk in animals following exposure to space radiation (protons and high-LET radiation). However, to date these studies have been able to use only a relatively small number of animals. More studies, with larger numbers of animals, are in progress, particularly using the resources of the NSRL, although little attention has yet been paid to gender differences. DOE conducted a series of large-scale animal studies from the 1960s to the1990s examining gender effects in mice and dogs exposed to high-LET neutrons (which have biological properties similar, although not identical, to those of heavy charged particles in space radiation and can be produced by interactions of charged particles in spacecraft and in astronauts' bodies). These studies also pointed to an increased risk among female animals and a difference in the spectrum of tumors observed between the sexes.[179–182] Recent studies[183] have also pointed to gender differences in cataract incidence in rats exposed to cosmic (high-LET) radiation. Therefore, defining gender differences in responses to space radiation exposure is an important knowledge gap.

Research Models and Platforms

Mathematical Models

Mathematical models, such as the digital astronaut program,[184] may be useful in predicting or verifying gender differences in physiological responses to spaceflight. However, more experimental data are needed to make such models useful.

Animal Models

Ground-based and flight studies should be used to evaluate effects of microgravity simulation on reproductive function and to draw correlations between altered reproductive hormones and changes in body composition and health status. Additional studies should be used to evaluate radiation risks because space exposures in general cannot provide the doses needed to test radiation effects in a reasonable sample size.

Bed Rest

Bed rest studies should be continued that allow direct gender comparisons and an understanding of the effects of microgravity simulation on reproductive function, and potential bone countermeasures (continuous estrogen/progesterone use, bisphosphonates, nutritional supplements, etc.).

Analog Environments

Analog environments (see Chapter 5) with small groups can be used to study single-gender and mixed-gender behavioral interactions.

Recommendations and Priorities

There is a critical lack of information about the effects of spaceflight on gonadal function and bone loss, as well as about effects of cosmic radiation on women's health. Basic information is needed about the effects of microgravity and circadian disruption on gonadotropin release and on estrogen and progesterone concentrations. In general, development of countermeasures should account for gender differences. Also, whenever feasible, modifications to the EVA suit design should accommodate smaller individuals or crew members with less upper body strength, to enhance mobility during EVA tasks.

Summary and Conclusions

The most significant gender issues that should be addressed in the next decade include an understanding of possible differences in bone loss and radiation risks and development of effective countermeasures. For crew selection, the most important consideration should be to choose the people most qualified to perform the required tasks. A critical issue is to select crews that work well together. Antarctic studies have found that mixed-gender teams may experience sexual jealousies and rivalries but often are more stable than all-male teams.[185,186]

Thermoregulation

The importance of maintaining body temperature is well understood by both clinicians and lay persons. The normal resting core temperature is tightly controlled around 37°C. For survival, the degree of overheating is more critical than overcooling. Although core temperature is tightly regulated, it fluctuates normally in a daily, 24-h circadian rhythm. This small, approximately 1°C, rise and fall in temperature is associated with hormonal cues that are important in regulating food intake, sleep, cognitive function, and immune surveillance.

Effects of Spaceflight

The effect of microgravity on thermoregulation usually is not viewed as an issue as critical as the effects of microgravity on the cardiovascular or musculoskeletal systems. During normal spaceflight, astronauts are enclosed in an artificial environment that maintains ambient conditions within a narrow temperature range to which they can easily adapt. However, to develop more efficient and effective environmental systems and to understand the consequences of environmental failures, the effects of spaceflight on human thermoregulation must be fully understood.

Physical heat exchange between the human body and the environment occurs through four main channels—radiation, conduction, convection, and evaporation—and the sum of heat transfer by all four must balance body heat production. In microgravity there may be significant alterations to each channel of heat transfer, compared with Earth environments.

Resting heat production for an average man is approximately 1,824 kcal/day and approximately 18 percent lower for a woman. Heat production measurements obtained during flights have shown that average metabolic rates are similar to preflight rates.[187,188] However, with longer flights it is possible that resting metabolic rates may be reduced if there are large decreases in metabolically active tissue (e.g., with muscle atrophy).

Thermal risk is particularly great during EVA, when radiant heat may be gained on the surface of the EVA suit exposed to the Sun or lost from the surface exposed to the lower temperature of open space (−270°C). Astronauts are protected from radiant energy fluctuations by the selection of a material and color (white) of the outer EVA suit that will insulate the crew member while reflecting radiant heat.

Conductive heat exchange occurs between two objects of different temperature that are in direct contact. During spaceflight, conductive heat loss becomes an issue when the insulative property of the EVA suit is insufficient to prevent heat exchange between the astronaut and a surface in contact with the astronaut. For example, due to insufficient glove insulation a crew member experienced frostbite while training for an EVA to repair the Hubble Space Telescope.[189]

Convective heat exchange refers to the exchange of thermal energy between hot and cold objects by the physical transfer of matter such as a liquid or gas. In humans, convective heat transfer occurs at two levels. First, body heat is transferred from the body core to the skin by dilating skin blood vessels to increase blood flow to the skin. This regulation of blood flow is under the control of the autonomic nervous system and has been found to be impaired after spaceflight or simulated microgravity.[190,191] Second, heat is lost by convection from the surface of the skin via the air circulating near the skin and from the respiratory tract to the expired air. At very low airflows ($<0.12\ m \cdot s^{-1}$) heat loss occurs by natural or free convection, in which the air nearest the skin surface becomes warmer and lighter than the surrounding air. In a 1-*g* environment, the warm air rises away from the skin, removing heat. However, in a microgravity environment, this warm air does not rise but remains close to the skin.[192] Novak and coworkers[193] confirmed that at airflows less than $2\ m \cdot s^{-1}$, convective heat loss from an artificially heated metal cylinder is impaired during spaceflight.

The effect of microgravity on evaporative heat loss is unclear at this time, but it appears that both sensible and insensible water loss are reduced Thermotolerance, which is the ability to tolerate a given level of core temperature without symptoms of heat illness, depends on the individual's level of aerobic conditioning and heat acclimation.[194] Before flight, most astronauts are relatively fit and have attained some level of heat acclimation. However, after spaceflight most crew members undergo a significant degree of aerobic deconditioning[195,196,197] and have become deacclimated to heat. This may result in a faster rise in core temperature during exercise or heat exposure, reducing productivity and increasing the risk of heat-related illness.

Another well-documented effect of spaceflight is a decrease in blood volume, due to a rapid, 10-20 percent loss of plasma and a more gradual decrease in red blood cell mass.[198] A loss of plasma volume of this magnitude on Earth severely compromises skin blood flow and sweating responses during body heating.[199] In two crew members after a 115-day spaceflight, a faster rise in core temperature during submaximal cycle exercise was attributed to a "reduced sensitivity" of both the skin blood flow and sweating responses.[200]

Crew members orbiting Earth experience complete day/night cycles approximately every 90 min. This change in light cycling or the overall lower lighting conditions during spaceflight have been postulated to produce a dampening and delay in circadian temperature fluctuations.[201] In ground-based studies, circadian desynchronization causes disruption of sleep and eating cycles, altered insulin regulation, and elevated levels of stress hormones such as cortisol and catecholamines.[202] During the first couple of weeks of spaceflight, circadian changes in body temperature have shown only minor changes in phase or amplitude,[203,204] although there is a reduced quality of sleep, which becomes shorter in duration and more disturbed. In one crew member during a long-duration spaceflight, circadian rhythm was well maintained during the first 100 days of flight, but then there was almost complete flattening of circadian temperature fluctuations, which was associated with sleep disruption.[205]

The thermoregulatory responses to cold during spaceflight have been even less studied than responses to heat. However, results from ground-based studies (see bed rest section below) suggest that physiological defenses against cold stress also are impaired, including delayed vasoconstriction and less-effective shivering response.[206] Altered vasomotor responses in the hands and feet may increase thermal discomfort and increase susceptibility to frostbite.

In the field of thermoregulation, technology always will be crucial for preventing thermoregulatory stress. Further refinement of environmental control systems, to perhaps include feedback and assessment of the thermal status of the crew members, may be needed as space vehicles and crew members are exposed to different and possibly more challenging environmental and work scenarios. Emergency procedures in the event of a loss of environmental control should be considered.

Countermeasures

Historically, thermoregulation issues during spaceflight have been addressed by engineering solutions. Under normal spaceflight conditions, cabin temperature is maintained in a thermoneutral range. On the ISS, for example, the environmental control system was designed to maintain cabin temperature between 18°C and 27°C and relative humidity from 25 to 70 percent.[207] However, there have been many incidents where the engineering solutions have failed and crew members were exposed to extreme thermal environments. For example, after the explosion of oxygen tanks during the Apollo 13 mission, the crew lived in the lunar module, which cooled rapidly to approximately 11°C. In another example, during the launch of Skylab a solar array and a meteoroid shield were lost and a piece of the shield inhibited opening of the second solar array. This led, among other things, to the temperature in the laboratory stabilizing at 58°C. When the first crew arrived, they had to rotate at intervals between the hot Skylab laboratory and their cooler command module until they could deploy a solar shade.

Since the *Challenger* accident, crew members are required to wear a pressure garment during space shuttle launch and landing that consists of an antigravity suit, a cooling garment, and an outer impermeable shell.[208] Despite the liquid cooling garment, crew members report feeling hot during re-entry;[209] this is a concern because increasing body temperature could reduce their orthostatic tolerance during re-entry.[210] Even mild elevations in core temperature during landing could result in cognitive and manual performance deficits.[211]

During EVA, crew members wear a pressurized suit to reduce the risk of decompression sickness and to sustain a pressurized environment essential for normal gas exchange through the lungs.

After observations of thermal discomfort during EVA in the Gemini program, a liquid cooling garment has been included in all EVA suit systems starting with Apollo. The current EVA suit has an upper heat-removal limit of 504 kcal/h (2,000 Btu/h) for 15 min or 252 kcal/h (1,000 Btu/h) for up to 7 h.[212] In a recent report estimating the heat removal requirements for a future lunar EVA in which the crew members would ambulate over the surface wearing their EVA suits, the average heat production while performing a 10-km "walkback" was 598 kcal/h (2,374 Btu/h). The subjects in this test unanimously reported inadequate cooling, and their core temperatures rose by an average of 1°C.[213] Thus, improvements in the cooling system of the EVA suit are required to prevent hyperthermia and possible impaired work performance of EVA crew members in future lunar or exploratory missions. This will require the development of smart EVA suits, with automatic regulation of cooling based on physiological input, a higher cooling capacity, and possibly an ability to modify the heat transfer characteristics of the spacesuit skin.

Gaps in Knowledge

Since the physiological consequences of long-term spaceflight for human thermoregulation are still unknown, engineering solutions to control thermal stress may be inadequate. The current thermal models to predict human thermoregulation during and after exposure to microgravity and partial gravity require validation for long-duration flights.

Evaporative cooling normally is the primary channel for humans to dissipate heat during heat exposure. Yet the effect of microgravity on evaporative heat loss is unknown.

Convective heat loss is a major channel for heat loss under resting conditions. Because of changes in blood flow distribution and increased peripheral blood flow after adaptation to microgravity, there may be an accumulation of heat near the surface of the skin and under clothing. Changes in convective heat exchange may alter thermal comfort and must be considered when choosing clothing and environmental conditions in the spacecraft or EVA suit.

The effects of adaptation to microgravity on heat and cold thermotolerance are not understood. It is likely that loss and redistribution of body fluids, decreased vascular reactivity, and reduced physical transfer of heat may limit the range of tolerable body temperature and increase risk of temperature-induced injuries and illnesses.

No data currently are available regarding the effects of spaceflight on cold thermoregulation. There is even a paucity of ground-based data on the effects of reduced-gravity exposure on cold response or tolerance to cold.

This is an important gap in knowledge for prevention of frostbite or other cold-related injuries during EVA or in the event of a prolonged power outage.

Finally, the consequences of altered thermoregulation on alertness, cognitive performance, food intake, sleep, and immune function require further study.

Research Models and Platforms

Mathematical Models of Thermoregulation

Mathematical models are used to predict human thermal responses during spaceflight.[214-217] However, none of the current models accurately incorporates physiological adaptations to microgravity. Fundamental research is required to understand the effects of microgravity on physical heat transfer, physiological responses, and possible changes in thermal tolerance. Multisystem models should be employed to probe the consequences of other microgravity-induced adaptations—for example, impaired baroreceptor function—on human tolerance for hot or cold environments.

Animal Models

Altered circadian control of body temperature during spaceflight has been demonstrated in primates, rats, beetles, and even fungi.[218] In constrained primates, spaceflight is associated with an early drop in core temperature,[219] a dampening of circadian temperature fluctuation, and desychronization of circadian rhythm.[220,221,222] Animal models in microgravity or simulated microgravity may be required to study physiological responses and molecular pathways that determine thermotolerance.

Bed Rest

Bed rest simulates cardiovascular changes and fluid shifts thought to impair thermoregulation in microgravity.[223] Thermoregulatory changes during bed rest concur with many of the changes reported during and after spaceflight, including a gradual reduction in resting metabolic rate,[224] lower resting skin temperatures, hyporeactive vasomotor responses to thermal stress, and changes in circadian rhythm. Bed rest is thus an excellent model to study specific mechanisms and potential countermeasures for the impaired thermoregulatory responses during a microgravity exposure. Bed rest impairs thermoregulatory responses to heat during exercise,[225] whereas supine exercise (for 90 min each day at 75 percent of pre-bed rest maximal heart rate) preserved the skin blood flow and sweating responses[226] and maximal capacities.[227] The effect of bed rest on thermoregulatory responses to cold was examined by immersing male subjects in cold water (28°C) for 100 min. After 35 days of bed rest, the fall in rectal temperature during a cold water immersion was more rapid than after 5 weeks of recovery from bed rest. Whether exercise countermeasures can prevent these changes in response to cold is unknown. Mekjavic and coworkers attributed the greater and more rapid fall in body temperature to a less pronounced vasoconstrictor response and a delayed-onset and less vigorous shivering response.[228] An impaired vasoconstrictor response also was reported after 14 days of bed rest, when a cold stimulus was applied by immersing one foot in ice water while the forearm blood flow was dilated by reactive hyperemia.[229]

Recommendations and Prioritization in Fundamental Research

1. Flight studies are needed to quantify the effects of microgravity on human thermoregulation. Overestimation of human thermal tolerance may result in an increased risk of heat injury during EVA, landing, or emergency egress. Such flight studies will be aided by the development of non-encumbering methods to accurately measure and record body temperatures over 72-h periods during spaceflight.

Thermoregulatory measures should be obtained in at least 10 astronauts during a prolonged spaceflight. Measurements should be obtained during rest and during exercise before flight, approximately every 30 days during flight, and post-flight. Measurements should include core and skin temperatures, heat production (indirect calorimetry), sweat loss, and blood flow distribution.

2. There have been few, if any, molecular studies of the effect of microgravity on thermotolerance. This gap can begin to be addressed by genomic studies to identify specific molecular pathways altered by microgravity

that might enhance or reduce thermal tolerance. This information may be useful for predicting which astronauts might be most susceptible to heat intolerance and for developing pharmacological countermeasures to prevent heat injuries and improve thermal tolerance.

Studies should be conducted in animals and humans during ground-based and flight studies. This objective may be met by obtaining blood, tissue, or saliva measurements during the studies outlined above and comparing specific molecular markers (e.g., heat shock proteins) against physiological responses.

3. Heat tolerance should be assessed in ground-based studies using animals and humans to assess the physiological mechanisms by which microgravity might alter thermal tolerance. This information may be used in the prevention and treatment of heat illness during a mission. Also, specific cooling countermeasures for use in microgravity can be evaluated.

Measurements should include assessment of gut permeability, endotoxin and inflammatory markers after an exertional heat stress in humans, and markers of tissue damage and inflammation in specific body tissues (skeletal muscle, brain, heart) in rats or mice.

4. Human circadian rhythm measurements are needed in a larger sample of humans during long-duration spaceflight, with continuous rather than intermittent measurement of core temperature. This information may be useful in preventing decrements in astronaut performance resulting from increased stress and sleep disruption caused by circadian dysrythmia.

Bed rest studies (with simulated flight-like light/dark cycling) and flight research should be undertaken to obtain core temperature, blood pressure and heart rate, stress (e.g., insulin, cortisol, urinary norepinephrine) and inflammatory markers (e.g., cytokines, C reactive protein), sleep, performance, and cognitive measurements to assess interactions between circadian dysfunction and overall health, performance, and cognitive function.

5. Cold studies should be continued to assess thermoregulatory responses to passive, prolonged cold exposure after bed rest or spaceflight. Studies should focus on susceptibility, awareness, and prevention of hypothermia and local tissue injuries (frostbite).

Measurements should include core and skin temperatures and vasoconstriction and shivering responses. Effects of prolonged bed rest with and without exercise countermeasures to maintain sympathetic, cardiovascular, and body fluid responses on cold tolerance should be explored. Effective methods for rewarming that could be applied in spaceflight could be developed and evaluated during these studies.

Summary and Conclusions

Maintaining a safe level of body temperature is critical for the health and productivity of the crew during spaceflight. Normally, humans thermoregulate within an approximately 5°C range of core temperature (35°C to 40°C). However, during spaceflight behavioral responses are limited, cooling and heating systems may fail, and it is suspected that thermotolerance to cold or hot conditions may be reduced. Fundamental research is required to understand the effects of spaceflight on human thermoregulation. Applied research is needed to develop countermeasures and emergency treatments for hyperthermia and hypothermia. Such applied research could lead to the development of rapid whole body cooling and heating methodologies that would be practical and effective in a microgravity environment. It could also enable the identification of personal cooling or heating methods that would permit crew members to perform their normal activities in a spacecraft or habitat in which the ambient temperature control is not functioning.

INCREASING TRANSLATIONAL RESEARCH IN THE SPACE/LIFE SCIENCES

Implementation of a Clinical and Translational Science Framework

Biomedical research is carried out in a continuum that begins at discovery and ends with documentation that an intervention improves health or function. Such a discovery may occur through research at the bench, in an animal laboratory, or at the clinical level, deriving from a physiological investigation, a controlled clinical trial, or from a serendipitous observation in a single subject or group of subjects. If gaps impede this process of moving from discovery to deliverable, this process may founder. In spite of much insightful basic and clinical space science research by scientists over many years, the extent to which the products of this discovery process have led to interventions or countermeasures that are now routinely employed has been limited.

Progress from Fundamental to Applied Knowledge

When an important biomedical discovery is made, it should ideally set in motion an orderly sequence of studies that take the discovery to achievement of a deliverable such as a drug, an intervention, or a countermeasure that addresses a health concern. The totality of this process is sometimes referred to as clinical and translational science,[230] which is best viewed as a way of organizing the research process. In 2006, the progression of these principles was developed at NIH and began to permeate the biomedical research community in both NIH intramural research and, with the inauguration of the Clinical and Translational Science Awards (CTSAs), the extramural research community in the academic health centers and universities. In some ways, elements of NASA's publicized philosophy of "better, faster, cheaper" influenced the CTSA conceptual development. After 4 years of deployment across 49 academic medical centers encompassing perhaps 8,000 individual research projects, with a mandate to expand to a system of 60 centers, this program has in fact become the largest functionally integrated research network in biomedicine. The CTSAs support Clinical Research Units (formerly called General Clinical Research Centers), which now represent the nation's major capability for intensive patient-oriented research—for example, as might be required in microgravity models such as bed rest, nutritional interventions, exercise interventions, energy balance, drug trials, and imaging capabilities that permit visualization of human physiology in real time.

Findings and Recommendations

The reorganization of the nation's biomedical research infrastructure into a clinical translational science network represents a substantial opportunity for NASA to emulate and to partner in this 21st century approach to applying new biomedical knowledge. Two major changes are recommended.

1. *Implementation of a clinical and translational framework for space life sciences research.* In spite of the large number of studies of countermeasures that have been carried out over the years, the results that have actually reached implementation are quite limited. This translation failure might best be addressed by systematic analysis of where in the translational process inefficiencies might be occurring. Overcoming or at least minimizing such gaps or obstacles has been a hallmark of the NIH CTSA program. Although this program is only entering its fifth, the longest existing awards have just gone through renewal review and were generally considered to be doing well. The kinds of interventions currently being undertaken to improve the process of clinical and translational research in the CTSA network should be generalizable to some aspects of NASA's research enterprise. Increased research investment by NASA in basic science and in the discovery phase of this process, including studies of animal models, might be a means to prime the process of NASA research.

One important aspect of this process is the deployment of informatics capabilities addressing all aspects of the research process so that captured information about all transactions—whether research, contractual, bureaucratic, laboratory, facility, etc.—are monitored to examine inefficiencies that may be delaying or in some cases sidelining a rapid and orderly discovery process. Hallmarks of this approach include:

A. Focusing on investigator needs first;
B. Collecting and analyzing metrics to assist in the evaluation of the success of projects;
C. Inclusion of auditing to ensure evaluation metrics are available to support program review and future prioritization; and
D. Leveraging existing resources whenever possible to avoid duplication of effort.

2. *A refocus on research as a priority to enable NASA to move forward with the best possible science.* To improve NASA's research enterprise, important changes will need to be implemented. Since the retirement of

Spacelab, in which many sophisticated experiments took place in the context of dedicated research missions implemented by a highly trained and supportive crew, the priority for research has been reduced to levels that compromise the research endeavor. For example, individual astronauts may elect not to participate in research projects for reasons such as concerns about confidentiality or simply a lack of interest, with no clear link to flight status or mission objectives. Communication between investigators and the actual subjects (the crew) is kept to a minimum, and misunderstandings and miscommunications about the goals, objectives, and risk/benefit ratios of specific studies are common. Mission managers, who are not uniformly well grounded in research, sometimes control crew availability, making decisions about crew scheduling that can compromise research studies. Even when non-advocate review committees provide outside consultation and peer review, their recommendations may not be followed, with attendant costs in terms of quality of research. To fix these systemic problems and improve NASA's research programs over the next decade, the panel offers the following suggestions.

A. *A change of attitude should be implemented concerning research within NASA.* Although it seems obvious that in order to improve research, research must be a priority, this truth is not always reflected in day-to-day NASA decisions. Every employee from management through crew needs to subscribe to the view that one of the key objectives of the organization is to support and conduct research. The designation of an ISS crew science officer is a positive step, but it does not achieve the goal in its entirety.

B. *Human research is an obligatory component of spaceflight operations.* Most crew members are fantastic advocates for research and go to extraordinary lengths to do a superb job for ground-based investigators. However, if a crew member does not want to participate, that individual should understand that not being assigned to a flight is the implication of that decision. These decisions should therefore be made before specific flight assignments as part of crew selection and indeed as part of the astronaut selection process from the beginning. The panel understands that this perspective might raise issues of compulsion to participate. Of course, no individual should ever be forced to participate in research against his or her will. However, it seems reasonable and ethical that, if participating in research is part of the job, then not participating may lead to a different type of assignment. This recommended philosophy is consistent with the occupational medicine model recommended in previous reports.[231] Difficulties inherent in this approach exist, but the panel encourages NASA to partner with other entities to find a workable solution. Furthermore, the panel recognizes that not all ethicists will agree with the prescriptive approach outlined above, and NASA should engage a variety of individuals with expertise in addressing ethical concerns to advise it on these issues.

The ISS and the U.S. National Laboratory as an Analog of a CTSA Clinical Research Unit

The ISS is now essentially complete and is beginning to realize the goal of serving as a U.S. National Laboratory.[232] This will present opportunities for NASA as well as for public, private, and governmental entities such as NIH. Such a laboratory can be of inestimable value in addressing scientific as well as practical issues in basic and human physiology, pathophysiology, pharmacology, and discovery-based translational science.

It is fortunate that the full ISS capability comes when NIH is rolling out its national CTSA network. There is, at one level, an analogy between the CTSA Clinical Research Units, which provide the capability for complex and intensive clinical studies of basic and clinical physiology and pathophysiology, and the U.S. National Laboratory on the ISS, with its goal of providing a facility where compelling biomedical research of the highest importance for understanding space physiology can be carried out.

However, there are enormous differences in logistical issues and costs of research in circumstances as disparate as the ISS and a Clinical Research Unit. Also, there are special issues of research opportunity on the ISS, given the necessarily circumscribed and targeted mission equipment. Nevertheless, comparison of the ISS's potential function in scientific discovery with a CTSA Clinical Research Unit should prompt discussions about how NASA and NIH might be able to cooperate in research selection and implementation. Indeed, NASA could interact with NIH and its National Center for Advancing Translational Science (which oversees the CTSAs) to explore some of the following ways to interact and strengthen relationships between these agencies:

1. *Knowledge sharing.* At the simplest level, knowledge sharing might take the form of comparative sharing of best practices and analysis of research metrics, but at a more substantive level it should include a genuine collaborative effort between NASA and NIH to jointly evaluate, fund, and place the most compelling biomedical research in the U.S. National Laboratory. NIH Program Announcement 09-120, "Biomedical Research on the International Space Station," is a notable example of such cooperation, but it is limited to cell biology and molecular biology experiments. Building on this model, the facilities of the CTSA Clinical Research Units could be made available to NASA as a research capability for bed rest studies and complex land-based protocols that could provide preliminary data for future research on the U.S. National Laboratory. With experience, cooperative evaluation and review of proposed research might enable even greater future collaboration.

2. *Establishment of standing research prioritization panels or committees that include flight medicine, NASA and NIH science staff, and the external research community.* The NIH style of peer review has withstood the test of time. Standing committee members are invited to defined terms of service, with an experienced panel chair and SRA (scientific review administrator). A specific NASA study section, perhaps created within the NIH Center for Scientific Review system, as opposed to the current system where every NASA review panel is constituted ad hoc, would create a cadre of experienced reviewers who could review proposals consistently. However, NASA and the National Space Biomedical Research Institute should retain the programmatic right to focus on research areas of high priority. This approach would help ensure that highly reviewed science would be implemented based on both scientific and programmatic priorities, simultaneously expanding knowledge sharing between the agencies.

3. *Collaboration in ground-based research.* The CTSA effort establish a network of Clinical Research Units that is the ideal infrastructure to increase opportunities for research on circadian rhythms and for postural and bed rest analog studies. With dedicated nursing and dietetic staff and a national coordinating network, a strong CTSA partnership could increase the number of NASA translational researchers and the annual participation rate in announced research opportunities in areas relevant to space biology and biomedicine.

Aspects of Clinical and Translational Science Applicable to Space Biomedical Research

Astronaut Health and Genomic Medicine

The Human Genome Project has accelerated scientific discovery. While variation in the human genome has long been the cornerstone of human genetics, the diverse set of novel molecular tools now available is radically altering how we think not only about genetics but also about the essence of disease and diagnosis. By comparing genetic differences at the nucleotide and haplotype level, coupled with sophisticated phenotyping and genome-wide association studies (GWAS), the complementary approach of medical resequencing for rare variants, and searching for somatic mutations, new insights into interindividual differences and the susceptibilities that distinguish individuals are emerging and are leading to the new discipline of personalized medicine.[233]

Known Effects and Gaps in Fundamental and Applied Knowledge

Personalized medicine has penetrated many fields, including oncology,[234] cardiovascular disease.[235] and sleep loss. With respect to sleep loss, the large phenotypic interindividual differences in neurobehavioral vulnerability to acute and chronic partial sleep deprivation[236] might compromise astronaut health and safety in spaceflight due to loss of circadian entrainment, abrupt sleep shifts for logistical reasons, high-tempo work schedules, and the effects of the microgravity environment.

Despite this progress elsewhere, the vision of personalized medicine has been delayed in the current NASA Human Research Program science portfolio. This delay may relate to concerns that genetic information would be used for job selection in the United States.

Findings and Recommendations

NASA can benefit from the rapid scientific advances in personalized medicine. An era is emerging when these unusually rapid advances are not only expanding our knowledge of human health but also altering the fundamental basis of the health care equation. Reorganizing NASA's approach by aligning it more with this new paradigm has important advantages for both science and the health of astronauts, although appropriate implementation may require a cultural change within NASA. A somewhat different issue, but one of considerable importance, is obtaining phenotypic analysis and physiological information on individual astronauts. Privacy issues have driven an understandable reluctance to make such data available to investigators. However encryption and de-identification methods for medical records are progressing rapidly, and NASA should explore recent advances in the de-identification of data such as the approach described in the synthetic derivative developed by Masys.[237]

Clinical Pharmacology in the Microgravity Environment

The body's absorption of and response to medications are generally perceived to be relatively unaffected by the microgravity environment. This view reflects the paucity of reports of serious deviations from expected drug effect in space. However a recent in-depth report[238] based on evidence from spaceflight experience calls attention to the possibility of alterations in clinical pharmacology in the microgravity environment that may have escaped previous recognition. Policies to better document deviations from land-based norms have been put in place.

Known Effects

There is evidence that, in some circumstances, there may be changes in the individual pharmacodynamic and/or pharmacokinetic profile of certain drugs caused by physiological adjustments or other factors in the microgravity environment.[239] Some of the effects observed have been related to drug-drug interactions, for example altered drug handling related to agents whose metabolism depends on the enzyme CYP2D6. A specific example has been the concomitant use of midodrine (the alpha agonist administered to increase vascular tone) and promethazine (an agent commonly used by astronauts for its antiemetic and sedative effect). While this specific interaction has received the greatest attention, there are 14 drugs in the current space shuttle and ISS formularies that have CYP2D6 effects. Other differences in drug effect may relate to less well studied phenomena such as alterations in gastrointestinal transit time and gastrointestinal absorption,[240] microgravity-induced alterations in blood and plasma volume,[241] and altered intestinal microflora (e.g., bacterial hyperproliferation).

Gaps in Fundamental and Applied Knowledge

A better understanding is needed of the consequences of physiological variables and how they alter drug absorption and handling at the kidney. Also it is important to understand if there are systematic alterations in the expression of key gene products of importance in drug disposition in the microgravity environment. These studies need to be conducted in the space microgravity environment, as well as in analog environments.

Findings and Recommendations

Efforts should be made using both ground-based microgravity analogs and experiments in the microgravity environment to determine if alterations in volume of distribution, drug metabolism, and hepatic and renal clearance occur in microgravity. If these effects occur and are of a magnitude that might have consequences for the health of astronauts, consideration of drugs with less extensive metabolism or more stable metabolism might be of value in selecting the drugs for future use in space.

OVERARCHING ISSUES AND GAPS IN THE KNOWLEDGE BASE

Educating the Next Generation of Space Translational Scientists

Only an estimated 14 percent of new scientific discoveries are successfully translated to routine clinical practice, and the time for those translations to occur has averaged 17 years.[242] If this translation rate remains similar for the life and microgravity science disciplines, then the research progress based on decadal survey recommendations will be very small. Returning to the analogy of the CTSA, that reorganization of the U.S. biomedical research enterprise expands the emphasis on education and training programs for target audiences ranging from practitioners to researchers to students as one way to expedite the translation of discoveries to practice. One ancillary goal of these programs is to increase the number and quality of collaborations among practitioners, scientists, patients and administrators.[243] This model of expanded education offers clues to the design of unique educational programs that improve translation in the space life sciences, including one or both of the following elements:

- A curriculum-based program focused on existing flight surgeons and physician-astronauts that will seek to expand their research knowledge and skill set (such programs are comparable to the NIH's K30 awards) and
- Mentored research training, similar to the NIH K12 awards.

An essential element of these programs is that they demand that a majority of the trainee's time be protected for instruction and research. When such expectations are not feasible or practical, CTSA institutions often provide continuing and professional education in specific areas of research. Such programs, if tailored to the biological concerns of extended-duration spaceflight, could provide meaningful opportunities for management personnel, physicians, and astronauts to expand their understanding of the research payloads for which they are responsible. Moreover, they could include combinations of virtual and traditional modes of instruction.

The Procurement Process and Its Effect on Flight Research

Procurements of flight hardware by NASA space life sciences have sometimes been characterized by non-competitive bidding by contractors, lack of consultation or involvement of experienced commercial developers of similar equipment, and inadequate ground-based testing prior to flight operations. Under these conditions, the design-development-testing-evaluation process is shortened or improperly completed, causing loss of data, extensive additional work, and increased costs to correct the problems. For example, the flight treadmill has undergone several iterations whereby the science requirements (ability to apply and quantify foot-ward loading, motorized control of treadmill speed, and speed range) have not been met. Once these treadmills were in orbit, the deficiencies were recognized, resulting in increased cost and development of a new treadmill by the same hardware developers. The interim resistance exercise device (iRED) is another example of the inefficiencies of this type of procurement processing.

When specialized equipment is needed that is commercially available, NASA should use or modify an off-the-shelf device. Such commercial hardware is usually thoroughly tested, often meets science requirements, and often can be modified for flight. When a commercial device is not appropriate for spaceflight, NASA should encourage collaboration between developers of commercial equipment and engineers who build flight hardware. There are several notable examples of integrative design strategies that result in hardware that functions well. Important features of these approaches include:

- Strong NASA reliance on science working groups whose members are potential customers of flight hardware;
- Regular inclusion of users (especially scientists and flight personnel) in the design process from the earliest planning stages through completion of the hardware;
- When possible, use of commercially available devices to develop and manufacture flight hardware;
- Rigorous ground-based testing to failure; and
- Sufficient time to iterate design cycles.

To ensure that flight hardware is robust and operates according to scientific or operational specifications, development timelines should not circumvent scientific feedback and re-engineering efforts.

International Collaboration Between Space Agencies

There is a need to reinvigorate international scientific cooperation, specifically the International Space Life Sciences Working Group. Such cooperation worked well in the decades before 2000 and will undoubtedly reduce costs to NASA. New partnerships with India, Russia, Australia, and China should be encouraged. Expansion of this working group and joint research announcements will aid discovery and internationalize space life sciences, offering opportunities for collaboration in ground studies prior to flight experiments.

SUMMARY

There is no lack of space life sciences research to be completed in the next decade. Critical to completing this work in the most useful fashion will be setting research priorities according to the most sensible sequence, the most frequent or severe risk to be mitigated, and the availability of resources. The following are the highest-priority research recommendations selected from the previous sections of this chapter.

Stress

Sensory-Motor

To ensure the safety of future commercial orbital and exploration crews, post-landing vertigo and orthostatic intolerance should be quantified in a sufficiently large sample of returning ISS crews, as part of the immediate post-flight medical exam. The effects of mission duration should be determined, and the causes of individual differences in egress performance identified. It is also important to assess the effectiveness and side effects of anti-motion sickness drugs on memory, cognition, sensory-motor adaptation, urinary retention, and sleep under operational conditions on the ISS. **(CC1)**

Artificial Gravity

Engineering design of exploration spacecraft for multiyear missions depends on whether artificial gravity (AG) is needed as a multisystem (bone, muscle, cardiovascular, vestibular, and immune systems) countermeasure and whether continuous large-radius AG is needed or intermittent exercise within lower-body negative pressure or short-radius AG is sufficient. If NASA determines that AG is required, human bed rest/centrifuge studies in ground laboratories are essential to establish dose-response relationships and determine what gravitational force level, gradient, rate of revolution, duration, and frequency are adequate. This research will require several years and should be resumed as soon as possible. The European Space Agency and the French Centre National d'Etudes Spatiales have active human and ground-based AG research programs, and NASA should forge a collaborative relationship with them. A short-radius centrifuge for rodents and, if possible, for humans on the ISS will ultimately be essential to validate effectiveness and operational acceptability. **(CC2)**

EVA

Because human space exploration involves significant EVA activity, the risk of decompression sickness must be mitigated. Doppler ultrasound studies on several human subjects in the ISS airlock and in ground chambers are needed to determine whether there is an effect of gravity on micronucleation and/or intrapulmonary shunting, or whether the unexpectedly low prevalence of decompression sickness on the space shuttle and the ISS is due to underreporting. Altitude chamber studies (1- to 2-year effort) are needed to determine operationally acceptable low suit pressure and hypobaric hypoxia limits. Lower habitat pressures could potentially eliminate prebreathe requirements, and lower suit pressure could improve glove and joint flexibility in advanced suits. However, use of such pressures would complicate the interpretation of scientific data. The decision to employ hypobaric and/or hypoxic habitat and suit environments thus involves significant scientific versus engineering tradeoffs and will depend on mission objectives. **(CC3)**

Food, Nutrition, and Energy Balance

Recent data indicate that crews on long-duration spaceflights are at risk for nutrient deficiency, particularly deficiencies in vitamins E and K, folate, and antioxidants. The next decade's work should focus on optimizing dietary strategies for crews (especially when UV exposure is limited) and food preservation strategies that will maintain bioavailability for 12 or more months. This work can be conducted in analog or spaceflight environments. **(CC4)**

If a crewed mission to Mars becomes a NASA priority, one of the first research programs that should be initiated is a robust food science program focused on preserving nutrient stability for 3 or more years. **(CC5)**

Because the energy requirements for long-duration exploration are still not fully defined, intervention trials (ground-based laboratory studies, ground-based spaceflight analog field studies, and spaceflight) should include food and energy intake as an outcome variable. **(CC6)**

Radiation Biology

The fundamental question in space radiation biology is, What will be the health consequences of exposure of astronauts to radiation in space? The radiation in question is high-LET radiation, which has been studied only infrequently to date. To understand the health risks associated with spaceflight, identification of the effects (short-term and long-term) of such radiation exposure is essential. The current studies in this field are predominantly at the molecular-cellular level and to a lesser extent at the whole-animal level. Continuation of these studies during the coming decade is essential for a full understanding of the consequences of exposure to radiation and a better estimate of radiation risk in space. Once risks have been determined, development of mitigators will be essential.

Human Studies

Longer-duration missions on the Moon or to Mars, where risks of exposure to solar particle events and galactic cosmic rays are higher, could result in exposures higher than astronauts have previously encountered. The only detrimental consequence of irradiation that has been shown in astronauts to date that is considered to be radiation-dose-dependent is the induction of cataracts.[244,245] The induction may be stochastic (with no dose threshold), and the pattern may be somewhat different from that with cataracts induced by lower-LET radiation (which usually begin in the subcapsular posterior region of the eye). Thus, continued study and follow-up of astronauts for cataract incidence, quality, and pathology related to radiation exposures is essential not only for understanding risk from cataracts but also for understanding radiation-induced late tissue toxicities in humans in general. The cataract studies should continue for the next 10 years and beyond in order to develop the largest database possible, since the sample size is so small. **(CC7)**

Animal Studies

Much of the ongoing work in the field of space radiation biology is being done at the molecular and cellular levels in order to identify end points that should be examined in whole animals. Major questions to be resolved include risks from cancer, cataracts (which show sex-specific differences in mice), cardiovascular disease, neurologic dysfunction, degenerative diseases, and acute toxicities such as fever, nausea, bone marrow suppression, and others. Animal studies inform an important knowledge base to extrapolate human risks since voluntary exposure of humans to radiation at such doses is neither possible nor prudent. While much is known about the effects of the low-LET radiation (gamma rays, x-rays) used in conventional radiotherapy, little is known about the consequences of exposure to high-LET space radiation. It is not likely that the needed information will be acquired in 10 years but, as studies continue, they are likely to influence decisions about risks posed by exposure to radiation in space. At present, most of the studies being conducted by NASA involve rodents (mice and rats), although additional animal models are being added for particular purposes. The next 10 years should expand the use of animal models that mimic human disease, including nonhuman primates, based on how well they mimic human disease. **(CC8)**

Cellular Studies

Continued cellular studies during the next 10 years are essential for developing end points and markers that can be used to define acute and late radiation toxicities. The work requires continued ground-based studies using radiation facilities that are able to mimic space radiation exposures, such as the NASA Space Radiation Laboratory at Brookhaven National Laboratory. Ground-based exposures are highly recommended because space exposures are of insufficient dose to induce detectable biological consequences. This constraint hampers studies of the combined effects of microgravity and radiation, which thus require either reliance on ground-based models (such as hindlimb unloading) or the installation of a radiation facility on the ISS.

Recently Brookhaven National Laboratory has done much to improve its ability to generate mixed field radiations that mimic solar particle events and is working to develop additional mixed field capabilities; studies using this facility or one comparable to it will be important for space radiobiology in the coming decade. **(CC9)**

Biological Sex/Gender

A renewed emphasis on expanding our understanding of gender differences in adaptation to the spaceflight environment should be part of future flight and ground-based research solicitations. Of particular interest are potential differences in bone, muscle, and cardiovascular function that might be amenable to targeted, gender-specific therapy.

Ground-based animal studies should be performed during the next 10 years to evaluate long-term radiation risks in male and female animals (mice, rats, and primates). Large-scale, multidecade animal studies, results from studies of atomic bomb casualties, and radiotherapy studies of patients exposed to low-LET radiation should report sex differences in the type and incidence of tumors following exposure to radiation. Future research should focus on the mechanisms responsible for sex-specific differences that may aid in the development and utilization of countermeasures for protection against radiation in the space environment. **(CC10)**

Thermoregulation

Human studies that systematically investigate the biophysical principles of thermal balance should be performed to determine whether microgravity reduces the threshold for thermal intolerance. This research area is highly relevant for safe spaceflight, especially during EVA and times of heavy work loads. Without an understanding of whether thermal tolerance is altered by microgravity, risk assessment is not possible. This research could be conducted in bed rest studies over the next 5 to 10 years, and may include rarified-gas environments to alter thermal exchange with the environment. Thermotolerance tests should be performed in male and female subjects before and after microgravity deconditioning and should include measurements of body temperature, gut permeability, blood endotoxins, and inflammatory and anti-inflammatory markers. Furthermore, methods to produce rapid, sustained, and physiologically appropriate cooling should be evaluated first during ground-based studies and then during parabolic flight, followed by spaceflight validation studies. **(CC11)**

Overarching Issue—Integrated Countermeasures Development

Research is needed to develop and test integrated countermeasures, including exercise, for periods exceeding 180 days in microgravity with and without artificial gravity. On-orbit measurements of loads and exercise histories are also important. This research is needed to maintain crew health and well-being during prolonged existence in microgravity and could require a decade to achieve a reasonable sample of human subjects.

Ground-based (e.g., bed rest) studies are needed to test and optimize integrated exercise countermeasures (both hardware and protocol), especially those that include an artificial gravity component. National and international cooperative studies utilizing the existing CTSA network may promote greater resource sharing and accelerate research. Effective use of the International Space Life Sciences Working Group and joint research announcements will promote efficient use of resources and internationalize space life sciences, offering opportunities for collaboration in ground studies prior to flight experiments.

REFERENCES

1. Gernhardt, M., NASA Exploration Technology Development Program. 2009. "Integrating Life Science, Engineering and Operations Research to Optimize Human Safety and Performance in Planetary Exploration," presentation at the Joint Meeting of Panels for the Decadal Survey on Biological and Physical Sciences in Space, August 20. National Research Council, Washington, D.C.
2. Behnke, A.R. 1942. Investigations concerned with problems of high altitude flying and deep diving: Application of certain findings pertaining to physical fitness to the general military service. *The Military Surgeon* 90:9-2.
3. Nishi, R.T., Brubakk, A.O., and Eftedal, O. 2003. Bubble detection. Pp. 501-529 in *Bennett and Elliot's Physiology and Medicine of Diving* (A.O. Brubakk and T.S. Neuman, eds.). Saunders Health Books, N.Y.
4. Lovering, A.T., Eldridge, M.W., and Stickland, M.K. 2009. Counterpoint: Exercise-induced intrapulmonary shunting is real. *Journal of Applied Physiology* 107:994-997.
5. Stepanek, J., and Webb, J.T. 2008. Physiology of decompressive stress. Chapter 3 in *Fundamentals of Aerospace Medicine*, 4th Edition (J.R. Davis, R. Johnson, J. Stepanek, and J.A. Fogarty, eds.). Lippincott Williams and Wilkins, Baltimore, Md.
6. Jersey, S.L., Baril, R.T., McCarty, R.D., and Millhouse, C.M. 2010. Severe neurological decompression sickness in a U-2 pilot. *Aviation, Space, and Environmental Medicine* 81(1):64-68.
7. Gernhardt, M.L., Conkin, J., Vann, R.D., Pollock, N.W., and Feiverson, A.H. 2004. DCS risks in ground-based hypobaric trials vs. extravehicular activity. *Undersea and Hyperbaric Medicine* 31(3):338.
8. Dujić, Z., Palada, I., Valic, Z., Duplancić, D., Obad, A., Wisløff, U., and Brubakk, A.O. 2006. Exogenous nitric oxide and bubble formation in divers. *Medicine and Science in Sports and Exercise* 38(8):1432-1435.
9. Dujić, Z., Palada, I., Duplancić, D., Marinovic-terzic, I., Bakovic, D., Ivancev, V., Valic, Z., Eterovic, D., Petri, N., Wisløff, U., and Brubakk, A.O. 2004. Aerobic exercise before diving reduces venous gas bubble formation in humans. *Journal of Physiology* 555:637-642.
10. Gernhardt, M.L. 1991. Development and evaluation of a decompression stress index based on tissue bubble dynamics [Ph.D. thesis]. University of Pennsylvania, Philadelphia, Pa.
11. Møllerløkken, A., Gutvik, C., Berge, V.J., Jørgensen, A., Løset, A., and Brubakk, A.O. 2007. Recompression during decompression and effects on bubble formation in the pig. *Aviation, Space, and Environmental Medicine* 78(6):557-560.
12. Gernhardt, M.L., Pollock, N.W., Vann, R.D., Natoli, M.J., Nishi, R.Y., Sullivan, P.J., Conkin, J., Dervay, J.P., and Moore, A.L. 2004. Development of an in-suit exercise prebreathe protocol supporting extravehicular activity in microgr. *Undersea and Hyperbaric Medicine* 31(3):338.
13. Buckey, J.C. 2006. *Space Physiology.* Oxford University Press, Oxford, U.K.
14. Stepanek, J., and Webb, J.T. 2008. Physiology of decompressive stress. Chapter 3 in *Fundamentals of Aerospace Medicine*, 4th Edition (J.R. Davis, R. Johnson, J. Stepanek, J.A. Fogarty, eds.). Lippincott Williams and Wilkins, Baltimore, Md.
15. Webb, J.T., Pilmanis, A.A., Prisk, G.K., and Muza, S.R. 2009. Prevention from decompression sickness using low pressure Moon and Mars habitats and pressure suits: Testing of environment for decompression sickness risk. White paper submitted to the Decadal Survey on Biological and Physical Sciences in Space, National Research Council, Washington, D.C.
16. NASA. NASA Education. 2009. Holding a Winning Hand. January 9. Available at http://www.nasa.gov/audience/foreducators/holding-a-winning-hand_prt.htm.
17. Speiss, B.D. 2009. Perfluorocarbon emulsions as a promising technology: A review of tissue and vascular gas dynamics. *Journal of Applied Physiology* 106:1444-1452.
18. National Research Council. 1997. *Advanced Technology for Human Support in Space.* National Academy Press, Washington, D.C., Chapter 4, pp. 82-96.
19. National Research Council. 2008. *A Constrained Space Exploration Technology Program: A Review of NASA's Exploration Technology Development Program.* The National Academies Press, Washington, D.C., pp. 39-40.
20. Clément, G., and Bukley, A., eds. 2007. *Artificial Gravity.* Springer, New York, N.Y.
21. Joosten, B.K. 2007. *Preliminary Assessment of Artificial Gravity Impacts to Deep-Space Vehicle Design.* NASA Johnson Space Center Document JSC-63743. NASA Johnson Space Center, Houston, Tex.
22. Hargens, A.R., Whalen, R.T., Watenpaugh, D.E., Schwandt, D.F. and Krock, L.P. 1991. Lower body negative pressure to provide load bearing in space. *Aviation, Space, and Environmental Medicine* 62:934-937.
23. Lee, S.M., Schneider, S.M., Boda, W.L., Watenpaugh, D.E., Macias, B.R., Meyer, R.S., and Hargens, A.R. 2009. LBNP exercise protects aerobic capacity and sprint speed of female twins during 30 days of bed rest. *Journal of Applied Physiology* 106:919-928.

24. Smith, S.M., Davis-Street, J.E., Fesperman, J.V., Calkins, D.S., Bawa, M., Macias, B.R., Meyer, R.S., and Hargens, A.R. 2003. Evaluation of treadmill exercise in a lower body negative pressure chamber as a countermeasure for weightlessness-induced bone loss: A bed rest study with identical twins. *Journal of Bone and Mineral Research* 18:2223-2230.
25. Zwart, S.R., Hargens, A.R., Lee, S.M., Macias, B.R., Watenpaugh, D.E., Tse, K., and Smith, S.M. 2007. Lower body negative pressure treadmill exercise as a countermeasure for bed rest-induced bone loss in female identical twins. *Bone* 40:529-537.
26. Cao, P., Kimura, S., Macias, B.R., Ueno, T., Watenpaugh, D.E., and Hargens, A.R. 2005. Exercise within lower body negative pressure partially counteracts lumbar spine deconditioning associated with 28-day bed rest. *Journal of Applied Physiology* 99:39-44.
27. Macias, B.R., Cao, P., Watenpaugh, D.E., and Hargens, A.R. 2007. LBNP treadmill exercise maintains spine function and muscle strength in identical twins during 28-days simulated microgravity. *Journal of Applied Physiology* 102:2274-2278.
28. Watenpaughl, D.E., O'Leary, D.D., Schneider, S.M., Lee, S.M., Macias, B.R., Tanaka, K., Hughson, R.L., and Hargens, A.R. 2007. Lower body negative pressure exercise plus brief postexercise lower body negative pressure improve post-bed rest orthostatic tolerance. *Journal of Applied Physiology* 103:1964-1972.
29. Guinet, P., Schneider, S.M., Macias, B.R., Watenpaugh, D.E., Hughson, R.L., Le Traon, A.P., Bansard, J.Y., and Hargens, A.R. 2009. WISE-2005: Effect of aerobic and resistive exercises on orthostatic tolerance during 60 days bed rest in women. *European Journal of Applied Physiology* 106:217-227.
30. Lee, S.M.C., Schneider, S.M., Boda, W.L., Watenpaugh, D.E., Macias, B.R., Meyer, R.S., and Hargens, A.R. 2007. Supine LBNP exercise maintains exercise capacity in male twins during 30-d bed rest. *Medicine and Science in Sports and Exercise* 39:1315-1326.
31. Schneider, S.M., Lee, S.M., Macias, B.R., Watenpaugh, D.E., and Hargens, A.R. 2009. WISE-2005: Exercise and nutrition countermeasures for upright VO_2pk during bed rest. *Medicine and Science in Sports and Exercise* 41:2165-2176.
32. Monga, M., Macias, B., Groppo, E., Kostelec, M., and Hargens, A. 2006. Renal stone risk in a simulated microgravity environment: Impact of treadmill exercise with lower body negative pressure. *Journal of Urology* 176:127-131.
33. Graybiel, A. 1975. Velocities, angular accelerations and Coriolis accelerations. Pp. 247-304 in *Foundations of Space Biology and Medicine* (M. Calvin and O.G. Gazenko, eds.). Volume II, Book I. NASA, Washington D.C.
34. Graybiel, A., Miller, E.F., and Homick, J.L. 1977. Experiment M131: Human vestibular function. Chapter 11 in *Biomedical Results of Skylab* (R.S. Johnston and L.F. Deitlein, eds.). NASA SP-377. U.S. Government Printing Office, Washington, D.C.
35. Lackner, J.R., and Graybiel, A. 1986. The effective intensity of Coriolis, cross-coupling stimulation is gravitoinertial force dependent: Implications for space motion sickness. *Aviation, Space, and Environmental Medicine* 57(3):229-235.
36. Lackner, J.R., and DiZio, P. 2000. Artificial gravity as a countermeasure in long-duration spaceflight. *Journal of Neuroscience Research* 62:169-176.
37. Young, L.R., Yajimia, K., and Paloski, W.F. 2009. *Artificial Gravity Research as a Tool in Space Exploration.* International Academy of Astronautics, Study Group 2.2, Final Report. International Academy of Astronautics, Paris, France.
38. Clement, G., and Buckley, A., eds. 2007. *Artificial Gravity.* Springer, New York, N.Y.
39. Young, L.R., Yajimia, K., and Paloski, W.F. 2009. *Artificial Gravity Research as a Tool in Space Exploration.* International Academy of Astronautics, Study Group 2.2, Final Report. International Academy of Astronautics, Paris, France.
40. Oman, C.M., Lichtenberg, B.K., Money, K.E., and McCoy, R.K. 1986. MIT/Canadian vestibular experiments on the Spacelab-1 mission: 4. Space motion sickness: Symptoms, stimuli, and predictability. *Experimental Brain Research* 64:316-334.
41. Oman, C.M., Lichtenberg, B.K., and Money, K.E. 1990. Symptoms and signs of space motion sickness on Spacelab-1. Pp. 217-246 in *Motion and Space Sickness* (G.H. Crampton, ed.). CRC Press, Boca Raton, Fla.
42. Davis, J.R., Vanderploeg, J.M., Santy, P.A., Jennings, R.T., and Stewart, D.F. 1988. Space motion sickness during the first 24 flights of the space shuttle. *Aviation, Space, and Environmental Medicine* 59:1185-1189.
43. Jennings, R.T. 1998. Managing space motion sickness. *Journal of Vestibular Research* 8(1):67-70.
44. Reason, J.T. 1978. Motion sickness adaptation: A neural mismatch model. *Journal of the Royal Society of Medicine* 71:819-829.
45. Oman, C.M. 1990. Motion sickness: A synthesis and evaluation of the sensory conflict theory. *Canadian Journal of Physiology and Pharmacology* 68:294-303.
46. Lackner, J.R., and DiZio, P. 2006. Space motion sickness. *Experimental Brain Research* 175:377-399.
47. Yates, B.J., Miller, A.D., and Lucot, J.B. 1998. Physiological basis and pharmacology of motion sickness: An update. *Brain Research Bulletin* 47:395-406.

48. Yates, B.J., Miller, A.D., and Lucot, J.B. 1998. Physiological basis and pharmacology of motion sickness: An update. *Brain Research Bulletin* 47:395-406.
49. NASA. 2010. *Human Integration Design Handbook.* NASA STD-23001. Section 6.5 Acceleration. Available at http://ston.jsc.nasa.gov/collections/TRS/_techrep/SP-2010-3407.pdf.
50. Adelstein, B.D., Beutter, B.R., Kaiser, M.K., McCann, R.S., Stone, L.S., Anderson, M.R., Renema, F., and Paloski, W.R. 2008. Influence of combined whole body vibration plus G-loading on visual performance. Report to the NASA Human Research Program, November. NASA Ames Research Center, Moffett Field, Calif.
51. West, J. 2005. Chapter 3.19, pp. 1-38 in *Mars Mission Concept Exploration and Refinement: Study Phase 2 Final Report.* Human Factors Engineering. CS Draper Laboratory, Cambridge, Mass.
52. Banks, R.D., Brinkley, J.W., Alnutt, R., and Harding, R.M. 2008. Human response to acceleration. Chapter 4, pp. 83-109 in *Fundamentals of Aerospace Medicine*, 4th Edition (J.R. Davis, R. Johnson, J. Stepanek, and J.A. Fogarty, eds.). Lippincott Williams and Wilkins, Baltimore, Md.
53. NASA. 2010. *Human Integration Design Handbook.* NASA STD 3001. Volume 2, Baseline edition, Section 6.5, January 27. Available at http://ston.jsc.nasa.gov/collections/TRS/_techrep/SP-2010-3407.pdf.
54. Smith, S.D., Goodman, J.R., and Grosveld, F.W. 2008. Vibration and acoustics. Chapter 5, pp. 110-141 in *Fundamentals of Aerospace Medicine*, 4th Edition (J.R. Davis, R. Johnson, J. Stepanek, and J.A. Fogarty, eds.). Lippincott Williams and Wilkins, Baltimore, Md.
55. Platts, S.H., Stenger, M., Phillips, T., Arzeno, N., and Summers, R. 2008. *Evidence Book: Risk of Orthostatic Intolerance During Re-exposure to Gravity.* NASA Human Research Program, Human Health Countermeasures Element. HRP-47060. NASA Johnson Space Center, Houston, Tex.
56. Heldt, T., Shim, E.B., Kamm, R.D., and Mark, R.G. 2002. Computational modeling of cardiovascular response to orthostatic stress. *Journal of Applied Physiology* 92:1239-1254.
57. Summers, R., Coleman, T., and Meck, J. 2008. Development of the Digital Astronaut Program for the analysis of the mechanisms of physiologic adaptation to microgravity: Validation of the cardiovascular system module. *Acta Astronautica* 63(7-10):758-762.
58. NASA. 2008. *Human Research Program Evidence Book.* NASA Johnson Space Center, Houston, Tex. Available at http://humanresearch.jsc.nasa.gov/elements/smo/hrp_evidence_book.asp.
59. NASA. 2008. *Human Research Program Evidence Book.* NASA Johnson Space Center, Houston, Tex. Available at http://humanresearch.jsc.nasa.gov/elements/smo/hrp_evidence_book.asp.
60. Smith, S., Zwart, S., Block, G., Rice, B., and Davis-Street, J. 2005. The nutritional status of astronauts is altered after long-term space flight aboard the International Space Station. *Journal of Nutrition* 135:437-443.
61. Smith, S., Wastney, M., Morukov, B., Larina, I.M., Nyquist, L.E., Abrams, S.A., Taran, E.N., Shih, C.Y., Nillen, J.L., Davis-Street, J.E., Rice, B.L., and Lane, H.W. 1999. Calcium metabolism before, during, and after a 3-mo spaceflight:kinetic and biochemical changes. *American Journal of Physiology* 277:R1-R10.
62. Smith, S., Davis-Street, J., Rice, B., Nillen, J., Gillman, P., and Block, G. 2001. Nutritional status assessment in semi-closed environments: Ground-based and space flight studies in humans. *Journal of Nutrition* 131:2053-2061.
63. Smith, S., Zwart, S., Block, G., Rice, B., and Davis-Street, J. 2005. The nutritional status of astronauts is altered after long-term space flight aboard the International Space Station. *Journal of Nutrition* 135:437-443.
64. Trivedi, D., Doll, R., and Khaw, K. 2003. Effect of four monthly oral vitamin D3 (cholecalciferol) supplementation on fractures and mortality in men and women living in the community: Randomized double blind controlled trial. *British Medical Journal* 326:469.
65. Smith, S., Zwart, S., Block, G., Rice, B., and Davis-Street, J. 2005. The nutritional status of astronauts is altered after long-term space flight aboard the International Space Station. *Journal of Nutrition* 135:437-443.
66. Smith, S., Zwart, S., Block, G., Rice, B., and Davis-Street, J. 2005.The nutritional status of astronauts is altered after long-term space flight aboard the International Space Station. *Journal of Nutrition* 135:437-443.
67. Vermeer, C., Wolf, J., Craciun, A., and Knapen, M. 1998. Bone markers during a 6-month space flight: Effects of vitamin K supplementation. *Journal of Gravitational Physics* 5:65-69.
68. Sato, Y., Tsuru, T., Oizumi, K., and Kaji, M. 1999. Vitamin K deficiency and osteopenia in disuse-affected limbs of vitamin D-deficient elderly stroke patients. *American Journal of Physical Medicine and Rehabilitation* 78:317-322.
69. Smith, S., Zwart, S., Block, G., Rice, B., and Davis-Street, J. 2005.The nutritional status of astronauts is altered after long-term space flight aboard the International Space Station. *Journal of Nutrition* 135:437-443.
70. Smith, S., Zwart, S., Block, G., Rice, B., and Davis-Street, J. 2005.The nutritional status of astronauts is altered after long-term space flight aboard the International Space Station. *Journal of Nutrition* 135:437-443.

71. Smith, S., Davis-Street, J., Rice, B., Nillen, J., Gillman, P., and Block, G. 2001. Nutritional status assessment in semi-closed environments: Ground-based and space flight studies in humans. *Journal of Nutrition* 131:2053-2061.
72. Stein, T., Leskiw, M., Schluter, M., Hoyt, R.W., Lane, H.W., Gretebeck, R.E., and LeBlanc AD. 1999. Energy expenditure and balance during spaceflight on the space shuttle. *American Journal of Physiology* 276:R1739-R1784.
73. Lane, H., Gretebeck, R., Schoeller, D., Davis-Street, J., Socki, R., and Gibson, E. 1997. Comparison of ground-based and space flight energy expenditure and water turnover in middle-aged healthy male US astronauts. *American Journal of Clinical Nutrition* 65:4-12.
74. LeBlanc, A., Rowe, R., and Schneider, V. 1995. Regional muscle loss after short duration spaceflight. *Aviation, Space, and Environmental Medicine* 66:1151-1154.
75. Ferrando, A. 2002. Alterations in protein metabolism during space flight and inactivity. *Nutrition* 18:837-841.
76. Ferrando, A., Lane, H., Stuart, C., and Wolfe, R. 1996. Prolonged bed rest decreases skeletal muscle and whole-body protein synthesis. *American Journal of Physiology (Endocrinology and Metabolism)* 270:E627-E633.
77. Smith, S., Wastney, M., O'Brien, K., Morukov, B.V., Larina, I.M., Abrams, S.A., Davis-Street, J.E., Oganov, V., and Shackelford, L.C. 2005. Bone markers, calcium metabolism, and calcium kinetics during extended-duration space flight on the Mir space station. *Journal of Bone and Mineral Research* 20:208-218.
78. NASA. 2008. *Human Research Program Evidence Book.* NASA Johnson Space Center, Houston, Tex. Available at http://humanresearch.jsc.nasa.gov/elements/smo/hrp_evidence_book.asp.
79. Zwart, S., Kloris, V., Perchonok, M., Braby, L., and Smith, S. 2009. Assessment of nutrient stability in foods from the space food system after long-duration spaceflight on the ISS. *Journal of Food Science* 74:H209-H217.
80. Smith, S., Zwart, S., Block, G., Rice, B., and Davis-Street, J. 2005.The nutritional status of astronauts is altered after long-term space flight aboard the International Space Station. *Journal of Nutrition* 135:437-443.
81. Vermeer, C., Wolf, J., Craciun, A., and Knapen, M. 1998. Bone markers during a 6-month space flight: Effects of vitamin K supplementation. *Journal of Gravitational Physics* 5:65-69.
82. Cornelli, U., Terranova, R., Luca, S., Cornelli, M., and Alberti, A. 2001. Bioavailability and antioxidant activity of some food supplements in men and women using the D-roms test as a marker of oxidative stress. *Journal of Nutrition* 131:3208-3211.
83. National Research Council. 2000. *Radiation and the International Space Station: Recommendations to Reduce Risk.* National Academy Press, Washington, D.C.
84. Institute of Medicine and the National Research Council. 2006. *A Risk Reduction Strategy for Human Exploration of Space: A Review of NASA's Bioastronautics Roadmap* (D.E. Longnecker and R.A. Molins, eds). The National Academies Press, Washington, D.C.
85. Institute of Medicine. 2007. *Review of NASA's Space Flight Health Standards-Setting Process: Letter Report* (R.S. Williams, ed.). The National Academies Press, Washington, D.C.
86. National Research Council. 2008. *Managing Space Radiation Risk in the New Era of Space Exploration.* The National Academies Press, Washington, D.C.
87. Particle Therapy Co-Operative Group. Particle Therapy Facilities in Operation (incl. patient statistics). Available at http://ptcog.web.psi.ch/ptcentres.html.
88. National Council on Radiation Protection and Measurements (NCRP). 2006. *Information Needed to Make Radiation Protection Recommendations for Space Missions Beyond Low-Earth Orbit.* NCRP Report 153. NCRP, Bethesda, Md.
89. NASA. Human Adaptation and Countermeasures Division. Space Radiation. Available at http://hacd.jsc.nasa.gov/projects/space_radiation_overview.cfm.
90. Hall, E.J., and Giaccia, J.A. 2005. *Radiobiology for the Radiologist.* Lippincott Williams and Wilkins, Baltimore, Md.
91. Paganetti, H., Niemierko, A., Ancukiewicz, M., Gerweck, L.E., Goitein, M., Loeffler, J.S., and Suit, H.D. 2002. Relative biological effectiveness (RBE) values for proton beam therapy. *International Journal of Radiation Oncology, Biology, Physics* 53(2):407-421.
92. NASA/Brookhaven National Laboratory. Space Radiation Biology Program. Available at http://www.bnl.gov/medical/NASA/.
93. Ritter, S., and Durante, M. 2010. Heavy-ion induced chromosomal aberrations: A review. *Mutation Research* 14:01(1):38-46. Review.
94. Held, K.D. 2009. Effects of low fluences of radiations found in space on cellular systems. *International Journal of Radiation Biology* 85(5):379-390. Review.
95. Roig, A.I., Hight, S.K., Minna, J.D., Shay, J.W., Rusek, A., and Story, M.D. 2010. DNA damage intensity in fibroblasts in a 3-dimensional collagen matrix correlates with the Bragg curve energy distribution of a high LET particle. *International Journal of Radiation Biology* 86(3):194-204.

96. Tian, J., Pecaut, M.J., Slater, J.M., and Gridley, D.S. 2010. Spaceflight modulates expression of extracellular matrix, adhesion, and profibrotic molecules in mouse lung. *Journal of Applied Physiology* 108(1):162-171.
97. Takahashi, A., Su, X., Suzuki, H., Omori, K., Seki, M., Hashizume, T., Shimazu, T., Ishioka, N., Iwasaki, T., and Ohnishi, T. 2010. p53-dependent adaptive responses in human cells exposed to space radiations. *International Journal of Radiation Oncology, Biology, Physics* 78(4):1171-1176.
98. Institute of Medicine and the National Research Council. 2006. *A Risk Reduction Strategy for Human Exploration of Space: A Review of NASA's Bioastronautics Roadmap* (D.E. Longnecker and R.A. Molins, eds). The National Academies Press, Washington, D.C.
99. National Research Council. 2006. *Health Risks from Exposure to Low Levels of Ionizing Radiation. BEIR VII Phase 2.* The National Academies Press, Washington, D.C.
100. Suzuki, T., Richards, S.M., Liu, S., Jensen, R.V., and Sullivan, D.A. 2009. Influence of sex on gene expression in human corneal epithelial cells. *Molecular Vision* 15:2554-2569.
101. Henderson, M.A., Valluri, S., DesRosiers, C., Lopez, J.T., Batuello, C.N., Caperell-Grant, A., Mendonca, M.S., Powers, E.M., Bigsby, R.M., and Dynlacht, J.R. 2009. Effect of gender on radiation-induced cataractogenesis. *Radiation Research* 172(1):129-133.
102. Henderson, M.A., Valluri, S., Garrett, J., Lopez, J.T., Caperell-Grant, A., Mendonca, M.S., Rusek, A., Bigsby, R.M., and Dynlacht, J.R. 2010. Effects of estrogen and gender on cataractogenesis induced by high-LET radiation. *Radiation Research* 173(2):191-196.
103. Miltenyi, Z., Keresztes, K., Garai, I., Edes, I., Galajda, Z., Toth, L., Illés, A. 2004. Radiation-induced coronary artery disease in Hodgkin's disease. *Cardiovascular Radiation Medicine* 5(1):38-43.
104. Heidenreich, P.A., Schnittger, I., Strauss, H.W., Vagelos, R.H., Lee, B.K., Mariscal, C.S., Tate, D.J., Horning, S.J., Hoppe, R.T., and Hancock, S.L. 2007. Screening for coronary artery disease after mediastinal irradiation for Hodgkin's disease. *Journal of Clinical Oncology* 25(1):43-49.
105. Virmani, R., Farb, A., Carter, A.J., and Jones, R.M. 1999. Pathology of radiation-induced coronary artery disease in human and pig. *Cardiovascular Radiation Medicine* 1(1):98-101.
106. Shimizu, Y., Kodama, K., Nishi, N., Kasagi, F., Suyama, A., Soda, M., Grant, E.J., Sugiyama, H., Sakata, R., Moriwaki, H., Hayashi, M., Konda, M., and Shore, R.E. 2010. Radiation exposure and circulatory disease risk: Hiroshima and Nagasaki atomic bomb survivor data, 1950-2003. *British Medical Journal* 340:b5349.
107. Preston, D.L., Shimizu, Y., Pierce, D.A., Suyama, A., and Mabuchi, K. 2003. Studies of mortality of atomic bomb survivors. Report 13: Solid cancer and noncancer disease mortality: 1950-1997. *Radiation Research* 160(4):381-407.
108. Institute of Medicine and the National Research Council. 2006. *A Risk Reduction Strategy for Human Exploration of Space: A Review of NASA's Bioastronautics Roadmap* (D.E. Longnecker and R.A. Molins, eds). The National Academies Press, Washington, D.C.
109. Shuryak, I., Hahnfeldt, P., Hlatky, L., Sachs, R.K., and Brenner, D.J. 2009. A new view of radiation-induced cancer: Integrating short- and long-term processes. Part II: Second cancer risk estimation. *Radiation and Environmental Biophysics* 48(3):275-286.
110. Sachs, R.K., Shuryak, I., Brenner, D., Fakir, H., Hlatky, L., and Hahnfeldt, P. 2007. Second cancers after fractionated radiotherapy: Stochastic population dynamics effects. *Journal of Theoretical Biology* 249(3):518-531.
111. Brengues, M., Paap, B., Bittner, M., Amundson, S., Seligmann, B., Korn, R., Lenigk, R., and Zenhausern, F. 2010. Biodosimetry on small blood volume using gene expression assay. *Health Physics* 98(2):179-185.
112. Booth, F.W., and Lees, J.J. 2007. Fundamental questions about genes, inactivity and chronic disease. *Physiological Genomics* 17:146-157.
113. Booth, F.W., Chakravarthy, N.V., and Sprangenburg, E.E. 2002. Exercise and gene expression: Physiological regulation of the human genome through physical activity. *Journal of Physiology* 543:399-411.
114. Hammond, T.G., Lewis, F.C., Goodwin, T.J., Linnehan, R.M., Wolf, D.A., Hire, K.P., Campbell, W.C., Benes, E., O'Reilly, K.C., Globus, R.K., and Kaysen, J.H. 1999. Gene expression in space. *Nature Medicine* 5(4):359.
115. Allen, D.L., Bandstra, E.R., Harrison, B.C., Thorng, S., Stodieck, L.S., Kostenuik, P.J., Morony, S., Lacey, D.L., Hammond, T.G., Leinwand, L.L., Argraves, W.S., Bateman, T.A., and Barth, J.L. 2009. Effects of spaceflight on murine skeletal muscle gene expression. *Journal of Applied Physiology* 106(2):582-595.
116. World Spaceflight. Astronaut/Cosmonaut Statistics. Available at http://www.worldspaceflight.com/bios/stats.php.
117. Encycopedia Astronautica. Women of Space. Available at http://www.astronautix.com/articles/womspace.htm.
118. NASA. Astronaut Biographies. Available at http://www.jsc.nasa.gov/Bios/astrobio.html.
119. CBS News. Spaceflight Statistics. Available at http://www.cbsnews.com/network/news/space/spacestats.html.

120. National Research Council. 1998. *A Strategy for Research in Space Biology and Medicine into the Next Century.* National Academy Press, Washington, D.C.
121. Mark, S. 2005. Sex- and gender-based medicine: Venus, Mars and beyond. *Gender Medicine* 2(3):131-136.
122. Mark, S. 2007. From Earth to Mars: Sex differences and their implications for musculoskeletal health. *Journal of American Academy of Orthopedic Surgeons* 1(15 Suppl.):S19-S21.
123. Harm, D.L., Jennings, R.T., Meck, J.V., Powell, M.R., Putcha, L., Sams, C.P., Schneider, S.M., Shackelford, L.C., Smith, S.M., and Whitson, P.A. 2001. Invited review: Gender issues related to spaceflight: A NASA perspective. *Journal of Applied Physiology* 91(5):2374-2383.
124. Guinet, P., Schneider, S.M., Macias, B.R., Watenpaugh, D.E., Hughson, R.L., Le Traon, A.P., Bansard, J.Y., and Hargens, A.R. 2009. WISE-2005: Effect of aerobic and resistive exercises on orthostatic tolerance during 60 days bed rest in women. *European Journal of Applied Physiology* 106(2):217-227.
125. Trappe, S., Creer, A., Slivka, D., Minchev, K., and Trappe, T. 2007. Single muscle fiber function with concurrent exercise or nutrition countermeasures during 60 days of bed rest in women. *Journal of Applied Physiology* 103(4):1242-1250.
126. Trudel, G., Payne, M., Madler, B., Ramachandran, N., Lecompte, M., Wade, C., Biolo, G., Blanc, S., Hughson, R., Bear, L., and Uhthoff, H.K. 2009. Bone marrow fat accumulation after 60 days of bed rest persisted 1 year after activities were resumed along with hemopoietic stimulation: The Women International Space simulation for Exploration Study. *Journal of Applied Physiology* 107(2):540-548.
127. Arbeille, P., Kerbeci, P., Greaves, D., Schneider, S., Hargens, A., and Hughson, R. 2007. Arterial and venous response to Tilt with LBNP test after a 60 day HDT bedrest (WISE study). *Journal of Gravitational Physiology* 14(1):P47-P48.
128. Maillet, A., Zaouali-Ajina, M., Vorobiev, D., Blanc, S., Pastouchkova, L., Reushkina, G., Morukov, B., Grigoriev, A.I., Gharib, C., and Gauquelin-Koch, G. 2000. Orthostatic tolerance and hormonal changes in women during 120 days of head-down bed rest. *Aviation, Space, and Environmental Medicine* 71(7):706-714.
129. Smith, S.M., Zwart, S.R., Heer, M., Lee, S.M., Baecker, N., Meuche, S., Macias, B.R., Shackelford, L.C., Schneider, S., and Hargens, A.R. 2008. WISE-2005: Supine treadmill exercise within lower body negative pressure and flywheel resistive exercise as a countermeasure to bed rest-induced bone loss in women during 60-day simulated microgravity. *Bone* 42(3):572-581.
130. Dorfman, T.A., Levine, B.D., Tillery, T., Peshock, R.M., Hastings, J.L., Schneider, S.M.., Macias, B.R., Biolo, G., and Hargens, A.R. 2007. Cardiac atrophy in women following bed rest. *Journal of Applied Physiology* 103(1):8-16.
131. Chopard, A., Lecunff, M., Danger, R., Lamirault, G., Bihouee, A., Teusan, R., Jasmin, B.J., Marini, J.F., and Leger, J.J. 2009. Large-scale mRNA analysis of female skeletal muscles during 60 days of bed rest with and without exercise or dietary protein supplementation as countermeasures. *Physiological Genomics* 38(3):291-302.
132. Lee, S.M.C., Schneider, S.M., Boda, W.L., Watenpaugh, D.E., Macias, B.R., Meyer, R.S., and Hargens, A.R. 2009. LBNP exercise protects aerobic capacity and sprint speed of female twins during 30 days of bed rest. *Journal of Applied Physiology* 106:919-928.
133. Rock, J.A., and Fortney, S.M. 1984. Medical and surgical considerations for women in spaceflight. *Obstetrical and Gynecological Survey* 39(8):525-535.
134. Becker, J.L. 1994. Women's health issues and space-based medical technologies. *Earth Space Review* 3(2):15-19.
135. Sandler, H., and Winters, D.L. 1978. *Physiological Responses of Women to Simulated Weightlessness. A Review of the Significant Findings of the First Female Bedrest Study.* NASA SP-430. NASA Scientific and Technical Information Office, Washington, D.C.
136. Institute of Medicine. 2001. *Exploring the Biological Contributions to Human Health: Does Sex Matter?* (T.M. Wizemann and M.L. Pardue, eds.) National Academy Press, Washington, D.C., pp. 45-78.
137. Institute of Medicine. 2001. *Exploring the Biological Contributions to Human Health: Does Sex Matter?* (T.M. Wizemann and M.L. Pardue, eds.) National Academy Press, Washington, D.C., pp. 13-27.
138. Fillingim, R.B., King, C.D., Ribeiro-Dasilva, M.C., Rahim-Williams, B., and Riley, J.L., III. 2009. Sex, gender and pain: A review of recent clinical and experimental findings. *The Journal of Pain* 10(5):447-485.
139. Baker, M.A., ed. 1987. Sensory functioning. Pp. 5-36 in *Sex Differences in Human Performance*. John Wiley and Sons, New York, N.Y.
140. Harm, D.L., Jennings, R.T., Meck, J.V., Powell, M.R., Putcha, L., Sams, C.P., Schneider, S.M., Shackelford, L.C., Smith, S.M., and Whitson, P.A. 2001. Invited review: Gender issues related to spaceflight: A NASA perspective. *Journal of Applied Physiology* 91(5):2374-2383.
141. Harm, D.L., Jennings, R.T., Meck, J.V., Powell, M.R., Putcha, L., Sams, C.P., Schneider, S.M., Shackelford, L.C., Smith, S.M., and Whitson, P.A. 2001. Invited review: Gender issues related to spaceflight: A NASA perspective. *Journal of Applied Physiology* 91(5):2374-2383.

142. Deschenes, M.R., McCoy, R.W., Holdren, A.N., and Eason, M.K. 2009. Gender influences neuromuscular adaptations to muscle unloading. *European Journal of Applied Physiology* 105(6):889-897.
143. Barr, Y.R., Bacal, K., Jones, J.A., and Hamilton, D.R. 2007. Breast cancer and spaceflight: Risk and management. *Aviation, Space, and Environmental Medicine* 78(4 Suppl.):A26-A37.
144. Henderson, M.A., Valluri, S., Garrett, J., Lopez, J.T., Caperell-Grant, A., Mendonca, M.S., Rusek, A., Bigsby, R.M., and Dynlacht, J.R. 2010. Effects of estrogen and gender on cataractogenesis induced by high-LET radiation. *Radiation Research* 173(2):191.
145. Mark, S., Cavanagh, P., and Rice, A.J. 2007. *Sex Differences in Bone Health: Bone Loss During Spaceflight.* Cleveland Clinic Press, Cleveland, Ohio.
146. Mark S. 2007. From Earth to Mars: Sex differences and their implications for musculoskeletal health. *Journal of the American Academy of Orthopedic Surgery* 15(Suppl. 1):S19-S21.
147. Jennings, R.T., and Santy, P.A. 1989. Reproduction in the space environment: Part II. Concerns for reproduction. *Review of Obstetrics and Gynecology* 45:7-17.
148. LeBlanc, A., Rowe, R., Schneider, SV., and Hendrick, T. 1995. Regional muscle loss after short-duration spaceflight. *Aviation, Space and Environmental Medicine* 66:1151-1154.
149. Trappe, S., Creer, A., Slivka, D., Minchev, K., and Trappe, T. 2007. Single muscle fiber function with concurrent exercise or nutrition countermeasures during 60 days of bed rest in women. *Journal of Applied Physiology* 103(4):1242-1250.
150. Buckey, J.C., Jr. 2006. Gender: Identifying and managing relevant differences. Pp. 207-221 in *Space Physiology*, 1st Edition. Oxford University Press, New York, N.Y.
151. Harm, D.L., Jennings, R.T., Meck, J.V., Powell, M.R., Putcha, L., Sams, C.P., Schneider, S.M., Shackelford, L.C., Smith, S.M., and Whitson, P.A. 2001. Invited review: Gender issues related to spaceflight: A NASA perspective. *Journal of Applied Physiology* 91(5):2374-2383.
152. Institute of Medicine. 2001. *Exploring the Biological Contributions to Human Health: Does Sex Matter?* (T.M. Wizemann and M.L. Pardue, eds.) National Academy Press, Washington, D.C., pp. 13-27.
153. Jennings, R.T., and Santy, P.A. 1989. Reproduction in the space environment: Part II. Concerns for reproduction. *Review of Obstetrics and Gynecology* 45:7-17.
154. Riley, D.A., and Wong-Riley, M.T.T. 2003. Neuromusclar development is altered by spaceflight. Pp. 95-104 in *The Neurolab Spacelab Mission: Neuroscience Research in Space* (J.C. Buckey and J.L. Homick, eds.). NASA Johnson Space Center, Houston, Tex.
155. Raymond, J., Dememes, D., Blanc. E., and Dechesne, C.J. 2003. Development of the vestibular system in microgravity. Pp. 143-150 in *The Neurolab Spacelab Mission: Neuroscience Research in Space* (J.C. Buckey and J.L. Homick, eds.). NASA Johnson Space Center, Houston, Tex.
156. Ronca, A.E., Fritzsch, B., Bruce, L.L., and Alberts, J.R. 2008. Orbital spaceflight during pregnancy shapes function of mammalian vestibular system. *Behavioral Neuroscience* 122(1):224-232.
157. Harm, D.L., Jennings, R.T., Meck, J.V., Powell, M.R., Putcha, L., Sams, C.P., Schneider, S.M., Shackelford, L.C., Smith, S.M., and Whitson, P.A. 2001. Invited review: Gender issues related to spaceflight: A NASA perspective. *Journal of Applied Physiology* 91(5):2374-2383.
158. Waters, W.W., Ziegler, M.G., and Meck, J.V. 2002. Postspaceflight orthostatic hypotension occurs mostly in women and is predicted by low vascular resistance. *Journal of Applied Physiology* 92(2):586-594.
159. Javis, S.S., Florian, J.P., Curren, M.J., and Pawelczyk, J.A. 2009. Sex differences in vasoconstriction reserve during 70 deg head-up tilt. *Experimental Physiology* 95:184-193.
160. Fu, Q., Arbab-Zadeh, A., Perhonen, M.A., Zhang, R., Zuckerman, J.H., and Levine, B.D. 2004. Hemodynamics of orthostatic intolerance: Implications for gender differences. *American Journal of Physiology. Heart and Circulatory Physiology* 286:H449-H457.
161. Heyward, V.H. 2010. *Advanced Fitness Assessment and Exercise Prescription*, 6th Edition. Human Kinetics, Champaign, Ill., p. 133, Table 6.2.
162. Harm, D.L., Jennings, R.T., Meck, J.V., Powell, M.R., Putcha, L., Sams, C.P., Schneider, S.M., Shackelford, L.C., Smith, S.M., and Whitson, P.A. 2001. Invited review: Gender issues related to spaceflight: A NASA perspective. *Journal of Applied Physiology* 91(5):2374-2383.
163. NASA. 2008. *Evidence Book: Risk of Compromised EVA Performance and Crew Health Due to Inadequate EVA Suit Systems.* NASA-JSC HRP-47060. NASA Johnson Space Center, Houston, Tex.
164. Tanaka, K., Danaher, P., Webb, P., and Hargens, A.R. 2009. Mobility of the elastic counterpressure suit glove. *Aviation, Space, and Environmental Medicine* 80:890-893.

165. Lee, S.M., Schneider, S.M., Boda, W.L., Watenpaugh, D.E., Macias, B.R., Meyer, R.S., and Hargens, A.R. 2009. LBNP exercise protects aerobic capacity and sprint speed of female twins during 30 days of bed rest. *Journal of Applied Physiology* 106:919-928.
166. Trappe, T.A., Burd, N.A., Louis, E.S., and Trappe. S.W. 2007. Influence of concurrent exercise or nutrition countermeasures on thigh and calf muscle size and function during 60 days of bed rest in women. *Acta Physiology* 191:147-159.
167. Zwart, S.R., Hargens, A.R., Lee, S.M., Macias, B.R., Watenpaugh, D.E., Tse, K., Smith, S.M. 2007. Lower body negative pressure treadmill exercise as a countermeasure for bed rest-induced bone loss in female identical twins. *Bone* 40:529-537.
168. National Institute of Allergy and Infectious Diseases. Radiation Countermeasures. Medical Countermeasures Against Radiological and Nuclear Threats. Program Highlights. Available at http://www.niaid.nih.gov/topics/radnuc/program/Pages/default.aspx.
169. Schneider, S.M., Lee, S.M.C., Macias, B.R., Watenpaugh, D.E., and Hargens, A.R. 2009. WISE-2005: Exercise and nutrition countermeasures for upright VO_2pk during bed rest. *Medicine and Science in Sports and Exercise* 41:2165-2176.
170. Clemons, M., and Goss, P. 2001. Estrogen and the risk of breast cancer. *New England Journal of Medicine* 344:276-285.
171. Barratt-Connor, E., Cauley, J.A., Kulkarni, P.M., Sashegyi, A., Cox, D.A., and Geiger, M.J. 2004. Risk-benefit profile for raloxifene: 4-years data from Multiple Outcomes of Raloxifene Evaluation (MORE) randomized trial. *Journal of Bone and Mineral Research* 29:1270-1275.
172. Mark, S. 2007. Sex differences in bone health. In *Bone Loss during Spaceflight.*1st Edition (P. Cavanaugh and A. Rice, eds.). Cleveland Clinic Press, Cleveland, Ohio.
173. NASA. International Space Station. Fact Sheet. Bisphosphonates as a Countermeasure to Spaceflight Induced Bone Loss (Bisphosphonates). Available at http://www.nasa.gov/mission_pages/station/research/experiments/Bisphosphonates.html.
174. Drake, M.T., Clarke, B.L., and Khosla, S. 2008. Bisphosphonates: Mechanism of action and role in clinical practice. *Mayo Clinic Proceedings* 83(9):1032-1045.
175. Bhuriya, R., Singh, M., Molnar, J., Arora, R., and Khosla, S. 2009. Bisphosphonate use in women and the risk of atrial fibrillation: A systematic review and meta-analysis. *International Journal of Cardiology* 142(3):213-217.
176. Whyte, M., Wenkert, D., Clements, K.L., McAlister, W.H., and Mumm, S. 2003. Bisphosphonate-induced osteopetrosis. *New England Journal of Medicine* 349:457-463.
177. Lyles, K.W., Colon-Emeric, C.S., Magaziner, J.S., Adachi, J.D., Pieper, C.F., Mautalen, C., Hyldstrup, L., Recknor, C., Nordsletten, L., Moore, K.A., Lavecchia, C., et al. 2007. Zoledronic acid and clinical fracture and mortality after hip fracture. *New England Journal of Medicine* 357:1799-1809.
178. National Research Council. 2006. *Health Risks from Exposure to Low Levels of Ionizing Radiation. BEIR VII Phase 2.* The National Academies Press, Washington, D.C.
179. Grahn, D., Lombard, L., and Carnes, B. 1992. The comparative tumorigenic effects of fission neutrons and cobalt-60 7 rays in the B6CF1 mouse. *Radiation Research* 129(1):19-36.
180. Thomson, J., Williamson, F., and Grahn, D. 1985. Life shortening in mice exposed to fission neutrons and γ rays: IV. Further studies with fractionated neutron exposures life shortening in mice exposed to fission neutrons and γ rays. *Radiation Research* 103(1):77-88.
181. Washington State University. National Radiobiology Archives. Available at http://www.ustur.wsu.edu/NRA/radioarchive.html.
182. Northwestern University. Janus Tissue Archive. Available at http://janus.northwestern.edu/.
183. Henderson, M.A., Valluri, S., Garrett, J., Lopez, J.T., Caperell-Grant, A., Mendonca, M.S., Rusek, A., Bigsby, R.M., and Dynlacht, J.R. 2010. Effects of estrogen and gender on cataractogenesis induced by high-LET radiation. *Radiation Research* 173(2):191.
184. NASA. Human Research Program. Digital Astronaut Simulates Human Body in Space. Available at http://spaceflight-systems.grc.nasa.gov/advanced/humanresearch/digital/.
185. Kanas, N., and Manzey, D. 2003. Human interactions. Pp. 75-106 in *Space Psychology and Psychiatry*, 1st Edition. Kluwer Academic Publishers, Boston, Mass.
186. Rosnet, E., Jurion, S., Cazes, G., and Bachelard, C. 2004. Mixed-gender groups: Coping strategies and factors of psychological adaptation in a polar environment. *Aviation, Space, and Environmental Medicine* 75(7 Suppl.):C10-C13.
187. Rambaut, P.C., Leach, C.S., and Leonard, J.J. 1977. Observations in energy balance in man during space flight. *American Journal of Physiology* 233:R208-R212.
188. Schoeller, D.A., and Gretebeck, R.J. 1999. Energy utilization and exercise in spaceflight. Pp. 97-115 in *Nutrition in Spaceflight and Weightlessness* (H.W. Lane and D.A. Schoeller, eds.). CRC Press LLC, Boca Raton, Fla.
189. NASA Engineering Network. Public Lessons Learned Entry 0876. Available at http://www.nasa.gov/offices.oce/llis/0876.html.

190. Novak, L. 1991. Our experience in the evaluation of the thermal comfort during space flight and in the simulated space environment. *Acta Astronautica* 23:179-186.
191. Winget, C.M., Vernikos-Danellis, J., Cronin, S.E., Leach, C.S., Rambaut, P.C., and Mack, P.B. 1972. Circadian rhythm asynchrony in man during hypokinesis. *Journal of Applied Physiology* 33:640-643.
192. Hardy, J.D. 1949. Heat transfer. Pp. 78-108 in *Physiology of Heat Regulation* (L.H. Newburgh, ed.). W.B. Saunders Co., Philadelphia, Pa.
193. Novak, L. 1991. Our experience in the evaluation of the thermal comfort during space flight and in the simulated space environment. *Acta Astronautica* 23:179-186.
194. Kregel, K.C. 2002. Heat shock proteins: Modifying factors in physiological stress responses and acquired thermotolerance. *Journal of Applied Physiology* 92:2177-2186.
195. Fortney, S.M., Mikhaylov, V., Lee, S.M.C., Kobzev, Y., Gonzalez, R.R., and Greenleaf, J.E. 1998. Body temperature and thermoregulation during submaximal exercise after 115-day spaceflight. *Aviation, Space, and Environmental Medicine* 69:137-141.
196. Fortney, S.M., and Greenleaf, J.E. 1996. The physiology of bed rest. Pp. 889-939 in *Handbook of Physiology* (M.J. Fregley and C.M. Blatteis, eds.). Environmental Physiology, Volume II. American Physiological Society, Oxford University Press, Oxford, U.K.
197. Shibasaki, M., Wilson, T.E., Cui, J., Levine, B.D., and Crandall, G.G. 2003. Exercise throughout 6 degrees head-down tilt bed rest preserves thermoregulatory responses. *Journal of Applied Physiology* 95:1817-1823.
198. Alfrey, C.P., Udden, M.M., Leach-Huntoon, C., Driscoll, T., and Pickett, M.H. 1996. Control of red blood cell mass in spaceflight. *Journal of Applied Physiology* 81:98-104.
199. Sawka, M.N., Young, A.J., Francesconi, R.P., Muza, S.R., and Pandolf, K.B. 1985. Thermoregulatory and blood responses during exercise at graded hypohydration levels. *Journal of Applied Physiology* 59:1394-1401.
200. Fortney, S.M., Mikhaylov, V., Lee, S.M.C., Kobzev, Y., Gonzalez, R.R., and Greenleaf, J.E. 1998. Body temperature and thermoregulation during submaximal exercise after 115-day spaceflight. *Aviation, Space, and Environmental Medicine* 69:137-141.
201. Dijk, D.J., Neri, D.F., Wyatt, J.K., Ronda, J.M., Riel, E., Ritz-De Cecco, A., Hughes, R.J., Elliott, A.R., Prisk, G.K., West, J.B., and Czeisler, C.A. 2001. Sleep, performance, circadian rhythms, and light-dark cycles during two space shuttle flights. *American Journal of Physiology. Regulatory, Integrative and Comparative Physiology* 281:R1647-R1664.
202. Scheer, F.A., Hilton, M.F., Mantzoros, C.S., and Shea, S.A. 2009. Adverse metabolic and cardiovascular consequences of circadian misalignment. *Proceedings of the National Academy of Sciences U.S.A.* 106(11):4453-4458.
203. Dijk, D.J., Neri, D.F., Wyatt, J.K., Ronda, J.M., Riel, E., Ritz-De Cecco, A., Hughes, R.J., Elliott, A.R., Prisk, G.K., West, J.B., and Czeisler, C.A. 2001. Sleep, performance, circadian rhythms, and light-dark cycles during two space shuttle flights. *American Journal of Physiology. Regulatory, Integrative and Comparative Physiology* 281:R1647-R1664.
204. Monk, T.H., Buysse, D, J., Billy, B.D., Kennedy, K.S., and Willrich, L.M. 1998. Sleep and circadian rhythms in four orbiting astronauts. *Journal of Biological Rhythms* 13:188-201.
205. Monk, T.H., Kennedy, K.S., Rose, L.R., and Lininger, J.M. 2001. Decreased human circardian pacemaker influence after 100 days in space: A case study. *Psychosomatic Medicine* 63:881-885.
206. Mekjavic, I.B., Golja, P., Tipton, M.J., and Eiken, O. 2005. Human thermoregulatory function during exercise and immersion after 35 days of horizontal bed-rest and recovery. *European Journal of Applied Physiology* 95:163-171.
207. Wieland, P.O. 1998. *Living Together in Space: The Design and Operation of the Life Support Systems on the International Space Station.* NASA TM-206956, Volume 1. NASA Marshall Space Flight Center, Huntsville, Ala.
208. Piscane, V.L., Kuznetz, L.H., Logan, J.S., Clark, J.B., and Wissler, E.H. 2007. *Use of Thermoregulatory Models to Enhance Space Shuttle and Space Station Operations and Review of Human Thermoregulatory Control.* NASA TM 0022493. NASA Johnson Space Center, Houston, Tex.
209. Rimmer, D.W., Dijk, D.J., Ronda, J.M., Hoyt, R., and Pawelczyk, J.A. 1999. Efficacy of liquid cooling garments to minimize heat strain during space shuttle deorbit and landing. *Medicine and Science in Sports and Exercise* 31:S305.
210. Allan, J.R., and Crossely, R.J. 1972. Effect of controlled elevation of body temperature on human tolerance to +Gz acceleration. *Journal of Applied Physiology* 33:418-420.
211. Pisacane, V.L., Kuznetz, L.H., Logan, J.S., Clark, J.B., and Wissler, E.H. 2007. Thermoregulatory models of space shuttle and space station activities. *Aviation, Space, and Environmental Medicine* 78:S48-S55.
212. Hamilton Sunstrand. 2006. *Extra Mobility Unit Design and Performance Requirements Specification.* Report SVHS7800, Release date January 23. Hamilton Sunstrand, Houston, Tex.
213. Norcross, J.R., Lee, L.R., Clowers, K.G., Morency, R.M., Desantis, L., De Witt, J.K., Jones, J.A., Vos, J.R., and Gernhardt, M.L. 2009. *Feasability of Performing a Suited 10-km Ambulation on the Moon—Final report of the EVA Walkback Test (EWT).* NASA/TP 2009-214796. NASA Johnson Space Center, Houston, Tex. October.

214. Piscane, V.L., Kuznetz, L.H., Logan, J.S., Clark, J.B., and Wissler, E.H. 2007. *Use of Thermoregulatory Models to Enhance Space Shuttle and Space Station Operations and Review of Human Thermoregulatory Control.* NASA TM 0022493. NASA Johnson Space Center, Houston, Tex.
215. Stolwijk, J.H.H., and Hardy, J.D. 1966. Temperature regulation in man—A theoretical study. *Pfluegers Archiv European Journal of Physiology* 291:129-162.
216. Kuznetz, L.H. 1968. *A Model for the Transient Metabolic Response of Man in Space.* Crew Systems Division. NASA-MSC MS-EC-R-68-4 and NASA-JSC N78-78393. NASA Johnson Space Center, Houston, Tex.
217. Wissler, E.H. 1964. A mathematical model of the human thermal system. *Bulletin of Mathematical Biophysics* 62:66-78.
218. Fuller, C.A., Hoban-Higgins, M.H., Klimovitsky, V.Y., Griffin, D.W., and Alpatov, A.M. 1996. Primate circadian rhythms during spaceflight: Results from Cosmos 2044 and 2229. *Journal of Applied Physiology* 81(1):188-193.
219. Klimovitskiy, V.Y., Alpatov, A.M., Sulzman, F.M., and Fuller, C.A. 1987. Inflight circadian rhythms and temperature homeostasis aboard Cosmos-1514 biosatellite. *Kosmicheskaya Biologiya I Aviakomicheskaya Meditsnia* 21(5):14-18.
220. Fuller, C.A., Hoban-Higgins, M.H., Klimovitsky, V.Y., Griffin, D.W., and Alpatov, A.M. 1996. Primate circadian rhythms during spaceflight: Results from Cosmos 2044 and 2229. *Journal of Applied Physiology* 81(1):188-193.
221. Sulzman, F.M., Ferraro, J.S., Fuller, C.A., Moore-Ede, M.C., Klimovitsky, V., Magedov, V., and Alpatov, A.M. 1992. Thermoregulatory responses of rhesus monkeys during spaceflight. *Physiology and Behavior* 51(3):585-591.
222. Alpatov A.M., Hobban-Higgins, T.M., Klimovitsky, V.Y., Turnurova, E.G., and Fuller, C.A. 2000. Circadian rhythms in macaca mulatta monkeys during Bion II flight. *Journal of Gravitational Physiology* 7(1):S119-S123.
223. Fortney, S.M., and Greenleaf, J.E. 1996. The physiology of bed rest. Pp. 889-939 in *Handbook of Physiology* (M.J. Fregley and C.M. Blatteis, eds.). Environmental Physiology, Volume II. American Physiological Society, Oxford University Press.
224. Winget, C.M., Vernikos-Danellis, J., Cronin, S.E., Leach, C.S., Rambaut, P.C. and Mack, P.B. 1972. Circadian rhythm asynchrony in man during hypokinesis. *Journal of Applied Physiology* 33:640-643.
225. Greenleaf, J.E., and Resse, R.D. 1980. Exercise thermoregulation after 14 days bed rest. *Journal of Applied Physiology* 48:72-78.
226. Shibasaki, M., Wilson, T.E., Cui, J., Levine, B.D., and Crandall, G.G. 2003. Exercise throughout 6 degrees head-down tilt bed rest preserves thermoregulatory responses. *Journal of Applied Physiology* 95:1817-1823.
227. Greenleaf, J.E., and Resse, R.D. 1980. Exercise thermoregulation after 14 days bed rest. *Journal of Applied Physiology* 48:72-78.
228. Mekjavic, I.B., Golja, P, Tipton, M.J., and Eiken, O. 2005. Human thermoregulatory function during exercise and immersion after 35 days of horizontal bed-rest and recovery. *European Journal of Applied Physiology* 95:163-171.
229. Shoemaker, J.K., Hogeman, C.S., Silber, D.H., Gray, K., Herr, M., and Sinoway, L.I. 1998. Head-down-tilt bed rest alters forearm vasodilator and vasoconstrictor responses. *Journal of Applied Physiology* 84:1756-1762.
230. Robertson, D., and Williams, G.H. 2009. *Clinical and Translational Science: Introduction to Human Research.* Elsevier, San Diego, Calif.
231. Institute of Medicine. 2001. *Safe Passage: Astronaut Care for Exploratory Missions.* The National Academies Press, Washington, D.C. Chapter 6, pp. 173-188.
232. Uhran, M.L. 2009. Progress toward establishing a US National Laboratory on the International Space Station. *Acta Astronautica* 66(1-2):149-156.
233. Ginsburg, G.S., and Willard, H.F. 2009. *Essentials of Genomic and Personalized Medicine.* Elsevier, Amsterdam, The Netherlands.
234. Collins, I., and Workman, P. 2006. New approaches to molecular cancer therapeutics. *Nature Chemical Biology* 2(12):689-700.
235. Watanabe, H., Darbar, D., Kaiser, D.W., Jiramongkolchai, K., Chopra, S., Donahue, B.S., Kannankeril, P.J., Roden, D.M. 2009. Mutations in sodium channel β1- and β2-subunits associated with atrial fibrillation. *Circulation: Arrhythmia and Electrophysiology* 2:268-275.
236. Goel, N., Banks, S., Mignot, E., and Dinges, D.F. 2009. *PER3* polymorphism predicts cumulative sleep homeostatic but not neurobehavioral changes to chronic partial sleep deprivation. *PLoS ONE* 4(6):e5874.
237. Roden, D.M., Pulley, J.M., Basford, M.A., Bernard, G.R., Clayton, E.W., Balser, J.R., and Masys, D.R. 2008. Development of a large-scale de-identified DNA biobank to enable personalized medicine. *Clinical Pharmacology and Therapeutics* 84:362-369.
238. NASA. 2008. *Evidence Book: Risk of Therapeutic Failure Due to Ineffectiveness of Medication.* HRP-47060. NASA Johnson Space Center, Houston, Tex. March.
239. Putcha, L., and Cintron, N.M. 1991. Pharmacokinetic consequences of spaceflight. *Annals of the New York Academy of Sciences* 618:615-618.

240. Amidon, G.L., DeBrincat, G.A., and Najib, N. 1991. Effects of gravity on gastric emptying, intestinal transit and drug absortion. *Journal of Clinical Pharmacology* 31:968-973.
241. Diedrich, A., Paranjape, S.Y., and Robertson, D. 2007. Plasma and blood volume in space. *American Journal of the Medical Sciences* 334:962-967.
242. Balas, E.A., and Boren, S.A. 2000. *Yearbook of Medical Informatics: Managing Clinical Knowledge for Health Care Improvement.* Schattauer Verlagsgesellschaft mbH, Stuttgart, Germany.
243. Westfall, J.M., Mold, J., and Fagan, L. 2007. Practice based research—"Blue Highways" on the NIH Roadmap. *Journal of the American Medical Association* 297:403-407.
244. National Council on Radiation Protection and Measurements (NCRP). 2006. *Information Needed to Make Radiation Protection Recommendations for Space Missions Beyond Low-Earth Orbit.* NCRP Report 153. NCRP, Bethesda, Md.
245. Durante, M., and Cucinotta, F.A. 2008. Heavy ion carcinogenesis and human space exploration. *Nature Reviews Cancer* 8:465-472.

8

Fundamental Physical Sciences in Space

Understanding the universe is a daunting task, yet our curiosity and wonder over centuries and civilizations has led physical scientists to seek answers to some of the most compelling questions of all. How did the universe come to be? What is it made of? What forces rule its behavior? Is there life elsewhere? In seeking answers to these questions, scientists search for the simplest laws that not only explain the universe but also predict behavior within it. Within the fundamental physical sciences activity at NASA, the panel identified two overarching quests that characterize the goals and motivations behind this compelling research: (1) to discover and explore the laws governing matter, space, and time and (2) to discover and understand the organizing principles of complex systems from which structure and dynamics emerge. A robust physical sciences program pursuing these quests is essential to NASA's effort to explore and develop space and promises societal benefits and technologies for improving life on Earth.

Discovery of fundamentally new knowledge and the subsequent development of engineered systems have advanced the human condition and supported the world's economy. Fundamental research across a wide range of disciplines and settings is both enabled by this rapid technological progress and helps to enable that progress. As part of this broad enterprise, fundamental physical sciences are both a customer of and a supplier in NASA's commitment to space exploration. For example, some of the most important questions in physics today can be answered only in the unique environment of space, and addressing them is enabled by NASA's commitment to exploration. But the results of investigations in the fundamental physical sciences also enable NASA's exploration mission by empowering the development of new materials and energy sources, time and frequency standards for navigation, and technologies that help humans adapt to the hostile conditions in space.

NASA-sponsored research in fundamental physical sciences must be far reaching. For example, discovery and exploration of physical laws can be pursued through efforts to detect and understand dark matter and dark energy, the search for gravitational waves (enabled by the long baselines available only in space for measuring small metric variations in space itself), and studies of the origins of the universe, mass, and time. In addition, NASA-sponsored research should address the complexity that is observed all around us, which emerges from simple physical laws of many particles acting cooperatively, and new organizing principles emerging as systems increase in size. We are just beginning to understand such complex systems, ranging from bacteria to galactic clusters, and to seize the great opportunity for profound discoveries and wide-ranging applications. The unique conditions of space, such as weightlessness and access to high vacuum, will also enable the development of powerful new technologies and scientific experiments—for example, space-based optical clocks for enhanced navigation on Earth and in space and the transmission of phase information between advanced clocks (which require microgravity) over large distances through the vacuum of space, where the lack of dispersion through a medium enables highly accurate relative timing and frequency information to test Lorentz variation at unprecedented limits.

To pursue these quests, NASA should support a comprehensive program providing regular access to space, complemented by a robust ground-based program of supporting investigations, flight-definition studies, and education of the next generation of scientists. Such a balanced program will foster a broad scientific community to ensure that NASA pursues the best science, both enabled by and enabling exploration. We know that traditionally this fundamental science mission is best accomplished through peer-reviewed selection processes that are responsive to the most compelling scientific ideas of our time. Neither the overall mission program nor specific scientific projects should be dictated during peer review or at any other stage in the planning process. Instead, areas of scientific thrust are discussed below where, historically, shared facilities have either already been developed or are likely to be available in the future.

In this chapter, four scientific "thrusts" are described that define the frontier of space-based fundamental physical science. Each of these thrusts is discussed in its own section, which provides technical background as well as some typical investigations that might form the basis of an initial program. Other important areas of physical inquiry, including fluid physics, materials, and combustion, have a fundamental component as well, but because they are covered in Chapter 9 of this report, they will not be discussed here.* At the end of this chapter, the panel's overall findings are discussed and recommendations for research in the fundamental physical sciences are provided, including statements about scientific content as well as platforms and facilities needed for success.

RESEARCH ISSUES

Thrust I: Soft-Condensed-Matter Physics and Complex Fluids

Complex fluids and soft condensed matter are materials with multiple levels of structure. That is, they are composed of objects that themselves contain many atoms or molecules. The field encompasses colloids, emulsions, foams, liquid crystals, dusty plasmas, and granular material. With large particles, slow dynamics, and controllable interactions, it is possible to use such systems as models for a wide variety of physical phenomena. Basic insights have been gained into diverse fields such as phase transitions, nucleation and growth of crystals, symmetry breaking, field theory, spinodal decomposition, and the development of the early universe,† ergodicity breaking and glass formation, turbulence, and chaos. The complexity of the basic building blocks and the variety of their interactions have led to the discovery of novel phases as well as interesting processes and dynamics.

Along with their utility for studying fundamental phenomena, complex fluids/soft materials are ubiquitous in the food, chemicals, petroleum, cosmetics, pharmaceutical, liquid-crystal display, and plastics industries. Granular and fluid flow and related processes are essential to present and emerging technologies. The direct contribution of these materials and processes amount to ~5 percent of the U.S. GDP and ~30 percent of the manufacturing output of the United States alone (>$1 trillion). They also play heavily in the construction, textile, printing, and electronics industries.[1]

The softness of the materials may be associated with the large size of the basic units. They are easily deformed and their statics and dynamics are governed by surface tension and entropic forces. On Earth these weak forces are typically dominated by gravity. Thus microgravity is required to probe the underlying properties of these

---

*Astronomy and astrophysics and fundamental physics overlap scientifically in many significant ways. This report has avoided duplication with those areas of fundamental physics (e.g., detection of gravitational waves using the Laser Interferometer Space Antenna) that have been carefully considered by the astronomy and astrophysics decadal survey in *New Worlds, New Horizons in Astronomy and Astrophysics* (National Research Council, The National Academies Press, Washington, D.C., 2010). Rather, this study has concentrated on experimental physics performed on small, self-contained space platforms that are typically designed and operated by small investigator teams, rather than the large observational observatories or experiments that are dealt with in *New Worlds, New Horizons*.

†The application of ϕ^4 field theories to understand spontaneous symmetry breaking led to research into the use of condensed matter systems to model cosmology. This has been the topic of theoretical work by Wojciech Hubert Zurek and experimental work by W.D. McCormick and others in Manchester, England. The subdiscipline is summarized in the book *The Universe in a Helium Droplet* by Grigory E. Volovik, The International Series of Monographs on Physics, Oxford University Press, 2003.

materials and to use them as models to explore other phenomena. Experiments on Earth are frequently hampered by sedimentation, flows, and suppression of thermodynamic fluctuations.

NASA realized the important role of microgravity research when the field of complex fluids was in its infancy. Within the complex fluids community, NASA's fostering of this developing area is well acknowledged. For almost two decades, important discoveries in the field were reported at the annual NASA complex fluids meeting. The broad support for ground-based research culminating in flight results led to important discoveries that bootstrapped the field and sparked major efforts in leading universities here and abroad. It also led to international collaboration on both ground-based and flight projects that inspired new initiatives together with the European Space Agency (ESA), and by ESA alone. For example, between 1998 and 2000, the research sponsored by the program produced several hundred papers that were published in internationally recognized journals. Of these papers, more than 120 were published in the *Journal of Fluid Mechanics* and *Physics of Fluids*, two prominent journals for fluid dynamics; 44 in *Physical Review Letters*, a leading physics journal; 8 in *Nature*; and 7 in *Science*, the last 2 of which are among the most prestigious scientific journals in the world.[2] This new field has developed into an important research area of physics and materials science and is now found in the science departments of every major university in the world. Complex fluids and soft matter are a key component of the microgravity research of space agencies internationally.

Practitioners of the field have gained some important experience in the conduct of microgravity experiments. An interesting aspect has been the active participation of the astronauts in conducting the research. In several instances unexpected discoveries resulted—for example, surprisingly large correlations in phase separations with spinodal decomposition, in the microgravity crystallization of glasses, and in the growth of dendritic (treelike) structure of crystals. In each case the astronauts, in contact with the principal investigators, were able to modify the equipment or improvise a new apparatus from stuff on the spacecraft to successfully record the discovery and make quantitative measurements. As a result, at least one much-cited paper was coauthored with astronauts.[3] There is, of course, the additional educational and motivational aspect of having graduate and postdoctoral students in live communication with their experiments and the astronauts during the flight.

Complex fluids and soft matter materials are excellent candidates for study in the microgravity laboratory. Colloids, polymer and colloidal gels, foams, emulsions, soap solutions, and the like are particularly susceptible to gravity because of the gradients that are formed in their properties under gravity. Hence, microgravity provides a unique opportunity to eliminate these gradients and to study the long-time dynamics of such systems free from such gravitational interference. Similar benefits accrue for colloids, gels, and dusty plasmas, whose density and morphology are height dependent under gravity. Similarly, in granular materials, stress chains and yield properties are height dependent and sensitive to the magnitude of gravity. While increased gravity can be effectively mimicked using a centrifuge, it is also important to explore under reduced gravity, which has been made available to researchers exclusively through NASA-sponsored microgravity research.

There are fundamental aspects of the issues discussed in Chapter 9 that should be supported by NASA. For example, the properties of granular materials are of ubiquitous concern in any mission to the Moon or Mars, crewed or robotic. (It is noteworthy in this connection that the Mars rover Spirit has been stuck in the martian soil since May 2010.) One highly relevant area of fundamental research is the development of robust constitutive equations that describe the strain-strain rate relationships for granular materials under reduced gravity. In fact the effect of reduced gravity on the properties of complex fluids in general provides a productive experimental environment to improve our fundamental understanding. Experiments in the range 0 to 1 g are most appropriately done on a microgravity platform.

As discussed in Chapter 9, there are also multiple issues surrounding spaceflight that need to be addressed, including processing, heating, and cooling of fluids. While applied NASA-sponsored research should be organized to have maximum impact on missions, targeted fundamental research would be beneficial as well, both for support of applied research and as possible opportunities for high-impact, space-based research projects.

Thrust II: Precision Measurements of Fundamental Forces and Symmetries

Space offers unique conditions to address important questions concerning the fundamental laws of nature and affords greater sensitivity than ground-based experiments in certain areas. In particular, high-precision measurements in space can test relativistic gravity and fundamental particle physics in ways that are not practical on Earth. Promising theoretical approaches to quantum gravity and physics beyond the currently accepted standard model of fundamental physics typically predict new forces, violations of fundamental symmetries, or time-varying physical constants. Such novel effects provide distinct signatures for precision experimental searches that are often best carried out in space. Two examples are discussed in more detail below; however, there are also many other opportunities for space-based precision measurement, each offering a unique opportunity for a major discovery in fundamental physics.[‡]

First, note that Einstein's theory of general relativity assumes an exact equivalence between gravitational mass and inertial mass. This equivalence principle (EP) states that all objects, no matter what they are made of, move under gravity in exactly the same way, depending only on their mass. Although both Newton's and Einstein's laws of physics assume that this principle holds exactly, the latest theories of modern physics usually predict that there should be small (less than one trillionth of a percent) violations of the EP at the fractional level of $\sim 10^{-13}$ to 10^{-19}. These predicted violations, while small, may be related to quantum gravity and to explanations of dark energy, which are among the most important topics of modern physics. Detecting these small but predicted violations of the EP would have a revolutionary impact on our understanding of basic physics. The EP is best tested in space, where (1) there is little or no friction, and no seismic or thermal activity or other sources of noise, and where (2) the dominant gravitational forces exerted by the sun, the planets, and other bodies of the solar system are easier to measure accurately. As a result, there are many promising space-based approaches to improved EP tests. There are two slightly different versions of the EP, known as the "weak" and "strong" versions, and both can be tested in space. Some, but not all, representative missions are briefly discussed here.

The MICROSCOPE (Micro Satellite à trainee Compensée pour l'Observation du Principe d'Equivalence) satellite mission under development by ESA and the French Centre National d'Etudes Spatiales is scheduled for launch in 2012. Its design goal is to achieve a differential acceleration accuracy to probe the weak EP at a sensitivity of 10^{-15}. The proposed Satellite Test of Equivalence Principle (STEP) mission will test the weak EP using cryogenically controlled test masses on a spacecraft orbiting Earth. STEP will search for a violation of the weak EP with a fractional accuracy of 10^{-18}, which is accurate enough to test some of the current leading theories that might go beyond Einstein's general theory of relativity. For testing of the strong EP, lunar laser ranging experiments—that is, experiments reflecting laser beams off retroreflector arrays placed on the Moon by the Apollo astronauts and by an uncrewed Soviet lander—set limits of $\sim 10^{-13}$ for any possible inequality in the ratios of the gravitational and inertial masses for Earth and the Moon. Although at present the Earth-Moon-Sun system is best for tests of the strong EP, over the next decade a major advance will come from interplanetary laser ranging, such as a retroreflector on a martian lander. Technology is available to conduct such measurements with a timing precision of a few picoseconds, which would lead to 100-fold improvements in tests of the strong EP.

Second, the standard model and general relativity, both of which are broad and powerful theories, are thought to be the effective low-energy limits of an underlying "ultimate theory" that unifies all of physics, including gravity and particle physics, at the so-called Planck scale. The Planck scale corresponds to enormous energies ($\sim 10^{19}$ GeV), which are not obtainable even in the most powerful particle supercollider that can be built. Recently it has been realized that the ultimate theory may well allow low-energy violations of a fundamental principle known as Lorentz symmetry (the symmetry of physics under rotations and boosts), as well as a related fundamental principle known as charge-parity-time (CPT) symmetry, which states that particle interactions should behave the same if one could simultaneously reverse the charge of the particles, their "parity" or handedness, and the direction of the flow of time. One could detect violations of Lorentz and CPT symmetry by finding, say, small variations in particle masses (Hughes-Drever effects), the speed of light (Michelson-Morley effects), and many other properties, as a function of orientation and boost in the universe, and as a function of the local gravitational potential. Precision searches for such Lorentz and CPT violations are undergoing intense experimental investigation across

[‡]For a more extensive discussion of the scientific opportunities described in this thrust, the panel refers the reader to recent review articles such as S.G. Turyshev, U.E. Israelsson, M. Shao, N. Yu, A. Kusenko, E.L. Wright, C.W.F. Everitt, M. Kasevich, J.A. Lipa, J.C. Mester, R.D. Reasenberg, et al., *International Journal of Modern Physics D* 16:1879-1925, 2007, and S.G. Turyshev, *European Physical Journal-Special Topics* 163:227-253, 2008.

many physics subfields. In some cases the experiments are sensitive to energies at the Planck scale, although as yet no violation has been observed in any system. This suite of experimental efforts is proceeding in concert with theoretical work used to interpret and compare different experiments.

Over the next decade, space-based experiments could improve the sensitivity to possible violations of Lorentz and CPT symmetry by several orders of magnitude. One important class of such experiments consists of clock comparison experiments, in which two or more highly stable space-based clocks are simultaneously operated and their clock rates compared and correlated with position and velocity in a gravitational potential. Einstein's general theory of relativity tells us that clock rates vary with velocity and gravitational potential but should not otherwise depend on position or orientation of the clock. The comparison of space-based clocks may improve Hughes-Drever tests by several orders of magnitude. Such major advances in sensitivity will arise from space-based operation because it offers (1) better stability and accuracy for clocks referenced to cold atoms and (2) access to a wider range of boosts, orientations, and gravity gradients in space than on Earth. In addition, the constancy and isotropy of the speed of light can also be tested by measuring the time it takes light to travel between a space-based clock and a ground clock. High-stability clocks orbiting Earth, combined with a sufficiently accurate time and frequency transfer link, could improve present sensitivity in this area by more than three orders of magnitude. The clocks might include microwave and optical clocks based on atomic transitions or stabilized cavities.

Space-based precision measurements may also enable NASA's current Exploration mission in the form of improved navigation and communication. In recent years there have been great advances in precision measurement technologies for fundamental physical properties such as time, gravity, and optical wavelength. For example, atomic clock performance (stability, accuracy) has improved by two orders of magnitude over the past decade, driven primarily by breakthroughs in fundamental physics in areas such as cold atoms, ion traps, and laser frequency combs. NASA, through its earlier Code U Fundamental Physics Program, supported research in this area. Advances in this research have been recognized in recent years with the award of Nobel prizes in physics to Eric Cornell, Wolfgang Ketterle, and Carl Weiman in 2001 and to Steven Chu, William Phillips, and Claude Cohen-Tannoudji in 1997.

Many precision measurement technologies could be readily adapted for space operation to enable both human and scientific exploration. For example, microwave atomic clocks with fractional frequency stability and accuracy better than 10^{-15} have been space qualified and are being prepared for 1 to 3 years of operation on the International Space Station (ISS) as part of the European mission known as ACES (Atomic Clock Ensemble in Space). Optical clocks based on optical transitions in cold atoms and laser frequency combs to allow counting of optical frequencies have already demonstrated fractional frequency stability and accuracy of $\sim 10^{-17}$ in ground-based labs. Operation in microgravity will allow the use of colder, denser atomic ensembles, with resulting advantages in clock stability and reduced systematic frequency shifts that could reach a stability and accuracy level of 10^{-18} to 10^{-19}. A network of space-based optical clocks could provide a universal high-precision time reference for space- and ground-based navigation, communication, and geodesy. This universal positioning system (UPS) could greatly improve Global Positioning System (GPS) performance and bring state-of-the-art navigation capabilities to space exploration.

Thrust III: Quantum Gases

When the temperature of a gas is decreased, the quantum, wavelike properties of the constituent atoms or molecules become more apparent. The gas becomes a "quantum gas" when the size of the individual particle's wavepacket becomes large compared to the length scale of interactions between the particles. In this limit, the wavelike properties of the particle motion and the indistinguishability of the particles become important, and collective quantum behavior begins to dominate the gas. On further cooling, the wavepacket size can become as large as or larger than the interparticle spacing, and the individual character of the particles is subsumed by a cooperative behavior, such that they become either a superconductor for a charged system or a superfluid for a neutral system.

One of the most dramatic developments in fundamental physics in the past two decades has been the realization of a superfluid Bose-Einstein condensate (BEC) in a dilute atomic gas.[4–7] The physics of the BEC connects the field of atomic, molecular, and optical physics to the field of condensed-matter physics, linking together two fundamental themes recommended for inclusion in NASA's Exploration Enterprise. The BEC has remarkable properties in common with the much denser phases discovered early in the 20th century—superfluidity in helium and superconductivity in certain metals—as well as with the matter in the core of a neutron star. The key to creating a BEC is to go beyond the already low temperatures achievable using laser cooling alone. Further cooling is achieved by evaporating atoms from a trap, typically created using magnetic fields, cooling the cloud somewhat as the coffee remaining in a cup is cooled when the hotter molecules evaporate and escape. For terrestrial experiments, the evaporation initially occurs from the surface of a three-dimensional trap; as the evaporation proceeds, however, gravitational compression in the trap causes the trapped gas to become almost two-dimensional, and cooling is restricted to a narrow ring and ceases. Typically this cooling limitation sets in at a few nanokelvin, at which point the size of the atomic wavepacket is a few tens of microns, or about the diameter of a human hair. In the absence of gravity, temperatures on the order of a picokelvin or less should be achievable, corresponding to atomic wavepacket sizes of nearly a millimeter! This is an astonishingly large size, since the wavelike properties of ordinary matter are normally limited to distances comparable to atomic sizes. But at the ultralow temperatures that may be achieved in space, the BEC wavepackets exist at a length scale observable by the unaided human eye. The temperature limitations imposed on quantum gases is not the only impact of gravity. Gravity makes the precise observation of freely expanding condensates difficult or impossible in Earth-based laboratories because it induces density stratification, which blurs and masks the system's underlying behavior. On Earth the trap that supports the BEC must be strong enough to provide a force to counter gravity, thus keeping the atoms or molecules within the trap. The strength of the trap perturbs the state of the particles and influences their collective behavior. In microgravity a trap that is 100,000 times weaker can contain the particles. This greatly reduces the experimental perturbations on the system, allowing its fundamental properties to be observed and systematically experimented with.

A remarkable range of physical phenomena can be investigated using BECs, but many of them only in space. Aspects of the formation of the BEC and its intrinsic quantum properties represent one rich class. For example, one fundamental excitation of a BEC is the quantum vortex (Figure 8.1). Research on vortex formation and relaxation can be used to probe phase-transition models of the early universe and can also give insight into the structure of neutron stars.

BECs can be contained in one-, two-, or three-dimensional lattices formed by precisely controllable optical standing waves. This configuration opens new windows onto the phases of quantum-dominated matter and can be used to simulate the properties of crystalline solids. In these systems, properties such as the shape and depth of the lattice potential can be varied continuously. Quantum phase transitions (one of which is the "superfluid-

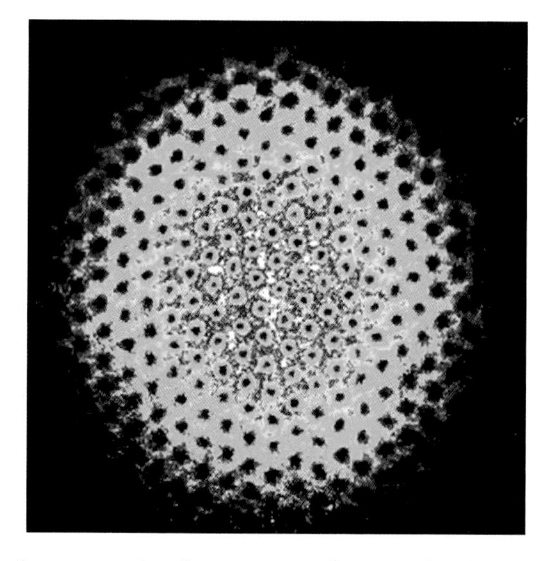

FIGURE 8.1 Data showing quantized vortices in a Bose-Einstein condensate, ranging from sparse to dense vortices in the left-hand figure. The right-hand figure shows a very dense, regular array of quantized vortices. SOURCE: *Left:* Courtesy of W. Ketterle, Massachusetts Institute of Technology. *Right:* Courtesy of E.A. Cornell, University of Colorado, Boulder.

to-Mott" insulator transition from a continuous quantum fluid to a discrete atomic lattice) can thereby be studied in a controlled, clear, and precise manner that is impossible in an ordinary solid where the chemical composition dictates the properties.[8] Furthermore, exact theoretical models can be developed and tested, revealing the key strengths and weaknesses of our basic understanding of whether a material conducts electricity or impedes it. In essence, this represents the use of the BEC to realize the kind of quantum simulation foreseen by the visionary physicist and Nobel laureate Richard Feynman.

As picokelvin BECs are realized in a space-based laboratory, scientists will be able to create and understand the competing forms of order that often govern the complex structure observed in the world around us. One of the remarkable aspects of quantum gases, and the BEC in particular, is the exquisite sensitivity to interparticle interactions. Because of the low thermal energies and the great size of the quantum wavepackets, the system becomes sensitive to tiny but important long-range forces and interactions. When combined with optical lattices, this allows replication and investigation of the building blocks of complicated matter such as magnetic and electric materials.

All particles can be classified as either "bosons" or "fermions." Unlike bosons, which tend to condense together, fermions tend to repel one another (more accurately, they cannot occupy the same quantum state). Fermionic matter is ubiquitous in the universe. It includes systems such as the electron gas that makes metals resilient, elastic, and conductive, and it is the source of forces that stabilize white dwarf stars against gravitational collapse. If the particles of a quantum gas are identical fermions, then another class of physics can be investigated. Fermions that interact via repulsive interactions are predicted to show a rich phase diagram when placed in an optical lattice, allowing us to test key theoretical models and amplify our understanding of a broad range of phenomena. Of practical importance is the unresolved mechanism of superconductivity in high-temperature superconductors. Of fundamental importance is the so-called color superfluidity of quarks in quantum chromodynamics, which describes nucleon and quark interactions at subnuclear length scales. (The term "color" here has nothing to do with color in the ordinary sense; it refers instead to quantum states of matter.) Analogs to both might be observable in cold Fermi gases in space.

The quest for the lowest energy quantum configuration of fermions in ultracold gases involves research on mixtures of ultracold bosons and fermions. On the practical side this is because the evaporative cooling used to achieve a BEC in a Bose gas does not work with fermions. The thermalization that allows the evaporating gas to cool relies on collisions between the atoms, which in turn are forbidden by the intrinsic exclusion exhibited by fermions for each other (the Pauli exclusion principle). To solve this, a Bose-Fermi mixture is used in which the bosons are evaporatively cooled and the fermions are "sympathetically" refrigerated by interaction with the bosons. One simple consequence is that the Fermi gas can only be made as cold as the companion BEC. However, heavier particles sink under the influence of gravity, so that as the gases become colder the species separate in space and cooling ceases. The solution to this problem is to remove gravity. Theoretical work on these mixtures has been prolific, and experimentation will yield exciting discoveries spanning the quantum-mechanical properties of extremely weakly interacting systems to strongly interacting ones.

Experiments with quantum gases in space will allow the study of matter in regimes not achievable on Earth. They will support new developments and applications of breakthrough technologies such as the atom laser, a bright source of coherent matter waves analogous to coherent light waves of the familiar laser. Another important impact of these systems will be in next-generation technologies and quantum sensors. Examples include ultraprecise atomic clocks and matter-wave interference devices with exquisite sensitivity to rotation and gravity. Space-based matter-wave interferometers can set new standards in inertial and gravitational sensing for basic research, navigation, geodesy, and geology.

An excellent example of a cold-atom quantum sensor is the cold-atom interferometer. As described above in connection with the Bose-Einstein condensate, when atoms are cooled, the length scale characterizing their quantum behavior increases. This allows the construction of "atom optical" devices in analogy with conventional optical devices.[9] One of these devices, the cold-atom interferometer, much resembles its optical counterpart. An input beam consisting of atoms of equal velocity is split into two parts. The two beams are made to propagate through different paths in space and are then recombined. If the two paths differ in length, there can be either constructive or destructive interference of the matter waves, and matter-wave fringes are observed. Such devices have already been tested as rotation sensors and as detectors to measure fundamental quantities such as photon momentum and the local force of gravity. Used for inertial navigation, these rotational sensors rival the best gyroscopes available and are potentially important for space navigation applications.

Thrust IV: Condensed Matter and Critical Phenomena

One of the great scientific successes enabled by the microgravity environment over the past two decades concerns better understanding of the behavior of materials under a special set of thermodynamic conditions known as a criticality.[10] If one maintains a system at its critical density ρ_c, then the observable liquid-vapor phase boundary in the pressure-versus-temperature phase diagram will end abruptly at a critical pressure P_c and a critical temperature T_c. At this critical point, the distinction between liquid and vapor phases disappears, creating a foglike critical state dominated by large fluctuations between the liquid and vapor phases. More than 130 years ago Johannes van der Waals observed that all fluids, when compared at the same reduced temperature and reduced pressure, have approximately the same compressibility factor, and that they all deviate from ideal gas behavior to about the same degree. This principle of corresponding states initiated the study of critical phenomena.[11] The 1982 Nobel prize in physics was awarded to Kenneth Wilson for his development of renormalization group techniques applied to critical phenomena. These techniques provide a powerful, systematic method of calculating the effect of critical fluctuations on the behaviors of many systems, leading to quantitative predictions of critical exponents and amplitude ratios and to the calculation of corrections to these predictions as the system is moved away from its critical point.[12,13]

Many other important materials, including superfluids, magnetic materials, and colloids, undergo transitions between ordered and disordered phases. Each of these systems has its own distinct physical property, called an "order parameter," that is zero in the disordered phase; each exhibits large fluctuations about its zero mean as the critical point is approached, and increases from zero as the ordered phase is entered.

Important advances in our understanding of critical phenomena came from the Lambda Point Experiment (LPE) that flew within the cargo bay of the space shuttle in 1992. That experiment provided a stringent test of advanced theories of static critical phenomena by measuring the heat capacity of ^{4}He near the superfluid critical point to within better than one part in 10^8 of the critical point temperature.[14] This experiment extended the precision by three orders of magnitude compared to what had been possible without access to the weightless laboratory of space.

A similar experiment, the Confined Helium Experiment (CHeX), flew in 1998 to extend these measurements to systems in two-dimensional confinement, again approaching the critical temperature with the same unprecedented level of precision as the LPE experiment. Comparisons of the data from these measurements provided a stringent test of the theory of finite size-scaling.[15]

With NASA support, experiments have been designed to elucidate critical phenomena in other classes of universality, to explore fundamentally new effects that are observed when a system near its critical point is driven away from equilibrium, both in the bulk and near boundaries.[16] These experiments, which have been promoted to the level of flight readiness, provide a near-term opportunity to obtain a well-defined science return from an available microgravity laboratory. The flight of these experiments would provide insight into the behavior of these critical systems that cannot be obtained on Earth. The Low-Temperature Microgravity Physics Facility (LTMPF) is a multiflight facility designed to attach to the Japanese Experiment Module/Exposed Facility of the ISS.[17] LTMPF has been engineered to support these and other experiments that test fundamental symmetries, such as the Lorentz invariance, discussed in Thrust II above, for many months in a well-controlled cryogenic environment. This facility is approximately 70 percent complete; once completed, it will facilitate these and other experiments in a high state of readiness for flight on the ISS and, possibly, on other platforms. LTMPF is able to support experiments in many thrusts, including superconducting oscillators for tests of relativity; superconducting proof masses for gravitational tests; and cold condensed-matter systems for studies of critical phenomena; and for the study of new ordered phases at low temperatures, as discussed below.

The elucidation of critical phenomena at accuracies that can be obtained only in space, as described above, not only would represent a major scientific advance but also would create the opportunity to apply these scientific results and flight engineering systems to advance the search for new phases and organizing principles of matter.

These important advances will be enabled by NASA's Exploration mission. These advances in turn promise to enable the Exploration mission through the development of new devices that use emerging measurement science that requires the microgravity laboratory.

Superfluid helium droplets (Figure 8.2) provide a convenient microscopic laboratory in which to study the structure and behavior of atoms and molecules, and to search for new phases of matter. Evidence of superfluidity has been observed in nanometer-sized small clusters of hydrogen in superfluid helium droplets using lasers as the experimental probe.[18] The microgravity environment may provide a laboratory where these and other interesting effects can be explored in a droplet that is stable without continuous intervention by external fields, as are required on Earth. External fields are often used on Earth to stabilize and suspend bubbles and drops, and these forces can readily mask interesting new ordered phases of matter and collective phenomena that could be studied if there were no such fields. The weightless laboratory is an important platform on which to explore new structures, interactions, and phases of matter because the systematic effects of gravity are removed and other biasing experimental effects can be controlled. It is important to provide a microgravity laboratory for these studies, since the understanding that is gained will advance our fundamental knowledge of physics, which could be important for future engineered systems.

To understand the importance of this basic knowledge for the development of new engineered systems, consider how basic research on superfluidity has led to the development of new inertial devices, such as new superfluid gyroscopes that operate on the superfluid Josephson effects in ^{3}He and in ^{4}He.[19] These devices may prove useful in future space exploration missions, and in some terrestrial applications. They use dynamical superfluid properties to detect rotations and may someday detect rotation rates that are far smaller than those that can be detected with conventional laser gyroscopes based on the Sagnac effect. Other devices, such as ultrastable blackbody devices that use the superfluid transition in ^{4}He as a fixed-point reference, may be useful in long-duration space radiometry measurements of the cosmic microwave background.[20] Finally, it may be possible to extend the technology developed to support the measurement of critical phenomena in space to enable many other space projects or missions. For example, charged-particle sensors and lightweight superconducting magnets may someday prove useful in detecting and deflecting the "prevailing wind" of dangerous cosmic radiation away from long-duration flight crews on the Moon or in transit to Mars or the outer planets. Such systems, if they can be engineered successfully, would effectively provide a substitute magnetosphere to protect flight crews from lethal charged particle flux once they are outside the protection of Earth's magnetosphere. Our understanding of how to contain and control the low-temperature environment in microgravity enables an entirely new class of superconducting sensors and

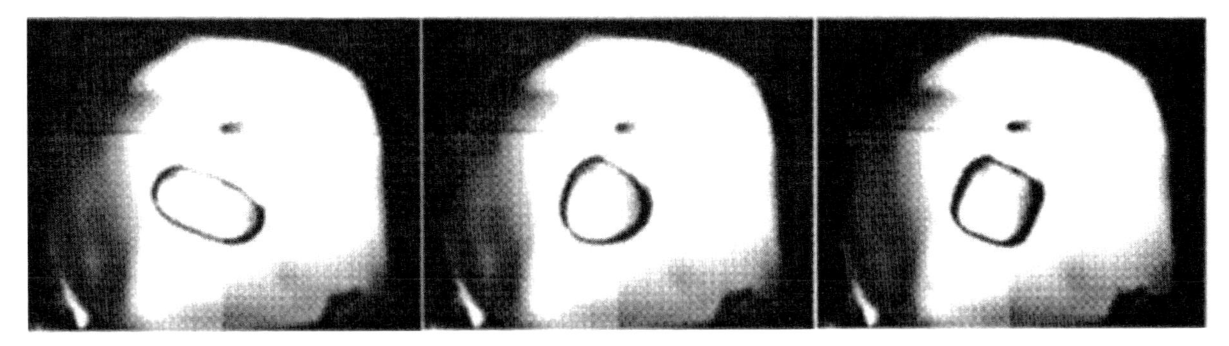

FIGURE 8.2 Levitated He droplet and a crystal boundary. Liquid helium is diamagnetic, so at sufficient values of the product of magnetic field and field gradient, BNB, the magnetostrictive force exceeds the force of gravity on the droplet, causing it to levitate. Once the droplet is levitated, phase change can be studied without the effects of the container, permitting the crystal boundaries between the body-centered cubic and hexagonal close-packed phases of solid helium, such as the boundary shown here, to be observed. SOURCE: With kind permission from Springer Science+Business Media: *Journal of Low Temperature Physics*, Oscillations of charged helium II drops, Volume 110(1/2), 1998, p. 177, D.L. Whitaker, M.A. Weilert, C.L. Vicente, H.J. Maris, and G.M. Seidel, Figure 5.3.

devices that may be used in many enabling applications. Cryogenic systems that operate at higher temperatures also provide methods for collection and preservation of samples from the planets and from the cosmic wind for planetary studies.

AVAILABLE AND NEEDED PLATFORMS

Fundamental physical science can benefit from experiments that are deployed on different platforms. In contrast to other subdisciplines of microgravity, where space-based research can be conducted on a particular platform, fundamental physics is so varied that all foreseeable platform modalities must be considered. For example, many energy-sensitive experiments will seek low Earth orbits on the ISS that record data only when the ISS is well outside the South Atlantic Anomaly. Other high-precision experiments will be ultrasensitive to vibration and other perturbations, requiring a free-flying platform. Still other experiments in complex fluids may benefit from reduced gravity but not zero gravity, making a lunar basing preferable. Yet other experiments that seek to test gravity theories through distant space ranging and tracking may be accommodated on future deep-space probes. Essentially all known space deployment platforms should remain on the table for fundamental physics experiments since their specific science-driven platform requirements cannot be reliably generalized.

Ground-Based Research

Ground-based research can, at low cost, answer fundamental scientific questions and enable space research and applications. It does so by identifying new opportunities for transformative space-based physical science; by resolving measurement and system feasibility issues before larger investments are made on space-based experimental platforms; and by serving as an active core community of experimental and theoretical scientists who carry out and support space-based experiments and interpret the results. Ground-based fundamental physics research in heat, mass, and momentum transport; materials physics; combustion; and granular materials can also help in the design of human flight systems and launch capabilities. As an example, the development of constitutive models for granular flow may enhance the performance of robotic planetary explorers and make them more robust. Another example might be the development of methods of ceramic processing that would combine recycled materials from the spacecraft with lunar regolith to build lunar habitats.

Aircraft and Drop Towers

Aircraft (parabolic zero-gravity flight) and drop towers, which provide a few seconds of microgravity conditions at a time, can test the feasibility and utility of microgravity and assess the prospects for experiments carried out under long-term microgravity or reduced gravity. Some experiments can be completed during a single drop or atmospheric flight. One example of experiments conducted in a transient microgravity environment is the "filament stretching" experiments that determine the behavior of polymeric fluids when they are rapidly stretched to fine filaments undisturbed by the effects of gravity. Another example is the use of the ZARM drop tower in Bremen, Germany, by ESA investigators to test BEC formation in a microgravity environment.

The International Space Station

The International Space Station and the space shuttle cargo bay have already enabled ground-breaking experiments on gelation and phase separation in colloidal suspensions and tests of critical phenomena in the Lambda Point Experiment and the CHeX described above. To make full use of the ISS and its delivery systems for fundamental physics studies, the LTMPF should be deployed. This would enable clock experiments, ultrasensitive measurements of gravitation, and critical point experiments to be carried out in microgravity. Certain aspects of fundamental physics experiments that have flown, or that have been prepared for spaceflight within the fundamental physical sciences programs, could be treated as facilities themselves, which guest investigators could use to multiply the science returns. This could also be done with the ongoing development of laser cooling and atomic physics experiments in space, as well as experiments that test gravity and fundamental symmetries. These user-based facilities could be developed by the laboratory groups that possess the expertise, and the NASA flight centers could anchor the programs and provide limited engineering support and project management for these efforts. Most of the technical development for these and future flight facilities could be done in the university laboratories that developed the measurement science and prototype flight systems.

Free-Flying Spacecraft

While the ISS provides a convenient, accessible laboratory for many microgravity experiments, the environment provided by free-flying spacecraft avoids the negative aspects of gravitational field perturbations from the movement of personnel, radio noise from a plethora of ISS electrical and electronic infrastructure, accelerations from orbital stabilization, conflicting requirements from concurrent experiments, and limitations on experimental parameters imposed by human safety requirements. Thus, free-flying spacecraft may be used when extremely low-noise and low-stray-acceleration environments are required, or when specific orbits are required to obtain the science return. An example of this is the Gravity Probe B spacecraft, which flew in a nearly polar orbit to perform high-resolution tests of Einstein's theory of general relativity. Many of the highest priority experiments in Thrust II (precision measurements of fundamental forces and symmetries), such as the MICROSCOPE and STEP missions, will require dedicated free-flying spacecraft. Free-flying deep-space probes may also be required when it is necessary to locate the experiment far from the Sun, either on a trajectory that is set to leave the solar system or to a Lagrange point.

Lunar or Martian Bases

Lunar or martian bases would be used for fundamental seismographic studies of the Moon or Mars, yielding insight into the interiors of these bodies and their geological history. The compositions of their regoliths, their magnetic fields, and their atmospheric phenomena (in the case of Mars) could be studied from such bases. In the longer term, such bases might also be used as platforms for large telescopes. The lunar regolith might be combined with waste aluminum metal to create new materials on the Moon, as discussed in Chapter 9. In short, a lunar or martian base might someday function as a stable, long-term laboratory for reduced gravity experimentation.

PROGRAM RECOMMENDATIONS FOR EXPERIMENT-SPECIFIC SUPPORT FACILITIES ON VARIOUS PLATFORMS

Fundamental physical science in space is enabled both by dedicated, single-experiment, free-flying platforms, such as Gravity Probe B and STEP, and by specially designed pieces of space hardware that allows researchers to experiment on different systems. Such systems include static and dynamic light-scattering facilities for the study of complex fluids and soft-condensed-matter physics; atomic clock ensembles such as ACES; and future optical-magnetic systems that permit the creation of quantum gases and the study of new physical phenomena in them. Other shared facilities that are important in this research include photographic systems and microscopy facilities incorporating confocal and laser tweezer capabilities. In the past NASA organized teams of researchers to help in the development of special hardware for fundamental physical science in space. In addition to producing useful flight hardware, these teams can also advance the overall state of terrestrial technology in their fields. Collaborations with similar groups in Europe have resulted in highly sophisticated facilities and have motivated further agreements to develop and share future flight facilities. NASA has also enabled the formation of international networks that produce state-of-the-art materials, particularly colloids that would otherwise not have been available.

The panel recommends that a ground-based program be reinstated to support technology development and ground-based science that will enable future flights. The program must continue during the time between major flight platforms in order to maintain the technology base and the intellectual community that is essential to the advancement of programs within NASA and that has historically contributed to the technological strength of the United States.

Flight facilities that currently support studying the physics of complex fluids and soft condensed matter should be continued, and LTMPF should be completed. Experiments that have flown or that have been prepared for spaceflight should be treated as facilities themselves, and guest investigators should be encouraged to use the same hardware, increasing the science return. This should also be done for the facilities being developed for laser cooling and atomic physics experiments in space as well as for testing gravity and fundamental symmetries. These user-based facilities should be developed by the laboratory groups that possess the expertise, and the NASA flight centers should anchor the programs and provide engineering support and project management for them. Most of the technical development efforts for these and future flight facilities should be done in the university laboratories where the measurement science and prototype flight systems were and are being developed.

Because of the close coupling between fundamental and applied aspects of fluids physics, NASA funding for applied and mission-enabling research, if restored, would be most fruitful if combined with well-targeted support for fundamental and mission-enabled research on fluids physics and complex fluids. The most important topics in complex fluids and fluids physics fundamental research are identical to those for which applied research is needed: multiphase flow, capillary-driven flow, and instabilities, especially in microgravity.

In general, NASA should support fundamental research that generates conceptual breakthroughs, that has a high impact on science, that can accelerate applied mission-enabling research, and that can increase public awareness of science generally and of NASA's missions and objectives in particular. To develop a few highly appropriate space-based fundamental experiments, a much larger repertoire of ground-based experiments (perhaps 100 to 150 ongoing studies) should be supported. The panel recommends that fundamental studies be carefully targeted and should meet four criteria: (1) the science involved should support experiments for which microgravity is required, (2) it should be relevant to NASA's missions, (3) it should encourage interactions with international partners, and (4) it should have an impact on education.

Finally, the fundamental physics program must remain agile enough that it can be pursued at low cost and be capable of generating interesting and unexpected new physical observations. Examples of such observations include the Pioneer anomaly and the continuing collection of tracking data from Voyager as it passes the heliopause and leaves the solar system, exploring and testing fundamental physical phenomena over a wide range of length scales. These particular efforts require only the resources of the Deep Space Network and a small effort within the ground-based program, and do not represent a major allocation of resources.

RESEARCH PROGRAM RECOMMENDATIONS

The panel found that the highest-priority areas of research in the area of fundamental physical sciences at NASA should be (1) soft-condensed-matter physics and complex fluids, (2) precision measurements of fundamental forces and symmetries, (3) quantum gases, and (4) critical phenomena. These areas embody important scientific objectives that can be studied only in the laboratory of space. NASA's new program in fundamental physical (FP) sciences in space should include these four areas, which the panel also calls thrust areas. Other important areas of physical science in space, including fluids physics, materials, and combustion, are described in Chapter 9, which covers the applied physical sciences.

Recommended Program Element 1: Research on Complex Fluids and Soft Matter (FP1)

Complex fluids and soft condensed matter are excellent candidates for study in the microgravity laboratory. They are materials with multiple levels of structure and their softness typically results from the large size of the basic units. They are easily deformed, and their statics and dynamics are governed by surface tension and entropic forces. On Earth, these weak forces are easily overwhelmed by gravity. Colloids, polymer and colloidal gels, foams, emulsions, soap solutions, and the like are particularly susceptible to gravity owing to the gradients that are formed in their properties under gravity. Microgravity provides a unique opportunity to eliminate these gradients and permit studying the long-term dynamics of such systems free from such gravitational interference. Microgravity is required as well to probe the basic properties of these materials and to use the materials as models to explore other phenomena. Experiments on Earth are hampered by sedimentation, flows, and the suppression of thermodynamic fluctuations. Similar issues emerge for colloids, gels, and dusty plasmas, whose density and morphology under gravity are height dependent. Similarly, the stress chains and yield properties of granular materials are dependent on height and sensitive to the magnitude of gravity.

Recommended Program Element 2: Research That Tests and Expands Understanding of the Fundamental Forces and Symmetries of Nature (FP2)

Space offers unique conditions to address important questions about the fundamental laws of nature, with sensitivity beyond that of ground-based experiments in many areas. In particular, high-precision measurements in space can test relativistic gravity and fundamental particle physics and related symmetries in ways that are not practical on Earth. Atomic clocks in space, probably optical but potentially microwave too, are useful in the study of time variation of the fundamental constants and have many more applications. Promising theoretical approaches to quantum gravity and physics beyond the currently accepted standard model of fundamental physics typically predict new forces, violations of fundamental symmetries, or time-varying physical constants. Such novel effects provide distinct signatures for precision experimental searches that are often best carried out in space.

Recommended Program Element 3: Research Related to the Physics and Applications of Quantum Gases (FP3)

A remarkable range of different physical phenomena can be investigated using quantum gases such as BECs and degenerate Fermi gases; many of these investigations can be done only in space. Aspects of the formation of BECs and their intrinsic quantum properties represent one rich class. Research on vortex formation and relaxation can be used to probe phase-transition models of the early universe and can also give insight into the structure of neutron stars. As picokelvin BECs are realized in a space-based laboratory, scientists will be able to create and understand the competing forms of order that often govern the complex structure observed in the world around us. One of the remarkable aspects of quantum gases, and BECs in particular, is their exquisite sensitivity to interparticle interactions. Because of the low thermal energies and the large size of their quantum wavepackets, the system becomes sensitive to tiny but important long-range forces and interactions. When combined with optical lattices, this sensitivity allows replication and investigation of the building blocks of complicated matter such as magnetic and electric materials.

Recommended Program Element 4: Investigations of Matter in the Vicinity of Critical Points (FP4)

Over that past two decades the microgravity environment has given us a better understanding of the behavior of materials in the vicinity of thermodynamically determined critical points. With NASA support, experiments have been designed to elucidate critical phenomena in other universality classes and to explore fundamentally new effects that are observed when a system near its critical point is driven away from equilibrium, both in the bulk and near its boundaries. These experiments have been designed and brought to the level of advanced flight readiness; this should allow obtaining a well-defined science return from an available microgravity laboratory. The flight of these experiments would provide insight into the behavior of these critical systems that cannot be obtained on Earth.

These four recommended program areas, or thrusts, share four important strengths: (1) they have significant potential to address some of the grand scientific challenges of our time, (2) they are synergistic with other NASA needs, (3) they have a great need for access to space, and (4) they have significant potential to affect the terrestrial research enterprise.

PROGRAMMATIC CONCLUSIONS, FINDINGS, AND RECOMMENDATIONS

A healthy and sustainable program of fundamental physical sciences in space will require a mix of multi-user and single-experiment space-based facilities and human-tended and free-flyer platforms. It also will require a strong ground-based program that is community driven and that allocates resources based on peer review, including the resources for flight experiments. The ground-based program will serve three essential functions. First, it will identify new opportunities for transformative, space-based physical science. Second, it will foster advances in instrumentation that are essential to accomplishing the ambitious objectives of future flight programs. Third, once the flight experiment is selected, the ground-based program will provide an active core community that can conduct competitive peer reviews to select the efforts, including the development of relevant theory, that might best support the flight experiment and ensure that fundamental understanding of physical sciences is advanced.

It is in the nature of fundamental research that its timescale is less well defined than the timescale for applied research. Nonetheless, achieving the program set out here will require that in the next 2 years NASA initiates and utilizes a peer-review system to select investigators for a ground-based research program encompassing each of the four fundamental physics thrusts identified in this chapter. In addition, experiments and facilities that could rapidly be made available for a possible return to flight program status need to be peer reviewed. NASA will need to continue to sponsor an international symposium series centered on the opportunities and viability of research missions within the fundamental physical sciences in space.

Looking ahead, a successful program 3 to 4 years from now will have NASA begin evaluating proposals for space-based fundamental physics—including those that use both free-flyer platforms and the ISS—to select compelling research that has demonstrated flight viability and a clear need for microgravity as demonstrated by the ground-based program. This effort will probably build on existing connections with space-research programs in other countries throughout the world and establish new ones. During the following 5 years it will be important to begin a process to reassess the direction of research, and to adjust the program priorities if necessary. This may be accomplished in an open and transparent manner through the international symposium series mentioned above.

Recommendation 1: A successful exploration program in the physical sciences necessitates first of all a ground-based fundamental physical sciences program. Such a ground-based program must also eventually support flight commitments in the fundamental physical sciences.

Recommendation 2: Flight experiments and facilities that could rapidly be made available for a return to flight should be peer reviewed. To justify this recommendation the panel points to the numerous existing experiments and supporting facilities that are at an advanced stage of flight readiness.

Recommendation 3: In funding projects, NASA should seek partnerships with other agencies and other nations. Research in fundamental physical science is supported by many federal agencies in the United States and is widely supported internationally.

Recommendation 4: NASA should build a program in fundamental physical sciences sufficiently large to attract prominent scientists, both flight- and ground-based, to create a vibrant ground-based program and to generate potential space-based missions.

Past experience in the NASA microgravity program suggests that a critical mass of 100 to 150 funded investigators would provide coverage of all the physical sciences of importance to NASA, engender synergy among investigators, and ensure spirited and regular meetings of the investigators. It would also provide a steady flow of projects for transitioning to flight. A program of this size is also consistent with that of other successful research programs in the physical sciences in the United States and other countries, where at least 100 to 150 investigators are needed to sustain a healthy and productive enterprise.

REFERENCES

1. Chaikin, P., and Nagel, S. 2003. Report on the NASA Soft and Complex Condensed Matter Workshop, NASA/CR-2003-212618. NASA, Washington, D.C.
2. National Research Council. 2003. *Assessment of Directions in Microgravity and Physical Sciences Research at NASA*. The National Academies Press, Washington, D.C.

3. Zhu, J.X., Li, M., Rogers, R., Meyer, W., Ottewill, R.H., Russell, W.B., and Chaikin, P.M. 1997. Crystallization of hard-sphere colloids in microgravity. *Nature* 387:883-885.
4. Anderson, M.H., Ensher, J.R., Matthews, M.R., Wieman, C.E., and Cornell, E.A. 1995. Observation of Bose-Einstein condensation in a dilute atomic vapor. *Science* 269(5221):198-201.
5. Davis, K.B., Mewes, M.-O., Andrews, M.R., van Druten, N.J., Durfee, D.S., Kurn, D.M., and Ketterle, W. 1995. Bose-Einstein condensation in a gas of sodium atoms. *Physical Review Letters* 75(22):3969-3973.
6. Cornell, E.A., and Wieman, C.E. 1998. The Bose-Einstein condensate. *Scientific American* 278(3):40-45.
7. Pitaevskii, L.P., and Stringari, S. 2003. *Bose-Einstein Condensation.* Clarendon Press, Oxford.
8. Greiner, M., Mandel, O., Esslinger, T., Hänsch, T.W., and Bloch, I. 2002. Quantum phase transition from a superfluid to a Mott insulator in a gas of ultracold atoms. *Nature* 415(6867):39-44.
9. Meystre, P. 2001. *Atom Optics.* Springer-Verlag, New York, N.Y.
10. Barmatz, M., Hahn, I., Lipa, J.A., and Duncan, R.V. 2007. Critical phenomena in microgravity: Past, present, and future. *Reviews of Modern Physics* 79:1-52.
11. Stanley, H.E. 1971. *Introduction to Phase Transitions and Critical Phenomena.* Oxford University Press, Oxford, U.K., and New York, N.Y.
12. Wilson, K.G. 1971. Renormalization group and critical phenomena. I. Renormalization group and the Kadanoff scaling picture. *Physical Review B* 4:3174-3183.
13. Wilson, K.G. 1971. Renormalization group and critical phenomena. II. Phase-space cell analysis of critical behavior. *Physical Review B* 4:3184- 3205.
14. Lipa, J.A., Nissan, J.A., Stricker, D.A., Swanson, D.R., and Chui, T.C.P. 2003. *Physical Review B* 68:174518.
15. Lipa, J., Swanson, D.R., Nissen, J.A., Geng, Z.K., Williamson, P.R., Strieker, D.A., Chui, T.C.P., Israelsson, U., and Larson, M. 2000. *Physical Review Letters* 84:4894.
16. Lammerzahl, C., Ahlers, G., Ashby, N., Barmatz, M., Biermann, P. L., Dittus, H., Dohm, V., Duncan, R., Gibble, K., Lipa, J., Lockerbie, N., Mulders, N., and Salomon, C. 2004. Review: Experiments in fundamental physics scheduled and in development for the ISS. *General Relativity and Gravitation* 36:615-649.
17. Larson, M., Croonquist, A., Dick, G.J., and Liu, Y.M. 2003. The science capability of the Low Temperature Microgravity Physics Facility. *Physica B: Physics of Condensed Matter* 329:1588-1589.
18. Toennies, J.P., and Vilesov, A.F. 2004. *Superfluid Helium Droplets: A Uniquely Cold Nanomatrix for Molecules and Molecular Complexes.* Wiley-VCH Verlag GmbH and Co. KGaA, Weinheim.
19. Simmonds, R.W., Marchenkov, A., Hoskinson, E., Davis, J.C., and Packard, R.E. 2001. Quantum interference of superfluid ^{3}He. *Nature* 412:55-58.
20. Green, C.J., Sergatskov, D.A., and Duncan, R.V. 2005. Demonstration of an ultra-stable temperature platform. *Journal of Low Temperature Physics*138:871-876.

9

Applied Physical Sciences

Applied physical sciences are central to many key exploration technologies. Many of the design challenges of new exploration technology systems are addressed by applied physical sciences research on fluid physics, combustion, and materials, as described in this chapter. This research will enable new exploration capabilities and yield new insights into a broad range of physical phenomena in space and on Earth. Applied physical sciences research will result in fundamental approaches for improved power generation, propulsion, life support, and safety.

A broad range of space-related technologies are advanced by key areas of research in the applied physical sciences:

- *Fluid physics.* Flows involving multiple phases (e.g., liquids and vapors) are present in many current and proposed space systems. Moreover, multiphase systems and thermal transport processes are enabling for proposed human exploration missions by NASA.[1] Complex fluids, including the granular physics associated with the flow and compaction of soil found on planetary bodies, present engineering challenges on topics ranging from habitat design to the degradation of life support systems because of dust.
- *Combustion.* Fire safety is integral to astronaut safety; fires behave differently on Earth than in space, in part because of differences in the ways that materials burn and gases flow. The understanding of material flammability in reduced gravity is incomplete. Concerns also exist about the effectiveness of fire detection and fire suppression systems designed for reduced gravity. An improved understanding of combustion in reduced gravity can lead to more efficient use of combustion processes on Earth, a cleaner environment, and better fire safety.
- *Materials science.* Lightweight materials, self-healing materials, and other new materials tailored to NASA's demanding missions are central to the success of future human and robotic exploration. Research in reduced gravity can lead to new insights about how various processing methods affect the internal structure and, ultimately, the properties of materials.

Reduced gravity provides a unique environment for fundamental research in fluid physics, combustion, and materials science, which are central to the robust performance of current and proposed space systems. On Earth, gravity induces fluid motions that may mask important phenomena under investigation. For example, multiphase flow regimes and the performance of heat transfer phase-change systems on Earth are very different than they are in the absence of gravity. Flame ignition, propagation, and quenching are also affected by gravity. If certain combustion processes related to the presence of gravity can be excluded, then it is possible to extract important, fundamental data and insights relevant to similar systems both on Earth and in space. This includes information on chemical reaction rates, diffusion coefficients, and radiation coefficients, as well as insights into flame structures, soot formation, and droplet combustion. Furthermore, the production and processing of new materials often involve a vapor or liquid. The examination of crystal growth and solidification processes in reduced gravity can provide new insights into the manner in which a liquid or vapor transforms into a crystal and the accompanying pattern formation processes, such as dendritic and cellular growth.

This chapter presents recommendations for applied physical sciences research that *enables* space exploration and *is enabled by* space exploration. These recommendations are based on the expertise of the panel, on approximately 40 white papers submitted by the scientific community, on briefings presented to the panel, and on other documents reviewed by the panel. Chief among these documents were two previous reports from the National Research Council (NRC). The first, *Microgravity Research in Support of Technologies for the Human Exploration and Development of Space and Planetary Bodies*,[2] known as the HEDS report, contains an extensive discussion of the central role that research in the applied physical sciences plays in enabling space exploration, as well as an encyclopedic survey of important exploration technologies and the physical phenomena underlying these technologies. It discusses many key technology areas, including power generation and storage, space propulsion, life support, hazard control (fire and radiation safety), and materials production and storage. Drawing on the many technologies in these broad areas, the HEDS report also presents extensive background on fluid physics, on topics such as interfacial phenomena, multiphase flow, and heat transfer; combustion and fire safety; and materials. The second report, *Assessment of Directions in Microgravity and Physical Sciences Research at NASA*,[3] focuses on research that is enabled by reduced-gravity research platforms supplied by the National Aeronautics and Space Administration (NASA). As with the HEDS report, this assessment contains an expansive discussion of background information and important research questions in fluid physics, combustion, and materials science. Both reports contain additional supporting information that, due to space limitations, is not included in this chapter.

Recommended research portfolios in fluid physics, combustion, and materials science are presented below. In general, to make the most out of each experimental opportunity, research should include a combination of reduced-gravity experiments, numerical simulation, and analysis. The timeline for each of these portfolios was constructed assuming that, over the next 10 years, the International Space Station (ISS) will be available with adequate crew time and up-mass and down-mass capabilities, current ground-based facilities will remain available, and funding will be available to support an expanded research program. As noted below, most of the recommended research should be structured to facilitate the development of related critical technologies, as detailed in Chapter 10, "Translation to Space Exploration Systems" (see Tables 10.3 and 10.4). The rest of the recommended research (i.e., in the areas of complex fluid physics, numerical simulation of combustion, and fundamental materials research) is broadly applicable. Although this research is generally applicable to research topics listed in Tables 10.3 and 10.4, it is not easily focused on the specific critical technologies associated with those topics.

This chapter concludes with a summary of the recommended research, a review of key facilities that will enable recommended research, and a discussion of programmatic recommendations.

FLUID PHYSICS

The panel considered many interesting NASA-related research issues in fluid mechanics (e.g., aerodynamics, hypersonic flows, and plasma dynamics). This section focuses on the gravity-related research issues of most crucial importance to NASA's future crewed and uncrewed missions. Areas of particular interest are reduced-gravity multiphase flows, cryogenics, and heat transfer: database and modeling; interfacial flows and phenomena; dynamic granular material behavior; granular subsurface geotechnics; dust mitigation; and fundamental research in complex fluid physics. Each of these fields encompasses a myriad of individual phenomena. The targeted topics, of highest priority for NASA, are both enabling to, and enabled by, NASA's access to reduced-gravity environments.

As discussed below and in Chapter 10, advances in multiphase flow and heat transfer provide enabling technology for many of NASA's proposed crewed missions.[4] A recent survey of NASA and industry identified high-priority gravity-related challenges such as the following:[5] (1) storage and handling of cryogens and other liquids,*

*"Other liquids" includes liquid oxygen, helium, and hydrogen, which are used for breathing, cooling, and propulsion, as well as non-cryogenic fuels such as hydrazine.

(2) life support, (3) power generation, and (4) thermal control. Specific examples include fluid management for both crewed and robotic missions; water processing for plants and people, including reclamation, recycling, hydrolysis, and hydration; mission-enabling phase-change technology for power production and thermal management; medical fluids; and food.

Experiments in both applied and fundamental fluid physics are needed. Key fluid physics issues include reduced-gravity multiphase flow, cryogenics and heat transfer, interfacial flows† and phenomena, dynamic granular material behavior, granular subsurface geotechnics, dust mitigation, and complex fluid physics. Research into all of these issues is uniquely enabled by NASA reduced-gravity facilities.

Space systems involving multiphase flows face significant design challenges due to the strong dependence of such flows on gravity.[6] In particular, reduced gravity poses unique challenges when weak forces that are often masked by Earth gravity dominate fluid behavior in unexpected ways.[7,8,9] For reasons detailed in the section entitled "Thermal Management" in Chapter 10, system designers have often avoided designs that use multiphase phenomena pending additional research and experimental demonstrations.[10] An improved understanding both of the key forces involved in multiphase flows and of the consequences of multiphase flows is necessary to prevent system failures and to enable designers to develop heat exchangers that employ multiphase flows (e.g., enhanced evaporation and condensing surfaces for evaporative and condensing heat exchange). Such an understanding could lead to better and more robust design principles and could dramatically enhance system performance and reliability while reducing system volume, mass, and cost. Attaining such knowledge is a high priority, as it is often mission enabling.[11] In addition, important research along these lines is uniquely enabled by access to reduced gravity (e.g., on the ISS and in spacecraft, sounding rockets, aircraft, and drop towers).

Robust phase separations by means of interfacial flows will be essential in numerous water-processing systems for long-duration life support in reduced gravity. A basic understanding of and the ability to predict the performance of multiphase, cryogenic flows and phase separation are essential to the design of the on-orbit propulsion fueling depots described in Chapter 10. Additionally, phase-change systems achieve competitive, if not mission-enabling, performance advantages for high power production (e.g., Rankine cycle phase separations) and thermal control systems (e.g., high-performance heat pipes), and they are inherent in space transportation systems for cryogenic propellants (e.g., ullage positioning, priming, and venting).[12] Although many classifications are possible,[13,14,15] the most challenging gravity-dependent vapor-liquid interfacial flows are as follows: (1) high-inertia multiphase flow and heat transfer and (2) interfacial-induced flows and phenomena.

The nature of gravity-dependent interfacial flows depends on the relative importance of inertia, acceleration, and fluid and system properties.[16] These flows may also be complicated by phase-change heat transfer, chemical and biological reactions, particulates, contaminates, and so on. System scale-up may be a challenge in partial gravity (on the Moon or Mars), but the microgravity environment on the ISS or other spacecraft is the most problematic. Many of these challenges may be dramatically reduced or eliminated for robotic missions (when no life support is necessary) or when an artificial-gravity countermeasure is used. In any event, the importance of research on multiphase flows and processes for space applications has been thoroughly, if not exhaustively, documented in previous studies and white papers.[17-21]

Fluid flow and heat transfer are interrelated processes in which gravity plays an important role.[22] During boiling heat transfer, for example, the formation, growth, and departure of a vapor bubble during nucleation on a heated surface induce motion in the host liquid in the presence of a gravitational field. The induced flow improves heat transfer from the solid surface. The relevant phase-change processes are boiling under pool and forced-flow conditions, evaporation at vapor-liquid interfaces, and condensation. Research literature, unfortunately, contains only very limited data on pool boiling in reduced gravity. Thus, available correlations and models are unable to provide reliable data on nucleate boiling and critical heat flux in reduced gravity.

Flow boiling occurs when liquid is pushed over the heated surface by external means—for example, with a pump. A few studies of flow boiling under reduced gravity have been performed.[23] In reduced gravity, vapor bubbles become much larger as local coalescences occur on the heater surface, and they rarely detach from the solid. Detailed data from systematic experimental studies are needed to validate numerical simulation models for the development of the flow regimes in a heated channel and prediction of pressure drops and heat-transfer rates.

---

†"Interfacial flows" are those in which an interface between two liquid or gas phases and the forces associated with the presence of the interface are important in determining the resulting flow.

During nucleate boiling, evaporation occurs from a thin liquid layer that forms between the vapor-liquid interface and the solid wall, as well as around the vapor-liquid interface of the vapor bubble located in the superheated liquid layer adjacent to the wall. Thin-film evaporation is associated with very high heat fluxes, and so an understanding of thin-film behavior in reduced gravity is extremely important for the development of mechanistic models for boiling and for other applications in propellant storage, life support systems, and dissipation of waste heat. Physical models for thin-film evaporation should include the impact of a wide variety of relevant forces (e.g., disjoining pressure, viscous force, inertia force, capillary force, gravity force, and recoil pressure). Capillary pressure gradients resulting from temperature and/or concentration gradients (Marangoni convection) and wettability of the surface also play an important role. The stability and rupture of thin films with or without external forces such as electric, magnetic, and acoustic forces should also be considered.

Condensation is important to thermal management, power systems, in situ resource utilization (ISRU), and propellant management, but little data on condensation in tubes in reduced gravity exist in the literature. Such data are badly needed to validate numerical simulation tools that could be used for the design of space-based two-phase heat rejection systems.

A better understanding of granular physics would have tremendous practical importance and is critical for enabling the human or robotic exploration of the Moon or Mars for the following reasons: (1) little is known about Moon and Mars regolith (soil) except that it is unlike soils found on Earth; (2) exploration of the Moon and Mars requires the direct interaction of human explorers and/or equipment with granular matter; (3) fundamental equations that define the influence of gravity on the compaction and flow of granular materials are lacking; (4) granular matter is critical in life-threatening situations ranging from the collapse of structures to lung disease caused by the inhalation of dust; (5) extraterrestrial exploration is enabled by habitat construction, ISRU, mining, and surface transportation, all of which involve granular matter; (6) technologies that handle granular materials, such as heavy construction equipment and foundation construction processes, cannot be directly transferred from existing terrestrial applications to future extraterrestrial applications; and (7) electrostatic forces can dominate the behavior of granular matter in reduced gravity and almost complete vacuum.[24,25]

Research in Support of NASA's Exploration Missions

Reduced-Gravity Multiphase Flows, Cryogenics, and Heat Transfer: Database and Modeling

In reduced gravity, the limitations on the empirically based predictive methods used on Earth for relatively high-speed multiphase flows do not allow NASA to exploit the advantages of using multiphase technology in space. This is because there is essentially no reliable database for flow regimes, void fraction, two-phase pressure drop, and wall heat transfer in reduced gravity. Therefore, a new predictive capability and design methodology need to be developed. In particular, physically based multiphase thermal-hydraulic models will be required to quantify accurately the effect of gravity (on Earth, Mars, the Moon, and in space). To be effective, such models must necessarily be developed with, and assessed against, appropriate small-scale, reduced-gravity data, and they must be capable of accurately scaling up these data to the relatively large multiphase systems and processes required on NASA's future human exploration missions.

In multiphase gas and liquid flows, the interfacial structures, flow regimes, and bubble sizes depend on the internal length scale mostly determined by the Taylor wavelength. In normal gravity flow, this length scale is on the order of several millimeters, and multiphase flow can easily reach near-equilibrium interfacial structures called flow regimes. Therefore, the use of the standard flow regime map and regime-dependent constitutive models has been quite successful. However, in reduced gravity, this length scale can be an order-of-magnitude larger than in normal gravity. Under such conditions the evolution of the interfacial structures is much more prolonged and complicated. These interfacial structures strongly affect the formation of flow regimes, phase separation, nucleation characteristics, bubble dynamics, vapor addition, and critical heat flux. A simple dynamical model of the interfacial structures would be a useful tool. This model should be compatible with standard computational multiphase fluid dynamic (CMFD) codes using two-fluid or multifield formulations.

Phase separation and distribution are the key gravity-dependent multiphase fluid mechanic phenomena that must be understood and accurately modeled to meet the needs of future systems. For forced fast flows beyond the natural wicking rates of capillary systems, separation and stratification of vapor and liquid phases occur when a multiphase mixture is accelerated. This property can be used to advantage when designing active and passive phase separators, but it can also cause problems in devices with complex geometries, such as parallel channel arrays, outlet plenums, and conduit fittings. Significantly, many phase separators are sensitive to gravity because of the phase distribution that enters the separators. In any event, the ability to predict phase distribution (i.e., flow regimes) accurately is essential for any new three-dimensional analytical or numerical models developed by NASA.

Pronounced lateral phase distribution occurs in multiphase conduit flows on Earth.[26,27] This phenomenon also occurs in microgravity,[28,29,30] but it behaves quite differently, strongly influencing phase separation, pressure drop, and phase-change heat transfer. Therefore, an understanding of the evolution of interfacial structure under reduced gravity is required.

Concerning heat transfer, the limited data available indicate that, for pool boiling, reduced gravity can enhance nucleate boiling heat transfer, but it can also significantly reduce critical heat flux.[31] Nevertheless, important scientific questions remain concerning the effect of gravity on boiling and condensation phenomena, including the surface nucleation of bubbles, bubble dynamics (bubble growth, merger, departure, and post-departure trajectory), phasic structure near a heated surface, rate of heat transfer, critical or dry-out heat flux, and both drop-wise and film-wise modes of condensation heat transfer. Also, as discussed below, related thin-film and interfacial phenomena, including surface wetting, need further study. Experiments to support model development will need access to reduced gravity.

As noted in Chapter 10, two-phase forced convective heat transfer is essential for high-power thermal management systems; NASA is likely to use this mode of heat transfer for energy production and utilization if it is available. In contrast to single-phase (gas or liquid) heat transfer, relatively little is currently known about the effect of gravity on heat transfer within two-phase systems during forced convective boiling and condensation. Detailed study of the effect of gravity on the forced convective condensation and boiling curves is particularly needed, especially with respect to the ebullition cycle (i.e., the formation and motion of vapor bubbles during surface boiling), critical heat flux, and the heat transfer during transition, film boiling, and quenching. In addition, the effectiveness of heat transfer enhancement devices (e.g., twisted ribbons) should be assessed in appropriate reduced-gravity experiments. The output of all of these heat transfer studies should be used to produce a new thermal design basis (i.e., models and correlations) that is valid in reduced gravity and for geometries suitable for NASA's future missions.[32,33]

In order to use multiphase technology extensively in space, NASA must develop reliable, physically based predictive capabilities with respect to effect of gravity on multiphase systems and associated transport processes. These models should be based on detailed numerical simulations and/or reduced-gravity data (which currently are almost nonexistent); they should be capable of accurately scaling up so that they can be used to design and analyze hardware of interest on NASA's spacecraft and extraterrestrial habitats. Physically based, three-dimensional, two-fluid models of multiphase flow and heat transfer have been developed by rigorously averaging the Navier-Stokes equations of fluid mechanics and the associated mass and energy conservation equations.[34,35] This results in a set of equations that can be efficiently integrated numerically. However, important physics is lost during the averaging process, and this information must be re-introduced into the two-fluid model through the use of mechanistic closure relations to describe the various interfacial and wall transfers mathematically.[36] Once this is done, the resultant multiphase model can be efficiently evaluated using a suitable CMFD solver such as NPHASE.[37] CMFD models of this type are widely used on Earth for transient and steady-state analysis of multiphase flow and heat-transfer phenomena in industrial systems and processes (e.g., nuclear reactors, chemical reactors, oil wells, etc.). However, these two-fluid models use closure relations that were developed for conditions on Earth; they are unreliable for applications in reduced gravity. Appropriate new closure relations can be developed, but this will require detailed measurements in reduced gravity along with numerical "data" from direct numerical simulation (DNS) or other numerical simulation techniques (e.g., lattice Boltzmann techniques) of multiphase flows.

Fortunately, recent advances in computational hardware (e.g., high-speed massively parallel processors) have enabled the detailed simulation of multiphase flow and heat-transfer phenomena using DNS.[38,39,40] This is an important new development, which allows near-first-principle predictions to be made. For example, in recent turbulent multiphase flow simulations, all interfaces were explicitly tracked, and the computational mesh was fine enough to resolve the turbulence structure as well as all significant mass, momentum, and energy transfers at the interfaces and at the wall of the conduit without the need for phenomenological closure relations.[41] Although DNS is numerically intensive, it is within the current state of the art and produces detailed three-dimensional results that quantify the effect of gravity on multiphase flow and heat transfer. Moreover, in principle, detailed numerical results can also be obtained using molecular simulation techniques,[42,43,44] but this approach has not yet been as thoroughly developed as DNS has for multiphase flow and heat-transfer applications. In any event, either DNS or molecular simulation results can be appropriately averaged and used as "data" in conjunction with physical data to help develop the closure relations needed by CMFD models.[45] CMFD will be a very effective tool for NASA to perform analysis and design studies for various engineering systems and technologies. However, for CMFD to predict multiphase flow behaviors effectively in reduced gravity, a dynamical model for the evolution of interfacial structures should be developed and implemented. The conventional approach using only a flow regime map should not be applied. The coupling of a CMFD approach to heat transfer and vapor generation at the wall is essential. Significantly, CMFD models require much less computational power than do DNS and molecular simulations, and thus CMFD models can be used more readily to design and analyze relevant multiphase systems and processes. Moreover, DNS or molecular simulations can also be used to analyze directly various phenomena that may be important in reduced gravity but not on Earth. For example, disjoining pressure-induced forces (i.e., van der Waals forces) at the contact line in a horizontal conduit leads to a remarkable stratified-to-annular flow regime change when going from Earth gravity to reduced gravity.[46] Thus, DNS or molecular simulations can also provide the insight needed to develop accurate three-dimensional multiscale CMFD models that are much more numerically efficient than DNS or molecular simulations and yet capture the same phenomena. This Grand Challenge type of research and development approach is consistent with the recommendations in previous NRC studies[47] and past international workshops on scientific issues in multiphase flow and heat transfer.[48,49]

Interfacial Flows and Phenomena

Numerous inadvertent or purposeful reduced-gravity multiphase flows are driven by gradients in surface tension caused by temperature, concentration, wetting conditions, and other factors, or by force fields involving pressure, shear forces, electric fields, magnetic fields, acoustics, acceleration, and so on. These are multiphase flows in which the influence of surface tension in the flow direction is strong, including phenomena relating to bubbles, drops, and nucleation. An extensive variety of flows and phenomena are represented here and either are present or find application in virtually *all* aspects of fluid systems aboard spacecraft. These systems include propellant management systems, phase distributions for power cycles, thermal control systems (e.g., heat pipes), water recycling for life support (for plants and crew), and other systems. Important challenges for research along these lines include the identification and assessment of key NASA application-specific phenomena. A prime example might be heat transfer due to temperature gradients near moving contact lines in cryotanks with small amounts of non-condensable gases.[50] Such studies contain elements of fundamental discovery, as well as inspiration for improved concepts in which concurrent increases in technology readiness level (TRL) should be expected. In any event, this research should be able to produce verified models and tools for advanced system design and analysis.

Spontaneous reduced-gravity interfacial flows, including wicking, can be exploited to control large quantities of liquids in reduced gravity. Such flows provide a passive means of liquid handling, enhancing possibilities for robust (no moving parts) primary or redundant solutions to reduced-gravity fluids management problems. Specific mission-enabling research that requires access to reduced gravity includes surface spreading (partial wetting and non-wetting), the coalescence of drops or bubbles, the rupturing of liquid films, local and global equilibriums, and stability, among other topics. Key challenges include passive phase separation and models for cryogenic fluid management and liquid handling for life support systems. Specific problems include complications associated with complex geometries and surfaces, heat and mass transfer, reactions, surfactants, contaminants, and moving contact-line boundary conditions, particularly for systems with partial wetting. With regard to the storage of cryogenic fuels, the impacts of length scale, fluid-structure interactions, and other considerations are important. Of practical concern, particularly for water processing, is the ability to predict system performance with confidence despite widely varying and perhaps poorly defined wetting conditions.[51]

Global multiphase system response phenomena are also important. In particular, it is well known that there can be global interactions among the various interconnected components in phase-change systems on Earth, which can lead to system-wide instabilities and failures. Limiting global instabilities include flow excursion instability, density-wave oscillations, and pressure-drop oscillations.[52] These are relatively low-frequency instabilities that can cause large-amplitude flow changes. These instability modes are relatively well understood for conditions on Earth,[53] but it is not yet clear how they will manifest themselves in multiphase systems in reduced gravity.[54] Preliminary analysis indicates that there may be a significant geometry-dependent effect of gravity,[55] but long-duration data in reduced gravity are needed to assess these results. In addition, development of new analytical models is needed for predicting and scaling up the observed system instabilities. The transient three-dimensional CMFD models previously discussed can be used for this purpose, but it appears that simpler, one-dimensional drift-flux models may be sufficient.[56] Other flow-stability problems that warrant further investigation include the following: freeze-thaw cycles, flow excursions, system start-up transients, and cross-contamination.

Dynamic Granular Material Behavior

During the landing and launching of a spacecraft, its rocket exhaust directly interacts with the regolith, causing a spray of material that can damage the spacecraft or nearby structures and reduce visibility.[57] Although the Apollo and Viking missions included considerable research concerning blast effects on planetary regolith, questions remain with regard to scaling laws for erosion craters, jet plumes in a vacuum, the extent of spray, and the interaction between an impinging body or jet and a heterogeneous regolith. Rovers for future surface exploration missions would likely be designed for a range of 100 km or more. The mobility of wheeled vehicles is a direct function of the wheel and terrain interactions. Challenges include improving the understanding of rolling wheel contact behavior along with multiscale modeling of terrain and wheel interaction under reduced-gravity conditions, the potential for vehicles to kick up loosely packed regolith that may contaminate or damage instruments or the vehicles themselves, environmental impact of vehicles on virgin regolith, designing for wheel contacts with various loads, and detailed boundary-traction models at reduced gravity and for unusual materials (both the wheel surface and the soil). An interesting inverse problem is that of extracting geotechnical information about the soil based simply on footprints and vehicle tracks.

Granular Subsurface Geotechnics

Temporary and/or permanent structures on or beneath the surface of the Moon and Mars will be needed for exploration and settlement. For example, buried structures could be used to provide radiation shielding. Lunar and terrestrial soils have different geologic origins: lunar soils have been "weathered" by meteoroid bombardment instead of wind or water. As a result, lunar soil particles are much sharper than their terrestrial counterparts and are composed of agglutinates (aggregates of smaller soil particles bonded together by melting during micrometeoroid impacts). These agglutinates can easily be crushed. Further, the lunar regolith has not been characterized deeper than a few meters below the surface.[58] Basic soil mechanics information for the reliable design of buried structures and safe mining for ISRU requires the sampling of soil at depth. Conventional terrestrial methods impart large material disturbances and require heavy machinery. Thus, research to develop novel in situ soil sampling and characterization techniques in the reduced gravity of the Moon and Mars is warranted.

Terrestrial sands[59] and simulated lunar soils[60] tested on the space shuttle have demonstrated very high strength and elastic moduli for the low effective stresses expected on the Moon or Mars. Similar characterizations are needed for the soils specific to the Moon and Mars, for which the ISS is an ideal platform. The low-strain deformational characteristics of lunar and martian soils are the key to avoiding differential settling of structures, particularly because the impact of reduced gravity on compaction is not clearly understood. In particular, regolith re-compaction (after mining or excavation) may be quite different on the Moon than for materials encountered on Earth because of the jagged and brittle nature of the lunar regolith in a dry, erosion-free vacuum environment. Even on Mars, the dusty soil has slowed exploration by rovers and could present problems in providing a firm foundation for structures.[61]

Exploration missions that rely on ISRU will require equipment for the excavating, mining, crushing, sieving, and conveying of lunar or martian soil. Currently, there is limited understanding of the behavior of granular materials in rough rolling or sliding contact for irregularly shaped particles such as jagged regolith. New multiscale models are needed for developing the understanding of and predicting the tangling of particles and particle interactions with wheels and abrasion on handling equipment.[62] Discrete element methods, in which interactions between many independent particles are modeled, may be useful in this regard, but further development to expand current capabilities beyond spherical particles to irregular-shaped particles is needed for lunar applications.[63] Upscaling to large, structural-scale systems will cause computational difficulties. Continuum models require the development of soil-specific constitutive relations, for which samples of lunar and martian soils will be needed, because simulated soils tested to date lack the ability to mimic accurately the effects of the crushable agglutinates.[64] Geologic variability hampers accurate prediction for even the most sophisticated terrestrial-based soil models. Thus the nature of spatial variability of lunar and martian soils needs to be assessed. Additional research would be needed to support the needs of asteroid surface missions.

Applications for ISRU, excavation, landslides, and other phenomena are related to the transition from static (jammed) to dynamic (flowing) states. Often gravity plays a key role in the initiation of flow. In granular flows, particles of different sizes tend to segregate because of percolation or buoyancy. Furthermore, granular flows can be influenced by interstitial gas.[65] Particle shape, electrostatic effects, frictional and shape properties, thermal cycling, and size distributions will also affect flow characteristics. A better understanding of the fundamental physical aspects of granular flow (e.g., pile, chute, and tumbler flows) is needed to enable designing for situations in which it is necessary to have granular flow (ISRU) and to prevent dangerous situations (e.g., landslides), and to contribute to a better understanding of the geomorphic features of the Moon and Mars. Although some progress has been made,[66] improved scaling laws that include the impact of gravity are needed to address many of the challenges described above.

Dust Mitigation

The fine regoliths on the surface of the Moon and Mars are essentially devoid of moisture and air. On the Moon, the dust that covers the surface is electrically charged, and thus it sticks to almost anything. The interaction of particles with solar radiation and the solar wind may result in dust lifting off the surface and falling back again, a process that is poorly understood and yet may be responsible for the streamers observed by Apollo-17 astronauts. Electrostatic charging of martian dust is also an open issue.[67]

Dust can interfere with many aspects of both human and robotic exploration. Concerns include the health effects of inhaling micron-sized jagged particles and/or silica dust, degradation of life support systems, obscuration of instruments, and damage to bearings, gears, and seals.[68] On Mars, dust storms with strong winds can last for weeks and can envelop the entire planet, and dust devils may create local hazards.[69,70] As with dust storms on Earth, wind-driven dust particles can damage equipment, foul filters, cause excessive wear, pose health risks, and reduce the amount of sunlight reaching solar panels. The interaction of dust with the solar wind, the thin atmosphere on Mars, and natural and artificial surfaces is unclear. Measurements of dust particle size and concentration are lacking. Surface electric fields cause dust to adhere to objects and drive dust transport, yet this process is poorly understood.

Fundamental Research in Complex Fluid Physics

As noted in Chapter 8, complex fluids are excellent candidates for study in reduced gravity. New experiments could greatly enhance the understanding of important reduced-gravity fluid physics phenomena and lead to new fundamental insights in fluid mechanics and transport in general. These experiments are enabled by NASA spaceflight, and they would address phenomena of broad interest to the science and engineering communities. This is particularly true with critical-point phenomena, rheology, and complex fluids such as biofluids, colloids, foams, nanoslurries, granular materials, plasmas, and liquid crystals.[71] The scientific community should design specific experiments to be conducted in response to research solicitations by NASA. Citing granular physics as an example, granular flows are typically driven by gravity, and experiments with granular flows at reduced gravity could reveal much about the influence of gravity on such flows. Other aspects of granular physics include particle clustering, self-assembly, and dissipation, all of which are altered by gravity under terrestrial conditions but could be studied free of sedimentation or gravity-induced stresses and friction in a microgravity environment. This enables, for example, the study of electrostatic effects, interstitial fluids, and the nature of flows involving particles of multiple sizes and/or shapes without the complication of buoyancy or gravitational settling.

Recommended Research in Fluid Physics

Recommended research in fluid physics is summarized below and in Table 9.1 toward the end of this chapter.

Reduced-Gravity Multiphase Flows, Cryogenics, and Heat Transfer: Database and Modeling

A detailed reduced-gravity database is essential for the development and assessment of reliable models for multiphase flow and heat transfer, but very little data is currently available. These data should include phase separation and distribution (i.e., flow regimes), heat transfer involving phase change (i.e., boiling and condensation), and/or advanced devices (e.g., twisted ribbons), pressure drop, and multiphase system stability. These data can be best acquired using a multipurpose phase-change test loop in the fluids integrated rack aboard the ISS. Direct numerical simulation or other numerical simulations (e.g., lattice Boltzmann techniques and molecular simulation) should also be performed to allow NASA to develop an understanding of mission-enabling phase distribution, separation, liquid management, and phase-change system phenomena. In conjunction with ISS data, the results from these detailed simulations can also be used as "data" to support the development of mechanistic three-dimensional CMFD models (e.g., to describe the required interfacial and wall closure relations), which are much more efficient than DNS or molecular simulation models and can be used for most design purposes. In addition, modeling efforts should include the following:

- One-dimensional drift flux models should be developed and used for phase-change system stability analysis.
- Modeling of the evolution of interfacial structures in reduced gravity, which NASA has initiated, should be completed.
- Bubble nucleation characteristics and interfacial structure evolution near a heating surface in terms of bubble departure size, departure frequency, and bubble motion should be analyzed and modeled for reduced-gravity boiling flow. Their relation to the occurrence of the critical heat flux also should be investigated.
- The effects of particular geometries on the evolution of the interfacial structures under reduced-gravity conditions should be modeled by a simple and effective approach.

All of the above models should be developed in a form that can be easily implemented into CMFD codes within the framework of the two-fluid model. This research could be performed in the next 10 years and beyond.

Research in this area should support the development of the following critical technologies described in Chapter 10 (see Tables 10.3 and 10.4): two-phase flow thermal management technologies; technologies to enable engine start after long quiescent periods, combustion stability at all gravity conditions, and deep throttle; supersonic retro propulsion systems; cryogenic fluid management technologies, including zero-boiloff propellant storage systems; regenerative fuel cells; thermoregulation technologies for lunar habitats, rovers, and space suits; and the development of fluid and air life support subsystems (e.g., to enable closed-loop air revitalization and closed-loop water recovery for extravehicular activity [EVA] and life support systems).

Interfacial Flows and Phenomena

Interfacial flows, which are affected by the forces associated with the presence of an interface between two liquid or gas phases, are central to spacecraft functioning. Flows of interest include storage and handling systems for cryogens and other liquids, life support systems, power generation, thermal control systems, and others. Research could lead to advanced systems that would be significantly more capable, more reliable, and more affordable than current systems. Relevant phenomena are strongly gravity-dependent and are highly enabling to and uniquely enabled by NASA missions. A wide variety of fundamental and applied fluid phenomena should be investigated, whenever possible employing multiuser facilities that develop mechanistic models for induced and/or spontaneous multiphase flows (with or without phase change). Research should include targeted experiments that expand core knowledge, improve designer options and confidence, and increase TRLs over the next 10 years and beyond.

Research in this area should support the development of the critical technologies described in Chapter 10 that are listed in the section above titled "Reduced-Gravity Multiphase Flows, Cryogenics, and Heat-Transfer: Database and Modeling."

Dynamic Granular Material Behavior and Granular Subsurface Geotechnics

Improved predictive capabilities related to the behavior of lunar and martian soils on the surface and at depth would enable advanced human and robotic planetary surface exploration and habitation. Surface operations such as wheel/track-soil interaction and cratering would benefit from the development of particle-scale and multiscale models and simulations of key dynamic interactions with soil, including the crushing and compaction of agglutinates. ISRU mining, the design of structural foundations and anchors, and berm/trench stability analysis would benefit from improved soil-specific computational models and methods for sampling planetary soil at depth. Model development can begin in the first part of the decade, but the refinement of site-specific models will likely require ground-based and ISS testing of actual lunar soils.

Research in this area should support the development of the following critical technologies described in Chapter 10 (see Tables 10.3 and 10.4): regolith- and dust-tolerant systems for planetary surface construction and teleoperated and autonomous construction.

Dust Mitigation

The development of fundamentals-based strategies and methods for dust mitigation would enable advanced human and robotic exploration of planetary bodies. Areas of interest include experimental methods, the understanding of the fundamental physics of dust accumulation and electrostatic interactions, and methods for modeling dust accumulation. Issues related to dust seals, environmental hazards, solar panel obscuration, and sensor fouling should also be addressed. Much of this work can be done with ISS and ground-based studies early in this decade.

Research in this area should support the development of the following critical technologies described in Chapter 10 (see Tables 10.3 and 10.4): dust mitigation technologies and systems for EVA and life support systems, for planetary surface construction, and for lunar water and oxygen extraction systems.

Complex Fluid Physics

Unique experiments for understanding complex fluid physics in microgravity are enabled by the ISS. Such experiments can unravel the behavior of complex fluids, including granular materials, colloids, foams, nanoslurries, biofluids, plasmas, non-Newtonian fluids, critical-point fluids, and liquid crystals, without the bias of gravity. The ISS microgravity environment further enables unique capabilities for fundamental experiments of complex flow systems that can explore fundamental fluid physics and geological systems with small-scale models. These studies could be accomplished with a combination of ground-based and ISS efforts in the next 10 years and beyond.

COMBUSTION

Combustion has evolved into an extremely multidisciplinary subject, as diverse as fire safety and astrophysics. In the most fundamental sense, combustion deals with the process and effects of energy released into a surrounding medium and the response and feedback of that background. Usually, the focus of combustion is on a reaction front, where reactants are converted into combustion products. Such fronts exhibit flames, deflagrations, detonations, and myriad intermediate states. Sometimes what seems to be the most esoteric combustion question in, for example, details of a flame structure or dynamics becomes a key issue for an application that is of critical importance to safety or a developing technology.

NASA's reduced-gravity combustion research program has had a number of broad objectives, ranging from increasing the fundamental understanding of the basic physical processes to preventing and controlling fires on spacecraft. A better understanding of combustion itself has evolved as NASA reduced-gravity facilities have been used to eliminate buoyancy effects, which often dominate terrestrial combustion, and to compare measurements with theory and numerical simulations. In addition, the program has tried to relate fundamental combustion principles to applications such as the simulation of fire growth in spacecraft and the assessment of actual fire risks. In this area, the understanding of combustion has provided underpinning information that enables a better understanding of material flammability and fire prevention measures that can improve fire safety in the future.[72,73]

NASA's reduced-gravity combustion research has led to enabling technologies for space exploration, and it has provided new insights into fundamental combustion processes. Both areas are addressed further in the recommendations.

Research in Support of NASA's Exploration Missions

Combustion research in support of NASA's exploration missions is addressed in Chapter 10, "Translation to Space Exploration Systems."

Fundamental Combustion Research

This section focuses on gravity-related combustion research issues of most crucial importance to NASA's future crewed and uncrewed missions. In particular, NASA should support fundamental combustion research in fire safety and combustion processes. Research in both of these areas would be facilitated by more capable numerical simulations of combustion. Because fire safety is so important both as a topic of fundamental combustion research and as an operational element of human space exploration missions, fire safety is addressed below and in Chapter 10.

Fire Safety

Fire safety includes fire prevention, detection, and suppression (all of which are discussed here and in Chapter 10) and post-fire recovery (which is discussed only in Chapter 10). Wherever a fuel and an oxidizer appear in reasonable proximity, there could be a scenario in which they meet by accident in the vicinity of an ignition source. To minimize the likelihood and impact of accidental fires in spacecraft, combustion needs to be better understood in all relevant environments (that is, on Earth and in the reduced-gravity environments of the space shuttle [and replacement vehicles], the ISS, the Moon, and Mars).

Dealing with accidental fires in reduced gravity is different from dealing with them on Earth in several important aspects. On Earth, buildings are designed to allow inhabitants to escape to a safe outside location. Spacecraft and habitats in reduced gravity, such as on the Moon or Mars, are enclosed by pressurized vessels with hostile outside environments, and so outside escape is generally not a viable option. Thus, every aspect of fire prevention, detection, and suppression is more critical in reduced gravity than in terrestrial scenarios.

Several fires have occurred in spacecraft on the ground or in space. As a result, new approaches to high-pressure oxygen atmospheres, design standards, flammability of materials, flammability testing, operational emergency procedures, quality control, and suppression strategies have been developed.[74,75]

From a fundamental science perspective, fluid transport in normal gravity is controlled by buoyancy. As a result, terrestrial systems are designed to deal with aspects of the combustion process that behave very differently in space, and an improved knowledge of combustion in reduced gravity is essential in order to adapt fire safety concepts and systems to the more stringent conditions of the space environment.

Fire Prevention

A major part of NASA's strategy for fire prevention is to improve the design and selection processes for materials used on spacecraft. This improvement involves determining acceptable materials, techniques to reduce material flammability, and ways to monitor and control ignition sources. Material screening for flammability is now based on empirical procedures that use standard tests performed in normal gravity.[76] For example, one test used for many solid materials considers upward ignition and flame growth. If the flame spreads more than six inches without self-extinguishing, the material fails the test (test 1 in NASA-STD-6001A). Fire-control measures may include, for example, installing firebreaks to limit the extent of flame spread or covering flammable materials with nonflammable materials. Despite these and many other measures, data taken in reduced gravity suggest that existing screening methods could be improved to provide a more scientifically based description of the material properties that enable combustible materials to sustain and grow a fire.[77,78,79] Future research should encompass existing space atmospheres and the new atmospheric conditions (enriched oxygen, reduced pressures) proposed for spacecraft cabins and surface habitats in future space missions.[80] Therefore, a more fundamental understanding of processes such as pyrolysis and combustion in microgravity are urgently needed to define the limitations of current screening tests and, if necessary, to propose improved tests. Experimental and theoretical methods are needed to improve the understanding of how screening-test results in normal gravity can be used to predict performance in reduced gravity.

Fire Detection

The earlier a fire is detected, the less the damage it can cause. At the same time, false alarms should be minimized. To be able to design and best exploit a fire detection system, it is essential to understand what is being detected and to capture adequate quantities of the effluent from the fire. The current ISS uses smoke or particle detectors that are sensitive to a given range of particle sizes. Key questions include the following:

- What should be sensed for faster and more reliable detection?
- If smoke particles are to be used, what ranges of particle sizes indicate incipient fires in reduced gravity?
- How can the detector signal be used best to initiate the fire suppression?

Early investigations suggested that smoke particles emerging from flames are larger in reduced gravity than in normal gravity. (This is attributed to the longer residence times that occur in low-speed flows in reduced gravity.[81]) Nonetheless, a more recent ISS experiment indicates that the particle sizes from pure pyrolysis processes (no fire) can actually be smaller than on Earth.[82] Since pyrolysis from the overheating of electrical insulation is of concern in space and an accumulation of pyrolysis products is a dangerous pre-fire situation, detectors should cover the entire size range for both fire and pyrolysis products. It is important to continue current efforts to characterize the chemical products of fires and to determine the particle-size distribution produced by the combustion of spacecraft materials in reduced gravity (to help define "fire signatures") and to understand the transport of combustion products from the fire to the detector.

Fire Suppression

Fires are typically suppressed by releasing a spray of gas or droplets. Two other methods, unique to space, are also available: cutting off air ventilation and depressurizing the spacecraft cabin. Early microgravity research suggested that most materials cannot maintain a flame when there is no convective airflow.[83,84] Even when combustion is sustained in a quiescent atmosphere (such as for a small droplet or a candle), flames that are supported by purely diffusive transport processes are relatively weak; thus they can be controlled and extinguished more easily. Shutting down spacecraft ventilation is now a standard procedure after a fire is detected. The second approach unique to space is to depressurize the cabin by opening an external vent. When the pressure is reduced enough, a fire is extinguished.[85,86] This approach requires astronauts to don space suits or to move to a different module. Also, the fire can briefly intensify during the depressurization process because of the induced flows caused by depressurization. Depressurization is normally considered a last resort.

Fire suppression agents used in spacecraft should be (1) extremely efficient, (2) nontoxic and causing little or no damage to the equipment, (3) easy to clean up with onboard resources, and (4) able to reach hard-to-access places, such as inside equipment racks. Types of suppression agents used in the past include portable aqueous gel or foam (Apollo, Skylab, Mir, and the ISS), bottled carbon dioxide (ISS), and Halon (space shuttle).[87,88] The most effective fire suppression agents and deployment methods have yet to be determined for specific space applications. In addition to the suppression agents currently used, water mists are promising candidates; they have demonstrated the ability to extinguish pool fires[89] and solid-fuel fires.[90,91]

There are several ways in which fire suppression agents extinguish a fire. One is by changing gas-phase processes so as to slow key reaction rates and to increase heat losses (thereby affecting the combustion reaction). For example, at high temperatures, Halon decomposes to release halogen atoms, such as bromine and fluorine, that combine with active hydrogen atoms to quench the flame propagation reaction. Alternately, fire suppression agents may cool or block the surface of a solid fuel, which reduces the pyrolysis rate (i.e., reduction of fuel supply). A few studies have examined gas-phase mechanisms for fire suppression in reduced gravity,[92,93] but little has been done to study mechanisms related directly to the fuel surface. New multidimensional simulations predict that to extinguish a flame most effectively, water mist must be injected differently under different gravity conditions. This is a multiphase combustion problem involving issues of gravity, buoyant and forced convection, and droplet sizing.

Verifying and understanding analytical and experimental results as well as better understanding the interactions among suppression agents, flames, and surfaces in reduced gravity will provide insight into the fundamental differences in balances between buoyancy and other forces and will contribute to the development of improved fire-extinguishing systems. In addition, methods for screening materials for use in spacecraft should consider factors related to fire suppression.

Combustion Processes

Studies of combustion in reduced gravity lead to a greater understanding of terrestrial combustion. Flames are controlled by energy released from exothermic chemical reactions and the interaction of this energy with the atmosphere. On Earth, the chemical reactions, energy, fluid dynamics, and gravity-induced buoyancy interact nonlinearly. Their effects are usually only separable through theoretical analysis and numerical simulation. By varying or eliminating the effects of gravity, one can extract fundamental data that are important for understanding combustion systems. Such data include parameters such as chemical reaction rates, diffusion coefficients, and radiation coefficients that strongly influence the ignition, propagation, and extinction of combustion waves. This approach has been implemented to some extent for the limited time and spatial scales in existing reduced-gravity platforms.

Besides gravity, there may be other factors specific to the space environment that differ from typical terrestrial environments. These include atmospheric conditions, local material compositions, stoichiometry, and ignition sources. The effects of these environmental changes on combustion in reduced gravity are unclear, and some range of relevant parameters should be investigated.

Data from reduced-gravity experiments, when combined with theory and computations, have "contributed to our fundamental knowledge of some of the most basic combustion phenomena, to the improvement of fire safety on present and future space missions, and to the advancement of knowledge about some of the most important practical problems in combustion on Earth."[94] These results include the discovery of new phenomena that challenged previous theories and demonstrated the actual behavior of previously proposed combustion states. The existence of theoretically predicted flame balls has been revealed, and their persistence and stability in reduced gravity have been demonstrated. Flame balls have been shown to evolve into flame strings as the combustion mixture becomes leaner, demonstrating the persistence of burning in very lean conditions. In addition, reduced-gravity experiments have demonstrated theoretically predicted pulsating flames, halos of soot rings around burning fuel droplets, greatly altered burning properties of metal wires, and the altered distribution of soot and particulate sizes in diffusion flames, which shows the inadequacy of smoke detectors used in space. A radiative quenching limit in low-straining or low-speed flows has been discovered, along with an upper size limit to spherical droplet flames. Experiments in reduced gravity have shown that flame spread over solid fuels changes because of both gravity and the nature of the background (enriched-oxygen) atmosphere, and in oxygen-starved conditions, flamelets and fingers appear before the final extinction occurs. Thin cellulose has been found capable of sustaining a flame in a lower-oxygen atmosphere in lunar gravity than in Earth gravity.

Much of the new information derived from combustion experiments in reduced gravity has been published in NASA reports, journal articles, conference proceedings,[95] a special issue of *Combustion and Flame*,[96] a special section of the *AIAA Journal*,[97] and a dedicated summary monograph on microgravity combustion.[98] The work has also left many unsolved or unexplained phenomena, as discussed below. The main point is that combustion phenomena behave differently as gravity changes from Earth gravity to zero gravity (or *g*-jitter). Some of this difference is related to the effects of gravity on the combustion front itself, and some of it is related to how gravity affects convection or the supply of fuel.

Gas-Phase Combustion

Gas-phase combustion, including ignition, persists as a fundamental research area that has important practical consequences in terms of fire safety. As a gaseous fuel and oxidizer mix, conditions for igniting fires (which might be detectable early) or explosions (which might not) can occur. Ignition limits for combustible gas-phase mixtures are often expressed in terms of the stoichiometry or the percentage of fuel; they are usually determined by specified laboratory tests. Other critical factors, some of which are seldom considered, are the method and rate of energy input into the reactants and the geometry and size of the entire system. Some work has been done, for example, to extend the ignition-limit concept and consider the power-energy relation required for hydrogen ignition in reduced gravity, but little has been done recently. It now appears that the ignition limits can change as a function of the size of the system. More work is required on gas-phase ignition in conditions representative of those found in specific microgravity environments and as a function of sources of input energy.[99,10]

It is very rare, on Earth and in space, for an area, cabin, or room to be completely filled with a stoichiometric, homogeneous mix of fuel and oxidizer. Instead, there are variations in mixture composition, stoichiometry, temperature, and/or other energy input. These result in gradients of chemical reactivity, but very little is known about the behavior of flames propagating through reactivity gradients. In any case, when obstacles are present, the situation becomes even more dangerous, as turbulence intensifies and flames accelerate. Reactivity gradients are important at all stages of a fire or an explosion, from ignition and propagation through to extinction. Reduced-gravity experiments can be used to learn more about flame ignition, propagation, and extinction in reactivity gradients. Such experiments could enable more accurate assessments of the risk and behavior of fires and explosions in space environments.

What is the flame structure near the flammability limits in microgravity? Previous studies of combustion in reduced gravity have shown that gaseous flammability limits may be extended beyond those measured on Earth.[101,102] It is now speculated that such limits might not exist at all—that a diffusive or hydrodynamic mechanism may cause extinction, or that flame balls sometimes evolve into flame strings, and, as the mixture becomes leaner, flame balls promote the persistence of burning in very lean conditions. To understand these issues, future research should investigate the nature of hydrogen and methane flame structures as the flame approaches the currently accepted flammability limits with respect to dilution or stoichiometry.

Most combustors and unwanted fires involve diffusion flames. There remain significant gaps in the understanding of these flames—particularly with regard to chemical kinetics, transport, radiation, soot formation, pollutant emissions, flame stability, and extinction—in part because long residence times possible in reduced gravity can result in unusual behavior. Gravity, which has a tremendous impact on most diffusion flames, is an important parameter to vary in experiments and models. Studies of laminar and turbulent diffusion flames in reduced gravity can yield insight into fundamental turbulent-flame structure. It can also provide important understanding of the limits and validity of the multidimensional computational models now commonly used.

Some of the most persistently uncertain input data in flame modeling are the molecular diffusion coefficients of lightweight species. They are often as important as the most important chemical reaction rate. Analyses have shown these to be particularly critical near the extinction limits. Sustained experiments in reduced gravity, in conjunction with measurements on Earth, can be used to determine accurate values of diffusion coefficients required for all models of flame behavior.

Droplet Combustion

Liquid fuels are normally burned as sprays or droplets. Reduced gravity enables the detailed study of the burning of individual droplets or droplet arrays to help improve the understanding of vaporization and combustion processes, including transient burning processes, multicomponent diffusion, extinction limits, and soot formation. Optimizing and controlling this form of combustion can lead to higher efficiency and cleaner combustion. Both pure liquid and multicomponent fuels, as well as synthetic fuels and biofuels, should be included.[103,104]

Solid Combustion and Smoldering

Gas-phase combustion from a solid fuel produces a special type of diffusion flame that is profoundly affected by gravity. Smoldering has many applications in material synthesis, but it is mostly encountered as a fire safety problem (e.g., with insulation or cables). Local diffusion flames anchored next to a region of solid fuel can spread and grow in size as the temperature of the solid increases. Differences in the types of convection (buoyant versus forced), flow-spread directions (with flame spread opposed to or concurrent with the flow directions), sample thickness, and charring characteristics affect the spread rate and extinction limits. Currently, the level of understanding is very low with respect to flame spread and growth with common solid fuels, such as cellulose and plastics, in a low-speed forced flow in reduced gravity. For complex solids (e.g., Nomex®, which is used extensively in space), there is even less understanding. Part of the difficulty is the lack of a reduced-gravity platform over an extended period of time for testing solids with realistic thicknesses. Because flame growth is a major concern in spacecraft fire safety, fundamental studies of solid-fuel flammability (ignition, extinction, and flame-spreading processes) are urgently needed. Validated numerical models are needed for conditions representative of the environment of future spacecraft and extraterrestrial habitats. The development of these models would require studies in which physical variables such as gravity, flow velocity, pressure, and oxygen percentage are varied. A detailed numerical model for solid-fuel combustion is also needed, along with simplified, phenomenological submodels that can be included in full-scale models of fires for large volumes, representing large portions of future space habitats.[105]

Smoldering, which is a surface reaction, is another mode of combustion unique to solid fuels. Smoldering is greatly affected by oxygen transport and heat loss. In reduced gravity there is less heat loss, but there is also a lower rate of oxygen transport. Thus, it is not clear whether the tendency to smolder is higher or lower in reduced gravity. Transitions between the flame and smoldering modes can also occur. Smoldering is a slow process, and so long-duration experiments are required to study it.

Other Combustion Regimes and Applications

Combustible mixtures, under supercritical conditions of increased pressure and low temperatures, are now being considered for liquid rocket fuels and for possible applications to solid-waste processing in reduced gravity.[106] Supercritical combustion supports many types of propagating flames, which are susceptible to a range of flame instabilities, including some influenced by gravity. Supercritical fuel droplets in background oxidizers represent another multiphase regime of combustion, and mechanisms for atomization and spray combustion may be different in these regimes.

Microscale combustion devices reduce spatial constraints and increase operational efficiency. A fundamental weakness of such devices is the large conductive heat loss resulting in low energy-conversion efficiency. By using a heat-recirculating design (i.e., a spiral counterflow or "Swiss roll" design), this liability is eliminated.[107] A number of interesting new phenomena or physical mechanisms have been proposed for use in reduced gravity. One example is a propulsion device using channels with diameters on the order of a few flame thicknesses. In this geometry, very large thrust is created by the interaction of a flame with boundary layers.[108] This capability presents possibilities for propulsion engines and thrusters in microgravity using almost any available fuel.

Recommended Research in Combustion

Recommended research in combustion is summarized below and in Table 9.1 toward the end of this chapter.

Fire Safety

Improved methods for screening materials in terms of flammability in space environments will enable safer space missions. Present tests, performed in normal gravity, are not adequate for reduced-gravity scenarios. By 2020, improvements can be achieved that would supplement current screening methods. Beyond 2020, new methods could be optimized and implemented.

Research in this area should support the development of materials standards related to ignition, flame spread, and toxic or corrosive gas generation in various environments and gravitational force fields, as described in Chapter 10 (see Table 10.3).

Combustion Processes

Space exploration is enabled by and enables extended combustion experiments in reduced gravity that have longer durations and larger scales and that include current and future fuels (e.g., biofuels and synthetic fuels), as well as practical aerospace materials, that NASA may consider for future missions. Previous microgravity experiments made new discoveries and raised new questions. There are now new fuels and new space exploration missions. In addition, most of the solid materials tested to date in reduced-gravity combustion experiments have been simple materials in the form of combustion samples. Future experiments should include more complex materials and shapes. Results from such experiments, along with a deeper understanding of fundamental combustion phenomena, will contribute to technologies for improving fire safety in space and terrestrial applications. By 2020, droplet-phase, gas-phase, and solid combustion experiments on the ISS could be completed, and the preparation and planning for larger-scale, longer-duration experiments could begin. Beyond 2020, larger-scale, longer-duration experiments would be conducted.

Research in this area should support the development of materials standards related to ignition, flame spread, and toxic or corrosive gas generation in various environments and gravitational force fields, as described in Chapter 10 (see Table 10.3).

Numerical Simulation of Combustion

Numerical models are powerful and necessary tools for studying combustion processes in reduced gravity, including issues related to fire safety in spacecraft. The development and validation of detailed single and multiphase numerical combustion models are needed to relate reduced-gravity and Earth-gravity tests. Because of limited access to space for experimentation, numerical simulations are necessary enabling tools for the prediction, design, and interpretation of data from experiments and missions. By 2020, selected numerical models could be validated with reduced-gravity experiments. Beyond 2020, these models could be integrated with experiments and design.

Research in this area should support the development of the following critical technologies, described in Chapter 10 (see Table 10.3): materials standards related to ignition, flame spread, and toxic or corrosive gas generation in various environments and *g* fields; particle detectors; and fire suppression systems.

MATERIALS SCIENCE

Materials science and engineering help shape elements of the modern world, from integrated circuits, to single-crystal turbine blades, to ceramic fuel cells. The study of seemingly disparate materials is unified by the paradigm that the synthesis and processing of a material affect its structure, which in turn governs its properties. Thus the design of materials for terrestrial or space applications, including in situ resource utilization, requires a deep understanding of the relationship between the properties of a material and its structure and of the manner in which synthesis and processing affect its structure. Materials research in reduced gravity has the potential to explore processes that govern materials production on the ground and produce new materials for use during spaceflight and the exploration of planetary bodies. These processes and materials may be critical for applications related to space exploration, and they may lead to new materials for various applications in space and on Earth.[‡]

Research in Support of NASA's Exploration Missions

Materials science and engineering are central to NASA's exploration mission, both crewed and uncrewed. This section focuses on gravity-related research issues in the development of new materials that (1) are most critical to meeting NASA's unique requirements and (2) would not be developed by other governmental agencies. Areas of particular interest include materials synthesis and processing, advanced materials, ISRU, and fundamental materials research. Enabling synthesis and processing techniques related to these topics are also of great importance. For

[‡]Table 13.3 in this report describes the potential for all recommended research to impact space exploration and to meet terrestrial needs.

example, computational materials science would contribute significantly to the creation of new, unique materials in an efficient and cost-effective manner. Advances in the fundamental understanding of materials synthesis, processing, computational materials science, and microstructural control, coupled with the development of new advanced materials, has the potential for creating novel materials that could improve weight and/or property factors—for example by providing reduced weight and increased operating temperatures or improving structural strength by 10 to 30 percent. An example of a new, unique property is self-healing capabilities after a meteoroid impact.

Materials Synthesis and Processing

Improvements to existing synthesis processes and the development of new processes are needed for the ground-based synthesis of a wide range of enabling materials, both traditional and advanced, for space exploration and for materials repair and production during missions. Applications of interest range from the structural components of spacecraft to nanoscale sensors that can be used to determine the health of astronauts. Many synthesis techniques of interest involve vapor- or liquid-phase transport and so can be influenced by convection. Thus research in this area is also be enabled by access to reduced-gravity platforms.

Combustion synthesis, or self-propagating high-temperature synthesis (SHS), is one example of a versatile reaction synthesis process that is ideally suited to operate in reduced gravity.[109,110] Past work has demonstrated that SHS can efficiently produce ceramics, intermetallics, glasses, functionally graded materials, and composites (metal-matrix composites, ceramic-matrix composites, and intermetallic-matrix composites) in a vacuum and with minimum energy input.[111–114] The raw materials for these SHS reactions can be fabricated from metals, elements, and compounds extracted from planetary regolith. SHS can be used to manufacture materials as well as for joining and repairing components.

Vapor-phase reactions involve reactants in a gaseous phase that participate in a gas-gas, gas-liquid, or gas-solid reaction to synthesize a product. The products can be ceramics, metals, or semiconductors in the form of nano- to microscale particles, films, or whiskers. These can be final products, as in the case of films or crystals, which have a variety of applications such as solar cells and high-temperature, wear- and corrosion-resistant coatings and lubricants.[115,116] Vapor-phase processing can produce bulk materials that are composed of nanoscale crystals. This unique structure gives rise to high-strength, low-weight materials that are important for NASA space applications. Vapor-phase processing can also produce carbon nanotubes and particulate reinforcements that can be used in other synthesis processes as growth initiation sites, raw materials for combustion reactions, and strengthening materials in metal-matrix and ceramic-matrix composites. In addition, semiconducting materials are normally produced using vapor-phase processing, and nanoscale semiconductors can be used for novel electronic devices and detectors. For example, the gas-detection properties of semiconductors in the form of nanowires and quantum dots (nanoscale cluster of atoms) can be used in sensors for liquid propellant tanks, for biomolecules of importance to monitoring astronaut health, or for toxic gases.

Previous materials science research in reduced gravity has utilized the Materials International Space Station Experiment (MISSE) in which a series of experimental trays were flown on the exterior of the ISS to measure space environment effects, such as atomic oxygen, vacuum, ultraviolet (UV) radiation, charged-particle radiation, thermal cycling, and debris and micrometeorite impact on spacecraft materials and components. These experiments have provided important information and a knowledge base on the degradation of materials, such as polymers, coatings, thin films, seals, solar cells, and paints.

With the advent of powerful computers and efficient algorithms, the virtual synthesis and processing of many materials are now possible. First-principle calculations at various levels are sufficiently developed to predict accurately, for example, crystal structure, surface reactivity, electronic band structure, and the mechanical strength of many materials using only atomic numbers of constituent elements as inputs. Thermodynamic and kinetic databases can be used to predict the phases that are present and transformation rates in complex multicomponent materials. Mesoscale simulations are used to predict solute segregation following solidification as well as precipitate size and spatial distributions in multicomponent alloys. On a larger scale, the three-dimensional computation of structural development of materials is now possible. It is thus possible to link conditions under which a material is processed and its resulting structure and properties with greater accuracy. However, despite significant improvements in these techniques, the complexity of materials used for spaceflight applications still require that computational methods be supplemented with experiments to verify results, to provide missing data, and to provide information on processes that cannot yet be modeled. An integrated computational and experimental effort (that is, the so-called integrated computational material engineering approach) reduces the number of prototype materials required. This approach drastically decreases the time and cost of developing new materials, and it should facilitate the development of new materials for NASA's unique applications. Research to develop new computational methods could further reduce the time required to design new materials.

Advanced Materials

Materials with novel mechanical properties and low density are needed for spaceflight components. Examples include aluminum, magnesium, and titanium alloys, as well as metal-matrix and ceramic-matrix composites. These materials may behave differently in the space environment (reduced gravity, high vacuum, atomic oxygen, etc.). Even so, very little, if any, work has been conducted on the effects of the space environment on the mechanical properties (strength, fatigue, fracture toughness, etc.) of some advanced intermetallic and composite structural materials. As new materials are developed—and as existing materials are adapted for new space applications—the effects of the space environment on these materials should be investigated.

"Smart" materials have the ability to monitor the condition of a structure and/or to adapt the structure and properties of materials to the environmental conditions. Smart materials may even have the capability of repairing (self-healing) a damaged structure. A vast array of smart materials can be employed in a variety of ways that will enable new space exploration capabilities. Shape-memory alloys and polymers are restored to their original shape or form a new shape within a certain temperature range, magnetic field, pH range, or light sensitivity.[117] These shape-memory materials can be used in small devices like actuators and also in the construction of buildings that can form large structures from compact structures. Self-healing materials possess the ability to repair cracks due to impact, fatigue, or wear. Many of these materials are now being developed and designed, but further, significant research must be performed in order to fully define their properties and capabilities in the space environment.

Bulk materials and surface-engineered coatings, primarily ceramics and intermetallics, that can withstand high temperatures (up to 1,200°C) and ultrahigh temperatures (1,200°C to 2,200°C) are crucial for heat shields, rocket nozzles, and other high-temperature applications. Present research consists of the synthesis of these materials as well as processing using sintering, reactive hot pressing, and a combination of combustion synthesis and hot pressing for bulk materials,[118,119,120] and vapor deposition techniques are used to produce high-temperature coatings.

New surface engineering and advanced coatings are needed to improve the ability of materials to perform in the harsh environments associated with space missions.[121] Various applications will need specific properties to be enhanced and will require multifunctional, multiproperty capabilities. The materials used in heat shields and rocket nozzles require a high-temperature coating to improve performance. Increased fatigue and wear lifetimes as well as improved fracture toughness will benefit components that experience repeated loads or friction.[122] New coatings could also give components multifunctional properties, such as a combination of wear and corrosion resistance, resistance to thermal fatigue and oxidation, and low coefficient of friction.[123] Smart coatings that provide information on how the surface of a component is performing (strain, wear, corrosion, oxidation, etc.) and/or provide a self-healing and repair function would also be useful in the harsh space environment.

In Situ Resource Utilization

Processes to acquire minerals from lunar or martian regolith by means of in situ separation and concentration will need to be developed in order to synthesize materials in space and to manufacture and, in some cases, to repair components in space. Conventional mineral processing, metal extraction, and refining processes have been used to extract concentrated minerals, metals, and other materials from simulated regolith. Hydro-, pyro-, and electro-metallurgical processes have been used to extract metals from regolith.[124-127] Fused salt electrolysis can be used to extract reactive metals such as aluminum, calcium, and silicon and also to generate oxygen from oxide regolith. Past research has demonstrated the ability to extract hydrogen from regolith using vacuum pyrolysis.[128]

Even so, oxygen-extraction processes need to be advanced to secure a source of oxygen for prolonged missions. In addition, each of these techniques requires additional research to develop the most energy-efficient extraction processes. Research should also investigate the possibility of extracting hydrogen from the solar wind.[129]

Materials synthesis methods that have been proposed can be designed to utilize materials extracted from regolith or the regolith itself.[130] It may also be feasible to use materials discarded from prior missions. The separation of oxides into oxygen and metallic elements takes place during oxygen extraction; the metallic elements can be a very important by-product. Silicon can be used for solar cells. Aluminum, titanium, and iron can be used for structural components. In addition, regolith compounds can be the raw materials for combustion synthesis reactions used for the fabrication of strategic components or welding/joining of structural components. Reactive elements, such as aluminum, that are extracted from the regolith or from scrapped spacecraft bodies and components can be used as a fuel for combustion synthesis reactions to fabricate and repair materials and components on the planetary surface. Alternatively, regolith separated into appropriate minerals could be fused into structural construction blocks using a solar concentrator or solar furnace.

The most accessible and abundant source of energy on the surface of the Moon or Mars is solar power. Since there is an abundance of silicate materials in the regolith, silicon-based photovoltaics and solar concentrators will be able to utilize solar energy for power, providing that essential processes, such as extracting silicon from the regolith, can be developed.[131] The efficiency of solar cells has increased rapidly in recent years, which has improved their ability to power space missions.[132] The current trade-off for space applications is weight for efficiency.[133] Efficient solar cells are not lightweight, and lightweight solar cells are not efficient. Investigations into organic photovoltaics are underway in an effort to achieve a better efficiency-to-weight ratio for space applications.[134]

It may be possible to fabricate materials for high-efficiency, low-operating-temperature solid-oxide fuel cells from the regolith, which would provide a supplementary planetary energy source. Much of the current research on fuel cells is very diverse in regard to the materials being investigated. Many of these systems may be applicable to space exploration missions, but the specific characteristics required for the rigors of space travel are not at present being investigated.[135] System performance at low temperatures and pressures should be studied in greater depth in order to produce a fuel cell capable of operating efficiently in the vacuum of space.

There is also a great need to find creative technologies that use the regolith to fabricate and, in some cases, repair components needed for life support, construction, energy generation and storage, forming tools, and other purposes.

Fundamental Materials Research

The relationship between the processing of a material and the resulting structure is, in many cases, not well understood. For example, many materials are processed using approaches that involve a change in phase, such as a transition from a liquid or vapor to a solid. Density changes and the influence of temperature and concentration on the density of a fluid or vapor lead to buoyancy-driven convection or sedimentation on Earth, which frequently masks processes under study. Thus the study of a wide variety of phenomena related to processing that involves transformation from a liquid or a vapor to a solid can benefit greatly from experiments in reduced gravity. The objective of the experiments and associated ground-based modeling would be to gain a deeper understanding of processes underlying transformations, which would thus improve processes used on Earth.[136] A second objective would be to provide insight into the effects of reduced gravity on processes of importance for ISRU applications and materials repair during spaceflight.

Materials Synthesis and Processing

The properties of many materials are directly related to the phases that nucleate from a liquid during processing. Thus, nucleation is an essential, but poorly understood, aspect of materials processing. Microgravity experiments, in which liquids can be held and solidified without a container, thus removing the effects of walls, can shed new light on the nucleation process. In addition, convection due to compositional inhomogeneities that accompany the formation of nuclei can be avoided in microgravity. It is thus possible to cool liquids far below their melting points.

The ability to undercool liquids in a controlled manner allows the study of the formation of stable and metastable phases, the formation of glasses, the relationship between liquid structure and the resulting crystal structure, and the thermophysical properties of deeply undercooled liquids. Containerless processing would also allow properties of liquids such as thermal conductivity and heat capacity to be determined at the very high temperatures at which crucible materials are not readily available.

Crystals are frequently grown from a liquid or vapor. On Earth, crystal properties are affected by the density differences between the phases, the temperature and composition dependence of the density of the parent phase, and variations in the surface tension of a vapor-liquid interfacial-induced convection. This convection leads to non-uniform compositions as well as defects in the resulting crystal. Microgravity allows these crystal growth phenomena to be studied without the confounding effects of gravitationally induced (buoyant) convection. The ability to grow crystals is important to the U.S. economy and research infrastructure; NASA funding has played a central role in supporting research in this important area, but recent reductions in funding by NASA (and other agencies) have had a large negative impact on crystal growth research.[137]

The synthesis of metal-ceramic composites[138] and ultra-lightweight porous materials[139,140] is also significantly affected by convection. Removing the gravity-driven buoyancy effects can result in a more uniform distribution and shape of the ceramic phase, the pores, and the ratio of open porosity to closed porosity. These uniform porosity structures have several important applications for space exploration that range from gas and liquid filters and separation membranes to drug delivery systems for astronauts.

Microstructure and Property Control

Microstructures are central to properties of materials as diverse as nanoscale precipitates in aluminum alloys and single-crystal turbine blades. These two structures also provide excellent examples of (1) pattern formation during solidification that involves dendrites and cells and (2) self-organization, which occurs during nucleation, growth, and Ostwald ripening of multiphase materials. The dendritic and cellular patterns that are formed during solidification result from the destabilizing effects of diffusion of impurities in the liquid and the stabilizing influence of surface energy and a temperature gradient. The effects of interfacial energy and processing conditions, such as the temperature gradient and growth velocity, remain an area of active investigation. Similarly, the two-phase microstructures that form during liquid-phase sintering and phase transformations, and which are governed by diffusion, interfacial energy, and kinetics, require further investigation. Unfortunately, buoyant convection or sedimentation makes it very difficult to study the physics that underlie processes such as dendritic and cellular solidification, liquid-phase sintering, and phase separation. The effects of interactions between individual dendrites or cells on spatial distribution and structure, the evolution of dendrites during transient heating or cooling, and the effects of noise and initial conditions on resulting patterns remain unclear. Interactions between dendrites are particularly important in the development of solid-liquid mixtures, so-called mushy zones, and resulting fluid instabilities need further investigation. Fluid flow in mushy zones can lead to especially deleterious casting defects in single-crystal turbine blades, and it remains an area of active research. In some cases, the effects of complex interactions among some combinations of factors also remain difficult to predict. Given this lack of understanding, in many cases it is difficult to identify processing conditions that will produce a desired set of material properties.

Computational materials science is an enabling technology at the mesoscale. With the advent of powerful computers and efficient numerical algorithms, such as the phase-field and level-set methods, in some cases it is possible to predict the evolution of interfaces or precipitates in solids in three dimensions on experimentally accessible timescales. Further development of algorithms and methods will allow, for example, thermal treatments to be designed for a given alloy that yield a certain set of properties and ultimately will eliminate the expensive and time-consuming trial-and-error approach to determining the processing path for a given alloy. Research in this area is also necessary to support the experimental microgravity effort by providing explanations for the observed behavior and suggesting future experiments.

Recommended Research in Materials Science

Recommended research in materials science is summarized below and in Table 9.1 in the next section.

Materials Synthesis and Processing and Control of Microstructure and Properties

The space environment enables the production of benchmark data for the development of microstructure, crystal growth, nucleation, and the synthesis of composites. These data, along with computational materials science, will improve the methods used for the synthesis and processing of existing and new materials on the ground. This would increase the ability to understand and predict formation of microstructures in a wide range of materials in both terrestrial and space environments. Results that will alter the fundamental understanding of these processes and improve terrestrial materials processing could be obtained by 2020, with longer-term efforts requiring new hardware.

Advanced Materials

New advanced materials would enable operations under increasingly harsh space environments and reduce the cost of human exploration. By 2020, advanced materials that meet new property requirements could be designed and developed using both current and novel materials synthesis and processing techniques and computational methods.

Research in this area should support the development of the following critical technologies described in Chapter 10 (see Table 10.4): inflatable aerodynamic decelerators, space nuclear propulsion, fission surface power, and radiation protection systems.

In Situ Resource Utilization

There is a strategic and critical need to utilize extraterrestrial resources for future space exploration and thereby extend human space exploration capabilities. Fundamental and applied research is required in developing technologies for the extraction, synthesis, and processing of minerals, metals, and other materials that are available on extraterrestrial surfaces. By 2020, a select group of strategic elements (e.g., oxygen and silicon), materials, and components could be identified and produced from simulated lunar and martian regolith both on Earth and in reduced gravity.

Research in this area should support the development of the following critical technologies described in Chapter 10 (see Table 10.4): ISRU capability planning and lunar water and oxygen extraction systems.

RESEARCH PRIORITIZATION AND RECOMMENDATIONS

The following criteria were used to develop a subset of high-priority topics from those presented above in this chapter.

- *Enabling research*

  1. The importance of the problem being addressed by this research.
  2. The degree of impact that the research will have on the problem being addressed.
  3. The likelihood that the research will be successful in addressing the problem. The risk of an investigation failing to reach a successful conclusion.
  4. A reasonable potential that needed resources such as crew time and research platforms could become available.
  5. The consumption of program resources compared to that by other potential investigations.
  6. The contribution to terrestrial value (medicine, economy, education, national security, etc.).

7. The efficiency of the proposed research in terms of addressing multiple questions in a single investigation.
8. The impact on furthering fundamental knowledge in relevant fields.

- *"Enabled-by" research*

1. The level of impact on fundamental knowledge.
2. The breadth of impact across a number of disciplines.
3. The likelihood that the research will produce a definitive answer.
4. A reasonable potential that needed resources such as crew time and research platforms could become available.
5. The consumption of program resources compared to what might be consumed by other potential investigations.
6. The contribution to terrestrial value.
7. The efficiency of the proposed research in terms of addressing multiple questions in a single investigation.
8. The impact on enabling space exploration.

The recommended high-priority research areas in applied physical sciences are listed below and in Table 9.1, which also lists current gaps, the specific research recommendations that cover a 20-year period, and the expected research outcomes.

Fluid Physics Recommendations

- *Reduced-gravity multiphase flows, cryogenics, and heat transfer: database and modeling*—NASA should create a detailed reduced-gravity database that includes phase separation and distribution (i.e., flow regimes), phase-change heat transfer, pressure drop, and multiphase system stability. In addition, NASA should support the development and use of direct numerical simulation and molecular simulation techniques (e.g., lattice Boltzmann methods) to improve the understanding of mission-enabling phase distribution, separation, liquid management, and phase-change system phenomena. **(AP1)**
- *Interfacial flows and phenomena*—NASA should investigate interfacial flows (including induced and spontaneous multiphase and cryogenic flows with or without phase change) relevant to storage and handling systems for cryogens and other liquids, life support systems, power generation, thermal control systems, and other important multiphase systems. **(AP2)**
- *Dynamic granular material behavior and granular subsurface geotechnics*—NASA should improve predictive capabilities related to the behavior of lunar and martian soils on the surface and at depth, with the ultimate goal of developing site-specific models. This would likely require both the ground-based and ISS testing of actual lunar soils. **(AP3)**
- *Dust mitigation*—NASA should develop fundamentals-based strategies and methods for dust mitigation during human and robotic exploration of planetary bodies. This should include experimental methods, the understanding of the fundamental physics of dust accumulation and electrostatic interactions, and methods for modeling dust accumulation. **(AP4)**
- *Complex fluid physics*—NASA should conduct experiments on the ISS leading to an understanding of complex fluid physics in microgravity, particularly with regard to the behavior of granular materials, colloids, foams, nanoslurries, biofluids, plasmas, non-Newtonian fluids, critical-point fluids, and liquid crystals. **(AP5)**

Combustion Recommendations

- *Fire safety*—NASA should develop improved methods for screening materials in terms of flammability and fire suppression in space environments. **(AP6)**
- *Combustion processes*—NASA should conduct droplet-phase, gas-phase, and solid combustion experiments in reduced gravity with longer durations, larger scales, new fuels (e.g., biofuels and synthetic fuels), and practical aerospace materials relevant to future missions. **(AP7)**
- *Numerical simulation of combustion*—NASA should develop and validate detailed single-phase and multiphase numerical combustion models to relate reduced-gravity and Earth-gravity tests, to interpret data from experiments and missions, and to facilitate experimental and design activities. **(AP8)**

Materials Science Recommendations

- *Materials synthesis and processing and the control of microstructure and properties*—NASA should support research in reduced gravity on the development of materials synthesis and processing and the control of microstructures to improve the properties of existing and new materials on the ground. **(AP9)**
- *Advanced materials*—NASA should support research to develop new and advanced materials that would enable operations in increasingly harsh space environments and reduce the cost of human exploration. **(AP10)**
- *In situ resource utilization*—NASA should support fundamental and applied research to develop technologies that would facilitate the extraction, synthesis, and processing of minerals, metals, and other materials that are available on extraterrestrial surfaces. **(AP11)**

FACILITIES

Currently, NASA has ground-based laboratories, drop towers, aircraft, the space shuttle, and the ISS for studying reduced-gravity phenomena. Reduced-gravity test durations range from several seconds to days.

The ISS creates significant possibilities for experimentation because of its microgravity environment and the ability to vent to an infinite high vacuum. The combustion integrated rack on the ISS has two inserts, the MDCA (Multi-user Droplet Combustion Assembly) and ACME (Advanced Combustion via Microgravity Experiments). The combustion integrated rack is currently configured for two sets of investigations using MDCA: the Flame Extinguishment Experiment (FLEX), which focuses on the efficiency of fire suppressants in microgravity; and FLEX-2, which focuses on more fundamental science issues. The fluids integrated rack is a multipurpose facility for fluid physics research in space. In addition, the microgravity glove box can be used for a wealth of experiments in the combustion, fluids, and materials areas. The Space Dynamically Responding Ultrasonic Matrix System (SpaceDRUMS) has recently been installed on the ISS to provide a facility for the containerless processing of materials using acoustic levitation.

Reduced-gravity experimental platforms are currently limited to aircraft (test duration up to 30 s with 10^{-2} *g*-jitter) and NASA's newly designed centrifuge in drop towers (test duration up to 5 s). The latter produces a very clean artificial gravitational level from 0 to 1 *g*, but it is subject to the Coriolis force and gravity gradients due to its small size. No reduced-gravity platforms capable of supporting longer test durations are currently available.

Other than the now-canceled ISS centrifuge, future possibilities include a rotating free-flyer (with or without a tether). This could include an emptied cargo vessel for long-duration experiments. For example, when cargo is delivered to the ISS, it comes in relatively large vessels (with volumes up to about 40 m^3), which are emptied and then ejected back into Earth's atmosphere where they burn up on re-entry. Before they are destroyed, however, these vessels could be used for relatively large-scale microgravity experiments lasting up to 2 h. The possibility of using these vessels for experiments is a potentially important opportunity for both basic science research and research in support of the exploration mission. For example, empty cargo vessels could be used for fire safety tests to assess whether the properties of a combustion system scale to larger sizes. The absence of *g*-jitter created by equipment and astronauts also makes them an ideal platform for crystal growth experiments that are particularly sensitive to vibrations.

TABLE 9.1 High-Priority Research Areas and Topics, Status, Recommended Research, and Outcomes for 2020 and Beyond

Research Area and Topic	Current Status	2010-2020	2020 and Beyond	Outcomes
Fluid Physics				
Reduced-gravity multiphase flows, cryogenics, and heat transfer: database and modeling: Phase separation and distribution, phase-change heat transfer, pressure drop, and multiphase system stability. (AP1)[a]	—Only very limited, mostly qualitative, reduced-gravity data exist, leading to insufficient designer confidence. —Few reliable detailed simulations of reduced-gravity multiphase phenomena exist.	—Design and build a multipurpose phase-change test loop for the fluids integrated rack aboard the International Space Station (ISS). —Acquire targeted database on phase distribution and separation, phase-change heat transfer (e.g., boiling and condensation), pressure drop, and system stability. —Perform detailed direct numerical simulations (DNSs) or molecular simulations of selected phase distribution, liquid management, cryogenics, and phase-change phenomena at reduced gravity.	—Acquire comprehensive, detailed three-dimensional data on phase distribution and separation and phase-change heat transfer. —Develop mechanistic, multiscale three-dimensional computational multiphase fluid dynamic models (using a reduced-gravity database and DNS or molecular simulation results). —Develop a one-dimensional drift-flux model based on a reduced-gravity database.	—A reliable database with which to develop and assess accurate models for the design and analysis of new and/or significantly improved systems for NASA (e.g., for power production and utilization, waste water recovery, on-orbit fueling, in situ resource utilization (ISRU) extraction of water from surface materials, etc.). —Reliable predictive capabilities for multiphase flow and heat transfer at reduced-gravity levels for system design, scale-up, and analysis.
Interfacial flows and phenomena: Acquisition of data and development of mechanistic models for induced and/or spontaneous multiphase and cryogenic flows with and without heat transfer. (AP2)	Reduced-gravity thermo-capillary and buoyancy-driven flows are reasonably well understood. In contrast, there is a poor understanding of problems dominated by moving contact lines with partial wetting.	Perform targeted experiments that expand core knowledge and improve designer options and confidence (multiuser facilities preferred).	Expand breadth of experiments to increase technology readiness level.	A reliable database, models, and experience sufficient to design and analyze mission-enabling spacecraft fluid systems with dramatically increased reliability. Terrestrial applications also expected.

TABLE 9.1 Continued

Research Area and Topic	Current Status	2010-2020	2020 and Beyond	Outcomes
Dynamic granular material behavior and granular subsurface geotechnics: Granular flow dynamics and geotechnics for Moon and Mars environments for human or robotic exploration, ISRU mining, and habitation. (AP3)	—Computational methods and models are limited to simple particle shapes; impact of gravity is uncertain. —Current characterization methods are based largely on empiricism specific to Earth; exploration and sampling techniques are unsuitable for extraterrestrial use. —No lunar soil data are available at depths below a few meters, where structures may be sited and mining for ISRU will occur.	—Develop computational methods and models for irregular-shaped particles and crushing/compaction of agglutinates. —Improve understanding of dynamic interactions with vehicle systems and cratering, including effects of gravity. —Develop methods for in situ sampling at depth. —Develop suitable models for stress-strain behavior of soils for foundations, berms, etc.	—Improve particle-scale models for irregular-shaped particles. —Improve multiscale models that include the effects of gravity. —Collect lunar soil samples for Earth-based characterization. —Develop methods for excavating and conveying materials for ISRU. —Develop methods for the design of below-grade structures and foundations.	—Predictive capability for interactions between vehicles and granular materials, cratering, excavating, and the jamming-flow transition for complex granular systems at reduced gravity. —Accurate and reliable predictive models of lunar and martian soil behavior for analysis and design of structural foundations, berms, slopes, and excavations. —Accurate and reliable computational models for the deformational and strength behavior of extraterrestrial soils.
Dust mitigation: Development of fundamentals-based strategies for dust mitigation on lunar and martian surfaces. (AP4)	—There is qualitative evidence of dust-related challenges from previous lunar missions and observed atmospheric dust-related phenomenon on Mars. —Minimal fundamental understanding of the physics of dust accumulation exists.	—Conduct ISS and ground-based experimental investigation of dust accumulation, specifically electrostatic effects. —Develop models/simulations of dust in extraterrestrial planetary environments.	—Extend approaches specific to lunar or martian environments. —Develop practical methods for mitigating adverse effects of dust on seals, sensors, and solar panels.	—Fundamental understanding of dust behavior in reduced gravity. —Practical approaches for mitigating impacts of dust on mechanical systems and sensors.
Complex fluid physics: Utilization of ISS microgravity environment to study the fundamental physics of complex fluids and flows. (AP5)	Many important microgravity experiments have been completed, but fundamental aspects of many issues have yet to be resolved.	—Conduct fundamental experiments on the ISS to unravel the complex behavior of granular material, colloids, foams, biofluids, plasmas, non-Newtonian fluids, critical-point fluids, and liquid crystals without gravitational bias. —Conduct fundamental experiments on complex flow systems otherwise biased by gravity.	Continue relevant research on the ISS.	A better understanding of the physics of complex fluids and flows.

continued

TABLE 9.1 Continued

Research Area and Topic	Current Status	2010-2020	2020 and Beyond	Outcomes
Combustion				
Fire safety: Screening methods for material flammability and fire suppression for space applications. (AP6)	Current terrestrial methods are inadequate.	Improve and supplement current materials screening methods.	Optimize and implement new methods.	Improved fire safety for astronauts.
Combustion processes: Combustion experiments in reduced gravity to cover longer durations, larger scales, new fuels, and practical aerospace materials relevant to future missions. (AP7)	Incomplete knowledge of basic processes and their response to reduced gravity existed. The importance of gravity in combustion has been demonstrated and new phenomena in reduced gravity have been discovered.	Complete droplet-phase, gas-phase, and solid experiments on the ISS. Begin preparations and planning for large-scale, long-duration experiment.	Conduct larger-scale, longer-duration experiments.	Deeper understanding of fundamental combustion phenomena. Some of the fundamental knowledge contributes to enabling technologies for fire safety. Other knowledge contributes to terrestrial applications.
Numerical simulation of combustion: Development and validation of single and multiphase numerical combustion models that relate reduced-gravity and Earth-gravity tests. (AP8)	Many computational tools are available; input data and boundary conditions are incomplete. Theoretical and numerical treatment of solid processes in combustion should be improved.	Develop and validate selected numerical models with reduced-gravity experiments.	Integrate models with experiment and design.	Validated numerical models to enable prediction, design, and interpretation of data from experiments and missions.
Materials Science				
Materials synthesis and processing and the control of microstructure and properties: Study of materials synthesis and processing and mircrostructure development that are affected by gravity. (AP9)	Very little research has been done in the past 8 years. Prior to this, extensive research was carried out, as outlined in prior National Research Council reports.[b,c]	Provide benchmark data for materials synthesis and processing and microstructural control using reduced gravity.	Employ new experimental facilities to address questions that cannot be answered with existing ISS facilities.	An increased ability (1) to understand and predict the formation of microstructure and properties of a wide range of materials in terrestrial and space environments and (2) to create new materials.
Advanced materials: Materials that enable the NASA mission. (AP10)	Very little fundamental research has been conducted on advanced materials for space exploration. NASA has relied on existing materials.	Develop novel, advanced materials using both experimental and computational experimental techniques and methods.	Develop novel materials that will significantly improve weight and property factors (e.g., decreased weight, increased operating temperature, and self-healing capabilities).	Improved spacecraft and mission capabilities at reduced cost.

TABLE 9.1 Continued

Research Area and Topic	Current Status	2010-2020	2020 and Beyond	Outcomes
In situ resource utilization: Fundamental studies of how to utilize in situ minerals and materials. (AP11)	Although the need has been recognized, little research has been conducted in this area.	Identify and produce a selected group of strategic elements (e.g., oxygen), materials, and components that enables space exploration and can be manufactured from extraterrestrial resources in both normal and reduced gravity.	Produce elements, materials, and/or components on the Moon, Mars, and/or asteroids.	Improved prospects for extended human exploration to extraterrestrial bodies.

[a]Recommendation identifiers are as listed with clarifying material in the main text of this chapter and also in Tables 13.1 and 13.2.
[b]National Research Council, *Microgravity Research in Support of Technologies for the Human Exploration and Development of Space and Planetary Bodies*, National Academy Press, Washington, D.C., 2000.
[c]National Research Council, *Assessment of Directions in Microgravity and Physical Sciences Research at NASA*, The National Academies Press, Washington, D.C., 2003.

PROGRAMMATIC RECOMMENDATIONS

NASA has long known (in some cases for many decades) about the need for research that enables the crewed exploration of space, but to date some needs have not been thoroughly addressed. As a consequence, for example, NASA and its contractors cannot reliably design and deploy large-scale multiphase systems and processes in space, fire safety research in reduced gravity is relatively immature, and spacecraft are designed using materials and design techniques that are generally available rather than tailored for a specific mission. To change this situation, NASA should alter the way that relevant applied research is solicited and funded. Moreover, the panel agrees with recent recommendations[141] that a new long-term space technology research program with realistic objectives and stable funding is urgently needed.

It appears unlikely that individual principal-investigator-driven research programs will satisfy the mission-oriented research needs of NASA. For example, in many cases well-coordinated research teams are needed. In others the direction of the research that satisfies NASA's needs may not be represented by the proposals received in response to a broad research announcement. The panel emphasizes the following pertinent observations made in a 2004 report prepared for NASA:[142]

> Industry and other government research agencies, such as the Defense Advanced Research Projects Agency and the Office of Naval Research, regularly engage in programmatic research. . . . These agencies typically request or invite the formation of appropriate research teams and solicit research proposals from one or more of these teams, [or they request mission-related proposals from individual principal investigators]. Research is initiated based on a careful internal or external review and assessment of the team's capabilities and responsiveness to the sponsor's needs of the proposed research. [Subsequent to funding, the] . . . sponsor or its designee continuously evaluates the performance of the research team or teams through programmatic meetings and the peer-reviewed technical publications that result from the research being performed. The process is a dynamic one where research directions are changed as required to accomplish [emerging] mission goals. The researchers who are involved in such programs normally find this an exciting and rewarding way to do research because they are all on the critical path, and peer pressure among team members encourages superior performance. (p. 15)

REFERENCES

1. National Research Council. 2000. *Microgravity Research in Support of Technologies for the Human Exploration and Development of Space and Planetary Bodies.* National Academy Press, Washington, D.C.
2. National Research Council. 2000. *Microgravity Research in Support of Technologies for the Human Exploration and Development of Space and Planetary Bodies.* National Academy Press, Washington, D.C.
3. National Research Council. 2003. *Assessment of Directions in Microgravity and Physical Sciences Research at NASA.* The National Academies Press, Washington, D.C.
4. National Research Council. 2000. *Microgravity Research in Support of Technologies for the Human Exploration and Development of Space and Planetary Bodies.* National Academy Press, Washington, D.C.
5. Weislogel, M.M. 2001. *Survey of Present and Future Challenges in Low-g Fluids Transport Processes.* NASA Contract Report C-74461-N. TDA Research, Wheat Ridge, Colo.
6. National Research Council. 2000. *Microgravity Research in Support of Technologies for the Human Exploration and Development of Space and Planetary Bodies.* National Academy Press, Washington, D.C.
7. Koster, J.N., and Sani, R.L., eds. 1990. *Low-Gravity Fluid Dynamics and Transport Phenomena.* Progress in Astronautics and Aeronautics, Volume 130. American Institute of Aeronautics and Astronautics, Reston, Va.
8. Dodge, F.T. 2000. *The New Dynamic Behavior of Liquids in Moving Containers.* Southwest Research Institute, San Antonio, Tex.
9. Langbein, D. 2002. *Capillary Surfaces: Shape-Stability-Dynamics, in Particular under Weightlessness.* Springer Tracts in Modern Physics, Volume 178. Springer-Verlag, Berlin, Heidelberg, New York.
10. National Research Council. 2000. *Microgravity Research in Support of Technologies for the Human Exploration and Development of Space and Planetary Bodies.* National Academy Press, Washington, D.C.
11. National Research Council. 2000. *Microgravity Research in Support of Technologies for the Human Exploration and Development of Space and Planetary Bodies.* National Academy Press, Washington, D.C.
12. National Research Council. 2000. *Microgravity Research in Support of Technologies for the Human Exploration and Development of Space and Planetary Bodies.* National Academy Press, Washington, D.C.
13. National Research Council. 2000. *Microgravity Research in Support of Technologies for the Human Exploration and Development of Space and Planetary Bodies.* National Academy Press, Washington, D.C.
14. Motil, B. 2000. *Workshop on Research Needs in Fluids Management for the Human Exploration of Space.* NASA-GRC Topical Report. NASA Glenn Research Center, Cleveland, Ohio. September 22.
15. McQuillen, J., Rame, E., Kassemi, M., Singh, B., and Motil, B. 2003. *Results of the Workshop on Two-Phase Flow, Fluid Stability and Dynamics: Issues in Power, Propulsion, and Advanced Life Support Systems.* NASA/TM 2003-212598, NASA Glenn Research Center, Cleveland, Ohio. May 15. Available at http://gltrs.grc.nasa.gov/reports/2003/TM-2003-212598.pdf.
16. Otto, E.W. 1966. Static and dynamic behaviors of the liquid-vapor interface during weightlessness. *A.I.Ch.E. Chemical Engineering Progress Symposium Series* 62:6:158-177.
17. National Research Council. 2000. *Microgravity Research in Support of Technologies for the Human Exploration and Development of Space and Planetary Bodies.* National Academy Press, Washington, D.C.
18. National Research Council. 2003. *Assessment of Directions in Microgravity and Physical Sciences Research at NASA.* The National Academies Press, Washington, D.C.
19. Motil, B. 2000. *Workshop on Research Needs in Fluids Management for the Human Exploration of Space.* NASA-GRC Topical Report. NASA Glenn Research Center, Cleveland, Ohio. September 22.
20. McQuillen, J., Rame, E., Kassemi, M., Singh, B., and Motil, B. 2003. *Results of the Workshop on Two-Phase Flow, Fluid Stability and Dynamics: Issues in Power, Propulsion, and Advanced Life Support Systems.* NASA/TM 2003-212598, NASA Glenn Research Center, Cleveland, Ohio. May 15. Available at http://gltrs.grc.nasa.gov/reports/2003/TM-2003-212598.pdf.
21. Chiaramonte, F., and Joshi, J. 2004. *Workshop on Critical Issues in Microgravity Fluids, Transport, and Reactor Processes in Advanced Human Support Technology.* NASA/TM 2004-212940. NASA Glenn Research Center, Cleveland, Ohio. February.
22. National Research Council. 2000. *Microgravity Research in Support of Technologies for the Human Exploration and Development of Space and Planetary Bodies.* National Academy Press, Washington, D.C.
23. National Research Council. 2003. *Assessment of Directions in Microgravity and Physical Sciences Research at NASA.* The National Academies Press, Washington, D.C.
24. National Research Council. 2000. *Microgravity Research in Support of Technologies for the Human Exploration and Development of Space and Planetary Bodies.* National Academy Press, Washington, D.C.

25. Wilkinson, R.A., Behringer, R.P., Jenkins, J.T., and Louge, M.Y. 2005. Granular materials and the risks they pose for success on the Moon and Mars. In *Space Technology and Applications International Forum-STAIF05* (M.S. El-Genk, ed.). AIP Conference Proceedings. American Institute of Physics, College Park, Md.
26. Lahey, R.T., Jr. 2005. The simulation of multidimensional multiphase flows. *Nuclear Engineering and Design* 235(10-12):1043-1060.
27. Serizawa, A. 1974. Fluid Dynamic Characteristics of Two-Phase Flow. Ph.D. Thesis, Kyoto University, Kyoto, Japan.
28. Colin, C., Fabre, J., and Dukler, A.E. 1991. Gas-liquid flow at microgravity conditions. I. Dispersed bubble and slug flow. *International Journal of Multiphase Flow* 17(4):533-544.
29. Dukler, A.E., Fabre, J.A., McQuillen, J.B., Vernon, R. 1988. Gas-liquid flow at microgravity conditions: Flow patterns and their transitions. *International Journal of Multiphase Flow* 14(4):389-400.
30. Heppner, D.B., King, C.D., and Libble, J.W. 1975. Zero-gravity experiments in two-phase fluid flow patterns. ASME Preprint IS-ENAS-24. American Society of Mechanical Engineers, New York, N.Y.
31. Lee, H.S., Merte, H., Jr., and Chiaramonte, F. 1997. Pool boiling curve in microgravity. *Journal of Thermophysics and Heat Transfer* 11:216-222.
32. G.W. Bush. 2004. "A Renewed Spirit of Discovery: The President's Vision for U.S. Space Exploration" in *The Vision for Space Exploration*. NP-2004-01-334-HQ. NASA, Washington, D.C.
33. Review of U.S. Human Spaceflight Plans Committee. 2009. *Seeking a Human Spaceflight Program Worthy of a Great Nation.* Office of Science and Technology Policy, Washington, D.C. October.
34. Drew, D.A., and Passman, S.L. 1998. *Theory of Multicomponent Fluids.* Applied Mathematical Series, Volume 135. Springer-Verlag, New York, N.Y.
35. Ishii, M. 1975. *Thermo-Fluid Dynamic Theory of Two-Phase Flow.* Collection de la Direction des Etudes et Recherches d'Electricite de France, No. 22. Editeur Eyrolles, Paris, France.
36. Lahey, R.T., Jr. 2005. The simulation of multidimensional multiphase flows. *Nuclear Engineering and Design* 235(10-12):1043-1060.
37. Antal, S.P., Ettorre, S.M., Kunz, R.F., and Podowski, M.Z. 2000. Development of a next generation computer code for the prediction of multicomponent, multiphase flow. In *Proceedings of the International Conference on Trends in Numerical and Physical Modeling for Industrial Multiphase Flow*. Cargese, France, September 27-29, 2000. Centre de Mathématiques et de Leurs Applications, Cachan cedex, France.
38. Dhir, V.K. 2001. Numerical simulations of pool boiling heat transfer. *AIChE Journal* 47:813-834.
39. Lahey, R.T., Jr., 2009. On the direct numerical simulation of two-phase flows. *Nuclear Engineering and Design* 239(5):867-879.
40. Tryggvason, G., Bunner, B., Esmaeeli, A., Juric, D., Al-Rawahi, N., Tauber, W., Han, J., Nas, S., and Jan, Y.-J. 2001. A front-tracking method for computations of multiphase flow. *Journal of Computational Physics* 169:708-759.
41. Lahey, R.T., Jr., 2009. On the direct numerical simulation of two-phase flows. *Nuclear Engineering and Design* 239(5):867-879.
42. Aidun, C.K., and Clausen, J.R. 2010. Lattice Boltzmann methods for complex flows. *Annual Review of Fluid Mechanics* 42:439-472.
43. Nourgaliev, R.R., Dhin, T.N., Theofanous, T.G., and Joseph, D. 2003. The Lattice Boltzmann equation method: Theoretical interpretation, numerics and implications. *International Journal of Multiphase Flow* 29:117-169.
44. Koplik, J., Yang, J.-X., and Banavar, J. 1991. Molecular dynamics of drop spreading on solid surfaces. *Physical Review Letters* 67:3539.
45. Lahey, R.T., Jr., 2009. On the direct numerical simulation of two-phase flows. *Nuclear Engineering and Design* 239(5):867-879.
46. National Research Council. 2000. *Microgravity Research in Support of Technologies for the Human Exploration and Development of Space and Planetary Bodies.* National Academy Press, Washington, D.C.
47. National Research Council. 2000. *Microgravity Research in Support of Technologies for the Human Exploration and Development of Space and Planetary Bodies.* National Academy Press, Washington, D.C.
48. Hanratty, T.J., Theofanous, T.G., Delhaye, J.-M., Eaton, J., McLaughlin, J., Prosperetti, A., Sundaresan, S., and Tryggvason, G. 2003. Workshop on Scientific Issues in Multiphase Flow—Workshop findings. *International Journal of Multiphase Flow* 29(7):1047-1059.
49. Prosperetti, A., and Tryggvason, G. 2003. Appendix-3: Report of the Study Group on Computational Physics. *International Journal of Multiphase Flow* 29(7):1089-1099.
50. NASA Glenn Research Center. Space Flight Systems. ISS Research Project. Zero Boil-Off Tank Experiment. Available at http://spaceflightsystems.grc.nasa.gov/Advanced/ISSResearch/MSG/ZBOT/.
51. Weislogel, M.M., Thomas, E.A., and Graf, J.C. 2009. A novel device addressing design challenges for passive fluid phase separations aboard spacecraft. *Microgravity Science and Technology* 21(3):257-268.

52. National Research Council. 2000. *Microgravity Research in Support of Technologies for the Human Exploration and Development of Space and Planetary Bodies.* National Academy Press, Washington, D.C.
53. Lahey, R.T., Jr., and Podowski, M.Z. 1989. On the analysis of various instabilities in two-phase flows. *Multiphase Science and Technology* 4:183-370.
54. Schlichting, W., Lahey, R.T., Jr., and Podowski, M.Z. 2010. The analysis of interacting instability modes in a phase change system. *Nuclear Engineering and Design* 240:3178-3201.
55. Achard, J.L., Drew, D.A., and Lahey, R.T., Jr. 1985. The analysis of nonlinear density-wave oscillations in boiling channels. *Journal of Fluid Mechanics* 155:213-232.
56. Lahey, R.T., Jr., and Dhir, V. 2004. *Research in Support of the Use of Rankine Cycle Energy Conversion Systems for Space Power and Propulsion.* NASA/CR-2004-213142. Prepared under cooperative agreement NCC3-975. NASA Glenn Research Center, Cleveland, Ohio. July.
57. Metzger, P.T., Immer, C.D., Donahue, C.M., Vu, B.T., Latta, R.C., and Deyo-Svendsen, M. 2009. Jet-induced cratering of a granular surface with application to lunar spaceports. *Journal of Aerospace Engineering* 22(1):24-32.
58. Colwell, J.E., Batiste, S., Horányi, M., Robertson, S., and Sture, S. 2007. Lunar surface: Dust dynamics and regolith mechanics. *Reviews of Geophysics* 45:RG2006.
59. Sture, S., Costes, N.C., Batiste, S.N., Lankton, M.R., Alshibli, K.A., Jeremic, B., Swanson, R.A., and Frank, M. 1998. Mechanics of granular materials at low effective stresses. *Journal of Aerospace Engineering* 11(3):67-72.
60. Alshibli, K.A., and Hasan, A. 2009. Strength properties of JSC-1A lunar regolith simulant. *Journal of Geotechnical and Geoenvironmental Engineering* 135(5):673-679.
61. Schafer, B., Gibbesch, A., Krenn, R., and Rebele, B. 2010. Planetary rover mobility simulation on soft and uneven terrain. *Vehicle System Dynamics* 48(1):149-169.
62. Andrade, J.E., and Tu, X. 2009. Multiscale framework for behavior prediction in granular media. *Mechanics of Materials* 41:652-669.
63. Cundall, P.A., and Strack, O.D.L. 1979. A discrete numerical model for granular assemblies. *Geotechnique* 29:47-65.
64. Colwell, J.E., Batiste, S., Horányi, M., Robertson, S., and Sture, S. 2007. Lunar surface: Dust dynamics and regolith mechanics. *Reviews of Geophysics* 45:RG2006.
65. Pak, H.K., van Doorn, E., and Behringer, R.P. 1995. Effects of ambient gases on granular materials under vertical vibration. *Physical Review Letters* 74:4643-4646.
66. Brucks, A., Arndt, T., Ottino, J.M., and Lueptow, R.M. 2007. Behavior of granular flow under variable *g*-levels, *Physical Review E* 75:032301.
67. Rover Team. 1997. Characterization of the martian surface deposits by the Mars Pathfinder Rover, Sojourner. *Science* 278(5344):1765-1768.
68. Slane, F.A., and Rodriguez, G. 2006. A layered architecture for mitigation of dust for manned and robotic space exploration. *Proceedings of the 10th Biennial ASCE Aerospace Division International Conference on Engineering, Construction, and Operations in Challenging Environments.* March 5-8, 2006, League City/Houston, Tex. (R.B. Malla, W.K. Binienda, and A.K. Maji, eds.). American Society of Civil Engineers, Reston, Va.
69. Kahre, M.A., Murphy, J.R., and Haberle, R.M. 2006. Modeling the martian dust cycle and surface dust reservoirs with the NASA Ames general circulation model. *Journal of Geophysical Research-Planets* 111:E06008.
70. Verba, C.A., Geissler, P.E., Titus, T.N., and Waller, D. 2010. Observations from the High Resolution Imaging Science Experiment (HiRISE): Martian dust devils in Gusev and Russell craters. *Journal of Geophysical Research* 115:E09002, doi:10.1029/2009JE003498.
71. National Research Council. 2003. *Assessment of Directions in Microgravity and Physical Sciences Research at NASA.* The National Academies Press, Washington, D.C.
72. Ross, H.D., ed. 2001. *Microgravity Combustion: Fire in Free Fall.* Academic Press, San Diego, Calif.
73. Urban, D.L., and King, M.K. 1999. NASA's microgravity combustion research program: Past and future. *Combustion and Flame* 116:319-320.
74. Ross, H.D., ed. 2001. *Microgravity Combustion: Fire in Free Fall.* Academic Press, San Diego, Calif.
75. NASA. 1967. *Report of Apollo 204 Review Board.* NASA Historical Reference Collection, NASA History Division, NASA, Washington, D.C.
76. NASA. 2008. *Flammability, Odor, Offgassing, and Compatibility Requirements and Test Procedures for Materials in Environments that Support Combustion.* NASA-STD-6001. Updated interim version of 6001A, dated April 21. Available at https://standards.nasa.gov/released/NASA/2008_04_21_ NASA-STD-_I_-6001A_InterimAPPROVED.pdf.
77. Maradey, J.F., T'ien, J.S., and Prahl, J.M. 1977. *The Upward and Downward Flame Propagation Limits of Rigid Polyurethane Foams.* CWRU Report FTAS/TR-77-131. Case Western Reserve University, Cleveland, Ohio.

78. Olson, S.L., Kashiwagi, T., Fujita, O., Kikuchi, M., and Ito, K. 2001. Experimental observations of spot radiative ignition and subsequent three-dimensional flame spread over thin cellulose fuels. *Combustion and Flame* 125(1-2):852-864.
79. Sacksteder, K.R., and T'ien, J.S. 1994. Buoyant downward diffusion flame spread and extinction in partial-gravity accelerations. *Proceedings of the Combustion Institute* 25:1685-1692.
80. Lange, K.E., Perka, A.T., Duffield, B.E., and Jeng, F.F. 2005. *Bounding the Spacecraft Atmosphere Desing Space for Future Exploration Missions.* NASA/CR-2005-213689. NASA Johnson Space Center, Houston, Tex.
81. Urban, D.L., Griffin, D., Ruff, G.A., Cleary, T., Yang, J., Mulholland, G., and Yuan, Z.G. 2005. Detection of smoke from microgravity fires. International Conference on Environmental Systems, Rome, Italy. Paper 2005-01-2930. July. SAE Transactions, Warrendale, Pa. pp. 375-384.
82. Urban, D.L., Ruff, G.A., Sheredy, W., Cleary, T., Yang, J., Mulholland, G., and Yuan, Z.G. 2009. AIAA Paper 2009-956. American Institute of Aeronautics and Astronautics, Reston, Va.
83. Olson, S.L., Ferkul, P.V., and T'ien, J.S. 1988. Near-limit flame spread over a thin solid fuel in microgravity. *Symposium (International) on Combustion* 22(1):1213-1222.
84. T'ien, J.S. 1986. Diffusion flame extinction at small stretch rate: The mechanisms of radiative loss. *Combustion and Flame* 65(1):31-34.
85. Goldmeer, J.S., T'ien, J.S., and Urban, D.L. 1999. Combustion and extinction of PMMA cylinders during depressurization in low gravity. *Fire Safety Journal* 32:61-88.
86. Kimzey, J.H. 1974. Skylab experiment M479 Zero Gravity Flammability, Skylab results. Pp. 115-130 in *Proceedings of the Third Space Processing Symposium*, Volume 1. NASA TMX-70252. NASA Marshall Space Flight Center, Huntsville, Ala.
87. Ruff, G.A., Urban, D.L., Pedley, M.D., and Johnson, P.T. 2009. Fire safety. Pp. 829-884 in *Safety Design for Space Systems* (G.E. Musgrave, A. Larsen, and T. Sgobba, eds.). Elsevier, Oxford, U.K.
88. Delplanque, J.P., Abbud-Madrid, A., McKinnon, J.T., Lewis, S.J., and Watson, J.D. 2004. Feasibility study of water mist for spacecraft fire suppression. Proceedings of the Halon Options Technical Working Conference (HOTWC-04), University of New Mexico, Albuquerque, N.M. May. Available at http://www.fire.nist.gov/bfrlpubs/fire04/art070.html.
89. Schwer, D., and Kailasanath, K. 2010. *Numerical Simulations of Water-Mist Suppression of Flames in Reduced-g Environments.* NRL-6410-020. Naval Research Laboratory, Washington, D.C.
90. Butz, J.R., and Abbud-Madrid, A. 2009. Advances in development of fine water mist portable extinguisher. SAE Paper 2009-01-2510. 39th International Conference on Environmental Systems, July 12-16, 2009, Savannah, Ga. SAE International, Warrendale, Pa.
91. Takahashi, F., and Katta, V.R. 2009. Extinguishment of diffusion flames around a cylinder in a coaxial air stream with dilution or water mist. *Proceedings of the Combustion Institute* 32:2615-2623.
92. Honda, L., and Ronney, P.D. 1998. Effects of ambient atmosphere on flame spread at microgravity. *Combustion Science and Technology* 133:267-291.
93. Takahashi, F., Linteris, G., and Katta, V.R. 2008. Extinguishment of methane diffusion flames by carbon dioxide in coflow air and oxygen-enriched microgravity environments. *Combustion and Flame* 155:37-53.
94. National Research Council. 2003. *Assessment of Directions in Microgravity and Physical Sciences Research at NASA.* The National Academies Press, Washington, D.C., p. 31.
95. NASA Conference Publications for various International Microgravity Combustion Workshops: Second (1993; NASA CP-10113, Center for AeroSpace Information, Hanover, Md.), Third (1995); NASA CP-10174, Center for AeroSpace Information, Hanover, Md.), Fourth (1997; NASA CP-1997-020547, http://www.archive.org/details/nasa_techdoc_19970020547), Fifth (1999; NASA CP-1999-208917, http://gltrs.grc.nasa.gov/reports/1999/CP-1999-208917.pdf), Sixth (2001; NASA CP-2001-210826, http://ncmr04610.cwru.edu/events/combustion2001/CP-2001-210826.pdf), and Seventh (2003, NASA CP-2003-212376, http://gltrs.grc.nasa.gov/reports/2003/CP-2003-212376-REV1.pdf).
96. Bowman, C.T., ed. 1999. Special Issue on Microgravity Combustion. *Combustion and Flame,* Volume 116.
97. Faeth, G.M., ed. 1998. Special Section on Microgravity Combustion. *AIAA Journal* 38:1337-1379.
98. Ross, H.D., ed. 2001. *Microgravity Combustion: Fire in Free Fall.* Academic Press, San Diego, Calif.
99. Faeth, G.M. 2001. Laminar and turbulent gaseous diffusion flames. Pp. 83-182 in *Microgravity Combustion: Fire in Free Fall* (H.D. Ross, ed.). Academic Press, San Diego, Calif.
100. Ronney, P.D. 2001. Premixed flames. Pp. 35-82 in *Microgravity Combustion: Fire in Free Fall* (H.D. Ross, ed.). Academic Press, San Diego, Calif.
101. Ronney, P.D. 1998. Understanding combustion processes through microgravity research. *Proceedings of the Combustion Institute* 27:2485-2506.

102. Williams, F.A., and Grcar, J.F. 2009. A hypothetical burning-velocity formula for very lean hydrogen-air mixtures. *Proceedings of the Combustion Institute* 32:1351-1357.
103. Choi, M.Y., and Dryer, F.L. 2001. Microgravity droplet combustion. Pp. 183-298 in *Microgravity Combustion: Fire in Free Fall* (H.D. Ross, ed.). Academic Press, San Diego, Calif.
104. Dietrich, D. L., Haggard, J.B., Jr., Dryer, F.L., Nayagam, V., Shaw, B.D., and Williams, F.A. 1996. Droplet combustion experiments in spacelab. *Proceedings of the Combustion Institute* 26:1201-1207.
105. T'ien, J.S., Shih, H.Y, Jiang, C.B., Ross, H., Miller, F.L., Fernandez-Pello, A.C., Torero, J.L., and Walther, D. 2001. Mechanisms of flame spread and smolder wave propagation. Pp. 299-418 in *Microgravity Combustion: Fire in Free Fall* (H.D. Ross, ed.). Academic Press, San Diego, Calif.
106. Sedej, M.M. 1985. Implementing supercritical water oxidation technology in a lunar base environmental control/life support system. Pp. 653-661 in *Lunar Bases and Space Activities of the 21st Century* (W.W. Mendell, ed.). Lunar and Planetary Institute, Houston, Tex.
107. Ronney, P.D. 2003. Analysis of non-adiabatic heat-recirculating combustors. *Combustion and Flame* 135:421-439.
108. Gamezo, V.N., and Oran, E.S. 2006. Flame acceleration in narrow tubes: Applications for micropropulsion in low-gravity environments. *AIAA Journal* 44:329-336.
109. Merzhanov, A. 2002. SHS processes in microgravity activities: First experiments in space. *Advances in Space Research* 29(4):487-495.
110. Munir, Z.A., and Anselmi-Tamburini, U. 1989. Self-propagating exothermic reactions: The synthesis of high-temperature materials by combustion. *Materials Science Reports* 3(6):279-365.
111. Anselmi-Tamburini, U., Spinolo, G., Flor, G., and Munir, Z.A. 1997. Combustion synthesis of Zr-Al intermetallic compounds. *Journal of Alloys and Compounds* 247(1-2):190-194.
112. Kunrath, A.O., Strohaecker, T.R., and Moore J.J. 1996. Combustion synthesis of metal matrix composites: Part I. *Scripta Materialia* 34(2):175-181.
113. Kunrath, A.O., Strohaecker, T.R., and Moore J.J. 1996. Combustion synthesis of metal matrix composites: Part II. *Scripta Materialia* 34(2):183-188.
114. Kunrath, A.O., Strohaecker, T.R., and Moore J.J. 1996. Combustion synthesis of metal matrix composites: Part III. *Scripta Materialia* 34(2):189-194.
115. Greene, J.E., Motooka, T., Sundgren, J.-E., Rockett, A., Gorbatkin, S., Lubben, D., and Barnett, S.A. 1986. A review of the present understanding of the role of ion/surface interactions and photo-induced reactions during vapor-phase crystal growth. *Journal of Crystal Growth* 79(1-3):19-32.
116. Pan, J., Cao, R., and Yuan, Y. 2006. A new approach to the mass production of titanium carbide, nitride and carbonitride whiskers by spouted bed chemical vapor deposition. *Materials Letters* 60(5):626-629.
117. Kiefer, B., Karaca, H.E., Lagoudas, D.C., and Karaman, I. 2007. Characterization and modeling of the magnetic field-induced strain and work output in Ni2MnGa magnetic shape memory alloys. *Journal of Magnetism and Magnetic Materials* 312(1):164-175.
118. Monteverde, F. 2005. Progress in the fabrication of ultra-high-temperature ceramics: In situ synthesis, microstructure and properties of a reactive hot-pressed HfB2-SiC composite. *Composites Science and Technology* 65(11-12):1869-1879.
119. Rangaraj, L., Divakar, C., and Jayaram, V. 2010. Fabrication and mechanisms of densification of ZrB2-based ultra high temperature ceramics by reactive hot pressing. *Journal of the European Ceramic Society* 30(1):129-138.
120. Tang, S., Deng, J., Wang, S., Liu, W., and Yang, K. 2007. Ablation behaviors of ultra-high temperature ceramic composites. *Materials Science and Engineering: A* 465(1-2):1-7.
121. Dworak, D., and Soucek, M. 2003. Protective space coatings: A ceramer approach for nanoscale materials. *Progress in Organic Coatings* 47(3-4):448-457.
122. Donnet, C., Fontaine, J., Le Mogne, T., Belin, M., Héau, C., Terrat, J.P., Vaux, F., and Pont, G. 1999. Diamond-like carbon-based functionally gradient coatings for space tribology. *Surface and Coatings Technology* 120-121:548-554.
123. Hogmark, S., Jacobson, S., and Larsson, M. 2000. Design and evaluation of tribological coatings. *Wear* 246(1-2):20-33.
124. Balasubramaniam, R., Gokoglu, S., and Hegde, U. 2010. The reduction of lunar regolith by carbothermal processing using methane. *International Journal of Mineral Processing* 96(1-4):54-61.
125. Faierson, E.J., Logan, K.V., Stewart, B.K., and Hunt, M.P. 2010. Demonstration of concept for fabrication of lunar physical assets utilizing lunar regolith simulant and a geothermite reaction. *Acta Astronautica* 67(1-2):38-45.
126. Faierson, E.J., Logan, K.V., Stewart, B.K., and Hunt, M.P. 2010. Demonstration of concept for fabrication of lunar physical assets utilizing lunar regolith simulant and a geothermite reaction. *Acta Astronautica* 67(1-2):38-45.
127. Johnson, S.W. 2001. Extraterrestrial facilities engineering. Pp. 727-757 in *Encyclopedia of Physical Science and Technology* (R.A. Meyers, ed.). Academic Press, New York, N.Y.

128. Carr, R.H., Bustin, R., and Gibson, E.K., Jr. 1987. A pyrolysis/gas chromatographic method for the determination of hydrogen in solid samples. *Analytica Chimica Acta* 202:251-256.
129. Ruiz, B., Diaz, J., Blair, B., and Duke, M.B. 2004. Is extraction of methane, hydrogen and oxygen from the lunar regolith economically feasible? *AIP Conference Proceedings* 699:984.
130. Robens, E., Bischoff, A., Schreiber, A., Dabrowski, A., and Unger, K.K. 2007. Investigation of surface properties of lunar regolith: Part I. *Applied Surface Science* 253(13):5709-5714.
131. Badescu, V. 2003. Model of a thermal energy storage device integrated into a solar assisted heat pump system for space heating. *Energy Conversion and Management* 44(10):1589-1604.
132. Ulrich, S., Veilleux, J., and Corbin, F. 2009. Power system design of ESMO. *Acta Astronautica* 64(2-3):244-255.
133. Reddy, M.R. 2003. Space solar cells—Tradeoff analysis. *Solar Energy Materials and Solar Cells* 77(2):175-208.
134. Reed, K., and Willenberg, H. 2009. Early commercial demonstration of space solar power using ultra-lightweight arrays. *Acta Astronautica* 65(9-10):1250-1260.
135. Sone, Y., Ueno, M., and Kuwajima, S. 2006. Fuel cell development for space applications: Fuel cell system in a closed environment. *Journal of Power Sources* 137(2):269-276.
136. National Research Council. 2009. *Frontiers in Crystalline Matter: From Discovery to Technology.* The National Academies Press, Washington, D.C.
137. National Research Council. 2009. *Frontiers in Crystalline Matter: From Discovery to Technology.* The National Academies Press, Washington, D.C.
138. Yi, H.C., Woodger, T.C., Moore, J.J., and Guigné, J.Y. 1998. The effect of gravity on the combustion synthesis of metal-ceramic composites. *Metallurgical and Materials Transactions* 29B:889-897.
139. Burkes, D.E., Moore, J.J., Yi, H.C., Gottoli, G., and Ayers, R.A. 2005. Effects of environmental gas on combustion synthesis and microstructure of Ni3-TiCx composites. *International Journal of Self-Propagating High-Temperature Synthesis* 14(4):293-304.
140. Yi, H.C., Guigné, J.Y., and Moore, J.J. 2005. Application of self-propagating high temperature (combustion) synthesis (SHS) in in-space fabrication and repair (ISFR) and in-situ resource utilization (ISRU). *International Journal of Self-Propagating High-Temperature Synthesis* 14(2):131.
141. Review of U.S. Human Spaceflight Plans Committee. 2009. *Seeking A Human Spaceflight Program Worthy of a Great Nation.* Office of Science and Technology Policy, Washington, D.C. October.
142. Lahey, R.T., Jr., and Dhir, V. 2004. *Research in Support of the Use of Rankine Cycle Energy Conversion Systems for Space Power and Propulsion.* NASA/CR-2004-213142. Prepared under cooperative agreement NCC3-975. NASA Glenn Research Center, Cleveland, Ohio. July.

10

Translation to Space Exploration Systems

This chapter identifies the technologies that enable exploration missions to the Moon, Mars, and elsewhere, as well as the foundational research in life and physical sciences. Science and technology development areas are recommended to support near-term objectives and operational systems (i.e., prior to 2020) and objectives and operational systems for the decade beyond 2020. While technologies and operational systems may be near term (prior to 2020) or longer term (2020 and beyond), it is anticipated that supporting research will be conducted in the coming decade. In addition to defining research in science, this chapter, where appropriate, includes discussions of establishing the technological know-how required to ensure the orderly transition of new technology into space exploration systems.

The National Aeronautics and Space Administration's (NASA's) future exploration missions are likely to include long durations, microgravity and partial-gravity environmental conditions, and extreme thermal and radiation environmental conditions. The specific environmental conditions needed to successfully perform the required research are identified. To provide information that can be utilized in the most flexible manner possible, the specific schedules or timetables by which these research objectives should be achieved are not specified; instead, the relative temporal sequences are described.

An initial assessment of science and technology needs is based on an update of topics found in the National Research Council's (NRC's) 2000 report, *Microgravity Research in Support of Technologies for Human Exploration and Development of Space and Planetary Bodies*.[1] Other major sources of information include NASA's project plans for the Exploration Technology Development Program,[2] NASA's report, *Technology Horizons: Game-Changing Technologies for the Lunar Architecture*,[3] and the space exploration options discussed in the final report of the Review of U.S. Human Spaceflight Plans Committee (also known as the Augustine Commission or Augustine Committee).[4] Current NASA planning documents, such as the lunar architecture, were used to illuminate the possible need of future exploration activities. Science and technology needs were categorized into seven topic areas: space power and thermal management; space propulsion; extravehicular activity (EVA); life support; fire safety; space resource extraction, processing, and utilization; and planetary surface construction.

This chapter describes necessary scientific research and technology development in each of these seven areas and categorizes the research recommendations in one of two time frames: either "Prior to 2020" or "2020 and Beyond." The two time periods are used to indicate when, given a best-case scenario, an operational system or exploration activity is likely to be implemented. Placing the implementation time either prior to or after 2020 can be stated with greater certainty than trying to establish when the research to support the implementation of these activities would need to commence. The research required to support implementation will have to be initiated well in advance of the desired implementation date; however, the exact time required to accomplish the needed research is uncertain due to technical reasons, as well as to budget and political factors, all of which are beyond the ability of the panel to predict with any precision. The strategic and tactical decisions regarding if and when a research program should be undertaken appropriately fall to those responsible for the implementation of the various missions under consideration. Since NASA's exploration mission schedule has been notional at best for decades, the "Prior to 2020" period should be viewed as representing near-term activities (e.g., near-Earth human exploration activities), while the "2020 and Beyond" period represents activities that enable longer-term exploration goals (e.g., human exploration of the lunar surface, planetary surfaces, or deep-space missions).

While there is always more that can be learned to further the understanding critical to enabling future exploration, prioritizing the various areas of potential study allows the critical needs to be translated into a plan that can be implemented successfully. Since NASA's missions, budgets, and priorities cannot be predicted, selecting one particular technology over another would be premature. Instead, this chapter describes the attributes of and development issues associated with viable technology options and explains the critical research gaps that need to be addressed if a particular option is taken. The chapter therefore divides technologies and their associated research challenges into two categories: "required" and "highly desirable." A required technology is one NASA needs to achieve an exploration objective. A highly desirable technology is one that offers a significant benefit in performance, efficiency, cost savings, or likelihood of mission success. The rating as either required or highly desirable applies directly to the technology or operational system; research challenges in the life and physical sciences are listed with the technology or system they enable. The committee's prioritization process took the following factors into account:

- The relative importance of the research to its topic area,
- The topic area's impact on overall exploration efforts,
- The interdependencies among topic areas and how knowledge in one area could be an enabler or prerequisite for advancing knowledge in another area, and
- Whether the topic area's knowledge needs were unique to NASA's exploration requirements such that they would be left unaddressed were NASA not to pursue them.

Table 10.3 at the end of the chapter summarizes the technologies (and their associated research challenges) required for implementation prior to 2020. Table 10.4 similarly summarizes the technologies (and their associated research challenges) required for implementation in 2020 and beyond.

The Integrative and Translational Research for the Human Systems Panel considered the realities and challenges of transitioning new technology to enable or greatly improve systems unique to NASA and critical to space exploration. The transition process requires that engineers understand and apply the research results of scientists and that research scientists work within the parameter space of specific mission categories. Successful transition from research results to implemented technology requires that program managers, engineering leaders, and research leaders create an environment where scientists interact with engineers on the specifics of system requirements. NASA leadership will know that a success-oriented environment has been achieved when they no longer hear the familiar refrains: "the engineers are not talking to the scientists," "the scientists are not working in the regimes of interest to the engineers," and "the program managers are risk-averse and not qualifying improved components/subsystems for flight." Chapter 12's section "Linking Science to Mission Capabilities Through Multidisciplinary Translational Programs" contains an in-depth discussion of the issues associated with creating organizations that can successfully translate research into technologies and technologies into exploration systems.

RESEARCH ISSUES AND TECHNOLOGY NEEDS

Space Power and Thermal Management

NASA's power generation, energy storage, and heat rejection technology needs in the coming decades are driven by three major and diverse categories of missions: (1) platforms for near-Earth science, resources (such as orbiting cryogenic propellant depots), and communications; (2) lunar and planetary surface missions; and (3) deep-space exploration probes. These categories give rise to a spectrum of future power and thermal management requirements ranging from a few watts (e.g., for microsatellites) to tens of kilowatts and perhaps megawatts (e.g., propulsion systems for exploration missions or for a permanent lunar or Mars presence), and durations ranging from a few hours or days (e.g., planetary rovers) to perhaps tens of years (e.g., permanent surface presence, deep-space missions). These power demands will be met by a variety of evolutionary and revolutionary technologies for providing prime energy sources, energy conversion, energy storage, thermal management and control, and heat rejection. Prime energy sources include insolation,* radioisotopes, and nuclear fission. Energy conversion technologies applicable to this broad range of power needs include those for both "direct" energy conversion (such as photovoltaic and thermoelectric systems, fuel cells, primary and secondary batteries, and/or alkali metal thermal-to-electric conversion approaches) and "indirect" conversion or heat engine approaches (Stirling, Brayton, or Rankine cycles).

Energy storage is accomplished principally by chemical or thermal energy storage (e.g., electrochemical batteries, fuel cells, or thermal energy reservoirs). Thermal management will be accomplished by thermal heat acquisition, transport, and rejection methods including radiators, evaporators, sublimators, and thermal storage techniques (e.g., thermal wadis†), as well as advanced environmental control refrigeration cycles. The mission environment also influences and in some cases dictates the technology selection. Deep-space and extrasolar missions (extending beyond the orbit of Jupiter, such as Voyager) can today only be performed using nuclear energy sources. Near-Earth missions, such as the International Space Station (ISS), can be conducted using photovoltaic, chemical, or thermal methods. Outer planet missions may combine photovoltaic, chemical, thermal, or nuclear sources. Extended lunar and planetary presence missions are greatly enhanced by nuclear power, often referred to in this context as fission surface power. The emerging area of in situ resource utilization (ISRU) will require more capable power systems than previously deployed. Novel power and thermal solutions that increase performance (e.g., increase efficiency) and/or reduce mass will enable human exploration on the surfaces of other solar system bodies. Figure 10.1 shows the power and duration-of-use regimes for different space power systems operating at 1 AU from the Sun. The region of photovoltaic operation shrinks as a mission moves farther from the Sun.

Although NASA faces a wide range of challenging power and thermal energy technology requirements, it does have partners in meeting some of these needs. The Department of Defense (DOD), the commercial satellite industry, and the international space community will need many of the same technologies that NASA will need. Partnering, formally and informally, will reduce NASA's research, development, and manufacturing production costs. Nevertheless, there remain NASA-unique power system technology needs for which NASA must bear the full cost, most notably nuclear technology—although the Department of Energy (DOE) would be a partner in the future development of nuclear systems, just as it is currently a partner in the program for radioisotope power systems (RPSs); the advanced Stirling radioisotope generator; and the fission surface power system technology development effort).‡

Prime power system mass for exploration missions to the Moon and Mars is a major fraction of the total mission mass to be transported from Earth. Thus, gains in efficiency and lifetime directly reduce mission cost. In

---

*Insolation is the solar radiation energy received on a given surface area in a given time. The term "insolation" is a contraction of "incoming solar radiation."

†Thermal wadis are engineered sources of stored solar energy using modified lunar regolith as a thermal storage mass.

‡DOE has a statutory responsibility "for the conduct of research and development activities relating to . . . production of atomic energy, including processes, materials, and devices related to such production" (Atomic Energy Act of 1954, as amended, Sec. 31). In addition, DOE, "as agent of and on behalf of the United States, shall be the exclusive owner of all [nuclear] production facilities" (Atomic Energy Act of 1954, as amended, Sec. 41). There are some exceptions regarding the ownership requirements, but they would not apply to the production of nuclear material to fuel space nuclear power systems. For example, under the existing memorandum of agreement between NASA and DOE regarding RPSs, DOE's responsibilities include the design, development, fabrication, evaluation, testing, and delivery of RPSs to meet NASA system-performance and schedule requirements. DOE also provides nuclear risk assessments; specifies minimum radiological, public-health, and safety criteria and procedures for the use of RPSs; provides safeguards and security guidance for NASA facilities and services; supports NASA operational plans, mission definition, environmental analysis, launch approval, and radiological contingency planning; affirms the flight readiness of RPSs with respect to nuclear safety; participates in the nuclear launch approval process; jointly investigates and reports nuclear incidents; and assumes legal liability for damage resulting from nuclear incidents and accidents involving RPSs.

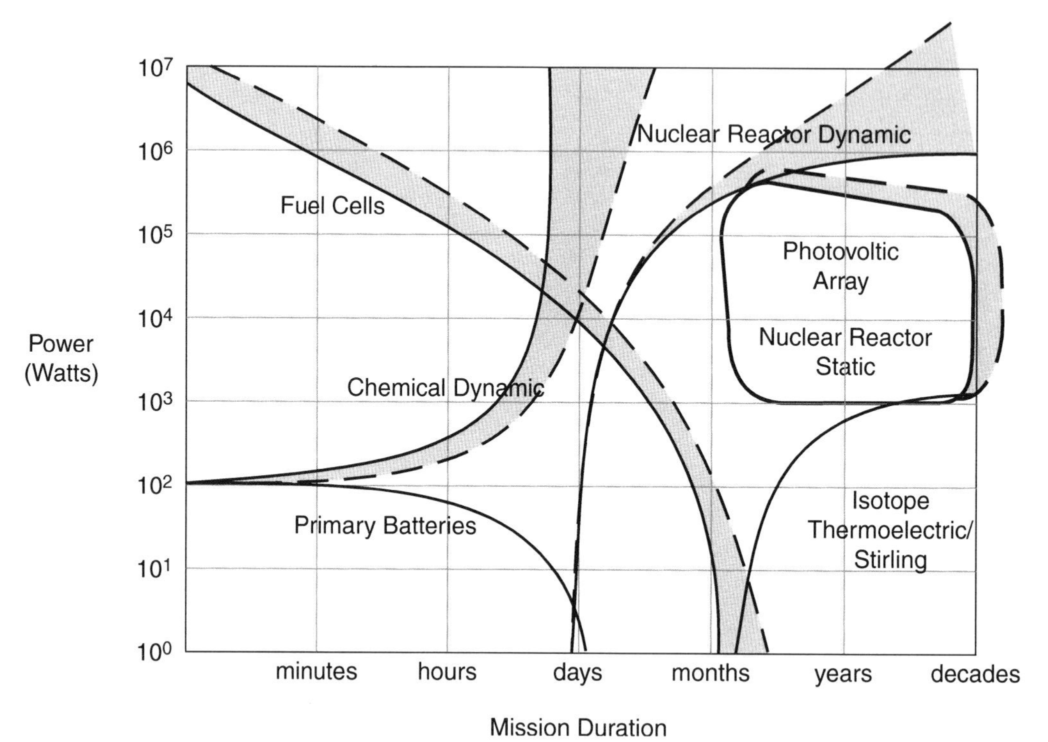

FIGURE 10.1 Diagram of the qualitative power and duration regimes for representative candidate space power systems. The gray areas are potential growth beyond the solid line in the next two decades. Generalizations are difficult since power requirements are highly mission-dependent. SOURCE: Anthony K. Hyder, University of Notre Dame, adapted from Figure 1.7 in A.K. Hyder, R.L. Wiley, G. Halpert, D.J. Flood, S. Sabripour, *Spacecraft Power Technologies*, Imperial College Press, London, 2000. © 2000 World Scientific.

addition, several of the enabling technologies for more affordable space exploration, such as small modular reactors, have the potential to transition to important and timely applications on Earth.

Power Generation Systems

Solar Power Systems

Photovoltaic power generation has an extensive heritage in terrestrial and space applications. Space-based systems ranging from a few watts ("nanosats") to many tens of kilowatts (space station) are currently in use.[5] Arrays have been developed for use in some of the most extreme temperature and radiation environments in space (e.g., the Mercury orbiter MESSENGER and the Jupiter orbiter Juno), although high radiation levels still degrade system performance over time and therefore limit mission lifetime.

Typical current spacecraft solar arrays achieve areal power densities of about 200-300 W/m^2 at specific powers ranging from 20 to 150 W/kg (using 27 percent efficient triple-junction solar cells). The efficiency of solar cells for space applications is projected to reach 37 percent by 2020. The ISS solar array produces 250 kW of power and is the largest space solar array to date. Concentrator arrays have been flown to distances of 1.5 AU from the Sun and have demonstrated 2.5 kW of output power (at 300 W/m^2 and 45 W/kg). Arrays capable of achieving up to 220 W/kg have been demonstrated on the ground.

Solar cells for terrestrial applications using concentrators have now demonstrated energy conversion efficiencies of 38.5 percent; efficiencies of 45 percent are on the near horizon and will enable further reductions in array area. Terrestrial systems capable of producing tens of megawatts are planned; such surface-based systems would have application to the exploration of planetary bodies with or without atmospheres.

TABLE 10.1 Solar Irradiance and Operating Temperatures for Different Environments

	Earth Orbit	Lunar Surface	Martian Surface
Solar irradiance (W/m^2)	1,368±41	1,368±41	590[a]
Temperature range (°C)	−100 to +100	−180 to +250	−140 to +20

[a]Average value from G.A. Landis and J. Appelbaum, Design considerations for Mars photovoltaic systems, pp. 1263-1270 in Conference Record of the Twenty First IEEE, Photovoltaic Specialists Conference, Kissimmee, Fla., May 21-25, 1990; actual irradiance varies with location and season on Mars.

Table 10.1 shows the solar irradiance and operating temperatures for Earth orbit and for lunar and martian surfaces; solar array size (and mass) requirements depend on these environmental parameters.

The total mass of a photovoltaic-based power generation system capable of producing tens of kilowatts on the lunar or martian surface will be driven by the energy storage components (e.g., batteries, regenerative fuel cells [RFCs],[§] or thermal energy reservoirs), power management electronics, and thermal management system, as well as the size of the solar array. The design of surface power systems will also need to consider the native dust environment, plasma arcing issues,[6] cosmic radiation, and the performance of associated thermal management components. Thermal management technologies should also be adapted to local environmental conditions.

Photovoltaic power generation on the surface of Mars requires an array approximately three times larger in area than needed in Earth orbit or on the lunar surface, due to the increased distance from the Sun. Energy storage requirements are far less demanding on Mars than on the Moon because of the much shorter martian "night" (12 h versus 2 weeks); a battery system on the surface of the Moon will require 28 times the capacity (and roughly the same increase in mass) as one on Mars to support the same load. But the effects of dust, dirt, and wind on the martian surface require additional mitigations that will increase the mass of any surface solar array. However, the net result is that photovoltaic power systems on Mars require much less mass to produce the same amount of steady-state power as a system on the Moon.

While the Juno mission will use photovoltaic power at Jupiter (at about 5.5 AU), the current practicality of using solar power diminishes with greater distances from the Sun, due to a combination of the fall-off in solar intensity and colder operating temperatures.[7]

Concentrating solar-thermal power systems are also in development for terrestrial applications and are potentially valid for space applications.[8] Typically, these consist of a solar concentrator such as a parabolic trough unit (which can produce heat at temperatures up to about 500°C) or a parabolic dish unit (which can produce heat at temperatures up to 1,000 to 2,000°C) combined with a heat engine (such as a Stirling cycle engine). For extraterrestrial surface applications, concentrating solar-thermal power systems can be manufactured from ultralight reflective foils creating deployable booms that can supply thermal or electricity power, or both, for ISRU.[9]

Lightweight solar arrays can be a low-mass, efficient source of power for near-Earth applications and can support future science missions that use solar electric power, such as visits to asteroids, cargo transport to the Moon or Mars, or possibly outer planet missions.

It is likely that the evolutionary development of multijunction solar cells will reach its practical achievable performance potential in the coming decade and additional modest performance gains will be found from technology development in concentrators. Next-generation photovoltaic technologies, such as nanotechnology-based solar cells or quantum-dot solar cells, have potential advantages of lower mass, lower cost, and/or higher efficiencies over existing photovoltaic technologies. Though these technologies are unlikely to be mass-competitive with RPSs for deep-space applications in the next 10 years, factors such as cost and availability may make them attractive alternatives for inner solar system missions in the future.

[§]All fuel cells utilize a fuel and an oxidizer. Fuel cells combine reactants to produce electrical power and waste heat. RFCs can either combine or produce reactants to produce or store power.

Nuclear Power Systems

Space nuclear power systems are among the NASA-unique technology needs. These include RPSs (Pu-238 energy source) and fission reactors as a primary energy source.

RPSs continue to be a high priority for NASA because they provide reliable, long-term power where solar power is not feasible. More than 26 NASA and DOD spacecraft have used radioisotope power since 1961. A recent NRC report[10] documents the current catastrophic lack of Pu-238, as well as lack of plans to produce a supply. The merits of Pu-238 include the low emitted-radiation shielding requirements, long half-life, and high energy density. Pu-238 production is a complex process including nuclear reactor irradiation and radiochemistry processing. Neither the United States nor any other country currently has the capability to produce Pu-238. Both the United States and Russia have small stockpiles of Pu-238; planned missions will consume it.¶ Russia has stopped selling Pu-238 to the United States.

Re-establishing Pu-238 production capability is critical to sustaining a deep-space mission capability and is a crosscutting enabler for research and development (R&D) as well as science missions. Congress has not approved DOE's fiscal year (FY) 2010 and 2011 budget requests to begin re-establishing a domestic Pu-238 production capability. While Pu-238 production is a technology/policy issue rather than an area of research, the issue is noted here because of its importance to deep-space science mission spacecraft.

Nuclear reactor power systems could supply substantially more electrical power than current prime power systems (solar cells or RPSs). NASA flew the System for Nuclear Auxiliary Power–10A (SNAP-10A), a 500-We thermoelectric fission reactor, in 1965 but has not flown a nuclear reactor since. The 1983 to 1994 100-kW SP-100 project and the last decade's Prometheus project were both terminated before reactor development could be completed.** Fission surface power systems are "an attractive power option for some lunar and Mars mission scenarios,"[11] and NASA has identified nuclear power reactors as one of 19 game-changing technologies for the lunar exploration architecture.[12] NASA has identified nuclear power as a high-priority technology because it releases many other exploration technologies from severe constraints on power use.[13] The recent confirmation of water at the lunar poles and on Mars underscores the need for ample power to enable ISRU for propellant production and life support.[14,15] The availability of nuclear reactor power systems would make it possible to relax stringent power constraints, thus reducing development costs across the entire lunar exploration architecture—except for the cost of the power system itself. Development of a nuclear power reactor for lunar missions will likely be a long and expensive effort. Looking back, efforts by both the SP-100 program and Prometheus Program to develop space nuclear reactors were terminated prematurely as support for these expensive projects dwindled over the years in the face of tight budgets and new agency priorities. Furthermore, even if space nuclear reactor systems were successfully developed, the cost of manufacturing each system (including the nuclear fuel) would be so high that such systems would be suitable only for very large missions that could afford a power system costing billions of dollars to develop.

NASA, in cooperation with DOE, is currently supporting a fission surface power system (FSPS) technology effort, but the magnitude, scope, and goals of this effort are quite modest in comparison to the total effort required to develop an operational reactor system. The current goal of the FSPS is to "generate the key products to allow Agency decision-makers to consider FSPS as a preferred option for flight development."[16] Continuation of NASA's FSPS program is desirable, and it would be advantageous to the project if the prototyping of the test unit were

---

¶The NRC study committee wrote, "The total amount of ^{238}Pu available for NASA is fixed, and essentially all of it is already dedicated to support several pending missions—the Mars Science Laboratory, Discovery 12, the Outer Planets Flagship 1 (OPF 1), and (perhaps) a small number of additional missions with a very small demand for ^{238}Pu. If the status quo persists, the United States will not be able to provide RPSs for any subsequent missions" (NRC, *Radioisotope Power Systems: An Imperative for Maintaining U.S. Leadership in Space Exploration*, The National Academies Press, Washington, D.C., 2009). The nominal launch date for OPF 1 is 2020 (see http://opfm.jpl.nasa.gov/), at which point the stockpile will be depleted.

**Project Prometheus was terminated in 2005 after it became clear that it would cost at least $4 billion to complete development of a spacecraft reactor module and at least $16 billion in total to develop the entire spacecraft and complete the mission, not counting the cost of the launch vehicle or any financial reserves to cover unexpected cost growth (Jet Propulsion Laboratory, *Project Prometheus Final Report*, 982-R120461, Jet Propulsion Laboratory, Pasadena, California, available at http://trs-new.jpl.nasa.gov/dspace/bitstream/2014/38185/1/05-3441.pdf, 2005, p. 178).

accelerated and carried out in parallel with testing to confirm the reactivity characteristics of the material assembly. Supporting physical science research includes high-temperature, low-weight materials for power conversion and materials for high-temperature radiators. New system geometries can be investigated further to take advantage of possible inherent shielding of the generators. Extending the current technology effort to develop an FSPS system could enable a power-rich lunar architecture, accelerating research on the lunar surface in all other areas. However, the hardware systems being assembled for testing do not include any nuclear fuel, and the current project does not include fuel development, which would be very costly (for DOE) and is unlikely to happen until NASA makes a firm commitment to deploy an FSPS. Once the current FSPS technology project is complete, NASA may decide to make fission surface power systems a priority, but that has not happened yet.

In conclusion, nuclear reactor systems have much to offer to the exploration of space, but they would also be very expensive to develop. History cautions against underestimating how difficult it would be to complete development of space nuclear power reactor systems on a scale large enough to support future U.S. space exploration missions.

Thermal Energy Conversion

Three dynamic energy conversion cycles are Stirling, Brayton, and Rankine. Stirling power conversion for space applications uses sealed gas/piston-linear alternator components that can operate at relatively high efficiency with comparatively small heat source-sink differential temperatures. Brayton power conversion uses a closed cycle version of gas turbine alternator technology. Rankine cycle systems use a space-adapted version of terrestrial two-phase steam power plants having a turbine and alternator. Stirling and Brayton systems, because they use gas as the working fluid, have gas heat transfer coefficients that are typically smaller than two-phase flow heat transfer coefficients. This necessitates larger heat transfer surfaces than Rankine cycle systems but eliminates issues related to zero gravity, freeze-thaw cycles, and two-phase flow. Thus, at higher powers, Rankine systems typically have smaller components than Stirling or Brayton systems because of the smaller heat transfer area required. Further, the Rankine cycle condenser (radiator) operates at a higher temperature and in an isothermal mode, and so the size/mass of the heat rejection system is much smaller.

Figure 10.2 compares various thermal energy conversion systems for producing electrical power as a function of specific mass (kilograms per kilowatt-electric) and power (kilowatts-electric). It shows that at low power, static conversion technologies such as thermoelectric and thermionic generally have lower specific mass, whereas at higher powers Stirling, Brayton, and finally, Rankine cycles have lower specific mass. Existing space power systems have been at relatively low power (e.g., the ISS, with a lifetime average electrical power of ~75 kWe produced from 260-kW solar arrays). However, future power systems for habitats, life support, ISRU, and propulsion all trend toward higher powers, including up to ~100 MWe. Thermoelectric and Stirling power conversion technologies are not mass competitive at these high power levels, as indicated in Figure 10.2. Thus, for NASA's projected high-power needs, the Brayton system may be a reasonable option to achieve required system mass performance. Figure 10.2 shows that there is little difference between a thermoelectric system and a Stirling system at low power levels. Therefore, at low powers, a new power conversion technology must be developed if performance gains are to be achieved. Thermophotovoltaic energy conversion, which operates at high temperatures and converts thermal power to electrical power using a photovoltaic-like technology, has the potential for reduced weight, high conversion efficiency, and operational simplicity.[17] However, it is currently an immature technology that would require significant R&D before it would be a viable option for NASA.

Energy Storage

Advanced energy storage technologies can offer an order-of-magnitude improvement over current technology. Lithium ion batteries offer a theoretical energy density of 700 W·h/kg and RFCs a theoretical energy density of 1,000 W·h/kg. Practical embodiments of these potential electrical energy storage devices will likely attain only about half the theoretical values, but this would nonetheless represent a 10-fold improvement over current

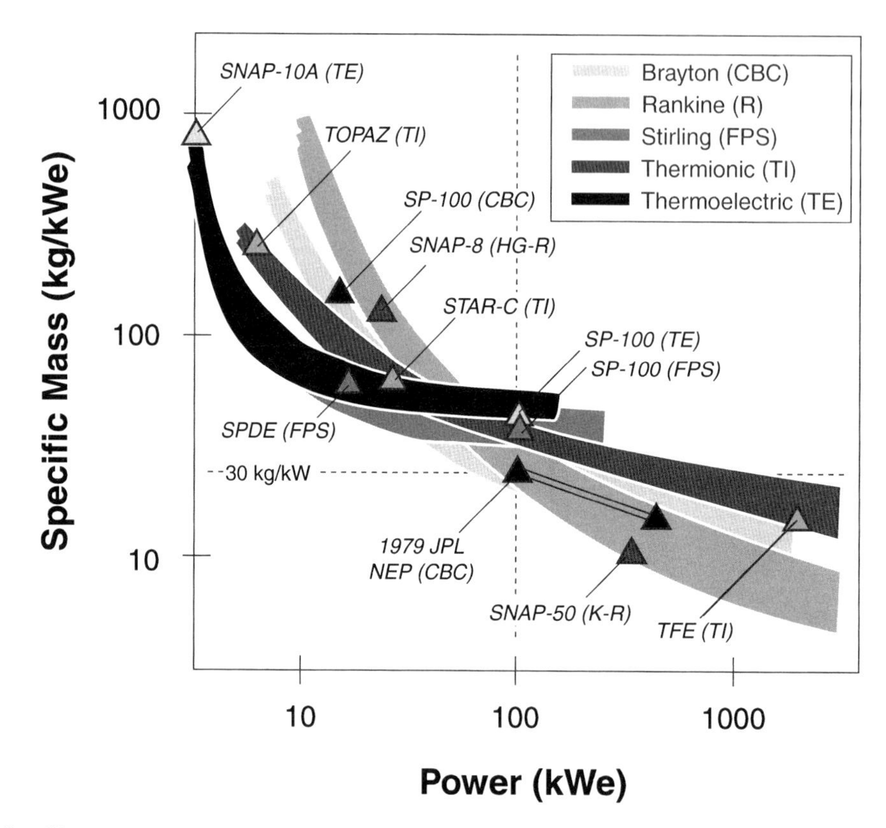

FIGURE 10.2 Specific mass versus power for various conversion technologies. SOURCE: L. Mason, Power Technology Options for Nuclear Electric Propulsion, Intersociety Engineering Conference on Energy Conversion Paper No. 20159, NASA Glenn Research Center, Cleveland, Ohio, 2002.

technology. Figure 10.3 shows the generally achievable energy and power densities for current and near-term electrochemical storage technologies.

NASA Glenn Research Center researchers and others have developed concepts for RFCs that would store energy on the ISS and on high-altitude balloons or high-altitude aircraft. They are now investigating RFCs for storing energy on the Moon or Mars.[18] A unitized RFC (Figure 10.4)[††] would use no electrolyzer; rather, it would regenerate water and store the hydrogen and oxygen as high-pressure gases directly through a single stack.

RFCs are considered viable options for solar energy storage in a wide variety of environments where the Sun eclipse period is several hours or longer. Reactant tankage size is proportional to the required stored energy (eclipse load power × eclipse time). For applications in which the stored energy requirement is modest, reactant tankage mass is also modest. However, for long eclipse time (14 days) and high eclipse power (tens of kilowatts or greater), such as for lunar surface power, the tankage mass becomes significant, perhaps as much as 5-10 percent of the overall mass. Thermal management, tankage mass optimization, and system mass versus operating pressure are also issues to consider. To use hydrogen and oxygen on the lunar surface or elsewhere on a large scale, safe, practical storage systems must be developed. Fuel cells, both primary and regenerative, have been in existence for years. However, the issues of dead-ended gas flow paths versus through-flow, cryogenic versus pressurized gas storage, thermal management, and reliable long-life operation (i.e., years) under reduced gravity and extreme temperatures all remain to be demonstrated for both planetary and orbital applications.

[††]Unitized RFCs are a particular packing geometry.

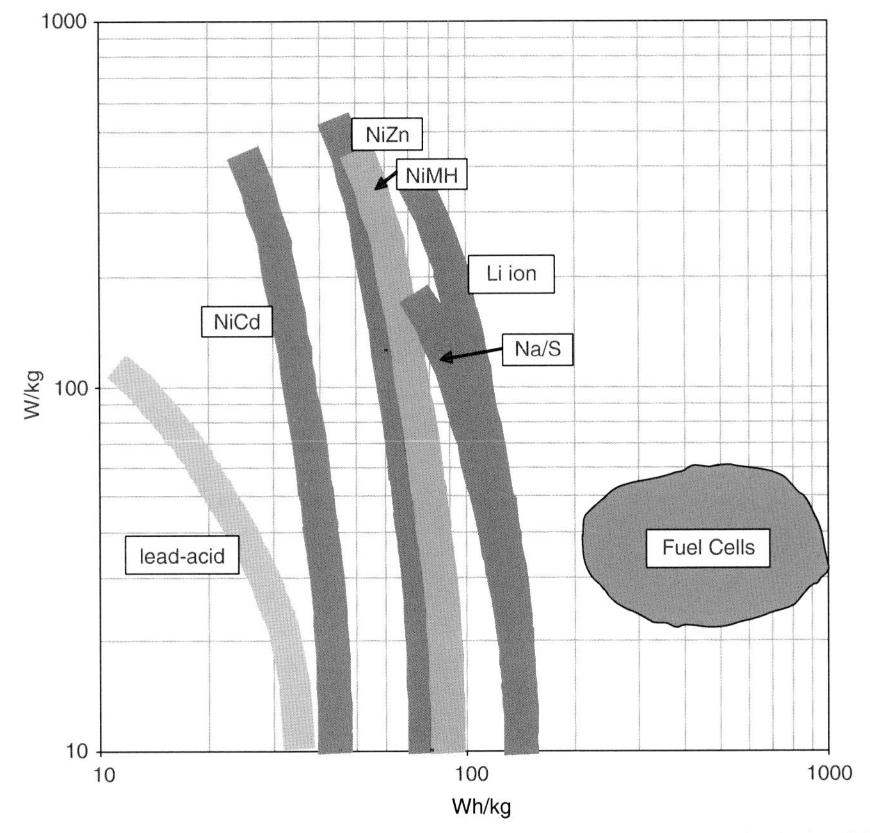

FIGURE 10.3 Generalized specific energy versus specific power map for several near-term technologies. SOURCE: Adapted from E.J. Cairns, Battery, Overview, Pp. 117-126 in *Encyclopedia of Energy* (C.J. Cleveland, ed.), Vol. 1, Elsevier, New York, Copyright 2004, with permission from Elsevier.

With few exceptions, NASA spacecraft and planned planetary systems all require electrical energy storage. The ISS is currently using nickel metal hydride batteries, but lithium ion batteries with the requisite energy storage density and cycle life are under development. Currently, nickel hydrogen batteries are the workhorse spacecraft energy storage approach but are being supplanted by rapidly evolving lithium ion approaches, which offer significant (>2 to 5 times) increases in energy density. Li-ion approaches are now baselined for most NASA robotic explorer missions under development in the Science Mission Directorate. If nuclear power is not used, RFCs and/or thermal energy storage must be developed in order to satisfy mass, power, and energy requirements for lunar and martian bases, especially when ISRU is incorporated into the base.

Thermal Management

The major design goals for any space thermal management system are high performance, reduced cost, reduced physical size, and high reliability. Earth-based system processes involving phase change and/or multiple phase flow have been shown to have the highest heat transfer coefficients.[19] The benefits of two-phase flows are illustrated in Figure 10.5, in which boiling heat transfer coefficients exceed single phase heat transfer coefficients by multiples in all cases.

While the thermal management technology requirements for NASA's different missions overlap, there are unique challenges posed by each environment. NASA requirements must ensure cooling for spacecraft on inner planet missions (which experience high insolation), on planetary surfaces (where dust from the lunar regolith

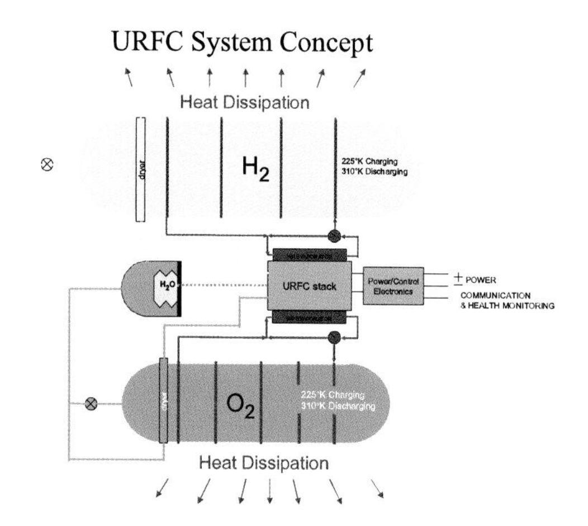

FIGURE 10.4 Unitized regenerative fuel cell system concept. SOURCE: K.A. Burke, *Unitized Regenerative Fuel Cell System Development*, NASA/TM-2003-212739, NASA Glenn Research Center, Cleveland, Ohio, 2003; available at http://ntrs.nasa.gov/archive/nasa/casi.ntrs.nasa.gov/ 20040027863_2004008361.pdf.

or martian soil degrades radiator surfaces), and on cis-lunar space missions. Further, high-power missions will ultimately require two-phase (single-component, vapor/liquid) thermal management technologies yet to be demonstrated in microgravity and in lunar and martian partial gravity. Gas and liquid phases in components such as boilers, condensers, and heat pipe radiators behave differently in other than Earth gravity due to the change in body force. An NRC report documented the research needs in two-phase thermal management technologies a decade ago,[20] but little has been done to advance the technology readiness level (TRL) of the component technologies. Two-phase flow technologies offer reduced system mass and size because their high heat transfer coefficients

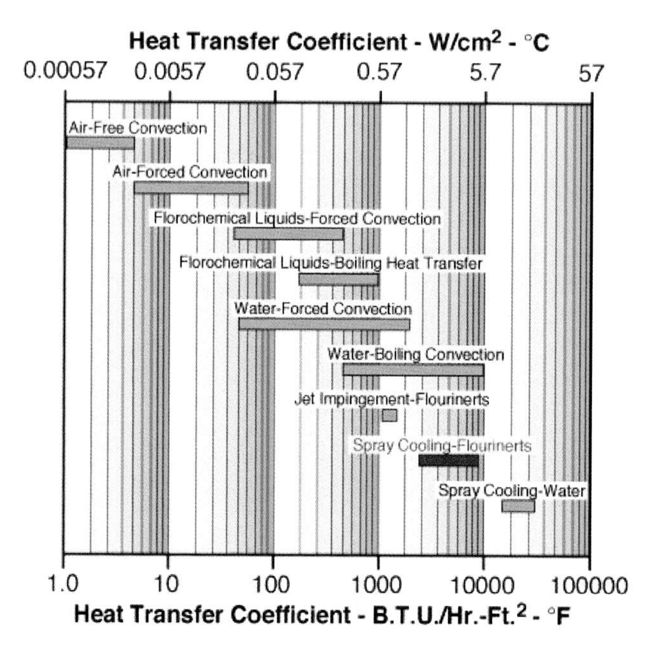

FIGURE 10.5 Comparison of heat transfer coefficient provided by different cooling technologies. SOURCE: K. Sienski and C. Culhane, Advanced system packaging for embedded high performance computing, pp. 211-214 in *Digest of Papers of the Government Microcircuit Applications Conference*, Volume 21, Orlando, Fla., March 1996.

require smaller and therefore lighter hardware than single-phase systems with the same capacity. Two-phase systems are naturally isothermal.

Active two-phase flow thermal management technologies operating in reduced-gravity conditions are essential to enable high-power thermal management systems.[21] However, such technologies require separation of the two phases of a single component. This need has been described in numerous NRC and NASA reports.[22,23] If missions requiring high power are to be carried out efficiently, this technology must be brought to TRL 8 and demonstrated at a system level; currently the technology is at perhaps TRL 6. Microchannel devices for heat exchange and phase separation are also under investigation. Component- and system-level tradeoffs among channel sizing, interchannel stability, pumping power, and overall system mass are being studied.

Gaps in the demonstration of two-phase technology have led NASA to avoid system designs based on multiphase flow. For example, the main cooling loops (ammonia and water) for the ISS are single-phase fluid loops, even though using two-phase flow would have resulted in reduced system mass and power requirements. NASA needs to be able to scale such systems confidently and predict their performance and system reliability by quantifying failure. Until these gaps are filled, NASA will not deploy two-phase flow systems, regardless of the cost and/or mass performance advantages they promise. Hence, more research is needed on two-phase systems subjected to reduced gravity.

Thermal Storage

Liquid-solid phase-change devices are commonly used in spacecraft thermal control to regulate temperature during peak heat generation periods and/or to stabilize payload component temperatures during Sun eclipse. A variety of waxes and hydrated salts have been employed. Their drawbacks include heat transfer rate limits (unless encapsulated in a high-conductivity porous matrix), relatively low effective heat capacity per unit mass, and freeze/thaw volume changes. Novel encapsulated thermal energy storage microcapsules mixed in a single-phase fluid have received some development funding from NASA but remain in an early stage of development. The effect of such a "slurry" working fluid is to both increase the heat transfer properties of the pumped liquid and increase its effective specific heat.

Thermal energy storage using modified surface material is a promising new thermal tool that can use in situ resources. Energy storage can be provided through the application of thermal energy reservoirs, with the thermal mass being provided by either materials brought from Earth (such as phase-change materials) or processed space resources. For example, lunar regolith can be modified through thermal process methods to yield a material whose thermal diffusivity[‡‡] is increased by approximately two orders of magnitude compared with untreated regolith. Thermal energy storage can also be directly integrated with solar concentrators for nighttime power generation. The placement of thermal wadis[24] has been proposed as an engineered source of heat (and power) for the protection of rovers and other exploration assets on the lunar surface.

Summary of Enabling Science and Technologies for Space Power and Thermal Management

Power and thermal management improvements will become increasingly important for future NASA mission needs, especially for missions that include ISRU. Important exploration technologies in space power and thermal management that would benefit from near-term R&D include the following:

[‡‡]A high thermal diffusivity is crucial for thermal energy storage because it allows a material's heat capacity to be utilized. Thermal diffusivity, which describes the rate at which heat flows through a material, is the thermal conductivity of that material divided by its volumetric heat capacity. In SI units, thermal diffusivity is measured in m^2/s.

Before 2020: Required

Two-Phase Flow Thermal Management Technologies. NASA would benefit from the potential advantages of low mass, small component size, and isothermality for future missions with high power requirements. Research should be conducted to address active two-phase flow questions relevant to thermal management. **(T1)**

2020 and Beyond: Required

Regenerative Fuel Cells. NASA would benefit from the enhanced flexibility in power and energy storage offered by regenerative fuel cells. The necessary research should be conducted to allow regenerative fuel cell technologies to be demonstrated in reduced-gravity environments, including research related to dead-ended gas flow paths versus through-flow, cryogenic versus pressurized gas storage, thermal management, and reliable long-life operation. **(T11)**

Energy Conversion Systems. NASA would benefit from additional energy conversion capabilities in the low- and high-power regimes, as shown in Figure 10.2. The development of thermal energy conversion technologies beyond the existing thermoelectric and Stirling systems is needed to enable higher-performance missions. Research should be done on high-temperature energy conversion cycles and devices coupled to essential working fluids, heat rejection systems, materials, etc. **(T12)**

Fission Surface Power. Fission surface power could be a valuable option to NASA in the future for missions with high power requirements. The continued development of supporting technologies and systems space nuclear reactor power would ensure that reactor power systems are a viable option for future space exploration missions. Areas of physical science research that enable the development of those systems include high-temperature, low-weight materials for power conversion and radiators. **(T13)**

Other

Radioisotope Production. While there are no underlying science or technology gaps, re-establishing domestic production of Pu-238 is necessary to ensure the continued viability of deep-space missions.

Space Propulsion

To support future space exploration missions, an evolutionary space transportation architecture will need to deliver humans, surface habitats, and transportation systems for the purposes of (1) exploration for science discovery and (2) maturing technology for the next exploration destination. In addition to supporting precursor missions in the near term, smaller scale and even micro propulsion options will be required in the far future, as exploration sensors and payloads become smaller and nanotechnology matures.

NASA's original Constellation space transportation architecture consisted of Earth-to-orbit launch vehicles for delivering humans and large cargos to orbit.[25] Proposed launch vehicles such as Ares I and Ares V were to use derivatives of the legacy large-thrust propulsion technology base including space shuttle solid rocket motors, Delta IV RS-68s, Apollo second and third stage J-2s, and Apollo/space shuttle aluminum cryogenic tanks. Thus, minimal technology development for launch vehicles was envisioned for missions to the Moon and Mars.

The Augustine Review Committee offered several other approaches in addition to the Constellation architecture, including the development of one launch vehicle for both humans and cargo, the use of current Evolved Expendable Launch Vehicles (EELVs), and space shuttle-derived concepts,[26] to improve affordability and involve commercial space enterprises. However, in all cases, little or no new propulsion technology was required for these launch vehicles because existing, or derivatives of existing, rocket engines were envisioned.

For small- to medium-lift launch vehicles (e.g., current EELVs), cryogenic propellant depots for on-orbit refueling could reduce cost, increase mission payload, and improve mission success.[27–30] Figure 10.6 illustrates the

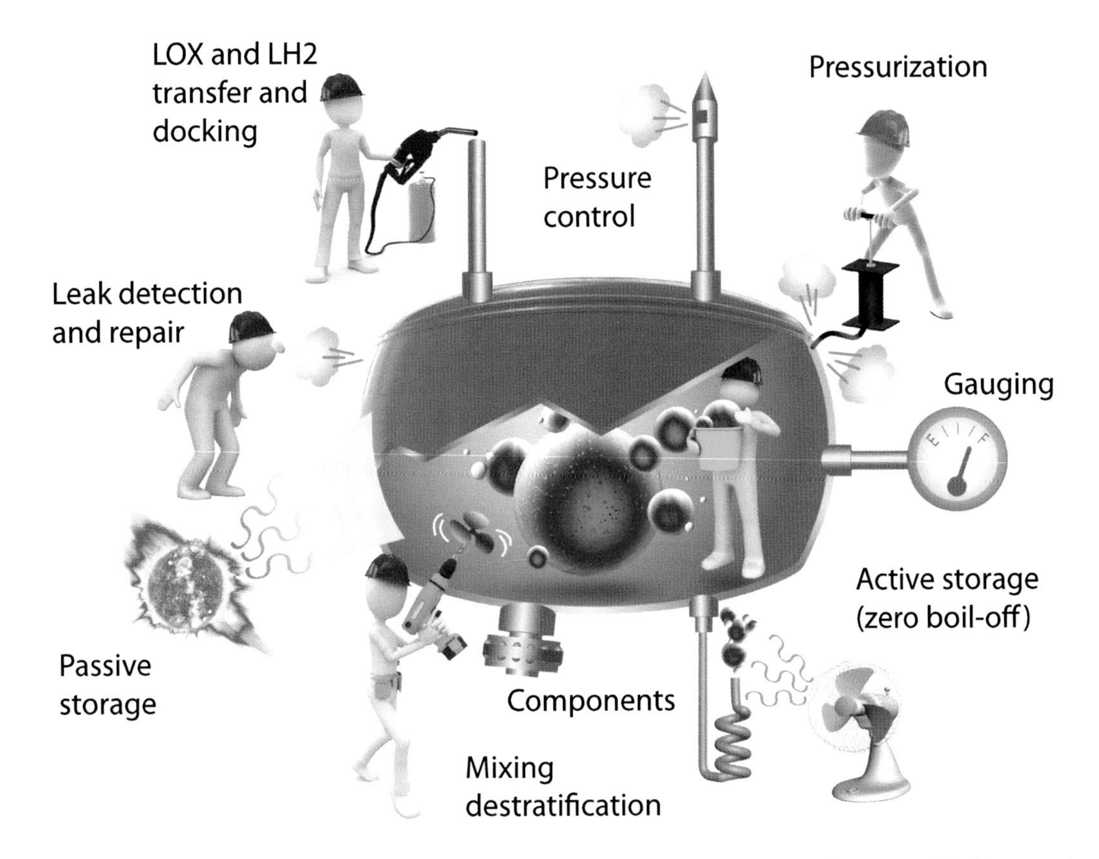

FIGURE 10.6 Guide to cryogenic propellant management for on-orbit fueling and vapor-free supercritical cryogenic propellant transfer. SOURCE: After D.J. Chato, Low Gravity Issues of Deep Space Refueling, paper presented at the 43rd AIAA Aerospace Sciences Meeting and Exhibit, Reno, Nev., January 2005; available at http://gltrs.grc.nasa.gov/reports/2005/TM-2005-213640.pdf.

principal operations of robotically managing cryogenic propellant storage and transfer. Propellant depots can be operated and filled through competitive commercial space launch companies, which, because of the cost advantage of even 4 to 6 EELVs over an equivalent heavy-lift launcher, would use multiple EELV launches to incrementally deliver propellant to a depot. This high-rate launch demand would provide a sustained competitive market for EELVs, with the potential to dramatically reduce the cost per kilogram to orbit compared with the alternative scenario of lifting all the propellant for a long-duration mission on a single launcher that also lifts the mission payload to orbit. Moreover, with depots, EELVs can deliver much heavier critical exploration transportation systems to orbit because only the inert mass of the systems would be lifted into orbit, improving ΔV[§§] capability by 75 percent over a vehicle delivered to orbit with all its propellant. For lunar missions, the Ares V could be eliminated, and for Mars missions, the majority of Ares V launches could be eliminated.[31] On-orbit refueling increases operational complexity and requires advanced cryogenic fluid transport and handling technology for reduced gravity, but 80 to 90 percent of the initial mass in low Earth orbit (IMLEO) propellant mass is decoupled from the delivery of the critical exploration transportation systems, providing an additional opportunity for tradeoffs to reduce costs.

A space exploration system's mass, power requirements, and cost are driven by the mission requirements (including safety) and destination. A long-term lunar mission, such as a 180-day mission to an outpost at Shackleton Crater near the south pole of the Moon, would dramatically increase the system requirements over those of

[§§] ΔV is the magnitude of the change in velocity.

the Apollo missions, which had only two-crew/several-day sorties near the Moon's equator. For every kilogram of mass delivered to the Moon, 600 kg of IMLEO transportation mass is required. Thus, advances in the in-space propulsion system, which is at least 80 percent of the IMLEO, have a dramatic compounding impact on total exploration architecture viability and sustainability.

NASA's existing plans for a lunar mission include the use of in-space propulsion based on derivatives of the Apollo oxygen/hydrogen J-2 for the Earth departure stage, Centaur oxygen/hydrogen RL-10 for the Altair lunar lander, and Delta II hypergolic storable AJ-10 for the lunar ascent stage and the Earth return service module. Higher-performance oxygen/methane engines were originally planned for the ascent stage and service module because of the synergy with the planned Mars descent and ascent modules,[32] but they were dropped because the technology was not sufficiently mature. In addition, the even more energetic oxygen/hydrogen engine was not considered because of the lack of technology for long-term (180 days for the Moon and 1 year for Mars) cryogenic storage and engine ignition concerns. Current cryogenic tank multilayer insulation has a liquid hydrogen boiloff rate of approximately 3 percent per month for the Lunar Module.[33] Thus, advanced active cooling techniques are desirable for lunar missions and necessary for the longer Mars missions or for long-term storage propellant depots for lunar missions.

For Mars missions in 2030 and beyond, the Mars Transfer Vehicle with Nuclear Thermal Rockets is projected to be two-thirds the IMLEO of a vehicle using oxygen-hydrogen propulsion, reducing the number of Ares V launches from 12 to 9.[34] Both the nuclear and the chemical propulsion systems can be developed using the existing technology base, although unique test facilities have to be developed for the nitrogen thermal rocket (NTR). For long-duration missions, such as a 900-day Mars human mission, zero-boiloff technologies would be required for cryogenic propellant storage. In addition, as described in earlier chapters, such a lengthy mission would affect system reliability and crew health in terms of physical/physiological deconditioning of the crew and exposure of both vehicle systems and the crew to radiation. For such long-duration missions, more than half the transits would likely be for cargo transfer, for which low-thrust, high-specific-impulse options like solar or nuclear electric propulsion (SEP or NEP) systems should be considered. However, the high cost of an NTR or a NEP system makes them suitable only for relatively long, heavy-payload missions that cannot easily be supported by non-nuclear power or propulsion systems.

For Mars orbit insertion and atmospheric descent, significant performance benefits can be achieved with proposed drag devices, which could reduce descent engine performance requirements by 60 to 80 percent. For Mars ascent vehicles, oxygen-methane engines have been proposed,[35] which could take advantage of ISRU of the carbon dioxide atmosphere and possibly the surface ice (water) of Mars to eliminate the need to supply ascent propellant from Earth.[36] Other options for ascent stages from the surface of the Moon and Mars, for which the architecture includes both crewed and uncrewed vehicles, include noncryogenic or Earth-storable propellants such as hydrazine or advanced storable propellants. Requirements for the high performance and high reliability of propellants following long-term exposure to the harsh environments in space and, for some systems, the surfaces of the Moon or Mars, will demand greater understanding of propellant storage and handling.

Cryogenic Fluid Management

Advances in in-space cryogenic fluid management technology can improve the affordability and performance of orbiting cryogenic propellant depots and hence the feasibility of long-duration exploration missions.[37] Research areas of particular interest include passive insulation and active cooling for zero boiloff; zero-gravity propellant transfer, including the automated coupling of cryogenic fluid lines; gauging the quantity of propellant in the tanks;[38] and the role of capillary forces in propellant management.[39] In addition, low-mass, cryogenic compatible, thermally insulated multifunctional materials for tank storage and fluid line transfer could potentially enhance safety and affordability over using multiple layers of material for each function. Micrometeroid protection for long-term propellant storage depots will require extensive materials interaction research to ensure that appropriate models are developed for depot replenishment schedules. Because some of the liquid could be dispersed as drops or globs floating in random parts of the container, the gauging of liquid quantity, both contained and transferred, in low gravity presents new, important problems. This issue is exacerbated by the low density of most cryogens, which will require large storage facilities.

Propellant boiloff mitigation can be achieved through the use of both active and passive systems. Passive techniques such as multilayer insulation and vapor-cooled shields are more mature than active systems, but they cannot achieve zero boiloff. The current state of the art in passive insulation limits propellant loss of liquid hydrogen to approximately 3 percent per month, a level not sufficient for long-term propellant storage. Active thermal management systems use a refrigeration system, called a cryo-cooler, to keep the propellant below its vaporization temperature. Research and demonstrations are required to assess refrigeration versus reliquefaction of the cryogens with respect to minimizing system mass and power requirements.

For in-space cryogenic transfer, new technologies are required for leak-free connect/disconnect and fluid transfer. For fluid transfer, a number of technologies could be applied including linear acceleration (where the fluid is pushed toward the feedline) or angular acceleration (where the fluid is forced to the outside of the tank for collection). The in-space depot will be required to continuously supply vapor-free supercritical cryogenic liquids to an orbital transfer vehicle at an acceptable flow rate and pressure drop. Research and development is needed to characterize and develop a design database for fundamental screen wicking characteristics, surface tension data, stratification, and screen channel outflow performance with cryogenic fluids.[40] A low-gravity mass gauging with an accuracy of better than 5 percent of fill tank will be required for the in-space depot. Although mass gauges available now lack the desired accuracy needed for low-gravity applications, two concepts—the compression mass gauge and the optical mass gauge—appear promising.

Cryogenic Propellants

As noted above, some future space architectures will include propulsion using liquid oxygen as oxidizer with both hydrogen and methane engines. Although the technology for these engines is well known for Earth applications, critical technologies are needed for lunar and planetary descent and ascent in zero or reduced gravity, including engine start, combustion stability, and deep throttle. These technologies require research in cryogenic fluid management, propellant ignition, flame stability, and active thermal control of the injectors and combustors. The associated physical processes[41] should be understood and predictable over the full range of gravitational variation, orientation, acceleration, temperatures, fluid phases, destratification, etc.

Noncryogenic (Earth Storable) Propellants

Earth-storable, hypergolic propellant engines for long-term placement of launch assets (uncrewed and crewed) on other moons and planets require a good understanding of the impacts of fluid migration, tribology for fluid handling devices, and chemistry (mixing or separation) in low-gravity environments. These technologies will be essential not only for propulsion but also for ISRU for relevant missions.

Supersonic Retro Propulsion and Aeroassist

For large-payload delivery to Mars and other planetary bodies with atmospheres, the use of large inflatable aerodynamic decelerators or rigid biconic/ellipsled aeroshells for atmospheric entry and descent deceleration, coupled with supersonic retro propulsion, could reduce the propulsive ΔV requirements by an order of magnitude relative to an all-propulsive descent system. The reduction in propulsive ΔV from aeroassist systems could significantly reduce IMLEO by factors of approximately 2 to 5 (depending on assumptions about system performance and mass), resulting in mission masses comparable to NEP or NTR systems.[42] Similar aeroassist technologies could also be used for planetary braking, eliminating the orbit insertion burns required for planetary orbit capture. Inflatable aerodynamic decelerators are projected to have a 35 percent mass advantage over rigid systems and an 80 percent mass advantage over all-propulsive systems.[43] However, the technology has not been demonstrated in a flight environment.

Current technology gaps include the materials and technologies to reduce the mass of inflatable and rigid aeroshells; the dynamic stability and thermal response of an inflatable structure at high Mach numbers; control requirements; the integrated precision landing performance including parachute performance and atmospheric uncertainties; and thermal protection system requirements, materials, and safety. Basic research is needed for the development of high-strength, high-temperature inflatable materials and for understanding the system's dynamic response in a planetary atmosphere. In addition, the flight environment is critical to the system design (including modeling and verification of stochastic atmospheres) and to the flow-field physics of the system (including chemical kinetics, boundary layer transition, and surface catalysis).

After aeroassist maneuvers decelerate the spacecraft from Mach 5 to Mach 2, supersonic retro propulsion completes the deceleration and landing. Unknowns include the interaction of the rocket plume with a planetary atmosphere at supersonic conditions; modeling of vehicle static and dynamic stability and control; the thrust and drag performance; the stability of the rocket nozzle; and the aerodynamic heating from the exhaust and flow field, including the radiation from the bow shock if one exists.

Nuclear Thermal Propulsion

NASA will benefit greatly from the development and demonstration of space nuclear reactors capable of supporting NTR[44] and NEP systems for missions beyond Mars and/or to enhance Mars exploration transportation capabilities. Although expensive to develop, the NTR is one of the leading propulsion system options for human Mars missions.[45] Such systems would have high thrust ($\cong$ 50,000 N) and high specific impulse (Isp $\cong$ 825-971 s),[46] a capability approximately twice that of today's oxygen/hydrogen chemical rocket engines. Hydrogen is the only working fluid for competitive NTR systems. Demonstrated in 20 rocket/reactor ground tests in the 1960s during the Rover/NERVA (Nuclear Engine for Rocket Vehicle Applications) Programs, the NTR uses heat from a fission reactor to directly heat liquid hydrogen propellant for rocket thrust.[47] NTR performance is limited by the temperature limits of the nuclear fuels and the regenerative cooling of the chamber.

Technology gaps include recapturing and updating the technology proven in the 1960s, ground tests, improving safety, reducing mass for affordability, and investigating long-life performance and reliability. Specific research and technologies include thermal control systems, efficient energy conversion and thermal transfer technologies, and lightweight/very high temperature thermal structures, along with safe and acceptable testing facilities. Enabling research includes the characterization of fissionable materials, development of a highly efficient, low-mass radiator, high-temperature superalloys and refractory metals for fuel cladding, and power conversion equipment to enable higher-temperature cycle operations and heat rejection. High-power (~0.5-MWe to multi-megawatt-class) helium-xenon Brayton or potassium Rankine power conversion units and efficient, long-life, high-power electric propulsion thrusters in the ~0.5- to 2-MWe size range are also desirable to contain the number of operational power conversion units and thrusters on the spacecraft. For bimodal systems providing both propulsion and modest (tens of kilowatts) power for life support and other subsystems, research and demonstration are required on the integration of a secondary closed He/Xe coolant loop into the engine design for the Brayton power cycle.

Nuclear Electric Propulsion

The NEP system is a low-thrust, high-specific-impulse option for transferring cargo for space exploration. It requires long transit times but significantly less propellant than systems with higher thrust. As a 2000 NRC report points out (at p. 41), "The energy source for NEP would be nuclear fission, with the generated heat transferred from the reactor to a suitable working fluid, rather than directly to a propellant gas as in the NTR. The working fluid would then be used to generate electric power through a thermodynamic cycle, and that electric power would drive a plasma or ion thruster. . . . [An NEP system would avoid] the temperature limitation on specific impulse that characterizes the NTR, and specific impulse can be thousands of seconds."[48] However, NEP systems would need high-burn-up/high-temperature nuclear fuels such as cermets, and high-temperature superalloys and refractory metals would be important for fuel cladding and power conversion equipment.

In addition to the nuclear technology needed to develop a high-power, long-life space nuclear power reactor, research gaps exist in the heat exchanger, thermal control, and scaling up of the electric thrusters in terms of power and longevity under continuous instead of intermittent operation. A Rankine cycle power conversion system would require fluid handling equipment such as piping, valves, pumps, turbines (with associated bearing systems), seals, and controls, all of which tend to be especially complex for the Rankine cycle, and design for high reliability will be correspondingly important. According to the 2000 NRC report (at p. 42), "Start-up and shutdown in orbit poses complex issues of design and operation, especially for a Rankine cycle with a liquid-metal working fluid and a nuclear heat source. Thawing . . . [in reduced gravity] of the nuclear reactor is required and the entire system must, in a controlled way, be brought to an equilibrium appropriate to steady-state operation. Transient, intermittent, or variable operations generally must also be considered carefully, taking account of system dynamics and possibilities for system instabilities . . . the propellant material, . . . [whether] liquid metal or noble gas, must be stored and maintained for especially long times," and the effects of NEP effluents on space design and operations must be defined.[49,50]

Several types of electric thrusters developed primarily by NASA have reached a high TRL and have been transferred successfully to commercial applications after decades of developments. These systems are used for small-scale intermittent applications (e.g., station keeping and orbit positioning).[51] Electric propulsion engine technology with power levels up to about 100 kW per thruster has been demonstrated at TRL 3 to 4; development of a megawatt-class NEP may need to include investigation of megawatt nuclear power sources and condensable propellant options. Development of megawatt nuclear power sources is required for either clusters of moderate power thrusters or the parallel development of megawatt power thrusters. In addition, condensable propellant options would pose unique issues for propellant transfer and management in very low gravity. Ground-based experiments, advanced modeling, and in-flight demonstrations would advance the understanding of fundamental processes involved in electric propulsion, leading to thrusters with greater performance and longer life, thereby enabling or enhancing some future exploration missions.

Solar Electric Propulsion

SEP is another low-thrust option appropriate for transferring cargo to the Moon or Mars. Unlike NEP, SEP can operate at a small scale, and a SEP system could also support small robotic exploration missions at an affordable cost. The feasibility of prospective SEP systems has been greatly enhanced by recent advances in solar cell efficiency and packaging. While low-power (kilowatt-class) SEP is currently widely applied on communications satellites, high-power systems would require significant additional development. A critical step for high-power SEP is the development of solar arrays with specific power in the 200-300 W/kg range. Based on recent demonstrations,[52] such arrays can be made operational during the next decade. They also can operate at very high voltages and are radiation resistant. Mission analysis studies[53] showed that a SEP freighter could deliver 10,000 to 20,000 kg to the lunar surface once per year for at least 5 years at a saving of >$2 billion compared to chemical propulsion systems. SEP is thus a viable and critical technology for future science missions.

A key issue presently unaddressed by supporting R&D is a space-based architecture using high-efficiency cargo transfer and significantly higher specific power and associated structural dimensions. Space-basing requires automated flight operations well beyond the state of the art, including formation flying, precision pointing, reducing physical disturbances in the arrays during continuous operation of the thrusters, and control of large space structures. The size, structural characteristics, and other attributes of the solar power array of a SEP system imply more complex operations than were being addressed[54] for the Orion element of the Constellation program, though they have much in common with current ISS and higher power geosynchronous satellite operations. The mass of the large structures would be minimized to increase payload capacity and/or reduce transit times. This would require validated knowledge of the structural behavior of very large, low-strength systems and innovative, likely dynamic, modes of control during flight.[55] With the present level of theoretical understanding, high-fidelity validation of flight operations and control for a high-power SEP will likely require high-fidelity ground testing as well as in-flight, ISS-captured, or free-flying demonstrations at an appropriate scale.

For robotic and specialized remote sensing applications, electric micropropulsion would be required. This would involve either scaled-down versions of thrusters or thrusters fabricated with micro-electromechanical systems. Micropropulsion could be used to enhance and enable microspacecraft (mass less than 100 kg) or pico-spacecraft (mass less than 10 kg).

High-power (megawatt) SEP is an alternative propulsion option to enable large payloads and short transit times for crewed space exploration missions.[56] Representative high-thrust SEP concepts include magnetoplasmadynamic and pulsed-inductive thrusters.[57] Research includes the scale-up from currently demonstrated kilowatt thrusters to megawatt class, with the associated requirements for lightweight power, efficient thermal control systems, long-life components, and high-efficiency/low-cost propellants similar to those for NEP systems.

Advanced Propulsion and Propellants

Magnetohydrodynamic (MHD) fluid accelerators are an example of a high-thrust, mid-TRL propulsion system. MHD fluid accelerators could potentially double or triple the specific-impulse performance of in-space cryogenic propellant rockets if adequate electrical power were available. This performance increase is accomplished by using the Lorentz force produced by crossed electric and magnetic fields to accelerate charged particles in the propellant. Conventionally, rocket performance is restricted to approximately 3,000 K because of material and active-cooling constraints; however, by direct infusion of power MHD accelerates the flow without increasing the temperature.[58,59,60]

The technology gaps for MHD are the large mass of the magnets and the high electrical power (and high currents) needed to drive the magnets. Research is required to develop highly conducting, low-mass magnets. Power could be supplied by nuclear fission or large photovoltaic collectors (for missions close enough to the Sun); however, the mass of the spacecraft power system could be greatly reduced through the use of microwave or direct concentrated solar beaming. Although the concept of power beaming to space has received attention since the 1980s, convincing flight tests or proof-of-principle demonstrations are yet to be conducted. Critical gaps still exist in advanced methods of systems dynamics and flexibility and the controls for precise alignment. At best, success requires an optimal location for a large power source with a stable line of sight to the spacecraft. The power requirements of MHD fluid accelerators may be comparable to the power requirements of high-power thrusters that would be needed for a large SEP or NEP system.

In addition to improving propulsion performance through power advancements, advanced propellants also have the potential of high specific impulse, ease of storage, and improved safety. Metallized-gelled propellants have shown dramatic payload improvement with large increases in bulk density and modest improvements in specific impulse.[61]

Summary of Enabling Science and Technologies for Space Propulsion

In-space propulsion systems represent 50 to 90 percent of the mass that has to be delivered to low Earth orbit, and therefore presents an opportunity for reducing the cost of the Exploration Program. Advances in propulsion performance (specific impulse, efficiency, thrust-to-weight ratio, propellant bulk density), reliability, thermal management, power generation and handling, and propellant storage and handling are key drivers to dramatically reduce mass, cost, and mission risk. The following recommendations address exploration technologies in space propulsion that would benefit from near-term R&D.

Before 2020: Required

Zero-Boiloff Propellant Storage. Research and technology is needed to enable zero-boiloff propellant storage for orbiting depots, on-orbit refueling, and long-duration exploration missions. To support this capability, physical sciences research should be conducted on advanced insulation materials, active cooling, multiphase flows, and capillary effectiveness. **(T2)**

Cryogenic Handling and Gauging. NASA would benefit from enhanced cryogenic fluid management for in-flight refueling, propellant depots, and long-duration planetary missions. Research in cryogenic handling and gauging under low-gravity conditions is needed. Research in in-space cryogenic fluid management includes active and passive storage, fluid transfer, gauging, pressurization, pressure control, leak detection, and mixing destratification. **(T3)**

2020 and Beyond: Required

Ascent and Descent Technologies. NASA will require lunar and planetary descent and ascent propulsion technologies, including engine start after long quiescent periods, combustion stability at all gravity conditions, and deep throttle. Areas of research include cryogenic fluid management, propellant ignition, combustion stability, and active thermal control of the injectors and combustors. **(T14)**

Inflatable Aerodynamic Decelerators for Bodies with Atmospheres. The availability of inflatable aerodynamic decelerators would reduce propulsive mass requirements. To develop these systems, physical science research is required in high-strength, low-density, high-temperature flexible materials; dynamics, stability and control; and aeroelasticity in the flight environment. **(T15)**

Supersonic Retro Propulsion System. A combination of a supersonic retro propulsion system with an aerodynamic decelerator completes the deceleration and landing and would reduce mass requirements. To enable development of such a system, physical sciences research is needed on flow-field interactions of the rocket plume with the atmosphere, aerothermodynamics of the flow, and the dynamic interactions and control of the vehicle. **(T16)**

Space Nuclear Reactors. The development and demonstration of space nuclear reactors capable of supporting nuclear thermal rockets (NTRs) are required for missions beyond Mars and/or to enhance Mars exploration transportation capabilities. Required technologies include thermal control systems, efficient energy conversion and thermal transfer technologies, and lightweight/very high temperature thermal structures, along with safe and acceptable testing facilities. Enabling physical science research has been identified in the section "Space Power and Thermal Management." Additional physical science research is required on liquid-metal cooling under reduced gravity, thawing under reduced gravity, and system dynamics. **(T17)**

Solar Electric Propulsion (SEP) Technologies. SEP is an important option for the efficient transfer of propellant and cargo to distant locations. To support the development of such systems, advances are needed in understanding the complex behavioral modes of very lightweight large space structures so as to enable development of innovative control methods for such structures during flight. In addition, research is needed in condensable propellants and in propellant transfer and management in very low gravity. **(T18)**

2020 and Beyond: Highly Desirable

Nuclear Electric Propulsion (NEP) Technologies. NEP will enable the very efficient transfer of propellant and other cargo to extended outposts on Mars and beyond. Areas of research, in addition to those summarized above under Space Nuclear Reactors (T17), include propellant management under reduced gravity and flow processes in electric thrusters.

Extravehicular Activity Systems

A new vision for EVA systems is emerging for exploration missions—one that encompasses spacesuits, rovers, and robotic assistants working collaboratively during mobile exploration sorties. In the past, EVA has enabled complex work outside a crewed space vehicle or lunar module, contributing a supporting operational role (repair, maintenance, observation, etc.). However, since the end of the Apollo missions, EVA has not typically served a primary mission role (excluding the EVAs on the Hubble Repair Missions). For space exploration missions to planetary surfaces (Moon/Mars/Flexible Path[¶¶]), EVA resumes a primary, required mission role that is critical to enabling successful mission operations, gaining new scientific knowledge, and gaining experience with exploration systems.

History of Extravehicular Activity in NASA Programs

NASA and Russian EVA systems documentation[62,63,64] provides necessary background, information on development, and knowledge about EVA capabilities. NASA's experience with EVA dates to the Gemini Program in 1965. As Jordan et al. have pointed out, "The capability for humans to work outside the spacecraft [has] proved invaluable time and again. For example, Skylab astronauts performed 12 contingency EVAs to fix unanticipated problems, repeatedly saving Skylab from abandonment. However, despite the advantages of EVA capability, early space shuttle designs did not include the means to perform EVA."[65]

Historically, an EVA system has consisted of the following components: a spacesuit; a portable life support system that provides the suit with a breathable atmosphere while removing carbon dioxide, water vapor, and trace contaminants; suit subsystems providing pressurization, mobility, temperature control, power, communications, and data systems, as well as protection from radiation and particle impacts; rovers and mobility aids; and tools (including robotic tools) that enable the EVA crew member to accomplish necessary mission tasks. NASA developed pressure protection during the Mercury program and took a clean-sheet approach to spacesuit design for each of the Gemini, Apollo, and space shuttle programs. An Apollo spacesuit was adapted for Skylab operations, and the space shuttle Extravehicular Mobility Unit (EMU) was enhanced and certified to meet the requirements for the ISS.

Because of indecision about developing a shuttle spacesuit, EMU development lagged behind the shuttle by approximately 4 years, which resulted in a very compressed development phase and certification and flight testing carried out in parallel with early shuttle flights. The most extensive information on the EMU can be found in Balinskas and Tepper's contractor report[66] and the Hamilton Sundstrand EMU Data Book,[67] as well as in conference proceedings.[68-78]

Existing NASA EVA systems include (1) a launch, entry, and abort (LEA) suit or an advanced crew escape suit (ACES), used inside a spacecraft or in emergencies, and (2) the EMU for microgravity EVA. The launch and entry suit, a partial-pressure suit, became operational for space shuttle flights after the *Challenger* accident and was used until 1998. The ACES, a full-pressure suit introduced in 1994, is the current space shuttle suit worn for launch and re-entry. The ACES technology derives from high-altitude pressure suits (used for advanced vehicles such as the SR-71, U-2, and X-15) and includes parachute bailout capability and a self-contained emergency oxygen system consisting of two 120-cubic-inch oxygen bottles that can provide at least 10 min of oxygen.[79]

The EMU is essentially a self-contained miniature spacecraft in the sense that it provides all of the functions necessary to sustain life in a human-sized, mobile form while enabling useful work in space. The EMU has undergone evolutionary development to meet the needs of microgravity spacewalks for the space shuttle as well as the ISS. The system is optimized for a microgravity environment and enables crew members to perform complex tasks safely. It incorporates a closed loop, portable life support system (PLSS). As Jordan et al. have observed, "That the EMU was initially designed as a limited capability suit to satisfy minimal mobility and operational requirements is astounding in light of the fact that it has subsequently been used to repair satellites, construct a massive space structure, and maintain the ISS. . . . [In fact] the EMU, like most complex engineering systems, has faced considerable uncertainty during its service life. Changes in the technical, political, and economic environments have often caused changes in requirements, which in turn necessitated design changes."[80] Initially, a new, advanced spacesuit to service the ISS and to serve as a test bed for planetary exploration technologies was planned. The 1989 decision not to build a space station-specific suit and instead to modify the EMU for construction and operation of the ISS resulted in 10 system requirements changes and 23 design/procedural changes.

[¶¶]The "flexible path" architecture to inner solar system locations, such as lunar orbit, Lagrange points, near-Earth objects, and the moons of Mars, followed by exploration of the lunar surface and/or martian surface, was put forward by the Augustine Commission (Review of U.S. Human Spaceflight Plans Committee, *Seeking A Human Spaceflight Program Worthy of a Great Nation*, Office of Science and Technology Policy, Washington, D.C., October 2009).

TABLE 10.2 Summary of U.S. Extravehicular Activity Duration by Program (1965 to 2009)

Program	Total EVA Duration (hours)	Suit Used
Gemini[a]	12:40	G-4C/G-5C
Apollo[b]	165:17	A7L/A7LB
Skylab	82:52	A7LB
Space shuttle[c,d,e]	1,894:09	EMU
International Space Station	835:02	EMU

[a]See http://history.nasa.gov/SP-4002/app1d.htm.
[b]D.S.F. Portree and R.C. Trevino, Walking to Olympus: and EVA Chronology, Monographs in Aerospace History Series #7, 1997.
[c]NASA space shuttle launch archive, available at http://science.ksc.nasa.gov/shuttle/missions/.
[d]Mission control center status reports, available at http://spaceflight1.nasa.gov/spacenews/reports/.
[e]Shuttle and ISS flight archive, available at http://www.astronautix.com/flights/.

Table 10.2 shows that, in the 40-year history of human spaceflight, NASA astronauts have logged approximately 3,000 h of total EVA time, the vast majority in EMUs on the space shuttle or the ISS. The total EVA duration for each new generation of spacesuit design has increased by more than an order of magnitude: 13 h for the G-4/G-5 series, 248 h for the A7L series, and 2,729 h for the EMU. A new paradigm is emerging for exploration missions in the next decades: with the "mountain of EVA" (Figure 10.7), another 10-fold increase in EVA operational hours is projected for exploration missions to the Moon or Mars. This large increase in EVA hours is relevant for any future mission to the Moon or Mars and is not dependent on a particular mission or architecture.

The literature and past scientific and technical recommendations regarding EVA systems[81-84] are assumed as baseline knowledge and are not recapitulated here. The focus here is on issues relevant to NASA in the next decade for achieving a translational portfolio to enable exploration missions to meet research and operational objectives in the life and physical sciences.

Future Extravehicular Activity Needs

The requirements for future EVA systems include (1) crew safety and mobility during LEA, (2) contingency microgravity EVA capability utilizing an umbilical, and (3) surface EVA capability.[85] The vision of mobile exploration EVA encompasses astronauts, rovers, robotic assistants, and possible mobile laboratories or bases in the next decade. Future EVA system concepts might include two spacesuit configurations. Suit 1 would provide an initial EVA capability for LEA and contingency microgravity EVA; Suit 2 would be designed for surface EVA capability. The two configurations would be modular designs incorporating many shared design elements and components. However, NASA has yet to develop a plan and processes to achieve this vision. NASA should further its interaction with systems engineering experts in industry, academia, and the DOD to leverage their knowledge and experience in fielding related systems with shared, modular components.

In addition to incremental design concepts that extend current EVA systems, bold design innovations should be encouraged, such as the Bio-Suit originally developed at the Massachusetts Institute of Technology under a NASA Institute for Advanced Concepts contract.[86,87]

EVA suit elements include the pressure garment subsystem; life support subsystem; and power, communications, avionics, and informatics (PCAI) subsystem. The capability to conduct EVA surface operations is a primary and critical component of future surface exploration missions, regardless of exploration destination. NASA has appropriately defined system capabilities and technologies for the Suit 1 concept, and the objectives for that concept are well understood. Current NASA and contractor knowledge and experience seem sufficient to implement the Suit 1 concept, which relies on existing and proven technologies; requirements definition and vehicle interface design reviews for it are ongoing. However, the panel did identify a research gap in NASA's Suit 1 concept in the area of joint mobility. EVA suits should have detailed and accurate specifications for mobility, joint torque, and joint range of motion requirements for both intravehicular activities and EVAs. Torque specifications based

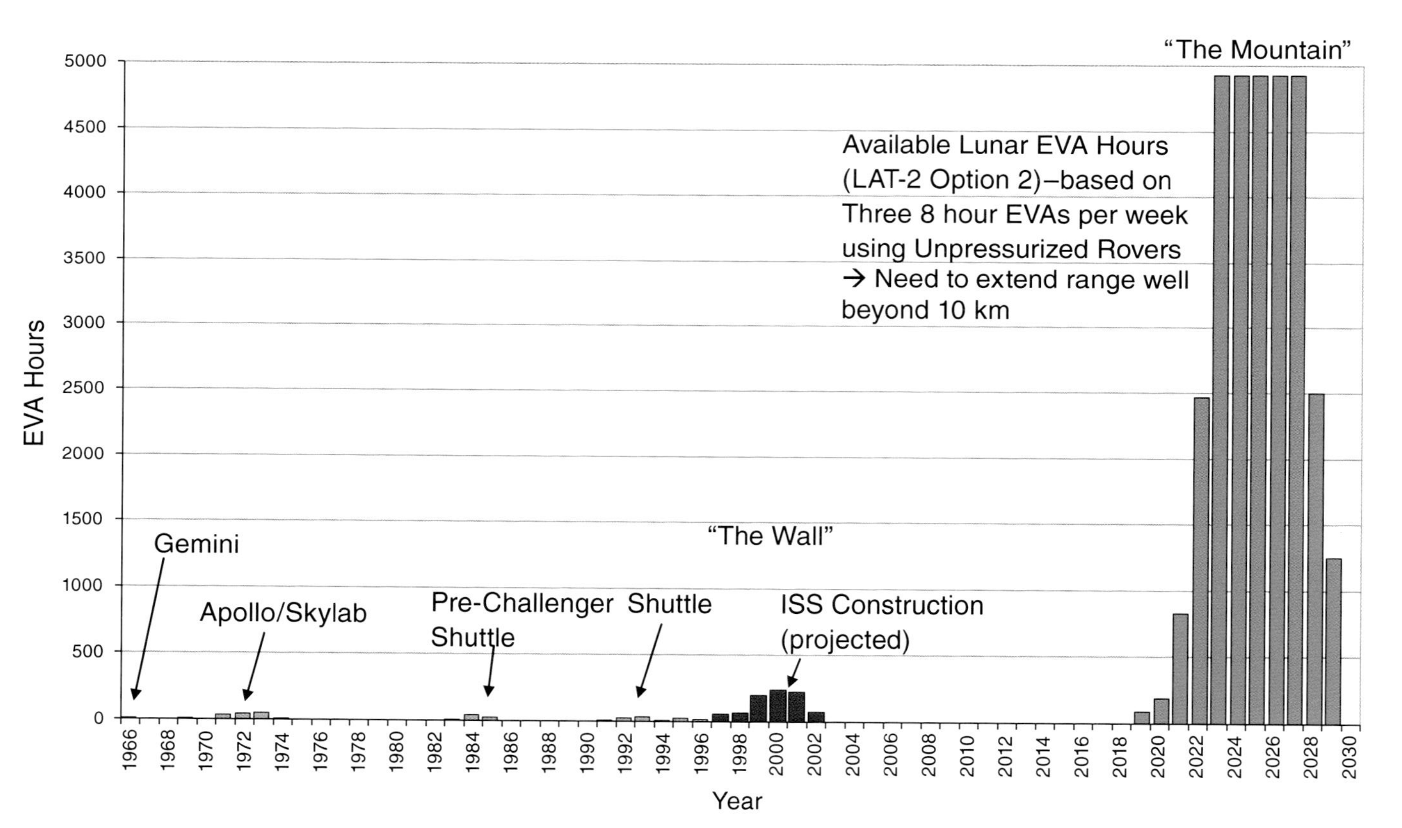

FIGURE 10.7 Annual cumulative hours of projected extravehicular activity (EVA), showing "mountain of EVA" for Exploration missions. SOURCE: M. Gernhardt, NASA Johnson Space Center, "Integrating Life Sciences, Engineering and Operations Research to Optimize Human Safety and Performance in Planetary Exploration," presentation to the NRC Decadal Survey on Biological and Physical Sciences in Space, August 20, 2009.

on prototype suits should be verified and refined using precise torque measurements that account for realistic, multijoint motions that an astronaut would naturally make if unrestricted by a suit. Further research is needed to improve EVA suit joint flexibility and conformal fit while considering anthropomorphic differences between men and women to prevent injuries to crew and enhance safety during lunar and planetary explorations.

The current EMU glove has caused problems with astronaut finger tip trauma. Improving space suit glove design is a suit-independent design challenge and represents a crosscutting, multidisciplinary research question.[88-91]

Developing a new EVA system to support anticipated surface operations on the Moon, Mars, or asteroids presents significant technical challenges, especially with regard to providing "mobility equal to that of an Earth-based geologist."[92] The engineering and biomedical requirements for pressure production, a breathable atmosphere, thermal control and ventilation, carbon dioxide removal, and waste management for surface EVA systems are understood and well specified in the current EVA system plans. However, critical elements of Suit 2 still require additional research and significant technological advances to provide enhanced exploration EVA capabilities. In particular, research and technology development is vital in the surface PLSS, the PCAI and information technology, and the interaction of the EVA suit with the surface environment.

A previous examination of the Exploration Technology Development Program stated that promising planetary PLSS technology enhancements include new heat-rejection technologies, variable pressure regulation, a rapid cycling amine swing bed for CO_2 control, and metabolic temperature swing absorption designs, even though some of these technologies are still at a low TRL.[93] The PCAI systems for surface operations will require additional capabilities. The DOD has made numerous advances in battery technology, displays, and speech and audio communications;[94] many of these technologies could be incorporated into or adapted to suit the EVA PCAI subsystem.

The aggregate risk from micrometeoroids and orbital debris during EVA over the life of the ISS has been doubled by the extension of the ISS mission to 2020. Thus, possible suit penetration risk due to micrometeoroids or orbital debris remains an ongoing concern.

For surface operations, protection from environmental hazards becomes a significant concern. Currently there are gaps in our understanding of and technological capabilities to mitigate harmful effects of continual dust exposure and possible micrometeoroid impacts on EVA systems. The effects of dust on the functionality and operational life of equipment are well known.[95,96] All of the Apollo missions were adversely affected by lunar dust. Significant effects included (1) visual obscuration from blowing dust while landing, (2) false instrument readings, (3) dust coating and contamination, (4) clogging of mechanisms, (5) abrasion, (6) thermal control problems, (7) seal failures, and (8) irritation from dust inhalation.[97] These problems probably could have been reduced if, early in the Apollo program, NASA had developed a better simulation environment including better simulants, higher vacuum, correlated simulations, and more realistic thermal and illumination environments.[98] Astronaut Richard Gordon reported:

> The cabin atmosphere was okay. On the way out, it was clean. On the way back, we got lunar dust in the command module. The system actually couldn't handle it; the system never did filter out the dust, and the dust was continuously run through the system and throughout the spacecraft without being removed.[99]

Knowledge gaps in dust mitigation still exist in the areas above.

Research in dust mitigation techniques and technologies will need to consider the specific requirements of the different surface environments. On the lunar surface, the dust transport phenomena are due to thermophoresis and electrostatics forces in vacuum, whereas on the martian surface, dust is transported primarily by CO_2 gas convection in the martian atmosphere. The strong martian winds, which are driven by the seasonal and diurnal insolation cycles combined with the effects of planetary rotation, cause the dust to suspend in the martian atmosphere for a long time.[100] The structure of dust particles is also different: martian dust particles are likely to be more rounded than lunar dust. The electrostatic charge is greater on lunar dust than on martian dust.

Requirements do not currently exist for thermal control and radiation protection for EVA systems, rovers, and habitats in the partial-gravity environment. The gaps in our understanding of thermal control and radiation biology are discussed in Chapter 7. Accumulation of surface electrical charge during long-duration EVAs has to be investigated within the context of Earth, the Moon, or martian environments. Plasma interactions with astronauts during EVAs in proximity to space structures with high-power, high-voltage solar arrays are also of particular importance.[101] Plasma interactions with suits should be further investigated.

For long planetary exploration missions, it will be important to reduce overall system mass (on-back and total mass, including expendables) beyond those attainable with current technologies and to expand the exploration radius from the lander/habitat by increasing system robustness and fault tolerance.

Finally, from a systems engineering perspective, it is important that EVA capabilities be integrated into the rest of systems and operations development from the earliest stages, in order to reduce life cycle and operational costs. Biomedical and EVA systems relationships and synergies should be identified for exploration missions. EVA technologies, human performance, and life and physical science phenomena in partial gravity need to be better understood. Design principles that include EVA systems with all other lunar/Mars surface systems and operations should be followed to produce an overall integrated mission architecture.

Summary of Enabling Science and Technologies for Extravehicular Activity

The following exploration technologies that are important to EVA would benefit from near-term R&D.

Before 2020: Required

EVA Suit Mobility Enhancements. NASA can enhance human performance and mobility in the next generation of EVA systems by minimizing astronaut injury and improving comfort. Significant mobility enhancements are also required. Research should be conducted to minimize joint torques, improve suit comfort, improve suit glove design, provide the equivalent of shirt-sleeve mobility, develop trauma countermeasures, and improve physical systems performance across the spectrum of gravity environments. Areas of biological and physical science research that will enable the development of these enhancements include suited astronaut computational modeling, biomechanics analysis for partial gravity, robot-human testing and quantification of advanced spacesuit joints and full body suits (joint torque characterization), and musculoskeletal modeling and suited range-of-motion studies as input for garments designed for comfort, protection, minimizing injury, and enhancing mobility. **(T4)**

Dust and Micrometeoroid Mitigation Systems. Benefits would arise from improved durability and maintainability of suits and EVA systems by developing technology to mitigate effects of dust and mitigate risks from micrometeoroids, orbital debris, and plasma. Physical science research should address the impact mechanics of particulates, the design of outer layer dust garments, advanced material and design concepts for micrometeoroid mitigation, possible magnetic repulsive technologies, and the quantification of plasma electrodynamic interactions with EVA systems. **(T5)**

2020 and Beyond: Required

Radiation Protection Systems. Define requirements for thermal control and radiation protection for EVA systems, rovers, and habitats and develop a plan for radiation shelters. **(T19)**

2020 and Beyond: Highly Desirable

Suit/Helmet Modular Information Technology Architecture. NASA would benefit from an enhanced architecture that maximizes exploration footprint and scientific return. Areas of biological and physical science research that enable the development of such an architecture include human factors engineering, display and information technology system design, and human-machine interactions with an emphasis on planning and real-time navigation.

PCAI Systems. NASA should leverage technology advances made in other agencies and in industry to improve its own PCAI capabilities. Advances in battery technology, displays, and speech and audio communications by DOD and industry should be incorporated into NASA's EVA PCAI subsystem. Areas of biological and physical sciences research that enable the development of these systems include advanced biomedical sensors, physiological monitoring, and health advising capabilities.

Planetary PLSS Technology. PLSS enhancements should include new heat-rejection technologies, variable pressure regulation, rapid cycling amine swing bed for CO_2 control, and metabolic temperature swing absorption designs.

Life Support Systems

A life support system (LSS) is necessary for space vehicles, rovers, and EVA systems. LSS functions include pressure control; atmosphere revitalization (removing carbon dioxide, water vapor, and trace contaminants); temperature and humidity control; waste collection; and fire prevention, detection, and suppression. Fire safety is discussed in the next section of this chapter. Dust mitigation and radiation protection are significant challenges as well. The literature and past scientific and technical recommendations regarding life support systems and technologies are assumed as baseline knowledge for this report. (Please see findings and recommendations on LSS in NRC reports published in 1997, 2000, and 2008.[102,103,104]) The focus here is on issues relevant for NASA in the next decade to achieve a translational portfolio to enable exploration missions to meet research and operational objectives in the life and physical sciences.

A range of atmosphere compositions are in use or proposed for various aspects of space exploration. Humans can survive only a few minutes without oxygen. Many biological responses are dependent on gas partial pressure, and total gas concentrations influence design tradeoffs in atmosphere selection, especially for space vehicles, habitats, rovers, and EVA systems.[105] At sea level on Earth, O_2 concentration is 21 percent and the normal O_2 partial pressure is 21 kPa. While human physiological needs set the requirement for partial pressure of O_2, fire risk increases as the absolute concentration of O_2 increases, and the level of engineering difficulty increases as the total pressure increases (for example, on joints and seals). Thus, the selection of the oxygen environment involves a tradeoff between engineering and fire safety issues. The atmosphere inside space vehicles and habitats can range from pure oxygen at low pressure to oxygen/nitrogen mixtures approximating Earth's atmosphere in both concentration and pressure. The current plan for future space vehicles is to use an oxygen-nitrogen atmosphere with up to 34 percent O_2. By contrast, the NASA EMU spacesuit supplies the astronaut with 100 percent oxygen, which is why the suit can be operated at a low pressure of 29.6 kPa (4.3 psi). Generation of O_2 can itself be a source of fire or explosion, as was the case in the Mir space station fire, so the method of O_2 generation should also consider fire safety implications.

Future Life Support System Needs

The LSS must provide adequate thermal control to maintain a suitable internal temperature regardless of internal activities (by astronauts and equipment) and the external environment. The LSS must also remove water vapor released by crew members and collect waste (fluids and solids). The space shuttle and the ISS provide a shirt-sleeve working environment for astronauts and various life support equipment. The ISS uses a pumped single-phase thermal bus to collect waste heat and transport it to heat rejection radiators.[106] For permanent settlements on the lunar or martian surface, environmental control of habitat, rovers, and in situ resource processing factories will be required, analogous to terrestrial HVAC systems, as well as a thermal bus.

NASA has appropriately defined LSS capabilities and technologies to meet the objectives for planned missions to the ISS and the Moon, including a new space vehicle, rovers, EVA systems, and surface habitats.[107] For missions to the ISS and initial short-duration (approximately 7-day) missions to the Moon, the LSS for space vehicles, including the lunar lander, would include a water loop to supply and store potable water; collection of human waste, which is then discarded; an oxygen loop to store oxygen; and a rapid cycling amine bed to collect CO_2, which is then vented overboard. The significantly enhanced LSS for a more permanent lunar presence would include continuous habitation for a crew of perhaps four. The current specifications and tradeoffs for this LSS include moving toward full closure of the water and oxygen loops, which necessitates wastewater recovery, brine recovery, and solid waste drying; O_2 generation from H_2O; and CO_2 reduction (to CH_4 or C). If lunar ISRU oxygen becomes available, full closure would no longer be necessary and the CO_2 reduction step might be eliminated. Capabilities and technologies that need to be further investigated and matured include electrolysis (for the Environmental Control and Life Support System, ISRU, and energy storage applications); CO_2 removal for habitats, pressurized rovers, and spacesuits; and compatibility with CO_2 reduction capabilities. Fluid system components (pumps, valves, sensors, etc.) need to be investigated and assessed for functionality in partial-gravity environments, in the context of system performance. Knowledge gaps in two-phase flow in partial gravity need to be filled.[108]

The Constellation Program's lunar architecture specifically emphasizes the goal of improving reliability and functionality of EVA and LSS.[109] Dust mitigation and thermal control on the lunar surface should be high priorities, but they had been omitted from NASA's prior lunar architecture. In the longer term, advanced bioregenerative life support technologies could support self-sufficient human habitation of space, as detailed in Chapter 4 of this report.

Environmental factors that need to be considered when designing an LSS for the lunar surface are (1) dust, (2) extreme temperatures and high temperature gradients, (3) radiation, (4) micrometeoroid impact, (5) low pressure, and (6) different gravity from Earth. As described in the EVA section above, significant knowledge gaps still exist in dust mitigation. Effects of extreme temperature and high temperature gradients on equipment, systems, and habitats are known.[110] Researchers[111] have suggested coupling ISRU mining and habitat construction as a way to control temperature swings and protect crew and equipment from cosmic radiation. Balasubramaniam et al.[112] performed computer simulations that show the advantage that thermal wadis can provide as a way to reduce the temperature swings. However, none of these technologies have been demonstrated in a relevant environment. Closed-loop air revitalization and closed-loop water recovery have been NASA goals since the 1990s. The developmental systems for neither capability have yet achieved a sufficiently high TRL.[113,114,115]

Cryogenic (<120 K) refrigeration of fluids stored in the liquid state and cryo-cooling of superconductors and infrared sensors will also be required. The development and adaption of cryogenic technology for life support systems would lead to more efficient storage of life support consumables (such as oxygen), provide superconductor technology for life support equipment, and lead to thermoelectric refrigeration for the storage of food supplies. These needs require adaptation of refrigeration and heat pump technology to the harsh surface temperatures, dust, and reduced gravity of exploration targets.

Finally, the panel notes that NASA should enhance its collaborations with ongoing international life support experiments and simulations (i.e., the European Space Agency, Japan, and Russia).

Summary of Enabling Science and Technologies for Life Support

NASA should undertake an incremental addition of advanced life support technologies and systems into existing designs in such a way as to establish a reliable performance track record before they are relied on as the primary LSS. The following exploration technologies that are important for LSS would benefit from near-term R&D.

Before 2020: Required

Fluid and Air Subsystems. NASA should undertake partial-gravity characterization and testing of fluid and air subsystems of new life support systems. Required research includes heat and mass transfer in porous media under low and microgravity conditions. **(T6)**

Before 2020: Highly Desirable

Environmental Atmospheric Sensors. Long-life atmosphere monitors should be developed and demonstrated to identify major atmosphere constituents, as well as trace contaminants that may be harmful to humans in a closed habitat.

2020 and Beyond: Required

Radiation Protection Systems. As described in Chapter 7, NASA would benefit from additional work in radiation shielding, countermeasures, and mitigators. NASA should continue to develop radiation protection and mitigation technologies and demonstrations. **(T19)**

Thermoregulation Technologies. NASA should develop and demonstrate technologies to support thermoregulation of habitats, rovers, and spacesuits on the lunar surface. Areas of biological and physical science research that could help enable the development of those technologies include understanding the human response characteristics to extreme physical conditions (temperature, pressure, oxygen level, etc.) and developing materials and methods to protect humans from these extreme conditions. **(T20)**

Closed-Loop Air Revitalization. NASA could achieve a significant savings in resupply/consumables by closing the air loop. Technology needs include CO_2 removal, recovery, and reduction; O_2 generation via electrolysis with high pressure capability; improved sorbents and catalysts for trace contaminant control; and atmosphere particulate control and monitoring. Fundamental physical sciences research to support the development of these technologies includes understanding the effects of gravity on the coupling of electrochemical systems with multiphase flow physics. **(T21)**

Closed-Loop Water Recovery. As with closed-loop air revitalization, NASA would benefit from significant savings in resupply/consumables with closed-loop water recovery. Technology needs include water recovery from wastewaters and brines, pretreatments, biocides, low expendable rates, and robustness. More fundamental work is also needed to assess the effect of variable gravity on multiphase flow systems (i.e., water management and recycling). **(T22)**

Dust Mitigation Technologies. Dust has a well-known adverse effect on life support systems. NASA should develop technologies to mitigate or eliminate the effects of dust on systems and hardware. Dust may coat solar energy harvesting systems, contaminate supplies, scuff equipment, create thermal control problems, and cause seal failures. NASA should develop, demonstrate, and test dust mitigation technologies needed to overcome these problems. Fundamental physical science research is needed to support the development of these technologies, specifically to understand the coupling of electrostatic fields and dust particulates dynamics in any dust mitigation technology. **(T23)**

2020 and Beyond: Highly Desirable

Solid Waste Treatment. NASA should develop and demonstrate long-duration waste stabilization and water recovery from solid wastes. As with closed-loop air revitalization, NASA would benefit from significant savings in resupply/consumables with recovery of water from solid waste.

Food Production and Bioregenerative Life Support Systems. As described in the recommendations in Chapter 4 of this report, NASA should develop an incremental approach toward the development of a bioregenerative life support system.

Fire Safety

Although they are relatively rare events, fires have occurred in space vehicles and habitats and will occur again. Fire safety is critical to any human space exploration because fires can have devastating consequences, including loss of life, loss of vehicle or habitat integrity, and mission failure. Historically, fire safety R&D has been treated as a subset of combustion research. While both basic and applied combustion research support fire safety research and it is essential to pursue deeper understanding of the combustion processes involved (see Chapter 9), many of the operational aspects of fire safety need to be defined in a manner that does not require a full understanding of the underlying basic principles of combustion. Because fire is a complex phenomenon that simultaneously involves many aspects of chemistry and physics, including kinetics, reaction dynamics, fluid mechanics, and heat and mass transfer in multiple phases, a comprehensive understanding of fire at the fundamental level does not currently exist. Instead, much of the current understanding of fire behavior is phenomenologically based, but with some reference to the underlying combustion physics and chemistry. As a result, many of the research and development needs for fire safety (fire safety R&D), especially at the translational stage, require the development of phenomenologically based models and correlations for materials response to fires, fire detection, fire suppression, and recovery from fire (and explosions). While both basic and applied combustion research are invaluable to spacecraft fire safety, there are other practical aspects of fire safety issues outside the scope of combustion science that are part of fire safety systems and thus are mission-enabling. In order to have a complete safety program, additional funding should be allocated to fire safety research separate from and in addition to funds allocated for combustion research, and the proposals for this research should be subject to end-user oriented reviews.

Fire safety for space exploration is predicated on the principles of prevention, detection, and suppression.[116] Little work has been done on post-fire clean-up or recovery, largely based on the false belief that fires are rare events and can be prevented, so clean-up and recovery from them are unimportant. However, although fires are rare, they are inevitable, and fires have occurred wherever human exploration has taken place. Furthermore, the potential for damage to mission-critical systems necessitates that strategies be developed and put in place for post-fire clean-up and recovery, especially for operations far from Earth, where terrestrial help for recovery is unavailable on any reasonable time frame. Unfortunately, much of fire safety to date has been predicated on assumptions about fire behavior in reduced gravity environments that have been called into question by recent research. This research has raised serious issues about the underlying assumptions for all three fire safety principles: prevention, detection, and suppression.

Materials Flammability and Toxicity

The cornerstone of fire safety prevention for space exploration since the tragic fire in Apollo 1 has been the qualification of materials used in space vehicles and proposed habitats. This qualification is based on NASA Standard 6001, which uses an upward flame spread test in 1 *g* as a screening test for materials flammability. This test is predicated on (1) the belief that upward flame spread represents a worst-case condition for any space exploration environment and (2) the desire to have a simple, 1-*g*-based test for materials flammability. Unfortunately, recently reported work has challenged the assumption that 1-*g* flame spread is the most severe. These results, although very limited, indicate that upward flame spread peaks somewhere in the range of 0.15 *g* to 0.5 *g*, i.e., the range of lunar and martian partial gravity.[117,118,119] As a result, there is a critical need going forward to reassess the validity of NASA Standard 6001 by developing substantial data on flame spread for various materials under relevant levels of gravity from 1 *g* down to microgravity levels, in order to develop more comprehensive screening tests for flammability of materials to be used in spacecraft and habitats. Furthermore, past research[120,121,122] has shown that the yield of toxic products of combustion can be many times higher in fire retardant materials than in non-fire retardant materials. Since toxic and/or corrosive agents can pose a serious threat to human life in space vehicles and habitats, control of the production of such agents should be included in a revised NASA 6001 Standard.

Finally, fire properties of various materials are affected by the oxygen environment. While the human need for oxygen depends on the partial pressure of oxygen, ignition and fire spread on materials depend on the absolute concentration of oxygen.[123] Recent moves to environments with higher concentrations of oxygen conflict with the current understanding of fire behavior in these atmospheres. Thus, more work is needed to establish the comprehensive definition of material flammability for spacecraft and the validity of material screening as a "preventive" measure. Furthermore, materials flammability should include defining the potential environment generated by a fire, in order to evaluate the tradeoffs between fire safety and human life support in various atmospheres and enable the design of an integrated human life support/fire safety strategy.

Fire Detection

Recent work has called into question some fundamental assumptions on current fire detection practices in space. Although the space shuttle uses ionization technology for fire detection, the ISS switched to use of photoelectric detection technology based on the belief that smoke particles in low-gravity environments would be large (>1 μm). This belief was based largely on fundamental combustion experiments carried out in microgravity, rather than on projects specifically designed to assess fire detection in reduced gravity. Since photoelectric detectors respond better than ionization detectors to particles over 1 μm in size, the assumption was that photoelectric detectors would provide earlier detection of fires in the ISS than ionization technology would. Recently reported results from Urban and coworkers[124] from work conducted on the space shuttle in microgravity show that particle sizes from smoldering materials may actually be too small (<0.3 μm) to be detected by photoelectric smoke detectors. Additional R&D is needed to validate this work and to better understand fire signatures, including particle sizes from both flaming and smoldering fires in reduced-gravity environments. More important, fire detection systems need to be tested in environments that approximate the actual environments in which they will be installed, thus reproducing the actual "fire signatures" that the detectors could see.

The assumption that where there are particles, there is smoke, and where there is smoke, there is fire, must also be supplanted by the development of multisignature fire detection schemes and systems. In low-gravity environments, relying on smoke as the harbinger of fire is not a reliable assumption. Smoke detectors rely on the detection of a certain density of particles in a particular size range. Particles generated from many other sources that are free to float about in a microgravity environment have the potential to generate false alarms from such smoke detectors. Detection of other fire signatures, including carbon monoxide, HCl, HCN, or soot precursors, can potentially shorten times to fire detection while reducing false alarms.

Given the fact that sensors are ubiquitous in crewed spacecraft, there could be value added in treating fire signatures as an input variable for detection system design. Not only can combinations of sensors provide faster and more reliable detection but they also can provide information that will enable astronauts to minimize the impact of the fire. Adequate management of this information requires the use and adaptation of computer-based fire models to integrate sensor data, speed computations, and reduce uncertainty.[125] Such a system would also likely require the study of improved training practices to enable more effective response.

Fire Suppression/Control

Fire suppression presents a significant problem for future crewed space missions. Fire suppression includes the deployment of a suppression agent and the interaction between the suppression agent and (1) the combustion reaction, (2) the fuel generating surfaces (pyrolysis), and (3) the non-pyrolyzing surfaces (and the associated potential cooling effect). The overall effect is that a suppression/control system needs to be addressed as an ensemble.

Currently, there is inconsistency between suppression agents used by different countries involved in the ISS. While NASA currently relies on CO_2 for the ISS (and halon for the space shuttle), the Russians rely on water-based agents.[126] Although there are advantages and disadvantages to various suppression agents, there is no qualification test for agents or for suppression systems that is specific to operation in reduced-gravity environments or high-O_2 atmospheres. While some recent work has shown the effectiveness of CO_2 in extinguishing gaseous diffusion flames,[127] NASA's reliance on gaseous CO_2 suppression systems for the ISS has been called into question by recent work from Sutula and coworkers.[128] They conducted a study of the delivery of extinguishing concentrations of CO_2 to areas similar to those found in electronic racks aboard the ISS. This research, which used Earth-based experiments coupled with low-gravity computer modeling, showed that it is unlikely that current CO_2 fire suppression systems can deliver a sufficient concentration of CO_2 to a fire for sufficient duration to extinguish the fire with no relight. Other recent work[129,130] has shown that water mist might be effective in various gravities, but many unanswered questions remain about how droplet size, size distribution, and injection method affect its efficacy under different gravity environments.

This work highlights the need not only for fundamental understanding of suppression agents but also for development of a methodology based on physical principles for qualifying the overall performance of fire suppression systems for relevant gravity levels and O_2 environments.

Computer Fire Modeling

Since large-scale fire tests in reduced gravity environments are currently impractical, computer-based fire modeling is of high importance to fire prevention, detection, and suppression for human space exploration. One of the important goals of applied fire research in reduced gravity should be to provide the data necessary to develop and validate computer fire models. Computer fire models are used extensively for design of terrestrial fire-safe structures. Furthermore, such modeling is already being used to assess the location and response of spacecraft fire detectors to various fire scenarios.[131] This modeling has already shown some weaknesses in the location of smoke detectors on the ISS and has identified problems associated with the effects of temporary storage of materials around the ISS. In addition, as discussed above, modeling by Sutula et al. has called into question the efficacy of CO_2 extinguishing systems on the ISS. Although the Sutula et al. findings were validated against 1-*g* data, there is no comparable reduced-gravity data to use in validating their results.

Work should be done to determine which existing fire models can be successfully used in reduced-gravity environments and the limitations of these uses. For example, one of the conclusions of the Sutula work is that large eddy simulation models do not work well for predicting fire suppression in rack geometries. Thus, the Fire Dynamics Simulator, a large eddy simulation model used for the smoke detector work of Urban et al. discussed above, requires further validation. Adaptation of computer fire models from terrestrial use to space exploration will require both an in-depth examination of how the physics is modeled and comparison with appropriate reduced-gravity data for validation. It is important to emphasize that fire models are required to resolve large volumes, and so the emphasis is not on the detailed resolution of the combustion chemistry but on the definition of simplified combustion, suppression, and detection models that are adequate for a coarse resolution grid. Thus the problem presented by the system defines the nature of the modeling strategy.

Post-fire Recovery

Fires during human space exploration have occurred and will occur again, since fires have followed every endeavor by mankind. The catastrophic fire of Apollo 1, a serious fire on Mir in 1997, and several documented incidents of smoldering or charred electrical components on the space shuttle underscore the potential for fire in spacecraft systems.[132,133] Thus, it is important that NASA not only prevent, detect, and suppress fires but also prepare for recovery from inevitable fires. Little research has been done to date on recovery from fires. Most of the efforts in fire safety have been directed toward prevention, detection, and suppression.

Some recent work has begun to examine what the toxic environment will be after a fire. One approach heats a composite disk (similar to a hockey puck) made of pellets of various materials that would be in a compartment, fused together in quantities representative of their proportions in the compartment.[134] The toxic gases that are generated by the smoldering of this disk are then analyzed. This approach, while novel, has never been validated even for terrestrial compartment fires and ignores the fact that materials interact during fires in ways that are different than is represented by their overall relative quantities.

As discussed above, determination of the toxic gas generation of materials in both smoldering and flaming fires should be part of the basic standard for qualifying materials to be used in space. Once that basic characterization is completed, research into how materials interact during fires will be required. All of this work should be combined with computer fire models to provide predictions of the kind of environments that would occur in a compartment during a specified fire. Such predictions should include generation of heat and toxic materials from both the fire and the extinguishing agent(s) and should include those materials that may deposit on surfaces.

Once the types of environments that can be expected from various fire scenarios are determined, strategies for clean-up should be developed. Such strategies might include isolating the compartment and venting it to space or using filtration canisters to absorb toxics from the environment. In any case, monitoring technology that can determine the state of the post-fire environment should be developed and tested. In particular, a strategy should be developed for determining when the atmosphere is breathable after a fire. One approach that bears research is whether a single species such as CO can be used as a surrogate for all toxics such that, when its level is below some threshold value, the air is sufficiently clear of all toxics to be deemed breathable again. This approach has been suggested by NASA and makes sense; however, substantial research and testing will be necessary to identify the right marker species and validate its critical level.

Work also needs to be done on assessing the impact of heat and products of combustion on mission-critical components and systems. For example, research should be conducted to determine the effects of heat and deposition of combustion products on electronic components and boards. This type of work will allow determination of what type of clean-up, if any, or replacement of electronics may be necessary following a fire to continue mission-critical operations.

Finally, since fire can cause a breach of a compartment, fire recovery should include strategies and equipment for repair of breaches in the integrity of a compartment. This might include patch kits or other materials and assemblies to seal a breach and to re-establish a habitable environment inside the compartment after a fire.

Explosion Mitigation and Recovery

Like fires, explosions can cause catastrophic loss of one or more compartments in a habitat or space vehicle. Hydrogen and oxygen will likely be on board any human exploration vehicle (and probably in habitats as well). High O_2 environments and reduced gravity affect flammability limits; therefore, materials such as various types of hydrocarbon liquids (e.g., oils and greases) and some metals can become explosive. So, as with fires, there is a need to develop sensors that can detect a potentially explosive atmosphere before explosive conditions are reached. Mitigation strategies should be developed once a potentially explosive environment is detected. Such strategies might include the introduction of an explosion suppression agent or a method to remove one of the components from the environment. Again, as in the fire case, strategies and equipment should be developed to repair breaches in compartments as the result of an explosion.

Summary of Enabling Science and Technologies for Fire Safety

Required by 2020

Materials Standards. NASA should develop and implement new testing standards to qualify materials for flight. Research is necessary in materials qualification for ignition, flame spread, and generation of toxic and/or corrosive gases in relevant atmospheres and reduced gravity levels. **(T7)**

Particle Detectors. NASA would benefit from improved particle (smoke) detectors. Research is necessary in characterizing particle sizes from smoldering and flaming fires under reduced gravity. **(T8)**

Fire Suppression Systems. NASA should develop and implement a standard methodology for qualifying suppression systems in relevant atmospheres and gravity levels and with various delivery systems. Research is needed to characterize the effectiveness of various fire suppression agents and systems under reduced gravity so that the qualification is based on physical principles. **(T9)**

Post-fire Environment Strategies. NASA would benefit from strategies to characterize a safe post-fire environment and to clean up a post-fire environment, including strategies and equipment to repair breaches in compartments as a result of fire or explosion, to the extent that such repairs are likely to be practical in a deployed spacecraft or habitat. Biological science research is needed to assess the toxicity of various fire products under reduced gravity. **(T10)**

Highly Desirable by 2020

Multisignature Fire Detection Technologies. Fire detection technologies that rely on other fire signatures in addition to smoke particles are likely to be faster and more reliable and can be used as information sources for effective response. To enable the development of such multisignature systems, research is needed to characterize fire signatures in addition to smoke particles in relevant atmospheres and at reduced gravity levels, as well as research on interpretation methods based on computational fire models.

Numerical Fire Modeling. NASA should develop and conduct experiments to provide validation data in relevant atmospheres and at reduced gravity levels.

Explosion Sensors. To identify a potentially explosive environment before explosive conditions are reached, sensors for incipient explosion detection need to be developed. Research to support this technology should address flammability limits under reduced gravity.

Explosion Mitigation Strategies. NASA should develop and test mitigation strategies to engage when potentially explosive conditions are reached. Research is needed on explosion suppression agents and/or methods to remove reactant components from a closed environment under reduced gravity.

Two other important recommendations for fire safety involve the organizational integration of fire safety R&D into operational fire safety for space vehicles and habitats. First, there appears to be no specific office in NASA that is responsible for fire safety per se. While all programs profess to have a high regard for fire safety, there do not appear to be specific personnel with the appropriate fire safety background who oversee the implementation of current fire safety knowledge in space vehicles and habitats. Second, there does not appear to be any specific mechanism for translation of fire safety R&D findings into the design and production of space vehicles and habitats. A NASA office specifically responsible for fire safety in vehicles and habitats would ensure that fire safety systems are based on the latest, best knowledge on fire prevention, detection, suppression, and mitigation. Furthermore, such an office would be able to inform the fire safety R&D community, both within NASA and outside, about specific needs for R&D on fire safety.

Space Resource Extraction, Processing, and Utilization

If humans are to undertake long-duration missions to the surfaces of other planetary bodies, beginning with the Moon and leading to Mars, utilization of local resources, or ISRU, will be an essential element. For reviews of many of the possibilities, see McKay et al.[135] and Duke et al.[136] For example, a number of modeling studies have demonstrated that, when all system elements are included (excavation, thermal and chemical processing, water electrolysis, product purification, power production, heat rejection, storage, etc.), the amount of oxygen that can be produced in a year from lunar resources exceeds by more than an order of magnitude the mass of equipment that must be brought from Earth to produce oxygen on the Moon. The system that has received the most study would use hydrogen brought from Earth to reduce lunar ilmenite ($FeTiO_3$) in a system in which the hydrogen is recycled and oxygen is the product.[137] These types of systems are beneficial to exploration missions because they offset the high cost of transporting equipment or propellant to planetary surfaces by using locally produced materials. They probably are essential for scenarios where permanent human presence is expected. The production of materials from resources found in the environments of the Moon, Mars, or other solid bodies can serve a wide variety of uses.

Initial ISRU applications would likely include propellants (H_2, O_2, CH_4, or other hydrocarbons), energy storage (H_2, O_2, and thermal mass materials), and life support consumables (H_2O, O_2, N_2). Unprocessed surface material will likely be used for radiation, meteoroid, or thermal shielding, particularly on the Moon. With experience gained from such efforts, more advanced ISRU systems in the future could potentially produce photovoltaic arrays for energy production,[138] solid materials for fabrication of spare parts and construction components, materials for radiation shielding, etc. In advanced applications, ISRU may play a role in the construction of pressurized structures for planetary surface operations. The benefits of ISRU derive in part from the fact that relatively small systems can work over long periods of time to produce relatively large amounts of product. The production and use of planetary resources can lead to modified interplanetary and surface architectures, reduced cost, and reduced risk for long-term space exploration. Many of the elements found in some ISRU production systems, such as reactors, electrolyzers, gas purification systems, and filters, are common or similar in ISRU and life support systems.

The benefits of ISRU are great; however, among the significant challenges are the following:

- No end-to-end ISRU system has been demonstrated, on Earth or in space. Not enough work has yet been done to determine which are the best candidates for flight systems.
- ISRU systems tend to require machinery with moving parts, raising problems of useful lifetime, maintenance, and repair.
- Complex robotic systems have yet to be demonstrated. Progress with rovers for Mars exploration is promising,[139] but ISRU systems will require more complex surface operations than moving over the surface and taking measurements.
- Excavating and processing materials from lunar or planetary surfaces will create a dusty environment, exacerbating issues with dust control and mitigation (as discussed above in this chapter).

The early results from recent robotic missions to the Moon are particularly encouraging for ISRU. From the Indian Chandrayaan-1 mission, we have learned that water and hydroxide molecules are mobile on the surface, with surface concentrations that vary through the lunar diurnal cycle.[140,141,142] From the Lunar Crater Observation and Sensing Satellite (LCROSS) impact mission, we have learned that there is substantial water ice, perhaps in concentrations of a few percent, and other volatiles present at least in the targeted cold trap, Cabeus Crater.[143] From the Diviner instrument on the Lunar Reconnaissance Orbiter (LRO),[144] we have learned that the polar regions are perhaps 20°C colder than previously expected. These results support the prospect that water and other volatiles may be present in useful concentrations at depths of about 1 m outside craters in the polar regions. If the other volatiles are confirmed to be the same constituents as might be expected due to cometary impacts (CH_4, other hydrocarbons, NH_3, etc.), then all of the elements necessary for life and eventual settlement or colonization may be present on the Moon.

There are a number of potential synergistic interactions between ISRU and other subsystems of a human planetary outpost. The production of propellant from in situ resources can offset the need to bring propellants from Earth and thereby drastically reduce the scale of a propulsion system designed for round-trip journeys. Production of propellants from local resources demands the development of reusable (restartable, at least) space transportation elements. Creation of cryogenic propellant depots in space, supplied from propellant sources on the Moon, can change dramatically the architecture and economic cost of exploration activities beyond low Earth orbit.[145]

For Mars, production of propellant from atmospheric carbon dioxide by reduction, using hydrogen brought from Earth, has been proposed as a straightforward way to produce methane, which significantly reduces the propellant to be brought from Earth. Water ice, which could be a direct source for hydrogen/oxygen or (with atmospheric CO_2) methane has been known for some time to exist at the martian poles.[146] Recently, Mars robotic missions, particularly the Mars Polar Orbiter, have given evidence of the widespread occurrence of large amounts of subsurface ice at least down to mid-latitudes.[147] These deposits may have formed during a previous wetter period on Mars and have been preserved by burial under dust or regolith. If such deposits can be found near places that are also targets for extended human exploration, they could become sources of propellant. The technology for extracting them would be similar to that required to mine lunar polar ice deposits (excavation, thermal extraction).

Some asteroids, including some near-Earth asteroids, have carbon-rich, water-rich compositions (up to 20 percent water by weight), judging from analyses of meteorites and remote spectroscopic analysis. Because it is unlikely that an asteroid would become a target for repeated or continuous human exploration, the development of asteroidal resources is likely only if a substantial, probably commercial, space products infrastructure develops. Lewis[148] has advocated the mining of iron-rich asteroids for platinum-group metals for commercial use on Earth. This would be a longer-term possibility.

ISRU systems designed to produce materials by thermochemical or electrochemical processes require significant amounts of power and could benefit from systems with high power output, such as nuclear fission power systems. As many production systems work best in continuous modes and may "freeze-up" if the process is terminated, continuous power generation is a significant benefit. Conversely, ISRU systems potentially allow for the production of photovoltaic solar power systems from local resources.[149] The production of H_2 and O_2 from local resources offers the potential to use them as fuel cell reactants. As another ISRU application to provide base power, the thermal energy storage media for nighttime power generation at a crewed outpost[150] could potentially be produced from locally available resources. There is also the potential for constructing thermal wadis: engineered sites providing heat and/or power, where rovers and other exploration hardware can hibernate while waiting out periods of darkness on a lunar or planetary surface.[151,152,153] These ISRU chemical and thermal energy storage options are of particular interest for the Moon, where long periods of darkness would require substantial energy storage for the protection of surface assets and for nighttime power generation, compared to a location in space or on Mars.

ISRU systems will likely be highly automated and, for the Moon, may be teleoperated from Earth even when humans are present at the exploration base. For Mars, they will have to be run with limited attention from Earth during robotic phases of exploration. Development of automated interfaces between elements of the ISRU system will be needed, particularly for mechanical elements such as between excavation and hauling rovers and material extraction elements or for transfer of cryogens between production and storage elements. Development of sensors that monitor and measure important aspects of the operation and provide information to control systems will be essential. The current NASA architecture for the Moon envisions the production of 1 metric ton of oxygen per year, for use in replenishing losses from LSSs and providing the larger reservoirs of oxygen and water that make a surface LSS more robust. Such a system would be a step toward, and could be viewed as a pilot-scale plant leading to, a subsequent propellant production facility. An oxygen production system might be based on the potential for extracting water ice from lunar polar regions or could alternately use one of the systems, now under development, for producing oxygen by chemical reaction of lunar regolith.[154]

The development of ISRU capabilities traditionally has been a commercial enterprise. If NASA's development of initial capability can solve key technical problems, commercial implementation of production systems could supply materials for continued government exploration and for other applications.

The development of ISRU capabilities would likely require a significant R&D thrust during the early stages of occupation of a lunar outpost. An ISRU laboratory module could focus on applications supporting outpost expansion, reducing operational risk and cost, and creating a more sustainable surface capability. These applications could include producing material for radiation and meteoroid shields, manufacturing spare parts, and constructing facilities using local materials (glass, composites, metals, concrete, etc.).

Recent NASA architectures for human exploration of Mars incorporate the use of martian oxygen and methane as propellant for lifting return flights from the surface of Mars.[155] Although human exploration of Mars is beyond the time frame of this decadal survey, an active robotic exploration program is ongoing. Opportunities should be sought to clarify the distribution of resources and feasible propellant extraction processes in conjunction with the robotic scientific exploration program.

Known Effects and Knowledge Gaps

For early applications, sufficient knowledge now exists on the distribution of potential resources on the Moon, except for the question of polar resources. The regolith everywhere on the Moon provides a source of oxygen (by reduction of metal oxides and silicates) and potentially hydrogen (hydrogen is generally present in the regolith, but at low levels: just 50 to 100 parts per million), but economical recovery has not been demonstrated. For polar regions, where there is remote-sensing evidence for enhanced hydrogen concentrations and recent reports of the existence of water (present in the LCROSS ejecta plume at apparent concentrations of 1 to 2 percent) and other volatiles,[156] surface exploration is needed to characterize the resource deposits (distribution, form, concentration, accessibility) before extraction techniques can be defined.

If it is determined that extraction of water from lunar polar regolith is feasible, three research questions become important: (1) How do the physical properties of icy regolith change between its ambient temperature and the temperature at which volatiles would be extracted (say, 0°C), and how do cold volatiles interact with higher-temperature environments? (2) How do flow processes work for granular solids at cryogenic temperatures in reduced gravity? (3) What are the behaviors of water-bearing granular materials at cryogenic temperatures, when mechanical friction heating is introduced? If sufficient water is located in shadowed polar regions, a substantial set of studies will be needed to learn to excavate and handle granular materials at cryogenic temperatures in the lunar vacuum.

For Mars, CO_2 in the atmosphere can be a source of carbon for the production of methane or other hydrocarbons, using hydrogen brought from Earth, and martian water ice can be a source of H_2O, H_2, and O_2. Although water ice is widespread on Mars,[157] little is known about its accessibility for ISRU uses. The behavior of cold, subsurface ices when exposed to a warmer surface environment needs to be understood.

NASA's ETDP for the past few years and its proposed ETDP project plan address production of oxygen on the Moon. Among the wide variety of potential ISRU processes that have been suggested,[158] research has focused on chemical extraction of O_2 by H_2 and CH_4 reduction[159] and molten regolith electrolysis.[160] Initial versions of oxygen extraction systems have been demonstrated in the field.[161] Lunar regolith excavation techniques have been brought to the level of field demonstration.[162] All unit operations associated with regolith excavation and processing have been demonstrated preliminarily at lunar analog test sites on Earth.[163] These systems are not yet flight-like versions and have not been integrated operationally. Several issues, such as system mass, power requirements and delivery, adaptation to the vacuum environment of the Moon, and long-term operation, will have to be solved to bring this technology to the level needed to produce the targeted quantities of oxygen. Including the current emphasis on the ETDP's laboratory and field development of lunar oxygen production, the following are key issues that must be addressed in a long-term exploration program that intends to incorporate ISRU into its strategy:

a. *Reliable, long-duration operations in a dusty environment.* The extraction techniques are operating on dusty materials, and so dust will be an ever-present problem. Because the techniques operate slowly over long periods of time, repeated or continuous feeding of dusty material from the lunar surface to and from reaction chambers will require development of new approaches to reliably and repeatedly seal the extraction systems. Problems that become research issues include the flow of solid materials in the vacuum environment, which will require a better understanding of how grain size, shape, and composition affect the transfer of granular materials between the surface and containment systems in reduced gravity, both in vacuum and in the pressurized volumes of extraction systems at elevated temperatures. Long-term behavior of materials from which excavation and extraction systems might be constructed are research concerns. See Chapter 9 for additional details on research issues in fluid physics, which applies to all areas of materials processing systems that involve the movement of materials on the lunar surface and within extraction systems.

b. *Gas or cryogen storage and transfer techniques.* Cryogenic storage of gases would provide higher performance (lighter tanks) than pressurized gas storage, but cryogenic storage is operationally more complex. Problems of reliably and repeatedly transferring O_2, H_2O, H_2, and CH_4 in dusty environments should be solved and long-term storage technology (e.g., zero-boiloff tanks) developed. To properly design equipment, the effects of reduced gravity on flow of cryogens should be understood.

c. End-to-end integrated tests should be carried out over suitable periods of time in thermo-vacuum chambers and in appropriate terrestrial field environments. It is especially important to demonstrate that oxygen of appropriate purity can be produced and its composition certified. It is also important to demonstrate that the system will continue to function through starts and stops that would be associated on the Moon with daytime operations and nighttime stand-downs, if nighttime power is not available.

d. An end-to-end robotic demonstration of oxygen production on the Moon (excavation, extraction, purification, storage, and transfer) is needed to validate operations of gravity-dependent processes in the 1/6-*g* lunar surface environment and to provide design requirements for a lunar outpost system. Physical modeling is needed of all elements of an end-to-end system that incorporates the relevant environments (vacuum, reduced gravity, static charging effects) on the flow of materials, thermal modifications of regolith materials (such as sintering), and heat transfer to support system design.

e. A wide variety of other materials could be accessible on the Moon as byproducts or via unique extraction pathways. Although ISRU processes for these materials are not included in early lunar outpost architectures, their development may improve the potential for development and sustainability of a long-duration lunar outpost and reduce the risks and cost for lunar-outpost growth phases. Some of these ISRU options could be developed into research tasks to be carried out by crew members in the lunar outpost. They might include production of consumables (C, H, N) from the lunar regolith; preparation of metals, ceramics, glasses, and concrete materials; and development and demonstration of forming and fabrication techniques. Research issues include identifying feasible processes and understanding the behavior of extraction systems in the lunar surface environment.

f. For ISRU on Mars, surface exploration will be necessary to define the properties of water ice deposits (location, extent, amounts of water contained, physical properties, etc.) that could be accessible for ISRU. This exploration could be carried out in conjunction with currently planned robotic missions or as a standalone Mars mission. A demonstration of CO_2 extraction from the martian atmosphere on a robotic exploration mission is needed to validate production techniques for CH_4 or other hydrocarbons.

ISRU Development and Test Environments

The environments on Earth in which materials production systems for ISRU are developed and tested must be chosen to allow all important variables to be tested before demonstrations are conducted on the Moon or Mars. For the most part, materials production, handling, and storage technologies (e.g., excavation, materials transportation and handling, chemical reactors) will look much like their terrestrial counterparts; significant advances in understanding systems for exploration applications have been gained by modeling based on terrestrial systems.[164] The principal physical differences that must be accounted for in the extraterrestrial ISRU systems are lower gravity, low external pressure (Mars) or vacuum (Moon), different raw materials, low temperatures (as low as 80 K on the Moon) and high temperatures (120°C on the Moon), and the absence of water for chemical processing. Fortunately, during system development, these physical conditions can to a great extent be tested in isolation. For example, subsystems for excavation, thermal extraction, or electrolysis can be tested and demonstrated in high vacuum and Earth gravity and also at 1/6 *g* (e.g., in parabolic aircraft flights) and room pressure. Simulants of lunar and martian materials can be prepared that mimic with sufficient fidelity the characteristics of lunar or martian starting materials. The following are relevant development and testing environments:

- *Laboratory bench-top experimentation.* Only a limited number of ISRU subsystem concepts have been verified in the laboratory. Laboratory-based experiments should be continued that define alternative, potentially higher-efficiency extraction techniques; extend the life expectancy of machinery with moving parts; and mitigate surface dust issues.
- *Terrestrial thermal-vacuum chambers to simulate lunar/Mars surface environments for end-to-end ISRU production tests.* Vacuum chambers should be dedicated to ISRU experimentation because of the need to carry out long-duration testing of materials production systems. Operation of a "dirty" thermal vacuum chamber is a significant challenge in itself.
- *Reduced-gravity aircraft flights.* These flights will generally be suitable for testing specific subsystems to verify functionality in lunar or martian gravity environments.
- *Analog field tests.* Analog testing environments are needed to demonstrate system interactions, evaluate repair and maintenance needs, and demonstrate long-lived operations. Terrestrial analog field tests are essential to demonstrate long-term reliability of candidate systems and to develop operational protocols.
- *Lunar robotic demonstrations to qualify systems and subsystems for lunar use; lunar robotic missions to explore polar regions for volatiles and nonvolatiles.* A vigorous program of robotic exploration of the Moon should be used to verify conditions near the lunar poles, develop resource maps, and demonstrate ISRU system end-to-end operations in the lunar environment. Because of the strategic importance of ISRU to exploration system architectures, robotic missions may be more important if NASA were to choose near-term exploration paths not principally focused on scientific exploration. Such missions might make use of landers and compact rovers that are in development by private teams competing to win the Google Lunar-X Prize*** and might make use of lunar assets, such as thermal wadis comprising regolith-derived thermal mass materials, as platforms that enable rovers and other exploration hardware to survive periods of cold and darkness on the lunar surface. In the architecture assumed for this study, lunar robotic missions, including demonstrations of ISRU, should be included within the next few years, if resource maps are to be developed and if the goal of producing 1 MT/year of oxygen is to be achieved by the program.

***The Google Lunar X PRIZE is a $30 million competition for the first privately funded team to send a robot to the Moon, travel 500 meters, and transmit video, images, and data back to Earth (see http://www.googlelunarxprize.org/).

- *Mars ISRU demonstrations.* Through Mars robotic missions, NASA should explore the potential for utilizing near-surface ice deposits. Likewise, robotic exploration missions to Mars should include capabilities, such as drills and surface geophysical sensors that can determine the amounts of ice and its accessibility to possible future human explorers. Demonstrating the feasibility of propellant production on Mars at an early stage in exploration can influence the later stages of robotic exploration. For example, robotic sample return missions could be made more effective if martian propellants were available.
- *Sample return missions to carbonaceous near-Earth asteroids to define their potential for resource extraction and utilization.* Prior to such missions, it is important to understand both the composition of the resources and the physical environment in which they exist.

Summary of Enabling Science and Technologies for Space Resource Extraction, Processing, and Utilization

The following exploration technologies that are important to space resource extraction, processing, and utilization would benefit from near-term R&D.

2020 and Beyond: Required

Lunar Oxygen Extraction System. A lunar oxygen extraction system would replenish life support system consumables and would be a critical first step in local resource utilization for larger-scale uses, such as a propellant production facility. Key supporting technologies should be developed to support an oxygen production system, including technologies to enable excavation, fluid handling, cryogenic transfer, and zero-boiloff cryogenic storage. Research will be required to develop techniques to mitigate environmental challenges (dust, repeated transfers to and from vacuum). A lunar oxygen extraction system will require an integrated research program to develop needed technologies, test them in lunar analog facilities (vacuum, field analog facilities), and conduct robotic demonstration missions to the Moon, potentially as piggy-back experiments on science exploration missions. **(T24)**

ISRU Capability Planning. Exploration missions should be conducted to the polar regions of the Moon and to Mars to delineate the resources that can be available at potential landing sites in order to properly plan ISRU capability development. The distribution and accessibility of water or hydrogen near the lunar poles and the delineation of ice on Mars as potential resources are important ISRU objectives and should be coordinated with scientific exploration missions to these targets. Water extraction from the lunar polar regolith will require fundamental research on physical properties and flow processes of the water-bearing material at cryogenic temperature. In addition, exploration for resources and development and demonstration of ISRU technology should be incorporated, as feasible, in relevant robotic missions. Two examples are (1) a Mars sample return mission that includes demonstration of the extraction of propellant from the martian atmosphere, and (2) robotic exploration missions to characterize near-Earth asteroids and return samples that would demonstrate their potential resource value. Finally, expansion of technical capabilities such as improved power production and storage and development of in-space propellant depots will improve the potential for utilizing off-Earth resources. Fundamental research is needed to provide a sound basis for how grain size, shape, and composition affect the transfer of cryogenic granular materials into continuous-process systems in reduced gravity and pressurized reactors. The uniqueness of the materials being processed and their low pressure, reduced gravity, and other special conditions will require new predictive physical models. **(T25)**

Planetary Surface Construction

This section covers the principal technologies and sciences necessary to plan, develop, deploy, construct, and maintain habitats, rovers, and other engineering systems related to construction on the surface of the Moon or Mars. Surface construction is a challenging task, given the harsh and isolated nature of the lunar and Mars environments, with their great temperature differentials, reduced gravity, partial or no atmospheric pressure, high doses of radiation, and potential micrometeorite bombardment. The equipment, tools, materials, technologies, and methods used on Earth will have to be modified—and some new ones will have to be created—to build habitats on the Moon and or Mars.

Construction

Most of the construction that will take place at a Moon or Mars base will likely be assembly and deployment of modules and equipment (unless and until ISRU capabilities advance to the point of providing construction materials). Although the ISS has provided extensive experience with assembling a large and complex structure in space, experience constructing and assembling structures on the surface of the Moon is lacking, with the exception of small experiments during the Apollo program. Despite the small size of those experiments, and even though all the tasks were completed successfully, some of the work proved to be very challenging, given the limitations of 1960s-era EVA suits and tools. Work done in simulated environments by the Desert RATS[†††] teams has provided some additional experience, and NASA has done several deployments of inflatable structures: one has been deployed in the Antarctic as a partial simulation of the extreme environment of the Moon or Mars.

Site Preparation

In most cases, some site preparation will be needed, particularly where a leveled field will be needed to install, deploy, or construct structures. Specific equipment will be needed to accomplish this. Although many conceptual studies have been proposed for these types of systems, few have been actually built and tested in simulated gravity fields and the extreme thermal and radiation environments of the Moon or Mars.

Equipment and Tools

Specific equipment and tools need to be developed to build, assemble, and deploy the different types of structures; these will be operated by robotic systems as well as by astronauts during EVA and intravehicular activities and in pressurized rovers. In some instances, these operations may be controlled from Earth. Although many tools and equipment were designed and fabricated for the ISS construction, few exist that were designed to operate on the surface of the Moon or Mars. Although the final construction equipment and tools will be designed specifically to the characteristics of the habitat and infrastructure to be deployed, an extensive list of possible tools and/or equipment (such as excavators, backhoe loaders, bulldozers, graders, or cranes) will need to be developed for lunar or martian habitat construction. From this list, a minimum set can be selected to perform the needed task(s). The issue of lunar and martian dust is of high priority when considering the survivability of construction systems and equipment. Research on materials and mechanisms to prevent equipment damage from planetary surface dust is critical.

Robotic systems will play a critical role in the construction of surface habitats. Robotics is a major gap not represented in the NRC's previous report, *Microgravity Research in Support of Technologies for the Human Exploration and Development of Space and Planetary Bodies*.[165] Autonomous, semi-autonomous, or teleoperated robots may prepare a site for construction, unload equipment from a lander, and provide transport to temporary or permanent sites. Robotic systems may also be used for assembling, deploying, and constructing many of the systems and infrastructure of a surface outpost. For robotic systems to be used effectively in surface outpost construction and operation, close coordination is needed between NASA's research programs in habitats and robotics.

Many robotic technologies and systems exist today in the manufacturing and commercial sectors that are applicable to planetary construction and assembly processes. Many of the technologies developed to operate the Mars rovers Spirit, Opportunity, and Phoenix are applicable to some construction-process robots (i.e., control

---

[†††]Desert Research and Technology Studies, or Desert RATS, is an annual field test led by NASA in collaboration with non-NASA research partners. The Desert RATS effort assesses preliminary exploration operational concepts for surface operation concepts, including rovers, EVA timelines, and ground support (see http://www.nasa.gov/exploration/analogs/).

systems, navigation cameras, hazcams,[‡‡‡] communications systems, warm electronics,[§§§] drives, motors, and mobility systems). Robotic systems that were designed for operations on Earth or on Mars will have to be adapted if they are to be applied to accomplishing tasks on the lunar surface.

Assembly

Initially, most of the construction activities will encompass assembling modules and elements prefabricated on Earth. Some of the equipment and tools listed in the previous subsection will be required to unload these elements from a lander, transport them to a significantly distant site, and install them in their final or temporary location.

No large structures have been assembled to date in partial-gravity or planetary surface simulators. Existing NASA facilities could be modified to accomplish such simulations. Some unloading and loading of modules has been simulated during the Desert RATS activities using the ATHLETE[¶¶¶] carrier. Most of the assembly knowledge to date is based on the ISS experience.

Precise mechanical berthing of modules with other modules, or with pressurized rovers, will be critical, given the proposed high frequency of these operations.

Structures

Continued research in advanced structures will be required to construct efficient, safe, and lightweight structures for habitats, rovers, and other surface infrastructure systems. Advanced structures will be required to reduce mass, enhance radiation protection, and operate in the extreme conditions of the Moon and Mars surfaces. Examples of such advanced structures and intelligent structures are self-healing structures, antimicrobial material coatings, carbon nanotube membranes for carbon dioxide capture, organic coatings, biomimetics structures, extreme temperature composites, and fabrics for inflatables. Different types of structures will be required to build the different elements of a surface outpost such as habitats, pressurized tanks for expendables, solar panels, radiators, rovers, robots, thermal insulation, and radiation and micrometeorite shields. The structures designed will be required to withstand the launch, landing, and operational loads of transportation, translation, and operation from Earth and on their final location on the planetary surface. In addition, they will have to endure the effects of radiation, extreme temperature differentials, and abrasiveness of the operational environment. Many types of structural systems are operating today in space. However, all these structures have been designed to operate in free space with minimal gravitational loads without the effects of landing, downloading, transportation, partial gravity, and dust present on the Moon or Mars. Planetary structures may include the following types:

- *Pressurized*—rigid, foldable, inflatable, metallic, composite, and hybrid;
- *Trusses*—space frames, foldable, telescopic, tensigrity, and quick assembly mechanisms;
- *Tents and shields*—frames, inflatable, rigid panels, fabrics, and films; and
- *Excavated*—walls, berms, trenches, caves, artificial craters, and foundations.

Habitats

It is uncertain at this time what size and type of habitats would be needed for a surface base on the Moon or Mars. The initial number of crew, resupply period, mission duration, and purposes of outposts have not yet been defined. These basic parameters are critical to the final design of any surface infrastructure that will support human life. The basic elements of a habitat, no matter its scale and purpose, will need to house and address the fundamental needs of humans to remain alive and productive. One of the latest and most ingenious design con-

---

[‡‡‡]Hazcams (short for Hazard Avoidance Cameras) are photographic cameras mounted on the front and rear of NASA's Spirit and Opportunity rover missions to Mars.

[§§§]The body of the Mars rover is called the warm electronics box, or "WEB" for short. The outer layer of the WEB protects the rover's computer, electronics, and batteries from temperature extremes.

[¶¶¶]The All-Terrain Hex-Legged Extra-Terrestrial Explorer (ATHLETE) is designed to roll over undulating terrain and "walk" over extremely rough or steep terrain so that robotic or human missions on the surface of the Moon can load, transport, manipulate, and deposit payloads to a range of desired sites (see http://www-robotics.jpl.nasa.gov/systems/).

cepts includes the ability to move the outpost and the entire related infrastructure to several distant sites during the design life cycle of the systems. This "mobility" concept adds a significant level of complexity to the habitats previously conceived as static structures.

Any human habitat, whether on a planetary surface or for deep-space exploration, will be required to house the following systems and subsystems:

- *Food*—preparation, delivery, consumption, and long-term storage;
- *Hygiene and waste management*—water dispensing, sinks, toilet, and personal toiletries;
- *Health maintenance*—exercise equipment, medical care equipment;
- *Sleeping accommodations*—horizontal bunks, privacy, crew personal equipment;
- *Operations center*—wardroom activities, communication and control;
- *Lighting*—general, task, emergency, EVA operations;
- *Furnishings*—seats, bunks, restraint systems, working surface, and scientific equipment;
- *Storage*—refrigerated, frozen, ambient, dry, and wet;
- *Acoustics*—surface materials, geometry layouts, equipment design, fireproof fabrics;
- *Airlocks*—egress and ingress, EVA suit donning, dust control;
- *Berthing ports*—connection between modules and rovers in a dust environment;
- *Dust control*—laminar flow air systems, materials, flooring systems, vacuums;
- *Radiation protection*—water, ice, polyethylene and other low-density and high-hydrogen-content materials, equipment layout, in situ materials; and
- *Mobility*—drives, wheels, suspension, winches, skids, etc.

Power and cooling for planetary habitats are considered in the "Space Power and Thermal Management" section of this chapter. Although most of these subsystems have been developed for the ISS, there is no experience with most of them on the Moon or Mars, with all the related effects of such sites. Some of the most critical areas that need further research are dust control, radiation protection, and mobility systems.

Maintenance

Almost all systems and subsystems of a habitat will have to be maintained, serviced, and/or replaced during any long-term mission. Therefore, they will have to be designed to minimize the impact of these activities with respect to time, complexity, mass, power, and safety. The actual number and types of spare parts will be determined by the final design of all the subsystems. Attributes such as robustness, modularity, commonality, interchangeability, simplicity, and protection/shielding will need to be applied to all operating systems. However, without specific design requirements for the subsystems, it is difficult to anticipate the maintenance schedules and spare parts required.

Rovers

Pressurized and unpressurized rovers greatly extend the range of exploration on planetary surfaces. Initially, uncrewed unpressurized rovers explore and survey sites, as Spirit and Opportunity have already done on Mars; in the future they will prepare sites for future human landings. Based on the latest designs of lunar surface systems architectures, pressurized rovers will play a prominent role as extensions to habitats and habitability support, as well as in mobility and exploration activities.

The Lunar Roving Vehicle, although unpressurized and very simple in design, gave the Apollo astronauts a wide range of mobility that would have been impossible without it. The Mars rovers have exceeded their design lives and primary mission requirements in terms of longevity and capabilities on the martian surface. However, many new technologies need to be advanced to deploy, operate, and maintain large habitable rovers for long-duration missions on the surface of the Moon or Mars. In particular, electromechanical systems are needed that can provide fine control and precise operation when performing complex simultaneous tasks.

Summary of Enabling Science and Technologies for Planetary Surface Construction

The following exploration technologies that are important to planetary surface construction would benefit from near-term R&D.

2020 and Beyond: Required

Construction: Teleoperated and Autonomous Construction. Robotic systems will play a critical role in the construction of surface habitats. An extensive, coordinated R&D program should be established to study human-robot interaction (including teleoperations) for the construction and operation of planetary surface habitats. **(T26)**

Construction: Regolith- and Dust-Tolerant Systems. Construction technologies are needed that include innovative designs for dust-tolerant mechanisms and fluid connectors, O-rings, and materials that can withstand the abrasive effects of regolith and provide a tight seal in its presence. To develop these technologies, a better understanding is needed of regolith mechanics, the behavior of ice-laden regolith, equipment traction, and the forces required to excavate, move, and compact regolith. There is a need to develop gravity-dependent soil models to better understand regolith strength, stiffness, and density on the Moon and Mars. **(T27)**

Habitats: Habitability Requirements. Surface habitability systems design requirements need to be developed. In particular, they should incorporate accurate lunar surface radiation modeling and simulation to predict crew dosage for long-duration missions within complex structures. **(T28)**

2020 and Beyond: Highly Desirable

Construction: Excavation Systems. NASA currently lacks excavation and earthmoving capabilities for the Moon or Mars. Research on human-controlled, semi-autonomous, and/or autonomous equipment to meet earthmoving and excavation requirements on the Moon or Mars is desired. Specific research on adaptation and downscaling of Earth-based subsystems is suggested.

Construction: Robots. The design and development of robots for the extreme environments of the Moon and Mars should include capabilities for the performance of the complex tasks required during construction.

Construction: Processes. Research is needed on construction and assembly processes under planetary simulated or scaled-down conditions.

Structures: Materials. NASA would benefit from structures and materials technology development for habitats, rovers, and other surface infrastructure systems that enable development of structures that have low mass and improved radiation protection capability and that can be deployed in extreme temperatures. Basic structural systems and materials research in reduced gravity and all other extreme planetary conditions is needed in several areas, including extreme temperature composites, alloys, fabrics for space suits, and inflatable structures.

Structures: Structural Systems. A number of structural systems will need to be developed and tested for the planetary environment, including pressurized structures, trusses, tents and shields, and excavated structures.

Maintenance: Subsystem Operations. Research should be conducted on maintenance strategies and operations of all subsystems and should be specific to the environment in which these systems will be operating.

Rovers. A number of technologies should be developed to enhance the use of rovers on a planetary surface, including mechanisms for mitigating the effects of extreme temperature (including highly reliable, lightweight thermal control, Sun shades, heat rejection systems, and radiator dust mitigation methods), mechanisms for mitigating the

TABLE 10.3 Current Research and Technologies Required to Support Objectives and Operational Systems up to 2020

Recommendation	Research Topic	Current Gap	Critical Technology	Enabling Research	Environmental Constraints	Crosscutting Applications
T1	Space power and thermal management	Inability to utilize multiphase flow systems to increase performance	Two-phase flow thermal management technologies	Harness ability to use active two-phase flow thermal management in reduced gravity fields	Partial and microgravity	Space and surface operations, propellant systems, EVA, life support, habitats, power, ISRU
T2	Space propulsion	Inability to limit boiloff of cryogenic propellants to extend storage	Zero-boiloff propellant storage systems	Research in such areas as advanced insulation materials, active cooling, multiphase flows, and capillary effectiveness	Full gravity range	Space and surface operations, ISRU
T3	Space propulsion	Lack of knowledge of cryogenic propellant flow, handling, and gauging in microgravity	Cryogenic fluid management technologies	Research to enable microgravity propellant flow, handling, and gauging	Partial and microgravity	Enables propellant depots, ISRU
T4	EVA	Inadequate mobility for suited crew	EVA suit mobility enhancements	Research in suit comfort, trauma countermeasures, and joint mobility to provide crew the mobility to perform tasks over extended periods without injury	Partial and microgravity	Space and surface operations
T5	EVA	Lack of suit durability in on-orbit, lunar, and martian environments	Dust and micrometeroid mitigation systems	Research and test beds to deal with durability and maintainability issues of suits stemming from micrometeoroid and orbital debris damage, dust exposure, and plasma	Partial and microgravity, temperature extremes	Space and surface operations
T6	Life support systems	Lack of understanding of partial-gravity effects on life support systems (fluid/air)	Fluid and air subsystems	Design, test, and operation of highly reliable life support fluid and air systems in reduced gravity environments	Partial gravity	Propellant systems, habitats and rovers, ISRU
T7	Fire safety	Lack of knowledge regarding materials flammability and toxicity in various atmospheres and gravity fields; lack of adequate standards to determine acceptable materials based on flammability and toxicity	Materials standards	Research to describe the flammability and toxicity of materials with respect to ignition, flame spread, and toxic/corrosive gas generation in various environments and gravity fields	Partial and microgravity, various O_2 atmospheres	Space and surface operations, space vehicles, habitats, rovers
T8	Fire safety	Current fire detection techniques lack reliability in reduced gravity fields	Particle detectors	Research to characterize particle sizes generated by smoldering and flaming fires; identification of other fire signatures that can facilitate fire detection	Partial and microgravity, various O_2 atmospheres	Space and surface operations, space vehicles, habitats, rovers

T9	Fire safety	Effectiveness of fire suppression systems in reduced gravity environment is not well understood	Fire suppression systems	Research to describe the effectiveness of fire suppression agents and systems against various types of fires in various spatial configurations and gravity fields	Partial and microgravity, various O_2 atmospheres	Surface and space operations, space vehicles, habitats, rovers
T10	Fire safety	Lack of knowledge of postfire environment	Post-fire environment strategies	Research to characterize post-fire environment and clean-up strategies including removal of toxic gases	Partial and microgravity, various O_2 atmospheres	Surface and space operations, space vehicles, habitats, rovers

TABLE 10.4 Current Research and Technologies Required to Support Objectives and Operational Systems for 2020 and Beyond

Recommendation	Research Topic	Current Gap	Critical Technology	Enabling Research	Environmental Constraints	Crosscutting Applications
T11	Space power and thermal management	Need energy storage density improvements (factor of 10) beyond current battery and fuel cells	Regenerative fuel cells (RFCs)	Research to enable RFC demonstration, including research related to dead-ended gas flow paths versus through-flow, cryogenic or pressurized gas storage, thermal management, and reliable, long-life operation	Partial and microgravity, extreme low temperature	Habitat and rovers, ISRU, surface operations, power, space vehicles
T12	Space power and thermal management	Current energy conversion systems are not efficient for all power regimes	Energy conversion technologies for low- and high-power regimes	Research in high-temperature energy conversion cycles and devices coupled to essential working fluids, heat rejection systems, materials, etc.	Partial and microgravity	Surface operations, ISRU, habitats
T13	Space power and thermal management	Photovoltaic and RPS systems not always adequate for very high power and high power density; lack of technology demonstration exists for fission surface power	Fission surface power	Research in high-temperature, low-mass materials for power conversion and radiators	Partial and microgravity	Space and surface operations
T14	Space propulsion	Lack of lunar and planetary descent and ascent propulsion capabilities	Technologies to enable engine start after long quiescent periods, combustion stability at all gravity conditions, and deep throttle	Research into cryogenic fluid management, propellant ignition, flame stability, and active thermal control of the injectors and combustors over the full range of gravities, orientations, fluid phases, etc.	Partial and microgravity, extreme low temperature	Propulsion for crew rescue and emergency maneuvers
T15	Space propulsion	Lack of aerodynamic decelerators to reduce propellant required for heavy payload entry to Mars	Inflatable aerodynamic decelerators	Research into flexible materials to enable the use of lightweight flexible aeroelastic re-entry devices	Partial and microgravity, Mars atmosphere	Subsystem for innovative crew return system
T16	Space propulsion	Lack of lunar and planetary descent propulsion capabilities	Supersonic retro propulsion system	Research on flow-field interactions of the rocket plume with atmosphere, aerothermodynamics of decelerator flows, dynamic interactions on the vehicle, and control for safe flight	Partial and microgravity, Mars atmosphere	Avoiding optical and sensor blinding during thrusting maneuvers
T17	Space propulsion	Lack of propulsion systems with high specific impulse	Space nuclear propulsion	Research in liquid-metal cooling under reduced gravity, thawing under reduced gravity, and system dynamics	Partial and microgravity	Nuclear power

T18	Space propulsion	Lack of efficient propulsion for very high specific impulse and high thrust for shorter trip times	Solar electric propulsion	Research in high specific power, multiphase thermal control, long-life engine components, control of lightweight space structures, and low-cost/efficient propellants	Partial gravity	Thermal management, space power
T19	EVA and life support systems	Inadequate protection from ionizing radiation exposure for EVA, rovers, habitats	Radiation protection systems	Research to enable crew to survive in the anticipated ionizing radiation environment	Full gravity range	Space and surface operations, habitat and rovers
T20	Life support systems	Inadequate protection from temperature extremes	Thermoregulation technologies	Research into thermoregulation of habitats, rovers, and spacesuits on the lunar surface	Temperature extremes	Space and surface operations, EVA, ISRU, propellant systems
T21	EVA and life support systems	Lack of closed loop air revitalization (oxygen loop closure)	Closed-loop air revitalization	Research in CO_2 removal, recovery, and reduction; O_2 generation via electrolysis with high pressure capability; improved sorbents and catalysts for trace contaminant control; and atmosphere particulate control and monitoring	Partial gravity	Thermal management, environmental control, habitats
T22	EVA and life support systems	Lack of closed loop water recovery	Closed-loop water recovery	Research in water recovery from wastewaters and brines, pretreatments, biocides, low expendable rates, and robustness; more research to assess the effect of lunar gravity on two-phase flow systems	Partial gravity	Thermal management, environmental control, habitats
T23	EVA and life support systems	Dust mitigation techniques and technologies do not exist	Dust mitigation technologies	Development and testing of dust countermeasures to mitigate the effects of dust coating, contamination, and abrasion and to prevent thermal control problems, seal failures, and inhalation/irritation	Partial gravity	Habitat and rovers, ISRU
T24	Space resource extraction, processing, and utilization	Current extraction process and mechanical systems operations are not suited to the lunar environment, e.g., cryogenic conditions at lunar poles	Lunar water and oxygen extraction system	Research to identify and adapt excavation, extraction, preparation, handling, and processing techniques; mitigate durability problems (dust, repeated operations); and demonstrate long-term operations in lunar environment	Partial gravity, cryogenic granular materials, vacuum, extreme temperatures	Surface operations, habitat construction, water management, power

continued

TABLE 10.4 Continued

Recommendation	Research Topic	Current Gap	Critical Technology	Enabling Research	Environmental Constraints	Crosscutting Applications
T25	Space resource extraction, processing, and utilization	Lack of knowledge regarding physical and handling properties of in situ resources	ISRU capability planning	Research (including remote assay and sampling) to characterize specific resources available at planned lunar and martian surface destinations available for ISRU planning and extraction	Partial gravity, cryogenic granular materials, extreme temperatures	Surface operations, habitat construction, propulsion, life support
T26	Planetary surface construction	Lack of understanding how to effectively integrate human and robotic operations	Teleoperated and autonomous construction	Research to determine how to best utilize human and robotic resources for construction and other surface operations	Partial gravity, extreme temperatures	Surface operations
T27	Planetary surface construction	Lack of information regarding regolith mechanics and properties	Regolith- and dust-tolerant systems	Research to describe the physical and mechanical properties of regolith to facilitate surface operations, construction, and ISRU	Partial gravity, extreme temperatures	Surface operations
T28	Planetary surface construction	Habitability requirements for partial-gravity operations unknown	Habitability requirements	Research to define partial-gravity habitability requirements for surface operations on the Moon and Mars	Partial gravity, extreme temperatures	Surface operations

abrasion on components and materials, suit docking/undocking systems, and dust-resistant seals. Improved modeling and simulation techniques for lunar thermal environments and dust characterization need to be developed.

Rovers: Windows. Large, lightweight windows with correct optical properties and protection from radiation/ultraviolet, micrometeoroid/orbital debris, blast effects, and scratching would enhance rover capabilities. Research is needed in new materials development, as well as in design.

Terrestrial Analogs. Several terrestrial analogs to space surface systems already exist and more are planned. These very useful facilities should be actively and systematically employed to assess designs, materials, and operation related to habitat design and construction.

SUMMARY AND CONCLUSIONS

The utility of a coherent plan that is appropriately resourced and consistently applied to enable exploration cannot be overemphasized. This is especially noteworthy in light of the frequent and large postponements and redirections that NASA's exploration-related goals have experienced over the past several decades. NASA's existing ETDP goals seem well aligned with the panel's recommendations, with augmentation as specified in this chapter.

Transition of technology on schedule and within budget to meet mission needs is an intellectual challenge worthy of the attention of our nation's best technologists. Usually it not treated as a job category. Rather, transition involves an ad hoc interplay among engineers transitioning the research findings, scientists continuing to advance the associated technology, and program managers assessing the risk, schedules, and budget. Seldom is the technological handoff a simple process. For example, advances in scientific understanding may be good enough to enable design and fabrication of a prototype of a new or improved major subsystem; nonetheless, research may continue on facets of the technology discovered during the prototyping. More attention should be given to understanding how to accomplish transition of technology within the NASA system. The goal is to reduce the uncertainty of the process for mission managers, thereby reducing unwarranted risk aversion and giving NASA the confidence to use tomorrow's technology sooner. Transitioning technology on schedule and within budget is integral to mission management. Attention should be given to improving NASA's confidence in predicting the transition of science to mission application, thereby improving projections of new systems to which NASA can aspire. When establishing major missions, NASA should ensure that program managers, engineers, and scientists will be true partners in transitioning the essential new technology. To improve the process, the specifics leading to successful transitions should be analyzed after the mission. In Chapter 12, the section "Linking Science to Mission Capabilities Through Multidisciplinary Translational Programs" describes more thoroughly the means to transition technology successfully within NASA.

The body of this chapter contains many recommendations. Tables 10.3 and 10.4 above summarize the research areas previously identified by the panel as required for prudent execution of the exploration program. Table 10.3 lists those topics for which information is required for activities that the exploration plan indicates will occur prior to 2020, and Table 10.4 lists research for the activities that are scheduled to occur in 2020 and beyond. Due to the uncertainty surrounding the funding that will be allocated to these various research topics, the panel did not factor in the lead time that would be needed for these research activities to provide answers to the questions they address. For example, planetary surface construction appears in the "2020 and Beyond" table, but it is essential that these activities be undertaken well in advance of 2020 to lead to operational systems and implementation in the 2020 time period. NASA can determine when to initiate a particular research project, based on the level of support and the state of knowledge that exists at the time the decision is made to pursue a future activity so that it will be ready at the appropriate time indicated in the tables. This approach implies that even topic areas listed in Table 10.4, "2020 and Beyond," might require initiation of the enabling research well before 2020.

Finally, in order for the efforts recommended here to yield the greatest benefit, NASA needs to ensure that explicit and robust organizational mechanisms and structures are in place that promote interdisciplinary collaboration and sharing of knowledge so that successful research is efficiently translated into applications.

REFERENCES

1. National Research Council. 2000. *Microgravity Research in Support of Technologies for Human Exploration and Development of Space and Planetary Bodies.* National Academy Press, Washington, D.C.
2. The following NASA Exploration Technology Development Program project plans were provided to the NRC Committee for the Decadal Survey on Biological and Physical Sciences in Space by the Exploration Technology Development Program, Advanced Capabilities Division, Exploration Systems Mission Directorate: FY10 Project Management Plan for Advanced Avionics and Processor Systems (AAPS), AAPS–PLAN-0001 (REV B), Document No. RHESE-PLAN-0001, October 08, 2009; Advanced Environmental Monitoring and Control Project Plan, Document No. ESTO FY10-01, Version 1.00, October 20, 2009; FY2010 Project Plan for Autonomous Landing and Hazard Avoidance Technology (ALHAT), Document No. ALHAT-1.0-001, September 21, 2009; FY 10 Project Plan for the Cryogenic Fluid Management (CFM) Project, October 1, 2009; Dust Management Project Plan, DUST-PLN-0001, Rev. B, October 27, 2009; Energy Storage Project Lithium-Ion Batteries and Fuel Cell Systems, Document ES08-105, Revision C, October 2, 2009; EVA Technology Development Project: Project Plan, CxP 72185, Annex 01, Rev. B, September 22, 2009; Exploration Life Support (ELS), Document No. JSC-65690 Rev B, October 15, 2009; Fire Prevention, Detection, and Suppression Project Plan, FPDS10-PP-001, Ver. 4.0, October 22, 2009; Project Plan (FY10) for Human-Robotics Systems (HRS), Document No. HRS1002, October 23, 2009; FY 2010 Project Plan for In-Situ Resource Utilization (ISRU), undated; Technology Development FY10 Project Plan Propulsion and Cryogenic Advanced Development (PCAD) Project, PCAD10_001, October 1, 2009; and FY 2010 Thermal Control System Development for Exploration Project Plan, October 20, 2009.
3. NASA. 2009. *Technology Horizons: Game-Changing Technologies for the Lunar Architecture.* NP-2010-01-237-LaRC. NASA Langley Research Center, Hampton, Va. September.
4. Review of U.S. Human Spaceflight Plans Committee. 2009. *Seeking a Human Spaceflight Program Worthy of a Great Nation.* Office of Science and Technology Policy, Washington, D.C. October.
5. Byers, D.C, and Dankanich, J.W. 2008. Geosynchronous-Earth-orbit communication satellite deliveries with integrated electric propulsion. *Journal of Propulsion and Power* 24(6):1369-1375.
6. Schneider, T., Mikellides, I.G., Jongeward, G.A., Peterson, T., Kerslake, T.W., Snyder D., and Ferguson, D. 2005. Solar arrays for direct-drive electric propulsion: Arcing at high voltages. *Journal of Spacecraft and Rockets* 42(3):543-550.
7. Benson, S.W., NASA Glenn Research Center. 2007. "Solar Power for Outer Planets Study," presentation to Outer Planets Assessment Group, November 8. Available at http://www.lpi.usra.edu/opag/nov_2007_meeting/presentations/solar_power.pdf.
8. Piszczor, M.F., Jr., O'Neill, M.J., Eskenazi, M.I., and Brandhorst, H.W., Jr., 2006. The Stretched Lens Array Square Rigger (SLASR) for Space Power. Presented at the 4th International Energy Conversion Engineering Conference and Exhibit (IECEC). AIAA Paper 2006-4137. American Institute of Aeronautics and Astronautics, Reston, Va.
9. Block, J., Straubel, M., and Wiedemann, M. 2010. Ultralight deployable booms for solar sails and other large gossamer structures in space. *Acta Astronautica* 68(7-8):984-992.
10. National Research Council. 2009. *Radioisotope Power Systems: An Imperative for Maintaining U.S. Leadership in Space Exploration.* The National Academies Press, Washington, D.C.
11. Warren, J., NASA Exploration Systems Mission Directorate. 2009. "Nuclear Power Systems for Exploration," presentation to the Panel on the Translation to Space Exploration Systems of the Decadal Survey on Biological and Physical Sciences in Space, November 12. National Research Council, Washington, D.C.
12. NASA. 2009. *Technology Horizons: Game-Changing Technologies for the Lunar Architecture.* NP-2010-01-237-LaRC. NASA Langley Research Center, Hampton, Va. September.
13. NASA. 2009. *Technology Horizons: Game-Changing Technologies for the Lunar Architecture.* NP-2010-01-237-LaRC. NASA Langley Research Center, Hampton, Va. September.
14. Craig, D., NASA Exploration Systems Mission Directorate. 2009. "Current Concepts for Lunar Outpost Habitation," presentation to the Panel on the Translation to Space Exploration Systems of the Decadal Survey on Biological and Physical Sciences in Space, November 12. National Research Council, Washington, D.C.
15. Warren, J., NASA Exploration Systems Mission Directorate. 2009. "Nuclear Power Systems for Exploration," presentation to the Panel on the Translation to Space Exploration Systems of the Decadal Survey on Biological and Physical Sciences in Space, November 12. National Research Council, Washington, D.C.
16. Warren, J., NASA Exploration Systems Mission Directorate. 2009. "Nuclear Power Systems for Exploration," presentation to the Panel on the Translation to Space Exploration Systems of the Decadal Survey on Biological and Physical Sciences in Space, November 12. National Research Council, Washington, D.C.
17. Kovacs, A., and Janhunen, P. 2010. Thermo-photovoltaic spacecraft electricity generation. *Astrophysics and Space Sciences Transactions* 6:19-26.

18. NASA Glenn Research Center. 2005. Fuel Cells: A Better Energy Source for Earth and Space. February 11. Available at http://www.nasa.gov/centers/glenn/technology/fuel_cells.html.
19. Sienski, K., Eden, R., and Schaefer, D. 1996. 3-D electronic interconnect packaging. *Aerospace Applications Conference Proceedings, IEEE* 1(3-10):363-373.
20. National Research Council. 2000. *Microgravity Research in Support of Technologies for Human Exploration and Development of Space and Planetary Bodies.* National Academy Press, Washington, D.C.
21. Lahey, R.T., Jr., and Dhir, V. 2004. *Research in Support of the Use of Rankine Cycle Energy Conversion Systems for Space Power and Propulsion.* NASA/CR-2004-213142. NASA Center for Aerospace Information, Hanover, Md. Available at http://gltrs.grc.nasa.gov/reports/2004/CR-2004-213142.pdf.
22. National Research Council. 2000. *Microgravity Research in Support of Technologies for Human Exploration and Development of Space and Planetary Bodies.* National Academy Press, Washington, D.C.
23. NASA. 2009. "Exploration Systems Mission Directorate: Life and Physical Sciences: Current Programmatic Content," briefing charts provided to Panel on the Translation to Space Exploration Systems of the Decadal Survey on Biological and Physical Sciences in Space, August 19. National Research Council, Washington, D.C.
24. Wegeng, R., Mankins, J.C., Balasubramaniam, R., Sacksteder, K., Gokoglu, S.A., Sanders, G.B., and Taylor, L.A. 2008. Thermal wadis in support of lunar science and exploration. Presented at the 6th International Energy Conversion Engineering Conference. AIAA Paper 2008-5632. American Institute of Aeronautics and Astronautics, Reston, Va.
25. Creech, S., and Sumrall, P. 2008. Ares V: Progress toward a heavy lift capability for the Moon and beyond. Presented at the AIAA SPACE 2008 Conference and Exposition, San Diego, Calif., September 9-11. AIAA Paper 2008-7775. American Institute of Aeronautics and Astronautics, Reston, Va.
26. Review of U.S. Human Spaceflight Plans Committee. 2009. *Seeking a Human Spaceflight Program Worthy of a Great Nation.* Office of Science and Technology Policy, Washington, D.C. October.
27. Review of U.S. Human Spaceflight Plans Committee. 2009. *Seeking a Human Spaceflight Program Worthy of a Great Nation.* Office of Science and Technology Policy, Washington, D.C. October.
28. NASA. 2005. *NASA's Exploration Systems Architecture Study.* Technical Memorandum NASA-TM-2005-214062. November. NASA, Washington, D.C.
29. Drake, B. 2009. *Human Exploration of Mars Design Reference Architecture 5.0 Addendum.* NASA/SP-2009-566-ADD. July. NASA, Washington, D.C.
30. National Research Council. 2000. *Microgravity Research in Support of Technologies for Human Exploration and Development of Space and Planetary Bodies.* National Academy Press, Washington, D.C.
31. Arney, D., and Wilhite, A. 2010. Orbital propellant depots enabling lunar architectures without heavy-lift launch vehicles. *Journal of Spacecraft and Rockets* 47(2):353-360.
32. NASA. 2005. *NASA's Exploration Systems Architecture Study.* NASA-TM-2005-214062. November. NASA, Washington, D.C.
33. Martin, J.J., and Hastings, L. 2001. *Large Scale Liquid Hydrogen Density Multilayer Insulation with a Foam Substrate.* NASA-TM-2001-211089. NASA, Washington, D.C., p. 33.
34. Drake, B. 2009. *Human Exploration of Mars Design Reference Architecture 5.0 Addendum.* NASA/SP-2009-566-ADD. July. NASA, Washington, D.C.
35. Sullivan, T.A., Linne, D., Bryant, L., and Kennedy, K. 1995. In-situ-produced methane and methane/carbon monoxide mixtures for return propulsion from Mars. *Journal of Propulsion and Power* 11(5):1056-1062.
36. Sanders, G.B., and Duke, M. 2005. *In-Situ Resource Utilization (ISRU) Capability Roadmap, NASA Workshop, Final Report,* May 19. NASA, Washington, D.C.
37. Griffin, J.W. 1986. Background and programmatic approach for the development of orbital fluid resupply. Presented at the AIAA/ASME/SAE/ASEE 22nd Joint Propulsion Conference, Huntsville, Ala. AIAA Paper 1986-1601. American Institute of Aeronautics and Astronautics, Reston, Va.
38. Review of U.S. Human Spaceflight Plans Committee. 2009. *Seeking a Human Spaceflight Program Worthy of a Great Nation.* Office of Science and Technology Policy, Washington, D.C. October.
39. Hastings, L.J., Tucker, S.P., Flachbart, R., Hedayat, A., and Nelson, S.L. 2005. Marshall Space Flight Center In-Space Cryogenic Fluid Management Program overview. Presented at the 41st AIAA/ASME/SAE/ASEE Joint Propulsion Conference and Exhibit, Tucson, Ariz. AIAA Paper 2005-3561. American Institute of Aeronautics and Astronautics, Reston, Va.
40. Howell, J.T., Mankins, J.C., and Fikes, J.C. 2006. In-Space Cryogenic Propellant Depot stepping stone. *Acta Astronautica* 59:230-235.

41. Chato, D.J. 2005. Low gravity issue of deep space refueling. Presented at the 43rd AIAA Aerospace Sciences Meeting and Exhibit, Reno, Nev. AIAA Paper 2005-1148. American Institute of Aeronautics and Astronautics, Reston, Va.
42. NASA. 2005. *NASA's Exploration Systems Architecture Study (ESAS) Final Report.* NASA-TM-2005-214062. NASA, Washington, D.C.
43. Drake, B. 2009. *Human Exploration of Mars Design Reference Architecture 5.0 Addendum.* NASA/SP-2009-566-ADD. July. NASA, Washington, D.C.
44. Borowski, S., and Schnitzler, B. NTR development strategy and key activities supporting NASA human Mars missions in the early-2030 timeframe. AIAA Paper 2010-6818. American Institute of Aeronautics and Astronautics, Reston, Va.
45. Review of U.S. Human Spaceflight Plans Committee. 2009. *Seeking a Human Spaceflight Program Worthy of a Great Nation.* Office of Science and Technology Policy, Washington, D.C. October, p. 102.
46. Humble, R.W., Henry, G.N., and Larson, W.J. 1995. *Space Propulsion, Analysis, and Design.* McGraw-Hill Companies, Inc., New York, p. 457.
47. Drake, B. 2009. *Human Exploration of Mars Design Reference Architecture 5.0 Addendum.* NASA/SP-2009-566-ADD. July 2009. NASA, Washington, D.C.
48. National Research Council. 2000. *Microgravity Research in Support of Technologies for Human Exploration and Development of Space and Planetary Bodies.* National Academy Press, Washington, D.C.
49. National Research Council. 2000. *Microgravity Research in Support of Technologies for Human Exploration and Development of Space and Planetary Bodies.* National Academy Press, Washington, D.C.
50. Mauk, B.H., Bythrow, P.F., and Gatsonis, N.A. 1993. Science plan for the Nuclear Electric Propulsion Space Test Program (NEPSTP). Presented at the 29th Joint Propulsion Conference, Monterey, CA, June 28-30. AIAA-93-1895. American Institute of Aeronautics and Astronautics, Reston, Va.
51. National Research Council. 2000. *Microgravity Research in Support of Technologies for Human Exploration and Development of Space and Planetary Bodies.* National Academy Press, Washington, D.C.
52. Brandhorst, H.W., Jr., Rodiek, J.A., O'Neill, M.J., and Eskenazi, M.I. 2006. Ultralight, compact, deployable, high-performance solar concentrator array for lunar surface power. Presented at the 4th International Energy Conversion Engineering Conference and Exhibit, San Diego, Calif., June 26-29. AIAA 2006-4104. American Institute of Aeronautics and Astronautics, Reston, Va.
53. Carpenter, C.B. 2006. Electrically propelled cargo spacecraft for sustained lunar supply operations. Presented at the 42nd AIAA/ASME/SAE/ASEE Joint Propulsion Conference and Exhibit, Sacramento, Calif., July 9-12. AIAA 2006-4435. American Institute of Aeronautics and Astronautics, Reston, Va.
54. National Research Council. 2008. *A Constrained Space Exploration Technology Program: A Review of NASA's Exploration Technology Program.* The National Academies Press, Washington, D.C.
55. Woodcock, G.R. 2004. Controllability of large SEP for Earth orbit raising. Presented at the 40th AIAA/ASME/SAE/ASEE Joint Propulsion Conference and Exhibit, Fort Lauderdale, Fla., July 11-14. AIAA 2004-3643. American Institute of Aeronautics and Astronautics, Reston, Va.
56. Chang Diaz, F.R. Squire, J.P., Bering III, E.A., George, J.A., Ilin, A.V., Petro, A.J., and Casady, L. 2001. The VASIMR engine approach to solar system exploration. Presented at the 39th AIAA Aerospace Sciences Meeting and Exhibit, January 8-11. AIAA 2001-0960. American Institute of Aeronautics and Astronautics, Reston, Va.
57. Kodys, A., and Choueiri, E. 2005. A review of the state-of-the-art in the performance of applied-field magnetoplasmadynamic thrusters. Presented at the 41st Joint Propulsion Conference, Tucson, Ariz., July 10-13. AIAA-2005-4247. American Institute of Aeronautics and Astronautics, Reston, Va.
58. Litchford, R.J., Cole, J.W., Lineberry, J.T., Chapman, J.N., Schmidt, H.J., and Lineberry, C.W. 2002. Magnetohydrodynamic augmented propulsion experiment: I. Performance analysis and design. Presented at the 33rd AIAA Plasmadynamics and Lasers Conference/14th International Conference on MHD Power Generation and High Temperature Technologies, Maui, Hawii, May 20-23. AIAA 2002-2184. American Institute of Aeronautics and Astronautics, Reston, Va.
59. Schulz, R.J., Chapman, J.N., and Rhodes, R.P. 1992. MHD augmented chemical rocket propulsion for space applications. Presented at the AIAA 23nd Plasmadynamics and Lasers Conference, Nashville, Tenn., July 6-8. AIAA Paper 92-3001. American Institute of Aeronautics and Astronautics, Reston, Va.
60. Litchford, R.J., Bitteker, L.J., and Jones, J.E. 2001. *Prospects for Nuclear Electric Propulsion Using Closed-Cycle Magnetohydrodynamic Energy Conversion.* NASA/TP-2001-211274. NASA Marshall Space Flight Center, Huntsville, Ala.
61. Palaszewski, B., and Powell, R. 1994. Launch vehicle propulsion using metallized propellants. *Journal of Propulsion and Power* 10(6):828-833.
62. Harris, G.L. 2001. *The Origins and Technology of the Advanced Extravehicular Space Suit.* AAS History Series, Volume 24. American Astronautical Society, San Diego, Calif.

63. Young, A. 2009. *Spacesuits: The Smithsonian National Air and Space Museum Collection*. PowerHouse Books, New York, N.Y.
64. Clark, P. 1988. *The Soviet Manned Space Program: An Illustrated History of the Men, the Missions, and the Spacecraft.* Orion Books, New York, N.Y.
65. Harris, G.L. 2001. *The Origins and Technology of the Advanced Extravehicular Space Suit.* AAS History Series, Volume 24. American Astronautical Society, San Diego, Calif.
66. Balinskas, R., and Tepper, E. 1994. *Extravehicular Mobility Unit Requirements Evolution.* Contract No. NAS 9-17873. Hamilton Sundstrand, Houston, Tex.
67. Hamilton Sundstrand. 2003. *NASA Extravehicular Mobility Unit LSS/SSA Data Book.* Revision J, February. Hamilton Sundstrand, Houston, Tex.
68. Wang, J., Yuan, W., and Yuan, X. 2009. Research progress of portable life support system for extravehicular activity space suit. *Space Medicine and Medical Engineering (Beijing)* 22(1):67-71.
69. Ross, A., Aitchison, L., and Daniel, B. 2008. Constellation space suit system development status. *Scientific and Technical Aerospace Report* 45(26): Abstract.
70. Chase, T.D., Splawn, K., and Christiansen, E.L. 2007. Extravehicular mobility unit penetration probability from micrometeoroids and orbital debris: Revised analytical model and potential space suit improvements. *Scientific and Technical Aerospace Report* 45(20): Abstract.
71. Jordan, N.C., Saleh, J.H., and Newman, D.J. 2006. The extravehicular mobility unit: A review of environment, requirements, and design changes in the US spacesuit. *Acta Astronautica* 59(12):1135-1145.
72. Jones, R.J., Graziosi, D., Ferl, J., Splawn, W.K., Cadogan, D., and Zetune, D. 2006. Micrometeoroid and orbital debris enhancements of shuttle extravehicular mobility unit thermal micrometeoroid garment. Presented at the International Conference on Environmental Systems, July 2006, Norfolk, Va. SAE Technical Paper 2006-01-2285. SAE International, Warrendale, Pa.
73. Bell, E.R., Jr., and Oswald, D.C. 2005. Past and present extravehicular mobility unit (EMU) operational requirements comparison for future space exploration. Presented at Space 2005, Long Beach, Calif. August 30-September 1, 2005. AIAA Paper 2005-6723. American Institute of Aeronautics and Astronautics, Reston, Va.
74. Akin, D.L. 2004. Robosuit: Robotic augmentations for future space suits. Presented at the International Conference on Environmental Systems, July 2004, Colorado Springs, Colo. SAE Technical Paper 2004-01-2292. SAE International, Warrendale, Pa.
75. Shavers, M.R., Saganti, P.B., Miller, J., and Cucinotta, F.A. 2003. *Radiation Protection Studies of International Space Station Extravehicular Activity Space Suits.* NASA/TP-2003-212051. NASA Johnson Space Center, Houston, Tex., Chapter 1, pp. 1-18.
76. Wilde, R.C., Baker, G.S., McBarron II, J.W., Graziosi, D., Persans, A., and Stein, J. 2000. Evolving EMU for space station: Status of current changes. IAA Paper 00-10103. International Academy of Astronautics, Paris, France.
77. Newman, D.J., Schmidt, P.B., and Rahn, D.B. 2000. Modeling the extravehicular mobility unit (EMU) space suit—Physiological implications for extravehicular activity (EVA). SAE Technical Paper 2000-01-2257. SAE International, Warrendale, Pa.
78. Graziosi, D., Stein, J., and Kearney, L. 2000. Space shuttle small EMU development. SAE Technical Paper 2000-01-2256. SAE International, Warrendale, Pa.
79. United Space Alliance. 2005. Crew Escape Systems 21002. Available at http://www.nasa.gov/centers/johnson/pdf/383443 main_crew_escape_workbook.pdf.
80. Jordan, N.C., Saleh, J.H., Newman, D.J. 2006. The extravehicular mobility unit: A review of environment, requirements, and design changes in the US spacesuit. *Acta Astronautica* 59(12):1135-1145.
81. National Research Council. 1997. *Advanced Technology for Human Support in Space.* National Academy Press, Washington, D.C.
82. National Research Council. 2000. *Microgravity Research in Support of Technologies for Human Exploration and Development of Space and Planetary Bodies.* National Academy Press, Washington, D.C.
83. National Research Council. 2003. *Assessment of Directions in Microgravity and Physical Sciences Research at NASA.* The National Academies Press, Washington, D.C.
84. National Research Council. 2008. *A Constrained Space Exploration Technology Program: A Review of NASA's Exploration Technology Program.* The National Academies Press, Washington, D.C.
85. Blanco, R., NASA. 2009. "Current and Future EVA Capabilities," presentation via teleconference call to Panel on the Translation to Space Exploration Systems of the Decadal Survey on Biological and Physical Sciences in Space, December 3. National Research Council, Washington, D.C.

86. Sim, Z., Bethke, K., Jordan, N., Dube, C., Hoffman, J., Brensinger, C., Trotti, G., and Newman, D. 2005. Implementation and testing of a mechanical counterpressure bio-suit system. Presented at the AIAA and SAE International Conference on Environmental Systems (ICES 2005), Rome, Italy, July. SAE Paper 2005-01-2968. SAE International, Warrendale, Pa.
87. Newman, D., Canina, M., and Trotti, G.L. 2007. Revolutionary design for astronaut exploration—Beyond the Bio-Suit. Presented at the Space Technology and Applications International Forum, STAIF-2007, Albuquerque, N.M., February 11-15. AIP Conference Proceedings 880. American Institute of Physics, College Park, Md.
88. Kosmo, J., Bassick, J., and Porter, K. 1988. Development of higher operating pressure extravehicular space-suit glove assemblies. Presented at the 18th SAE Intersociety Conference on Environmental Systems, San Francisco, Calif. July 11-13. SAE Technical Paper 881102. SAE International, Warrendale, Pa.
89. Wright, H.C. 1985. Enhancement of space suit glove performance. Presented at the 15th AIAA, SAE, ASME, AIChE, and ASMA Intersociety Conference on Environmental Systems, San Francisco, Calif., July 15-17. SAE Technical Paper 851335. SAE International, Warrendale, Pa.
90. Tanaka, K., Danaher, P., Webb, P., and Hargens, A.R. 2009. Mobility of the elastic counterpressure space suit glove. *Aviation, Space, and Environmental Medicine* 80(10):890-893.
91. Danaher, P., Tanaka, K., and Hargens, A.R. 2005. Mechanical counter-pressure vs. gas-pressurized spacesuit gloves: Grip and sensitivity. *Aviation* 76(4):381-384.
92. Blanco, R., NASA. 2009. "Current and Future EVA Capabilities," presentation via teleconference call to Panel on the Translation to Space Exploration Systems of the Decadal Survey on Biological and Physical Sciences in Space, December 3. National Research Council, Washington, D.C.
93. National Research Council. 2008. *A Constrained Space Exploration Technology Program: A Review of NASA's Exploration Technology Program.* The National Academies Press, Washington, D.C., p. 40.
94. National Research Council. 2008. *A Constrained Space Exploration Technology Program: A Review of NASA's Exploration Technology Program.* The National Academies Press, Washington, D.C., p. 40.
95. Taylor, L.A., Schmitt, H., Carrier, W., and Nakagawa, M. 2005. The lunar dust problem: From liability to asset. Presented at the AIAA 1st Space Exploration Conference, Orlando, Fla., January 30-February 1. AIAA Paper 2005-2510. American Institute of Aeronautics and Astronautics, Reston, Va.
96. Gaier, J.R. 2005. *The Effects of Lunar Dust on EVA Systems during the Apollo Missions.* NASA/TM-2005-213610. March. NASA Glenn Research Center, Cleveland, Ohio.
97. Gaier, J.R. 2005. *The Effects of Lunar Dust on EVA Systems during the Apollo Missions.* NASA/TM-2005-213610. March. NASA Glenn Research Center, Cleveland, Ohio.
98. Gaier, J.R. 2005. *The Effects of Lunar Dust on EVA Systems during the Apollo Missions.* NASA/TM-2005-213610. March. NASA Glenn Research Center, Cleveland, Ohio.
99. Gaier, J.R. 2005. *The Effects of Lunar Dust on EVA Systems during the Apollo Missions.* NASA/TM-2005-213610. March. NASA Glenn Research Center, Cleveland, Ohio.
100. Forget, F. 2004. Alien weather at the poles of Mars. *Science* 306:1298-1299.
101. Koontz, S., Valentine, M., Keeping, T., Edeen, M., Spetch, W., and Dalton, P. 2003. Assessment and Control of Spacecraft Charging Risks on the International Space Station, *Proceedings of the 8th Spacecraft Charging Technology Conference*, October 20-24. Available at http://dev.spis.org/projects/spine/home/tools/sctc/VIIIth.
102. National Research Council. 1997. *Advanced Technology for Human Support in Space.* National Academy Press, Washington, D.C.
103. National Research Council. 2000. *Microgravity Research in Support of Technologies for Human Exploration and Development of Space and Planetary Bodies.* National Academy Press, Washington, D.C.
104. National Research Council. 2008. *A Constrained Space Exploration Technology Program: A Review of NASA's Exploration Technology Program.* The National Academies Press, Washington, D.C.
105. Campbell, P.D. 2006. *Recommendations for Exploration Spacecraft Internal Atmospheres: The Final Report of the NASA Exploration Atmospheres Working Group.* NASA JSC-63309. January. NASA Johnson Space Center, Houston, Tex.
106. Science@NASA Headline News. 2001. Staying Cool on the ISS. March 21. Available at http://science.nasa.gov/headlines/y2001/ast21mar_1.htm.
107. Bagdigian, R., NASA Marshall Space Flight Center. 2009. "Environmental Control and Life Support in the Constellation Program," presentation to Panel on the Plant and Microbial Biology of the Decadal Survey on Biological and Physical Sciences in Space, October 8. National Research Council, Washington, D.C.

108. Balasubramaniam, E.R., Kizito, J., and Kassemi, M. 2006. *Two Phase Flow Modeling: Summary of Flow Regimes and Pressure Drop Correlations in Reduced and Partial Gravity.* NASA/CR-2006-214085. National Center for Space Exploration Research, Cleveland, Ohio.
109. Bagdigian, R., NASA Marshall Space Flight Center. 2009. "Environmental Control and Life Support in the Constellation Program," presentation to Panel on the Plant and Microbial Biology of the Decadal Survey on Biological and Physical Sciences in Space, October 8. National Research Council, Washington, D.C.
110. Thornton, J., Whittaker, W., Jones, H., Mackin, M., Barsa, R., and Gump, D. 2010. Thermal strategies for long duration mobile lunar surface missions. Presented at the 48th AIAA Aerospace Sciences Meeting Including the New Horizons Forum and Aerospace Exposition, Orlando, Fla., January 4-7. AIAA-2010-0798. American Institute of Aeronautics and Astronautics, Reston, Va.
111. Baiden, G., Grenier, L., and Blair, B. 2010. Lunar underground mining and construction: A terrestrial vision enabling space exploration and commerce. Presented at the 48th AIAA Aerospace Sciences Meeting Including the New Horizons Forum and Aerospace Exposition, Orlando, Fla., January 4-7. AIAA 2010-1548. American Institute of Aeronautics and Astronautics, Reston, Va.
112. Balasubramanian, R., Gokoglu, S., Sacksteder, K., Wegeng, R., and Suzuki, N. 2010. An extension of analysis of solar-heated thermal wadis to support extended-duration lunar exploration. Presented at the 48th AIAA Aerospace Sciences Meeting Including the New Horizons Forum and Aerospace Exposition, Orlando, Fla., January 4-7. AIAA 2010-0797. American Institute of Aeronautics and Astronautics, Reston, Va.
113. Bockstahler, K., Funke, H., Lucas. J., Witt, J., and Hovland, S. 2009. Design status of the closed-loop air revitalization system ARES for accommodation on the ISS. *SAE International Journal of Aerospace* 1(1):543-555.
114. Tomes, K., Long, D., Carter, L., and Flynn, M. 2007. Assessment of the Vapor Phase Catalytic Ammonia Removal (VPCAR) technology at the MSFC ECLS Test Facility. Presented at the SAE International Conference on Environmental Systems, July 9-12. SAE Technical Paper 2007-01-3036. SAE International, Warrendale, Pa.
115. NASA Ames Research Center. Ames Technology Capabilities and Facilities. Advanced Life Support. Available at http://www.nasa.gov/centers/ames/research/technology-onepagers/advanced-life-support_prt.htm.
116. Ruff, G.A., Urban, D.L., Pedley, M.D., and Johnson, P.T. 2009. Fire safety. Chapter 27 in *Safety Design for Space Systems* (G.E. Musgrave, A. Larsen, and T. Sgobba, eds.) Elsevier Press, Oxford, U.K.
117. Ferkul, P., and Olson, S. 2010. Zero-gravity centrifuge used for the evaluation of material flammability in lunar-gravity. Presented at the AIAA 40th International Conference on Environmental Systems. July 11-15. AIAA 2010-6260. American Institute of Aeronautics and Astronautics, Reston, Va.
118. Sacksteder, K., Ferkul, P.V., Feier, I.I., Kumar, A., and T'ien, J.S. 2003. *Upward and Downward Flame Spreading and Extinction in Partial Gravity Environments.* NASA/CP-2003-212376/REV1. Available at http://ntrs.nasa.gov/archive/nasa/casi.ntrs.nasa.gov/20040053567_2004055227.pdf.
119. Sacksteder, K., and T'ien, J. 1994. Buoyant downward diffusion flame spread and extinction in partial-gravity accelerations. *Symposium (International) on Combustion* 25:1685-1692.
120. Braun, E., Levin, B.C., Paabo, M., Gurman, J., Holt, T.H., and Steel, J.S. 1987. *Fire Toxicity Scaling.* NBSIR 87-3510. U.S. Dept. of Commerce, National Bureau of Standards, National Technical Information, Gaithersburg, Md.
121. Hshieh, F.-Y., and Beeson, H.D. 1995. Flammability testing of pure and flame retardant-treated cotton fabrics. *Fire and Materials* 19(5):233-239.
122. Levin, B.C., Paabo, M., Fultz, M.L., and Bailey, C.S. 1985. Generation of hydrogen cyanide from flexible polyurethane foam decomposed under different combustion conditions. *Fire and Materials* 9(3):125-134.
123. Lange, K.E., Perka, A.T., Duffield, B.E., and Jeng F.F. 2005. *Bounding the Spacecraft Atmosphere Design Space for Future Exploration Missions.* NASA/CR-2005-213689. NASA Johnson Space Center, Houston, Tex.
124. Urban, D.L., Ruff, G., Sheredy. W., Cleary, T., Yang, J., Mulholland, G., and Yuan, Z. 2009. Properties of smoke from overheated materials in low gravity. Presented at the 47th AIAA Aerospace Sciences Meeting, Orlando, Fla., January 5-8. AIAA-2009-0956. American Institute of Aeronautics and Astronautics, Reston, Va.
125. Cowlard, A., Jahn, W., Abecassis-Empis, C., Rein, G., and Torero, J.L. 2008. Sensor assisted fire fighting. *Fire Technology* 46:719-741.
126. Ruff, G.A., Urban, D.L., Pedley, M.D., and Johnson, P.T. 2009. Fire safety. Chapter 27 in *Safety Design for Space Systems* (G.E. Musgrave, A. Larsen, and T. Sgobba, eds.). Elsevier Press, Oxford, U.K.
127. Takahashi, F., Linteris, G., and Katta, V.R. 2008. Extinguishment of methane diffusion flames by carbon dioxide in coflow air and oxygen-enriched microgravity environments. *Combustion and Flame* 155:37-53.
128. Sutula, J.A. 2008. *Towards a Methodology for the Prediction of Flame Extinction and Suppression in Three-Dimensional Normal and Microgravity Environments*, Ph.D. Dissertation, University of Edinburgh, Scotland, U.K.

129. Butz, J.R., and Abbud-Madrid, A. 2009. Advances in development of fine water mist portable extinguisher. Presented at the 39th International Conference on Environmental Systems, Savannah, Ga., July 12-16. SAE Technical Paper 2009-01-2510. SAE International, Warrendale, Pa.
130. Takahashi, F., and Katta, V.R. 2009. Extinguishment of diffusion flames around a cylinder in a coaxial air stream with dilution or water mist. *Proceedings of the Combustion Institute* 32:2615-2623.
131. Urban, D.L., Ruff, G., Brooker, J., Cleary, T., Yang, J., Mulholland, G., and Yuan, Z. 2008. Spacecraft fire detection: Smoke properties and transport in low-gravity. Presented at the 46th AIAA Aerospace Sciences Meeting, Reno, Nev., January 7-10. AIAA-2008-0806. American Institute of Aeronautics and Astronautics, Reston, Va.
132. NASA. 1997. Small Fire Extinguished on Mir. News Release. Available at http://spaceflight.nasa.gov/history/shuttle-mir/history/h-f-linenger-fire.htm.
133. Friedman, R. 1996. *Risk and Issues in Fire Safety on the Space Station.* NASA/TM 106430. NASA Glenn Research Center, Cleveland, Ohio.
134. Urban, D.L., Ruff, G., Sheredy. W., Cleary, T., Yang, J., Mulholland, G., and Yuan, Z. 2009. Properties of smoke from overheated materials in low gravity. Presented at the 47th AIAA Aerospace Sciences Meeting, Orlando, Fla., January 5-8. AIAA-2009-0956. American Institute of Aeronautics and Astronautics, Reston, Va.
135. McKay, M.F., McKay, D.S. and Duke, M.B. 1992. *Space Resources.* NASA SP-509. NASA, Washington, D.C.
136. Duke, M.B., Gaddis, L.R., Taylor, G.J., and Schmitt, H.H. 2006. Developing the Moon. Pp. 597-655 in *New Views of the Moon* (B.L. Jolliff, M.A. Wieczorek, C.K. Shearer, and C.R. Neal, eds). Reviews of Mineralogy and Geochemistry, Volume 60. Mineralogical Society of America, Chantilly, Va.
137. Eagle Engineering. 1988. *Conceptual Design of a Lunar Oxygen Pilot Plant.* NASA/CR EEI 88-182. Contract NAS9-17878. NASA Johnson Space Center. Available at http://www.isruinfo.com//docs/LDEM_Draft4-updated.pdf.
138. Ignatiev, A., Kubricht, T., and Freundlich, A. 1998. Solar cell development on the surface of the Moon. IAA-98-IAA.13.2.03. International Astronautical Federation.
139. PBS News Hour. Mars Exploration Rovers. Available at http://www.pbs.org/newshour/indepth_coverage/science/mars rover/archive.html.
140. Pieters, C.M., Goswami, J.N., Clark, R.N., Annadurai, M., Boardman, J., Buratti, B., Combe, J.-P., Dyar, M.D., Green, R., Head, J.W., Hibbitts, C., et al. 2009. Character and spatial distribution of OH/H2O on the surface of the Moon seen by M3 on Chandrayaan-1. *Science* 326(5952):568-572.
141. Clark, R.N. 2009. Detection of adsorbed water and hydroxyl on the Moon. *Science* 326(5952):562-564.
142. Sunshine, J.M., Farnham, T.L., Feaga, L.M., Groussin, O., Merlin, F., Milliken, R.E., and A'Hearn, M.F. 2009. Temporal and spatial variability of lunar hydration as observed by the Deep Impact spacecraft. *Science* 326(5952):565-568.
143. Colaprete, K., Ennico, K., Wooden, D., Shirley, M., Heldmann, J., Marshall, W., Sollitt, L., Asphaug, E., Korycansky, D., Schultz, P., Hermalyn, B., et al. 2010. Water and more: An overview of LCROSS impact results. Presented at the 41st Lunar and Planetary Science Conference, The Woodlands, Tex., March 1-5. Abstract #2335. Lunar and Planetary Institute, Houston, Tex.
144. Paige, D.A., Greenhagen, B.T., Vasavada, A.R., Allen, C., Bandfield, J.L., Bowles, N.E., Calcutt, S.B., DeJong, E.M., Elphic, R.C., Foote, E.J., Foote, M.C., et al. 2010. DIVINER Lunar Radiometer Experiment: Early mapping mission results. Presented at the 41st Lunar and Planetary Science Conference, The Woodlands, Tex., March 1-5. Abstract #2267. Lunar and Planetary Institute, Houston, Tex.
145. Blair, B.R., Diaz, J., Duke, M.B., Lamassoure, E., Easter, R., Oderman, M., and Vaucher, M. 2002. *Space Resource Economic Tool Kit: The Case for Commercial Lunar Ice Mining.* Final Report to the NASA Exploration Team. Jet Propulsion Laboratory, Pasadena, Calif.
146. Beaty, D., Buxbaum, K., Meyer, M., Barlow, N., Boynton, W., Clark, B., Deming, J., Doran, P.T., Edgett, K., Hancock, S., Head, J., et al. 2006. Findings of the Mars Special Regions Science Analysis Group. *Astrobiology* 6(4):677-732.
147. Plaut, J.J., Holt, J.W., Head III, J.W., Gim, Y., Choudhary, P., Baker, D.M., Kress, A., and the SHARAD Team. 2010. Thick ice deposits in Deuteronilus Mensae, Mars: Regional Distribution from radar sounding. Presented at the 41st Lunar and Planetary Science Conference, March 1-5. Abstract #2454. Lunar and Planetary Institute, Houston, Tex.
148. Lewis, J.S. 1996. *Mining the Sky: Untold Riches from the Asteroids, Comets, and Planets.* Helix Books, Reading, Mass.
149. Ignatiev, A., Kubricht, T., and Freundlich, A. 1998. Solar cell development on the surface of the Moon. Presented at the 49th International Astronautical Congress. IAA-98-IAA.13.2.03. International Astronautical Federation, Paris, France.
150. Wegeng, R.S., Humble, P.H., Saunders, J.H., Feier, I.I., and Pestak, C.J. 2009. Thermal energy storage and power generation for the manned outpost using processed lunar regolith as thermal mass materials. Presented at the AIAA Space 2009 Conference, Pasadena, Calif., September 14-17. AIAA-2009-6534. American Institute of Aeronautics and Astronautics, Reston, Va.

151. Wegeng, R.S., Mankins, J., Taylor, L., and Sanders, G. 2007. Thermal energy reservoirs from processed lunar regolith. Presented at the AIAA 5th International Energy Conversion Engineering Conference, St. Louis, Mo., June 25-27. AIAA-2007-4821. American Institute of Aeronautics and Astronautics, Reston, Va.
152. Wegeng, R.S., Mankins, J., Balasubramaniam, R., Sacksteder, K., Gokoglu, S., Sanders, G., and Taylor, L. 2008. Thermal wadis in support of lunar science and exploration. Presented at the AIAA 6th International Energy Conversion Engineering Conference, Cleveland, Ohio, July 28-30. AIAA-2008-5632. American Institute of Aeronautics and Astronautics, Reston, Va.
153. Balasubramaniam, R., Gokoglu, S., Sacksteder, K., Wegeng, R., and Suzuki, N. 2009. Analysis of solar-heated thermal wadi to support extended-duration lunar exploration. Presented at the AIAA Aerospace Sciences Meeting, Orlando, Fla., January 5-8. AIAA-2009-1339. American Institute of Aeronautics and Astronautics, Reston, Va.
154. Larson, W.E., Sanders, G.B., Sacksteder, K.R., Simon, T.M., and Linne, D.L. 2008. NASA's in-situ resource utilization project: A path to sustainable exploration. Presented at the 59th International Astronautical Congress. IAC-08-A3.2.B13. International Astronautical Federation, Paris, France.
155. Drake, B. 2009. *Human Exploration of Mars Design Reference Architecture 5.0 Addendum.* NASA/SP-2009-566-ADD. July. NASA, Washington, D.C.
156. Colaprete, K., Ennico, K., Wooden, D., Shirley, M., Heldmann, J., Marshall, W., Sollitt, L., Asphaug, E., Korycansky, D., Schultz, P., Hermalyn, B., et al. 2010. Water and more: An overview of LCROSS impact results. Presented at the 41st Lunar and Planetary Science Conference, The Woodlands, Tex., March 1-5. Abstract #2335. Lunar and Planetary Institute, Houston, Tex.
157. Boynton, W.V., Feldman, W.C., Squyres, S.W., Prettyman, T.H., Brückner, J., Evans, L.G., Reedy, R.C., Starr, R., Arnold, J.R., Drake, D.M., Englert, P.A.J., et al. 2002. Distribution of hydrogen in the near-surface of Mars: Evidence for subsurface ice deposits. *Science* 297:81.
158. Taylor, L.A., and Carrier III, W.D. 1994. Oxygen production on the Moon: An overview and evaluation. Pp. 69-108 in *Resources of Near Earth Space* (J.S. Lewis, M.S. Matthews, M.L. Guerrieri, eds.). University of Arizona Press, Tucson, Ariz.
159. Larson, W.E., Sanders, G.B., Sacksteder, K.R., Simon, T.M., and Linne, D.L. 2008. NASA's in-situ resource utilization project: A path to sustainable exploration. Presented at the 59th International Astronautical Congress. IAC-08-A3.2.B13, International Astronautical Federation, Paris, France.
160. Larson, W.E., Sanders, G.B., Sacksteder, K.R., Simon, T.M., and Linne, D.L. 2008. NASA's in-situ resource utilization project: A path to sustainable exploration. Presented at the 59th International Astronautical Congress. IAC-08-A3.2.B13, International Astronautical Federation, Paris, France.
161. Carlson, R.R., Bland, D., Fox, R., Hamilton, J., and Schowengerdt, F. 2009. Lunar surface equipment testing and demonstrations at the PISCES Lunar Analog Facilities. Presented at the 27th International Symposium on Space Technology and Science, Tsukuba City, Ibaraki Prefecture, Japan. ISTS 2009-f-04. International Symposium on Space Technology and Science, Tokyo, Japan.
162. Carlson, R.R., Bland, D., Fox, R., Hamilton, J., and Schowengerdt, F. 2009. Lunar surface equipment testing and demonstrations at the PISCES Lunar Analog Facilities. Presented at the 27th International Symposium on Space Technology and Science, Tsukuba City, Ibaraki Prefecture, Japan. ISTS 2009-f-04. International Symposium on Space Technology and Science, Tokyo, Japan.
163. Carlson, R.R., Bland, D., Fox, R., Hamilton, J., and Schowengerdt, F. 2009. Lunar surface equipment testing and demonstrations at the PISCES Lunar Analog Facilities. Presented at the 27th International Symposium on Space Technology and Science, Tsukuba City, Ibaraki Prefecture, Japan. ISTS 2009-f-04. International Symposium on Space Technology and Science, Tokyo, Japan.
164. Eagle Engineering.1988. *Conceptual Design of a Lunar Oxygen Pilot Plant.* NASA/CR EEI 88-182. Contract NAS9-17878. NASA Johnson Space Center. Available at http://www.isruinfo.com//docs/LDEM_Draft4-updated.pdf.
165. National Research Council. 2000. *Microgravity Research in Support of Technologies for Human Exploration and Development of Space and Planetary Bodies.* National Academy Press, Washington, D.C.

11

The Role of the International Space Station

UNIQUE STATUS AND CAPABILITIES

Spanning a construction period of more than a decade and involving the coordinated efforts of many nations, the International Space Station (ISS) represents a stunning achievement of human engineering. After years of continual redesign, development, and assembly, the ISS is poised to begin fulfilling its intended role as a world-class scientific laboratory for studying biological and physical processes in the near absence of gravity. However, the ISS of today lacks a number of important research facilities, such as the 3-meter centrifuge, planned during earlier stages of its design.

The assembly of major U.S., European, and Japanese components of the ISS will be completed in 2011—13 years after the launch of the first ISS component, the Russian Zarya module in 1998. (The Russians may launch their own pressurized laboratory to the station in the 2012-2013 time frame, but those plans are not yet finalized.) The ISS reached its full crew complement of six in May 2009 and should continue to hold that many until at least 2020, and perhaps beyond. Although limited by crew and equipment availability, significant science research was conducted during the construction phase of the ISS, with small observational experiments carried out shortly after the initial launch and more meaningful work beginning after the arrival of the Expedition 1 crew in late 2000.

Flight research is generally part of a continuum of efforts that extend from laboratories and analog environments on the ground, through other low-gravity platforms as needed and available, and eventually into extended-duration flight. Like any process of scientific discovery this effort is iterative, and further cycles of integrated ground-based and flight research are likely to be warranted as understanding of the system under study evolves. Although research on the ISS is only one component of this endeavor, the capabilities provided by the ISS are vital to answering many of the most important research questions detailed in this report. The ISS provides a unique platform for research, and past studies of the National Research Council (NRC) have noted the critical importance of the ISS's capabilities to support the goal of long-term human exploration in space.* These capabilities include the ability to perform experiments of extended duration, the ability to continually revise experiment parameters on the basis of previous results, the flexibility in experimental design provided by human operators, and the availability of sophisticated experimental facilities with significant power and data resources. The ISS is the only

*See, for example, National Research Council, *Review of NASA Plans for the International Space Station*, The National Academies Press, Washington, D.C., 2006.

existing and available platform of its kind, and it is essential that its presence and dedication to research for the life and physical sciences be fully employed in the decade ahead.

Before the 2010 budget announcement, the research plan of the National Aeronautics and Space Administration (NASA) for ISS utilization was expected to focus on objectives required for lunar and Mars missions in support of Constellation program timelines. Participation by the United States in the ISS was expected to end in 2016. There is now a de-emphasis at NASA on lunar missions and an extension of the ISS mission to 2020. The change in focus strengthens the need for a permanent research laboratory in microgravity devoted to scientific research in space focused both on fundamental questions and on questions posed in response to the envisioned needs of future space missions.

AREAS OF RESEARCH ON THE INTERNATIONAL SPACE STATION

Each of the panel chapters (Chapters 4 through 10) in this report describes critical research questions, most of which will need to progress through the use of more than one research platform, including ground-based laboratories and facilities such as drop towers or parabolic flights, to use of the ISS. The platforms and facilities required for each research area are discussed in the individual chapters, but it can be noted that for the majority of investigations, the ISS will provide the most advantageous research platform once the investigations transition to flight. In many cases, the ISS will be the *only* platform capable of meeting the requirements of investigations, and the ISS is the only platform that can provide a very long duration microgravity environment. Summarized in the following sections are examples of areas of past and future life and physical sciences research benefiting from, or requiring, the capabilities of the ISS.

Life Sciences Research on the ISS

Although it is impossible to list all the various biological research projects that were conducted on the ISS prior to the current era, insights from a 2008 report from NASA indicate a spectrum that, for plants, ranges from investigating the influences of gravity on the molecular changes in *Arabidopsis thaliana* to studying the mechanisms of photosynthesis, phototropism, and gravity sensing.[1] Cellular biology studies included investigating gene expression changes in *Streptococcus pneumonia* and select microbes, exploring mechanisms of fungal pathogenesis and tumorgenesis, and observing changes in the responses of monocytes in cell culture, blood vessel development, and wound healing to the space environment. Also investigated were the chromosomal aberrations in the blood lymphocytes of astronauts and the effect of spaceflight on the reactivation of latent Epstein-Barr virus. Such analyses have revealed notable gaps in knowledge. For example, there has not been a comprehensive program dedicated to analyzing microbial populations and responses to spaceflight, yet microbes play significant roles in positive and negative aspects of human health and in the degradation of their environment through, for example, food spoilage and biofouling of equipment.

The final report of the Review of U.S. Human Spaceflight Plans Committee (also known as the Augustine Commission or Augustine Committee)[2] has emphasized that future astronauts will face three unique stressors: (1) prolonged exposure to solar and galactic radiation, (2) prolonged periods of exposure to microgravity, and (3) confinement in close, relatively austere quarters along with a small number of other crew members with whom the astronaut will have to live and work effectively for many months while having limited contact with family and friends. All of these stressors are present in the ISS environment. Accordingly, ISS research studies could profitably determine mission-specific effects of these and other relevant stressors, alone and in combination, on the general psychological and physical well-being of astronauts and on their ability to perform mission-related tasks. Aspects pertaining to crew member interactions and the behavioral aspects of isolation and confinement have been examined on the ISS,[3] but research with the full crew complement of six and prolonged mission durations is needed to address critical mission issues, such as the importance of sleep for astronaut performance and how best to maximize interpersonal behavior and maintain cognitive function so that the crew can function at its optimal level.

Experiments related to human physiology on the ISS have examined the effects of spaceflight on the central nervous system and spinal excitability, skeletal muscle, bone maintenance and loss, cardiovascular control, pulmonary function, locomotor dysfunction, ground reaction forces, and nutritional status. Investigations have included the effects of radiation, the influence of light on the sleep-awake cycle, the risk of renal stones, and the advantages of select pharmaceutical drugs.[4] This research has revealed important limitations on the current understanding of how humans react to the spaceflight environment and so to current approaches to maintaining astronaut health under these conditions. For example, recent findings for ISS astronauts who have been in space for 6 months or longer and who performed the recommended exercise regimens indicated that these countermeasures were unable to prevent loss of muscle mass or decrement in muscle performance.[5]

It is clear that further research is essential for attaining an understanding of how and why human physiology is altered in space and for the design of effective countermeasures that will help to maintain the necessary functional homeostasis when humans are faced with altered gravitation on future missions. The ISS provides a unique opportunity to carry out both fundamental and translational research on organ and systemic function in the absence of the gravity variable necessary to meet these goals. The presence of humans in the space laboratory for up to 6 months enables the development of the much-needed databases for the various physiological systems as well as a thorough evaluation of select countermeasures such as exercise and pharmacological agents. Insights can be gleaned, for example, concerning the effect of radiation on coronary heart disease and pharmacological interventions to reduce bone resorption within a given tour.

The prolonged access to space afforded by the ISS will also allow the probing of fundamental questions about animal biology not directly related to human health, such as the role of gravity in developmental biology, by examining how animals grow, develop, mature, and age. Notably absent from the 2008 report from NASA on ISS research accomplishments[6] and in subsequent reports were references to animal research being conducted in ISS modules, even though the capability exists. Because of budgetary constraints and policy decisions, this essential component of microgravity research has not been implemented. Also eliminated was a provision for a small-animal centrifuge that the Russians had previously demonstrated as an effective countermeasure to the effects of microgravity in Cosmos 936.[7] Thus, although there are some current facilities on the ISS to allow the inclusion of experiments on, for example, fruit flies or nematodes, the inclusion of an animal facility capable of housing rodents will be necessary. Without this capability, it will not be possible to conduct future experiments essential to advancing the basic understanding of animal physiology in space and to providing animal models for probing changes affecting the health of astronauts and for the development of suitable countermeasures. Thus, the availability of animals in the ISS National Laboratory would facilitate fundamental research on the effects of microgravity on inadequately studied systems, such as the immune, endocrine, reproductive, and nervous systems, while expanding knowledge of the mechanisms responsible for cellular and molecular changes in skeletal, muscular, and connective tissue systems. With the availability of "knock out" and "disease" animal models, new insights on how microgravity affects physiological mechanisms can be secured from space experiments. In addition, there is the potential to gain new data on tissue healing (especially fractures) and on the growth and development of animals over multiple generations.

One further element of research enabled by access to the ISS is its use as a test bed to facilitate studies on plant and microbial components of a bioregenerative life support system. Such research would allow exploration of the possibility of self-sufficiency for food production, water recycling, and regeneration of the craft's atmosphere for extended crewed missions, obviating the need for costly resupply. Establishing the robust elements of such a bioregenerative life support system, which will likely incorporate a combination of biological systems and physico-chemical technologies, requires extended research now that carefully integrates ground- and ISS-based work. Levels and quality of light, atmospheric composition, nutrient levels, and availability of water are all critical elements shaping plant growth in space; each of the elements needs to be optimized in a rigorously tested technology platform designed to maximize performance during spaceflight. Although such a research program will be enabled by access to the unique environment of the ISS, it is fundamentally aimed at enabling a long-term human presence in space. Developing a sustained research program combining the ground- and ISS-based design and validation of components will be critical to establishing the dynamic, integrated intramural and extramural research community necessary to support this area.

These examples highlight the ISS as an essential and integral component of any implementation of the life sciences research outlined in this decadal survey.

Physical Sciences Research on the ISS

The reduced-gravity platforms currently available to researchers in the physical sciences are aircraft, drop towers, sounding rockets (in Europe and Japan), the space shuttle, and the ISS. Aircraft provide partial gravity for 20 to 25 s, with a *g*-jitter of 10^{-2} *g*. Drop towers allow microgravity for a few seconds, and sounding rockets for a few minutes. In the space shuttle, gravity levels on the order of 10^{-4} *g* can be sustained for long periods of time. However, the ISS is a very long duration experiment platform providing acceleration levels on the order of 10^{-5} to 10^{-6} *g* under the right conditions.[8] The premise of the ISS has been that it will serve as a laboratory for research and for the development and testing of technologies that facilitate space exploration. It also provides a platform for basic and applied research in biological and physical sciences aimed at enhancing a fundamental understanding of phenomena and processes with eventual space and terrestrial applications.

The facilities available on the ISS for U.S. researchers in the physical sciences include the Microgravity Science Glovebox, the Combustion Integrated Rack, the Fluids Integrated Rack, the Materials Science Research Rack, the Space Dynamically Responding Ultrasound Matrix System (Space DRUMS), and several multiuser EXPRESS Racks.[9] In addition, the European Space Agency has the Fluid Science Laboratory, and the Japanese Aerospace Exploration Agency has the Ryutai and Kobairo Racks for fluid physics and materials science research. Through international collaboration, all of these facilities can be used to advance research in the physical sciences.

As of 2009, NASA had carried out 20 expeditions to the ISS. These expeditions have led to 52 experiments in the physical sciences, and 15 more were planned for expeditions 21 and 22. Some of the recent experiments that have been conducted in the basic fluid physics area include gelation and phase separation in colloidal suspensions, critical phenomena, crystallization of glasses, growth of dendritic crystals, properties of magneto-rheological fluids, properties of particle growth in liquid-metal mixtures, and stress/strain response in polymeric liquids under shearing. In the area of combustion and fire safety, investigations have included smoke and aerosol measurements and the study of soot emission from gas-jet flames.

The fundamental and applied microgravity research in the physical sciences for which the ISS can serve as a laboratory is described in detail elsewhere in this report (see Chapters 8, 9, and 10). Here only a brief summary is presented. In the fundamental physics area, the topics of interest are soft matter and complex fluids (materials with multiple levels of structure), including colloids, polymer and colloidal gels, foams, emulsions, liquid crystals, dusty plasmas, and granular materials. Because of the gradients that develop in their properties under gravity, the microgravity environment provides ideal conditions for understanding the dynamic behavior of such materials, allowing the testing of ideas about fundamental physical processes—varying from examination of the constitutive equations that describe the strain-rate relationships for granular materials through to analysis of crystal growth—without the confounding effects of convection-based imperfections in material deposition. Precision measurements of fundamental forces and symmetries are another area that can benefit greatly from the microgravity environment of the ISS. Some of the subtopics of interest are the study of the equivalence principle and theories behind the standard model and general relativity to ask whether different kinds of matter interact with gravity in the same way. The study of quantum gases can lead to a range of new technologies and understanding, from developing ultraprecise atomic clocks and quantum sensors to resolving the mechanism of superconductivity in high-temperature superconductors. Major advances in the understanding of phenomena near the critical point can be achieved through well-conceived experiments conducted on the ISS. The completion of the Low Temperature Microgravity Physics Facility would significantly enhance the capability of the ISS to support experiments in fundamental physics.

Applied physical sciences include fluid physics and heat transfer, combustion, and materials science. In the fluid physics area, multiphase flow phenomena and associated heat transfer have been identified as a critical area that would benefit greatly from experiments in the long-duration microgravity environment of the ISS. Experiments on pool boiling; forced-flow boiling, including phase separation and flow stability; closure relations for interfacial and wall heat, mass and momentum transfer; condensation; and capillary-driven flows would provide significant knowledge and a database with which computer models could be validated and systems could be designed. In addition, research aimed at increasing the efficiency and lifetime of power-generation and energy-storage systems would reduce costs by reducing mass and redundancy. All of these systems would benefit from research, prototyping, and testing on the ISS. For example, key advantages of spaceborne power systems based on the Rankine cycle are higher power and small component size, because two-phase-flow and heat-transfer coefficients are much larger than those for gas in non-two-phase systems. Unfortunately, little is known at present about the behavior of two-phase flows and associated heat transfer in a reduced-gravity environment. The possibility of developing a multipurpose, multiuser facility for multiphase flow research on the ISS should therefore be considered. Such a facility would also act as a catalyst for bringing together national and international researchers to address the problem in a cost-effective and comprehensive manner. Significant insights into the static and dynamic behavior of granular materials and dusts could be gained through experiments on the ISS. Such an understanding would be of value for applications on Earth and for human and robotic exploration of the Moon and Mars. (It is noteworthy that the Mars rover Spirit has been stuck in the martian soil, a granular material, since May 2009.)

Advances in propulsion performance (specific impulse, efficiency, thrust to weight, propellant bulk density), reliability, thermal management, power generation and handling, propellant storage and handling, and strategies for refueling on orbit are all key drivers for dramatically reducing mass, cost, and mission risk. The ISS provides unique opportunities for advances in a number of these areas through research on processes such as cryogenic two-phase fluid management, propellant transfer, engine starts, flame stability, active thermal control of injectors and combustors, and cryogenic fluid management.

In summary, the ISS platform is an essential and integral component of any implementation of the physical sciences research outlined in this decadal survey.

Utilizing the ISS for Research

The decadal survey committee strongly recommends that NASA intensify the utilization of the ISS as a world-class research laboratory engaged in both basic and applied research that enables space exploration and is enabled by the microgravity environment of the ISS. The goal should be to maximize the utilization of existing facilities and to engage world-class scientists and engineers to carry out research that leads to the development of space-related technologies. Ground-based experimental and theoretical work should form a significant component of the overall activity.

Cross-disciplinary research should be emphasized, and a research portfolio with prioritization should be developed and shared with the technical community.[10] To develop a vibrant research community that is committed to space-related research, NASA should have a firm plan for sustaining the research by providing adequate resources. Aside from benefiting directly from the research, NASA would be contributing to the creation of the workforce for the future. The process, from the acceptance of a proposal to preparation of the flight experiment to conduct of experiments on the ISS, should be streamlined, with a reduction in time from start to finish. This is essential to keep graduate students and other researchers engaged in the research activity. Some of the experimental rigs that have already been flown on the ISS can serve as facilities for future research investigations in the physical sciences. NASA should reconsider the placement of a centrifuge on the ISS so that long-duration partial-gravity experiments can be conducted. NASA should also strengthen and expand its collaborations with international partners. This would allow access to the facilities of the partner countries, avoid a duplication of research, and allow U.S. researchers to accomplish much more than they could otherwise.

Caveats

Discussed throughout this report are various topics within each of the areas described above. The committee reiterates, however, that although the ISS is a key component of the research infrastructure to be utilized by a biological and physical sciences research program, it is only one component of a healthy program. Other platforms will play an important role and, in particular, research on the ISS will have to be supported by other platforms, including a parallel ground-based program, to be scientifically credible.

REFERENCES

1. Evans, C.A., Robinson, J.A., Tate-Brown, J., Thumm, T., Crespo-Richey, J., Baumann, D., and Rhatigan, J. 2009. *International Space Station Science Research Accomplishments During the Assembly Years: An Analysis of Results from 2000-2008.* NASA/TP-2009-23146-Revison A. NASA Johnson Space Center, Houston, Tex., p. 1.
2. Review of U.S. Human Spaceflight Plans Committee. 2009. *Seeking a Human Spaceflight Program Worthy of a Great Nation.* Final Report. Available at http://www.nasa.gov/pdf/396093main_HSF_Cmte_FinalReport.pdf, p. 113.
3. Evans, C.A., Robinson, J.A., Tate-Brown, J., Thumm, T., Crespo-Richey, J., Baumann, D., and Rhatigan, J. 2009. *International Space Station Science Research Accomplishments During the Assembly Years: An Analysis of Results from 2000-2008.* NASA/TP-2009-23146-Revison A. NASA Johnson Space Center, Houston, Tex., p. 1.
4. Evans, C.A., Robinson, J.A., Tate-Brown, J., Thumm, T., Crespo-Richey, J., Baumann, D., and Rhatigan, J. 2009. *International Space Station Science Research Accomplishments During the Assembly Years: An Analysis of Results from 2000-2008.* NASA/TP-2009-23146-Revison A. NASA Johnson Space Center, Houston, Tex., p. 1.
5. Trappe, S., Costill, D., Gallagher, C., Creer, A., Peters, J.R., Evans, H., Riley, D.A., and Fitts, R.H. 2009. Exercise in space: Human skeletal muscle after 6 months aboard the International Space Station. *Journal of Applied Physiology* 106:1159-1168.
6. Evans, C.A., Robinson, J.A., Tate-Brown, J., Thumm, T., Crespo-Richey, J., Baumann, D., and Rhatigan, J. 2009. *International Space Station Science Research Accomplishments During the Assembly Years: An Analysis of Results from 2000-2008.* NASA/TP-2009-23146-Revison A. NASA Johnson Space Center, Houston, Tex., p. 1.
7. Kotovskaya, A.R., Ilyin, E.A., Korolkov, V.I., and Shipov, A.A. 1980. Artificial gravity in spaceflight. *Physiologist* 23(Suppl. 6):S27-S29.
8. DeLombard, R., Hrovat, K., Kelly, E., and McPherson, K. 2004. Microgravity environment on the International Space Station. AIAA Paper 2004-0125. NASA/TM-2004-213039. NASA Glenn Research Center, Cleveland, Ohio.
9. Robinson, J.A., NASA. 2009. "International Space Station," presentation to the Committee for the Decadal Survey on Biological and Physical Sciences in Space, August 19. National Research Council, Washington, D.C.
10. National Research Council. 2003. *Factors Affecting the Utilization of the International Space Station for Research in the Biological and Physical Sciences.* The National Academies Press, Washington, D.C.

12

Establishing a Life and Physical Sciences Research Program: Programmatic Issues

The National Aeronautics and Space Administration (NASA) faces numerous challenges in carrying out the aspirations of the United States to advance its space exploration mission. Over its 50-year history, the successful progress of NASA in space exploration has depended on the ability to address a wide range of biomedical, engineering, physical science, and related obstacles. The partnership of NASA with the research community reflects the original mandate from Congress in 1958 to promote science and technology, which requires an active and vibrant research program. In the past, NASA's commitments to life and physical sciences research activities for space exploration have been strong and productive. As highlighted in the 2003 National Research Council (NRC) report *Assessment of Directions in Microgravity and Physical Sciences Research at NASA*,[1] NASA's Physical Sciences Division had a strong outreach program in the early 1990s that engaged scientists in the microgravity disciplines, including combustion, fluid physics, fundamental physics, and materials science. Similarly, as emphasized in Chapters 4, 5, and 6 of the current report, there were great advances in scientific discovery in past decades as a result of NASA's strong commitment to life sciences research activities. The committee acknowledges the many achievements of NASA, a feat all the more remarkable given budgetary challenges and changes of overall emphasis within the agency. However, the committee is deeply concerned about the current state of NASA life and physical sciences research.

The scientific community engaged in space exploration research has dwindled in the past decade as a result of marked reductions in budget funding levels, from approximately $500 million shared equally between life and physical sciences in 2002 to the 2010 level of less than $200 million, with most of the latter going to the Human Research Program and only $47 million going to International Space Station (ISS) life and physical sciences research. There has been a corresponding reduction in the ISS research portfolio, from 966 investigations in 2002 to 285 in 2008.[2] NASA has acknowledged the decline in research activity but had also projected that the decline would be temporary. For example, in describing the run-up to full ISS utilization, the authors of a 2000 NASA report wrote:

> The high level of space life sciences research activity seen in the 1991-1995 period continued through 1996 and then began to taper off. This decline in the number of life sciences payloads is attributable to several factors: the close out of the Cosmos/Bion program in 1997, the end of the planned NASA/Mir collaboration in 1998, the retirement of Spacelab, and the requirement for Space Shuttle flights to conduct assembly of the International Space Station (ISS) beginning in late 1998. Flight experimentation should again pick up as ISS assembly reaches completion in the first few years of the twenty-first century.[3]

It is now time for NASA to return to a high level of programmatic vision and dedication to life and physical sciences research, to ensure that the considerable obstacles to long-duration human exploration missions in space can be resolved. As has always been the case, achievement of these goals will depend on a steady input of innovative high-quality research results. In turn, high-quality research will depend on NASA embracing life and physical sciences research as part of its core exploration mission and re-energizing a community of life and physical scientists and engineers focused on both exploration-enabling research and scientific discovery (i.e., fundamental research enabled by space exploration).

The committee concluded that considerable programmatic efforts will be required to overcome current obstacles, secure the trust of the scientific community, and restore the life and physical sciences research program to a committed, comprehensive, and highly visible organizational resource that effectively promotes research to meet the national space exploration agenda. Issues that are important in achieving these goals are discussed in the following sections.

PROGRAMMATIC ISSUES FOR STRENGTHENING THE RESEARCH ENTERPRISE

Elevating the Priority of Life and Physical Sciences Research in Space Exploration

As the nation and NASA prepare for the next decade of space exploration, numerous challenges must be met to ensure successful results. Among these challenges are the developments needed to reduce risks and costs, which will come from a deeper understanding of the performance of people, animals, microbes, plants, materials, and engineered systems in the environments of space. To meet these challenges, which span life and physical sciences, it is essential to develop a long-term strategic research plan, firmly anchored both within NASA and in a broad and diverse extramural research community. For such a plan to become a reality, research must be central to NASA's exploration mission and be supported throughout the agency as an essential means to achieve future space exploration goals. Feedback, associated with this decadal survey received from numerous interviews, town hall meetings, and white paper submissions, indicated that a large proportion of the research community does not see such an environment for life and physical sciences within the current exploration programs at NASA.

NASA has overcome a number of obstacles in fulfilling the original objectives identified by Congress. It has been a challenge from the outset to organize and manage the life and physical sciences research program within the overall NASA administrative infrastructure. Some of the organizational challenges included the ability to select and prioritize the most meritorious research projects, the provision of adequate and sustained support for such research projects, and the ability to attract a community of researchers with the necessary skills and experiences to conduct these studies and to create a new generation of scientists and engineers focused on research to answer questions relevant for space exploration missions. To continue to meet such challenges, it is of paramount importance that the life and physical sciences research portfolio supported by NASA, both extramurally and intramurally, receive appropriate attention and that its organizational structure be optimally designed to meet NASA's needs. *The utility of a coherent research plan that is appropriately resourced and consistently applied to enable exploration cannot be overemphasized.* This is especially noteworthy in light of the frequent and large postponements that NASA's exploration-related goals have experienced over the past several decades.

The NASA exploration research enterprise will be improved only if it is promoted and embraced horizontally and vertically throughout the organizational structure of NASA. Multiple factors have resulted in NASA life and physical sciences research being relegated to a very low priority, with many areas virtually eliminated. Since retirement of the Spacelab (in 1999) and the completion of the International Neurolab project (mission conducted in 1998), during which many sophisticated experiments took place in the context of dedicated research missions implemented by a highly trained and intellectually engaged crew, the priority for research has been reduced to levels that compromise the research endeavor and the success of future exploration missions. The perception that research is optional, rather than essential, is reflected in the attitudes of flight and ground personnel toward crew participation in research projects and appears to be driven by NASA's overall expectations and reward system for flight missions. Currently, astronauts can opt out of their participation in approved and manifested research projects, both in terms of serving as a subject and acting as a surrogate investigator for a research project. For example, in one ongoing extramural project, only 11 of 18 crew members agreed to participate in a project that involves noninvasive imaging (echo and magnetic resonance imaging) and ambulatory monitoring (Holter and blood pressure) of cardiovascular function; one participant dropped out 2 weeks before the flight. Similarly, in an ongoing intramural project that is evaluating a new evidence-based exercise prescription to minimize loss of muscle, bone, and cardiovascular function during ISS missions, only 3 of the first 6 crew members invited to participate were enrolled. Mission managers, who often have limited research background and are not incentivized to place a priority on research, control crew availability and make decisions concerning crew scheduling that can compromise research studies and outcomes, even when acceptable alternatives to these competing activities are available.

To address these systemic problems and restore the high priority of NASA's life and physical sciences research program over the next decade, the following steps are important.

- *Recognition of the need for a change of attitude and commitment to life and physical sciences research throughout the agency is essential.* To reflect a vision that life and physical sciences research is central to NASA's space exploration mission, research must be viewed as a priority. It is essential that every employee, from management through crew, subscribe to the view that a key objective of the organization is to support and conduct life and physical sciences research as an essential translational step in the execution of space exploration missions.
- *Acknowledging life and physical sciences research as an integral component of spaceflight operations.* For research to become a central component of exploration programs, it is necessary to develop a culture in which participation in research, both as a subject of investigation and as a surrogate investigator, is viewed as a fundamental element of the astronaut mission. Many crew members already display this attitude, and they frequently go to extraordinary lengths to participate in research studies in partnership with the extramural research community. However, the level of autonomy given to astronauts to choose whether or not to participate in research activities is surprising and inappropriate, given the scarce opportunities for human research in space. In addition, many types of experiments require specialized scientific or technical expertise to make knowledgeable observations, measurements, and judgments. It is important to optimize very limited opportunities to gain a better understanding of the effects of the space environment on human health, safety, and performance, and on systems that reduce risk and optimize performance, because such information will define the future limits of space exploration.

One possible solution is to include not only scientific and technical expertise but also willingness to participate in research that enables future space exploration as part of crew selection in the planning of specific mission assignments, and perhaps even as part of the astronaut selection process. Such an approach would define research participation as one of the responsibilities of an astronaut. The priority of research can also be reinforced during the training of ground support personnel (e.g., flight directors, mission controllers, training managers, and instructors).

Requiring astronauts to participate in research as part of a mission may generate questions about such issues as coercion and privacy rights. However, it is reasonable and ethical that, if research participation is defined as a component of being an astronaut, then the consequence of choosing not to participate would be understood to result in a different assignment. Astronauts would have to accept the risk of loss of confidentiality, with respect to data collected for research purposes, along with the many other risks associated with spaceflight. This approach would remain aligned with the Federal Policy for the Protection of Human Subjects.[4] Because the NASA exploration mission is of national importance, research opportunities to advance this agenda become part of strategic decisions. This philosophy is consistent with that embodied in the National Aeronautics and Space Act of 1958,[5] the Institute of Medicine (IOM) report *Safe Passage*,[6] and the NRC report *A Strategy for Research in Space Biology and Medicine in the New Century*.[7]

- *The collection and analysis of a broad array of physiological and psychological data from astronauts before, during, and after a mission is necessary to advance knowledge of the effects of the space environment on human health and to improve the safety of space exploration.* As discussed below (see "Improved Access to Samples and Data From Astronauts"), ensuring the health, safety, and performance of astronauts during future space exploration will depend on the existence of comprehensive databases on astronaut health and performance. In addition to data collected from distinct research projects, such databases would ideally be populated with operational data collected during missions and with medical data obtained before, during, and after exposure to the space environment; extensive follow-up data will be necessary to understand the potential long-term effects of spaceflight on health. The 2001 IOM report *Safe Passage*[8] addressed the issue of confidentiality of astronaut health data. That report indicated that medical information that is not part of a defined research protocol has been regarded by NASA as confidential, to be known only to the astronaut's flight surgeon and the astronaut. One finding of that report was as follows:

> Because of concerns about astronaut privacy, data and biological specimens that might ensure the health and safety of the astronaut corps for long-duration missions have not been analyzed. If these data and other data to be accumulated in the future are to be used to facilitate medical planning for the unique sets of pressures and extreme environments that astronauts will experience on long-duration space missions, the ethical concerns about astronaut privacy must be appropriately modified. (p. 177)

The report argued, and this committee concurs, that the emphasis on the privacy of astronaut health data has resulted in lost opportunities to advance the understanding of the risks to humans in the space environment. The IOM report recommended that NASA develop and use an occupational health model for the collection and analysis of astronaut health data. Under such a paradigm, the importance of understanding the risks of the space environment for future astronauts is at least as important as maintaining the confidentiality of individual medical information. The conclusion of the IOM committee was that the collection of individual medical data before, during, and after a mission should be expected by all astronauts who participate in space missions. Such an obligation is reasonable in return for the unique opportunity to travel in space as a representative of the United States. It is important that there be regular, planned analyses of these data to identify critical areas of research to enable future space exploration. The privacy of individual astronauts should be protected to the extent possible (e.g., by de-indentifying data as much as possible) but should not outweigh the need to understand the risks of long-duration space travel.

If there are legal concerns regarding the conclusion that participation in research that enables future space exploration should be part of the job of an astronaut, even if the confidentiality of the data generated cannot be ensured, NASA could bring the matter to the attention of the Secretary's Advisory Committee on Human Research Protections (SACHRP) of the Department of Health and Human Services. The SACHRP provides expert advice and recommendations to the secretary and to the assistant secretary for health on issues and topics pertaining to or associated with the protection of human research subjects. Specific topics include, but are not limited to, special populations and populations in which there are individually identifiable samples, data, or information.[9]

Conclusions

- The success of future space exploration depends on life and physical sciences research being central to NASA's exploration mission and being embraced throughout the agency as an essential translational step in the execution of space exploration missions.
- A successful life and physical sciences program will depend on research being an integral component of spaceflight operations and on astronauts' participation in these endeavors being viewed as a component of each mission.
- The collection and analysis of a broad array of physiological and psychological data from astronauts before, during, and after a mission are necessary for advancing knowledge of the effects of the space environment on human health and for improving the safety of human space exploration. If there are legal concerns about implementing this approach, they could be addressed by the Department of Health and Human Services SACHRP.

Establishing a Stable and Sufficient Funding Base

A renewed funding base for fundamental and applied life and physical sciences research is essential for attracting the scientific community that is needed to meet prioritized research objectives. Researchers must have a reasonable level of confidence in the sustainability of research funding, if they are expected to direct their laboratories, staff, and students on research issues relevant to space exploration. Given the time frame required for completion of the types and scales of experiments necessary for space exploration, grant funding mechanisms must typically span multiple years, with contingencies for delays in flight experiments. A stable research funding level is essential for re-invigorating a scientific community that will not only conduct the research to enable future space exploration but also advance scientific discoveries that are enabled by space exploration. Feedback from interviews, town hall meetings, and white paper submissions associated with this decadal survey suggested that a significant portion of the scientific community remains hopeful that NASA will restore and expand its sponsorship of life and physical sciences research. However, levels of skepticism are high. It is important for NASA to recognize that a failure to make good faith efforts at responding to the recommendations of this decadal survey could result in a loss of interest and support from the scientific community.

It is critical for a healthy space exploration research program, including a program of ground-based research, to support an appropriately balanced portfolio of intramural and extramural research (similar to the support of intramural and extramural research at the National Institutes of Health (NIH), where intramural research is approximately 10 percent of the budget). Although an intramural program is essential to ensure that there are timely and ongoing research efforts focused on barriers that limit space exploration, the current research portfolio is heavily weighted toward intramural projects. An extramural program increases the intellectual wealth and breadth of innovative crosscutting ideas to stimulate advances in both space exploration capabilities and fundamental scientific discoveries. The committee did not develop specific conclusions concerning the allocation of research funds between the intramural and extramural program. Its general conclusion was that the extramural budget has to be sufficiently large (e.g., about 75 percent of the total research budget) to support a robust extramural research program and to ensure that there will be a stable community of scientists who are prepared to lead future space exploration research and to train future researchers.

Research productivity will be optimized by strongly encouraging the collaboration of intramural scientists not only with extramural scientists but also with other agencies. It would be important that such opportunities for collaboration with other agencies be extended to both senior- and junior-level intramural laboratory personnel, with allowance for the release time and travel funds needed to support these activities. Life and physical sciences research for space exploration can potentially be supported by many federal agencies but, to date, efforts to align and coordinate research programs between such agencies have been only marginally successful. This may be due, in part, to the different missions of the respective agencies. However, there is a growing need to synergize multiagency efforts (as discussed below). This will become increasingly important following the retirement of the space shuttle from providing transportation services for research. An increased coordination among agencies would be expected to harness and leverage existing resources. Possible mechanisms for encouraging interagency collaborations include:

- Dedicated interagency funds for research that has applications on Earth and in space;
- Interagency strategic resource planning;
- Similar review process;
- Continued use of interagency workshops and symposia;
- Interagency dual-use technology pilot grant programs;
- Interagency, interdisciplinary mentored training programs; and
- Use of applicable mechanisms that already exist in other agencies (e.g., the Department of Energy).

Success for interagency initiatives will depend on support of such initiatives across the agencies, creation of a spirit of collaboration, and development of new partnerships leading to novel research teams. A comprehensive interagency team effort will serve as a creative scientific resource for implementation of a comprehensive space exploration research program.

However, success from such interactions will depend on the degree to which the collaboration is embraced by all stakeholders and seen as important to the mission of the specific agency. For example, the NIH may have little interest in supporting a project if the research funds are directed to the development of flight hardware for an experiment. It is promising that there are model collaborations from existing interdisciplinary programs in place, such as the NIH Clinical and Translational Science Award (CTSA) program, Ecology of Infectious Diseases (National Science Foundation [NSF] and NIH), National Plant Genome Initiative (NSF, NIH, the U.S. Department of Agriculture, DOE, the U.S. Agency for International Development, the Office of Science and Technology Policy, the Office of Management and Budget) and the U.K. Engineering and Physical Sciences Research Program. An advantage of interdisciplinary programs is a shared contribution of several agencies to the funding needed for ambitious and expensive projects that likely will be necessary to enable the space mission. It would be valuable to strengthen and sustain the historical collaborations of NASA with agencies such as the NIH in life sciences, and to expand them in physical sciences to other agencies such as DOE and the Department of Defense (DOD). These collaborations can build on such efforts as the *Memorandum of Understanding between the National Institutes of Health and the National Aeronautics and Space Administration for Cooperation in Space-Related Health Research*, which went into effect in September 2007.[10,11] This agreement established a framework of cooperation between the NIH and NASA to encourage (1) communication and interaction between the NIH and NASA research communities to facilitate space-related research and to integrate results from that research into an improved understanding of human physiology and human health; (2) exchange of ideas, information, and data arising from their respective research efforts; (3) development of biomedical research approaches and clinical technologies for use on Earth and in space; and (4) research in Earth- and space-based facilities that could improve human health on Earth and in space. In the physical sciences area there have been and are ongoing collaborations between NASA and other space agencies, such as the European Space Agency (ESA) and the German Space Agency (DLR). Examples include a joint experiment on the ISS with DLR on capillary channel flow, a microheater array boiling experiment with ESA, an advanced colloids experiment with ESA, and in the non-Newtonian fluids area, the Observation and Analysis of Smectic Islands in Space experiment with DLR. NASA would benefit from further expanding such collaborations in the future.

The United States has enjoyed a leadership position in space exploration due to its long and successful history of space missions. However, during the coming decade, it is likely that significant efforts in this area will be initiated by other nations. Because of unresolved problems in ensuring safe and successful long-duration missions that affect all nations attempting human spaceflight, a convergence of efforts would likely be of universal benefit. Similar steps as discussed above regarding interagency collaboration within the United States seem logical to explore in support of international scientific projects designed to resolve issues relevant to technological challenges and to astronaut health, safety, and performance. The research community that deals with life and physical sciences in space has remained quite robust internationally, even as NASA has reduced its support in this area. Investigations in Japan, Europe, and Russia have continued, with new results being published regularly.

To regain stature as the leader of the global scientific team in life and physical sciences in space, NASA needs to increase international scientific activities through interactions with such organizations as, but not limited to, the International Space Life Sciences Working Group (ISLSWG). Such cooperation worked well in the decades before 2000 and will undoubtedly reduce subsequent costs to NASA. As noted, in the physical sciences area, there are existing collaborations between NASA and such agencies as ESA, and these could be expanded. New partnerships, such as with India, Australia, and China, are possible. Strong interactions with groups such as the ISLSWG and the offering of joint research announcements with international partner agencies will aid discovery and internationalize space life and physical sciences, offering opportunities for collaboration in ground-based and flight experiments.

Conclusions

- In accord with elevating the priority of life and physical sciences research, it is important that the budget to support research be sufficient, sustained, and appropriately balanced between intramural and extramural activities. As a general conclusion regarding the allocation of funds, an extramural budget would need to support a sufficiently robust extramural research program to ensure that there will be a stable community of scientists and engineers prepared to lead future space exploration research and train the next generation of scientists and engineers.
- Research productivity and efficiency will be enhanced if the historical collaborations of NASA with other sponsoring agencies, such as the NIH, are sustained, strengthened, and expanded to include other agencies.

Improving the Process for Solicitation and Review of High-Quality Research

Scientists who plan to compete for major research grants typically conduct preliminary studies well in advance of submitting a grant application. Thus, familiarity with and predictability of the research solicitation process is critical for enabling researchers to plan and carry out activities in their laboratories that enable them to prepare high-quality research proposals. A regular frequency of solicitations, ideally with multiple solicitations per year, would serve to maintain a community of investigators focused on life and physical sciences research areas relevant to NASA, thereby creating a sustainable research network. This approach is used successfully by other major granting agencies including the NIH and NSF.

An important goal for any funding agency is to ensure that supported research projects are aligned with agency needs. To meet NASA's future space exploration goals, complex scientific and engineering problems need to be solved. Many of these problems will require team-based solutions bridging multiple scientific domains, which can be difficult to organize and sustain. To specifically incentivize this approach, research solicitations need to target not only individual principal investigator-driven applications but also team-driven research involving investigators with complementary interests. Further, solicitations should include both broad research announcements to encourage a wide array of highly innovative research grant applications and targeted research announcements to ensure that high-priority mission-oriented goals are met. This multifaceted solicitation approach would be expected to attract cutting-edge life and physical sciences research that both enables and is enabled by space exploration.

The rejuvenation of a life and physical sciences research program at NASA will depend on effective communications with the scientific community regarding research opportunities. Other major funding organizations (e.g., NIH, NSF) have established web-based links on their home pages for the dissemination of research information. In contrast, the NASA home page has no obvious link for scientists seeking information on research opportunities. Updating and expanding the Life Sciences Data Archive home page to include information on physical sciences research and providing a direct link to this site from the NASA home page would facilitate better communication with the life and physical sciences communities.

The process of peer review is firmly established as a mechanism for identifying the most meritorious research in any scientific area. The concept is embraced by the global research community and viewed as a guarantor for a transparent, fair, and equitable process that results in significant scientific progress. Life and physical scientists are familiar with the peer-review process utilized by many federal agencies, and the research community as a whole has extensive experience in navigating such processes. The legitimacy of the peer-review process is highly dependent on the adoption of review-panel recommendations by the respective funding agencies. Failure to follow recommendations raises the risk of alienating the research community, unless it is performed in a transparent and legitimate manner. The committee believes that NASA has a well-designed peer-review system for the evaluation of extramural research applications. The fluid physics program is one example. Several of the proposals from this program have gone on to become flight experiments. However, the committee also believes that the standards for Non-Advocate Review of intramural research could be elevated by ensuring that the review process, including actions taken by NASA as a result of review recommendations, is more transparent and includes clear rationales for prioritizing both intramural and extramural investigations. Past NRC reports[12] have noted that spaceflight opportunities should unconditionally give maximal access to peer-reviewed experiments that have a strong basis in ground tests and spaceflight performance verifications. Given the severe limitations of actual spaceflight access, it is important that spaceflights do not carry science that has not been deemed of high merit by peer review and prioritized by NASA or formal NASA science partnerships. For example, including experts in space medicine on peer-review panels for space life sciences would help achieve this goal. As part of this effort, it is also important that there be coordination among agency assets, commercial payload developers, and flight systems developers in a manner that serves the best science and reduces the end-to-end time for flight experiments.

The transparency of the process by which intramural and extramural research projects are selected for support after peer review for scientific merit could be ensured if NASA assembled a research advisory committee, composed of 10 to 15 independent life and physical scientists, to oversee and endorse the process. This committee would be charged with advising and making recommendations to the leadership of the life and physical sciences program on matters relating to research activities. This could include an evaluation of the cost versus scientific benefit of proposed space-based research. A similar approach is used by the NIH, where advisory councils within each of the institutes and centers provide a second-tier review of research under consideration for funding. A NASA Research Advisory Committee for Life and Physical Sciences would ensure that the very limited opportunities and funds for space-based research are extended to the best combination of high-quality intramural and extramural studies and that appropriate progress toward research goals is achieved after protocols are in place. The creation of such an advisory body would be expected to re-establish a solid bridge with the scientific community by giving it a voice in guiding NASA life and physical sciences research.

Conclusions

- Regularly issued solicitations for NASA-sponsored life and physical sciences research are necessary to attract investigators to research that enables or is enabled by space exploration. Effective solicitations should include broad research announcements to encourage a wide array of highly innovative applications, targeted research announcements to ensure that high-priority mission-oriented goals are met, and team research announcements that specifically foster multidisciplinary translational research.
- The legitimacy of NASA's peer-review systems for extramural and intramural research hinges on the assurance that the review process, including the actions taken by NASA as a result of review recommendations, is transparent and incorporates a clear rationale for prioritizing intramural and extramural investigations.
- The quality of NASA-supported research and the interactions with the scientific community would be enhanced by the assembly of a research advisory committee, composed of 10 to 15 independent life and physical scientists, to oversee and endorse the process by which intramural and extramural research projects are selected for support after peer review of their scientific merit. Such a committee would be charged with advising and making recommendations to the leadership of the life and physical sciences program on matters relating to research activities.

Rejuvenating a Strong Pipeline of Intellectual Capital Through Training and Mentoring Programs

Realizing the growing challenge of the need to rapidly translate basic findings to applications, the biomedical research enterprise in the United States has reorganized to expand the emphasis on education and training programs for target audiences, ranging from practitioners to researchers to students, as one way to expedite the translation of discoveries to practice. One successful vehicle for these efforts is the recently launched, NIH-funded Clinical and Translational Science Awards. An ancillary goal of the CTSA program is to increase the number and quality of collaborations among practitioners, scientists, patients, and administrators.

The NIH model of research education, which provides training for predoctoral and postdoctoral trainees (F and T awards) and junior faculty investigators (K awards), offers clues to how NASA could design unique educational programs that improve translation in the life and physical sciences, including some or all of the following elements:

- *A curriculum-based program for flight surgeons and physician-astronauts to expand their research knowledge and skill set.* Such programs are comparable to the NIH team-based K30 awards (clinical research curriculum awards).
- *Mentored research training of junior faculty in biomedical sciences, similar to the NIH K (career development) awards.* An essential element of these programs is that they demand that a majority of the trainee's time be protected for instruction and research. When such expectations are not feasible or practical, CTSAs provide continuing and professional education in specific areas of research.
- *Career enhancement awards for junior faculty in nonmedical life sciences, physical sciences, and engineering.* In the physical sciences and engineering, and in many nonmedically related disciplines in the life sciences, junior faculty are expected to develop independent research portfolios upon their academic appointment. Thus, to attract new talent, it is essential to create funding mechanisms specifically targeted for junior faculty. These

mechanisms typically provide significant multiyear support and are accompanied by recognition and prestige. Examples from which NASA could model such an award program include the NSF CAREER Award, the Office of Naval Research Young Investigator Program, the Air Force Office of Scientific Research Young Investigator Program, the Army Research Office Young Investigator Program, the Defense Advanced Research Projects Agency (DARPA) Young Faculty Award, and DOE Early Career Research Program.

- *Graduate student training opportunities.* NASA should consider multiple mechanisms to attract graduate students to space research. Opportunities could include small student research awards, such as the one supported by NASA through the American College of Sports Medicine; research internship opportunities in NASA laboratories; and predoctoral training stipends, such as the NIH predoctoral National Research Service Award.

In addition to the NIH model, the National Space Biomedical Research Institute has a very successful training program[13] that could be expanded. It includes (1) summer internships for undergraduate, graduate, and medical students; (2) graduate education opportunities, which are currently offered through Texas A&M University and Massachusetts Institute of Technology; and (3) postdoctoral research fellowships. Similar programs, if tailored to the concerns of extended-duration spaceflight, could also provide meaningful opportunities for management personnel, engineers, physicians, and astronauts to expand their understanding of the research payloads for which they are responsible. Moreover, such programs could include combinations of virtual and traditional modes of instruction.

A strong pipeline of intellectual capital can be developed by modeling a training and mentoring program after other successful programs in the life and physical sciences. A critical number of investigators is required to sustain a healthy and productive scientific community. Building a program in life and physical sciences would benefit from ensuring that an adequate number of investigators, including flight and ground-based investigators, are participating in research that will enable future space exploration.

Conclusion

- Educational programs and training opportunities effectively expand the pool of graduate students, scientists, and engineers who will be prepared to improve the translational application of fundamental and applied life and physical sciences research to space exploration needs.

Linking Science to Mission Capabilities Through Multidisciplinary Translational Programs

Complex systems problems of the type that crewed missions will increasingly encounter will need to be solved with integrated teams that are likely to include scientists from a number of disciplines, as well as engineers, mission analysts, and technology developers. The interplay between and among the life and physical sciences and engineering, along with a strong focus on cost-effectiveness, will require multidisciplinary approaches. Multidisciplinary translational programs can link the science to the gaps in mission capabilities through planned and implemented data collection mechanisms.

Research Team Approach

A life and physical sciences research program capable of addressing the complex problems posed by space exploration must include both horizontal integration across multiple disciplines and vertical translation of fundamental discoveries to practical application. The dependencies across research objectives, particularly across the broad array of research within the life and physical sciences, can best be defined and addressed through a team research approach. As an example, the loss of bone and muscle tissue will remain a barrier to prolonged spaceflight until effective countermeasures are developed. Although microgravity per se triggers bone and muscle tissue loss because of the reduction in mechanical loading forces, losses are likely exacerbated by additional factors in the space environment (e.g., nutrition, hormonal disruptions, psychological stress). Thus, the development of effective countermeasure strategies requires input from experts across multiple disciplines (e.g., basic bone and muscle biologists, cardiovascular physiologists, endocrinologists, exercise physiologists, nutritionists, biomechanists, behavioralists). Further, physical scientists and engineers must work side by side with life scientists to ensure that countermeasures developed and tested in ground-based studies can be implemented in the space environment.

In this regard, it is not surprising that exercise countermeasures for the preservation of bone and muscle mass on the ISS have been ineffective to date. Although new exercise equipment was recently positioned on the ISS, the previous generation of equipment did not have the capacity to provide an adequate stimulus intensity and, because of vibration issues, could not be used in a manner that would generate the desired high strain rates. If the development and deployment of such equipment is not driven by research needs, the quality of the research is compromised by the use of inferior technology. Such problems could seemingly be avoided or minimized by having life scientists, physical scientists, and engineers working together. An example of vertical integration in physical sciences would be collaboration among researchers with complementary talents and interests to develop a knowledge base that leads to design, development, and testing of the physical components (e.g., heat pipe radiator) of a system.

Beyond the need to provide scientific underpinnings to fulfill future space exploration goals, the space research community represents an ideal foundation where life and physical scientists and engineers can coalesce around common goals. Because scientific advances can occur as a result of serendipity, it is important to have life scientists, physical scientists, and engineers working side by side to take full advantage of both planned and serendipitous discoveries. Examples of the benefits of multidisciplinary interactions already exist. For instance, the development of protective gear and issues related to temperature and environment control during extravehicular activity have engaged both the life and physical sciences communities. Despite such examples, at this time a broad-based, multidisciplinary, integrative approach to conducting space exploration research has not been formally implemented. Because such research is more challenging to organize and conduct than research by individual investigators, scientists must be incentivized to participate in the former. Funding opportunities that require multidisciplinary research teams would provide the appropriate incentive. A long-term strategic plan to maximize team research opportunities and initiatives would be expected to lead to more-efficient solutions to the complex problems associated with space exploration. Implementing this initiative would require forming integrative teams of intramural and extramural scientists, with representation across the life and physical sciences, as well as across funding agencies, to assist in this crucial planning process. Team research models used in physical sciences and engineering by DARPA, with tangible outcomes at the end of the project, should be assessed and considered.

Translational Research—Advancing Research Discoveries to Mission Needs

To meet the demands for new scientific knowledge to guide future space missions, there is a strong need to improve the trajectory of research productivity. This might best be addressed by a systematic analysis of where inefficiencies might be occurring in the translational process. In the physical sciences, the gap between basic research studies and successful commercial or government applications has been referred to as "the valley of death."[14] Clinical and translational scientists in recent years have also begun to define particularly problematic gaps that commonly prove to be the "valley of death" for new drugs, devices, or interventions.

Overcoming, or at least minimizing, these gaps has been a hallmark of the CTSA program launched by NIH. The kinds of interventions currently being undertaken to improve the process of clinical and translational research in the CTSA network should be generalizable to various aspects of NASA's research enterprise.

One important aspect of the CTSA program is the deployment of informatics capabilities addressing all aspects of the research process so that information captured about all transactions (e.g., research, contractual, bureaucratic, laboratory, facility, etc.) are monitored to examine inefficiencies that may be delaying or, in some cases, sidelining a rapid and orderly discovery process. As discussed above, this process also provides important tools for increasing communication among scientists and between NASA and the extramural research community of investigators in space life and physical sciences. Hallmarks of the CTSA approach include (1) focusing on investigator needs, (2) collecting and analyzing metrics to assist in the evaluation of the success of projects, (3) inclusion of auditing to insure evaluation metrics are available to support program review and future prioritization, (4) leveraging existing resources whenever possible to avoid duplication of effort, and (5) building a well-connected community of investigators.

Centralized Information Networks

Centralized information networks based on NASA-sponsored research that are accessible to intramural and extramural investigators would be a valuable research tool. Modern analytical techniques offer a tremendous opportunity to understand the effects of spaceflight on life and physical science systems. High throughput techniques (e.g., genomics, proteomics, metabolomics, transcriptomics, etc.) generate vast amounts of data that can be mined and analyzed by multiple researchers. An example is the National Human Genome Research Institute's Encyclopedia of DNA Elements (ENCODE) project.[15] The creation of a formalized program to promote the sharing and analysis of such data would greatly enhance the science derived from flight opportunities. Elements of such a program would include guidelines on data sharing and community access, with a focus on rapid updating of datasets for shared access while respecting the rights of the principal investigators and confidentiality of participants in experiments. A program of analysis grants dedicated to the spaceflight-derived datasets, including operational medical data, would provide value-added interpretation while ensuring that all data are maximally mined for information. Larger scale multiple investigator experiments, with related science objectives, methods, and common data products would result in the production of large datasets and would emphasize analysis over implementation. This type of dataset, which is similar to those used by the space and Earth sciences within NASA, would likely be a tremendous resource for student research.

Improved Access to Samples and Data from Astronauts

The medical and scientific communities interested in human health, safety, and performance during long-duration spaceflight have been consistent in their requests for greater access to biological samples and other data collected from astronauts before, during, and after space missions.[16–19] The rights of astronauts to privacy have, at times, appeared to conflict with the need for access to valuable data to benefit future space travelers. One conclusion from the 2001 IOM report *Safe Passage*[20] was that

> NASA has devoted insufficient resources to developing and assessing the fundamental clinical information necessary for the safety of humans on long-duration missions beyond Earth orbit. Although humans have flown in space for nearly four decades, a paucity of useful clinical data have been collected and analyzed.

There are probably multiple reasons for the failure to advance this fundamental knowledge, including inadequate funding, competing mission priorities, lack of attention to research, and restricted access to data and biological samples. The IOM report included suggestions for resolving this conflict. Among the recommendations were that NASA should (1) establish a comprehensive health information system for astronauts, for the purpose of collecting and analyzing data, and (2) develop a strategic research plan designed to increase the knowledge base about the risks to astronaut health. Some of these goals have been met, but much remains to be completed to provide more widespread scientific access to such data.

There are potentially three different types of research that are important for advancing knowledge of the biological effects of the space environment, each with unique strategies for collecting biomedical data. The first is hypothesis-driven experiments designed for space that generate data to address specific biological questions. Such data may already be available through the Life Sciences Data Archive,[21] which according to its webpage is an active archive that provides information and data from spaceflight experiments funded by NASA. However, this archive does not appear to be current; some of the "New Experiments" described on the homepage are decades old. Although there is a mechanism to search the database, this approach yields only summary data from the experiments. The second type of research capitalizes on routinely captured operational data that could be used to address biomedical research questions. For example, NASA already has extensive operational data (e.g., the precision of navigation maneuvers) that could be linked to data on crew characteristics and made available for data mining. The availability of operational data would enable scientists to address such questions as whether astronauts who perform with less precision in space are characterized by potential mediators of performance (e.g., sleep disruption, motion sickness). The third type of research is on the acute and persistent effects of spaceflight on health. This approach requires the creation of a long-term astronaut health information registry to learn what, if any, chronic health problems are encountered by flight crews, including long after their mission. There is relatively little knowledge of long-term changes in the health of astronauts who have flown on long-duration space missions. As discussed above, policies on the availability of and access to astronaut health records have severely limited knowledge of the effects of the space environment on health, while hindering the development of effective countermeasures.

A strategy that would benefit all three of these research approaches would be the creation of robust databases that could be used by extramural scientists to address research questions. The databases could be populated retrospectively, with currently archived data from NASA-sponsored projects in the Life Sciences Data Archive, archived data from flight medicine, and available long-term follow-up health data such as the Longitudinal Study of Astronaut Health,[22] with plans to expand the databases prospectively. The databases could be generated through a formal research announcement to attract experienced independent investigators who have established similar population databases (e.g., Nurses' Health Study, the Framingham Study, the Women's Health Initiative). Coupling the database with a genetic bank and repository of astronaut samples would ensure the availability of the maximal amount of data to address future investigations. Because of the limited number of humans who undergo exposure to the space environment, maintaining an extensive and well-organized database and managing it as a resource to be shared with the scientific community has long been viewed an essential step for scientific discovery. At the same time, because few humans undergo exposure to the space environment, it is recognized that even a de-identified database may not fully protect the confidentiality of data.[23] As discussed above, the need to understand the risks of long-duration space exposure should be viewed as at least as high a priority as the protection of individual data privacy because the future of space exploration hinges on a solid knowledge of health risks. The informed consent process for astronauts can clearly specify the sources and types of data, including data collected from activities other than predefined research projects, that could be used for research purposes.

Conclusions

- A long-term strategic plan to maximize team research opportunities and initiatives would accelerate the trajectory of research discoveries and improve the efficiency of translating those discoveries to solutions for the complex problems associated with space exploration.
- Improved central information networks would facilitate data sharing with and analysis by the life and physical sciences communities and would enhance the science results derived from flight opportunities.
- Improving the access of the scientific community to samples and data collected from astronauts via central information networks would advance knowledge of the effects of the space environment on human health and improve the safety of space exploration. Any concerns regarding the confidentiality of data could best be addressed by the Department of Health and Human Services SACHRP.

Developing Commercial Sector Interactions to Advance Science, Technology, and Economic Growth

It is important that NASA's commercial sector interactions be as conducive to the advancement of science, technology, and economic growth as possible. As an example, contract specifications for commercial flight providers may hinder research unless they are formulated with specific requirements to accommodate science needs. Because up-mass to and down-mass from the ISS may be delivered by commercial contractors in the future, it is important that the contract specifications for vehicles include adequate capacity for transporting biological and/or inorganic samples and test apparatus. Conditioned down-mass is of particular importance because there are limited facilities on the ISS for storage of samples. Unless suitable down-mass transportation is made available, only the relatively simple analyses that can be conducted on the ISS will be feasible.

Broad, multidisciplinary teams will be necessary to coordinate and integrate activities across the commercial sector, the space medicine community, and the space operations community. Issues related to the control of intellectual property, technology transfer, conflicts of interest, and data integrity will also have to be addressed. In 2009 NASA issued an opportunity for non-U.S. government entities to propose uses of the ISS,[24] and in 2010, NASA issued a draft Cooperative Agreement Notice for management of the U.S. portion of the ISS for non-U.S. government users.[25] In both cases, private-sector entities are specifically targeted.

Commercial suborbital spaceflight is on the horizon and has the potential to provide a platform for the study in reduced gravity of rapidly occurring processes such as combustion phenomena. Several companies are now developing the hardware and procedures for suborbital flights.[26] The flights will be available to the public, initially at high cost but becoming more affordable as operations continue. While flight trajectories, and therefore dynamics, differ among the companies and remain to some extent proprietary, it is reasonable to expect flights to reach peak altitudes in the range of 100 km and flight times of about 15 min. Hypergravity launch and landing phases will surround a free-fall or near weightless (milli-gravity) phase of 4- to 5-min duration. One aspect that needs to be addressed is how to make flight opportunities available to the research community. A typical NIH or NSF grant, for example, will likely support only one or two such flights, which obviates much of their appeal. One approach is for funding agencies (NASA, NIH, NSF) to pool resources and purchase a set of flights to be dedicated to life and physical sciences or to research in general.

Vehicles under development by commercial suborbital companies, such as Virgin Galactic, Armadillo Aerospace, Blue Origin, Masten Space Systems, and XCOR Aerospace, will allow unprecedented access to the space environment and a new way to engage scientists, university researchers, and students.[27] The scientific community has reacted enthusiastically to the promise of these vehicles, with more than 200 scientists from around the country participating in a series of workshops with suborbital vehicle developers, a distinguished group of scientists coming together to form the Suborbital Applications Researchers Group, and a conference on next-generation suborbital research. NASA also quickly recognized the potential of commercial suborbital spacecraft and has formed the NASA Commercial Re-usable Suborbital Research Program at NASA Ames Research Center.[28] It is important that these types of educational networking opportunities be fostered to help catalyze research interactions among commercial developers, the scientific community, and NASA.

Conclusion

- With the retirement of the space shuttle pending, it will be important for NASA to foster interactions with the commercial sector, particularly commercial flight providers, in a manner that addresses research needs, with attention to such issues as control of intellectual property, technology transfer, conflicts of interest, and data integrity.

Synergies with Other National and International Agencies

NASA has the opportunity to leverage scientific advances in the life and physical sciences funded by other agencies and by other countries and to develop partnerships that produce research results that have genuine benefits to both partners.

Domestic Examples of Potential Synergies in the Life Sciences

- *Commercial companies.* Personalized medicine, which is being funded for use in many biomedical fields, including oncology, cardiovascular disease, and predicting responses to certain therapeutic drugs, could prove enormously beneficial to NASA.
- *Human health.* NIH is developing the methodology necessary to provide for the healthcare needs of today's civilian populations. Astronauts and other human beings in the future who may wish to travel and live in space vehicles and habitats can benefit from and contribute to research across multiple domains (e.g., bone, muscle, sleep, behavior). The unique environments in microgravity may provide a set of physical stimuli to humans that can lead to insights into human health and biology unachievable through any other research.
- *Radiation biology and health physics.* DOE and DOD have requirements to understand the effects of radia-

tion on humans. There may be joint-funding opportunities where radiation health issues overlap and results are of mutual interest. DOE has a low-dose radiation program, and NASA cofunds selected projects that it believes will be of scientific benefit. Further, because the NASA Space Radiation Laboratory at Brookhaven National Laboratory is the only site in the United States where biology research with energetic charged particles other than protons can be conducted, and because interest is growing in other countries in the development of charged particles for use in cancer therapy, answers to questions of interest to both NASA and the National Cancer Institute at the NIH might be pursued in an economical manner.

Domestic Examples in the Physical Sciences

- *DOE and NSF.* As an example of the kinds of joint opportunities that could be developed here, there is a growing community of researchers worldwide who are interested in performing carefully conducted laboratory physics experiments that address select obstacles that physics faces today and that will exploit the benefits of a space environment. DOE, NASA, and NSF have jointly funded the recent report of the NRC Committee on Atomic, Molecular, and Optical Sciences, which emphasized the significant discovery potential of future space-based experiments using new technologies and laboratory techniques that have the ability to probe the fundamental laws of nature at the highest levels of accuracy.[29]
- *DOD.* Multiagency support coordinated by DARPA and formulated through joint workshops could lead to significant progress in developing, for example, quantum technologies for space applications that benefit the entire discipline of space-based research in fundamental physics.
- *Federal Aviation Administration.* The Federal Aviation Administration will soon issue an award for a Center of Excellence for Commercial Space Transportation. As a result, there may be numerous opportunities for synergistic research projects.

Opportunities for International Collaborations

Because opportunities available for space-based experiments are extremely limited, significant collaboration with various international partners could avoid duplication of experimental capabilities in proposed experiments and ensure facilities are used to the maximum extent possible at the lowest cost. Collaboration with international agencies such as ESA and the Japanese Aerospace Exploration Agency (JAXA) could provide a coherent, defensible, research program that maximizes the experimental, analytical, and numerical capabilities of researchers worldwide.

As just one example of a successful international collaboration, the recent Super Critical Water Oxidation research at NASA Glenn Research Center has provided the basis of a new flight investigation on the ISS. A team of scientists at Glenn Research Center has partnered with scientists from the Centre National d'Etudes Spatiales and the Institute of Condensed Matter Chemistry at Bordeaux of the Centre National de la Recherche Scientifique in an investigation entitled the Supercritical Water Mixture experiment. The experiment is scheduled to be performed in the DECLIC2 facility on the ISS early in calendar year 2011.

Information on international project collaborations between NASA and ESA to design and build unique pieces of equipment for skeletal muscle and sensory-motor function testing has been produced by the Muscle Atrophy Research and Exercise System project. Several bed rest studies were sponsored by both ESA and NASA, including the Women's International Space Simulation Exploration study.

Promoting National and International Synergies

The following are examples of strategies that could be used to promote multinational efforts and synergies in biological and physical sciences space research.

- Hold joint workshops, including webinars.
- Have members of other agencies on peer-review teams.

- Develop joint working groups.
- Encourage through solicitations the development of joint proposal efforts.

As an example of the third strategy, NASA Advanced Life Support representatives participate annually with the International Advanced Life Support Working Group. Members include personnel from NASA Headquarters, Johnson Space Center, Kennedy Space Center, and Ames Research Center. Other participating agencies include ESA, JAXA, the Russian Institute of Biophysics, and the Russian Institute for Biomedical Problems. Through fellowships sponsored by the Japanese Society for Promotion of Science, a number of visits and collaborations were facilitated between NASA and Japanese scientists.

As an example of the fourth strategy, the Space-Time Asymmetry Research (STAR) project, a jointly proposed concept by the NASA Ames Research Center, Stanford University, and international partners from Saudi Arabia, Germany, and the United Kingdom, will test isotropy and symmetry of space-time at unprecedented precision. The STAR program is predicated upon building a series of focused, small-satellite missions. STAR will take an incremental mission approach, flying instruments with progressively increased precision and measurement scope in each of five flights. The program is designed specifically to attract extensive research participation and leadership by university students. STAR will challenge curious, young minds and train the next generation of space scientists and engineers.

ADMINISTRATIVE OVERSIGHT OF LIFE AND PHYSICAL SCIENCES RESEARCH

Currently, life and physical science endeavors focused on understanding phenomena in low-gravity environments have no clear institutional home at NASA. As determined by the committee from an examination of the highly varied history of these programs and as discussed in the final report of the Review of U.S. Human Spaceflight Plans Committee (also known as the Augustine Commission or Augustine Committee),[30] administratively embedding crucial forward-looking elements (such as the Life and Physical Sciences Research program) in larger or operationally focused organizations virtually guarantees that the resources for such elements will be depleted by other needs.

This chapter has focused on the essential needs for a successful renewed research endeavor in life and physical sciences. In the context of a programmatic home for an integrated research agenda, program leadership and execution are likely to be productive only if aggregated under a single management structure and housed in a NASA directorate or key organization that understands both the value of science and its potential application in future exploration missions.

Conclusions

- Leadership with both true scientific gravitas and a sufficiently high level in the overall organizational structure at NASA is needed to ensure that there will be a "voice at the table" when the agency engages in difficult deliberations about prioritizing resources and engaging in new activities.
- The successful renewal of a life and physical sciences research program will depend on strong leadership with a unique authority over a dedicated and enduring research funding stream.
- It is important that the positioning of leadership within the agency allows both the conduct of the necessary research programs as well as interactions, integration, and influence within the mission-planning elements that develop new exploration options.

SUMMARY

The committee recognized that the withdrawal of NASA support of life and physical sciences research over the past decade occurred for multiple reasons, many of which were unavoidable. Now that the assembly of the ISS is complete, it is time for NASA to turn to its goal of re-establishing support for life and physical sciences research. Building a "new" NASA Life and Physical Sciences Research Program will pose great challenges but will also present the opportunity to adopt modern-day approaches to solving the complex problems inherent to space exploration. The programmatic conclusions in this report are intended as a guide to restore research activities to a level of excellence that will ensure that NASA remains the undisputed international leader of life and physical sciences research that enables future space exploration and advances fundamental scientific discovery.

REFERENCES

1. National Research Council. 2003. *Assessment of Directions in Microgravity and Physical Sciences Research at NASA.* The National Academies Press, Washington D.C.
2. Tomko, D.L., NASA. 2009. "History of Life and Physical Sciences Research Programs at NASA," presentation to the Committee for the Decadal Survey on Biological and Physical Sciences in Space, May 6. National Research Council, Washington, D.C.
3. Souza, K., G. Etheridge, and P.X. Callahan, eds. 2000. Post-1995 missions and payloads. Chapter 5 in *Life into Space: Space Life Sciences Experiments. Ames Research Center, Kennedy Space Center, 1991-1998.* NASA SP-2000-534. NASA, Washington D.C.
4. Department of Health and Human Services. Office for Human Research Protections. Federal Policy for the Protection of Human Subjects. Available at http://www.hhs.gov/ohrp/policy/common.html.
5. National Aeronautics and Space Act of 1958.
6. Institute of Medicine. 2001. *Safe Passage: Astronaut Care for Exploration Missions.* National Academy Press, Washington D.C.
7. National Research Council. 1998. *A Strategy for Research in Space Biology and Medicine in the New Century.* National Academy Press, Washington D.C.
8. Institute of Medicine. 2001. *Safe Passage: Astronaut Care for Exploration Missions.* National Academy Press, Washington D.C.
9. For more information see Department of Health and Human Services, Office for Human Research Protections, Secretary's Advisory Committee on Human Research Protections, Charter, available at http://www.hhs.gov/ohrp/sachrp/charter/index.html.
10. *NIH Record.* 2007. NIH, NASA Partner for Health Research in Space. Volume LIX, No. 21, October 19. Available at http://nihrecord.od.nih.gov/newsletters/2007/10_19_2007/story1.htm.
11. Memorandum of Understanding between the National Institutes of Health and the National Aeronautics and Space Administration for Cooperation in Space-Related Health Research. See http://www.niams.nih.gov/News_and_Events/NIH_NASA_Activities/nih_nasa_mou.asp.
12. National Research Council. 1998. *A Strategy for Research in Space Biology and Medicine in the New Century.* National Academy Press, Washington D.C.
13. NASA/Brookhaven National Laboratory Space Radiation Program. Space Radiobiology. Available at http://www.bnl.gov/medical/nasa/LTSF.asp.
14. National Research Council. 2005. *Accelerating Technology Transition: Bridging the Valley of Death for Materials and Processes in Defense Systems.* The National Academies Press, Washington, D.C.
15. National Institutes of Health. National Human Genome Research Institute. Research Funding. The ENCODE Project: Encyclopedia of DNA Elements. Available at http://www.genome.gov/ENCODE/.
16. National Research Council. 1998. *A Strategy for Research in Space Biology and Medicine in the New Century.* National Academy Press, Washington, D.C.
17. National Research Council. 2000. *Review of NASA's Biomedical Research Program.* National Academy Press, Washington, D.C.
18. National Research Council and National Academy of Public Administration. 2003. *Factors Affecting the Utilization of the International Space Station for Research in the Biological and Physical Sciences.* The National Academies Press, Washington, D.C.
19. Institute of Medicine. 2007. *Review of NASA's Space Flight Health Standards-Setting Process: Letter Report.* The National Academies Press, Washington, D.C.
20. Institute of Medicine. *Safe Passage: Astronaut Care for Exploration Missions.* National Academy Press, Washington D.C., p. 71.
21. NASA Johnson Space Center. Life Sciences Data Repositories. NASA Human Research Program. Available at http://lsda.jsc.nasa.gov/lsda_home.cfm.

22. The Longitudinal Study of Astronaut Health was reviewed in the 2004 Institute of Medicine report *Review of NASA's Longitudinal Study of Astronaut Health* (The National Academies Press, Washington, D.C.).
23. National Institutes of Health. Educational Materials. HIPAA Privacy Rule. Information for Researchers. Research Repositories, Databases, and the HIPAA Privacy Rule. Available at http://privacyruleandresearch.nih.gov/research_repositories.asp.
24. NASA. 2009. National Lab Opportunity. NNH09CAO003O. Issued August 6, 2009.
25. NASA. 2010. Cooperative Agreement Notice, draft, NNH11SOMD002C.
26. Sanderson, K. 2010. Science lines up for seat to space. *Nature* 463:716-717.
27. Commercial Spaceflight Federation. 2009. "Using Next-Generation Suborbital Spacecraft for Research and Education Missions in the Biological and Physical Sciences," presentation to the Decadal Survey on Biological and Physical Sciences in Space, October. National Research Council, Washington, D.C.
28. See the NASA Office of the Chief Technologist Flight Opportunities Program website, available at http://suborbitalex.arc.nasa.gov/, for more details on the NASA CRuSR Program.
29. National Research Council. 2007. *Controlling the Quantum World: The Science of Atoms, Molecules, and Photons.* The National Academies Press, Washington, D.C.
30. Review of U.S. Human Spaceflight Plans Committee. 2009. *Seeking a Human Spaceflight Program Worthy of a Great Nation.* Office of Science and Technology Policy, Washington, D.C., October, p. 113. Available at http://legislative.nasa.gov/396093main_HSF_Cmte_FinalReport.pdf.

13

Establishing a Life and Physical Sciences Research Program: An Integrated Microgravity Research Portfolio

NASA has a strong and successful track record in human spaceflight made possible by a backbone of scientific and engineering research accomplishments. At this time, the United States finds itself at a stage where future decisions regarding space exploration and activities will depend on the generation of new knowledge in the life and physical sciences to ensure successful implementation of the human exploration options chosen. This decadal survey identifies a number of research questions that need to be addressed to provide a sound basis for any future crewed space program, as well as research questions that can be addressed uniquely using the space environment. The relative urgency of resolving these questions will depend on policy decisions about the future direction of the space exploration program. For example, conducting an extended crewed mission to the Moon or beyond will require a focus on one set of priorities, whereas capitalizing on space assets to resolve terrestrial scientific challenges will involve a different set of priorities. Irrespective of such policy decisions, however, the committee concluded that a number of fundamental questions and research areas will have to be addressed as part of an integrated approach that allows sufficient flexibility for policymakers to choose viable, cost-effective paths for the U.S. crewed space program in the future. Some of these research areas relate to understanding the impacts of extended exposure to microgravity conditions and how to mitigate those impacts. Other fundamental research areas address the need for technological advances that can reduce the cost of space exploration, as well as the challenges posed for humans by extended space travel with only very remote possibilities for logistical support and replenishment. These questions are identified and priorities discussed in the preceding chapters, with Chapters 4 through 10 each focusing on a number of specific research disciplines in a related area.

In this chapter, the committee presents an integrated research portfolio that synthesizes the highest-priority recommendations developed in Chapters 4 through 10 by the panels and the committee that reflects broad priorities that cut across all the discipline areas. The committee believes that the research questions crystallized through the individual panels and summarized in this chapter go to the core of challenges that need to be resolved to advance future human space exploration. It is understood that as a result of scientific gains, additional questions and issues may arise that will need novel solutions beyond the priorities and recommendations listed in this report. However, the committee believes that a plan with sufficient flexibility with regard to an order of implementation is provided here to serve as a blueprint, when guided by overall directions and goals identified by informed policymakers, for NASA to expand and refine its space exploration program.

It is also the committee's strong belief that undertaking an integrated portfolio of research will require programmatic reforms at NASA to establish a strong life and physical sciences enterprise, as discussed in Chapter 12.

BOX 13.1
Summary of Support for NASA's Robust Life and Physical Sciences Research Program, 1996-2001

In fiscal year 1996 the budget for NASA's Office of Life and Microgravity Science and Applications covered a portfolio that mirrors much of the set of integrated recommendations presented in this report as well as the development of a great deal of hardware for the conduct of that portfolio on the International Space Station (ISS). By 2001, some hardware was still being developed, but hardware expenditures had dropped off significantly, allowing the number of funded tasks to begin to increase. By 2010, however, the breadth of the portfolio had shrunk considerably, and the number of tasks had dropped by about two-thirds. Currently there is no single source for obtaining a full accounting of all ground- and space-based life and microgravity science research conducted by NASA, but by any measure both the content of the funded current portfolio and the sum of supported tasks are considerably lower than in 1996-2001.

Fiscal Year	Number of Tasks[a]	Budget (million $)	Program Contents
1996	872	~500	Technology and applications for space research and human support in space, environmental health (microbiology, toxicology, barophysiology, and radiobiology), advanced life support, space human factors, advanced space suits, space biology research, plant biology, combustion science, materials science, fluids, fundamental physics, and supporting orbital operations and research
2001	1,014	~300	Advanced human support, biomedical countermeasures, gravitational biology and ecology, microgravity research, materials science, environmental health, tissue engineering, telescience, human factors, radiation research
2010	364	~150	Research supporting human exploration and ISS life and physical sciences research, including the Human Research Program and the small portion of research within the Exploration Technology Demonstration Program that is related to life and physical sciences research

NOTE: Numbers obtained from NASA task books and presentations to the Committee for the Decadal Survey on Biological and Physical Sciences in Space.
[a]Correlates closely with number of principal investigators.

Such an enterprise will serve as a necessary foundation for the agency to build a solid, robust, and transparent research base shaped by the recommendations from this decadal survey coupled with future policy directions.

The committee points out that a large integrated portfolio of research similar to the complete set of research recommendations contained in this study was supported by NASA in the mid-1990s through the early 2000s (Box 13.1).

PRIORITIZING RESEARCH

In assembling the recommended integrated portfolio of research, the committee has mapped the chapters' highest-priority recommendations against eight prioritization criteria that it believes are relevant to broadly informing policy decisions with regard to future space program options (Box 13.2). The recommendations address unanswered questions related to the health and welfare of humans undertaking extended space missions; to technologies needed to support such missions; and to logistical issues potentially affecting the health of space travel-

BOX 13.2
Criteria Used for Categorization of Research Recommendations

In its categorization of research, whether basic, applied, or translational, the committee used the following eight prioritization criteria developed to capture the potential value of the results of research (information, engineered systems, publications, or new concepts).

- Prioritization Criterion 1: The extent to which the results of the research will reduce uncertainty about both the benefits and the risks of space exploration (*Positive Impact on Exploration Efforts, Improved Access to Data or to Samples, Risk Reduction*)
- Prioritization Criterion 2: The extent to which the results of the research will reduce the costs of space exploration (*Potential to Enhance Mission Options or to Reduce Mission Costs*)
- Prioritization Criterion 3: The extent to which the results of the research may lead to entirely new options for exploration missions (*Positive Impact on Exploration Efforts, Improved Access to Data or to Samples*)
- Prioritization Criterion 4: The extent to which the results of the research will provide full or partial answers to grand science challenges that the space environment provides a unique means to address (*Relative Impact Within Research Field*)
- Prioritization Criterion 5: The extent to which the results of the research are uniquely needed by NASA, as opposed to any other agencies (*Needs Unique to NASA Exploration Programs*)
- Prioritization Criterion 6: The extent to which the results of the research can be synergistic with other agencies' needs (*Research Programs That Could Be Dual-Use*)
- Prioritization Criterion 7: The extent to which the research must use the space environment to achieve useful knowledge (*Research Value of Using Reduced-Gravity Environment*)
- Prioritization Criterion 8: The extent to which the results of the research could lead to either faster or better solutions to terrestrial problems or to terrestrial economic benefit (*Ability to Translate Results to Terrestrial Needs*)

The committee did not weight these criteria, a step that would require assumptions about policy decisions not yet made, or subject to change in the future. The criteria and priorities outlined in this chapter, based on clear metrics, provide a basis for a complete, transparent, and robust research program, which the committee believes is required to fully address NASA's future needs for the life and physical sciences research essential to successful space exploration.

ers, such as adequate nutrition, exposure to radiation, thermoregulation, immune function, stress, and behavioral aspects. The eight criteria listed in Box 13.2 are also offered as a tool that will allow further down-selection to a focused research program that can support any future policy decisions and the associated technology development or knowledge requirements.

Although suggestions are provided below for further prioritization of recommendations already identified as being of the highest priority for specific research areas, none of these high-priority recommendations should be interpreted as being unnecessary. Recognizing that the relative order in which the recommendations will be addressed is likely to depend on the future directions of NASA's exploration and research programs, the committee underscores that all of the recommendations individually are of high merit and collectively constitute important components of an integrated research portfolio. Adoption of such a portfolio will serve as a foundation for the success of future U.S. space exploration efforts, which will require integration, diverse teams, and a translational scientific approach as discussed in Chapter 12.

In addition to the recommendations forming an integrated research portfolio, most of the discipline panels identified a set of important priorities more extensive than what is summarized in this chapter. Although the subset of recommendations provided here should form the core of a renewed physical and life sciences research program, rebuilding an integrated program commensurate with the scale of past microgravity work by NASA will require that the larger set of priorities identified by the panels also be considered as priorities for implementation. Some of the key issues to be addressed in the integrated research portfolio are the effects of the space environment on life support components, the management of the risk of infections to humans, behavior having an impact on individual and group functioning, risks and effects of space missions on human physiological systems, fundamental physical challenges, applied fluid physics and fire safety, and finally, translational challenges arising at the interface bridging basic and applied research in both the life and physical sciences.

Chapters 4 through 10 identify research questions important both to successful space exploration and to advances in fundamental physics and biology enabled by access to space. These two, very connected concepts—the science enabled by exploration and the science that enables exploration—speak strongly to the powerful role of science within the human spaceflight endeavor. Each recommendation listed in Table 13.1 is identified by the committee as either enabling or enabled by exploration. (Some of the recommended research fits both categories.) Further, the research recommendations are also dependent on and define the resources needed to accomplish identified goals. Those resources include hardware and flight opportunities together with robust ground-based programs that place highly evolved experiments in the best position for success upon access to spaceflight.

Ultimately, the research recommended in this decadal study must be further prioritized based on future policy developments, a task that the information summarized in Tables 13.2 and 13.3 is meant to facilitate. Examples of how these tables can be used to develop a research portfolio for a mission-focused policy decision and a knowledge-focused policy decision (see Boxes 13.3 and 13.4 below in this chapter) are meant to indicate a possible approach, and not to be prescriptive.

FACILITY AND PLATFORM REQUIREMENTS

Microgravity research facilities can be divided into two classes: space-based and ground-based. The International Space Station (ISS), discussed in Chapters 3 and 11, is the only space-based facility providing a long-term environment for scientists worldwide to carry out microgravity experiments. Short-term space-based facilities are free-flyers and satellites. Ground-based facilities include parabolic flights, drop towers and sounding rockets, bed rest facilities, accelerators, and medical clinics. A research portfolio that draws on communities of investigators using model organisms, robust technology, and all available ground and flight platforms will greatly facilitate this endeavor. Such an approach will allow solidifying critical new discoveries, decrease the time from selection to flight, shorten the discovery confirmation process, and enhance the outcomes of mission-driven life and physical sciences research.

Ground-Based Research Platforms

Ground-based research provides the basis for the design of flight-based research and can, at low cost, address fundamental scientific questions that enable space research and applications by resolving measurement and system feasibility issues. Ground-based fundamental physics research in heat, mass, and momentum transport, materials physics, combustion, and granular materials supports the design of human flight systems and launch capabilities. Space radiation in particular can be simulated well in ground-based laboratories. Accelerators at the NASA Space Radiation Laboratory at Brookhaven National Laboratory produce both high-energy protons and the energetic nuclei of heavier elements, allowing focused, mechanistic studies of the biological consequences for mammalian cells and other relevant model systems (plants, microbes, etc.) of exposure to radiation. Continued availability of space radiation facilities to NASA investigators is critical, as is broad access provided in a timely fashion to meet agency needs.

Aircraft (parabolic zero-gravity flight) and drop towers, which provide a few seconds of microgravity conditions at a time, can enable tests of technical feasibility and also serve as platforms for experiments that can be completed during a single drop or atmospheric flight. For translational programs such as in situ resource utilization (ISRU), analog field tests can be used to demonstrate system interactions, to evaluate repair and maintenance needs,

TABLE 13.1 Summary of Highest-Priority Recommendations Made in Chapters 4 Through 10

Recommendation Identifier[a]	Recommendation	Enabled by (EB) and/or Enabling (E) Space Exploration
Plant and Microbial Biology (Chapter 4)		
P1	Establish a microbial observatory program on the ISS to conduct long-term, multigenerational studies of microbial population dynamics.	EB
P2	Establish a robust spaceflight program of research analyzing plant and microbial growth and physiological responses to the multiple stimuli encountered in spaceflight environments.	EB
P3	Develop a research program aimed at demonstrating the roles of microbial-plant systems in long-term life support systems.	EB/E
Behavior and Mental Health (Chapter 5)		
B1	Develop sensitive, meaningful, and valid measures of mission-relevant performance for both astronauts and mission control personnel.	E
B2	Conduct integrated translational research in which long-duration missions are simulated specifically for the purpose of studying the interrelationships among individual functioning, cognitive performance, sleep, and group dynamics.	E
B3	Determine the genetic, physiological, and psychological underpinnings of individual differences in resilience to stressors during extended space missions, with development of an individualized medicine approach to sustaining astronauts during such missions.	E
B4	Conduct research to enhance cohesiveness, team performance, and effectiveness of multinational crews, especially under conditions of extreme isolation and autonomy.	EB/E
Animal and Human Biology (Chapter 6)		
AH1	The efficacy of bisphosphonates should be tested in an adequate population of astronauts on the ISS during a 6-month mission.	EB/E
AH2	The preservation/reversibility of bone structure/strength should be evaluated when assessing countermeasures.	EB/E
AH3	Bone loss studies of genetically altered mice exposed to weightlessness are strongly recommended.	EB
AH4	New osteoporosis drugs under clinical development should be tested in animal models of weightlessness.	EB
AH5	Conduct studies to identify underlying mechanisms regulating net skeletal muscle protein balance and protein turnover during states of unloading and recovery.	EB/E
AH6	Conduct studies to develop and test new prototype exercise devices and to optimize physical activity paradigms/prescriptions targeting multisystem countermeasures.	EB/E
AH7	Determine the daily levels and pattern of recruitment of flexor and extensor muscles of the neck, trunk, arms, and legs at 1 *g* and after being in a novel gravitational environment for up to 6 months.	EB
AH8	Determine the basic mechanisms, adaptations, and clinical significance of changes in regional vascular/interstitial pressures (Starling forces) during long-duration space missions.	EB/E
AH9	Investigate the effects of prolonged periods of microgravity and partial gravity (3/8 or 1/6 *g*) on the determinants of task-specific, enabling levels of work capacity.	EB/E
AH10	Determine the integrative mechanisms of orthostatic intolerance after restoration of gravitational gradients (both 1 *g* and 3/8 *g*).	EB/E

continued

TABLE 13.1 Continued

Recommendation Identifier[a]	Recommendation	Enabled by (EB) and/or Enabling (E) Space Exploration
AH11	Collaborative studies among flight medicine and cardiovascular epidemiologists are recommended to determine the best screening strategies to avoid flying astronauts with subclinical coronary heart disease that could become manifest during a long-duration exploration-class mission (3 years).	EB/E
AH12	Determine the amount and site of the deposition of aerosols of different sizes in the lungs of humans and animals in microgravity.	EB/E
AH13	Multiple parameters of T cell activation in cells should be obtained from astronauts before and after re-entry to establish which parameters are altered during flight.	EB
AH14	Both to address the mechanism(s) of the changes in the immune system and to develop measures to limit the changes, data from multiple organ/system-based studies need to be integrated.	EB/E
AH15	Perform mouse studies of immunization and challenge on the ISS, using immune samples acquired both prior to and immediately upon re-entry, to establish the biological relevance of the changes observed in the immune system. Parameters examined need to be aligned with those in humans influenced by flight.	EB
AH16	Studies should be conducted on transmission across generations of structural and functional changes induced by exposure to space during development. Ground-based studies should be conducted to develop specialized habitats to support reproducing and developing rodents in space.	EB
Crosscutting Issues for Humans in the Space Environment (Chapter 7)		
CC1	To ensure the safety of future commercial orbital and exploration crews, quantify post-landing vertigo and orthostatic intolerance in a sufficiently large sample of returning ISS crews, as part of the immediate post-flight medical exam.	EB/E
CC2	Determine whether artificial gravity (AG) is needed as a multisystem countermeasure and whether continuous large-radius AG is needed or intermittent exercise within lower-body negative pressure or short-radius AG is sufficient. Human studies in ground laboratories are essential to establish dose-response relationships, and what gravity level, gradient, rotations per minute, duration, and frequency are adequate.	E
CC3	Conduct studies on humans to determine whether there is an effect of gravity on micronucleation and/or intrapulmonary shunting or whether the unexpectedly low prevalence of decompression sickness on the space shuttle/ISS is due to underreporting. Conduct studies to determine operationally acceptable low suit pressure and hypobaric hypoxia limits.	E
CC4	Determine optimal dietary strategies for crews and food preservation strategies that will maintain bioavailability for 12 or more months.	E
CC5	Initiate a robust food science program focused on preserving nutrient stability for 3 or more years.	E
CC6	Include food and energy intake as an outcome variable in dietary intervention trials in humans.	EB/E
CC7	Conduct longitudinal studies of astronauts for cataract incidence, quality, and pathology related to radiation exposures to understand both cataract risk and radiation-induced late tissue toxicities in humans.	E
CC8	Expand the use of animal studies to assess space radiation risks to humans from cancer, cataracts, cardiovascular disease, neurologic dysfunction, degenerative diseases, and acute toxicities such as fever, nausea, bone marrow suppression, and others.	E

TABLE 13.1 Continued

Recommendation Identifier[a]	Recommendation	Enabled by (EB) and/or Enabling (E) Space Exploration
CC9	Continue ground-based cellular studies to develop end points and markers for acute and late radiation toxicities, using radiation facilities that are able to mimic space radiation exposures.	E
CC10	Expand understanding of gender differences in adaptation to the spaceflight environment through flight- and ground-based research, particularly potential differences in bone, muscle, and cardiovascular function and long-term radiation risks.	EB/E
CC11	Investigate the biophysical principles of thermal balance to determine whether microgravity reduces the threshold for thermal intolerance.	EB/E
Fundamental Physical Sciences in Space (Chapter 8)		
FP1	Research on complex fluids and soft matter. Microgravity provides a unique opportunity to study structures and forces important to the properties of these materials without the interference caused by Earth-strength gravity.	EB/E
FP2	Understanding of the fundamental forces and symmetries of nature. High-precision measurements in space can test relativistic gravity, fundamental high-energy physics, and related symmetries in ways that are not practical on Earth. Novel effects predicted by new theoretical approaches provide distinct signatures for precision experimental searches that are often best carried out in space.	EB
FP3	Research related to the physics and applications of quantum gases. The space environment enables many investigations, not feasible on Earth, of the remarkably unusual properties of quantum gases and degenerate Fermi gases.	EB/E
FP4	Investigations of matter near a critical phase transition. Experiments that have already been designed and brought to a level of flight readiness can elucidate how materials behave in the vicinity of thermodynamically determined critical points. These experiments, which require a microgravity environment, will provide insights into new effects observable when such systems are driven away from equilibrium conditions.	EB
Applied Physical Sciences in Space (Chapter 9)		
AP1	Reduced-gravity multiphase flows, cryogenics and heat transfer database and modeling, including phase separation and distribution (i.e., flow regimes), phase-change heat transfer, pressure drop, and multiphase system stability.	EB/E
AP2	Interfacial flows and phenomena (including induced and spontaneous multiphase flows with or without phase change) relevant to storage and handling systems for cryogens and other liquids, life support systems, power generation, thermal control systems, and other important multiphase systems.	EB/E
AP3	Dynamic granular material behavior and subsurface geotechnics to improve predictions and site-specific models of lunar and martian soil behavior.	E
AP4	Development of fundamentals-based strategies and methods for dust mitigation during advanced human and robotic exploration of planetary bodies.	E
AP5	Experiments on the ISS to understand complex fluid physics in microgravity, including fluid behavior of granular materials, colloids and foams, biofluids, non-Newtonian and critical point fluids, etc.	EB
AP6	Fire safety research to improve methods for screening materials for flammability and fire suppression in space environments.	E
AP7	Combustion processes research, including reduced-gravity experiments with longer durations, larger scales, new fuels, and practical aerospace materials relevant to future missions.	EB/E

continued

TABLE 13.1 Continued

Recommendation Identifier[a]	Recommendation	Enabled by (EB) and/or Enabling (E) Space Exploration
AP8	Research on numerical simulation of combustion to develop and validate detailed single phase and multiphase combustion models for interpreting and facilitating combustion experiments and tests.	E
AP9	Reduced-gravity research on materials synthesis and processing and control of microstructure and properties, to improve the properties of existing and new materials on the ground.	EB/E
AP10	Development of new and advanced materials that enable operations in harsh space environments and reduce the cost of human space exploration.	E
AP11	Fundamental and applied research to develop technologies that facilitate extraction, synthesis, and processing of minerals, metals, and other materials available on extraterrestrial surfaces.	EB/E
Translation to Space Exploration Systems (Chapter 10)		
TSES1	Conduct research to address issues for active two-phase flow relevant to thermal management. (T1)	E
TSES2	To support zero-boiloff propellant storage and cryogenic fluid management technologies, conduct research on advanced insulation materials research, active cooling, multiphase flows, and capillary effectiveness (T2), as well as active and passive storage, fluid transfer, gauging, pressurization, pressure control, leak detection, and mixing destratification (T3).	E
TSES3	NASA should enhance surface mobility; relevant research includes suited astronaut computational modeling, biomechanics analysis for partial gravity, robot-human testing of advanced spacesuit joints and full body suits, and musculoskeletal modeling and suited range-of-motion studies (T4), plus studies of human-robot interaction (including teleoperations) for the construction and operation of planetary surface habitats (T26).	E
TSES4	NASA should develop and demonstrate technologies to mitigate the effects of dust on extravehicular activity (EVA) systems and suits, life support systems, and surface construction systems. Supporting research includes impact mechanics of particulates, design of outer-layer dust garments, advanced material and design concepts for micrometeoroid mitigation, magnetic repulsive technologies, and the quantification of plasma electrodynamic interactions with EVA systems (T5); dynamics of electrostatic field coupling with dust (T23); and regolith mechanics and gravity-dependent soil models (T27).	E
TSES5	NASA should define requirements for thermal control, micrometeoroid and orbital debris impact and protection, and radiation protection for EVA systems, rovers, and habitats and develop a plan for radiation shelters. (T19)	E
TSES6	NASA should conduct research for the development and demonstration of closed-loop life support systems and supporting technologies. Fundamental research includes heat and mass transfer in porous media under partial gravity and microgravity conditions (T6) and understanding the effect of variable gravity on multiphase flow systems. (T21, T22)	E
TSES7	NASA should develop and demonstrate technologies to support thermoregulation of habitats, rovers, and spacesuits on the lunar surface. (T20)	E
TSES8	NASA should perform critical fire safety research to develop new standards to qualify materials for flight and to improve fire and particle detectors. Supporting research is necessary in materials qualification for ignition, flame spread, and generation of toxic and/or corrosive gases (T7) and in characterizing particle sizes from smoldering and flaming fires under reduced gravity (T8).	E

TABLE 13.1 Continued

Recommendation Identifier[a]	Recommendation	Enabled by (EB) and/or Enabling (E) Space Exploration
TSES9	NASA should develop a standard methodology for qualifying fire suppression systems in relevant atmospheres and gravity levels and would benefit from strategies for safe post-fire recovery. Specific research is needed to characterize the effectiveness of fire suppression agents and systems under reduced gravity (T9) and to assess the toxicity of various fire products (T10).	E
TSES10	Research should be conducted to allow regenerative fuel cell technologies to be demonstrated in reduced-gravity environments. (T11)	E
TSES11	To support the development of new energy conversion technologies, research should be done on high-temperature energy conversion cycles, device coupling to essential working fluids, heat rejection systems, materials, etc. (T12). Research is also required on more efficient surface-base primary power and on the technologies to enable solar electric propulsion as an option to transfer large masses of propellant and cargo to distant locations (T18).	E
TSES12	To make fission surface power systems a viable option, research is needed on high-temperature, low-weight materials for power conversion and radiators and on other supporting technologies. (T13)	E
TSES13	Development and demonstration of ascent and descent system technologies are needed, including ascent/descent propulsion technologies, inflatable aerodynamic decelerators, and supersonic retro propulsion systems. The required research includes propellant ignition, flame stability, and active thermal control (T14); lightweight flexible materials (T15); and rocket plume aerothermodynamics and vehicle dynamics and control (T16).	E
TSES14	Research is required to support the development and demonstration of space nuclear propulsion systems, including liquid-metal cooling under reduced gravity, thawing under reduced gravity, and system dynamics. (T17)	E
TSES15	Research is needed to identify and adapt excavation, extraction, preparation, handling, and processing techniques for a lunar water/oxygen extraction system. (T24)	E
TSES16	NASA should establish plans for surface operations, particularly ISRU capability development and surface habitats. Research is needed to characterize resources available at lunar and martian surface destinations (T25) and to define surface habitability systems design requirements (T28).	E

[a]Identifiers correspond to the identifiers given to the highest-priority recommendations listed at the ends of Chapters 4 through 10, which provide context and clarifying discussion.

TABLE 13.2 Highest-Priority Recommendations That Provide **High** Support in Meeting Each of Eight Specific Prioritization Criteria

	<----------Prioritization Criteria---------->							
	(1) Positive Impact on Exploration Efforts, Improved Access to Data or to Samples, Risk Reduction	**(2)** Potential to Enhance Mission Options or to Reduce Mission Costs	**(3)** Positive Impact on Exploration Efforts, Improved Access to Data or to Samples	**(4)** Relative Impact Within Research Field	**(5)** Needs Unique to NASA Exploration Programs	**(6)** Research Programs That Could Be Dual-Use	**(7)** Research Value of Using Reduced-Gravity Environment	**(8)** Ability to Translate Results to Terrestrial Needs
Life Sciences	P2, P3, B1, B2, B3, B4, AH1, AH2, AH3, AH5, AH6, AH7, AH8, AH9, AH10, AH11	P3, B1, B2, B3, B4, AH6, AH9, AH10, AH11	P3, B4, AH1, AH2, AH3, AH5, AH6, AH7, AH8, AH9, AH10, AH11	P1, P2, B3, B4, AH9, AH10, AH11, AH16	P1, P2, P3, AH1, AH2, AH3, AH4, AH5, AH6, AH7, AH8, AH9, AH10, AH11, AH16	B1, B2, B3, B4, AH1, AH2, AH3, AH4, AH5, AH6, AH7, AH9, AH10	P1, B1, B4, AH12, AH16	B1, B2, B3, B4, AH1, AH2, AH3, AH4, AH5, AH6, AH7
Translational Life Sciences	CCH2, CCH4, CCH7	CCH2, CCH4, CCH6, CCH7	CCH2, CCH4, CCH6, CCH7, CCH8	CCH2, CCH6	CCH1, CCH2, CHH3, CCH6, CCH7, CCH8		CCH1, CHH2, CHH3, CCH7, CCH11	
Physical Sciences	AP1, AP4, AP6, AP8, AP11	AP1, AP2, AP10, AP11	AP1, AP2, AP3, AP10, AP11	FP1, FP2, FP3, AP5, AP7, AP8, AP9	AP1, AP2, AP3, AP4, AP6, AP11	AP7, AP8, AP9, AP10	FP1, FP2, FP3, FP4, AP1, AP2, AP5, AP6, AP7, AP9	AP1, AP2, AP7, AP8, AP9
Translational Physical Sciences	TSES1, TSES2, TSES3, TSES14	TSES1, TSES3, TSES5, TSES10	TSES14		TSES2, TSES3, TSES4, TSES5, TSES6, TSES7, TSES12, TSES13, TSES14, TSES 16	TSES10, TSES11, TSES12	TSES1, TSES2, TSES3, TSES4, TSES5, TSES12, TSES13, TSES14, TSES15, TSES16	TSES10

NOTE: Identifiers are as listed in Table 13.1 and correspond with the recommendations listed there and also presented with clarifying discussion in Chapters 4 through 10.

TABLE 13.3 Level of Support Provided by High-Priority Recommendations for Each of Eight Prioritization Criteria

Recommendation Identifier[a] Within Suggested Program Elements	(1) Positive Impact on Exploration Efforts, Improved Access to Data or to Samples, Risk Reduction	(2) Potential to Enhance Mission Options or to Reduce Mission Costs	(3) Positive Impact on Exploration Efforts, Improved Access to Data or to Samples	(4) Relative Impact Within Research Field	(5) Needs Unique to NASA Exploration Programs	(6) Research Programs That Could Be Dual-Use	(7) Research Value of Using Reduced-Gravity Environment	(8) Ability to Translate Results to Terrestrial Needs
	<-------------------- Prioritization Criteria							-------------------->
				Plant and Microbial Biology Research				
P1	Medium	Low	Low	**High**	**High**	Medium	**High**	Medium
P2	**High**	Medium	Medium	**High**	**High**	Medium	Medium	Medium
P3	**High**	**High**	**High**	Low	**High**	Medium	Medium	Medium
				Human Behavior and Mental Health Research				
B1	**High**	**High**	Low	Medium	Low	**High**	**High**	**High**
B2	**High**	**High**	Low	Medium	Low	**High**	Low	**High**
B3	**High**	**High**	Medium	**High**	Low	**High**	Low	**High**
B4	**High**	**High**	**High**	**High**	Medium	**High**	**High**	**High**
				Animal and Human Biological Research				
AH1	**High**	Medium	**High**	Medium	**High**	**High**	Medium	**High**
AH2	**High**	Medium	**High**	Medium	**High**	**High**	Medium	**High**
AH3	**High**	Medium	**High**	Medium	**High**	**High**	Medium	**High**
AH4	Medium	Medium	Medium	Medium	**High**	**High**	Medium	**High**
AH5	**High**	Medium	**High**	Medium	**High**	**High**	Medium	**High**
AH6	**High**	**High**	**High**	Medium	**High**	**High**	Medium	**High**
AH7	**High**	Medium	**High**	Medium	**High**	**High**	Medium	**High**
AH8	**High**	Medium	**High**	Medium	**High**	Medium	Medium	Medium
AH9	**High**	**High**	**High**	**High**	**High**	**High**	Medium	Medium
AH10	**High**	**High**	**High**	**High**	**High**	**High**	Medium	Medium
AH11	**High**	**High**	**High**	**High**	**High**	Medium	Medium	Medium
AH12	Medium	Medium	Medium	Medium	Medium	Low	**High**	Medium
AH13	Medium	Low	Medium	Medium	Medium	Medium	Medium	Medium
AH14	Medium	Low	Medium	Medium	Medium	Medium	Medium	Medium
AH15	Medium/Low	Low	Medium	Medium	Medium	Medium	Medium	Medium
AH16	Medium/Low	Medium/Low	Medium/Low	**High**	**High**	Low	**High**	Medium

continued

TABLE 13.3 Continued

Recommendation Identifier[a] Within Suggested Program Elements	<------------Prioritization Criteria------------>							
	(1) Positive Impact on Exploration Efforts, Improved Access to Data or to Samples, Risk Reduction	**(2)** Potential to Enhance Mission Options or to Reduce Mission Costs	**(3)** Positive Impact on Exploration Efforts, Improved Access to Data or to Samples	**(4)** Relative Impact Within Research Field	**(5)** Needs Unique to NASA Exploration Programs	**(6)** Research Programs That Could Be Dual-Use	**(7)** Research Value of Using Reduced-Gravity Environment	**(8)** Ability to Translate Results to Terrestrial Needs
				Crosscutting Research for the Human System				
CC1	Medium	Low	Low	**Low**	**High**	Low	**High**	Medium
CC2	**High**	**High**	**High**	**High**	**High**	Low	**High**	Low
CC3	Medium	Medium	Medium	Low	**High**	Low	**High**	Low
CC4	**High**	**High**	**High**	Medium	Medium	Medium	**Medium**	Medium
CC5	Medium	Medium	Medium	Medium	Medium	Medium	Medium	Medium
CC6	Medium	**High**	**High**	**High**	**High**	Medium	Low	Medium
CC7	**High**	**High**	**High**	Low	**High**	Low	**High**	Low
CC8	Medium	Medium	**High**	Low	**High**	Low	Low	Low
CC9	Medium	Low	Low	Low	Medium	Low	Low	Low
CC10	Medium	Medium/Low	Medium	Low	Medium	Medium	Low	Medium
CC11	Medium	Medium/Low	Medium	Low	Medium	Medium/Low	High/Medium	Medium
				Fundamental Physical Sciences Research				
FP1	Low	Low	Medium	**High**	Low	Medium	**High**	Medium
FP2	Low	Low	Low	**High**	Low	Medium	**High**	Medium
FP3	Low	Low	Medium	**High**	Low	Medium	**High**	Medium
FP4	Low	Low	Low	**Medium**	Low	Medium	**High**	Medium
				Applied Physical Sciences Research				
AP1	**High**	**High**	**High**	Medium	**High**	Low	**High**	**High**
AP2	Medium	**High**	**High**	Medium	**High**	Medium	**High**	**High**
AP3	Medium	Medium	**High**	Low	**High**	N/A	Low	Low
AP4	**High**	Medium	Medium	Low	**High**	N/A	Medium	Low
AP5	Low	Low	Medium	**High**	Low	Medium	**High**	Medium
AP6	**High**	Medium	Low	Low	**High**	Low	**High**	Medium
AP7	Medium	N/A	N/A	**High**	Medium	**High**	**High**	**High**
AP8	**High**	Medium	Low	**High**	Medium	**High**	N/A	**High**
AP9	N/A	N/A	Low	**High**	Low	**High**	**High**	**High**
AP10	Low	**High**	**High**	Medium	Medium	**High**	Low	Medium
AP11	**High**	**High**	**High**	Low	**High**	N/A	Medium	N/A

Translation to Space Exploration Systems Research								
TSES1	**High**	**High**	Low	Low	Medium	Medium	**High**	Low
TSES2	**High**	High	Medium	Low	**High**	Medium	**High**	Medium
TSES3	**High**	**High**	**High**	Low	**High**	Medium	**High**	Medium
TSES4	Medium	Medium	**High**	Low	**High**	Low	**High**	Low
TSES5	Medium	Medium	**High**	Low	**High**	Low	Medium	Low
TSES6	Medium	Medium	Medium	Low	**High**	Low	Medium	Low
TSES7	Medium	Medium	**High**	Low	**High**	Medium	Medium	Medium
TSES8	Low	Low	Low	Low	Medium	Medium	**High**	Medium
TSES9	Low	Low	Low	Low	Medium	Medium	**High**	Medium
TSES10	Medium	**High**	Low	Low	Medium	**High**	Medium	Medium
TSES11	Medium	Medium	Low	Low	Medium	**High**	Low	Medium
TSES12	Medium	Medium	Low	Low	**High**	**High**	**High**	Medium
TSES13	Medium	Medium	Low	Low	**High**	Medium	**High**	Medium
TSES14	**High**	Medium	**High**	Medium	**High**	Medium	**High**	Medium
TSES15	Medium	Medium	Low	Low	Medium	Low	**High**	Low
TSES16	Medium	Medium	Low	Low	**High**	Low	**High**	Low

[a]Identifiers are listed in Table 13.1 and correspond with the recommendations listed there and also presented the ends of Chapters 4 through 10, which provide context and clarifying discussion.

and to demonstrate long-lived operations. Terrestrial analog field tests are essential to demonstrate the long-term reliability of candidate systems and to develop operational protocols.

Ground-based research is also important for the development of exercise countermeasures, including bed rest studies. Findings from animal models have generated fundamental knowledge concerning the effects of microgravity on muscle and bone physiology. Further, new avenues of animal research can unfold in the areas of epigenetics of gene expression and protein turnover in response to unloading stimuli. Such ground-based studies will benefit from shared specimens and data from space experiments using new technological approaches such as transcript profiling.

Analog Environments

Analog environments (e.g., the ISS as an analog for remote and low-gravity planetary surfaces; polar and undersea research facilities) and rigorously designed experimental simulations (e.g., long-duration chamber studies) that faithfully mirror actual mission parameters (e.g., isolation, confinement, workload, long and uncertain time duration, communication delays, disruption of diurnal sleep-wake cycles) can help to support a balanced research portfolio. Analog opportunities offered through the ISS are discussed in Chapter 11.

Flight Platforms

Uncrewed flight opportunities on free-flyers provide a venue to conduct short-duration experiments, ideally with an animal centrifuge available to provide proper 1-*g* controls for animal specimens and to address the impact of microgravity on biological systems. Free-flyers are well suited for experiments involving virulent organisms or toxic, radioactive, or otherwise dangerous materials that pose a risk to humans. Suborbital platforms and parabolic flights are key in providing a short-duration microgravity environment for biological and physical sciences studies of phenomena and behaviors that may show significant effects during the transitions between 1 *g* and microgravity that will occur in planetary arrivals and departures.

Free-flying spacecraft can also be used for fundamental physics experiments that require an extremely low-noise and low-stray-acceleration environment or a specific orbit. Future possibilities include a rotating free-flyer (with or without a tether), perhaps with an emptied cargo vessel for long-duration experiments. Before ISS cargo vessels are destroyed, they can potentially be used for relatively large-scale microgravity experiments, such as fire safety tests. The absence of *g*-jitter also makes them an ideal platform for crystal growth experiments that are particularly sensitive to vibrations.

Planetary or Lunar Surfaces as Platforms

Many biological processes are compromised in microgravity, and the gravity threshold for restoring proper function is unknown. Availability of lunar bases for carrying out biological experimentation and for testing bio-regenerative life support systems would allow assessment of whether biological functions will be normal (similar to those in 1 *g*) in partial gravity. Lunar or martian bases would also be useful for conducting planetary research described in other studies,[1,2] such as fundamental seismographic studies, yielding insight into planets' interiors and their geological history, as well as allowing studies of their regolith compositions, magnetic fields, and atmospheric phenomena (in the case of Mars) that are relevant to human exploration. In the longer term, such bases might also be used as platforms for large telescopes and provide a stable, long-term laboratory setting for reduced-gravity experimentation. Robotic exploration of the Moon could verify conditions near the lunar poles, develop resource maps, and demonstrate ISRU system end-to-end operations in the lunar environment. Robotic missions may be of particular importance for near-term exploration paths not directly focused on lunar exploration that could use landers or compact rovers. Lunar assets, such as thermal wadis comprising regolith-derived thermal mass materials, could serve as platforms that enable rovers and other exploration hardware to survive periods of cold and darkness.

Space Platforms for Research Beyond 2020

Although most of the recommendations in this report address the current decade, the committee recognizes the long time constant inherent in the implementation of some of the recommendations and thus the importance of planning for the period 2020-2029. The efforts for that decade include extension of research findings from the 2010-2019 decade and completion of remaining gaps. Although specific gaps are challenging to predict, it can be expected that some projects started in the 2010-2019 decade will not reach maturity in that period. Likely to be available in 2020-2029, for example, are new transport vehicles capable of carrying astronauts well beyond low Earth orbit—emphasizing the need for research leading to compact low-power yet highly effective devices that will provide countermeasures for changes in multiple human systems during long voyages in microgravity. Further, the role of partial gravity in preventing deterioration in important physiological systems will have to be clearly understood and countermeasures developed, if necessary, to mitigate those effects. NASA should therefore consider a flexible infrastructure of experimental facilities that could be upgraded to novel exploration systems.

A lunar outpost established as a key national scientific resource could prove to be an important research platform for ongoing studies in partial gravity, providing, among other benefits, sustainable research laboratories for biological research on model systems addressing key scientific areas related to microgravity.

HIGHEST-PRIORITY RESEARCH AREAS AND OBJECTIVES

Table 13.1 *summarizes*, by discipline, the research elements selected by the panels, in close coordination with the committee, as having the highest priority, and which this survey recommends for inclusion in NASA's new portfolio of biological and physical sciences research. The committee concluded that the elements listed in Table 13.1 are important in the creation of a compelling program of life and physical sciences research that can address both fundamental scientific goals and exploration technology needs. These research elements are not described in detail here; instead, unique identifiers listed in Table 13.1 allow locating related full descriptions in Chapters 4 through 10 (where each identifier is listed after a recommendation selected as having highest priority). These identifiers are also shown in Tables 13.2 and 13.3, which map the research elements to the eight prioritization criteria used by the committee. The committee believes that these recommended research areas are the most critical to advancing the national space research program, and that these elements collectively constitute the *core* of an integrated research portfolio in microgravity. It should be kept in mind that this list of recommendations represents the distillation of priorities from an exceptionally large number of disciplines that have in the past typically been treated in separate, more narrowly focused studies. Most of the panel chapters contain additional recommended research—important to a program in that discipline—that was not selected for the integrated portfolio.

RESEARCH PORTFOLIO SELECTION OPTIONS

In Table 13.2, the committee maps the highest-priority recommendations (each indicated by the unique identifier listed in Table 13.1) from Chapters 4 through 10 to the eight prioritization criteria defined in Box 13.2. The research areas listed under a given criterion in Table 13.2 are those categorized in Table 13.3 as providing "high" support for that particular criterion. This mapping is intended to help provide a basis for policy-related ordering of an integrated research portfolio, depending on future policy decisions.

As examples of how the information in Table 13.2 might be used, consider two bounding policy options that could drive a research portfolio. The first is a decision to send humans to Mars (Box 13.3). Clearly Prioritization Criteria 1 and 2 would be the most important for prioritizing the research to support this policy, and supporting the associated recommended research areas in an integrated program with clear translational end points would be essential. These translational end points must enable realization of specific design goals that would be unachievable without successful research. In this first example Prioritization Criteria 3 and 5 would also have to be taken into consideration when selecting the science necessary to achieve this policy goal.

The second sample policy option is a decision to hold off on advanced human missions until a new base of capability is developed and to focus instead in the near term on advancing leading-edge science (Box 13.4) and

BOX 13.3
Sample Bounding Policy Option One

Goal: Send Humans to Mars

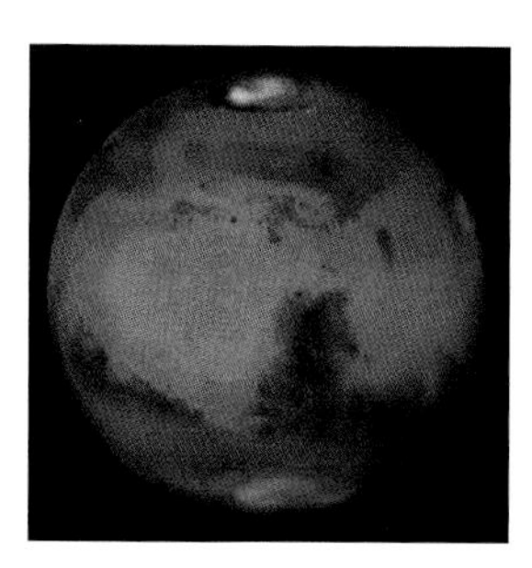

Prioritization Criteria 1 and 2 will be the most important functions in prioritizing research to support the goal of sending humans to Mars, and a way must be found to support the recommendations associated with these priorities in an integrated program with clear translational end points. Prioritization Criteria 3 and 5 will also have to be taken into consideration to achieve the science necessary to achieve this policy goal.

Criterion 1. The extent to which the results of the research will reduce uncertainty about both the benefits and the risks of space exploration (*Positive Impact on Exploration Efforts, Improved Access to Data or to Samples, Risk Reduction*)

→ **Relevant research recommendations**

- Life sciences: P2, P3, B1, B2, B3, B4, AH1, AH2, AH3, AH5, AH6, AH7, AH8, AH9, AH10, AH11
- Life sciences translational: CCH2, CCH4, CCH7
- Physical sciences: AP1, AP4, AP6, AP8
- Physical sciences translational: TSES1, TSES2, TSES3, TSES14

(AH1:) The efficacy of bisphosphonates should be tested in an adequate population of astronauts on the ISS during a 6-month mission.

(TSES2:) Research should be conducted in support of zero-boiloff propellant storage and cryogenic fluid management. Physical sciences research includes studies of advanced insulation materials, active cooling, multiphase flows, and capillary effectiveness (T2), as well as active and passive storage, fluid transfer, gauging, pressurization, pressure control, leak detection, and mixing destratification (T3).

Criterion 2. The extent to which the results of the research will reduce the costs of space exploration (*Potential to Enhance Mission Options or to Reduce Mission Costs*)

→ **Relevant research recommendations**

Criterion 3. The extent to which the results of the research may lead to entirely new options for exploration missions (*Positive Impact on Exploration Efforts, Improved Access to Data or to Samples*)

→ **Relevant research recommendations**

Criterion 5. The extent to which the results of research are uniquely needed by NASA, as opposed to any other agencies (*Needs Unique to NASA Exploration Programs*)

→ **Relevant research recommendations**

BOX 13.4
Sample Bounding Policy Option Two

Goal: Develop New Capabilities by Advancing Leading-Edge Science

The goal of developing new capabilities by advancing leading-edge science represents a decision to postpone an advanced human mission until a new base of capability is developed with which to plan. The focus in the near term will be on advancing leading-edge science and the value of space assets to terrestrial needs. In this case, Prioritization Criteria 4, 5, and 8 will have primary importance, and Prioritization Criteria 6 and 7 may also be of importance in building the integrated research portfolio that best supports this policy goal.

Criterion 4. The extent to which the results of the research will fully or partially answer grand science challenges that the space environment provides a unique means to address (*Relative Impact Within Research Field*)

Relevant research recommendations

- Life sciences: P1, P2, B3, B4, AH9, AH10, AH11, AH16
- Life sciences translational: CCH2, CCH6
- Physical sciences: FP1, FP2, FP3, AP5, AP7, AP8, AP9
- Physical sciences translational: None

Establish a "microbial observatory" program on the ISS to conduct long-term multi-generational studies of microbial population dynamics.

Microgravity provides a unique opportunity to study long time dynamics of colloids, polymer and colloidal gels, foams, emulsions, and soap solutions free from gravitational interference.

Criterion 5. The extent to which the results of the research are uniquely needed by NASA, as opposed to any other agencies (*Needs Unique to NASA Exploration Programs*)

Relevant research recommendations

Criterion 8. The extent to which the results of the research could lead to either faster or better solutions to terrestrial problems or to terrestrial economic benefit (*Ability to Translate Results to Terrestrial Needs*)

Relevant research recommendations

Criterion 6. The extent to which the results of the research can be synergistic with other agencies' needs (*Research Programs That Could Be Dual-Use*)

Relevant research recommendations

Criterion 7. The extent to which the research must use the space environment to achieve useful knowledge (*Research Value of Using Reduced-Gravity Environment*)

Relevant research recommendations

the value of our space assets to terrestrial needs. In this case, Prioritization Criteria 4, 5, and 8 would have primary importance, and Prioritization Criteria 6 and 7 might also be of importance in building the integrated research portfolio that best supports this policy goal.

In addition to providing a basis for prioritization, Table 13.2 also illustrates the interdependence among the different individual research recommendations, none of which, as pointed out above, should be seen in isolation. Although an exact order of dependency among the individual recommendations is not specified, their grouping clearly indicates their interdependence and underscores the importance of an integrated approach.

TIMELINE FOR THE CONDUCT OF RESEARCH

The committee was tasked with developing a timeline for the conduct of its recommended research, and except where indicated otherwise the panel chapters contain rough estimates—based on assumptions of robust programmatic support and reasonable access to flight opportunities—of time frames for the individual research areas. The committee identified priority areas and questions that need to be addressed during the present decade (2010-2020), as well as more overarching areas going beyond 2020. It refrained from suggesting a detailed timeline for the overall research portfolio, because this will depend to a major extent on future policy and funding decisions. It is the committee's belief and hope that the high-priority recommended research and its categorization according to eight prioritization criteria will serve to inform policymakers about knowledge needed irrespective of decisions that might favor long-term human space exploration, planetary surface habitation and presence, or more basic and fundamental research. In the committee's view, all of these endeavors will require a science portfolio integrated so as to enable NASA to derive optimal benefits and science return from its investments in research, as well as from support provided by other government agencies and/or commercial sources.

An integrated research portfolio can also enable the identification and execution of radical new options to reduce cost and risk for the U.S. space program. Specifically, new options that offer significant reductions in cost and/or risk can best be conceived and developed in the context of integrated solutions to science and engineering challenges and inclusion of translational end points.

Many of the thematic chapters include information on the current status of research and what would be reasonable expectations with regard to accomplishments for the decade 2010-2019 versus 2020 and beyond. Much of this estimation is based on the time required to conduct experiments and on the near-term expected availability of platforms for conducting research. For a detailed summary of the rationale for and the respective targets of research for the decades 2010-2019 and 2020-2029, the reader is referred to each of the thematic chapters (4 through 10). In addition, the mapping of research areas to prioritization criteria presented in Table 13.2 offers an approach to considering timelines for research, as does Table 13.3, in which the disciplinary panels have further classified each high-priority recommendation as being of high, medium, or low applicability with respect to each of the eight prioritization criteria.

The committee chose this tabular presentation to avoid redundancy and to provide a ready means for NASA to identify specific components of an integrated research portfolio judged most likely to contribute to capability and flexibility for achieving space exploration program goals, as represented by the eight prioritization criteria shown. Thus, for example, in considering a martian exploration mission (see Box 13.3), each recommendation can be seen in Table 13.2 as ranked at a finer granularity with regard to its importance in addressing that specific goal. If NASA were to decide to increase synergism with other agencies in building its research program, the recommendations most relevant to addressing this priority would be found under Prioritization Criteria 6 in Table 13.2, and the relative importance of all identified high-priority recommendations for this specific action item would be as indicated in Table 13.3.

The committee anticipates that the categorization offered in Table 13.3 will guide NASA's decision making on timeline and urgency issues. The committee realizes that a careful assessment of timeline goals will require a comprehensive and broad overview of space-relevant research and will require a strong life and physical sciences research organization in the agency. Hence, the programmatic focus and recommendations summarized in Chapter 12 will be a key mechanism to ensure that specific, thematic committee recommendations can be adapted to a flexible timeline responsive to NASA's overarching goals.

IMPACT OF SCIENCE ON DEFINING U.S. SPACE EXPLORATION POLICY

Implicit in this report are integrative visions of the science advances necessary to underpin and enable major new components, revolutionary systems, and bold exploration architectures for human space exploration. Essential to achieving affordable, safe, and productive space exploration systems, such advances are central to the U.S. space exploration policy and agenda. Their system-level aspects are fully addressed in the technical literature cited in Chapters 4 through 10. The panels drew on their collective knowledge of science and technology and both the references and their associated issues to define the scientific barriers, unit-processes, and physical challenges worthy of inclusion in the recommendations in this report.

Impediments to revitalizing the U.S. space exploration agenda include costs, past inability to accurately predict costs and schedule, and uncertainties about mission and crew risk. The technical communities recognize their obligations to deal with those impediments. Indeed, typical flow-downs from science as discussed in this report include improvements in function and efficiency, subsequent reductions in mass, and direct or implied reductions in cost. The starting point for much of the life sciences research is reducing mission and crew risk, an undertaking for which new understanding is required to make safe human passage possible to, for example, Mars. Better scientific understanding will also greatly improve the fidelity of overall cost and schedule predictions associated with development of new systems.

A few examples from preceding chapters of this report illustrate these points. One revolutionary and mission architecture-changing system involves on-orbit depots for cryogenic rocket fuels. The scientific foundations required to make this Apollo-era notion a reality are specified in the report. For some lunar missions, such a depot could produce the major cost savings of an Ares 1 launch system replacing the Ares 5. The highly publicized collection or production of large amounts of water from the Moon or Mars will require scientific understanding of how to retrieve and refine water-bearing materials from the extremely cold, rugged regions on those bodies. Once produced, that water could be transported to surface bases or to orbiting facilities for conversion into liquid oxygen and hydrogen by innovative solar-powered cryogenic processing systems and then stored in the on-orbit depots. All of these hardware and systems implementations require or will be enhanced by new scientific understanding. Such advances point the way to a new era in defining space exploration.

Part of gaining support for crewed Mars missions is being able to address with confidence the questions of protecting the health, safety, and job performance capabilities of crew members during the months-long transits to and from Mars. The life sciences research portfolio recommended in this report constitutes an integrated complex of scientific pursuits pertaining to multiple different biological systems and aimed at reducing to a minimum the health hazards of space explorers, thereby providing quantitative answers to the questions associated with visiting Mars. In other words, sustained research successes are required before humans can safely go to Mars and return.

Thus, this report is much more than a catalog of research recommendations; it identifies the scientific resources and provides tools to help in defining and developing with greater confidence the future of U.S. space exploration and scientific discovery.

REFERENCES

1. National Research Council. 2003. *New Frontiers in the Solar System: An Integrated Exploration Strategy.* The National Academies Press, Washington, D.C.
2. National Research Council. 2007. *The Scientific Context for Exploration of the Moon.* The National Academies Press, Washington, D.C.

Appendixes

A

Statement of Task

BACKGROUND

Consistent with U.S. Space Exploration policy, NASA intends to conduct a series of robotic and crewed exploration missions over the next decades. These include missions to the International Space Station (ISS) and other missions to low Earth orbit and missions to the Moon. These missions will involve a combination of factors such as reduced gravity level, radiation, life support and extended-duration confinement. In addition, these missions present multidisciplinary scientific and engineering challenges and opportunities that are both fundamental and applied in nature. Meeting these scientific challenges will require an understanding of biological and physical sciences and their intersections in partial and microgravity environments.

Previous congressional language in the NASA Authorization Act of 2005 had reserved a portion of space station research funding for fundamental research in life and microgravity sciences. More recently, Congress provided additional direction regarding life and microgravity research (Explanatory Statement accompanying the FY 2008 Omnibus Appropriations Act (P.L. 110-161)) by stating: "Achieving the goals of the Exploration Initiative will require a greater understanding of life and physical sciences phenomena in microgravity as well as in the partial gravity environments of the Moon and Mars. Therefore, the Administrator is directed to enter into an arrangement with the National Research Council to conduct a "decadal survey" of life and physical sciences research in microgravity and partial gravity to establish priorities for research for the 2010-2020 decade."

In early 2010, guidance was provided to NASA in the fiscal year (FY) 2011 Presidential Budget request which would extend the lifetime of ISS to 2020—considerably altering both the research capacity and role of ISS in any proposed program of life and microgravity research. Additional changes initiated by the budget request would greatly affect both the organization and likely scale of these programs at NASA. In order to ensure that the committee could both provide timely input to these organizational changes, and incorporate consideration of their possible impact into its final report recommendations, the committee will develop an interim report focused on key near term issues, followed by a reassessment of its portfolio assumptions and recommendations prior to completion of the final full report.

STATEMENT OF TASK

Consistent with the direction in the Explanatory Statement accompanying the FY 2008 Omnibus Appropriations Act (P.L. 110-161), the National Research Council will organize a decadal survey to establish priorities and provide recommendations for life and physical sciences research in microgravity and partial gravity for the 2010-2020 decade. The committee will develop criteria for the prioritization.

The decadal survey will define research areas, recommend a research portfolio and a timeline for conducting that research, identify facility and platform requirements as appropriate, provide rationales for suggested program elements, define dependencies between research objectives, identify terrestrial benefits, and specify whether the research product directly enables exploration or produces fundamental new knowledge. These areas will be categorized as either those that are required to enable exploration missions or those that are enabled or facilitated because of exploration missions.

The decadal survey should:

- Define research areas that enable exploration missions or that are enabled by exploration missions;
- For each of the two categories above, define and prioritize an integrated life and physical sciences research portfolio and associated objectives;
- Develop a timeline for the next decade for these research objectives and identify dependencies between the objectives;
- Identify terrestrial, airborne, and space-based platforms and facilities that could most effectively achieve the objectives;
- Explain how the objectives could enable exploration activities, produce knowledge, or provide benefits to space and other applications;
- Identify potential research synergies between NASA and other U.S. government agencies, as well as with commercial entities and international partners; and
- Identify potential research objectives beyond 2020.

The results of the decadal survey will assist in defining and aligning life and physical sciences research to meet the needs of exploration missions. The recommendations regarding the timeline and sequence of research are intended to allow NASA to develop an implementation plan that will impact future exploration missions. The survey should focus on the aforementioned tasks and should not recommend budgetary levels. This decadal survey should build upon the findings and recommendations of previous National Academies' studies conducted in this area.

Prior to the publication of the final report, a brief interim report will be developed that is intended to address near term challenges faced by NASA as it reorganizes its programs to comply with directions to NASA in the President's FY 2011 Budget that substantially affect the conduct of ISS science in particular, and life and microgravity science in general. The interim report will focus on issues identified by the committee that relate to:

1. ISS as a platform for conducting life and physical sciences research, and
2. Programmatic support of a healthy and sustainable life and physical sciences research program at NASA.

The interim report will identify programmatic needs and issues to guide near-term decisions that are critical to strengthening the organization and management of life and physical sciences research at NASA. The interim report will also identify a number of broad topics that represent near-term opportunities for ISS research. These areas, along with research more suited to other platforms, will be discussed in greater detail in the final report. In addition to any relevant findings, the interim report may include recommendations to the extent that they are useful and that adequate justification for them can be provided in this short report.

B

Glossary and Selected Acronyms

25-OH vitamin D3	25-hydroxycholecalciferol, a metabolically active form of vitamin D
ACES	Atomic Clock Ensemble in Space, a European space mission; advanced crew escape suit
ACME	Advanced Combustion via Microgravity Experiments
ACTH	adrenocorticotrophic hormone
actin	protein that forms one component of the internal skeleton (cytoskeleton) of the cell
ADH	antidiuretic hormone
AG	artificial gravity generated by devices creating centrifugal forces
AGE	arterial gas emboli
agglutinates	in lunar regolith, easily crushable aggregates of smaller soil particles that have been bonded by melting during micrometeoroid impacts
AHB	Animal and Human Biology (Panel)
amyloplast	plant organelle that contains starch; because of its high density it moves within the cell in response to the direction of gravity
ANP	atrial natriuretic peptide
applied physical sciences	the study of physical sciences with particular applications in mind; in this report, the applied physical sciences of particular interest are fluid physics, combustion, and materials science
ARED	advanced resistive exercise device
ATV	Automated Transfer Vehicle

auxin	plant hormone thought to play a key role in regulating plant growth responses to gravity; in addition it regulates many other developmental processes in plants
AVP	arginine vasopressin
BDCF	Baseline Data Collection Facility (at NASA Kennedy Space Center)
BEC	Bose-Einstein condensate, at a temperature near absolute zero, atoms behave as a "superatom"
biofilm	complex aggregation of different microbes growing on a surface, generally living within a matrix of secreted compounds
biofuel	gas or liquid fuel produced from biomass, the biological materials produced by living organisms
biomolecule	chemical compound found in living organisms
Bion	Russian space capsule that can support animals (e.g., monkeys, rats) and insects in orbit for up to 3 weeks
bioregenerative life support	life support system based on biological components designed to regenerate air and water and produce food to sustain crew members on extended missions
bisphosphonate	a pharmaceutical drug to prevent bone loss
BMD	bone mineral density
BNL	Brookhaven National Laboratory
boiling curve	plot of heat flux versus the difference between (a) the temperature of the wall where heat is being added to a boiling liquid and (b) the temperature of the liquid
Brayton cycle	a thermodynamic cycle used for power generation that features high conversion efficiency and a single-phase working fluid but with the drawback of relatively low heat-rejection temperatures, requiring relatively large and massive radiators
buoyant convection	a form of convection in which the movement of the working fluid (gas or liquid) is caused by density differences at different points in the fluid; also referred as natural or free convection; see also *forced convection*
CADMOS	Centre d'Aide au Développement des activités en Micro-pesanteur et des Operations Spatiales (operated by CNES)
carbon nanotubes	hollow tubes that are made of pure carbon and are just a few nanometers in diameter
CCDev	Commercial Crew Development (NASA contract)
CD4/CD8	subgroups of immune cells used to fight infection
cellular solidification	a mode of solidification that (1) forms a fine-grained material (compared, for example, to dendritic solidification) and (2) facilitates close control of the microstructures within the material, in part because the direction of growth is determined by the direction of heat flow within the material as it solidifies and not by crystallographic properties of the material

CELSS	Closed Ecological Life Support System; Controlled Environment Life Support System
ceramic-matrix composite	a composite material that uses a ceramic material (i.e., a ceramic matrix) to bind together the strengthening agent embedded in the matrix
CHeX	Confined Helium Experiment
chute flow	a flow of granular material down the inclined surface of a chute
CIR	Combustion Integrated Rack (on the ISS)
closed porosity	a measure of the void spaces in a material (e.g., as a percentage of the total volume of a material) that considers only those void spaces that are sealed off from the external surface of the material; total porosity is the sum of open and closed porosities
closure relations	small scale models that provide data on important physics phenomena that are lost when physical data or DNS results are averaged; these relations are necessary for two-fluid CMFD models to close (that is, to define all the unknowns in the model), and the accuracy of these two-fluid models is limited by the accuracy of the closure models upon which they rely
CMFD	computational multiphase fluid dynamics: a numerical approach using high-speed computers for evaluating the conservation equations that describe multiphase flows
CNES	Centre National d'Études Spatiales, the French government space agency
colloid	any gas, solid, or liquid in a fine state of subdivision, with particles too small to be seen in an ordinary microscope, that is dispersed in a continuous gaseous, liquid, or solid medium and either does not settle or settles very slowly
combustion synthesis	a technique for synthesizing materials that uses highly exothermic, self-sustaining reactions
complex fluids	fluids that are homogeneous at macroscopic scales but have a complex structure at microscopic scales; common examples include colloidal suspensions of solid particles in liquid (e.g., paint or ink); emulsions of two immiscible liquids such as oil and water (e.g., milk or mayonnaise); foams, which are a mixture of liquid and gas (e.g., shaving cream); and liquid crystals
composite	a combination of two or more materials that (1) have significantly different physical or chemical properties and (2) remain separate and distinct on a macroscopic level within the finished product; in a typical composite, one material (the matrix) is used to bind together a strengthening agent, which may take the form of filaments, foils, flakes, or other particles
condensation curve	plot of heat flux versus the difference between (1) the temperature of the wall where heat is being removed from a gas that is being condensed and (2) the temperature of the gas
constant gravity stimulus	the natural force of attraction exerted by celestial bodies, e.g., Earth
convection	the transfer of energy and mass in a fluid (liquid or gas) caused by the physical movement of molecules within the fluid; see also *buoyant convection* and *forced convection*
countermeasure	a physiological intervention to maintain normal organ and/or systemic function

CPT symmetry	the concept that the universe should behave the same if one could simultaneously reverse the charge of all particles, their "parity" or handedness, and the direction of the flow of time
critical heat flux	the maximum rate of heat transfer that occurs before a breakdown in the boiling process; during nucleate boiling, this occurs when the boiling process makes a transition to film boiling
critical point	the temperature and pressure above which the liquid and gas forms of a material no longer exist as distinct phases because the material takes the form of a supercritical fluid; the critical point for water is 705°F and 3,200 psi
CTSA	Clinical and Translational Science Award
CVP	central venous pressure
cytokinin	plant hormone that plays important roles in regulating development
cytoskeleton	internal protein skeleton of the cell; made of microtubules, microfilaments, and in animals, intermediate filaments
Damec	Danish Medical Centre of Research
DARPA	Defense Advanced Research Projects Agency
DCI	decompression illness (see DCS)
DCS	decompression sickness
DDREF	dose and dose rate effectiveness factor
deflagration	vigorous burning with subsonic flame propagation
dendrite	tree-like crystal that forms during solidification from a liquid
Desert RATS	Desert Research and Technology Studies
detonation	explosive combustion that spreads supersonically via shock compression
diffusion flame	a flame in which the oxidizer combines with the fuel (by diffusion) and burns simultaneously; in most combustion systems or fires, fuel and air are initially unmixed, resulting in the formation of diffusion flames, which typically have a distinct edge that defines the limits of the region where combustion is occurring; the alternative is premixed flames, which occur when fuel and oxidizer are mixed before they burn
DLR	German Aerospace Center (Deutsches Zentrum für Luft- und Raumfahrt)
DNS	direct numerical simulation: a simulation in computational fluid dynamics in which the Navier-Stokes equations for turbulent flows are numerically solved
DOD	U.S. Department of Defense
DOE	U.S. Department of Energy
DOF	degrees of freedom
down mass	capacity to transfer payload from a location in space, such as low Earth orbit, to Earth

drift-flux model	a computational approach for predicting the performance of a multiphase fluid that considers the performance of the fluid as a whole, rather than assessing different phases individually; although less sophisticated than CMFD models, the simplicity of drift-flux models is advantageous for engineering tasks where the sophistication of a CMFD model is not needed
drop tower	a facility, which may be above or below ground, in which experiments are subjected to free-fall for a few seconds to create conditions of weightlessness
DTH	delayed-type hypersensitivity
EBV/VZV	latent herpes viruses
EDMP	experiment data management plan
EDS	Emergency Detection System
EELV	Evolved Expendable Launch Vehicle
elastic modulus	the relative stiffness of a material within the elastic range, which can be calculated as the ratio of stress to strain
electrolysis	passing a direct electric current through an ion-containing solution to produce chemical changes at the electrodes
electrometallurgical	related to the use of electricity and electrolysis to extract metals from ore, regolith, or other materials
EMG	electromyographic activity (as measured in skeletal muscle)
EMU	Extravehicular Mobility Unit (space shuttle EVA suit)
endodermis	specialized layer of cells enclosing the transport tissues (vasculature) of the plant
EP	the general relativity equivalence principle that all objects, regardless of their composition, move under gravity in exactly the same way, depending only on their mass
epigenetics	factors modulating genetic expression without altering DNA sequences
EPM	European Physiology Module (of the ISS)
ESA	European Space Agency
ETDP	Exploration Technology Development Program
ethylene	a lightweight hydrocarbon, C_2H_4; ethylene is used by plants as a growth signal
eukaryote	a cell in which the genetic information is enclosed in a membrane-bounded structure called the nucleus; eukaryotes generally contain many other membrane-bounded regions of specialized function, called organelles
EVA	extravehicular activity, for example space walks performed by astronauts outside of the ISS
excursion	see *flow excursion*
EXPRESS	expedite the processing of experiments to space station; a standardized rack configuration used on the ISS

extinction limit	the minimum conditions necessary to sustain combustion of a flowing gas; for example, in some combustors, the extinction limit is the minimum time that a point in the flow stream (of mixed air and fuel) must spend in the combustor to sustain continuous combustion for a given set of conditions, such as fuel type, fuel/air mix, pressure, and temperature
fermions	a class of fundamental particles that includes systems such as the electron gas that makes metals resilient, elastic, and conductive and that is the source of forces that stabilize white dwarf stars against collapse
FIR	Fluids Integrated Rack (on the ISS)
flammability limit	the limiting conditions under which combustion of a given type can be sustained in a given environment; for example, in gas-phase combustion, flammability limits are primarily a function of the fuel type, the total pressure, the concentrations of fuel and oxygen, and the temperature; flammability limits typically describe upper and lower bounds (e.g., the maximum and minimum limits) on fuel concentration
FLEX	Flame Extinguishment Experiment
flow excursion	an event in which a two-phase system goes from one operating state to another but does not return to the original state
fMRI	functional magnetic resonance imaging
forced convection	a form of convection in which the movement of the working fluid (gas or liquid) is externally imposed, for example, by a blower or pump; see also *buoyant convection*
free-flyer	a satellite that can be used for automated microgravity research in both biological and physical sciences, such as growing bacteria in space or exposing materials to the space environment, among many other uses; mission durations, satellite bus and payload sizes, and mission purposes vary widely; free-flyers can operate either with or without human interaction and may or may not return samples or data back to Earth autonomously; some free-flyers will only transmit data back to Earth and are not designed for re-entry
FSB	Fundamental Space Biology (NASA program)
FSL	Fluid Science Laboratory (on the ISS)
FSPS	Fission Surface Power System (joint NASA/DOE technology effort)
fuel cell	device that converts chemical energy of a fuel directly into electrical energy
functional residual capacity	amount of air in the lungs after exhaling
FY	fiscal year
genome	the entire genetic information of an organism
geomorphic	relating to the surface features of a landscape and the forces that shaped them
geotechnics	practical application of geological science to mining, civil engineering, etc.
GeV	billion (giga) electron volts: unit of measure for high-energy particles

GH	growth hormone
g-jitter	gravity-jitter: small fluctuations in acceleration that are present in a spacecraft environment and are caused by machinery, rocket firings, astronauts in motion, etc.
global equilibrium	state in which intensive properties of a system are homogeneous and constant throughout the system
gravitaxis	the swimming of an organism in a direction determined by the gravity vector
gravitropism	directional growth response of plant stems and roots to the force of gravity
gravity	a force per unit mass experienced by a physical body as a result of mutual attraction with all other bodies, independent of electromagnetic or other forces
green revolution	a range of research and development advances applied to crop plants, such as reducing the height of some cereals, that greatly increased worldwide agricultural yields from the 1940s to 1970s
GWAS	genome-wide association study(ies)
Gy	gray: the SI unit of absorbed dose of ionizing radiation
halon	any of a group of compounds used as fire suppression agents; they are created by replacing the hydrogen atoms of a hydrocarbon with halogen atoms, such as bromine or fluorine; for example, Halon 1301, used on the space shuttle, is bromotrifluoromethane: CF_3Br
heat exchanger	device that facilitates the transfer of heat from a hot source to a cold sink
heat pipe	a container of two-phase fluid used to transfer heat efficiently
heat sink	a reservoir to absorb thermal energy
HEDS	abbreviated name of the NRC report titled *Microgravity Research in Support of Technologies for the Human Exploration and Development of Space and Planetary Bodies* (2000)
heliopause	the theoretical boundary of the solar system where the Sun's solar wind is stopped by the interstellar medium; the heliopause is at a distance of about 140 AU from the Sun
Henry Gauer reflex	a head-ward movement of fluid occurring during spaceflight
hot pressing	a process in which the particles of a powder are welded together by the simultaneous application of pressure and heat; hot pressing is also known as pressure sintering (see also *sintering* and *liquid-phase sintering*)
hPa	hectopascal; a unit of measure commonly used for barometric pressure; 1,013 hPa is the barometric pressure equivalent to 760 mm Hg (1 atmosphere), the nominal atmospheric pressure at Earth's surface
HRF	Human Research Facility; either of two facilities, HRF-1 and HRF-2, on the ISS
HTV	H-II Transfer Vehicle; a Japanese launch vehicle
HU	hindlimb unloading [model]; a rodent model for unloading skeletal muscle in vivo in ground-based experiments

hydrometallurgical	related to the use of chemical processes involving water-based solutions to extract metals from ore, regolith, or other materials
hypobaric pressure	pressure less than 1 atmosphere
hypoxia	a condition wherein an organism receives insufficient oxygen to support its metabolism
HZE particles	high-energy particles such as iron nuclei present in cosmic rays; HZE particles have an energy range of about 10^2 to 10^3 MeV per nucleon
ICP	intracranial pressure
ignition limit	the minimum conditions that must be present for combustion to start in a given environment in the presence of a spark; for example, with gas phase combustion, ignition limits are primarily a function of the fuel type, the total pressure, the concentrations of fuel and oxygen, and the temperature
IL	interleukin
IMLEO	initial mass in low Earth orbit
insolation	the solar radiation energy received on a given surface area in a given time
intensive property	physical property of a system or material that does not depend on the size of the system or the amount of the material; examples include pressure, temperature, density, viscosity, and boiling point, but not mass, energy, volume, or stiffness
interfacial phenomena	material behaviors associated with the boundaries (faces) between different phases, including those between similar phases of different materials
interstitial gas	gas that may be present in the openings or pore spaces in rock or soil
IOM	Institute of Medicine
iRED	interim resistive exercise device
ISLSWG	International Space Life Sciences Working Group
ISPR	International Standard Payload Rack (on the ISS)
ISRU	in situ resource utilization; the proposed use of resources found or manufactured on the Moon, Mars, or other planetary bodies to further the goals of a space mission
ISS	International Space Station
IVGEN	Intravenous Fluid Generation for Exploration (project)
JAXA	Japan Aerospace Exploration Agency
Josephson effect	a phenomenon of electric current across two weakly coupled superconductors separated by a very thin insulating barrier (a Josephson junction); Josephson effects in ^{3}He and ^{4}He have applications in advanced technology such as new superfluid gyroscopes
JSC	Johnson Space Center

laminar flame	flame that occurs in an environment where fluid flow is laminar rather than turbulent (that is, the flow is smooth and orderly, with little mixing between adjacent fluid layers); laminar flames are impractical because of the low rate of mixing of fuel and air
LBNP	lower body negative pressure
LCROSS	Lunar Crater Observation and Sensing Satellite
LEA	launch, entry, and abort (suit)
LED	light-emitting diode
LEO	low Earth orbit; approximately 100 to 1,200 miles above Earth's surface
LET	linear energy transfer; the amount of energy deposited per unit distance that a charged particle travels; high-LET radiation includes the heavier-than-protons charged-particle radiation found in galactic cosmic rays; the biological concerns are that such radiation is more damaging than is low-LET radiation such as the x-rays, gamma rays, or protons used in clinical/medical applications
lignification	the production of the polymer lignin in plant cell walls; leads to extremely strong support tissues within the plant
liquid-phase sintering	a sintering process that occurs in the presence of a liquid that coexists with the powder being sintered at the sintering temperature; the liquid phase increases the bonding rate because the capillary forces associated with the presence of the liquid are equivalent to very large external pressures (see also *sintering* and *hot pressing*)
lodging	the bending over of plant stems in response to extreme weather such as wind and rain; in cereal crops, lodging can lead to poor grain formation and problems with harvesting; lodged plants can often right themselves through the gravitropic response of their stems
Lorentz symmetry	a symmetry of physics under rotations and boosts
low-shear modeled microgravity	a fluid-based microbial culture environment using a rotating vessel, for which the very low shear forces generated have been shown to mimic some of the effects of microgravity
LPE	Lambda Point Experiment
LRO	Lunar Reconnaissance Orbiter
LSS	life support system
LTMPF	Low-Temperature Microgravity Physics Facility; a multiflight facility designed to attach to the Japanese Experiment Module/Exposed Facility of the ISS
LVEDV	left ventricular end diastolic volume
MARES	Muscle Atrophy Research and Exercise System (on the ISS)
MASER	Material Science Experiment Rocket
MDCA	Multi-user Droplet Combustion Assembly
MEDES	Institute for Space Medicine and Physiology, Toulouse, France
MELFI	Minus Eighty Degree Laboratory Freezer for the ISS

MELiSSA	Micro-Ecological Life Support System Alternative; a project from a consortium of European and Canadian research laboratories and universities that is managed by the European Space Agency; investigates artificial microbe/plant ecosystems with an aim to develop elements of a bioregenerative life support system
mesoscale	of intermediate size; in materials science, of a size ranging from approximately 10 microns to 1 millimeter
metabolomics	an analytical technique that comprehensively catalogs the small-molecule metabolites present in an organism
metagenomics	the study of the multiple genomes found in environmental samples
metal-ceramic composite	a composite with both metal and ceramic components, such as ceramic particles dispersed in a metal matrix or metal filaments embedded in a ceramic matrix; see also *composite*
metal-matrix composite	a composite material that uses a metallic substance (i.e., a metal matrix) to bind together the strengthening agent embedded in the matrix; see also, *composite*
MHC	myosin heavy chain; the motor protein regulating muscle contraction
MHD	magnetohydrodynamics
microarray	an analytical technique by which the levels of expression of thousands of genes can be assayed simultaneously
microgravity	an environment in which there is very little net gravitational force, such as in free-fall or in orbit
MICROSCOPE	a room-temperature weak equivalence principle experiment in space relying on electrostatic differential accelerometers
MISSE	Materials International Space Station Experiment
mixed fields	mixtures of protons with heavier charged particles or of a variety of heavy particles
model system	an organism that is particularly tractable to study and for which there is a large body of information about its development and response systems, and that is used to infer how other similar biological systems may respond or develop; examples: for bacteria, *Escherichia coli*; for animals, the nematode *Caenorhabditis elegans* and the fruit fly *Drosophila melanogaster*; and for plants, thale cress (*Arabidopsis thaliana*) and rice (*Oryza sativa*)
motor unit	a group of muscle fibers of similar properties innervated by a common neuron
MPLM	Multi-Purpose Logistics Module (on the ISS)
MRM1	Mini-Research Module-1 (on the ISS)
MSG	Microgravity Science Glovebox (on the ISS)
MSL	Materials Science Laboratory (on the ISS)
MSL-1	Microgravity Science Lab (space shuttle mission)
MSNA	muscle sympathetic nerve activity
MSRR-1	Materials Science Research Rack-1

multiphase	any process involving a mixture of two or more phases (solid, liquid, and gas); a glass of ice water is a multiphase system
myostatin	an antigrowth factor protein that impacts bone and muscle formation
nanoslurry	a mixture of nanoscale particles and a liquid
NASA	National Aeronautics and Space Administration
Navier-Stokes equations	the equations of motion for a viscous fluid in terms of pressure, density, external force, fluid velocity, and viscosity
NE	norepinephrine
NEP	nuclear electric propulsion
NGF	nerve growth factor
NIAID	National Institute of Allergy and Infectious Diseases
NIH	National Institutes of Health
Nomex®	an artificial heat- and fire-resistant fabric manufactured by the DuPont Corporation
NRC	National Research Council
NSBRI	National Space Biomedical Research Institute
NSF	National Science Foundation
NSRL	NASA Space Radiation Laboratory at Brookhaven National Laboratory; a facility able to generate the spectrum of radiation types to which astronauts are likely to be exposed in space
NTR	nuclear thermal rocket
nucleate boiling	in pool boiling, the boiling that occurs when individual bubbles of gas appear on the heat transfer surface (that is heating the fluid) and then rise to the surface, as opposed to film boiling, which occurs when the bubbles of gas are formed so rapidly that they combine to form a gas film that covers the heat transfer surface
open porosity	a measure of the void spaces in a material (e.g., as a percentage of the total volume of a material) that considers only those void spaces that are connected to the external surface of the material; total porosity is the sum of open and closed porosities
order parameter	a parameter of a system that is zero in the disordered phase, exhibits large fluctuations about its zero mean as the critical point is approached, and grows from zero to larger values as the ordered phase is entered
organelle	a membrane-bounded structure that is found within a cell and is a site of specialized function
orthostatic intolerance	inability to maintain normal blood pressure while standing
osmotic force	the driving force of water movement across the membrane of a cell

Ostwald ripening	tendency for a particle dispersion to grow in diameter over time as smaller particles (with higher solubility) dissolve preferentially, with subsequent crystallization onto larger particles, making them even larger
PCAI	power, communications, avionics, and informatics
PHA	polyhydroxyalkanoate
phase	a homogeneous and physically distinct state of aggregation of a substance, e.g., solid, liquid, or vapor phase
phase separation	separation of a mixture of phases into individual component phases
physically based model	a model of system behavior based on fundamental physical principles (e.g., thermodynamic laws) and the appropriate physical mechanisms (e.g., heat transfer, capillary flow), as opposed to an empirical model, which is based primarily on experimental measurements and incorporates only a limited theoretical understanding of the system
pile flow	a flow of granular material along the inclined surface of a stationary pile
Planck scale	corresponds to energies of $\sim 10^{19}$ GeV
PLSS	portable life support system
pO_2 (or PO_2)	partial pressure of oxygen
pool boiling	boiling that occurs when the heating surface is submerged in a relatively large body of still liquid (there is no liquid movement except that which arises naturally from buoyant convection currents and from agitation by bubbles of gas that form during the boiling process)
PRA	plasma renin activity
protein balance	the net status of protein content in a muscle fiber; if the protein balance is negative, the fiber atrophies
protein turnover	the process in a cell by which any given protein stock undergoes simultaneous processes of synthesis and degradation
proteomics	an analytical approach for the large-scale identification of the proteins present within an organism; can be used to monitor how the spectrum of proteins changes with environmental changes
psi	pounds per square inch
PTSD	post-traumatic stress disorder
pyrolysis	decomposition of a material or compound due to heating without combustion, which is prevented by the absence of oxygen or any other oxidizing reagents
pyrometallurgical	related to the use of heat-based processes, such as smelting, to extract metals from ore, regolith, or other materials
QT interval	in cardiology, a measure of time that represents the interval between electrical depolarization and repolarization of the left and right ventricles of the heart

quantum gas	a system of particles in which the size of an individual particle's quantum wavelength becomes large compared to the length scale of interactions between the particles in the system
quantum phase transition	transition from a continuous quantum fluid to a discrete atomic lattice, such as the "superfluid-to-Mott" insulator transition
quorum sensing	the coordination of responses from bacterial populations through the exchange of small signaling molecules; quorum sensing allows bacteria to respond to their own population levels

R&D	research and development
radiation	anything propagated as rays, waves, or a stream of particles, but especially light and other electromagnetic waves or the emission from radioactive substances
radioisotope	a radioactive isotope of an element
Rankine cycle	a thermodynamic cycle for power generation that uses separate boilers and condensers with two-phase (liquid/vapor) mixtures with high conversion efficiencies and high heat-rejection temperatures, allowing reduced radiator mass and areas
RANKL	an orthoclase-stimulating peptide that induces bone loss
RE	resistance exercise
reaction wood	strengthening tissue that forms upon mechanical stress of woody plants, such as occurs from wind, snow build up, or the weight of the plant
reactive hot pressing	a hot pressing process in which powders are mixed and an exothermic chemical reaction occurs
reactive oxygen species	highly reactive molecules derived from oxygen, such as superoxide or hydrogen peroxide; reactive oxygen species are produced during normal metabolism, but they can be deleterious to the cell; they are also widely used as signaling molecules that regulate organism function
reduced gravity	gravity levels less than 1 *g*
REM	rapid eye movement
residence time	the length of time that combustion gases are in the combuster; it is larger for larger combusters and shorter for systems with higher gas velocities
resorption	the process of losing bone material
RFC	regenerative fuel cell
regolith	surface rock, especially used to describe the lunar surface soil
rheology	the science of the deformation and flow of liquids and solids
rpm	revolutions per minute
RPS	radioisotope power system
RVLM	rostral ventrolateral medulla; a brain region

SACHRP	Secretary's Advisory Committee on Human Research Protections (of the U.S. Department of Health and Human Services)
SARG	Suborbital Applications Researchers Group
sclerostin	a bone factor gene stimulating bone growth
self-healing material	polymer composite designed to automatically repair cracks within the material that may be caused by impact, fatigue, or wear
SEP	solar electric propulsion
SHS	self-propagating high-temperature synthesis: see *combustion synthesis*
sintering	a process in which the particles of a packed powder are bonded to each other by heating to a high temperature below the melting temperature (but generally above one-half the absolute melting temperature); this process generally takes place without external pressure (see also *hot pressing* and *liquid-phase sintering*)
SLS	Space Life Sciences (as referred to in STS space shuttle science missions, e.g., SLS-1)
solar particle event	flux of energetic ions and/or electrons of solar origin
sounding rocket	uncrewed rocket used for short, non-orbital flights; the most common uses are to study Earth's atmosphere and to conduct microgravity research
Spacelab	Spacelab was a reusable laboratory module flown in the space shuttle's cargo bay and used for microgravity experiments that were operated and/or monitored by astronauts. Spacelab had four main components: a pressurized laboratory module with a shirt-sleeve working environment; a tunnel for gaining access to the module; one or more pallets for exposing materials and equipment to space; and an instrument pointing system for astronomical, solar, and/or Earth observations, along with other targets. A memorandum of understanding was signed in 1973 between the European Space Agency (then the European Space Research Organization) and NASA (with Marshall Space Flight Center as the lead NASA center) to design and develop the laboratory. The 10-foot-long pressurized modules were built by an industrial consortium and flew on all five space shuttle vehicles between 1983 and 1998.
SpaceX	Space Exploration Technologies Corporation
specific impulse	efficiency of rocket engines expressed as thrust per unit mass of flow rate produced by burning rocket propellant
spinodal decomposition	a mechanism by which a solution of two or more components can separate into distinct regions (or phases) with distinctly different chemical compositions and physical properties
STAR	Space-Time Asymmetry Research (project)
Starling Landis equation	an equation designed to estimate pressures in the capillary beds of the circulatory system
STEP	Satellite Test of Equivalence Principle; a space mission to test the weak equivalence principle using cryogenically controlled test masses on a spacecraft orbiting Earth

Stirling cycle	a method of power conversion that utilizes sealed gas/piston-linear alternator components and can operate at relatively high efficiency with comparatively small heat source-sink differential temperatures
stoichiometric mix	a "perfect" mix of a combustible gas and air, such that there is just enough oxygen to support combustion of all the fuel present
stoichiometry	the proportions in which chemical elements combine or are produced and the weight relations in a chemical reaction, such as combustion
strain	deformation of a body in response to an external force
stress	external force per unit area acting on a body
STS	Space Transport System; formal name for the U.S. space shuttles; used with a number to designate a specific space shuttle flight, e.g., STS-17
superfluid	a fluid, such as a liquid form of helium, exhibiting a frictionless flow at temperatures close to absolute zero
surface engineered coating	advanced coating consisting of multiple, thin layers designed to improve the performance of a given component in a particular application by improving the mechanical, physical, and/or chemical properties of that component
surface spreading	the phenomenon observed when a relatively insoluble liquid is placed on the clean surface of another liquid (or when a liquid is placed on the smooth surface of a solid)
surfactant	surface-active agent; also known as a wetting agent, a surfactant lowers the surface tension of a liquid, allowing easier spreading
SWAB	Surface, Water, and Air Biocharacterization (program); an environmental sampling program established by NASA to document the microbes found in the water and air supply and on the surfaces of the ISS
T3	triiodothyronine (the active form of thyroid hormone)
T4	thyroxine
tensegrity	a structure in which compression and tension forces are balanced throughout a network; in a cell, tensegrity is thought to reside in the rigid and flexible components of the cytoskeleton that are connected together and so can rapidly transmit mechanical forces throughout this network
thermal wadi	an engineered source of stored solar energy using modified lunar regolith as a thermal storage mass
thermophotovoltaic	the selective emission and conversion to electrical energy of thermally produced photons
thermophysical	related to physical properties that are affected by temperature
TKSC	[JAXA] Tsukuba Space Center; located in Tsukuba, Japan
TNF-α	tumor necrosis factor-alpha; a cytokine that induces inflammatory responses
transcriptional profiling	the use of approaches such as microarray analysis to catalog the expression/activity of a wide range of genes in an organism

transcriptome	the spectrum of genes that are being actively expressed at any moment in time; the transcriptome can change as an organism experiences new stimuli and changes the genes it is expressing in response to those stimuli
tree of life	a depiction of the interrelatedness of the various kingdoms of life as branches on a tree, with the trunk reflecting their common ancestry; DNA sequencing has been used in recent years to more closely define these relationships and so locate organisms more precisely within this tree representation of ancestry
TRL	technology readiness level; one of a set of nine graded definitions/descriptions (TRL-1 to TRL-9) of stages of technology maturity; for example TRL-1 indicates that a basic principle has been observed and reported, TRL-8 indicates a design qualified for spaceflight
TSH	thyroid stimulating hormone
tubulin	a protein that forms one component of the internal skeleton, the cytoskeleton, of cells
tumbler flow	a flow of granular material in a rotating drum
turbulent flame	flame that occurs in an environment where fluid flow is turbulent rather than laminar (that is, the flow is chaotic and disorganized, with substantial mixing between adjacent fluid layers); all practical combustion systems with liquid or gas fuels use turbulent flow to provide adequate mixing of fuel and air
twisted ribbons	twisted metal strips placed in the water-filled tubes of a boiler to increase the boiling rate by providing additional nucleation sites for the formation of gas bubbles
unitized RFC	a regenerative fuel cell with a particular packing geometry
up-mass	capacity to transfer payload from Earth to a location in space, such as low Earth orbit
USOC	user support and operation center (ESA centers)
UV	ultraviolet, a portion of the electromagnetic spectrum
Van der Waals forces	a group of relatively weak and temporary intermolecular interactions that generally result when a molecule or group of molecules become polarized into a magnetic dipole, most often because of uneven or shifting distributions within the electron clouds of the atoms
vasculature (plant)	specialized tissue that transports water, mineral nutrients, and sugars produced by photosynthesis around the plant; consists of two specialized cell types: xylem that principally transports water and phloem, which is largely responsible for the movements of sugars
VGE	venous gas emboli
wetting	the ability of a liquid to maintain contact with a solid surface; the degree of wetting is determined by a force balance between adhesive and cohesive forces
WORF	Window Observational Research Facility (on the ISS)

ZARM	Zentrum für angewandte Raumfahrttechnologie und Mikrogravitation in Bremen, Germany
zero gravity	an environment in which the net vector of all gravitational and accelerative forces acting on a body is essentially zero; see *microgravity*

C

Committee, Panel, and Staff Biographical Information

COMMITTEE ON DECADAL SURVEY ON BIOLOGICAL AND PHYSICAL SCIENCES IN SPACE

ELIZABETH R. CANTWELL, *Co-Chair*, is the director for mission development in the Engineering Directorate at the Lawrence Livermore National Laboratory. Until August 2010, she was the deputy associate laboratory director in the National Security Directorate of the Oak Ridge National Laboratory (ORNL). Prior to joining ORNL, she was the deputy division leader for science and technology in the International Space and Response (ISR) Division at the Los Alamos National Laboratory (LANL). As division leader for ISR, she was responsible for the execution of projects from small, principal investigator (PI)-driven basic science projects through the delivery of large satellites and instruments into the space environment or other field deployments. Until June 2005, she served as the section leader for the Micro and Nanotechnology Center at Lawrence Livermore's Engineering Research Center for fabricating small sensors and devices. She began her career building life support systems for human spaceflight missions with NASA and later went on to serve as a program manager in the Life Sciences Division at NASA Headquarters. Dr. Cantwell earned her Ph.D. in mechanical engineering from the University of California, Berkeley. Her National Research Council (NRC) experience includes past membership on the Committee on NASA's Bioastronautics Critical Path Roadmap, the Space Station Panel of the Review of NASA Strategic Roadmaps, the Committee on Technology for Human/Robotic Exploration and Development of Space, and the Committee on Advanced Technology for Human Support in Space.

WENDY M. KOHRT, *Co-Chair*, is a professor of medicine in the Division of Geriatric Medicine at the University of Colorado, Denver, Anschutz Medical Campus, and an adjunct professor of integrative physiology at the University of Colorado, Boulder. Her research interests are aging, exercise, regional adiposity, energy metabolism, and the effects of changes in the endocrine system on human physiology. She has written articles on increasing bone mineral density through exercise and hormone therapy, the preservation of bone health through physical activity, lower-body adiposity and metabolic protection in postmenopausal women, and protection of bone mass by estrogens and raloxifene during exercise-induced weight loss. She is currently a consultant to NASA's Johnson Space Center (JSC) for the Exercise Countermeasures Program Investigator Team working on optimization of the exercise prescription for the preservation of musculoskeletal and cardiovascular health on the International Space Station (ISS). She is the PI of a clinical trial, COX Inhibition and Musculoskeletal Responses to Exercise, funded by the National Institute on Aging. Another focus of Dr. Kohrt's research is bone health in aging and the extent to which lifestyle behaviors can protect against bone loss. She is a member of the American College of Sports

Medicine (ACSM), the American Society for Bone and Mineral Research, and the Endocrine Society, among other professional societies. Dr. Kohrt received her B.S. in physical education and mathematics from the University of Wisconsin, Stevens Point, and her M.S. and Ph.D. in exercise science from Arizona State University; she completed postdoctoral research training in applied physiology and gerontology at Washington University School of Medicine in St. Louis. She has extensive advisory committee experience as both member and chair, including service on NASA program reviews and such high-profile committees as the U.S. Physical Activity Guidelines Advisory Committee.

LARS BERGLUND is a professor of medicine, the associate dean for research, and the director of the National Institutes of Health (NIH)-funded Clinical and Translational Science Center (CTSC) at the University of California, Davis (UC Davis); he also serves as a physician at the Sacramento VA [Department of Veterans Affairs] Medical Center. He received his Ph.D. in 1977 and his M.D. in 1981, both from Uppsala University, Sweden. His internship and residency in internal medicine and clinical chemistry were completed at the Karolinska Institute, Stockholm, Sweden, where he served as a faculty member in the Department of Clinical Chemistry (1986-1993). Dr. Berglund was recruited to Columbia University as a Florence Irving Associate Professor of Medicine in 1993 and became professor of medicine in 2000. He served as the associate director for the Columbia University General Clinical Research Center (GCRC) from 1997. In 2002, he was recruited to UC Davis, and in 2004 he became the first program director of the UC Davis GCRC. Dr. Berglund became the first assistant dean of clinical research at UC Davis in 2004 and the associate dean of clinical and translational research in 2006. Also in 2006 he became the first director of the NIH-funded UC Davis CTSC. In 2009, Dr. Berglund assumed the position of associate dean for research in the UC Davis School of Medicine. As CTSC director, Dr. Berglund ensures that administrative, patient care, and research reporting procedures are carried out in conformity with NIH, UC Davis, and VA policies. In addition, he sets goals and standards for the CTSC, encourages investigators to utilize the CTSC, and fosters collaborations between clinical and basic science investigators. He serves on several Clinical and Translational Science Awards (CTSA) consortium committees and was a co-chair for the CTSA Consortium Oversight Committee (2006-2008). Dr. Berglund's research focus is in the area of lipoprotein metabolism and cardiovascular disease, and his research is funded by the National Heart, Lung, and Blood Institute. He has published more than 190 peer-reviewed papers and is a member of the editorial board of seven journals, including *Arteriosclerosis, Thrombosis and Vascular Biology*, the *Journal of Clinical Endocrinology and Metabolism*, and *Clinical and Translational Science*. He serves on numerous advisory boards and is a member of the American Heart Association Peer Review Committee and of the Clinical Guidelines subcommittee of the Endocrine Society, and he serves as chair of the NIH AIDS, Clinical Research and Epidemiology Study Section.

NICHOLAS P. BIGELOW is the Lee A. DuBridge Professor of Physics and Optics, the chair of the Department of Physics and Astronomy, and a senior scientist at the Laboratory for Laser Energetics at the University of Rochester. Dr. Bigelow's research interests are in the areas of quantum optics and quantum physics. His recent work has focused on the creation and study of ultracold quantum gases, the manipulation and control of atomic motion using light pressure forces, the laser cooling and trapping of atoms and molecules, Bose-Einstein condensation, and the basic quantum nature of the atom-photon interaction. Prior to joining the faculty at the University of Rochester, Dr. Bigelow was a member of the technical staff at AT&T Bell Laboratories. He then went to the Ecole Normale Superieure in Paris, France, where he worked in the Laboratoire Kastler-Brossel. In addition to receiving numerous awards, including a Young Investigator Award from the National Science Foundation (NSF) and a Packard Foundation Fellowship, Dr. Bigelow was the chair of the Fundamental Physics Discipline Working Group in the NASA Microgravity Physics Program. He received his B.S. in engineering physics and in electrical engineering from Lehigh University, and his M.S. and Ph.D. in physics from Cornell University. Dr. Bigelow has served on numerous advisory committees for organizations including the National Research Council, NASA, NSF, and the Department of Energy (DOE).

LEONARD H. CAVENY is an aerospace consultant and former director of science and technology for the Ballistic Missile Defense Organization. His previous experience also includes service as the deputy director of innovative science and technology for the Strategic Defense Initiative Organization, staff specialist for the Office of the Deputy Undersecretary for Research and Advanced Technology for the Department of Defense (DOD), and program manager for energy conversion for the Air Force Office of Scientific Research (AFOSR). From 1969 to 1980, as a senior professional staff of Princeton University's Aerospace and Mechanical Sciences Department, he guided graduate student research and served as principal investigator. Dr. Caveny's expertise includes solid rocket propulsion, aerothermochemistry flight experiments, electric propulsion, space solar power, diagnostics of reacting flows, combustion, propellants, refractory materials, and aeroacoustics. He earned his B.S. and M.S. in mechanical engineering from the Georgia Institute of Technology and his Ph.D. in mechanical engineering from the University of Alabama. He is a fellow of the American Institute of Aeronautics and Astronautics (AIAA). He served on the NRC's Committee for the Review of NASA's Pioneering Revolutionary Technology Program and as chair of the NRC Panel to Review Air Force Office of Scientific Research Proposals in Propulsion.

VIJAY K. DHIR is a professor and the dean of the Henry Samueli School of Engineering and Applied Science at the University of California, Los Angeles (UCLA). He was previously the chair of the UCLA Department of Mechanical and Aerospace Engineering. His research focuses on two-phase heat transfer, boiling and condensation, thermal and hydrodynamic stability, thermal hydraulics of nuclear reactors, microgravity heat transfer, and soil remediation. In addition to his work at UCLA, for the past 30 years Dr. Dhir has been a consultant for numerous organizations, including General Electric Corporation, Rockwell International, the Nuclear Regulatory Commission, LANL, and the Brookhaven National Laboratory. He was elected to the National Academy of Engineering (NAE) for his work on boiling heat transfer and nuclear reactor thermal hydraulics and safety. Dr. Dhir is also a fellow of the American Society of Mechanical Engineers (ASME) and the American Nuclear Society, a recipient of ASME's Heat Transfer Memorial Award, and the senior technical editor of ASME's *Journal of Heat Transfer*. Since 1999, a team of researchers led by Dr. Dhir has been taking part in a NASA research program to examine the effects of boiling in space. He received his B.S. in mechanical engineering from Punjab University in India, his M.Tech. in mechanical engineering from the Indian Institute of Technology, and his Ph.D. in mechanical engineering from the University of Kentucky.

JOEL E. DIMSDALE is a distinguished professor emeritus and research professor of psychiatry at the University of California, San Diego (UCSD), School of Medicine. Dr. Dimsdale's major research interests include sympathetic nervous system physiology as it relates to stress, blood pressure, and sleep; cultural factors in illness; and quality of life; his clinical subspecialty is consultation psychiatry. He is an active investigator, a former career awardee of the American Heart Association, and a past president of the Academy of Behavioral Medicine Research, the American Psychosomatic Society, and the Society of Behavioral Medicine. Dr. Dimsdale serves on numerous editorial boards and is editor-in-chief emeritus of *Psychosomatic Medicine*, a previous guest editor of *Circulation*, and former chair of the Sleep Research Society's Committee on Research. He has been a consultant to the President's Commission on Mental Health and the Institute of Medicine (IOM) and is a long-time reviewer for NIH. Dr. Dimsdale is the former chair of the UCSD Academic Senate. He heads the Translational Research Scholars Program for UCSD's Clinical and Translational Research Institute. Dr. Dimsdale received his B.A. in biology from Carleton College and his M.A. in sociology and M.D. from Stanford University. In 1980 he served on the advisory committee and was vice chair of the clinical panel for the IOM Conference on Bio-behavioral Approaches to Sudden Death.

NIKOLAOS A. GATSONIS is a professor in the Mechanical Engineering Department and the director of the Aerospace Engineering Program at Worcester Polytechnic Institute (WPI). From 1991 to 1993 he was a postdoctoral fellow at the Space Department of the Johns Hopkins University (JHU) Applied Physics Laboratory (APL). His research interests include simulation methods and modeling of macro- to nanoscale fluid and plasma transport processes and development of plasma diagnostics. Dr. Gatsonis's research in spacecraft/environment interactions, spacecraft electric propulsion, and micropropulsion involved participation in several spaceflight and ground-based experiments. His research has been supported by AFOSR, JHU APL, NASA, and NSF and through industrial collaborations. In addition to receiving numerous teaching awards, Dr. Gatsonis received the WPI Trustees Award for Outstanding Research and Creative Scholarship (2004) and the George I. Alden Chair in Engineering (2007-2010).

He was an associate editor of the AIAA *Journal of Spacecraft and Rockets* (2003-2006), a member of the AIAA Electric Propulsion Technical Committee (1998-2003), and a member of the AIAA Space Science Technical Committee (1992-1996). He received his undergraduate degree in physics from the Aristotle University of Thessaloniki in Greece, an M.S. in atmospheric science from the University of Michigan, and M.S. and Ph.D. degrees in the Department of Aeronautics and Astronautics of the Massachusetts Institute of Technology (MIT).

SIMON GILROY is a professor of botany in the Botany Department at the University of Wisconsin-Madison. Dr. Gilroy's research utilizes a multidisciplinary approach to study the interaction of environmental sensing and development in plants. One area of focus in his work is on understanding the molecules involved in the signals that allow plants to monitor and adapt to their environment—specifically, how these signals are perceived and translated to the development and control of a plant's growth. Dr. Gilroy and his team have investigated the cellular basis for gravity and mechano-signaling in the growing root and, in one approach, have mapped the sensory cells in the root using laser ablation and investigated the signaling events in these cells in response to gravity and touch stimulation. Another major project is focused on defining the signaling pathways responsible for plant hormone action. Among his recent publications in this field is *Plant Tropisms* (edited with P.H. Masson), a comprehensive review of the current state of knowledge on the molecular and cell biological processes that govern plant tropisms. Dr. Gilroy received his Ph.D. in botany from the University of Edinburgh. He served as a board member of the American Society for Gravitational and Space Biology (ASGSB).

BENJAMIN D. LEVINE is a professor of medicine and cardiology and holds a distinguished professorship in exercise science at the University of Texas Southwestern (UT Southwestern) Medical Center at Dallas. He is the director for the Institute for Exercise and Environmental Medicine (IEEM) at Texas Health Presbyterian Dallas where he also holds the S. Finley Ewing Jr. Chair for Wellness and the Harry S. Moss Heart Chair for Cardiovascular Research. Dr. Levine founded the IEEM in 1992; it has become one of the premier laboratories in the world for the study of human integrative physiology. His global research interests center on the adaptive capacity of the circulation in response to exercise training, deconditioning, aging, and environmental stimuli such as spaceflight and high altitude. He is a fellow of the American College of Cardiology and of the American College of Sports Medicine and is on the board of trustees/board of directors of the ACSM (for which he is currently vice president), the American Autonomic Society, and the International Hypoxia Symposium. A Henry Luce Foundation and Fulbright Scholar, he received the Peter van Handel Award from the United States Olympic Committee (for outstanding research), the Research Award from the Wilderness Medical Society, the Honor Award from the Texas Chapter of ACSM, and the Citation Award from the National ACSM. He was elected to the Association of University Cardiologists, received the Michael J. Joyner International Teaching Award from the Danish Cardiovascular Research Academy, and has been selected as one of the "Best Doctors" for cardiovascular medicine in Dallas and America by his peers. Dr. Levine has an extensive background in space medicine, serving as a co-investigator on four Spacelab missions (Spacelab Life Sciences [SLS]-1, SLS-2, D-2, and Neurolab) and the MIR space station; he is currently the PI of a large, cardiovascular experiment on the ISS called the ICV, or Integrated Cardiovascular experiment. He has completed multiple bed-rest studies with a long, sustained track record of funding by NASA and the National Space Biomedical Research Institute (NSBRI), for which he became team leader of the Cardiovascular Section in 2007. His many other leadership roles for NASA and NSBRI have included serving on the first Board of Scientific Counselors for NSBRI, directing the Cardiovascular Unit of the UT Southwestern NASA SCORT in integrative physiology, and advising NASA's flight surgeons on cardiovascular medical issues. Dr. Levine earned his B.A. magna cum laude in human biology from Brown University and his M.D. from Harvard Medical School. He completed his internship and residency in internal medicine at Stanford University Medical Center, followed by a cardiology fellowship at UT Southwestern where he trained under the renowned gravitational physiologist C. Gunnar Blomqivst.

RODOLFO R. LLINAS* is the Thomas and Suzanne Murphy Professor of Neuroscience and chair of the Department of Physiology and Neuroscience at New York University Medical Center. His research pertains mostly to neuroscience from the molecular to the cognitive level. Dr. Llinas focuses on the intrinsic electrophysiological properties of mammalian neurons in vitro. In particular, he studies the ionic channels that generate some of the sodium and calcium currents responsible for the electrophysiological properties of neurons and their distribution in different cell types. Dr. Llinas also looks at the role of calcium conductance in synaptic transmissions and at the concept of calcium microdomains; examines the cerebellar control of movement and thalamocortical connectivity; and is mapping the human brain using noninvasive magnetoencephalography. He received his M.D. from the Universidad Javeriana in Bogotá, Colombia, and his Ph.D. in neuroscience from the Australian National University in Canberra. Dr. Llinas is a member of the National Academy of Sciences (NAS). He served on the NRC's U.S. National Committee for the International Brain Research Organization and on the steering group for a Workshop on Bionics and Space Exploration.

KATHRYN V. LOGAN is the director of the Center for Multifunctional Aerospace Materials and the Samuel P. Langley Professor in the Department of Materials Science and Engineering at the Virginia Polytechnic Institute and State University. Her research interests are advanced synthesis and processing, design of materials, high-temperature solid-state diffusion, refractory material development, analytical materials characterization, and mechanical properties of materials. At the Center, Dr. Logan is responsible for high-performance, multifunction aerospace materials research and has overseen the development of a variety of new materials and structures. She is interested in creating unique materials that help in the human exploration of space. In addition to her materials work, she and her students are building a large radio-frequency induction press that will be capable of forming large-surface-area materials for space exploration programs. Once complete, this apparatus will be able to form unique components and structures not yet possible using standard techniques. Dr. Logan is a fellow and past president of the American Ceramic Society and the National Institute of Ceramic Engineers and is on the external advisory board for Clemson University's Department of Materials Science and Engineering. She has served on the NRC's Board on Army Science and Technology.

PHILIPPA MARRACK† is an investigator at the Howard Hughes Medical Institute, a senior faculty member in the Integrated Department of Immunology at National Jewish Health, and a professor in the Department of Biochemistry and Molecular Biology and the Department of Immunology and Medicine at the University of Colorado Health Sciences Center in Denver. Dr. Marrack's research interests include the creation, specificity, survival, and activation of T cells; cellular and molecular immunology; microbial pathogenesis; mammalian development; cell biology; and pathogenicity. She did much of the pioneering research on T cells, including the discovery that T cells have receptors to distinguish between dangerous microbes and a molecule called MHC. Dr. Marrack's current research focuses on how the body realizes that it has been injected with alum, a precipitate of aluminum salts. Her work is partially supported by funds from NIH and by fellowships from the Leukemia and Lymphoma Society. Dr. Marrack earned her B.A. in biochemistry and her Ph.D. in biological sciences from Cambridge University. In addition her membership in the National Academy of Sciences and the Institute of Medicine, she is a member of the Royal Society of the United Kingdom, the American Academy of Arts and Sciences, the American Association of Immunology, and the British Society for Immunology.

GABOR A. SOMORJAI is a professor in the Department of Chemistry at the University of California, Berkeley, and a faculty senior scientist in the Materials Sciences Division and a group leader of the Surface Science and Catalysis Program at the Lawrence Berkeley National Laboratory. Dr. Somorjai's research interests are in the field of surface science. His group is studying the structure, bonding, and reactivity at solid surfaces on the molecular scale; this knowledge then contributes to the understanding of macroscopic surface phenomena, adsorption, heterogeneous catalysis, and biocompatibility on the molecular level. To this end, he also develops instruments

*Rodolfo R. Llinas was a member of the committee through mid-December 2009.

†Philippa Marrack was a member of the committee through mid-May 2010.

for nanoscale characterization of surfaces, including sum frequency generation surface vibrational spectroscopy, high-pressure scanning tunneling microscopy, and high-pressure x-ray photoelectron spectroscopy. Dr. Somorjai received his Ph.D. in chemistry from the University of California, Berkeley. He is a member of the National Academy of Sciences.

CHARLES M. TIPTON is professor emeritus of physiology at the University of Arizona. He retired after 35 years of directing an exercise physiology laboratory that employed animal models to investigate mechanisms associated with acute and chronic exercise; the laboratory was continuously supported by federal, state, and private funds, including support from NIH and NASA. During his career, Dr. Tipton held appointments or joint appointments with departments of physical education, physiology and biophysics, biomedical engineering, orthopaedic surgery, and surgery, as well as exercise and sport sciences. In addition, he was a visiting senior scientist at the NASA Ames Research Center. Dr. Tipton is a former president of the American College of Sports Medicine, editor of *Medicine and Science in Sports and Exercise*, associate editor of the *Journal of Applied Physiology*, and Councilor of the American Physiological Society (APS); he received honor awards for research both from the American College of Sports Medicine and from the Environmental and Exercise Section of APS, and he received the Founders Award from the American Society for Gravitational and Space Biology. Besides being chair of the NIH Applied Physiology and Bioengineering Study Section, he has served on numerous space-related panels, including the NASA Review Panel on Space Medicine and Countermeasures, the NASA-IDI Cardiopulmonary Physiology Review Panel, and AIBS panels for microgravity research, and on the NASA-Bion Biospecimen Peer Review Panel. Currently, he is a member of External Advisory Committee to the National Space Biomedical Research Institute. After receiving his B.S. in physical education from Springfield College and an M.S. in physical education from the University of Illinois, Urbana-Champaign, he taught science, biology, and physical education in select high schools in Illinois. Later, he returned to the University of Illinois and received his Ph.D. in physiology with minors in biochemistry and anatomy.

JOSE L. TORERO is the Building Research Establishment (BRE)/Royal Academy of Engineering Professor of Fire Safety Engineering, the director of the BRE Centre for Fire Safety Engineering, and the head of the Institute for Infrastructure and Environment at the University of Edinburgh, Scotland. Prior to taking the helm at the Centre, Dr. Torero was an associate professor in the Department of Fire Protection Engineering and an affiliate associate professor in the Department of Aerospace Engineering at the University of Maryland. He is a fellow of the U.K. Royal Academy of Engineering and Royal Society of Edinburgh. His research is primarily in the areas of fire dynamics, smoke detection and management, protection and suppression systems, fire-induced skin burns, and the behavior of structures in the event of a fire—in particular, fire behavior in complex environments like spacecraft. Dr. Torero is a member of numerous organizations, including the International Association for Fire Safety Science for which he serves as vice chair and the Fire Safety Committee of the International Council for Tall Buildings and Urban Habitat for which he serves as chair. He also served on the AIAA Microgravity and Space Processes Technical Committee, the ASME K-11 Fire and Combustion Committee, and NASA's Mars or Bust and Fire Safety Committee. He is the editor-in-chief of *Fire Safety Journal*, associate editor of *Combustion Science and Technology*, and a member of the editorial boards of *Progress in Energy and Combustion Science*, *Fire Technology*, and *Fire Science and Technology*. His academic distinctions include the Society of Fire Protection Engineers' Arthur B. Guise Medal for eminent contributions to fire science and the Tam Dalyell Medal for excellence in engaging the public with science. He received his B.Sc. from the Catholic University of Peru and his M.Sc. and Ph.D. from the University of California, Berkeley.

ROBERT WEGENG is a chief engineer in the Energy and Efficiency Division at Battelle's Pacific Northwest National Laboratory. During his more than two decades of employment with Battelle, Mr. Wegeng has contributed as an engineer and project manager to projects supported by the federal government—for the Defense Advanced Research Projects Agency, DOD, DOE, and NASA—and by commercial organizations in the energy, aerospace, and chemical process industries. He was vice chair of the 2nd and 4th International Conferences on Microreaction Technology, a conference held jointly by Battelle, the Institute of Microtechnology, the German Society for Chemical Apparatus, Chemical Engineering and Biotechnology, and the American Institute for Chemical Engineers. Mr. Wegeng has been involved in numerous projects dealing with alternative energy and with human exploration architecture, including in situ resource utilization (ISRU). He has written in the latter area, outlining microchemical and thermal systems for ISRU, and on a microchannel in situ propellant system as an enabling technology for Mars architecture concepts. Mr. Wegeng's experience in solar thermochemical fuel production demonstrates a strong foundation in chemistry, which he applies to his technical papers on ISRU. In collaboration with researchers at NASA, he has developed a means to keep robotic systems warm enough to operate in harsh extraterrestrial environments. Mr. Wegeng is the co-recipient of two R&D 100 Awards and he has registered 88 patents (U.S. and foreign).

GAYLE E. WOLOSCHAK is a professor of radiation oncology and of radiology and cell and molecular biology at the Northwestern University Feinberg School of Medicine. Her research is focused on nanocomposites and molecular consequences of radiation exposure. Dr. Woloschak's work is oriented toward function use of nanocomposites for intracellular manipulation, imaging, and gene silencing. Her work on motor neuron disease is designed to lead to an understanding of the molecular basis for the combined abnormalities from a molecular-cellular perspective. She received her Ph.D. in medical sciences (microbiology) from the Medical College of Ohio. Dr. Woloschak served on the NRC Committee on Evaluation of Radiation Shielding for Space Exploration and the Committee to Assess Potential Health Effects from Exposures to PAVE PAWS Low-Level Phased-Array Radiofrequency Energy. She has served on review panels for NIH, NASA, DOE, DOD, and other organizations and has chaired several international workshops on biological and medical applications of microprobes.

ANIMAL AND HUMAN BIOLOGY PANEL

KENNETH M. BALDWIN, *Chair*, is a professor of physiology and biophysics at the University of California, Irvine, and School of Medicine. Dr. Baldwin's laboratory research focuses on the impact of activity patterns or exercise regimens on the biochemical and physiologic properties of cardiac and skeletal muscle in mammals. His research has demonstrated that muscle systems are in a dynamic state of biological adaptation, referred to as plasticity. Various subcellular components and proteins can be changed both qualitatively and quantitatively in accordance with how the muscle system is continually stressed (or unstressed) by activities such as chronic locomotion, muscle loading, and muscle unloading such as during chronic bed rest. Of primary interest is how the effects of these various activities are translated into biochemical events that lead to alterations in protein expression in muscle. Because the role of myosin is that of both a structural and regulatory protein involved in muscle contraction, work in Dr. Baldwin's laboratory focuses on factors that influence the expression of different isoforms of myosin in both cardiac and skeletal muscle. As a corollary to these experiments, Dr. Baldwin's group, in conjunction with NASA, recently sent rats on several space shuttle missions to study the effects of weightlessness on skeletal muscle. Dr. Baldwin was appointed chair of the NASA Life and Microgravity Sciences Advisory Committee in 1998 and was appointed to the NASA Advisory Council in 1999. His academic distinctions include the 1998 APS Edward Adolph Award, the 1998 ACSM Southwest Chapter Achievement Award, and the 1999 NASA Public Service Award. Recently, he received the ACSM prestigious Honor Award for his research in muscle plasticity. He received his Ph.D. from the University of Iowa.

FRANÇOIS M. ABBOUD is the Edith King Pearson Chair in Cardiovascular Research, a professor of medicine and molecular physiology and biophysics, the director of the Cardiovascular Research Center, and the associate vice president for research at the University of Iowa. He was chair of the Department of Internal Medicine from 1976 through 2002. His NIH Program Project Grant on Integrative Neurobiology of Cardiovascular Regulation has been supported since 1971. His human studies have focused on the integrated control of sympathetic activity in physiological and pathological states (e.g., sleep apnea and hypertension). Dr. Abboud has received a number of awards, including the ASPET (American Society for Pharmacology and Experimental Therapeutics) Award for Experimental Therapeutics; the Wiggers Award, the Ludwig Award, and the Walter B. Cannon Award Lectureship of the American Physiological Society; the Research Achievement Award, the Gold Heart Award, and the Distinguished Scientist Award of the American Heart Association; and the CIBA Award and Medal for Hypertension Research of the Council for High Blood Pressure Research. Most recently Dr. Abboud received the Kober Medal of the Association of American Physicians. He was the editor-in-chief of *Circulation Research* and co-editor of the *Handbook of Physiology on Peripheral Circulation and Organ Blood Flow*. He served on the Advisory Council of the National Heart, Lung, and Blood Institute and is on the Editorial Advisory Board of *Clinical Autonomic Research* and on the NRC's Sleep Medicine and Research Committee. He is a member of the Institute of Medicine and of the American Academy of Arts and Sciences.

PETER R. CAVANAGH is a professor and the endowed chair in Women's Sports Medicine and Lifetime Fitness at the University of Washington School of Medicine, where he is building a research and education program on the bone and joint health of active women. His other research interests include lower-extremity biomechanics, athletic footwear, bone loss during long-duration spaceflight, bone health in women on Earth, and the foot complications of diabetes. Dr. Cavanagh is the principal investigator of an experiment that was recently completed onboard the ISS. His latest book, *Bone Loss During Spaceflight*, was published in 2007. He has authored, co-authored, or edited more than 400 papers, abstracts, chapters, and books and has mentored more than 70 graduate students and postdoctoral fellows. He is a member of the American College of Sports Medicine, the American Diabetes Association, the American Orthopedic Foot and Ankle Society, the American Society of Biomechanics, and the International Society of Biomechanics. His more recent honors include the 2007 Laurence R. Young Space Biomedical Research Award from NASA/NSBRI and the 2009 Edward James Olmos Award for Advocacy in Amputation Prevention. After completing undergraduate studies at Loughborough College at the University of Nottingham, United Kingdom, Dr. Cavanagh received his Ph.D. in anatomy and human biomechanics at the Royal Free Medical School at the University of London. He also received a D.Sc. degree from the Faculty of Medicine at the University of London.

V. REGGIE EDGERTON is a professor of physiological science in the Department of Physiological Sciences at the University of California, Los Angeles. Previously he served as the director of the Brain Research Institute. His research interests include neural control of movement and neuromuscular plasticity. Dr. Edgerton's laboratory focuses on two main research questions: how and to what extent the nervous system controls protein expression in skeletal muscle fibers, and how the neural networks in the lumbar spinal cord of mammals, including humans, control stepping, including the question of how this stepping pattern becomes modified by chronically imposing specific motor tasks on the limbs after complete spinal cord injury. He is also studying how to develop robotic devices that can help laboratory animals and humans with neuromuscular deficits to walk. Such a device is being developed for use by crew members in maintaining a critical level of control of locomotion in variable gravitational environments. Dr. Edgerton is a member of the American Physiological Society. He received his Ph.D. in exercise physiology from Michigan State University.

DONNA MURASKO is the dean of the College of Arts and Sciences and a professor of biology and a professor of microbiology and immunology in the College of Medicine at Drexel University. She also served as vice provost and chair of the Department of Microbiology and Immunology at Drexel. Although Dr. Murasko's initial training was in tumor immunology, the focus of her research for more than 20 years has been the changes that occur in immune response with increasing age. Utilizing both mouse models and human samples, she has focused on the immune response to viruses. She is a member of numerous professional societies and a fellow of the American Association for the Advancement of Science (AAAS). She received her B.A. in bacteriology from the Douglass College of Rutgers and her Ph.D. in microbiology from the M.S. Hershey Medical Center at Pennsylvania State University.

JOHN T. POTTS, JR., is the Jackson Distinguished Professor of Clinical Medicine at Harvard Medical School. He served as the director of research and physician-in-chief emeritus at Massachusetts General Hospital (MGH). He completed his internship and residency at MGH from 1957 to 1959 before moving on to NIH. Dr. Potts remained at NIH from 1959 to 1968, when he returned to MGH as the chief of endocrinology. He also served as the chair of the Department of Medicine and physician-in-chief. In his role as director of research, Dr. Potts was responsible for developing policies and strategies for preserving and strengthening the extensive scientific research effort at

MGH. Dr. Potts is a director of ReceptorBase, Inc., and Zeltiq Aesthetics; a founder of Radius Health, Inc.; and a member of the scientific advisory boards of MPMP Capital and HealthCare Ventures, as well as the medical advisory board of Cell Genesys. Dr. Potts received his B.A. from LaSalle College and his M.D. from the University of Pennsylvania. He is a member of the National Academy of Sciences, the Institute of Medicine, and the American Academy of Arts and Sciences. He is the author or co-author of more than 500 scientific publications. Dr. Potts has served as a member of the NRC Committee on Non-heart-Beating Organ Transplantation II: The Scientific and Ethical Basis for Practice and Protocols, the Project on Medical and Ethical Issues in Maintaining the Viability of Organs for Transplantation (for which he was the principal investigator), and the Board on Health Sciences Policy.

APRIL E. RONCA is a professor of obstetrics and gynecology and (jointly) neurobiology and anatomy and molecular medicine/translational science at Wake Forest University School of Medicine. She is also the director of the Women's Health Center of Excellence Research Program. Dr. Ronca previously spent 6 years as the director of the Developmental Neurobiology and Behavior Laboratory at the NASA Ames Research Center. The main focus of Dr. Ronca's research is mammalian pregnancy, birth, and the transition from prenatal to postnatal life, with an emphasis on sensory development and neurodevelopmental disorders. She was an investigator on two NASA space shuttle experiments examining gravitational influences on pregnancy and prenatal development. Dr. Ronca has received numerous research awards from NIH and NASA and has published more than 60 papers and book chapters. She is the 2004 recipient of the NASA Exceptional Achievement Medal and the Thora Halstead Young Investigator's Award from the American Society for Gravitational and Space Biology (ASGSB). Dr. Ronca serves on the ASGSB board of directors, the editorial boards for ASGSB and *Reproductive Biology and Endocrinology* and is a member of the NIH Biobehavioral Regulation Learning and Ethology Study Section. She has served on numerous NASA advisory and review panels. Dr. Ronca received her B.S. in psychology and her Ph.D. in neuroscience from the Ohio State University as a presidential fellow.

CHARLES M. TIPTON. *See the committee listing above.*

CHARLES H. TURNER[‡] was the Chancellor's Professor of Biomedical Engineering and Orthopedic Surgery at Indiana University-Purdue University, Indianapolis (IUPUI). Dr. Turner was also the director of orthopedic research in the Department of Orthopedic Surgery and the associate chair for biomedical engineering at IUPUI. Prior to joining the faculty at Indiana University in 1991, he spent 4 years with the Osteoporosis Research Center at Creighton University. The main focus of Dr. Turner's research was treatments for the bone disease osteoporosis. His recent research focused on molecular genetics using transgenic and congenic mice to identify new ways to make bone stronger. He won numerous awards for his research in musculoskeletal biomechanics and bone biology, including grants from NIH, the Fuller Albright Award from the American Society for Bone and Mineral Research, and the Outstanding Young Investigator Awards from the Whitaker Foundation and the Health Future Foundation. In 2002, Dr. Turner was a fellow of the American Institute of Medical and Biological Engineers. He served as a consultant in biomechanics and orthopedic science for NIH, NSF, NASA, the Food and Drug Administration, the Canadian Institutes of Health Research, the Swiss National Science Foundation, the Austrian Science Fund, the Israel Science Foundation, and the Wellcome Trust (England). He published more than 400 scientific papers and abstracts on topics in biomechanics, bone biology, and orthopedic science, and he has given more than 100 invited presentations on research topics in musculoskeletal biomechanics worldwide. Dr. Turner received a B.S. in mechanical engineering from Texas Tech University and his Ph.D. in biomedical engineering from Tulane University.

JOHN B. WEST is a distinguished professor of medicine and physiology in the School of Medicine, University of California, San Diego. Dr. West's research interests are in respiratory physiology, including environmental and exercise physiology. He has had a long association with NASA, spending a sabbatical year at the NASA Ames Research Center in 1967-1968 and subsequently being a principal investigator of a large project to measure pulmonary function in astronauts during spaceflight. A number of his experiments were carried out on SpaceLab in the

[‡]Deceased July 2010.

1990s. Dr. West has long had an interest in high-altitude physiology and led the 1981 American Medical Research Expedition to Everest, during which the first measurements of human physiology were obtained on the summit. He is the author or editor of 22 books and more than 400 publications. Dr. West received his M.D. from Adelaide University, Australia, in 1958 and his Ph.D. from London University, United Kingdom, in 1960. Dr. West was elected to the Institute of Medicine in 2004. He has served on numerous NRC or IOM committees, including the Committee on Advanced Space Technology and the Committee on Space Biology and Medicine. He is currently a member of the NRC Committee on Aerospace Medicine and the Medicine of Extreme Environments.

APPLIED PHYSICAL SCIENCES PANEL

PETER W. VOORHEES, *Chair*, is the Frank C. Engelhart Professor and chair of the Department of Materials Science and Engineering at Northwestern University. He was a member of the Metallurgy Division at the National Institute of Standards and Technology (NIST) until joining the Department of Materials Science and Engineering at Northwestern University in 1988. He has published more than 160 papers in the area of the thermodynamics and kinetics of phase transformations. He has flown experiments investigating the dynamics of coarsening processes on the space shuttle and, more recently, on the ISS. Professor Voorhees's research interests include the dynamics of coarsening processes, nanowire growth, solid-oxide fuel cells, and the three-dimensional morphology of interfaces in materials. He received both his B.S. and his Ph.D. in materials engineering from Rensselaer Polytechnic Institute (RPI). Dr. Voorhees last served as the chair of the NRC Committee on Microgravity Research (2000-2003), and he was a member of the Space Studies Board (1998-2003).

NIKOLAOS A. GATSONIS. *See the committee listing above.*

RICHARD T. LAHEY, JR., is the Edward E. Hood Professor Emeritus of Engineering at Rensselaer Polytechnic Institute and a founding director of PJM, LLC, the largest supplier and market operator of wholesale electricity in the United States. He was formerly the director of the Center for Multiphase Research and the dean of engineering at RPI. He also served as chair of the Department of Nuclear Engineering and Science and the faculty senate at RPI. Prior to joining Rensselaer's faculty in 1975, he held several technical and managerial positions with the General Electric Company, including overall responsibility for all domestic and foreign research and development programs associated with boiling water nuclear reactor thermal-hydraulic and safety technology. His research has focused on multiphase flow and heat transfer technology. He has received numerous honors and awards, including the E.O. Lawrence Memorial Award of the Department of Energy, the ANS Seaborg Medal, the AIChE Kern Award, and the Glenn Murphy Award of the American Society of Engineering Education. He was elected to the National Academy of Engineering for contributions to the field of multiphase flow and heat transfer and nuclear reactor safety technology. He received his B.S. in marine engineering from the U.S. Merchant Marine Academy, his M.S. in mechanical engineering from RPI, his M.E. in engineering mechanics from Columbia University, and his Ph.D. in mechanical engineering from Stanford University. Dr. Lahey served as a member of the NRC Committee on Microgravity Research (1997-2000), the Electric Power/Energy Systems Engineering Peer Committee (1999-2002), and the Committee on Safety and Security of Commercial Spent Nuclear Fuel Storage (2004-2005).

RICHARD M. LUEPTOW is a professor of mechanical engineering, the senior associate dean of the McCormick School of Engineering and Applied Science, co-director of the Master of Product Development Program, and formerly Charles Deering McCormick Professor of Teaching Excellence at Northwestern University. Prior to joining the Northwestern faculty, he was a senior research engineer at the Haemonetics Corporation in Braintree, Massachusetts. Dr. Lueptow's research interests and expertise range from fundamental flow physics, to water purification on manned spacecraft, to planetary acoustics. He has studied rotating filtration, dry granular flows and granular slurries, the fundamental physics of circular Couette flow, filtration systems, fire suppression sprays, and acoustic gas composition sensors. Dr. Lueptow has received numerous awards, including the NIH National Research Service Award (1978-1980), the American Society for Engineering Education Dow Outstanding Young Faculty Award (1993), the William M. Carey Award from the National Fire Protection Research Foundation (2002,

2003), and the Space Act Award from the NASA Inventions and Contributions Board for Rotating Reverse Osmosis (2004). Dr. Lueptow received his B.S. in engineering from Michigan Technological University and his Sc.D. in mechanical engineering from MIT. He has frequent visiting appointments at the University of Marseille and is a fellow of the American Physical Society.

JOHN J. MOORE is a materials scientist who holds the position of Trustees' Professor and head of the Department of Metallurgical and Materials Engineering at the Colorado School of Mines (CSM). Dr. Moore is also the director of the interdisciplinary graduate program in materials science and the director of the Advanced Coatings and Surface Engineering Laboratory/Advanced Combustion Synthesis and Engineering Laboratory (ACSEL) at CSM. ACSEL is a national and international leader in research on advanced coatings, surface engineering, and advanced materials synthesis and processing. Prior appointments held by Dr. Moore include those as head of the Department of Metallurgical and Materials Engineering at CSM; professor and head of the Department of Chemical and Materials Engineering at the University of Auckland, New Zealand; professor of metallurgical engineering at the University of Minnesota; and manager of industrial engineering and production control at Birmingham Aluminium Castings in the United Kingdom. Dr. Moore's research interests fall into two main areas: advanced coatings and surface engineering and self-propagating high-temperature (combustion) synthesis (SHS) of advanced materials. Both of these research areas are operated through ACSEL. The main objective of ACSEL is to perform fundamental research in advanced physical vapor deposition and chemical vapor deposition systems that will aid the thin films, coatings, and surface engineering industry, and the application of SHS in the development of new processes for net shaped metal matrix, ceramic matrix, and inetermetallic matrix composite materials. Dr. Moore has published more than 600 papers in international journals on materials processing and has been awarded 14 patents. He was awarded a B.Sc. in materials science and engineering from the University of Surrey, United Kingdom, and a Ph.D. and a D.Eng. in industrial metallurgy from the University of Birmingham, U.K. Dr. Moore is the chair of the Scientific Advisory Board of XsunX, a start-up photovoltaics company in Oregon; a member of the board of directors of Hazen Research, Inc., Golden, Colorado; and the chief scientific officer and company secretary of Advanced Materials Solutions, Inc., a high-tech materials company based in Golden, Colorado.

ELAINE S. ORAN is the senior scientist for reactive flow physics at the Naval Research Laboratory; an adjunct professor in the Department of Aerospace Engineering at the University of Michigan, Ann Arbor; and a visiting professor at the University of Leeds, United Kingdom. Dr. Oran designs numerical methods for simulating complex fluid dynamic processes and then uses these methods to solve a wide variety of scientific and engineering problems. Her recent research interests include combustion and propulsion, rarefied gases and microfluidics, fluid turbulence, materials engineering, high-performance computing and parallel architectures, computational science and numerical analysis, biophysical fluid dynamics, wave equations, and astrophysical phenomena such as supernova explosions. Dr. Oran is a member of the National Academy of Engineering; a fellow of the AIAA, the American Physical Society, and the Society for Industrial and Applied Mathematics; and a member of the Combustion Institute, the American Society of Mechanical Engineering, and the Society of Women Engineers. Her numerous awards include the Presidential Rank Award of Distinguished Senior Professional, the degree of Docteur Honoris Causa from Ecole Centrale de Lyon, and the Zeldovich Gold Medal from the Combustion Institute. Dr. Oran received her B.A. in chemistry and physics from Bryn Mawr College and her M.Ph. in physics and Ph.D. in engineering and applied sciences from Yale University. She is currently a member of the NRC's Aeronautics and Space Engineering Board and has served on many NAE committees, including the Aerospace Engineering Peer Committee, the Audit Committee (2005-2008), and the Bernard M. Gordon Prize Committee (2004-2006).

AMY L. RECHENMACHER is an assistant professor in the Sonny Astani Department of Civil and Environmental Engineering at the University of Southern California (USC). Prior to joining the faculty of USC, she was an assistant professor at the Johns Hopkins University, where she taught soil mechanics and digital imaging geotechnical engineering courses and methods. She also worked briefly as a soil specialist on an archaeological excavation of the ancient Mut Temple ruins in Egypt. Dr. Rechenmacher studies the mechanics of dense granular flows, in particular the behavior of local flows, and micro- and mesoscale and thermodynamic aspects of granular material deformation in three contexts: in shear bands or in fractures (geometrically constrained zones of intense shear) in sands, in simulated fault gouge, and in idealized dense granular assemblies. A purpose of her work is to help researchers understand and quantify the thermodynamics of dense granular flows. Her pioneering applications of experimental imaging in soils have yielded some of the first-ever nondestructive characterizations of grain-scale motion in real granular materials. Dr. Rechenmacher is a member of two national engineering honor societies, Tau Beta Pi and Chi Epsilon, and of numerous professional engineering organizations, including the American Society of Civil Engineers; she is also a member of the American Geophysical Union, the Geo-Institute, and the International Association of Foundation Drilling. Dr. Rechenmacher received the 2008 NSF Early Career Award for her work in the kinematics of granular material behavior and geotechnical engineering applications. She earned her B.S. in civil engineering from Iowa State University, her M.A.S. in civil engineering from Cornell University, and her Ph.D. in geotechnical engineering from Northwestern University's McCormick School of Engineering.

JAMES S. T'IEN is the Leonard Case Jr. Professor of Engineering in the Department of Mechanical and Aerospace Engineering at Case Western Reserve University. Dr. T'ien's research interests are in the areas of combustion, propulsion, fire research, and chemically reacting flows. He served as the chief scientist in combustion at the National Center for Microgravity Research on Fluid and Combustion and at the National Center for Space Exploration on Fluid and Combustion (1998-2006). Current funding for his projects comes from NASA, NIST, and the Department of Homeland Security. He has received numerous awards, including a Daniel and Florence Guggenheim Fellowship in Jet Propulsion and a Public Service Medal from NASA (2000). In addition, he is an honorable member of the Combustion Institute's Chinese Taipei Section and a member of ASME and AIAA. He has served on numerous advisory committees and working groups, including NASA's Microgravity Combustion Science Discipline Working Group and AIAA's Microgravity Science and Space Processes Committee. Dr. T'ien received his B.S. in mechanical engineering from the National Taiwan University, his M.S. in engineering science from Purdue University, and his Ph.D. in aerospace and mechanical sciences from Princeton University. He served on the NRC Committee to Identify Innovative Research Needs to Foster Improved Fire Safety in the United States (2001-2003).

MARK M. WEISLOGEL is a professor in the Thermal and Fluid Sciences Group in the Maseeh College of Engineering and Computer Science at Portland State University (PSU). Dr. Weislogel has research experience from government and private institutions. While employed by NASA, he proposed and conducted experiments relating to microgravity fluid mechanics. This unique subtopic area within fluid mechanics provides significant challenges for designers of fluids management systems for aerospace applications as well as exciting learning opportunities for students. He continues to make extensive use of NASA ground-based low-gravity facilities such as drop towers and low-gravity aircraft and has completed experiments on the space shuttle, the Russian Mir space station, and the ISS. While in the private sector, Dr. Weislogel served as PI for applied research projects concerning high-performance heat transport systems, micrometeorite-safe space-based radiators, microscale cooling systems, emergency oxygen supply systems, and astronaut sleep stations. His research at PSU includes passive noncapillary cooling cycles for satellite thermal control and capillary fluidics at both micro- and macro-length scales. His teaching interests include heat transfer, fluid mechanics, and applied mathematics as it relates to these subjects. Dr. Weislogel has written more than 50 publications. He has B.S. and M.S. degrees from Washington State University, and he earned a Ph.D. from Northwestern University in 1996.

FUNDAMENTAL PHYSICS PANEL

ROBERT V. DUNCAN, *Chair*, is the vice chancellor for research at the University of Missouri. Prior to assuming his current position, Dr. Duncan served as a professor of physics and astronomy at the University of New Mexico (UNM), as a visiting associate on the physics faculty of the California Institute of Technology (Caltech), as a joint associate professor of electrical and computer engineering at UNM, and as the associate dean for research in the College of Arts and Sciences at UNM. His expertise is in low-temperature physics, and he has served as PI on a fundamental physics research program for NASA. As the director of the New Mexico Consortium's Institute in the

Los Alamos National Laboratory, he worked to fund major conferences and summer schools in quantitative biology, information science and technology, energy and environment, and astrophysics and cosmology. Dr. Duncan is a fellow and life member of the American Physical Society. He was named the Gordon and Betty Moore Distinguished Scholar in the Division of Physics, Mathematics, and Astronomy at Caltech in 2004 and has recently served as chair both for the American Physical Society's Topical Group on Instrumentation and Measurement and for its International Symposium on Quantum Fluids and Solids. He received his B.S. in physics from the Massachusetts Institute of Technology in 1982 and his Ph.D. in physics from the University of California, Santa Barbara, in 1988.

NICHOLAS P. BIGELOW. *See the committee listing above.*

PAUL M. CHAIKIN is the Silver Professor of Physics at New York University (NYU). Prior to joining the faculty of NYU, Dr. Chaikin was a professor at the University of California, Los Angeles; University de Paris-Sud, Orsay; the University of Pennsylvania; and Princeton University. He also worked as a research associate and consultant for Exxon Research and Engineering Company and the NEC Institute. He has published more than 300 articles in journals such as *Physics Review*, *Journal of Materials Science*, *Physical Review Letters*, *Journal of Physics, Nature,* and *Science*, and he co-authored the book *Principles of Condensed Matter Physics* with T.C. Lubensky (1995). Dr. Chaikin has served on the editorial boards of numerous refereed journals and has lectured at more than 200 meetings and at more than 200 universities and laboratories. He has also registered two patents. He received a Sloan Foundation Fellowship and a John Simon Guggenheim Foundation Fellowship. He is a member of the National Academy of Sciences and the American Academy of Arts and Sciences. He received his B.S. from Caltech in 1966 and his Ph.D. in physics from the University of Pennsylvania in 1971. Dr. Chaikin served as a member of the NRC Committee on Opportunities in High Magnetic Field Science (2003-2005) and the Committee on an Assessment of and Outlook for NSF's Material Research Laboratory Program (2005-2007), and he currently serves as a member of the Solid State Sciences Committee (2007-2010).

RONALD G. LARSON is the George Granger Brown Professor of Chemical Engineering and the former chair of the Chemical Engineering Department at the University of Michigan, Ann Arbor. Prior to joining the faculty of the University of Michigan, Dr. Larson worked for Bell Laboratories. He served as president of the Society of Rheology from 1998 to 1999 and is currently chair of the Division of Polymer Science in the American Physical Society. His research interests include complex fluids, polymers, fluid mechanics, surfactants, biomolecules, transport theory, rheology, instabilities, and constitutive theory. Dr. Larson has received numerous awards, including the Bingham Medal from the Society of Rheology (2002), the Alpha Chi Sigma Award from the American Institute of Chemical Engineers (2000), the Publication Award from the *Journal of Rheology* (1999), and the Excellence Award from the Chemical Engineering Department at the University of Michigan (1998). He earned his B.S. and Ph.D. in chemical engineering from the University of Minnesota in 1975 and 1980, respectively. Dr. Larson is a member of the National Academy of Engineering. In addition, he has served on the NRC Engineering Review Panel and on the Committee to Review Proposals for the 2007 State of Ohio Wright Centers of Innovation and the Research and Commercialization Program in Engineering and Physical Sciences.

W. CARL LINEBERGER is the E.U. Condon Distinguished Professor of Chemistry at the University of Colorado, Boulder, and a fellow at JILA, a joint University of Colorado-NIST facility. Prior to joining the faculty at the University of Colorado, Dr. Lineberger held various positions including that of research physicist at the U.S. Army Ballistic Research Laboratory and chair of the Joint Institute for Laboratory Astrophysics. His work is primarily experimental, using a wide variety of laser-based techniques to study the structure and reactivity of gas phase ions. Dr. Lineberger is a member of the National Academy of Sciences, the American Academy of Arts and Sciences, Sigma Xi, the American Chemical Society (ACS), and the Optical Society of America. He is a fellow of the American Physical Society and of AAAS. He has received numerous awards, including the ACS Peter Debye Award in Physical Chemistry, the ACS Irving Langmuir Prize in Chemical Physics, the Optical Society of America William F. Meggers Prize, and the Bomem-Michelson Prize. He is also on the editorial advisory board of *Chemical Physics Letters*. He received his B.E.E., M.S.E.E., and Ph.D. from the Georgia Institute of Technology. Dr. Lineberger has served on numerous NRC committees, and he is currently an active member of the NRC Report Review Committee.

RONALD WALSWORTH is a senior lecturer in the Department of Physics at Harvard University and a senior physicist at the Smithsonian Astrophysical Observatory. His research group pursues a wide range of experimental investigations, including the development of atomic clocks; precise tests of physical laws and symmetries; laser frequency combs as improved optical wavelength calibrators for astrophysical spectroscopy, with applications to exoplanet searches; sensitive magnetometry with nitrogen vacancy (NV) color centers in diamond; and the development of new bioimaging tools, with a focus on pulmonary physiology and brain science. Dr. Walsworth has received numerous awards, including the Francis M. Pipkin Award on precision measurements from the American Physical Society (2005). He received his B.S. in physics from Duke University and his Ph.D. in physics from Harvard University. Dr. Walsworth currently serves on the NRC's Committee on Atomic, Molecular, and Optical Sciences.

HUMAN BEHAVIOR AND MENTAL HEALTH PANEL

THOMAS J. BALKIN, *Chair*, is the chief of the Department of Behavioral Biology at the Walter Reed Army Institute of Research where he has served since 1985. Dr. Balkin first worked as a research psychologist in the Human Psychopharmacology Branch of the Department of Behavioral Biology and has been in his current position since 1995. He is also a co-director of the Sleep Disorders Center at Howard County (Maryland) General Hospital. He is the author of numerous articles and studies on sleep issues, including those involving alertness, fatigue, and performance. He has been affiliated with several sleep disorders centers and sleep centers since 1981 and has served on boards and committees focused on issues of sleep and performance for organizations such as the North Atlantic Treaty Organization, NIH, and other government agencies. Dr. Balkin received his B.S. from Syracuse University; his M.S. in experimental psychology from the State University of New York, College at Cortland; and his Ph.D. in experimental psychology from Bowling Green State University.

JOEL E. DIMSDALE. *See the committee listing above.*

NICK KANAS is a professor of psychiatry at the University of California, San Francisco. Dr. Kanas's research interests include the psychological interactions of people under stress and ways that they can cope better with stressors in their environment. For more than 16 years, he has studied astronauts living and working in space. He has been the PI of two large NASA-funded international studies involving the Mir space station and the ISS, and he is currently the PI of a NASA-funded study aimed at examining the effects of increased crew anatomy in space. In 1999, Dr. Kanas received the Aerospace Medical Association Raymond F. Longacre Award for Outstanding Accomplishment in the Psychological and Psychiatric Aspects of Aerospace Medicine, and in 2008 he received the International Academy of Astronautics (IAA) Life Sciences Award. In 2003 he received the J. Elliott Royer Award for excellence in academic psychiatry. Dr. Kanas is the co-author of the book *Space Psychology and Psychiatry*, which was the recipient of the 2004 IAA Life Sciences Book Award. He served as a member of the NRC Committee on Space Biology and Medicine Panel on Human Behavior.

GLORIA R. LEON, Professor Emerita of Psychology at the University of Minnesota, received her Ph.D. in mental health psychology from the University of Maryland. Her first academic appointment was in the Department of Psychology at Rutgers University, and in 1974 she was appointed assistant professor in the Department of Psychology at the University of Minnesota. At Minnesota, she was a member of both the clinical and personality doctoral programs, serving for 10 years as the director of the clinical psychology graduate program. She has conducted extensive research on the assessment of personality and behavioral functioning after traumatic situations including the Holocaust, Vietnam combat, disasters, and living in the Chernobyl area. She has studied polar expeditions with teams composed of single-gender, mixed-gender, and cross-national members and continues research in this area as an analog for space exploration. Dr. Leon was co-principal investigator on a NASA-funded program of research focused on developing more effective protective clothing for astronauts during extravehicular activity (EVA) using principles of physiological design, and technologies for more accurately monitoring the thermal status of astronauts during extended EVA. She has been a member of a number of NASA committees and workshops focused on behavioral health and human performance in space, NASA peer-review panels, the International Astronautics Association psychosocial committee, and program committees for various space-related congresses. She was a member of the external advisory committee of the National Space Biomedical Research Institute, Neurobehavioral and Psychosocial team (2004-2007). Dr. Leon is currently a member of the Institute of Medicine's Committee on Medicine in Extreme Environments and of the NASA Human Research Program's Behavioral Health and Performance Standing Review Panel.

LAWRENCE A. PALINKAS is the Albert G. and Frances Lomas Feldman Professor of Social Policy and Health at the University of Southern California and an adjunct professor of medicine and family and preventive medicine at the University of California, San Diego. A medical anthropologist, his primary areas of expertise lie in preventive medicine, cross-cultural medicine, and health services research. He has conducted extensive research on human adaptation to isolated, confined, and extreme environments, most notably in Antarctica and the high Arctic. Among his scholarly achievements are receiving the Antarctic Service Medal awarded by the National Science Foundation and the U.S. Navy in 1989, serving as the deputy chief officer of the Life Sciences Standing Committee on Antarctic Research in 2002, and serving as the chair of the NSBRI External Advisory Council, 2003-2005. He is an elected fellow of the American Anthropological Association and of the Society for Applied Anthropology and the author of more than 250 publications. Dr. Palinkas received his B.A. from the University of Chicago and both his M.A. and Ph.D. from UCSD. He served as a member of the NRC Committee on Space Biology and Medicine, as chair of the Panel on Human Behavior, and as a member of the Scientific Committee on Antarctic Research and the Committee on NASA's Bioastronautics Critical Path Roadmap.

MRIGANKA SUR, Massachusetts Institute of Technology.[§]

INTEGRATIVE AND TRANSLATIONAL RESEARCH FOR THE HUMAN SYSTEMS PANEL

JAMES A. PAWELCZYK, *Chair*, is an associate professor of physiology and kinesiology at Pennsylvania State University. Dr. Pawelczyk served as a payload specialist on STS-90 Neurolab. During the 16-day Spacelab flight, the seven-person crew aboard NASA space shuttle *Columbia* served as both experiment subjects and operators for 26 individual life sciences experiments focusing on the effects of microgravity on the brain and nervous system. Dr. Pawelczyk's primary research interests include the neural control of circulation, particularly skeletal muscle blood flow, as it is affected by exercise or spaceflight. Dr. Pawelczyk is a member of the American Heart Association, the American Physiological Society, the American College of Sports Medicine, and the Society for Neuroscience. He has won numerous awards, including the Young Investigator Award from the Life Sciences Project Division of the NASA Office of Life and Microgravity Science Applications (1994) and the NASA Space Flight Medal (1998). He earned two B.A. degrees, in biology and psychology, from the University of Rochester in 1982, an M.S. in physiology from Pennsylvania State University in 1985, and a Ph.D. in biology (physiology) from the University of North Texas in 1982. Dr. Pawelczyk has previously served as a member of the NRC's Review of NASA Strategic Roadmaps: Space Station Panel (2005), Committee on NASA's Bioastronautics Critical Path Roadmap (2004-2006), Committee to Review NASA's Space Flight Standards (2006-2007), Planning Committee for the Issues in Space Science and Technology Workshop Series (2007-2008), and Committee on NASA's Research on Human Health Risks (2007-2008). He currently serves as a member of the NRC Committee on Aerospace Medicine and Medicine of Extreme Environments (2006-2009) and as a member of the Space Studies Board (2006-2010).

ALAN R. HARGENS is a professor in and the director of the Clinical Physiology Laboratory at the University of California, San Diego. He previously served as the chief of the Space Physiology Branch, a senior research physiologist and space station project scientist at the NASA Ames Research Center, and a consulting professor

[§]Mriganka Sur was a member of the panel through mid-December 2009.

of human biology at Stanford University. Dr. Hargens's laboratory at UCSD focuses on orthopaedic and clinical physiology, with recent research concerning gravity effects on the cardiovascular and musculoskeletal systems of humans and animals. Dr. Hargens also investigates countermeasures to challenge the cardiovascular and musculoskeletal systems and to provide artificial gravity for spaceflight and Moon/Mars habitation. This research also is translated to aid in the postsurgical treatment and rehabilitation of orthopaedic patients and to improve the performance of athletes. In addition to his research on gravitational stress, Dr. Hargens studies tissue fluid and osmotic pressures, including those in giraffes to understand how they prevent dependent edema, those in skeletal muscle to diagnose compartment syndromes, and those in intervertebral discs to help understand low-back pain. Dr. Hargens's laboratory has performed numerous bed rest studies, ranging from 1 to 60 days in length, and has investigated spinal adaptations to microgravity in rats (Cosmos 1887), mice (STS-131 and STS-133), and astronauts (IML-2 and ISS). He served as chair of NASA's Science Working Group for the Space Station Centrifuge Facility, as a member of the NASA Cardiopulmonary Discipline Working Group, as chair of the NASA Panel for Human Health from Earth to Space, as a member of NASA Committee on Development of Countermeasures for Long Duration Space Flight, as a member of the International Multidisciplinary Artificial Gravity Project Review, as a co-chair of the NASA Human Health Countermeasure Element Standing Review Panel, and as a member of the NASA Human Research Program Cardiovascular Risks Panel. Dr. Hargens is a fellow of the American College of Sports Medicine and a fellow of the Aerospace Medical Association. He is associate editor of the *Journal of Gravitational Physiology*. He is the recipient of an NIH Research Career Development Award, the Elizabeth Winston Lanier Award, the American Physiology Society Recognition Award, and two NASA honor awards. He is a past president of the International Society of Adaptive Medicine. He received his Ph.D. in marine biology from the Scripps Institution of Oceanography at UCSD in 1971.

ROBERT L. HELMREICH is a retired professor of psychology, having taught at the University of Texas, Austin, from 1966 to 2006. He was PI of the University of Texas Human Factors Research Project, which studied individual and team performance, human error, and the influence of culture on behavior in aviation and medicine. He was a member of the Space Life Sciences Committee for NASA's University Space Research Association, and he is a fellow of the Royal Aeronautical Society and the American Psychological Association. Dr. Helmreich received the Flight Safety Foundation Distinguished Service Award for his contributions to aviation safety through the study and development of team training techniques (Crew Resource Management) for flight crews. He was also awarded Laurels from *Aviation Week and Space Technology* for his research related to human factors in aviation. He received the Distinguished Service Award of the Flight Safety Foundation and the Human Factors Award of Airbus Industrie in 2004. Dr. Helmreich is the 2005 recipient of the Public Service Award of the American Association of Anesthesia Nurses and of the University of Texas College of Liberal Arts Pro Bene Meritis Award, and he received the David S. Sheridan Award from Albany Medical College in 1997. He has written more than 200 papers, chapters, and scientific reports, and he is the author (with Ashleigh Merritt) of the 1998 book *Culture at Work in Aviation and Medicine: National, Organizational, and Professional Influences*. Dr. Helmreich received his B.S. and Ph.D. from Yale University in 1959 and 1966, respectively. He served on the NRC Panel on Human Behavior, the Panel on Workload Transition, the Committee on Space Biology and Medicine, the Committee on Human-Systems Integration, the Committee on Human Factors, and the Panel on Human Factors in Air Traffic Control Automation.

JOANNE R. LUPTON is a Distinguished Professor, Regent's Professor, and University Faculty Fellow at Texas A&M University, College Station, and holder of the William W. Allen Endowed Chair in Human Nutrition. Dr. Lupton's research is on the effects of diet on colon physiology and colon cancer, with a particular focus on dietary fiber and n-3 fatty acids. She translates basic research on diet and colon physiology to science-based public policy and has consulted with individuals in Japan, South Korea, China, Taiwan, and elsewhere on the definition of dietary fiber and on establishing dietary guidance systems in those countries. She is the PI on a training grant to train Ph.D. students in space life sciences and is the former team leader for nutrition and physical fitness for the NSBRI. Her research is supported by grants from the NIH/National Cancer Institute, NASA, and NSBRI. Dr. Lupton chaired the Macronutrients Panel for Dietary Reference Intakes for the Food and Nutrition Board of the

NRC, which determined the intake values for protein, carbohydrates, fats, fiber, and energy. She also chaired the NRC panel to determine the definition of dietary fiber. She was a member of the 2005 Dietary Guidelines Committee. Dr. Lupton spent 1 year at the Food and Drug Administration helping to develop levels of scientific evidence required for health claims. She is a past president of the American Society for Nutrition (ASN) and a member of the Institute of Medicine. Dr. Lupton has mentored more than 50 M.S. and Ph.D. students while at Texas A&M and received the Dannon/ASN mentoring award in 2004. In 2007 she received the Texas A&M Distinguished Achievement Award for Research. Her undergraduate degree is from Mount Holyoke College, and her Ph.D. in nutrition is from the University of California, Davis.

CHARLES M. OMAN is a senior research engineer, senior lecturer, and director of the Man-Vehicle Laboratory in the Department of Aeronautics and Astronautics at the Massachusetts Institute of Technology. Dr. Oman's group studies the physiological and cognitive limitations of humans in aircraft and spacecraft and tries to develop new ways of improving human-vehicle effectiveness and safety. The laboratory takes an interdisciplinary approach, utilizing techniques from manual and supervisory control theory, estimation, signal processing, biomechanics, cognitive, computational and physiological neuroscience, sensory-motor physiology, human factors, and biostatistics. Dr. Oman flew experiments on visual and vestibular function in spatial orientation on nine space shuttle missions, including six Spacelab flights. Since 1997 he has led the Sensorimotor Adaptation Research Team of the NSBRI. Dr. Oman served on the NASA Advisory Council's Biological and Physical Research Advisory Committee and the NRC Panel on Robotic Access and Human Planetary Landing Systems. He chaired the NASA Space Station Utilization Advisory Subcommittee (2004-2005). He is a member of the International Academy of Astronautics. Dr. Oman received his B.S.E. from Princeton University and his Ph.D. from the Massachusetts Institute of Technology.

DAVID ROBERTSON is the Elton Yates Professor of Medicine, Pharmacology, and Neurology at Vanderbilt University. He directs the Clinical and Translational Research Center of the Vanderbilt CTSA and the Vanderbilt Center for Space Physiology and Medicine. He established the Vanderbilt Autonomic Dysfunction Center in 1978 as the first international facility for research, education, and patient care of autonomic nervous system diseases. Dr. Robertson was founding president of the American Autonomic Society in 1990 and was founding president of the Association for Patient Oriented Research in 1998. He directed the Vanderbilt Medical Scientist Training Program for 10 years and the Division of Movement Disorders in the Department of Neurology for 8 years. Dr. Robertson is PI for the NIH Autonomic Rare Diseases Clinical Research Consortium, which coordinates national clinical trials and natural history studies in multiple system atrophy and other autonomic diseases. He served as a PI on the NASA Neurolab Mission aboard the space shuttle *Columbia* in 1998. He is the editor of the *Primer on the Autonomic Nervous System* and *Clinical and Translational Science: Introduction to Human Research*, the first textbook of the new discipline of clinical and translational research. In 2003, he received the inaugural Distinguished Educator Award of the Association for Clinical Research Training. Dr. Robertson received his B.A. in 1969 and his M.D. in 1973 from Vanderbilt University and received postdoctoral medical training at the Johns Hopkins Hospital.

SUZANNE M. SCHNEIDER is an associate professor in the Department of Health, Exercise and Sports Sciences at the University of New Mexico. Prior to joining the UNM faculty in 2002, Dr. Schneider was a research physiologist at NASA's Johnson Space Center (1989-2001) and a project scientist at the Human Research Facility of the ISS (1994-2001). Her research focuses on thermal physiology and microgravity, including the thermal responses of females to microgravity. Dr. Schneider is on the editorial board of the *American Journal of Physiology and Aviation* and *Space and Environmental Medicine*, and she has presented at meetings of the Federation of American Scientists for Experimental Biology, the American College of Sports Medicine, and the Aerospace Medical Association (ASMA). She received her B.A. in biology from the University of Missouri and her Ph.D. in physiology from St. Louis University. She completed her postdoctoral fellowship at the John B. Pierce Foundation, with an appointment with the Department of Epidemiology at Yale University.

GAYLE E. WOLOSCHAK. *See the committee listing above.*

PLANT AND MICROBIAL BIOLOGY PANEL

TERRI L. LOMAX, *Chair*, is the vice chancellor for research and graduate studies and a professor of plant biology at North Carolina State University. She previously served as dean of the university's graduate school, where she led the development and implementation of a strategic plan for graduate education. Prior to joining the faculty at North Carolina, Dr. Lomax served at NASA Headquarters in Washington, D.C., first as the division director of the fundamental space biology program and later as the acting deputy associate administrator for research for the Exploration Systems Mission Directorate. She was also a professor of botany and plant pathology at Oregon State University, where she directed the program for the analysis of biotechnology issues. Dr. Lomax's research is focused on plant development biology, specifically on how multiple hormones interact to regulate plant growth and responses to the environment. Dr. Lomax was a foundation board member and treasurer for the American Society for Plant Biologists (ASPB; 1997-2000) and a member of the executive committee of ASPB for 8 years. She served as a member of the board of governors of the American Society for Gravitational and Space Biology (1995-1998) and has been on the editorial advisory board of *The Plant Journal* since 1994. Her numerous awards include a Fulbright Fellowship, the Savery Award for Outstanding Young Faculty at the Oregon State University College of Agricultural Sciences, and the NASA Institute for Advanced Concepts Fellow. Dr. Lomax received her B.S. in botany from the University of Washington, her M.S. in botany/biology from San Diego State University, and her Ph.D. in biological sciences from Stanford University.

PAUL BLOUNT is an associate professor of physiology in the Graduate School of Biomedical Sciences at the University of Texas Southwestern Medical Center at Dallas. Prior to joining the faculty at UT Southwestern, Dr. Blount was a fellow at Washington University in St. Louis, studying G-protein coupled receptors in the Neuroscience Department and at the University of Wisconsin-Madison, studying bacterial mechanosensitive channels in biophysics. At UT Southwestern, Dr. Blount performs research that is aimed at determining how organisms detect mechanical force. This ability is required for the senses of touch, hearing, and balance, as well as for the determination of arterial pressures and osmotic gradients across cellular envelopes. Dr. Blount's laboratory utilizes microbes and microbial sensors to explore the general functional principles of biological mechanosensors. Because of the interdisciplinary nature of this basic research, the interests of Dr. Blount and the members of his laboratory range from microbial homeostasis to understanding the molecular mechanisms of ion channel gating. He is a member of the American Society for Microbiology and a member of the Biophysical Society. Dr. Blount received his B.A. in microbiology from the University of California, San Diego, and his Ph.D. in biological sciences (neurosciences) from the Washington University School of Medicine.

ROBERT J. FERL is a professor at and the director of the Interdisciplinary Center for Biotechnology Research at the University of Florida (UF). Dr. Ferl's research agenda includes analysis of the fundamental biological processes involved in plant adaptations to environments, with an emphasis on the particular environments and opportunities presented by the space exploration life sciences. He is an expert in the area of plant gene responses and adaptations to environmental stresses and the signal transduction processes that control environmental responses. The fundamental issues driving his research program include the recognition of environmental stress, the signal-transduction mechanisms that convert the recognition of stress into biochemical activity, and the gene activation that ultimately leads to response and adaptation to environmental stress. Most recently these studies have led to the examination of protein interactions as fundamental mechanisms for metabolic regulation of plant biochemistry. Dr. Ferl has been funded continuously for more than 25 years, most recently with grants from NASA, NSF, and the U.S. Department of Agriculture (USDA). He has frequently been asked to serve as a scientific adviser and reviewer for national agencies including NASA, NSF, and USDA. Recent articles that he has written have been chosen as cover articles for the journals *Plant Physiology* and *Molecular Biology of the Cell*. His application of basic science to the questions of advanced life support for NASA's exploration initiative has captured great interest in his research. Dr. Ferl has recently served as the developer and director of the virtual center for Exploration Life Sciences, a joint academic research and education venture between UF/Institute of Food and Agricultural Sciences (IFAS), and the NASA Kennedy Space Center (KSC). In that role he has been responsible for research and academic program development at KSC, and he has facilitated the recruitment of faculty and research programs from UF/IFAS to be located at KSC. Many articles, news spots, and online stories have covered Dr. Ferl's research on plant adaptations to spaceflight and extraterrestrial environments. He has been asked to serve on the Science Council of the Universities Space Research Association, and he was appointed by NASA as the only plant molecular biologist to be a member of the Lunar Exploration Analysis Group, which will outline scientific research priorities for the return of humans to the Moon.

SIMON GILROY. *See the committee listing above.*

E. PETER GREENBERG is a professor of microbiology at the University of Washington School of Medicine. After completing a postdoctoral fellowship at Harvard University, Dr. Greenberg was on the faculty at Cornell University and then the University of Iowa College of Medicine before moving to the University of Washington in 2005. The research in Dr. Greenberg's laboratory is focused on the emerging field of sociomicrobiology, and he is widely credited as a founder of the quorum sensing field. He has concentrated much of his effort on *Pseudomonas aeruginosa*, an opportunistic pathogenic bacterium that can cause both acute and persistent biofilm infection. Quorum sensing allows certain bacterial species to monitor their own population density and respond by activating transcription of specific sets of genes. Dr. Greenberg is a member of the National Academy of Sciences, the American Academy of Arts and Sciences, AAAS, and the American Academy of Microbiology. He holds a B.A. in biology from Western Washington University, an M.S. in microbiology from the University of Iowa, and a Ph.D. in microbiology from the University of Massachusetts. Dr. Greenberg is a member of the editorial board of *Proceedings of the National Academy of Sciences*. He was also a member of the NRC 2007 Selman A. Waksman Award in Microbiology Selection Committee.

TRANSLATION TO SPACE EXPLORATION SYSTEMS PANEL

JAMES P. BAGIAN, *Chair*, is the director of the Center for Health Engineering and Patient Safety and a professor in the Medical School and the College of Engineering at the University of Michigan. Previously, he served as the first director of the VA National Center for Patient Safety (NCPS) and the first chief patient safety officer for the Department of Veterans Affairs from 1999 to 2010. There he developed numerous patient safety-related tools and programs that have been adopted nationally and internationally. Dr. Bagian served as a NASA astronaut; he is a veteran of two space shuttle missions, which included his service as the lead mission specialist for the first dedicated Life Sciences Spacelab mission. Currently his primary interest and expertise involve the development and implementation of multidisciplinary programs and projects that involve the integration of engineering, medical/life sciences, and human factor disciplines. At present he is applying the majority of his attention to the application of systems engineering approaches to the analysis of medical adverse events and the development and implementation of suitable corrective actions that will enhance patient safety, primarily through preventive means. He received his B.S. in mechanical engineering from Drexel University and his M.D. from Jefferson Medical College at Thomas Jefferson University. Dr. Bagian is a member of both the National Academy of Engineering and the Institute of Medicine. He has served on numerous NRC committees, including the Task Group on Research on the International Space Station (2001-2003), the Committee on Space Biology and Medicine (2000-2003), the Review of NASA Strategic Roadmaps: Space Station Panel (2005), the Committee on NASA's Bioastronautics Critical Path Roadmap (2004-2006), and the Committee on Optimizing Graduate Medical Trainee (Resident) Hours and Work Schedules to Improve Patient Safety (2007-2009).

FREDERICK R. BEST is an associate professor in the Nuclear Engineering Department at Texas A&M University, College Station, and the director of the Space Engineering Research Center and the Interphase Transport Phenomena Laboratory. Prior to joining Texas A&M in 1983, Dr. Best was a visiting professor at the Massachusetts Institute of Technology and a technical coordinator in the nuclear area of the MIT Electric Utility Research Program. His research is focused on zero gravity two-phase flow, reactor thermal hydraulics, and interphase transport phenomena. He earned his B.S. in mechanical engineering in 1968 from Manhattan College in New York City, and he received his M.S. in nuclear engineering in 1969 and his Ph.D. in nuclear engineering in 1980, both from MIT. Dr. Best has served on the NASA BIO-PLEX Review Panel (2000-2001); the NASA Peer Review Panel (chair, 2000-2001); the NASA Advanced Life Support, Science, and Technology Working Group (1999-2001); and the Air Force Science Advisory Board (1995-1996). He is currently serving on the NASA/JPL Technology Review Board for the New Millennium Program ST-8.

DAVID C. BYERS, Independent consultant, Torrance, California.¶

LEONARD H. CAVENY. *See the committee listing above.*

MICHAEL B. DUKE is a planetary geologist who recently retired as the director of the Center for Commercial Applications of Combustion in Space at the Colorado School of Mines (CSM). His principal research focuses on the general area of study that relates to the use of in situ resources to support human exploration missions to the Moon and Mars. His planetary science interests relate to the mineralogy and petrology of meteorites and lunar materials. Dr. Duke worked at the NASA Johnson Space Center for 25 years prior to accepting the position at CSM in 1998. He has also been a research scientist at the U.S. Geological Survey (1963-1970) and the curator of NASA's lunar sample collection (1970-1977). Dr. Duke received the NASA Exceptional Scientific Achievement Award and the AIAA's Space Science Medal, and he was a distinguished federal executive. Dr. Duke received his B.S. and Ph.D. from the California Institute of Technology. He served on the NRC's Committee on the Scientific Context for the Exploration of the Moon and the Panel on Solar System Exploration.

JOHN P. KIZITO is an assistant professor of mechanical engineering at North Carolina Agricultural and Technical State University (A&T). Prior to joining the faculty of A&T, Dr. Kizito was a research assistant, research scientist, and adjunct assistant professor at Case Western Reserve University from 1988 to 2007 and a staff scientist at the National Center for Space Exploration Research at the NASA Glenn Research Center from 1999 to 2007. His research focuses on developing thermal management methods, high heat flux control, energy storage (phase-change materials), water management in polymer electrolyte membrane fuel cells, innovative computational fluid dynamics techniques, gravity-driven transport phenomena, medical devices and instrumentation, microgravity research, in situ resource utilization, and lunar and Mars exploration equipment. He is a member of ASME, AIAA, and ASGSB. Dr. Kizito was awarded a 1988 Fulbright Award, a 1988 Rice-Cullimore Award, and five NASA Team Achievement Awards between the years of 2001 and 2003. He received his B.S. in mechanical engineering from Makere University and his M.S. and Ph.D. in mechanical engineering from Case Western Reserve University.

DAVID Y. KUSNIERKIEWICZ is the chief engineer of the Space Department of the Johns Hopkins University Applied Physics Laboratory, where he has worked since 1983. He has an extensive background in designing, integrating, and testing power system electronics for spacecraft. Mr. Kusnierkiewicz was the mission system engineer for the NASA New Horizons Pluto-Kuiper-Belt Mission, and he is the mission and spacecraft system engineer for the NASA Thermosphere, Ionosphere, Mesosphere, Energetics and Dynamics (TIMED) program, which launched in December 2001. He has served on numerous review boards for NASA missions, including Lunar Reconnaissance Orbiter and Lunar Robotic Explorer, as well as for the missions Dawn, Juno, Radiation Belt Storm Probes, and ST-8 (part of the New Millennium Program). He has received three NASA Group Achievement Awards. Mr. Kusnierkiewicz received his B.S. and M.S. in electrical engineering from the University of Michigan, Ann Arbor. He has also served on the Mitigation Panel of the NRC Committee to Review Near-Earth Object Surveys and Hazard Mitigation Strategies (2009-2010) and is a Corresponding Member of the International Academy of Astronautics.

E. THOMAS MAHEFKEY, JR., is a consultant with Heat Transfer Technology Consultants, serving several firms in the areas of heat transfer and energy conversion. He retired from the Air Force Wright Laboratory in 1995 after 33 years as an engineer, scientist, and research manager. Before retiring, he was the deputy division chief for technol-

¶David C. Byers was a member of the panel through mid-December 2009.

ogy in the Jet Propulsion Laboratory's Aerospace Power Division. Dr. Mahefkey was instrumental in establishing the Thermal Energy/Heat Pipe and the Thermionics Laboratories in the Air Force Wright Laboratory. He is also an experienced educator, having held the rank of adjunct professor of mechanical engineering at the University of Dayton, Wright State University, the University of Kentucky, Ohio State University, and the Air Force Institute of Technology. Dr. Mahefkey's areas of expertise include thermionics, energy conversion, and heat transfer, and he has published extensively in these areas. He received his B.S. in aerospace engineering from St. Louis University and his M.S. in physics and his Ph.D. in mechanical engineering from the University of Dayton. He previously served as chair of the NRC Committee on Thermionic Research and Technology.

DAVA J. NEWMAN is a professor of aeronautics and astronautics and engineering systems, the director of the Technology and Policy Program, MacVicar Faculty Fellow at the Massachusetts Institute of Technology, and a Harvard-MIT Health Sciences and Technology Program affiliate faculty. She leads the MIT-Portugal Program's Bioengineering Systems effort. Dr. Newman's research focuses on the mechanics and energetic requirements of human performance across the continuum of gravity—from microgravity to lunar and martian gravity levels to hypergravity—research combining aerospace bioengineering, human-in-the-loop dynamics and control modeling, biomechanics, human-robotic cooperation, and bioastronautics. She was PI for the space shuttle Dynamic Load Sensors (DLS) experiment that measured astronaut-induced disturbances of the microgravity environment on mission STS-62. An advanced system, the Enhanced DLS experiment, flew onboard the Russian Mir space station (1996-1998). Dr. Newman was a co-investigator on the Mental Workload and Performance Experiment that flew on STS-42 to measure astronaut mental workload and fine motor control in microgravity. She is the author of more than 150 publications. She received her B.S. in aerospace engineering from the University of Notre Dame and an S.M. in aeronautics and astronautics, an S.M. in technology and policy, and her Ph.D. in aerospace biomedical engineering from MIT. Dr. Newman previously served as a member of the NRC Committee on Advanced Technology for Human Support in Space, the Committee on Engineering Challenges to the Long-Term Operation of the International Space Station, the Aeronautics and Space Engineering Board, the Steering Committee for Workshops on Issues of Technology Development for Human and Robotic Exploration and Development of Space, and the Committee on Full System Testing and Evaluation of Personal Protection Equipment Ensembles in Simulated Chemical and Biological Warfare Environments.

RICHARD J. ROBY is the president and technical director of Combustion Science and Engineering, Inc. (CSE). In addition to his role at CSE, he is an adjunct professor in the Department of Fire Protection Engineering at the University of Maryland, College Park, and served as an assistant and associate professor in the Mechanical Engineering Department of Virginia Polytechnic Institute and State University (1986-1992). Dr. Roby also serves as the chief executive officer/manager of LPP Combustion, LLC, a clean technology start-up company that specializes in clean combustion of conventional and renewable liquid fuels, and as chair of SafeAwake LLC, a fire safety company that produces secondary alert systems for those who are deaf or hard of hearing. He has more than 30 years of professional experience and has skills in chemical, mechanical, and fire protection engineering. Prior to joining CSE, Dr. Roby served as the director of combustion research at Hughes Associates, Inc. (1992-1998) and as a research assistant and research engineer at Stanford University (1983-1986), the Ford Motor Company Scientific Research Laboratories (1979-1983), and Cornell University (1977-1979). Dr. Roby serves as project manager for a variety of experimental and analytical combustion and fire science research and development projects, which include modeling of blow-off and flashback in gas turbine combustors, determining the effects of fuel constituents on combustor performance, developing and incorporating reduced kinetics mechanisms in computational fluid dynamics codes, creating innovative fire detection devices, and developing new fire models. He is a fellow of the Society of Fire Protection Engineers and a member of several other professional societies, including the Combustion Institute, the Society of Automotive Engineers, the American Society of Mechanical Engineers, the American Institute of Aeronautics and Astronautics, and the International Association for Fire Safety Science. He won the Outstanding Faculty Award from the Society of Automotive Engineers (1999), the National Fire Protection Research Foundation's Harry Bigglestone Award for Excellence in Communication of Fire Protection Concepts (1999, 2005, 2006, and 2007), and the William M. Carey Award at the 5th Fire and Detection Research

Application Symposium (2001). Dr. Roby holds more than 10 patents related to fire safety and combustion system design. He received his A.B. in chemistry and B.S. in chemical engineering in 1977 from Cornell University, his M.S. in mechanical engineering in 1980 from Cornell University, and his Ph.D. in mechanical engineering from Stanford University in 1988.

GUILLERMO TROTTI is currently the president of Trotti and Associates, Inc. (TAI), a firm that he founded in 1993 in Cambridge, Massachusetts. He is an internationally recognized architect and industrial designer with more than 30 years of experience designing space habitats and structures, architectural projects for the hospitality, entertainment, and education sectors. Previously, he was the president of Bell and Trotti, Inc. (BTI), a design and fabrication studio that specialized in high-technology architecture, exhibit design, industrial design, and space architecture. From its inception, BTI had a key role in designing diverse elements of the ISS for NASA and leading aerospace companies. TAI has worked with NASA's Institute of Advanced Concepts on revolutionary mission architecture concepts for exploring the Moon with habitable rovers. The Extreme Expeditionary Architecture: Mobile, Adaptable Systems for Space and Earth Exploration project proposes a revolutionary way for humans and machines to explore the Moon. TAI is also working with MIT on the Biosuit project, an advanced mechanical counterpressure spacesuit for lunar and Mars surface exploration. Mr. Trotti also has more than 25 years' experience teaching design in architecture and industrial design at the University of Houston and at the Rhode Island School of Design, respectively. He received his M.A. in architecture from Rice University in 1976. Mr. Trotti served on the NRC Committee to Review NASA's Exploration Technology Development Programs.

ALAN WILHITE is the Langley Distinguished Professor in the School of Aerospace Engineering at the Georgia Institute of Technology, and he also serves as the co-director of the Georgia Institute of Technology Center for Aerospace Systems Engineering. He currently resides at the National Institute of Aerospace teaching graduate classes and conducting research at the NASA Langley Research Center. He teaches and supervises research in systems engineering and aerospace systems design. He has numerous published articles and several book chapters in these areas. Dr. Wilhite has served as a researcher, systems program manager, and senior executive involved in the design and development of NASA space and aeronautic systems. He is an AIAA associate fellow and has served on several AIAA technical committees, such as space systems, space transportation, and computer-aided design. He is a member of the International Astronautical Federation and a member of its Systems Engineering Committee. He conducts research in system-of-systems architecture design, robust design, aerodynamics, propulsion, MDO, operations, cost, systems engineering, and risk. He has served as NASA's external chair for systems engineering, and he conducts research supporting NASA's vision in space exploration. Dr. Wilhite received his B.S. in aerospace engineering from North Carolina State University, his M.S. in flight systems from George Washington University, and his Ph.D. in aerospace engineering from North Carolina State University.

STAFF

SANDRA J. GRAHAM, *Study Director*, has been a senior program officer at the National Research Council's Space Studies Board (SSB) since 1994. During that time Dr. Graham has directed a large number of major studies, many of them focused on space research in biological and physical sciences and technology. More recent studies include an assessment of servicing options for the Hubble Space Telescope, a study of the societal impacts of severe space weather, and a review of NASA's Space Communications Program while on loan to the NRC's Aeronautics and Space Engineering Board (ASEB). Prior to joining the SSB, Dr. Graham held the position of senior scientist at the Bionetics Corporation, where she provided technical and science management support for NASA's Microgravity Science and Applications Division. She received her Ph.D. in inorganic chemistry from Duke University, where her research focused primarily on topics in bioinorganic chemistry, such as rate modeling and reaction chemistry of biological metal complexes and their analogs.

ALAN C. ANGLEMAN has been a senior program officer for the ASEB since 1993, directing studies on a wide variety of aerospace issues. Previously, Mr. Angleman worked for consulting firms in the Washington, D.C., area, providing engineering support services to the Department of Defense and NASA Headquarters. His professional career began with the U.S. Navy, where he served for 9 years as a nuclear-trained submarine officer. He has a B.S. in engineering physics from the U.S. Naval Academy and an M.S. in applied physics from the Johns Hopkins University.

IAN W. PRYKE retired from the European Space Agency (ESA) in 2003. He is currently a senior program officer (part time) with the SSB and a senior fellow and assistant professor at the Center for Aerospace Policy Research in the School of Public Policy of George Mason University; he also operates as an independent consultant. Mr. Pryke joined the European Space Research Organisation (later ESA) in 1969, working in the areas of data processing and satellite communications. In 1976 he transferred to ESA's Earth Observation Programme Office, where he was involved in the formulation of the Remote Sensing Programme. In 1979 he moved to the ESA Washington Office, where he was engaged in liaison work with both government and industry in the United States and Canada, becoming the head of the office in 1983. He holds a B.Sc. in physics from the University of London and an M.Sc. in space electronics and communications from the University of Kent.

ROBERT L. RIEMER joined the staff of the NRC in 1985. He served as senior program officer for the two most recent decadal surveys of astronomy and astrophysics and has worked on studies in many areas of physics and astronomy for the Board on Physics and Astronomy (for which he served as associate director, 1988-2000) and the SSB. Prior to joining the NRC, Dr. Riemer was a senior project geophysicist with Chevron Corporation. He received his Ph.D. in experimental high-energy physics from the University of Kansas-Lawrence and his B.S. in physics and astrophysics from the University of Wisconsin-Madison.

MAUREEN MELLODY has been a program officer with the ASEB since 2002, where she has worked on studies related to NASA's aeronautics research and development program, servicing options for the Hubble Space Telescope, and many other projects in space and aeronautics. Previously, she served as the 2001-2002 AIP Congressional Science Fellow in the office of Congressman Howard L. Berman (D-Calif.), focusing on intellectual property and technology transfer. Dr. Mellody also worked as a postdoctoral research scientist at the University of Michigan in 2001. Dr. Mellody received her Ph.D. in applied physics from the University of Michigan, her M.S. in applied physics from the University of Michigan, and her B.S. in physics from Virginia Polytechnic Institute and State University.

REGINA NORTH, consultant, has specialized throughout her research career in the area of human behavior in isolated and confined environments, including Arctic and Antarctic stations, offshore oil platforms, and the Mir space station and the International Space Station, and she has conducted field research in these environments. At the NASA Johnson Space Center, Ms. North worked as a senior research scientist and research program analyst for the ISS chief scientist, the chief of advanced programs, and the director of biological sciences and applications managing the ISS multidisciplinary research portfolio onboard the ISS. Previously, for the JSC Mission Operations Directorate, she was a certified instructor training astronaut crews to perform science and technology multidisciplinary experiments on the ISS. Ms. North has also worked as a researcher for the Man-Vehicle Laboratory at the Massachusetts Institute of Technology and the Arctic Research Institute of the Centre National de la Recherche Scientifique (CNRS), in Paris, France. She attended the graduate program in social sciences at the University of Sao Paulo, Brazil; the Ecole des Hautes Etudes en Sciences Sociales (School for Advanced Studies in the Social Sciences) at CNRS; and the International Space University at MIT. She served as president of the International Space University Alumni Association from 2003 to 2006. Ms. North is fluent in Portuguese, English, French, Spanish, and Italian and has a working knowledge of Russian.

CATHERINE A. GRUBER, editor, joined the Space Studies Board as a senior program assistant in 1995. Ms. Gruber first came to the NRC in 1988 as a senior secretary for the Computer Science and Telecommunications Board and also worked as an outreach assistant for the National Science Resources Center. She was a research assistant (chemist) in the National Institute of Mental Health's Laboratory of Cell Biology for 2 years. She has a B.A. in natural science from St. Mary's College of Maryland.

LEWIS GROSWALD, a research associate, joined the Space Studies Board as the Autumn 2008 Lloyd V. Berkner Space Policy Intern. Mr. Groswald is a graduate of George Washington University, where he received a master's degree in international science and technology policy and a bachelor's degree in international affairs, with a double concentration in conflict and security and Europe and Eurasia. Following his work with the National Space Society during his senior year as an undergraduate, Mr. Groswald decided to pursue a career in space policy, with a focus on educating the public on space issues and formulating policy.

DANIELLE JOHNSON-BLAND joined the Division on Behavioral and Social Sciences and Education as a senior program assistant in 2008. She was first assigned to the Center for Economic Governance, and International Studies, but shortly thereafter received additional assignments with the Committee on Law and Justice, the Committee on Population, and the Space Studies Board. Mrs. Johnson-Bland's interests include youth rehabilitation, criminal justice management, and juvenile justice reform. She holds a B.S. in social science from University of Maryland, University College.

LAURA TOTH is a senior program assistant for the National Materials Advisory Board, the Board on Manufacturing and Engineering Design, and the Board on Infrastructure and the Constructed Environment. She has been with the NRC since 2002 and has also worked with the Transportation Research Board and the Space Studies Board. Before joining the NRC, Ms. Toth worked in retail management for 15 years.

LINDA M. WALKER, a senior project assistant, has been with the NRC since 2007. Before her assignment with the Space Studies Board, she was on assignment with the National Academies Press. Prior to working at the NRC, she was with the Association for Healthcare Philanthropy in Falls Church, Virginia. Ms. Walker has 28 years of administrative experience.

ERIC WHITAKER is a senior program assistant at the NRC's Computer Science and Telecommunications Board (CSTB). Prior to joining the CSTB, he was a realtor with Long and Foster Real Estate, Inc., in the Washington, D.C. metropolitan area. Before that, he spent several years with the Public Broadcasting Service in Alexandria, Virginia, as an associate in the Corporate Support Department. He has a B.A. in communication and theater arts from Hampton University.

MICHAEL H. MOLONEY is the director of the SSB and the Aeronautics and Space Engineering Board at the NRC. Since joining the NRC in 2001, Dr. Moloney has served as a study director at the National Materials Advisory Board, the Board on Physics and Astronomy (BPA), the Board on Manufacturing and Engineering Design, and the Center for Economic, Governance, and International Studies. Before joining the SSB and ASEB in April 2010, he was associate director of the BPA and study director for the Astro2010 decadal survey for astronomy and astrophysics. In addition to his professional experience at the NRC, Dr. Moloney has more than 7 years' experience as a foreign-service officer for the Irish government and served in that capacity at the Embassy of Ireland in Washington, D.C., the Mission of Ireland to the United Nations in New York, and the Department of Foreign Affairs in Dublin, Ireland. A physicist, Dr. Moloney did his graduate Ph.D. work at Trinity College Dublin in Ireland. He received his undergraduate degree in experimental physics at University College Dublin, where he was awarded the Nevin Medal for Physics.